Scottish Life and Society

Scotland's Domestic Life

Publications of the European Ethnological
Research Centre

Scottish Life and Society:A Compendium of Scottish Ethnology
(14 Volumes).

Already published:
Volume 3 *Scotland's Buildings*
Volume 6 *Scotland's Domestic Life*
Volume 9 *The Individual and Community Life in Scotland*
Volume 11 *Institutions of Scotland: Education*
Volume 12 *Religion*
Volume 14 *Bibliography*

GENERAL EDITOR:
Alexander Fenton

Scottish Life
and Society

A COMPENDIUM OF

SCOTTISH ETHNOLOGY

SCOTLAND'S DOMESTIC LIFE

Edited by
Susan Storrier

JOHN DONALD
in association with
THE EUROPEAN ETHNOLOGICAL RESEARCH CENTRE
and
THE NATIONAL MUSEUMS OF SCOTLAND

First published in Great Britain in 2006 by
John Donald, an imprint of Birlinn Ltd
West Newington House
10 Newington Road
Edinburgh

www.birlinn.co.uk

COPYRIGHT © The European Ethnological Research Centre 2006

All rights reserved

ISBN 0 85976 649 7

All rights reserved

British Library Cataloguing-in-Publication Data
A catalogue record is available on request

We gratefully acknowledge the support of the following bodies:

The Scotland Inheritance Fund
The Russell Trust
The Binks Trust

Typeset by Carnegie Publishing Ltd, Lancaster
Printed and bound in Great Britain by The Cromwell Press

Contents

List of Figures   *viii*
List of Contributors   *xviii*
Foreword   *xxi*
Abbreviations   *xxiii*
Glossary   *xxiv*

Introduction   1

PART ONE: THEMES IN SCOTTISH DOMESTIC LIFE

1. Shelter from the Elements   7
   Geoffrey Stell
2. Security   23
   Sabina Strachan
3. Privacy   42
   Hannah Avis
4. Storage   55
   David Jones
5. Lighting   69
   Alexander Fenton
6. Heating and Sleeping   83
   Alexander Fenton
7. Rest   108
   Naomi E A Tarrant
8. Music   120
   Gary J West
9. Needlework   134
   Naomi E A Tarrant
10. Television, Video and Computing   144
    James Stewart and Kathy Buckner
11. Games and Sport   163
    John Burnett
12. Artistic Expression and Interiors   174
    Elizabeth Cumming
13. Spirituality   197
    David Reid
14. Clothing   223
    Naomi E A Tarrant and Alexander Fenton

15.	Toilette *Susan Storrier*	243
16.	Personal Hygiene *Miles K Oglethorpe* *with a section by Alexander Fenton*	277
17.	Housework *Una A Robertson*	307
18.	Auxiliary Work in the Rural Home *Alexander Fenton*	330
19.	Working from Home *Christopher A Whatley*	338
20.	Work in the Homes of Others *Jean Aitchison*	357
21.	Reading and Study *Heather Holmes*	377
22.	Gardens *Sally A Butler*	400
23.	Food *Una A Robertson*	418
24.	Childbirth *Lindsay Reid*	440
25.	Childcare and Children's Leisure *Lindsay Reid*	458
26.	Infirmity *Lindsay Reid*	482
27.	Medicine *Joyce Miller*	500
28.	Hierarchy and Authority *Lynn Jamieson*	515
29.	Status *Crissie W H White*	542
30.	Custom *Gary J West*	562
31.	Animals *Catherine Smith*	581
32.	Paying for Dwellings *Peter A Kemp*	597

PART TWO: LESS COMMON TYPES OF SCOTTISH HOUSEHOLD AND HOME

33.	'Time Out' Work *Roger Leitch*	617
34.	'Time Out' Leisure *Roger Leitch*	637

35.	Prisons *Rod MacCowan*	654
36.	Hospices and Nursing Homes *Derek B Murray*	672
37.	Children's Homes *Joe Francis*	683
38.	Residential Schools and Student Residences *Ian Morris*	701
39.	Shared Flats and Houses *Susan Storrier*	714
40.	Communes *David Riches*	737
41.	Religious Communities *Mark Dilworth*	749
42.	Fishing and Homes *Reginald Byron*	762
43.	Living Alone *Sinéad Power*	777
44.	Communities on the Move *Hugh Gentleman*	788
45.	Domestic Life in Temporary and Self-Built Dwellings *Hugh Gentleman*	818
46.	Idiosyncratic Homes *Deborah Mays*	833

PART THREE: HOMELESSNESS

47.	Homelessness *Peter A Kemp*	851
	Index	865

List of Figures

1.1	Northern pier, Portpatrick Harbour, Wigtownshire, 1985.	8
1.2	Steading, Burghmuir, Linlithgow, West Lothian, 2003.	10
1.3	Caisteal Uisdein (Hugh's Castle), Snizort, Skye.	11
1.4	Cottage and Byre, Caolas, Tiree, 1973.	13
1.5	Cottage and Byre, Caolas, Tiree, 1973.	14
2.1	Argyle Park Sheltered Housing, Edinburgh, 2005.	26
2.2	Wooden lock from St Kilda.	28
2.3	The White Caterthun, Angus, 2003.	31
2.4	Neighbourhood-watch sign, Edinburgh, 2005.	33
2.5	Red Road Development, Glasgow, 1987.	35
3.1	Exterior shot of Edinburgh tenement, 2005.	45
3.2	The lounge of a one-bedroom flat off Easter Road, Edinburgh, 2000.	46
4.1	Walter Geikie: Farmhouse kitchen interior, Cousland, Midlothian, c 1830.	56
4.2	Meal girnal in the form of a Scotch chest.	57
4.3	Provision cupboard belonging to William Munro, farmer, Clashnabuil, Easter Ross.	59
4.4	Stage-top sideboard, attributed to James Japp, Montrose, Angus, c 1810.	63
4.5	Meal barrels made at the Seaton Pottery in Aberdeen.	66
5.1	A bouet, in this case three-sided.	70
5.2	A peerman.	73
5.3	A cruisie, possibly from Angus.	74
5.4	Gas-light fitting in a house in Edinburgh, c 1900.	77
5.5	'A Woman's Place'. Perthshire interior of about 1850.	78
6.1	The kitchen range at Smithy Croft, Pitglassie, Auchterless, Aberdeenshire, early 1960s.	87
6.2	The central hearth in a blackhouse at Calbost, Isle of Lewis, 1964.	90
6.3	Hanging lum, Ardoyne, Insch, Aberdeenshire, 1961.	92
6.4	A photograph, probably of the late nineteenth century, showing the kitchen at Eastfield Farm, Lanarkshire.	93
6.5	A cast-iron stove, built into a formerly open hearth, at Estabin, Orkney, 1968.	95

6.6	A blackhouse or 'byre dwelling' in St Kilda, with a byre at one end and living space at the other.	98
6.7	The positions of the box beds in two small Aberdeenshire crofts.	101
7.1	The Bothy at Gagie, Angus, late nineteenth to early twentieth century.	110
7.2	'The Wake', by Alexander Carse, early nineteenth century.	112
7.3	Blue Room, Montgomerie House, Ayrshire, 1948.	113
7.4	Shetland chairs, with and without arms.	117
8.1	Waulking Tweed, Eriskay, 1934.	124
8.2	Informal music session, Forrest Glen, Dalry, 1985.	126
9.1	Smith family sampler, Lintmill, Aberdeenshire, late nineteenth century.	135
9.2	Folding and packing hand-woven and hand-knitted garments in polythene bags prior to their dispatch to customers, 1963.	136
9.3	Mrs Belle Ann McAngus of Hilton, Fearn, Ross-shire, knitting in 1975.	139
10.1	Table of households with home access to Internet.	146
10.2	Mr and Mrs Brimfield, Arbroath, Angus, relaxing at home by their radiogram.	147
10.3	George Hastie in his sitting room, Inverkeithing, Fife, 1960s.	147
10.4	Concealing the PC.	154
10.5	PC surrounded by a mix of documents and writing tools.	157
11.1	The Stuart family, probably father and daughters, playing chess at their house at Roslin, Midlothian, *c* 1900.	165
11.2	Playing cards on Lewis, probably *c* 1970.	168
12.1	Doorstep designs, Dunsyre, Lanarkshire.	177
12.2	Late sixteenth-century ceiling decoration at Prestongrange, Prestonpans, East Lothian, dated 1581.	178
12.3	Saloon, the House of Dun, Angus, designed by William Adam with stucco work by Joseph Enzer, 1742.	179
12.4	The drawing room, Holmwood House, Cathcart, Glasgow, designed by Alexander Thomson, 1857.	181
12.5	Mrs Blackie's bedroom, Hill House, Helensburgh, Dunbartonshire, designed and furnished by Charles Rennie Mackintosh, 1902–4.	183
12.6	Drawing room, Corriemulzie Cottage, Deeside, in 1863.	187
12.7	Parlour mantelpiece display at 111 New Street, Fisherrow, Midlothian.	188
12.8	The 'but' end of a South Uist croft with dresser made by Archibald MacPherson and display of household ceramics.	192

13.1	Crucifix on the wall above the mantelpiece at Sanna, Ardnamurchan, Argyll, 1930s.	203
13.2	Selection of personal items, including a wooden rosary, recovered from the sixteenth-century English vessel, the *Mary Rose*.	204
13.3	'Presbyterian Catechising', by John Philip, RA (1817–67).	207
13.4	'Family Devotions', household prayer, 1870s.	210
13.5	'He Walks a Portion wi' Judicious Care', by G Paul Chalmers, RSA.	214
13.6	'A Scottish Christening', by John Philip, RA (1817–67).	217
14.1	'The Arrival of the Country Cousins', by Alexander Carse, 1812.	225
14.2	'Highland Dance', by David Allan, about 1780.	229
14.3	A German Gypsy family, about 1906.	234
15.1	Evelyn and Alexa Thompson at home in Academy Street, Ayr, *c* 1900.	250
15.2	Tanning centre, Edinburgh, 2005.	261
15.3	Tillycorthie House, Udny, Aberdeenshire, 1920s.	257
15.4	Dressing table in a house in Ayr, 1979.	258
15.5	'The Golden Fleece', by Keith Henderson.	259
15.6	Young people on an outing, Ness, Lewis, probably the 1950s.	255
15.7	By 1900 beards were generally no longer favoured.	267
16.1	The old way in which miners took their bath.	282
16.2	The new way in which miners took their bath.	283
16.3 & 16.4	A typical modern bathroom in working-class housing.	289
16.5	Example of a bath-shower combination, appearing in the 1912 Shanks catalogue, *Appareils Sanitaire* of 1912, produced for the French and Belgian markets.	292
16.6	Example of a pedestal lavatory basin from the J & M Craig catalogue of 1914.	293
16.7	An example of an old, soiled pan closet, illustrated by Hellyer, 1886.	294
16.8	An example of a valve WC at Bielside House in East Lothian.	295
16.9	Shanks of Barrhead patent water closet, with blue floral pattern, and matching cistern.	296
17.1	An old woman sweeps with a besom made by binding a bundle of twigs or heather roots around a wooden pole.	312
17.2	A stone sink with enamel pails below.	313
17.3	The latest in technology from the 1930s, a Hoover vacuum cleaner complete with its box of attachments for specialised cleaning jobs.	314
17.4	Two boys selling sand from a cart.	315

17.5	A housemaid's box containing the tools and cleaning materials needed for her work around the house.	316
17.6	Town dwellers could enlist the services of a water caddy.	319
17.7	A girl uses a 'dolly' or 'posser' to push washing through the water.	322
17.8	A selection of implements including flat, box and charcoal irons; an iron stand and goffering iron; egg, polishing and Italian irons; a laundry stove for heating the irons; and a linen press with drawers.	324
17.9	Proprietary cleaning materials and washing powders including Oxydol, Rinso and several soaps.	326
18.1	Rivlin from Shetland.	331
18.2	Making a caisie in Orkney.	333
18.3	Winding simmons in Orkney.	334
18.4	Horsehair rope and winder 'crocan goisid', South Uist, 1965.	335
19.1	Woman spinning, Harris, 1937.	345
19.2	Weaving at Kirriemuir, Angus.	347
19.3	Weaving at Newcastleton, Roxburghshire, 1907.	348
19.4	Broom seller, Fife, c1850.	350
19.5	Basket making, Fife, 1950s.	351
20.1	Peeling the tatties.	358
20.2	The maid at Sanna, Argyll, trampling the washing, c1925.	359
20.3	The testament of Issobell Broun, servant girl, 1592.	360
20.4	A table of the Assessed Taxes chargeable to employers of male servants, 1854.	365
20.5	The sewing class at the Glasgow 'Dough' School, c1931.	371
20.6	Cooking and baking class at the Glasgow 'Dough' School, c1931.	372
21.1	Books used as decorative objects in the drawing room of Thomas Bonnar II, Edinburgh, c1880.	379
21.2	Book cases in William Reid's home, Edinburgh, 1894–6.	380
21.3	Flat pack in action. Shelving in a study, Burntisland, Fife, 2003.	381
21.4	Reading by the fire. Advertisement, 1861.	386
21.5	A man reads his newspaper on the left of a central fire with stone-built back at Kirbister, Birsay, Orkney, in the 1960s.	389
21.6	Forms of Reading.	391
21.7	Types of Experience	391
22.1	The gardens of Culross, Fife.	404
22.2	The Hermitage Folly, Dunkeld, Perthshire, built in 1758 by the 2nd Duke of Atholl.	405
22.3	Plants found in late eighteenth- and early nineteenth-century cottage gardens.	407

22.4	Edinodunensis Tabulam.	408
22.5	Keeping gardens was a patriotic duty during the Wars.	411
22.6	Two-storeyed terraced houses at Meadowbank in Ladywell, Livingston, West Lothian, 1968.	412
22.7	Working in the school garden at Lilliesleaf, Roxburghshire during World War I.	414
23.1	A selection of cookery books published in Scotland.	422
23.2	Numerous traders walked the streets selling their wares.	423
23.3	This gas cooker of the 1890s, incorporating a tank for hot water, is recognisably a link between the coal-burning range and modern-day appliances.	428
23.4	The *batterie de cuisine* of the larger kitchen.	429
23.5	Tin openers came in about the same time as corned beef from Argentina and often took the form of a bull's head with the bull's tail constituting the handle.	430
23.6	The cooling unit of this 1927 refrigerator was so bulky that it sat prominently on the top of the unit.	431
23.7	The suggested layout for a breakfast table of the 1890s.	432
23.8	In a menu '*à la russe*' one type of food comprised a course in the style still familiar on formal occasions today.	433
24.1	Graph showing the decline in home births in Scotland in the twentieth century.	441
24.2	Living room prepared for a homebirth in the 1980s.	446
24.3	The christening party of Donald Ban MacNeill, Oronsay, Argyll, c 1930.	450
24.4	Homebirth.	452
24.5	Homebirth.	453
25.1	Helping with what may be pet lambs, Shetland, 1950s.	459
25.2	Edith and Andrew Inkster, Shetland, c 1907.	462
25.3	Andrew Inkster with his sailor suit on, Shetland, c 1911.	463
25.4	200-year-old pram, photographed in Stornoway, Lewis.	465
25.5	Coming home from minnowing, Perthshire, late 1940s.	470
25.6	Children Highland dancing at Banchory Games, Kincardineshire, c 1952.	472
26.1	'La Chaumière Ecossaise' (from d'Hardiville, *Souvenirs des Highlands*, 1832, Paris, 1835).	483
26.2	Andrew Minto, Coulterhaugh, Biggar, Lanarkshire, being cared for at home, c 1910.	484
26.3	Nurse Mary Morrison and Annie MacLeod at Mrs MacLeod's door.	488
26.4	'Milan' Day Centre for Older Asians, Pilrig Street, Edinburgh, 1992.	492

27.1	Amulets.	505
27.2	The Lee Penny from Lanarkshire.	506
27.3	The Toothache Stone, Islay.	507
27.4	Chalybeate Well, West Linton.	508
27.5	Healing stones kept at Killin, Perthshire.	509
28.1	Group family portraits sometimes depict family hierarchies as well as particular occasions.	516
28.2	Golden Wedding, Turriff, Aberdeenshire.	516
28.3	Cottage interior with figures, by William Shiels, 1841.	517
28.4	Husband and wife on an afternoon jaunt to Cove, near Cockburnspath, Berwickshire, c 1900.	523
28.5	Dressing up, Dalhousie, Midlothian, 1888.	530
28.6	Games, Corstorphine, Edinburgh.	535
28.7	Charles (Chic) Rooney and his father Charles Rooney senior (and dog), from Dundee, on excursion to the Carse of Gowrie, Perthshire/Angus, 1936/7.	536
29.1	Blackhouse, corrugated-iron dwelling and stone-and-lime-mortared house together.	545
29.2	An 'own-door' flat in a council block of the 1960s.	548
29.3	Late eighteenth-century Georgian marine villa with Victorian update, in c 1925.	549
29.4	The sale of furniture from Seafield House, Lamlash, Isle of Arran, on 26 May 1885 lists the furnishings of the middle class at that time.	550
29.5	The speculative bungalow.	550
29.6	Victorian suburban villa of 1887.	554
29.7	Melfort House, Glasgow, 'office' on the rear boundary wall.	555
30.1	People of One Fire.	571
30.2	Guisers, Iochdar, South Uist, c 1983.	572
31.1	Micky the budgie, with Charles Rooney senior, Annie Rooney and granddaughter Catherine, 1955, Wedderburn Street, Dundee.	589
31.2	Kim, with the NCR Wildfowling Club, c 1965.	590
31.3	Kim, retrieving pheasant in snow, c 1965.	590
31.4	Dundee NCR Wildfowling Club, c 1965.	591
31.5	Kim with the Rooney sisters, Dundee.	592
31.6	Robin Spark and his cat greeting each other, Bruntsfield, Edinburgh, late 1990s.	593
32.1	House flitting, Edinburgh, 1934.	600
32.2	Rent Strike photograph.	602
33.1	Remains of a circular shieling hut, Uig, Lewis, c 1987.	619

33.2	Enjoying a break and a 'piece' at the berry picking, Aberlemno, Angus, 1957.	620
33.3	Former fisher girl, Isa Ritchie, aged 91, outside her home in Whitehills, Banffshire, in May 1991.	622
33.4	Late nineteenth-century Pitmillie salmon fishers' bothy, Hillhead, by Boarhills, on the east coast of Fife, with its pantile roof intact.	625
33.5	Salmon fishers' bothy, Hillhead, by Boarhills, Fife.	626
33.6	Feeing Saturday, Arbroath High Street, Angus, in 1918.	630
33.7	Joe Tindal, born 1918, Kirkton of Monikie, Angus.	632
34.1	The frontage of Dalnacardoch Inn, Perthshire, in 1994, since restored.	638
34.2	Brewing magnate Lord Burton (of Bass) in 1903, then owner of the 25,000 acre Glenquoich deer forest whose shooting guests included the King and the Tsar of Russia.	639
34.3	An isolated, windowless bothy in Upper Glen Almond, Perthshire, left to decline to discourage poachers.	642
34.4	The well-known climber, writer and poet, Sydney Scroggie, pictured at home in 1997 prior to becoming an Honorary Doctor of Laws from the University of Dundee.	643
34.5	Interior of Glen Affric hostel, Inverness-shire, under natural light with Iain Smart sitting to the left of the fire, where a sack of coal had been generously left for hillwalkers and hostellers in winter.	645
34.6	One of the sea-side pleasures for the very young. A one-man band, 1930s.	649
35.1	HMP Edinburgh Gatehouse.	655
35.2	Prison population, 8 March 2002.	657
35.3	Exterior of Glenesk House Accommodation Hall, HMP Edinburgh.	660
35.4	Cell, HMP Edinburgh.	661
35.5	Residential Hall daily routines for remand and convicted prisoners, HM P Edinburgh.	662–3
35.6	'A' Hall west elevation, HMP Edinburgh.	663
35.7	A typical week's menu.	665
35.8	Interior of the chapel, HMP Edinburgh.	668
35.9	The Visitor Centre interior, HMP Edinburgh.	669
36.1	Day Hospice – a homely place. St Columba's Hospice, Edinburgh.	673
36.2	St Columba's Hospice exterior view.	674
37.1	East Park Home for Infirm Children.	685
37.2	'Feeding the Children'; children in a workhouse in London.	687

37.3	Aberlour Home.	690
37.4	Aberlour Home.	690
37.5	Quarrier's Children's Home.	691
37.6	Homelea and Paisley Home.	691
38.1	This beautiful building is McNabb's House (named after Dollar's founder), Dollar Academy, Clackmannanshire.	704
38.2	Chatting in a bedroom, promotional material, Dollar Academy, Clackmannanshire.	705
38.3	Female pupils and others at a barbeque, Dollar Academy, Clackmannanshire.	706
38.4	Male pupils at a formal social event, Dollar Academy, Clackmannanshire.	706
39.1	Tenement in South Edinburgh, near the University of Edinburgh.	721
39.2	Livingroom of a shared flat.	722
39.3	Flatmates enjoying a barbeque in the drying green of a tenement, Edinburgh, 1990s.	729
40.1	A treasured icon of the Findhorn commune is the caravan in which Peter and Eileen Caddy arrived with their family in 1962.	739
40.2	'Today, the Community has diversified into more than 30 different businesses and initiatives, and provides a valuable insight into sustainable living. It is a place where everyday life is transformed by living, working and learning with spiritual values.'	740
40.3	The former toilet block now houses one of the thriving 'alternative' businesses which have been set up within the Findhorn commune's precincts.	741
40.4	One of the few communal areas not accessible to the general public, the Community Centre, is the focal point for casual relaxation within the Findhorn precinct.	742
40.5	Recently constructed accommodation for guests attending courses and conferences at the Findhorn commune is ecologically friendly in design, including turf-roofed houses for super-efficient insulation.	743
40.6	One from the cluster of whisky-barrel houses which are a distinctive feature of the Findhorn architectural landscape.	743
40.7	One of the several 'sanctuary' buildings within the Findhorn commune, which is sited in a peaceful garden.	744
41.1	St Andrews priory church and precincts, Fife, as they might have been about 1550, drawn by Alan Sorrel in 1965.	751
41.2	A typical monastic plan, based on Dundrennan (Cistercian), Kirkcudbrightshire.	753
41.3	Bird's-eye view, St Hugh's Charterhouse, Parkminster, Sussex.	754

41.4	Outline plan of Fort Augustus, Inverness-shire.	755
41.5	Fort Augustus Abbey, Inverness-shire, aerial view.	755
41.6	Fort Augustus Abbey, Inverness-shire, ground plan	756
42.1	Harbour at Anstruther and Cellardyke, 1880s.	767
42.2	Net mending, Ferryden, Angus.	768
42.3	Two women (at the left) mend, while the men tie the individual herring nets to the common 'buss rope' to form a 'fleet' ready to be set at sea. St Monans, Fife.	769
43.1	Increase in single-adult and single-pensioner households, Scotland, 1971–2001.	778
43.2	Single-adult and single-pensioner household type by local authority grouping.	779
43.3	Mrs Nettie (Annette) Kellock in her Rodney Street home, Edinburgh, 1973.	781
44.1	Changes in Traveller dwellings over the years.	792
44.2	Travellers camped illegally on a former bing in Bellshill, Lanarkshire in 1969.	793
44.3	Interior of a Traveller family's large bow tent in Kintyre, Argyll in 1970.	793
44.4	Continuity of traditional construction styles; paired bender tents.	794
44.5	A horse-drawn Traveller caravan near Wormit, Fife in 1953.	796
44.6	Interior of a comfortable modern Traveller trailer in Lanarkshire in 1969.	797
44.7	Children form an important part of Traveller life, as do horses. Camping on an old farm in Lanarkshire in 1970.	798
44.8	A new council site for Travelling People in Scotland in the late 1990s.	802
44.9	An older-style Show family's trailer in Scotland.	808
44.10	All modern comforts. A modern trailer at a Scottish fair in the late 1990s.	809
44.11	The exterior of a Show family's luxurious trailer at a Scottish fair in the present day.	810
45.1	A large mobile home site on the edge of a Scottish town in 1976.	819
45.2	Modern twin-unit mobile homes on a Scottish site in 1976.	820
45.3	Part of a small Scottish mobile-home site in 1976.	820
45.4	A large mobile-home site developed for an incoming workforce in northern Scotland in 1976.	821
45.5	A medium-sized hut site in the Scottish Borders in 1999.	825
45.6	Assorted huts on a large, long-established site in Angus in 1999.	825
45.7	Huts on a small site in Stirlingshire in 1999.	826
45.8	Huts on a small site near Loch Lomond in 1999.	826

46.1	Hospitalfield, Arbroath, Angus.	836
46.2	The Retreat, Berwickshire.	838
46.3	Dalkeith's Water Tower, Midlothian.	840
46.4	Crauchie Steading, East Lothian.	842
46.5	Alderston Coach House, Haddington, East Lothian.	843
46.6	The Old School, Dean Village, Edinburgh.	844
46.7	The Pineapple, Dunmore, Stirlingshire.	845
47.1	Woman in bed in a dormitory, common lodging house, Glasgow.	853
47.2	Upper and lower bunk, common lodging house, Glasgow.	854
47.3	Men in a common lodging house dining room, Glasgow.	855
47.4	Number of Households Applying to Local Authorities under the Homelessness Legislation.	859

List of Contributors

JEAN AITCHISON
Freelance writer and volunteer worker, Ayrshire Archives.

HANNAH AVIS
Lecturer, Geography, University of Edinburgh.

KATHY BUCKNER
Senior Lecturer, School of Computing, Napier University, Edinburgh.

JOHN BURNETT
Curator of Scottish Ethnology, National Museums of Scotland, Edinburgh.

SALLY A BUTLER
Lecturer, School of Landscape Architecture, Edinburgh College of Art.

PROFESSOR REGINALD BYRON
Professor of Anthropology and Sociology, University of Wales, Swansea.

DR ELIZABETH CUMMING
Freelance art historian, Edinburgh.

RIGHT REV MARK DILWORTH OSB†
Formerly Abbot, Fort Augustus, Inverness-shire.

PROFESSOR ALEXANDER FENTON
Director, European Ethnological Research Centre, Edinburgh.

DR JOE FRANCIS
Lecturer, Social Work, University of Edinburgh.

HUGH GENTLEMAN
Independent research consultant and formerly of the Central Research Unit of The Scottish Office, Edinburgh.

DR HEATHER HOLMES
The Scottish Executive and freelance writer, West Lothian.

PROFESSOR LYNN JAMIESON
Professor of Sociology, University of Edinburgh and Co-Director, Centre for Research on Families and Relationships, Edinburgh.

DAVID JONES
Lecturer, School of Art History, University of St Andrews and Editor of *Regional Furniture*.

PROFESSOR PETER A KEMP
Director, Social Policy Research Unit, University of York.

DR ROGER LEITCH
Freelance writer, Dundee.

ROD MacCOWAN
H M Deputy Chief Inspector of Prisons and formerly Governor, H M Prison, Edinburgh.

DR DEBORAH MAYS
Principal Inspector, Head of Listing, Historic Scotland, Edinburgh.

DR JOYCE MILLER
Witchcraft Survey, Scottish History, University of Edinburgh.

DR IAN MORRIS
Freelance writer, Edinburgh, and formerly H M Chief Inspector of Schools.

REV DR DEREK B MURRAY
Retired Baptist minister and formerly Chaplain, St Columba's Hospice, Edinburgh.

DR MILES K OGLETHORPE
Industrial Survey, Royal Commission on the Ancient and Historical Monuments of Scotland, Edinburgh.

SINÉAD POWER
Research student, Geography, University of Edinburgh.

REV DAVID REID
Minister and freelance historian, Fife.

DR LINDSAY REID
Freelance medical historian and research midwife, Fife.

DR DAVID RICHES
Senior Lecturer, Social Anthropology, University of St Andrews.

UNA A ROBERTSON
Freelance historian, Edinburgh.

CATHERINE SMITH
Archaeozoologist, SUAT (Scottish Urban Archaeological Trust) Ltd, Perth.

GEOFFREY STELL
Formerly Head of Architecture, Royal Commission on the Ancient and Historical Monuments of Scotland, Edinburgh.

DR JAMES STEWART
Research Fellow, Research Centre for Social Sciences, University of Edinburgh.

SUSAN STORRIER
Formerly Deputy Director, European Ethnological Research Centre, Edinburgh.

SABINA STRACHAN
Assistant Inspector of Ancient Monuments, Historic Scotland, Edinburgh.

NAOMI E A TARRANT
Formerly Curator of Costume and Textiles, National Museums of Scotland, Edinburgh.

DR GARY J WEST
Lecturer in Scottish Ethnology, Celtic and Scottish Studies, University of Edinburgh.

PROFESSOR CHRISTOPHER A WHATLEY
Bonar Chair of Modern History, University of Dundee.

CRISSIE W H WHITE
Senior Research Fellow, School of Design, Glasgow School of Art.

† = deceased contributor.

Foreword

The volume on *The Scottish Home*, edited by Annette Carruthers and published by the National Museums of Scotland in 1996, lavishly illustrated in black and white and in colour, is in many ways a precursor to this present volume. It might have seemed that it provided the last word on the subject, and that there was little further room for more on 'Scotland's Domestic Life'. Here, then, was a problem. A comparable volume had already been planned for the *Compendium* series. Could we now find enough material to fill it, without duplication?

In a partial answer to this seeming difficulty, in a review of *The Scottish Home* in *Folk Life* 35 (1996–97), 102, I noted that: 'In general the authors show themselves to be more at home in elucidating the details of the homes of the middle and higher classes of society, though country living in small houses is not ignored'. Yet the great bulk of the population lived in more humble circumstances, in which function rather than decoration and display was the primary factor, in respect of use and sharing of space within the home, the forms and nature of furniture and furnishings, tools and equipment, and the conduct of social life and communal living in general. In the planning of this present substantial volume, Susan Storrier has assembled a strong supporting cast of authors who have more than made good the deficiency and have proved that there was no shortage of material, including a good deal that is new.

Nevertheless it is not simply a question of élite versus everyday. The first person to receive a PhD in folklore in the United States of America, at Indiana University, was Warren E Roberts, who developed a special expertise in the study of buildings, furniture, crafts and associated tools. In his book, *Viewpoints on Folklife. Looking at the Overlooked* (Ann Arbor, London, 1988), he notes that it 'is sad but true that the great mass of humanity is simply overlooked when most people write or talk about the past and what life was like then' (xiii). He therefore sees it as his task to study the 'old, traditional way of life of the great mass of people who lived before the Industrial Revolution' (xiv), and in his studies of material culture, he excludes from consideration, inter alia, the twentieth-century industrial era and the élite culture of the pre-industrial era (16–17). Such exclusions do not now apply in ethnological studies, which take – or ought to take – into account the processes of historical change and social transformation that have been wrought over time in the living patterns of humanity at all levels. Annette Carruthers' *Scottish Home*, therefore, is to be seen at one level as an essential complement to *Scotland's Domestic Life*. Equally, other volumes in this *Compendium* series deal with aspects of the industrial world, and with the spiritual life and community organisation of the people(s) of Scotland. All of these are intended to complement each other, to provide as holistic a picture as it is possible to produce of a nation at the early years of a new millennium.

In the planning and editing of this volume, Susan Storrier has established a wide context for the study of Scotland's domestic life. The material culture of the home is complemented by articles on everyday work and leisure activities, whether individual or communal, on the administrative framework of provision for matters such as health, discipline and education, and, importantly, on the welfare problems of homelessness, and of living alone. There are answers to many questions but many questions are also identified here that point the way to future research. The care taken by the editor to cover or draw attention to the wide implications of what constitutes domestic life from birth onwards, and its influence on the attitudes of individuals as they grow up and join wider worlds, often nowadays in different countries, makes this an important book which should be widely welcomed.

Alexander Fenton

Abbreviations

CBA	Council for British Archaeology
DOST	*Dictionary of the Older Scottish Tongue*
EERC	European Ethnological Research Centre
NAS	National Archives of Scotland
NGS	National Galleries of Scotland
NHS	National Health Service
NLS	National Library of Scotland
NMRS	National Monuments Record of Scotland
NMS	National Museums of Scotland
NSA	*New (Second) Statistical Account of Scotland, 1845*
NTS	National Trust for Scotland
OSA	*The Statistical Account of Scotland, 1791–99*
PP	Parliamentary Papers
PSAS	*Proceedings of the Society of Antiquaries of Scotland*
RCAHMS	Royal Commission on the Ancient and Historical Monuments of Scotland
ROSC	*Review of Scottish Culture*
SA	Sound archive
SLA	Scottish Life Archive, NMS
SND	*Scottish National Dictionary*
SRO	Scottish Record Office
SS	*Scottish Studies*
SSS	School of Scottish Studies, University of Edinburgh
SVBWG	Scottish Vernacular Buildings Working Group
TCM	Town Council Minutes
TGSI	*Transactions of the Gaelic Society of Inverness*

Glossary

acanthus palmette, decorative, often freestanding carving in wood or stone resembling the foliage of the acanthus plant. Used by the Greeks and Romans and first revived in the Renaissance.
Act of Uniformity, Act of 1662 in which clergymen who refused to conform to the rules laid down in the *New Prayer Book* and to take an oath of loyalty to Church and king were excluded from holding any place in the Church of England.
Aga, solid-fuel cooker invented in the 1920s by physicist Nils Gustav Dalen. It is thermostatically controlled and provides hot water that can be circulated around the dwelling as central heating.
allopathy, method of treating disease by inducing a condition different from the cause of the disease. On this modern Western medical treatment is based. It contrasts with homeopathy, which is the method of treating disease by using a small amount of drug in order to produce symptoms similar to those of the disease being treated.
arles, sum of money paid to a servant upon engagement to bind the contract.
Art Deco, an inter-War style of architecture which takes its name from the *Exposition Internationale des Arts Décoratifs et Industriels Modernes* held in Paris in 1925.
autopiano, automated instrument which produced music by means of holes punched in thick card, linen or paper, c 1909.
bannock, round flat cake, usually of oatmeal or barley or pease meal, baked on a girdle.
barmkin, high-walled courtyard of a **tower house**, for the protection of cattle.
barricade, older-style winter tent used by Travelling People, characterised by having sleeping quarters off a central dome-roofed living area heated by an open fire built on the ground.
batten, in building construction, a strip, bar or plank of timber normally less than 4 inches (10 cm) in thickness.
bejant, university freshman.
ben, ben end, the 'room', as opposed to the kitchen, in a two-roomed house.
Berlin woolwork, form of canvas work using coloured wools and a squared design chart or, by the late Victorian period, a printed canvas.

billy, tin can with wire handle for making a brew of tea, etc.
blacklead, form of graphite or plumbago which was useful in the domestic context for polishing iron grates and similar surfaces.
bonnet laird, minor landowner who farmed his own land.
booth, in Shetland, a one or two-storeyed building for the storage of fishing goods and gear, often including mercantile accommodation. If multifunctional, separate entrances were the norm.
borax, naturally occurring mineral salt which, for domestic purposes, is in the form of white powder or crystals. It has both medicinal and water-softening properties.
brander, a) gridded frame on which items can be laid to dry above a fire b) gridiron.
braxy, bacterial disease of sheep.
bretasche, timber platform of limited length or a small, isolated covered chamber projecting from a castle wall, originally for defence and later for purely domestic purposes.
broch, late prehistoric structure (predominantly seventh century BC to second century AD), found mainly in the Northern Isles, the Western Isles, Caithness and Sutherland, consisting of a large round tower and hollow stone-built walls.
bull's-eye design, abstract design of concentric lines.
burgh, settlement established from the early medieval period as either a Royal Burgh or a Burgh of Barony and granted trading privileges by the Crown. In the later medieval period burghs were given other responsibilities. After 1832 some had elected councils and new Parliamentary burghs were created.
bylie up, to 'drum up' or make a wayside brew, usually of tea.

camphor, vegetable oil obtained from Camphora officinarum. Best known as a moth repellent but it has had other domestic uses and also has medicinal properties.
Central Midwives' Board for Scotland, statutory body for the regulation of midwives in Scotland from 1916 to 1983.
chalmer, chamber, room.
chapman, itinerant petty merchant or dealer.
chipcarving, method of woodcarving in which the wood is cut away in layers to expose an incised pattern.
cloot, clout, cloth, patch, rag or shred of fabric.
cran, a) iron support over a grate for pots b) tripod or support for a pot over hot embers c) see **swee**.
crannog, lake dwelling, usually defined as an artificial islet or an augmented natural islet in a loch.
credenza, serving table and sideboard frequently ending in curved cabinets designed for plate display.
crinothene, olefin synthetic fabric which imitates the stiffness of crinoline and is suited to pleating, folding and embroidery.

critical mass, the number of users necessary in order for use of a technology to spread within a particular market or community, dependent on factors such as economies of scale in production and distribution, development of uses and the network of connected users.

cross stitch, stitch worked on canvas or another type of fabric where the thread can be counted. A St Andrew's cross (x), as opposed to a St George's cross (+).

cruisie, simple open lamp of two iron boat-shaped shells, usually with a wick of the pith of rushes.

curer, individual whose business focused on the organisation of labour relating to the processing and preservation of fish by salting, pickling or smoking.

deck access, access to front doors of flats in high-rise blocks via long balconies from open stairwells.

deer watcher, usually a retired deerstalker living in bothy accommodation and engaged by an estate to stop the movement of deer onto other land.

digital divide, the supposed social and economic division between those people with access to information and communication technologies and the skills to make use of them, and those without, based on the theory that full participation in society and economy is dependent on use of these technologies.

dixhuitième, general term for the eighteenth-century revivalist styles popular in the Edwardian period.

dojon, the main residential tower of the castle, usually circular in plan.

droving, the seventeenth- to nineteenth-century custom of walking cattle raised in the Highlands and Islands to fairs and markets further south to supply urban centres in Scotland and England.

drugget, woven and felted coarse woollen fabric.

engobe, intermediate layer between the body of a ceramic article and the glaze, often giving the item the appearance of a superior clay, or providing a better base for decoration. Examples include a white engobe on a piece of earthenware or fireclay sanitary ware.

feeing fair, fair or market held at or near **Whitsunday** and **Martinmas**, where farmers engaged servants for the coming term.

Five Finger, see **Maw**.

florists' societies, groups of people with an interest in the breeding and showing of flowers who come together to share enthusiasms and compete in shows.

forestair, outside staircase on the front of a house giving access to the principal apartments.

fuller's earth, type of clay used in the manufacture of cloth. It is highly efficient at absorbing water, grease, oil and colouring material. It was used in the home to remove grease spots from fabrics, marble and other surfaces. It could also form part of a mixture to restore whiteness to linen.

garden city, idealised form of social living conceived by E Howard at the turn of the twentieth century. Garden suburbs were planned and developed first in England, before the concept spread elsewhere.

gelly, traditional form of Travellers' tent made of boughs and often covered with sailcloth or tarpaulin. Characterised by an internal tank fire and pipe through which the smoke escaped.

ghillie, estate employee who attends a sportsman concerned with deerstalking or taking salmon by reel.

glengarry, Highlanders' cap of thick milled wool, generally going to a point in front, with ribbons dangling behind.

Green Lady, Glasgow Municipal Midwife, so called because of the dark green uniform worn.

hanging lum, hingin lum, a wooden canopy chimney fixed against a gable wall.

heckle, to comb raw flax.

heids an thraws, arrangement of people sharing a bed whereby they lie in alternate pattern up and down the bed.

hillfort, Bronze Age or Iron Age settlement positioned upon a strategic summit and consisting of individual dwellings within ramparts and ditches.

housing pathways, (also referred to as housing careers or housing histories) a person's housing experience over the life course. As well as relating to the sequence of dwellings which an individual occupies, and to changes in tenure, it also embraces the changing meanings which a person attaches to his or her housing.

howdie, a) midwife b) untrained sick nurse c) woman who laid out the dead.

howf, a) meeting place, frequently a public house, sometimes implying a place of disrepute b) rough shelter or refuge, such as a cave.

hurlie bed, bed on wheels, normally stored below another bed.

ice house, underground chamber for the storage of ice formed during the winter and also certain foodstuffs.

interaction design, designing the user experience in interactive products including functionality, information content and access.

kebbuck heel, hard end piece of a cheese.

kie, cattle.

Kilner jar, patent glass jar in which fruit, etc, could be preserved.

kinetoscope, popular device of the 1890s which showed a series of consecutive images made to move past a viewing aperture when a handle was turned. Sets of images were available to purchase.

kist, chest, often used for clothes or meal.

kisting, **coffining**, **chesting**, the ceremonial procedure of placing a body into a coffin.
knot garden, originally an element of a garden planted in the form of the twisted pattern of a knot, made from miniature hedges. It expressed continuity or infinity in the garden. By the mid seventeenth century 'knot garden' had come to mean almost any box-edged garden of intricate patterns and flowers.
landscape mode, arrangement of window panes with a horizontal emphasis.
lighter, large open boat designed for ease of loading and unloading.
longhouse, single-storeyed building for human and livestock cohabitation. Also known as a byre dwelling.
lum, chimney.

macramé, knotted threadwork, normally using heavy-duty cotton or wool.
magic lantern, early form (1880s) of photographic projector. Sets of transparencies telling a story could be purchased.
Martinmas, 11 November. Originally the feast of St Martin de Tours. On this date **feeing fairs** were held.
Maw or **Five Finger**, card game in which players are each dealt five cards, and then play to win tricks. One of its oddities is that in black suits the lowest cards count highest.
meat, food in general, including butcher meat.
meerschaum, fine light, whitish clay, sometimes used for tobacco pipes.
ménage, means of saving in which a small sum paid regularly by an individual creates a fund which can be withdrawn for a specific purpose, such as Christmas.
merino, a) fine dress fabric, originally of merino (Spanish breed of sheep) wool b) knitted goods of merino wool, latterly mixed with cotton.
messan, lap dog.
Modern, **Modernism**, a style of architecture associated with the Modern Movement (1930s–1950s) which is characterised by a strictly rational, functional and non-traditional approach to design, construction and materials.
moiré wallpaper, heavy wallpaper manufactured by pressing silk or synthetic fabrics between engraved cylinders to emboss a waved or watered-effect design.
monthly nurse, midwife or nurse employed privately for a month to attend to a mother in labour and postnatally, usually under the supervision of a general practitioner.

narrowboat, slender freight vessel designed for canal and river use with scant accommodation.
nuclear fort, Early Historic defensive hilltop settlement consisting of a series of terraces within which different activities and hierarchies may have prevailed.

orraman, farm hand who undertook odd jobs and sometimes drove a pair of horses.
outerworks, series of ramparts and ditches, gateways and passages surrounding a settlement. They tend to be defensive in their form and can include relatively elaborate devices such as wooden stakes.
overmantel, ornamental stand in wood or (gilded or painted) plaster, often framing a mirror, placed immediately over a mantelpiece to complement its form.
ox gall, bile, the contents of the gallbladder, of oxen or bullocks. Used in a diluted form to clean carpets and mixed with lye to scrub floorboards in flea-infested dwellings. It was also used to clean cloth without removing the colour.

palliasse, straw-filled mattress.
pend, open-ended passageway. In urban settings a pend would lead from the street to the 'backlands', the rear properties, land or garden behind the front house/tenement.
petit point, diagonal stitch worked on canvas where the thread can be counted. One half (/ or \) of **cross stitch**.
petticoat tails, enriched shortbread made into a circular shape and divided into eight triangles.
philabeg, **filibeg** [Gaelic **feileadh beag**], kilt.
phonograph, predecessor of the gramophone, invented by T A Edison in 1877. Used for recording and reproducing music and other forms of sound.
Picturesque, visually pleasing man-adapted landscape, as opposed to wilderness where man has had little or no hand in its evolution.
piece, a) portable snack. For example, snack lunch b) snack to eat between meals, as in 'play piece'.
point block, **Modernist** multi-storeyed tower block of compact plan.
pokerwork, the art of decorative burning to create patterns in the tops or legs of wooden tables, chairs, stools and flatware.
polyphon, music box, a means of reproducing music not dissimilar to an **autopiano** but functioning by the agency of holes cut into brass discs.
ponyman, person, usually an estate worker, in charge of the pony or ponies during deerstalking activities.
praxinoscope, method of showing pictures, similar to the **kinetoscope**, but in this instance the images are drawn on mirrors.

quadrivium, the four branches of mathematics in medieval education; arithmetic, geometry, astronomy and music.
quarter day, the date which marks one of the quarters of the 'Celtic' year; 1 February (Imbolc/St Bride's Day), 1 May (Bealtainn/Beltane), 1 August (Lunasdal/Lammas), 1 November (Samhainn).

quarterer, farm worker who did miscellaneous jobs and errands and resided on the farm.
quine, unmarried woman or girl.
quoin, dressed corner stone.

raffia, the leaf bast of the Raphia palm tree, used as a strong alternative to straw for woven or plaited domestic artefacts.
real/royal tennis, tennis played in a closed court in which the ball can be bounced off the walls and other surfaces.
repoussé, relief work on metal materials produced by hammering or pressing on the reverse.
ride, wide avenue cut through woodlands to allow hunting parties to penetrate into the forest.
ring-ditch house, Iron Age roundhouse built of a base of stones with a timber superstructure and an outer ditch, found in the eastern Lowlands and elsewhere.
rivlin, a shoe of untanned cow hide or sheep skin worn hair side out, formerly widespread in Scotland and known from the early fifteenth century.
roller organ, device for producing music automatically by hand cranking a drum pierced with holes.
round-about fireside, circular **hingin lum**, hanging from the roof.
roundhouse, circular-plan house, often with thick dry-stone walls, of the Iron Age, for example a broch or wheelhouse.
rusticated, classical form of dressing masonry to create v-jointed or channelled joints.

sarking, boards laid on rafters as a base for the roof covering.
sgeulachd [Gaelic], tradition, legendary lore.
shift, a) change of clothing b) item of clothing that can be readily changed, especially underclothes c) women's undergarment.
sightline, vista to and from a specific location.
single end, home within a tenement of single-room proportions.
slab block, **Modernist** multi-storeyed tower block of elongated plan.
smoor, to cover or smother a fire at night so that it is still smouldering in the morning.
SO-grams, conversational prop used by informants to describe in a graphical way their social networks. Level of significance of acquaintances, frequency of contact, and type of media used in making contact are recorded.
souterrain, subterranean stone-lined structures dating from a number of periods found in various parts of Scotland. Souterrains range in size and plan form and may have sometimes functioned as grain stores. They are unlikely to represent a single phenomenon.
sowans, husks or seeds of oats, together with some fine meal, steeped in water for about a week until the mixture turns sour, then strained and the husks thoroughly squeezed to extract all the meal, when the

jelly-like liquor is left for a further period to ferment and separate, the solid glutinous matter which sinks to the bottom being *sowans*, and the liquid, *swats*.

springhouse, large byre, particular to the West Highlands where the cattle trade was intensive, located outside the 'baile' (village) and used in spring until the pasture improved.

stereoscope, device which involved two slightly different images set a few inches apart which were viewed simultaneously. It achieved its greatest popularity in the 1850s.

streakin boord, strykin beuird, streikin/straiking board, board on which a body was laid before **coffining**.

stucco, relief plaster decoration often modelled and moulded in lime and aggregate mixes, sometimes including marble dust.

swee, swey, pivoting iron bar built into one corner of a hearth for hanging pots, kettles, etc.

tangled-threid design, symmetrical linear pattern encapsulating the energy of a tangled thread.

technology lifecycle, process of development, adoption, consumer penetration and decline of a technology.

telework, work undertaken away from the normal workplace, often in the home, supported by information and communication technologies.

The Glorious Twelfth, the first day of the grouse-shooting season, usually the twelfth of August.

time shifting, recording television programmes for watching at another time, often later in the evening, rather than for keeping as a record or for repeat watching.

tower house, laird's house of the later medieval period using defensible architectural features from the preceding castle era.

trivium, the group of liberal arts first studied in medieval schools; grammar, rhetoric and logic.

trou-madame, game played on a specially made board in which the object is to roll balls into holes which give different scores.

truck system, system whereby workers are paid for their labour in goods whose cost is controlled by their employer.

tumbler lock, wooden lock with a simple mechanism which can be opened using a wooden key reminiscent of a lightly toothed comb.

vermiculated, literally 'worm eaten', a classical form of dressing masonry with the appearance of worm tracks.

wally, general term for fired clay, glazed or unglazed, as in wally close, wally dug (dog), wally teeth.

water caddy, person employed to carry water for others. Their services were redundant after a supply of water was piped into every dwelling.

wheelhouse, type of roundhouse with the ground plan featuring subdivisions radiating from the centre in a spoke-like formation.

Whitsunday, Whit Sunday, 28 May. On this date **feeing fairs** were held and many contracts and leases came to an end.

wright, wricht, vricht, carpenter, joiner.

Introduction

SUSAN STORRIER

In academic as well as popular discourse domestic life often collocates with 'house', 'hearth', and 'family', yet, while the importance of these terms cannot be denied, the story of home life – in any nation – is in reality a far wider one, in terms of the structures used as dwellings as well as the arrangements and households found within them. It is an impression of this variety as it has occurred in Scotland past and present, at a range of social levels and from the point of view of individuals of differing ages and cultural backgrounds which it is hoped this volume will provide.

In comparison with some other areas of Scottish experience, there has been little published on domestic life. There are notable exceptions of course,[1] but much remains to be both investigated and documented. This might seem curious. The natural world and most aspects of human existence have an impact on the dwelling, directly or otherwise. Work, war, politics, economics, community and all the other major features of the public world interact in some way with our homes and home life, with the dwelling providing a frame in which these vast entities can be studied in juxtaposition and perhaps on a more manageable scale. But dwellings and domestic life have profound importance beyond translation of the wider world. Most of us spend the greater part of our lives in our homes, whether at rest or active. This is true even for many people who work or study for long hours beyond the home. For others – for example those who work primarily in the home, the very young, many older people who are retired and many of those who are unemployed or in some way infirm (together making up a very substantial part of the population) – the dwelling can have a still greater role to play in life. It can be fairly said, then, that to study domestic life is to explore both the centre and the larger part of human experience.

Given this importance, why should domestic life have been relatively neglected as a topic for study? It is true that certain areas have received more substantial academic attention in Scotland, although in each of these there remains plenty of work still to be done; the furnishing and decoration of grand houses, for example, and at the other end of the spectrum, domestic life in one- or two-roomed urban tenements and some types of small rural home. Yet many other important areas have been essentially overlooked as a focus for detailed academic investigation, while it is the case that researchers and writers in general have more often sought to recreate the past than investigate the present.[2] Non-academic investigations of Scottish domestic life tend to follow a similar pattern. Some topics receive relatively high exposure – food, for example – while other themes are largely passed over and there is at times a tendency towards maintaining comforting images of the domestic past rather than reporting on the complexity and the shade and light of both past and present.

It may be that the patchy and oftentimes inadequate study of Scottish domestic life has been in part a reaction to notions of status. There is a long history of viewing the domestic arena as inferior to the public and dwellings have often been associated, correctly or not, with groups somehow regarded at times as being of lower status – women, children, older people, the unwell and the disabled. Thus home life may on occasion be considered a less prestigious focus of academic endeavour.

But there is more. Where there is a desire to research there is often not the means to do so, at least in a straightforward manner. The evidence for domestic life is limited.[3] For some topics there is scarcely any source material available, while for some others a picture must be painstakingly created from oblique fragments of information. Not only that, but some phenomena associated with the home, despite being reasonably well documented in particular instances, are difficult to define and discuss in more general or abstract terms, while even the intrinsically straightforward may lie in complex relation to other matters. The biggest challenge, though, comes from the sheer size of the story; to describe accurately and analyse the activities in even one household during, say, twenty-four hours – the actors, their experiences and roles, the processes contained within the home and those making connections with the wider world, and the manifestations of these, covert and overt, and much more – would make for a very substantial exercise indeed. Infinitely more is involved in an examination of the variety of Scottish home life.

Given this, it is clear that no single volume can pretend to be comprehensive in this context. This collection aspires to indicate the range of Scottish domestic experience. Even so, there has been much that has had to be omitted, owing not only to the constraints of the volume size and the lack of evidence and research in some areas, but also the demands placed upon the time of the experts in certain fields. It is for such reasons that it was not possible to include, for example, chapters on the themes of leaving and moving home and sexuality in the home, although the topics are at times touched on in other contributions.[4] Of those topics dealt with as chapters, some are presented as broad overviews while others form more specific studies, depending primarily on the evidence readily available.

Scotland's Domestic Life aims not only to inform and act as both an ethnological text and an ethnological tool, but to stimulate thought and further research, and where possible authors have stated potentially fruitful areas of investigation. The great advantage of studying domestic life is that it is something to which everyone can relate, no matter what his or her background. Most of us are in some way already ethnologists of the home. It is also hoped that this book will complement other recent works on domestic themes. It was felt that there remained a need for an up-to-date book shaped primarily around matters other than the physical divisions of the home and for this reason the volume, in its first section, contains chapters focused on major concepts and processes common to all or most homes, while the second section examines the interaction of these phenomena in a selection of less-common household and dwelling types. A few of the topics are dealt with for virtually the first time in an academic setting, while some others have not hitherto been presented in an overtly ethnological context. The volume, by its more exploratory nature, thus differs somewhat from many others in the *Compendium* series.[5]

The 36 authors who contributed to this book deserve wholehearted thanks for making time, often against a background of very much more pressing concerns, to share their expertise and frame it in a broadly ethnological perspective. The editorial challenge was to further this process and shape material originating from many disciplines and viewpoints and of different styles into an ethnological whole. It is the writers who are responsible for the success of this, by responding so patiently and kindly to requests for revisions and additional material.

Also to be very much thanked are the staff of the Scottish Life Archive of the National Museums of Scotland, Mark Mulhern of the European Ethnological Research Centre, the Series Editor, Professor Alexander Fenton, and the readers: Fionna Ashmore, Dr Angus Bancroft, John Beech, Hugh Cheape, Professor Edward J Cowan, John Crompton, George Dalgleish, Dr Ruth Emond, Dr Miles Glendinning, John G Harrison, Stephen Jackson, Dorothy Kidd, David Lynn, Iseabail Macleod, Ruth Madigan, Dr Morrice McCrae, Mary McGeady, R Ross Noble, Professor Lindsay Paterson, Dr Madhu Satsangi, Helena Scott, Dr Kirsten Stalker, Dr Steven Sutcliffe, Dr Kenneth Veitch, Dr Jeremy Weston.

Thanks are due also to those who offered feedback and suggestions, and those, too many to be listed but all very much appreciated, who selflessly helped in the search for authors.

NOTES

1. One of the most significant books of recent years is Carruthers, A ed. *The Scottish Home*, Edinburgh, 1996, on which this volume is intended to build.
2. The paucity of research in some areas is currently being addressed. The Centre for Research on Families and Relationships, Edinburgh, for example, is active in promoting and disseminating research into many hitherto unexplored or under-researched topics relating to contemporary Scottish domestic life. <www.crfr.ac.uk>
3. See Carruthers, A. Studying the Scottish home. In Carruthers, 1996, 11–36, for a detailed discussion of the available evidence.
4. For ease of access to material the index is supplemented by cross referencing between certain chapters.
5. Readers may wish to extend and frame their reading of domestic matters by consulting the other volumes of the *Compendium* series, of which Volumes Three, Seven and Nine (*Scotland's Buildings*; *Craft and Service: The Working Life of the Scots*; *The Individual and Community Life in Scotland*) have especially close connection.

PART ONE

•

Themes in Scottish Domestic Life

1 Shelter from the Elements

GEOFFREY STELL

INTRODUCTION

Physically and culturally Scotland is a land of contrasts and changeable moods. The background to human settlement is a landscape which ranges from flat or gently rolling and fertile lowlands to rugged, exposed and relatively barren Highlands and Islands. This contrast is reflected in a pattern of human settlement which likewise ranges between densely populated towns and cities and dispersed and isolated rural communities. Lying as it does in the track both of cyclonic depressions and of the more benign Gulf Stream flowing across the North Atlantic, the country is subject to a changeable climate which by turns may be either wet, windy and cold, or dry, mild and warm, although it is one which is generally temperate and only rarely subject to the seasonal extremes of cold and heat experienced in continental Europe.[1] Particularly along the western seaboard of Scotland settings for habitation may thus vary widely and sometimes dramatically – often within the same locality – from those estate and urban centres which are well sheltered and almost sub-tropical to those houses and cottages – past and present – which have clung to the less hospitable slopes of hills and glens or huddled around exposed coastlines (Fig. 1.1). Under pressure of population expansion in the eighteenth and nineteenth centuries, life on these marginal lands was the harsh lot of many, perhaps most, rural tenants and sub-tenants who, like their urban counterparts, had little or no choice in the location of their homes and workplaces.

An environment of such variation has long demanded, and given rise to, a corresponding variety of forms and methods of domestic building. Whatever their status or period, if houses were to provide effective shelter and to have a reasonable chance of survival they would have to be located, designed or adapted in a manner which enabled them to cope with ambient conditions. Traditionally, such structures would make best use of whatever materials and resources were available and would also require regular maintenance. With an annual rainfall and relative humidity which are high by continental-European standards, Scotland's weather has been a factor which designers, builders and those responsible for housing maintenance have ignored or underestimated at their peril. On the European continent the main technical challenge in modern times has generally been to provide insulation against intense but relatively dry cold and heat. In Scotland, as in

Figure 1.1 Northern pier, Portpatrick Harbour, Wigtownshire, 1985. Remains of the outer harbour redesigned by John Rennie (1814–18) and left unfinished in the aftermath of a ferocious storm in 1839. Crown Copyright: RCAHMS.

England, Wales and Ireland, the main need has been to counter problems of penetrating and rising damp. Alongside problems of overcrowding and lack of sanitation, all-pervading dampness was one of the major physical blights afflicting the notorious slums of nineteenth- and early twentieth-century urban Scotland. Whatever its causes – whether attributable to inadequately designed roofs, drains and structural ventilation systems, inappropriate and porous materials, or simply poor maintenance – the critical need to provide more effective means of water shedding, drainage and damp proofing has underlain and informed many significant advances in the modern Scottish house-building industry.

In an age which is acutely aware of the effects of global warming, possible links between climate history and building location and design clearly have a special fascination. Illustrations of possible connections between the physical environment and habitation appear to be readily found, for example, among those rural house types, of semi-subterranean or low, ground-hugging form, which are aerodynamically and materially suited to the windswept conditions which have long prevailed in the Western and Northern Isles. However, the exploration of indirect clues to Scotland's changing weather patterns through the ages is a much broader theme which lies beyond the immediate scope of this short chapter, although brief reference may here be made to a couple of examples. Firstly, the archaeological evidence of medieval settlement and agriculture in the Lammermuirs, in the south east, suggests that dwellings were then occupied and the associated ground ploughed at altitudes higher than would be comfortable and practicable today.[2] Secondly, a late seventeenth-century fashion for steeply pitched roofs, which would certainly have been better capable of effective snow shedding, may conceivably have been one architectural side effect of the 'Little Ice Age'[3] which intensified in the latter half of the seventeenth century. However, any possible correlation has yet to be tested more thoroughly than by casual observation of random examples, given that so little research has yet been conducted even into the basic identification of pre-1750 roofs in Scotland, let alone into the likely determinants of their construction and pitch.

Many traditional buildings, perhaps the majority, are unselfconscious and passive reflectors of their physical surroundings, usually in terms of their location and construction and, prior to the twentieth century, building materials were among the clearest tell-tale indicators of local or regional origins (see Fig. 29.1). Scotland as a whole has always enjoyed a natural abundance of workable stone, slate and lime, and the ways in which building stone has been expertly and ingeniously handled over the centuries is evident throughout much of the country. The ancient ancestry and relative sophistication of this masonry tradition is well attested by structures such as prehistoric brochs and highly developed late medieval tower houses, whose exteriors often belie the complexities of their design and construction. It is thus no coincidence, firstly, that many of the country's first or greatest architects and engineers came to their profession through a close association with building stone and masoncraft,[4] and, secondly, as described below, that by the nineteenth century building in stone had come to represent what now might be termed the socially inclusive norm across Scotland.[5]

By comparison, and certainly by the exceptional standards of medieval and later England, timber has generally been a less accessible resource and its historical use has accordingly been less extensive and less accomplished, though the scope and nature of Scottish timber construction and the development of the attendant carpentry and joinery skills should not be underestimated.[6] The extent and ancestry of Scottish timber building – and the use of native woodlands – are only now beginning to be fully explored and appreciated, and the performance of exposed timber in the Scottish environment to be technically and scientifically assessed.[7]

Where slate or tile was not available, which may have been the case with the bulk of Scottish rural housing before about 1800, thatch of various materials, including marram grass ('bents'), heather and bracken as well as straw and potato shaws, was applied as a roof covering over a groundwork of timber and turf. However, as modern experiments[8] and detailed surveys have confirmed, such basic structures of rubble stone, timber, turf and thatch, which now survive largely as ruins, necessitated a frequent – probably annual or at least biennial – round of repairs and reconstruction to keep them securely wind and watertight. Roofs of slate nailed to close-jointed sarking is regarded as the traditional Scottish norm, but in eastern Scotland, as in the rest of eastern Britain, the use of tiles, specifically S-shaped pantiles, as a roof covering over open frameworks of battens has remained almost equally prevalent, a distribution pattern which may in part relate to their sourcing and manufacture but also to the appropriateness of their use, in 80 to 100-year renewable cycles, among weather conditions which are generally drier than the west. Many such tiled roofs have been combined with courses of slates to form an apron or cloaking course at the eaves, a fairly widespread modification which has presumably been found to improve their water-shedding and wind-resistance properties (Fig. 1.2).

By contrast with these examples, there are forms of Scottish building which almost defiantly declare that the physical environment has had little or no influence on their setting and design. Scotland's lofty urban tenements and tower houses, for instance, are among the tallest late medieval and early modern buildings in the British Isles, the veritable skyscrapers of their age,[9] and many of them stand on prominent hilltops or crag-and-tail urban sites, buffeted by every gale and storm that blows (Fig. 1.3). But the general picture is undoubtedly dominated by modern buildings, many of which appear to take little heed of the fact that they have been planted in Scotland. Since the 1960s, and particularly in and around the cities of Glasgow, Edinburgh, Dundee and Aberdeen, Scotland has been host to numerous and massive

Figure 1.2 Steading, Burghmuir, Linlithgow, West Lothian, 2003. Detail of pantiled roof and slated eaves 'apron'. Courtesy of Geoffrey Stell.

Figure 1.3 Caisteal Uisdein (Hugh's Castle), Snizort, Skye. Artistic impression by David L. Roberts of a sixteenth-century residence on an exposed promontory site. From the original painting by David L. Roberts (1931–97). © *Dualchas* Museums and Heritage Service, The Highland Council.

multi-storeyed tower blocks of steel-and-concrete construction, comparable in height and bulk with any in Britain,[10] (see Fig. 2.5) while domestic architectural fantasies in *Art Deco* or Modern style, more suited to a gentle, Mediterranean-style climate, have also made their way into inter-War and post-War Scotland in limited but significant numbers.[11]

With this balance of qualified considerations in mind, this chapter briefly explores a little further how far the provision of shelter from the elements may be deduced from general considerations of location and setting and from aspects of building design, construction and adaptation, especially in relation to those dwelling types where the structural evidence is relatively clear and unadorned.

LOCATION AND SETTING

In Scotland, as elsewhere, the sites of many medieval and sub-medieval lairds' houses appear to have been selected to give them a view of the most valuable natural assets over which they would wish to exercise greatest vigilance and protection, namely, their estate demesne or table lands. Before extensive agricultural drainage schemes were undertaken in the eighteenth and nineteenth centuries, such lands generally meant gently sloping ground, south facing for preference, on which ploughing assisted the natural drainage and the creation of arable fields and gardens. The setting of

Torthorwald Castle, midway up the eastern slopes above the Lochar Moss, east of Dumfries, well exemplifies such locations. They are also demonstrated collectively by the group of lairds' houses set into the slopes around the rim and on the 'islands' (inches) of the Carse of Gowrie, a heavy, low-lying tract of ground in Perthshire and Angus, which, like the Howe of Fife and the Laigh of Moray, only later realised its class-1 agricultural potential through improved drainage schemes.[12]

Compared to hilltop, promontory or island sites which were unambiguously suited to thoroughgoing fortifications, such settings appear to represent a compromise between the provision of a view which overlooked the policies and some degree of protection from the elements. Indeed, in later times and probably for similar reasons, views of the landscape from what Sir John Clerk of Penicuik (1676–1755) characterised as 'cold, wild ground' commanding a 'dismal prospect of the moors',[13] continued to be deliberately eschewed by the builders of Scottish country houses. Such practices drew the contempt of English commentators such as Dorothy Wordsworth who entertained much more romantic notions of a house's setting in the landscape. She noted, for example, that Lord Breadalbane's Taymouth Castle, Perthshire, 'is in a hollow, without prospect ... seeing nothing and adorning nothing'.[14] In Scotland, elevated sites risked physical exposure, particularly in winter, and in order to mitigate the effects of their positions, houses and farms on estates which colonised the uplands during the era of agricultural improvement were frequently accompanied by plantations of trees, particularly in the form of linear shelter belts.[15] Providing some degree of protection for locations which are prone to exposure, not just by reason of their altitude, is indeed evident across a range of house types. It seems likely, for example, that those cottages in east-coast fishing villages which characteristically present their narrow and minimally-lit gabled walls, not their main frontages, to the waterside or foreshore, do so partly because they may thereby reduce potential everyday problems inherent in their exposed situations.

In general, as in modern times, sand, bog and other unstable or made ground were best avoided as house sites but if they had to be used, appropriate structural foundations, including timber rafts or branders, were laid.[16] The risk of inundation on the plains of the more powerful rivers has long been recognised – and experienced – in places such as Perth, where the River Tay has demonstrated its tendency to create such problems since at least the twelfth century. Indeed, in the long catalogue of disasters which have occasionally beset Scottish house dwellers, storms and natural calamities induced by extremes of weather appear to have been about equal to fire and accidents as recorded agencies of destruction and damage, and both categories have almost certainly outweighed the limited and localised effects of war.[17] Remarkably, this has remained Scotland's experience in modern times. Although the advent of aerial bombardment in World War II led to a few devastating attacks such as the Clydebank 'Blitz' in 1941, these were very much the exception rather than the rule, and, overall, the towns and cities of Scotland did not suffer to anything like the same destructive extent as those of most other combatant nations in Europe.[18]

Within much stricter ranges of choice than were available to the proprietors, the sites of townships and cot towns of the tenantry and sub-tenantry also made use of naturally draining slopes wherever possible and practicable. In both pre- and post-improvement times, buildings were generally grouped or formally laid out over thinner, less cultivable soils so that the best ground was not taken up. The buildings were also arranged in a manner which afforded shelter but made good functional use of the elements in, for example, traditional and improved barn design where opposed doors provided a means of controlling the natural application of wind to winnowing processes and wattled or louvred panels generally assisted the drying and storage facilities. Strong westerly winds have been an ever-present fact of life and habitation over much of Scotland, and on the relatively flat island of Tiree in the Inner Hebrides they have undoubtedly been a major factor in determining the fairly regular east-west alignment and sturdy low profile of the numerous nineteenth-century croft houses of traditional blackhouse form.[19] Local Gaelic proverbs affirm the wisdom of having a house that is set east-west (An iar 's an ear an dachaidh as fhèarr) and with its back to the wind and its face to the sun (Cùl ri gaoith agus aghaidh ri gréin) (Figs 1.4 and 1.5).

On Tiree, such winds are generally constant and 'fair', unpredictable only in their sheer force, but an area in the lee of Ben Hough, at 390 feet (119 m) in height one of only two hill masses of any significance on the island, is known to be subject to the damaging effects of an eddying or 'foul' wind, hence the relative paucity of existing buildings in this area. A similar gap is perceptible in the pre-crofting settlement pattern depicted on James Turnbull's map of 1768–9, which also shows a proliferation of small farms and a sizeable township at Hough itself on the windward side of Ben Hough. Here, strong winds were responsible for the devastating sand blows which later overwhelmed these buildings, leaving only the ruins of an isolated medieval chapel locked in sand dunes at Kilkenneth as testimony

Figure 1.4 Cottage and Byre, Caolas, Tiree, 1973. Front (east) wall. Crown Copyright: RCAHMS.

Figure 1.5 Cottage and Byre, Caolas, Tiree, 1973. Rear (west) wall. Crown Copyright: RCAHMS.

to the one-time densely settled communities on the island's western coastal fringe.[20]

For those who became caught up in the great displacements of the rural population from the late eighteenth century onwards, drawn to the coasts by successive booms in kelp manufacture and herring fishing, or driven from the hills and glens by the requirements of sheep farming and, later, of sporting estates, improved shelter from the elements might, at best, be solidly provided in the new, well-ordered planned villages; at worst, it might be afforded in only the most basic of temporary structures and bothies, sometimes in communal accommodation,[21] at times in what effectively became squatter settlements utilising whatever materials were to hand and located in marginal, exposed and dangerous locations. The Highland Clearances acquired much notoriety, and few clearance settlements were more notorious than that of the cliff-top site of Badbea on the coast of south-eastern Caithness, where children of the displaced Langwell tenantry were allegedly tethered to stakes like goats to stop them straying and being blown into the sea.[22]

In the centres of the towns and cities, for all but the wealthiest home-builders, the choice of location was generally restricted by considerations of space and questions of 'neighbourhood'. In Edinburgh and the Canongate as early as the seventeenth century, the clustering of tall properties in confined spaces is known to have generated amenity issues such as the limiting of sunlight and creation of shadows across neighbouring properties. However, it is questionable whether, as has recently been contended,[23] such tightly packed communities may have also been consciously striving to create a specifically 'weather-protected' environment for themselves. Whilst Renaissance arcades were being translated from fashionable European exemplars into a Scottish idiom to provide shelter for shops and market areas, any more general protection which might have been afforded by the alignment or re-alignment of the streets themselves appears to have been casual and purely adventitious. Indeed, when an opportunity for a comprehensive re-configuration of an urban layout arose, as it did in the case of Edinburgh New Town from the 1760s onwards, the streets were laid out deliberately on an open, spacious plan, knowingly giving even greater freedom to the wind and the rain.[24] To a greater or lesser extent, that degree of openness and indifference to the elements has tended to persist among schemes of urban housing – and among individual or communal house designs – ever since, in marked contrast to the commercial architecture of enclosed markets and shopping malls.

BUILDING DESIGN, CONSTRUCTION AND ADAPTATION

As adaptations of natural rock shelters, caves are among the most basic forms of habitation to provide protection from the elements, usually requiring only the minimum of man-made intervention. Exceptionally for Scotland, though not dissimilar in principle from the house-fronted caves of the Loire valley in France, the extensive complex of three caves beneath Culzean Castle, Ayrshire, has an architectural frontispiece of about 1600, and, reminiscent of troglodyte dwellings, Gilmerton Cove[25] on the southern outskirts of Edinburgh is a complete subterranean house of eighteenth-century creation comprising passages and rooms carved entirely out of the local sandstone. Underground structures or hypogea are found in the Western Isles, and semi-subterranean structures of sub-circular plan and turf-covered corbelled construction are among the oldest surviving standing structures of domestic origin on remote St Kilda.[26]

Above ground, the traditional Hebridean blackhouse, designed to make economical and effective use of the local Lewisian gneiss and scarce supplies of timber, had double walls, often up to six-feet six-inches (2 m) in thickness and as broad as they were tall; a hipped and thatched roof rested on the inner edge of a broad wall head, originally draining into the core between an inner wall which was mortared for insulation and an outer wall which was drystone built to assist drainage and ventilation.[27] Once characteristic of the Outer Isles, modified versions of the type also still survive on Tiree, where a traveller in 1802 noted the distinguishing fact that:

> ... the walls of the huts in Tirii are double, and the space between them is ... filled with earth and sand to make them the stronger, to bear the western blast, which sometimes makes dreadful havoc. Most commonly every hut has two doors, that when the wind blows hard, the one to the windward may be shut, and that to leeward opened.[28]

This last observation is not borne out by the almost universal evidence of single east-facing doors in the surviving croft houses of traditional blackhouse form. Their general effectiveness has, however, evidently remained above question, judging by the testimony which was later provided by an Argyll County Medical Officer of Health. In 1893 he wrote that:

> ... a Tiree dwelling will stand a hurricane without the least injury. The whistle of the wind is no more heard from within than in the interior of Ben Cruachan ... The wind strikes against the walls, and shoots over the roof without scarcely touching it.[29]

In traditional small farmsteads and later crofts, some degree of shelter was afforded by the parallel juxtaposition of house and outbuildings, creating a relatively narrow 'close' or passage between the ranges. This was the layout of the older style of houses of tenant farmers in the uplands of Moray and Nairn as described by one early nineteenth-century commentator:

> In the more stormy quarter of the district the house and offices were arranged in two lines, or so contrived as to have the doors mutually sheltered by the opposite building, from the penetrating blast, or the drifting snow; but in the low part of Moray, the turf hovels were placed in all the irregularity which chance might exhibit.[30]

One conspicuous feature of the 'improved' rural and urban built environment of Scotland from the later eighteenth century onwards was the increasing availability of cut stone for everyday constructional use as walling for permanent habitations. Having been a signifier of relatively high-status work in the medieval and early modern periods, stone masonry building practices extended across the social and architectural spectrum to become, by the nineteenth century, the preferred and predominant walling material for virtually all house types throughout the country, from the humblest cottages upwards. Masonry walls, ranging in treatment from dressed stone or ashlar to coursed or random rubble, often covered by a protective coat of render or harl,[31] rapidly displaced or reduced other materials – including brick, which became ubiquitous in England at about the same time – to isolated regional pockets or to internal structural use. Technical improvements, such as the creation of sub-floor ventilation systems, accompanied these changes, and porches of stone, timber and glass became an almost standard adjunct to the main entrances of many of these improved farmhouses and cottages across all parts of the country, providing not only extra shelter from the wind but also a system of double doors which minimised draughts and heat loss.[32] More recently, advances in the technology and mass production of glass and timber extensions have facilitated and fed an equally widespread fashion for double-glazed conservatories which, since their first introduction into nineteenth-century country houses, have permitted a sense of outdoor living in virtually all weathers. Viewing the elements in relative comfort has clearly become a feature of modern living which seems likely to persist, and, in Scotland as elsewhere within the last quarter of the twentieth century, the widespread application of energy-efficient forms of central heating, insulation, draught proofing and double glazing, designed for new houses and adapted for old, has created an expectation or 'norm' of shelter and comfort that is already several degrees – literally – above the levels tolerated by most households in the 1950s.

So far as less permanent habitations are concerned, apart from the relatively makeshift tents or 'benders' associated with nomadic lifestyles,[33] shieling huts and their derivative cabin forms constitute the smallest and most primitive category of traditional dwelling. They were usually designed and built to provide temporary or seasonal accommodation for families or migrant workers, with only the minimum of shelter created out of wood, stone or, latterly, sheet metal.[34] Tented or other temporary and basic communal accommodation has also long been a feature of military life, though permanent barrack blocks, including some built of timber at Fort William, Inverness-shire,[35] came to be associated with the forts and garrison posts established in Scotland from the seventeenth century onwards. The two World Wars of the twentieth century generated a variety of standard hut types, all of utilitarian design and employing materials and methods of construction capable of mass production and rapid assembly, the best known being the semicircular, steel-sheeted Nissen, the brainchild in 1915 of one Lieutenant Colonel Peter Nissen (1871–1930), a Canadian of Norwegian origin.[36] As in the rest of the United Kingdom, military camp sites proliferated all over Scotland, many associated with defences and installations on exposed hill and coastal sites to which, judging from oral testimonies, the thin-walled huts were scarcely equal, especially in winter. However, weatherproofed and well maintained, a number of such huts, even including some of World War I vintage, have managed to survive to this day, usually but not invariably converted to non-domestic, agricultural purposes and only rarely occupying their original wartime sites.

Factory-based mass-production and prefabrication techniques which had been honed in wartime were applied with equal vigour to the emergency programme of house building, accelerating pre-War urban slum-clearance schemes and making good any damage that had resulted from aerial bombardment. Under the Housing (Temporary Accommodation) Act of 1944, more than 32,000 'prefabs' were erected in Scotland between 1945 and 1949. They were of differing degrees of prefabrication and employed a variety of materials, the most notable on both counts being the aluminium 'AIROH' bungalow.[37] As built, they had a life expectancy of merely ten years, but here and there, re-clad and refitted, a number of them have survived in use as comfortable dwellings for more than half a century.

Non-traditional materials such as steel and pre-cast concrete were increasingly applied to post-War house construction, particularly public-sector cottages, three or four-storeyed tenements, and from 1949–50 onwards, multi-storeyed tower blocks of eight or more floors.[38] The creation of massive point blocks and slab blocks reached its zenith in the late 1960s when giant steel-framed towers of up to 31 storeys came to dominate the skyline of Glasgow and its suburbs. In the United Kingdom as a whole this fashion for high flats lasted longest in Aberdeen where ten- and 11-storeyed point blocks were still being erected as old people's sheltered housing as late as the mid 1980s.[39] By that date tower blocks elsewhere had become objects of controversy and stigma, beset by the combined effects of increasing domestic fuel costs and delinquent social behaviour which compounded design deficiencies in, for example, the inadequate provision of lifts. Other, more familiar problems also persisted, and flats in multi-storeyed and deck-access blocks appear to have been little better than their predecessors in coping with the cold and damp, though any assessment of their capabilities is distorted by a confused web of social-management and building-design issues. The fact that some tenants felt impelled to block ventilation outlets, for example, may well have reflected a level of draughtiness and discomfort attributable to the design, but this ad hoc 'remedy' itself almost certainly exacerbated the problems of condensation which the outlets had been specifically designed to obviate. In such circumstances, building performance in any one of a number of key areas was almost inevitably going to be regarded as inadequate. In a condition review of council houses undertaken by the City of Glasgow District Council in 1989, over 20 per cent of all dwellings erected since 1964 were found to be subject to condensation and over 14 per cent showed the effects of rain penetration. Overall, the total housing stock of 165,057 dwellings in 1989 included 46,022 (27.9 per cent) affected by condensation and 28,056 (17 per cent) by penetrating damp; a less uncomfortable finding was that rising damp afflicted only 3,109 (1.9 per cent) properties and these were mainly pre-1964 tenements.[40]

CONCLUSION

This brief general conspectus of a huge subject has perhaps served to demonstrate that, on the one hand, environmental conditions and the basic requirements of shelter have undoubtedly been of importance in many aspects of house location and design in Scotland, particularly among older and more traditional forms of building. On the other hand, such considerations have also been shown to have been of partial and limited impact, particularly in modern times when advances in building technology may have encouraged strong beliefs in the all-round technical capability and performance of materials and methods, irrespective of setting or ambient conditions. Indeed, a windy hilltop setting was deliberately chosen for Cumbernauld, one of Scotland's four major post-War new towns. In these respects, such buildings and housing developments are probably more the products of their political, social and technological background than of their physical environment, although any such determinism must itself be qualified. Like any person, a building may not be wholly explicable in terms of its origins and context, and in its performance it can expect to be tested as much by the vagaries of people as by those of the elements.

ACKNOWLEDGEMENT

The author is grateful to Dr Miles Glendinning, for reading and commenting on a draft of this paper.

NOTES

1. Pearce, Smith, 1990, 331–445 and especially 373, 376–7.
2. Parry, 1975; Parry, 1976; Parry, 1978; Smith, Parry, 1981, 3–16.
3. For the 'Little Ice Age', see Lamb, 1982, 201–30 at 209–14 and references cited; Whyte, 1981.
4. For example, Thomas Telford (1757–1834), Skempton, et al, 2002, 682–97.
5. McMillan, Gillanders, 1997; see also Fisher, 1976; Naismith, 1985, 76–86.
6. Hay, 1976; Stell, 2004.
7. Davies, Walker, Pendlebury, 2002; Smout, 1997.
8. Noble, 1983–4; Noble, 2000; Noble, 2003.
9. Stell, 1980; Stell, 1988.
10. Glendinning, Muthesius, 1994.
11. McKean, 1987.
12. Stell, 1985, 198 and notes 25–6.
13. Gray, 1892, 148.
14. Wordsworth, 1874, 93–4. Cited by Slater, 1980, 226–7.
15. Caird, 1980, 216.
16. Stell, 1984.
17. Stell, 1986.
18. MacPhail, 1974; War damage in Scotland receives no mention in either Richards, 1947; or Lambourne, 2001.
19. Boyd, 1986; Souness, 1992; see also, RCAHMS, 1980, 41, 244–6 (nos 370–2).
20. RCAHMS, 1980, 146–7 (no. 298).
21. Leitch, 2003; Holmes, 2003.
22. Gunn, 1972, 138, 142; Close-Brooks, 1995, 79–80 (no. 28).
23. McKean, 1999.
24. Cosh, 2003, graphically describes how Edinburgh's windy weather found even greater freedom to play across the broad, open streets of the New Town.
25. Coles, 1910–11, 265–71; see also Chapter 46 Idiosyncratic Homes.
26. Thomas, 1870; Stell, Harman, 1988, 21–3, 27–8.
27. Sinclair, 1953, 18–32; Fenton, 1989; Walker, McGregor, 1996; A stimulating, but now largely discredited, thesis that building customs in the Western and Northern Isles were derived from Norse ethnic origins is set out in Roussell, 1934.
28. Murray, 1803. Cited in RCAHMS, 1980, 41.
29. Cited in Souness, 1992, 82.
30. Leslie, 1813, 59; For traditional farm layouts in Orkney comprising closely-set parallel ranges see Fenton, 1978, 116–30.
31. Fisher, 1976, 17 and 2 (26), records the first use of the term in 1572. Harling has proved itself an effective and durable means of weatherproofing random rubble masonry
32. Naismith, 1985, 120–6.
33. See Chapter 44 Communities on the Move.
34. Jones, 1996, 37–43.
35. Tabraham, Grove, 1995, 39–42; See also NLS, MS 1646 Z.02/30a-b for 1746 drawings of additional timber-built barracks to be built at the fort; For a general review of barrack design, see Douet, 1998.

36. McCosh, 1997; For a summary of military hut typology see Lowry, 1996, 18–25.
37. Glendinning, Watters, 1999, 228–30; Horsey, 1990, 29.
38. Damer, 1991, for conditions in Glasgow council houses 1919–65.
39. Glendinning, Muthesius, 1994, 322.
40. Strathclyde Regional Council Education Department, npd, 74–6; For tenant commentaries on physical conditions in new housing schemes and high-rise blocks, see, for example, Castlemilk People's History Group, 1990; Glendinning, Muthesius, 1994, 322–3.

BIBLIOGRAPHY

Boyd, A. *Seann Taighean Tirisdeach [Old Tiree Houses]* (Càirdean nan Taighean Tugha [Friends of Thatched Houses]), Lochgilphead, 1986.
Caird, J B. The reshaped agricultural landscape. In Parry, M L, Slater, T R, eds, *The Making of the Scottish Countryside*, London and Montreal, 1980, 203–22.
Castlemilk People's History Group. *The Big Flit: Castlemilk's First Tenants*, Glasgow, 1990.
Close-Brooks, J. *The Highlands* (Exploring Scotland's Heritage), Edinburgh, 1995.
Coles, F R. Notices of rock-hewn caves in the valley of the Esk and other parts of Scotland, *PSAS*, 45 (1910–11), 265–301.
Cosh, M. *Edinburgh: The Golden Age*, Edinburgh, 2003.
Damer, S. A Social History of Glasgow Council Housing, 1919–65, typescript report for the Economic and Social Research Council, Centre for Housing Research, University of Glasgow, Glasgow, 1991.
Davies, I, Walker, B, Pendlebury, J. *Timber Cladding in Scotland*, Edinburgh, 2002.
Douet, J. *British Barracks, 1600–1914*, London, 1998.
Fenton, A. *The Northern Isles: Orkney and Shetland*, East Linton, 1978.
Fenton, A. *The Island Blackhouse*, Edinburgh, 1989.
Fisher, I. Building stone and slate: Some regional aspects of Scottish quarrying. In Fenton, A, Walker, B, Stell, G, eds, *Building Construction in Scotland: Some historical and regional aspects* (SVBWG), Dundee, 1976, 16–27.
Glendinning, M, Muthesius, S. *Tower Block: Modern Public Housing in England, Scotland, Wales, and Northern Ireland*, New Haven and London, 1994.
Glendinning, M, Watters, D M, eds. *Home Builders: Mactaggart and Mickel and the Scottish Housebuilding Industry* (RCAHMS), Edinburgh, 1999.
Gray, J M, ed. *Memoirs of the Life of Sir John Clerk of Penicuik* (Scottish History Society, 13), Edinburgh, 1892.
Gunn, M. Since the 'Forty-Five'. In Omand, D, ed, *The Caithness Book*, Inverness, 1972, 137–52.
Hay, G D. Some aspects of timber construction in Scotland. In Fenton, A, Walker, B, Stell, G, eds, *Building Construction in Scotland: Some historical and regional aspects* (SVBWG), Dundee, 1976, 28–38.
Holmes, H. Housing for seasonal agricultural workers. In Stell, G, Shaw, J, Storrier, S, eds, *Scotland's Buildings, Scottish Life and Society: A Compendium of Scottish Ethnology, Vol 3*, East Linton, 2003, 236–48.
Horsey, M. *Tenements and Towers: Glasgow Working-Class Housing, 1890–1990* (RCAHMS), Edinburgh, 1990.
Jones, D. Living in one or two rooms in the country. In Carruthers, A, ed, *The Scottish Home*, Edinburgh, 1996, 37–58.
Kleboe, J, ed. *Timber and the Built Environment Conference*, Edinburgh, 2004.
Lamb, H H. *Climate, History and the Modern World*, London, 1982.
Lambourne, N. *War Damage in Western Europe*, Edinburgh, 2001.

Leitch, R. Seasonal and temporary dwellings. In Stell, G, Shaw, J, Storrier, S, eds, *Scotland's Buildings, Scottish Life and Society: A Compendium of Scottish Ethnology, Vol 3*, East Linton, 2003, 224–35.
Leslie, W. *General View of the Agriculture of Nairn and Moray*, London, 1813.
Lowry, B, ed. *Twentieth-Century Defences in Britain: An Introductory Guide* (CBA), York, 1996.
MacPhail, I M M. *The Clydebank Blitz*, Clydebank, 1974.
McCosh, F W J. *Nissen of the Huts*, Bourne End, 1997.
McKean, C. *The Scottish Thirties*, Edinburgh, 1987.
McKean, C. The evolution of the weather-protected city. In Dennison, E P, ed, *Conservation and Change in Historic Towns* (CBA Research Report, 122), York, 1999, 24–38.
McMillan, A A, Gillanders, R J. *Quarries of Scotland* (Technical Conservation, Research and Education, Historic Scotland, Technical Advice Note 12), Edinburgh, 1997.
Murray, S. *A Companion, and Useful Guide to the Beauties in the Western Highlands of Scotland, and in the Hebrides*, London, 1803.
Naismith, R J. *Buildings of the Scottish Countryside*, London, 1985.
Noble, R R. Turf-walled houses of the Central Highlands: An experiment in reconstruction, *Folk Life*, 22 (1983–4), 68–83.
Noble, R R. Creel houses of the Scottish Highlands. In Owen, T M, ed, *From Corrib to Cultra: Folk essays in honour of Alan Gailey*, Belfast, 2000, 82–94.
Noble, R R. Earth building in the Central Highlands: Research and reconstruction. In Govan, S, ed, *Medieval or Later Rural Settlement in Scotland: Ten Years On*, Edinburgh, 2003, 45–51.
Parry, M L. Secular climatic change and marginal land, *Transactions of the Institute of British Geographers*, 64 (1975), 1–14.
Parry, M L. The abandonment of upland settlement in Southern Scotland, *Scottish Geographical Magazine*, 92 (1976), 50–60.
Parry, M L. *Climatic Change, Agriculture and Settlement*, Folkestone, 1978.
Pearce, E A, Smith, C G, eds. *The World Weather Guide*, London, 1990.
RCAHMS. *Inventory of Argyll*, 3, Edinburgh, 1980.
Richards, J M, ed. *The Bombed Buildings of Britain*, London, 2nd edn, 1947.
Riches, A, Stell, G, eds. *Materials and Traditions in Scottish Building: Essays in memory of Sonia Hackett* (SVBWG), Edinburgh, 1992.
Roussell, A. *Norse Building Customs in the Scottish Isles*, Copenhagen and London, 1934.
Sinclair, C. *The Thatched Houses of the Old Highlands*, Edinburgh, 1953.
Skempton, A W, Chrimes, M M, Cox, R C, Cross-Rudkin, P S M, Rennison, R W, Ruddock, E C, eds. *A Biographical Dictionary of Civil Engineers in Great Britain and Ireland, Vol 1, 1500–1830*, London, 2002.
Slater, T R. The mansion and policy. In Parry, M L, Slater, T R, eds, *The Making of the Scottish Countryside*, London and Montreal, 1980, 223–47.
Smith, C D, Parry, M L, eds. *Consequences of Climate Change*, Nottingham, 1981.
Smout, T C, ed. *Scottish Woodland History*, Dalkeith, 1997.
Souness, J R. Taighean tugha Tirisdeach [The thatched houses of Tiree]. In Riches, Stell, 1992, 81–94.
Stell, G P. Scottish burgh houses, 1560–1707. In Simpson, A T, Stevenson, S. *Townhouses and Structures in Medieval Scotland: A Seminar*, Glasgow, 1980, 1–31.
Stell, G P. Stone buildings with timber foundations: Some unanswered questions, *PSAS*, 114 (1984), 584–5
Stell, G P. The Scottish medieval castle: Form, function and 'evolution'. In Stringer, K J, ed, *Essays on the Nobility of Medieval Scotland*, Edinburgh, 1985, 195–209.
Stell, G P. Destruction, damage and decay: The collapse of Scottish medieval buildings, *ROSC*, 2 (1986), 59–69.
Stell, G P. Urban buildings. In Lynch, M, Spearman, M, Stell, G, eds, *The Scottish Medieval Town*, Edinburgh, 1988, 60–80.

Stell, G P. Timber in Scottish historic buildings. In Kleboe, 2004, 19–26.
Stell, G P, Harman, M. *Buildings of St Kilda* (RCAHMS), Edinburgh, 1988.
Strathclyde Regional Council Education Department. *Housing in Twentieth-Century Glasgow: Documents 1914 to 1990*, Glasgow, npd.
Tabraham, C, Grove, D. *Fortress Scotland and the Jacobites*, London, 1995.
Thomas, F L W. On the primitive dwellings and hypogea of the Outer Hebrides, *PSAS*, 7 (1870), 153–95.
Walker, B, McGregor, C. *The Hebridean Blackhouse: A Guide to Materials, Construction and Maintenance* (Technical Conservation, Research and Education, Historic Scotland, Technical Advice Note 5), Edinburgh, 1996.
Whyte, I D. Human response to short and long-term climatic fluctuations: The example of early Scotland. In Smith, Parry, 1981, 17–29.
Wordsworth, D. *Recollections of a Tour Made in Scotland in AD 1803*, Edinburgh, 1874.

2 Security

SABINA STRACHAN

SECURITY AND THE CULTURE OF DWELLING

To be secure is to be protected from danger and without care or anxiety. Throughout human history the search for security in its various forms – physical, psychological, emotional, spiritual or economic – has been a constant, at all social levels, across space and through time.

Most of the activities of human life touch upon the home. The dwelling is the setting which provides the primary context for life, of being in the world, and is often seen as the place of definitive security.[1] Amongst other things, the home seeks to offer continuity, the structure which best engenders security, as well as shelter, a form of security from the elements, thereby contributing to a secure social environment.[2] These basic requirements for human life interject into the built form, and the consideration of security shapes all dwellings to some extent, often in the subtlest of ways.

Security in the dwelling can be expressed at a number of levels, in relation to the individual, household or community or it may be focused upon property. There is, however, much overlap and similar processes and recurrent problems may be identified such as the elusive nature of security, for despite its undoubted importance to wellbeing security can be difficult to attain. It is the examination of such paradoxes which can be the most revealing.

SECURITY AND THE INDIVIDUAL

Fear and threat
There are many forms of threat to which people may be subject and, similarly, many ways in which a specific threat can be perceived by any one person. What we fear, and its expression, is intensely personal across space and time.[3] Perception of danger depends not only on one's general susceptibility to fear but also on the ability or willingness to defend oneself and how one expects to cope with the consequences of violation. Fear can be modified in a number of ways, if an individual has actually experienced or witnessed danger for example, and this leads to yet more complexity. Nuances such as these make generalisations difficult, though perceived threats and the resultant fears of groups tend to be less varied and allow for broad discussion.

Although some individuals do not believe it possible to arrive at a state of stability, perceiving life to be ever-changing and uncertain[4] and hence perceptually insecure, most of us aspire to achieve some level of security, especially personal security; that is to be comfortable with ourselves, to be stable and assured and in possession of a clear identity. Alienation and loneliness in any form will usually have an adverse effect upon personal security, as can, for some individuals, disability and long-term sickness. Personality also plays a role; by being intensely introverted, certain individuals struggle to form relationships[5] from which a sense of personal security may emerge.

The issue of security within one's home may be particularly important for all those who, no matter what their character, are socially marginalised in some way. People who live alone may feel especially vulnerable, and this can also be the case for some who live with others, for example, those in children's or old people's homes. It is also possible to include those individuals who do not have an adequate support network, who are unable to rely upon the emotional support of friends and family, or those who are excluded from society on the basis of mental or physical health problems or disability.

The home is sanctuary for many people beyond working age, especially in urban or physically isolated settings. As one older person remarked, 'Of course if it was safe I would go out. I'm 75. So I don't go out at all after dark. In winter, as soon as it's dark I stay in.'[6] There has been a steady growth in the average human lifespan in the post-War period and also in the number of single-person households.[7] These demographic changes tally with increased social isolation and 'urban paranoia' trends.[8] Some immigrants may also possess a heightened sense of the dwelling as a major form of protection. In times of community stress the risk of danger for such people may increase and also exacerbate feelings of not belonging and isolation.

The daily experience of home life is intrinsically related to gender, a factor which also predetermines to some degree that which we fear and our coping mechanisms. Women tend to feel more vulnerable and less able to counteract a physical attack. They are also more likely to fulfil the role of carer of children or the infirm, and their personal security may be compromised in response to feelings of isolation. Furthermore, home is cell and captivity, hell rather than haven, for all those who suffer any form of abuse within its walls.[9]

Physical threat is higher in economically or socially deprived areas – high crime figures dominate a number of the 1960s and 1970s schemes of social housing, for example – although reporting of incidents and forms of protection are greater in more affluent areas. Particularly vulnerable households tend to be allocated the poorest housing stock, and the disreputable surroundings may heighten their members' insecurity.

The rise in owner occupation of the last decades has led to an increased feeling of security within and emotional attachment to homes, dwellings in this category usually being imbued with heightened properties of comfort and relaxation. The new residential status of many Scots today touches upon their personal wellbeing, not least because the individual may participate more actively in their local community or feel the need to maintain or improve the appearance and ambience of their surroundings. Whilst a relatively high proportion of Scots remain tenants compared to the rest of the UK, expectations have changed and homeownership is often seen as desirable.[10]

Inevitably we compare ourselves and our sense of security with the past or, more often, an imagined version of this. Many people register a generally heightened feeling of insecurity in modern times. However, the level of security for members of earlier societies is a matter for conjecture. Death could more frequently strike prematurely. At the same time the spirituality and ritual surrounding death often had great importance for both the individual and the community in ways different from and sometimes more profound than now, when much has become 'sanitised' and de-personalised and belief systems have changed.[11] Symbolism in general in the past was often potent, the supernatural was frequently perceived as a reality, and, 'hearth, home and firelight afford[ed] some protection' from nature and the elements.[12]

Means of obtaining personal security
Survival is our primary target, the emphasis on which increases as our level of security decreases.[13] Beyond that, personal security is, to a considerable degree, born of a sense of belonging to place and/or people. The dwelling is usually a place which endures, while life within the home is commonly a shared experience, and the comfort found in the presence of other people is an important form of emotional security. In the 'search for group support', boundaries are mutually delineated and selected from unending possibilities, leading to the distinctiveness of individual households.[14]

A sense of security may also stem from one's religious belief system, or from one's physical surroundings if they are imbued with a sense of cosiness. Likewise, by surrounding oneself with familiar and significant objects, including photographs or religious artefacts or, in the distant past, urns or interments, one may increase a feeling of both identity and security. Over one third of the pits discovered beneath the floor level of the wheelhouse at Sollas, North Uist, for example, contained votive offerings including some animal interments.[15]

The sense of security an individual possesses is usually greater at higher levels of society. Status clearly affects the level of security which one feels entitled to and receives, and the risk to which one is exposed. The defence of the nuclear fort of Dunadd, Argyll, occupied between the sixth and ninth centuries AD, provides a clear illustration of this. The fort, on an Iron Age site, is situated on a naturally defensible summit with excellent sightlines. Each of the three terraces is ringed by thick walls. The wall surrounding the first of these is particularly substantial, and the initial approach is through a steep-sided ravine which once had timber outerworks.[16] At the

Figure 2.1 Argyle Park Sheltered Housing, Edinburgh, 2005. Source: M Mulhern.

heart of these extensive measures was the stronghold of the Dalriadian élite. The lower terraces seem to have accommodated different groups, perhaps their position and relative safety, and conversely their discomfort, dictated by rank. The defensive capabilities of such strongholds could be seen from miles around – their formidable form a deterrent in itself.

Protection of élites is often carried out by bodyguards or soldiers. In castles, the personal servant slept in his or her master's or mistress's bed chamber or in the corridor outside (Fig. 1.3). Such proximity was maintained until strife lessened and issues of fashion and privacy superseded those of security. Within castle walls, the second line of defence centred on the lord's apartments, the 'dojon', as the lavish example of Bothwell Castle, Lanarkshire, demonstrates. Bothwell was surrounded by a moat and drawbridge, and intruders faced missiles from the arrowslits and bretasche before negotiating the portcullis and snaking passageway. At Kildrummy Castle, Aberdeenshire, a drawbar secured the door between the Earl of Mar's great hall and personal chamber and the windows were protected by grilles and shutters with their own drawbars.[17]

For those nearer the periphery of society, other measures have been employed to address associated insecurity. Physically impaired and frail older people may feel more secure with the knowledge that help is nearby. The role of non-disabled neighbours in the sense of security of some disabled or older individuals is marked. Practical measures include the fitting of types of locks and latches which do not hinder access in case of emergency but still leave the occupant feeling safe from attack and with an acceptable degree of control over his or her own household matters. Direct lines to the emergency services can be installed in the homes of those whom local authorities feel are at risk and in sheltered housing schemes bell pushes or cord pulls alert wardens[18] (Fig. 2.1). However, such measures may at the same time provide a constant reminder of personal circumstance and therefore decrease psychological security. Likewise, people may feel their dwelling is sanitised and regimented to some degree and no longer entirely 'homely'.

In modern homes of all sorts, the use of protocol has increased. New measures have been created to enable us to deal with strangers who call either physically or electronically, and these include the use of such items as identity cards. These however are easily forged and, like many other phenomena, may create a false sense of security.

The individual within a household

Household members may need to be protected not only from the outside world but also from each other and themselves. Placing locks on internal doors is one method by which people's movements are restricted. The intention of latches and primitive wooden tumbler locks opened by simple devices[19] was perhaps not to prevent outsiders from entering but to ensure instead that vulnerable members of the household stayed inside, whether children or animals (Fig. 2.2). Making the home 'childproof' or 'child friendly', with for example stair barriers, socket covers and child-safety devices on cupboard doors, is modern expression of such thought. Protection may also be more than physical; for example, the creation of 'safe' areas within the home, essentially the formation of boundaries, will have the effect of providing emotional as well as practical security for both the parent and child.

The creation of social barriers between those living within a dwelling may result in an erosion of personal security as greater restriction leads to feelings of isolation and loss of independence. Physical segregation of people within the home is not unknown. Domestic space has been subdivided in many societies, for example in Neolithic Orkney, where archaeologists have hypothesised that areas within the home may have been defined by gender, age and status.[20] The security of individuals is usually interlinked with their relative status within the household. Distinctions can result in unequal access to safety resources, such as community interaction, knowledge, education and freedom and may help create or exacerbate interdependent relationships of fear and domination.[21] In most homes responsibility for the security of the household rests more on some members than others, usually parents, and sometimes fathers especially.

Today, technology, in the form of Internet home shopping and banking, for example, has lessened the need for face-to-face human interaction and thus the opportunity for physical threat. However, the same media

Figure 2.2 Wooden lock from St Kilda. Such simple devices may have been designed more to keep the vulnerable inside a building than to prevent intruders entering. Source: SSS 1530, C4/64. Photo: Ian Whitaker, May, 1957.

have created new threats, for example through the use of chat rooms as well as email and information security lapses.

SECURITY AND THE HOUSEHOLD

Threats to households and their perceptions of threat
Although many households are composed of one person, most constitute small communities, and the implication is that security can be related to household size. Some households are subject to greater threat from outside than others. Such threat may be occasional, intermittent or perpetual. Increased threat may be related to household type, building type or residential area. Soon after their construction, the form of tower blocks, for example, imposed as it was by designers, came to be viewed as provoking problems such as vandalism and delinquency,[22] in the same way as the tenement was abhorred by nineteenth-century commentators as the cause of deprivation and immorality.

Perception again plays an important role. Perceptions of the threat to household security are dependent upon a number of factors. These include the incidence of negative events, such as burglary, and individuals' reactions to the experience. The religious, ethnic and cultural nature of households may also be significant. As mentioned before, the sense of home as intensely private and protected space can be heightened for members of certain recent immigrant groups. Some Asian households, for example, where social life can on occasion be focused within the home to a greater extent than elsewhere, might regard threat to the safety of the household as especially critical.[23] Such households, and those of other 'minority' groups, may be more subject to threat and close community links can result in greater than average awareness of others being attacked.

The perception of threat is also shaped by whether or not any householder is present, specifically which members, and at what time of day. If the house is empty, alarms are activated and dead locks are positioned. Security provision lessens once people return, that is until they go to bed when the measures employed include having a phone in more than one room, or a mobile phone, or, in extreme cases, the provision of a weapon.

Means of protection
Many ways of protecting the household have been employed in Scotland. 'Superstition' plays a part, more so in the past or in some minority belief systems. The stranger is always seen as a potential threat; therefore the behaviour of the individuals at either side of the threshold between the home and the world is important and relates to custom.[24] Recently many physical barriers within the home have been dissolving with the popularity of open-plan living areas and there is increasingly little distinction between semi-public and private spaces. Similarly, the outside of homes have witnessed a blurring of public, semi-public and private space with the creation of 'open-plan' estates. What barriers remain, both inside and outside the dwelling, have as a consequence become more pertinent and more closely regulated by cultural nuances, such as the use of the back door only by particular people, or knocking at a teenager's bedroom door or the door of a home office before entering.

Insecurities regarding accidents at home have increased in recent decades, perhaps because some other types of insecurity have diminished. Greater personal financial security has become a reality for many and relative economic stability has characterised the country as a whole. The prevailing thought in the post-War period has been that accidents in the home are 'the modern epidemic plague' and can be prevented by designing ergonomic interiors.[25] Contributing to such interiors are, for example, breakfast bars devoid of sharp edges, bath grab rails, and carefully determined shelf heights and locations of light switches. In addition, potential health risks can be monitored electronically with infant alarms, smoke alarms and carbon-monoxide detectors commonplace in many homes. Household economics are a determining factor as to whether these safety measures are applied as a matter of priority, and also affect risk in terms of the standard of the dwelling's fabric, the quality of repair and the use of more or less-combustible furnishings.

Economic security
Tenancies and their conditions greatly affect the sense of security within renting households. Greater general economic security in the later sixteenth century, coupled with heritable tenure for the upper echelons of society, often fostered extravagant investment in houses and land. In later centuries some of these people's tenants had to drain and then cultivate new tracts of moorland each year, which often resulted in the pioneers' bankruptcy and the creation of improved land for a set of incoming tenants.[26] The Crofter's Act of 1886 gave tenants in the 'Crofting Counties' (Argyll, Caithness, Inverness-shire, Orkney, Ross and Cromarty, Shetland and Sutherland) greater rights. However, most crofters had to provide a house, which was not owned and therefore could not be sold or used as security. The net result was the widespread retention of impermanent building forms, in terms of materials and ease of construction and dismantling.[27] The houses of some crofting households were, however, built by the estate, and therefore were of a more permanent form. The Crofting Reform Act of 1976 at last made it easier for tenants to acquire their crofts.

Yearly, monthly, and to a lesser extent weekly tenancies were the norm for the Victorian working class, whose members came in large numbers to urban centres searching for employment. The shorter tenancies were often for subdivided buildings, occupied by those with the lowest incomes. In these a dwelling could consist of a single room, a 'single end', which in the worst cases might be shared with another family.[28] Such accommodation was 'characterised by a cycle of poor-quality short lets and insecurity of tenure'. The situation was not alleviated by local-government schemes in the first half of the twentieth century. Housing stock was redistributed with the cheapest associated with even poorer conditions. There was no expenditure on maintenance and the standard of the stock was not addressed. Eviction was rare although the bailiffs were an ever-present threat.[29]

Today, homeownership does not necessarily increase security for those buyers with lesser incomes who need to pay substantial interest on borrowed money and may struggle to meet monthly repayments, both on the building and some of its contents. Stress levels for these people may be higher than for those renting on a long-term basis, especially from non-private landlords such as the housing associations which succeeded local authority responsibility for much social housing. Private renting can be expensive and ownership is seen as desirable as mortgage repayments tally with, or better, private rent prices in competitive urban centres.

SECURITY AND THE COMMUNITY

Sense of place, community and possession is likely to have become more important in Scotland from the late Bronze Age into the Iron Age, when farming practices developed and settlement was therefore more permanent. Social hierarchies, perhaps greatly influenced by environmental conditions, were such that large enclosures containing numbers of dwellings were prevalent at both ends of the dating spectrum c1700 BC–AD 200, whereas smaller territorial groups dominated the intervening period. The most defensible summits within a given locality were not necessarily favoured for hillforts, therefore defensibility does not seem to have been the main priority. Multiple entrances through ramparts created numerous weak spots, such as those at the Brown and White Caterthuns in Angus (Fig. 2.3). Security of the territorial group therefore seems to have derived from displays of status rather than defensive capabilities. However in times of threat, fortifications, such as those erected about AD 400 at Traprain Law, East Lothian, were intensified, in this case in response to the Roman advance.[30] The scale of settlements was much smaller for groups which most often were based upon the extended family. Such peoples built monumental roundhouses, the most conspicuous of which are the stone-built brochs. Some brochs were surrounded by extensive complexes which would have sustained larger communities, such as those at Gurness and Midhowe in Orkney. MacSween's analysis of the distribution of brochs in Skye concludes that sites near territorial boundaries were favoured. Two hypotheses emerge: firstly that close observation allowed for rapid response if hostility occurred; and secondly that co-operation between groups was facilitated for hunting, agriculture or communal defence.[31]

Threat to community and its perception
Security within a human population may disappear suddenly as unpredictable and often uncontrollable natural factors, such as disease and climatic

Figure 2.3 The White Caterthun, Angus, 2003. A hillfort with spectacular ramparts yet numerous weak spots in these created a structure better suited to displays of status than defence. Source: S Storrier.

extremes, do not respect settlement boundaries nor the communities which reside within them. Most defendable threats to communities, however, are human in origin.

As before, the stranger can be regarded as a threat. Though strangers may arrive aggressively in one's community, as in war, a sense of threat is present at other times. Even in 'settled' communities differences may be exaggerated and innocents may be seen as enemies to such a degree that their homes may be denigrated and regarded as out of bounds. In the past such victimisation was inflicted upon 'tinker' camps, and modern parallels occur with the mobile villages of New Age travellers on the fringes of rural settlements[32] or asylum seekers who are placed in certain areas of poor urban housing. Community can feel particularly threatened by immigrant and migrant cultures when the traditions of the incoming people become influential. However it is unease such as this, or 'crisis in urban life',[33] which periodically fosters 'communality' at a time when centralisation has eroded communal confidence.

Neighbouring communities can also reduce our sense of security. Those living near to dangerous areas in cities may feel unsafe. Conversely the boundaries of areas regarded as deprived are often maintained and veiled by adjacent communities who subordinate others by prejudice and cloak and protect their own way of life.

Often the greatest threat to our safety comes from within those communities in which we live. There are spoils to be had from increasing combined human effort and interaction but at the same time the potential for disruption rises with group size. Modern urban Scots especially may belong to a number of largely discrete communities – of work, leisure, political or religious allegiance, etc, as well as neighbourhood. In any of these disputes may occur. For example, issues of trespass and rights-of-way have become more acute over time, as common gardens are divided and pends gated, and as concern over personal security and privacy has become more measured and boundaries firmly delineated.

The interference of authorities can greatly affect community security as institutional policies can sometimes reflect misconceptions and assumptions regarding continuity and way of life. Inhabitants of more economically and socially deprived areas are those at greatest risk from violent crime; however they are often the worst served in terms of policing.

Another source of threat comes from matters of general cultural and economic change. Trends in regional, national and international culture affect every individual's sense of security. One of the most important alterations in recent decades has been the lessening of community feeling and general interaction between people. Dispossession is 'the loss of collective and individual security [which] carries with it the constant threat of isolation as well as meaninglessness ... Man is no longer "at home" in society, in the cosmos, or ultimately with himself.'[34] Fear is greatest for dispossessed people who previously relied upon community, in which strong dependent relationships developed, and who have suddenly found themselves without this support.

In Scotland, as natural threat was more successfully overcome by technology, individual responsibility for defence decreased, while the impetus for forging communal ties lessened. Similar parallels can be seen with regard to the decline in religious and other belief.

Means of protecting community
Domain is fundamentally important to communities; any breach of this will threaten group dynamics and power structures, dependent upon the perceived severity of the threat. Rituals for defining community domain are commonplace; common ridings still take place, in the Border towns for example, in which mounted burghers travel the extent of the burgh common lands, today as part of a festival, formerly annually to define the extent of the townsfolk's combined holdings. Other means of binding members of a community together include shared popular knowledge or religious belief, whether displayed through public festivals or privately observed customs.

Initiatives such as neighbourhood-watch schemes were first adopted in the 1980s, often by wealthier communities where perceived threat was disproportionate to actual risk (Fig. 2.4). Consultative measures in 'deprived areas' with neighbourhood groups and agencies and the use of closed-circuit television have at once created an increased sense of security. But reporting and evidence have increased whilst actual crime levels remain static, thereby engendering additional worry. Preoccupation with preventative measures such as lighting timers, intruder alarms and window locks also contain this paradox and in addition these devices may affect people's daily routines.[35]

Figure 2.4 Neighbourhood-watch sign, Edinburgh, 2005. Source: Alexander Fenton.

Through community interaction both the individual and the household can feel more secure. The optimum conditions for wellbeing are an environment which provides sustenance and economic security, a lack of exploitation and a sense of community, through co-operation and concern for the welfare of those within the group, in addition to a feeling of belonging to the group and place. Culture is the framework which informs all of these elements and is unendingly variable. In large urban settings community becomes less 'manageable' and our interdependence upon others is therefore compromised and care for other individuals lessened. Infrastructure and amenity become central to community feeling or perceptions of security.

The tenement and its successors
Our view of the past is shaped to some extent by nostalgia. Nostalgia has had a large part to play in the societal representations associated with the tenement, a building form of particular importance in Scotland and one often associated with the positive aspects of community. Historically there has been little mass housing for the poorest echelons of society. Instead filtration takes place when homes begin to deteriorate or areas or types become less fashionable. Stone-built eighteenth- and nineteenth-century townhouses and tenements became subdivided and the lower classes gained access to these properties as the medieval dwellings which some formerly inhabited were cleared. The living conditions in tenements inevitably strengthened a sense of community, with strong familial and neighbourly relationships impinging upon wellbeing, self-worth and, often, sustenance. Support networks created for and by such residents were vitally important as 'a realistic response to low incomes, economic adversity and unpredictable domestic crisis. In the absence of state support for the relief in the home of illness, old age or unemployment, the "safety net" for most families was the neighbourhood itself.'[36]

One of the original intentions of tower-block design and the sky corridors of deck-access schemes[37] was to reinforce the community spirit of families displaced from condemned tenements (Fig. 2.5). These buildings were rapidly created in areas often dependent on heavy and manufacturing industry and within peripheral schemes where there was a lack of amenity. A surplus led to a high proportion of empty units and the ensuing problems were compounded by poor management in the form of lack of maintenance, inability to respond quickly to problems and an unsuitable mix of tenants. Too much control could also have a negative effect. Such difficulties, combined with the stark change of physical environment, resulted in the failure of many schemes and stigmatisation of the tenants and caused some people to return to their former, dilapidated tenement homes.[38] As a consequence some proposed housing schemes were redesigned prior to building and other projects were shelved when the drawbacks of this type of housing became apparent. However, in a number of instances good practice (intensive management, a strict letting policy and high-quality finishes), as in

Figure 2.5 Red Road Development, Glasgow, 1987. General view from the west. One of the aims of creating such buildings was to reinforce the community spirit of families displaced from condemned tenements.

Aberdeen, prevented many problems and the desired sense of community did indeed flourish.[39]

The impetus for improvements in working-class housing conditions was often directly associated with the concerns of decision makers and their peers. Hygiene was the primary concern of nineteenth-century commentators. It was their belief that the 'immoral' slum dweller harboured pestilence and disease which then spread to other classes. Satellite New Towns, following the Garden City ethos, were favoured by the planning profession of the mid twentieth century. One of their intentions was to avoid 'mixing the classes in too close proximity'.[40] The siting of large estates of council housing at the peripheries of cities had, in part, the same aim, but also attracted complaints from suburban neighbours who felt their safety compromised and saw property prices affected.

SECURITY AND POSSESSIONS

Objects and identity, place and order
To dwell is to a great extent to orient and identify oneself with one's surroundings, based upon the 'spatial irrelationships of objects of man's

identification ... to gain a world through the understanding of things'.[41] Familiarity is an important element in the concept of home and our routines usually breed contentment and provide security and comfort. The dwelling is akin to a safe into which we deposit items of meaning, our 'things'. The objects with which we grow up and which we choose to bring into our homes are central to creating both a sense of place and a sense of self.

Insecurity can arise when things in the home are felt to be 'out of place'. This creates anxiety, especially for those individuals who are predisposed to maintaining a high degree of control and order. However, for some, domestic untidiness can be as psychologically important, and is as jealously guarded. As disposable wealth increases, consumerism irrevocably alters the priorities of daily life and the desire to organise ever more complicated lives may be magnified.

Protecting the produce of the land
The associative status, or value, which is apportioned to a belonging will determine both the precautions and the actions taken to protect it in a given situation. In Scotland, cattle, across various levels of society, aside from providing a staple of the diet, signified degree of wealth and thereby status. Livestock therefore offered a high degree of economic security and were afforded an equally high degree of protection. This factor has done more than most to shape the physical form of the Scottish house across history and social standing.

In the medieval period, high walls formed barmkins, courtyards associated with tower houses, into which the prime livestock could be herded for protection in times of threat. Raids by neighbouring kin-based groups were as much rites of passage as banditry and they remained commonplace into the eighteenth century in some parts of the country.

Agricultural, rather than pastoral, settlement required carefully defined boundaries and led to the creation of more enduring rights. These could result in increased investment in the form of labour, capital and commitment from the household. However, the pastoral occupancy of marginal land predetermined the dominance of moveable wealth and the impermanence of buildings. Humans and cattle were over-wintered in the single-storeyed longhouses or 'byre dwellings' typical of the vast majority of the tenant class (see Fig. 29.1). The security of the animals was assured by the close proximity of their owners as warmth and care would be vital if they were to survive the lean winter. The economic importance of livestock for households increased as tenants were obliged to rear black cattle for export in response to the rising demand for beef which developed in tandem with urban growth. Droving of livestock to markets in fact became a mainstay of many Highland communities. Springhouses, essentially large byres, were a development in areas where stock rearing continued into the twentieth century; these offered additional shelter in early spring until both the overnight temperature and quality of grazing improved.

Cattle provided vital nourishment for Scotland's peoples, primarily in the form of milk and milk products and sometimes blood. In agricultural terms, bere, an older form of barley, was the dominant grain and its preservation required secure storage. During the Iron Age, souterrains were an integral part of many communities' buildings. The function of souterrains has been the cause of considerable debate but present opinion favours the view that, in the case of at least some examples, their cold, dry conditions were created for the storage of cereals, dairy produce, meat and fish. Large kidney-shaped examples proliferate in the eastern Lowlands, such as that at Pitcur in Perthshire. Orkney appears to have possessed smaller, oval souterrains, whereas narrow, lozenge-shaped examples are found in the Inner Hebrides.[42] Their scale appears to have been dependent on the amount of arable surplus associated with a particular environment.

A naturally occurring feature in shallow loch waters could be artificially augmented to create a level base for one or more buildings within an enclosure, creating a crannog. Crannogs have long been considered defensive structures, deterring primarily animal predators. Some crannogs were accessed by causeways, others by wooden drawbridges, while those at a greater distance from the shore were reached by boat. These islets were occupied from the Bronze Age until early modern times, with most dating to the Iron-Age and medieval periods.[43] Despite their long popularity, crannogs were not fully secure. Some would have been subject to periodic flooding and even the most prestigious examples, such as that at Finlaggan, Islay – the administrative centre of the Lordship of the Isles in the medieval period – were easily overcome by fire.

Feu duties were annual payments to landowners, who themselves owed dues to their nobles. These duties were paid in kind, therefore storerooms for perishables made up a considerable portion of the buildings within castle walls, and cellars or outbuildings were necessary as part of lairdly dwellings. Food items which required storage space tended to consist of salted or pickled fish or meat, cheese, butter and meal, often packed in kists (wooden boxes), barrels or earthenware containers. These commodities were periodically subject to theft, as court records testify.[44] As with bastles, the ground floor of a laird's house could be reserved for storage and the main living quarters of some lairds' and merchants' families would be on the first floor. Those dwellings associated with fishing rents, such as in Shetland, could store cured fish as well as nets, oil, etc, behind heavy doors in the thick-walled ground floor of their booths.

Protecting other possessions within the home

What we value – either monetarily, symbolically or sentimentally – has altered significantly as the way of life in Scotland has changed. Not only has what we choose to protect changed, but also the measures we undertake. Domestic belongings of any sort, and the house itself, can be protected by both tangible means – the use of manpower, concealment and warning and deterrent measures – and non-tangible phenomena -belief of various sorts, including religion and superstition. Tea, a luxury good until after World War I, was kept in locked caddies, and safes were installed for jewellery and documents in homes, to the extent that some large country houses had 'safe rooms'. Some items of extreme value, such as wills, were removed from the home and kept elsewhere, for example in the bank safety-deposit box or the lawyer's office. Concealment, in the form of secret compartments in furniture or hatches in floors, was the key to the security of sensitive material such as love letters, that relating to blackmail, and even contraband. Secret rooms existed where prohibited activities, ranging from illicit distilling to Roman Catholic worship, took place.

With regard to non-tangible means of protection, particular attention is paid to those apertures which breach boundaries. These may be 'guarded' by lucky horseshoes for example. Animal and human burials beneath the floor levels of homes in the prehistoric period may have performed a similar, though perhaps intensified, function. Two female burials were excavated in one of the buildings at the Neolithic settlement of Skara Brae in Orkney, whilst carved stones were concentrated at the low entranceways within the complex.[45]

Today, a major consideration of many house purchasers is a range of factors which reduce the likelihood of the dwelling being targeted by burglars or intruders. These may include whether or not the house is overlooked, ease of rear access, street lighting, clearly defined boundaries and the nature of the existing walling or fencing. Additional measures to deter unwanted entry range from gravel paths to dummy alarms, window roll shutters and guard dogs. Such measures can either provide comfort or provoke insecurity on the part of the occupants. Postcode marking and insuring items of value does not assure the safety of possessions, only that they can be traced or their monetary equivalent obtained should theft take place. The same applies to insurance against damage resulting from natural phenomena or accidents. The binding nature of lease agreements and the conditions of mortgages can result in the loss of the home also.

CONCLUSION

The emphases upon security have changed considerably over time and vary between class, age and gender. However, despite the heterogeneity surrounding security, some frequently occurring emotional responses regarding our perceptions, trust and self-worth are revealed. Individual, household and communal security, whether physical, emotional, mental, spiritual or economic, depends upon factors such as the level of threat at any given time, and, more particularly, perception of threat and resultant reactions to it.

Security within the dwelling underpins self-identity and is inevitably influenced by our particular cultural setting. One of the fundamental roles of a dwelling is the creation of a feeling of belonging and comfort which translates into our everyday lives and informs our future decisions. Our continuity is constantly reassessed as we strive to hold on to our homes and our home lives by degrees of protection and control.

'The combination of security, physical or psychological, with shelter is the essence of the "home".'[46] However, the individualism of modern life does little to bolster personal security as human interaction is increasingly replaced by social isolation and the value placed upon our material wealth outweighs consideration of our mental health. It is this modern threat which should be addressed and warrants defensive action to safeguard the 'essence of the home'.

NOTES

1. Hediger, 1977. Cited in Ingold, 1995, 73.
2. Carruthers, 1993, 30; Carruthers, 1996, 26; see also Chapter 1 Shelter from the Elements.
3. Tuan, 1979, 4.
4. Tuan, 1979, 203–4.
5. Varwell, 1981, 60–2.
6. Hanmar, 1984, 65. Cited in Crow, Allan, 1994, 164.
7. See Chapter 43 Living Alone.
8. Miles, 1988, 109–11.
9. Madigan, et al, 1990. Cited in Crow, Allan, 1994, 123–4, 128.
10. Crow, Allan, 1994, 121.
11. See Chapter 13 Spirituality; Chapter 26 Infirmity; Chapter 30 Custom.
12. Roberts, 1996, 67.
13. Scherer, 1972, 51.
14. Varwell, 1981, 9.
15. Armit, 1996, 153, 158.
16. Armit, 1998, 130–1.
17. Breeze, 1986, 14, 24.
18. Scottish Development Department, 1979, 4.1.2, 5.2.7, 5.7.15, 5.9.5.
19. Fenton, 1997, 147, 165.
20. Knap of Howar, Papa Westray, Orkney. Armit, 1998, 26, 28.
21. Berreman, 1975, 160, 176, 179.
22. Coleman, 1985, 12. Cited in Glendinning, Muthesius, 1994, 319.
23. Crow, Allan, 1994, 104.
24. Rapaport, 1969, 80. Cited in Roberts, 1996, 69.
25. Royal College of Surgeons, 1963, 9. Cited in Department of the Environment, 1974, 54–5.
26. Fenton, 1985, 58–9.
27. See Chapter 32 Paying for Dwellings.
28. Beaton, 1997, 49–50; Fenton, 1985, 58–9, 79, 89.
29. Horsey, 1990, 2, 5–8, 10, 12.
30. Armit, 1998, 17–19, 78–81.
31. MacSween, 1985, 32.
32. See Chapter 44 Communities on the Move.
33. Durrant, 1939, 119. Cited in Glendinning, Muthesius, 1994, 102.
34. Berger, 1977, 39. Cited in Varwell, 1981, 82.
35. Crow, Allan, 1994, 102, 163.
36. Bulmer, 1986, 92. Cited in Scherer, 1972, 41.
37. The decks were intended to represent the street, however, this reflected the terraces

of English towns and cities rather than the tradition of flatted living in Scotland. Horsey, 1990, 57.
38. Glendinning, Muthesius, 1994, 95–6, 319–22; Horsey, 1990, 20.
39. In Aberdeen demand was maintained and the last block was built in 1985 at Jasmine Place. Glendinning, Muthesuis, 1994, 322.
40. Horsey, 1990, 28, 32.
41. Norberg-Schulz, 1985, 15, 17.
42. Armit, 1998, 19, 84–6, 109.
43. Morrison, 1985.
44. Donaldson, 1954, 4–5, 8, 11, 14, 20, 33, 70, 73, 75–6. Cited in Fenton, 1997, 160.
45. Armit, 1996, 31–2.
46. Carruthers, 1996, 26.

BIBLIOGRAPHY

Armit, I. *The Archaeology of Skye and the Western Isles*, Edinburgh, 1996.
Armit, I. *Scotland's Hidden History*, Stroud, 1998.
Beaton, E. *Scotland's Traditional Houses: From Cottage to Tower House* (Discovering Historic Scotland series), Edinburgh, 1997.
Ben-Amos, D. Towards a definition of folklore in context. In Hammond, P B, ed, *Cultural and Social Anthropology: Introductory Readings in Ethnology*, London, 1975, 358–67.
Berger, P L. *Pyramids of Sacrifice: Political Ethics and Social Change*, Harmondsworth, 1977.
Berreman, G D. Race, caste, and other invidious distinctions. In Hammond, P B, ed, *Cultural and Social Anthropology: Introductory Readings in Ethnology*, London, 1975, 158–80.
Breeze, D J, ed. *Scottish Castles and Fortifications*, Edinburgh, 1986, reprinted 1990.
Bulmer, M. *Neighbours: The Work of Philip Abrams*, Cambridge, 1986.
Carruthers, A. Form and function in the Scottish home, 1600–1950, *ROSC*, 8 (1993), 29–34.
Carruthers, A. Studying the Scottish home. In Carruthers, A, ed, *The Scottish Home*, Edinburgh, 1996, 11–36.
Coleman, A. *Utopia on Trial*, London, 1985.
Condry, E. *Scottish Ethnology* (Association for Scottish Ethnology, monograph no 1), Edinburgh, 1981.
Crow, G, Allan, G. *Community Life: An Introduction to Local Social Relations*, Hemel Hempstead, 1994.
Department of the Environment. *Housing the Family Design Bulletin* (1966–71), Lancaster, 1974.
Donaldson, G, ed. *The Court Book of Shetland, 1602–04*, Edinburgh, 1954.
Durrant, R. *Watling: A Survey of Social Life on a New Housing Estate*, London, 1939.
Elliot, L, Atkinson, D. *The Age of Insecurity*, London, 1998.
Fenton, A. *The Shape of the Past 1: Essays in Scottish Ethnology*, Edinburgh, 1985.
Fenton, A. *The Northern Isles: Orkney and Shetland*, East Linton, 1997.
Foster, S. Analysis of spatial patterns in buildings as an insight into social structure, *Antiquity*, 63 (1989), 40–50.
Glendinning, M, Muthesuis, S. *Tower Block: Modern Public Housing in England, Scotland, Wales and Northern Ireland*, London, 1994.
Gosden, C. *Anthropology and Archaeology: A Changing Relationship*, London, 1999.
Hanmar, J, Saunders, S. *Well-Founded Fear: A Community Study of Violence to Women*, London, 1984.
Hediger, H. Nest and home, *Folia Primatologia*, 28 (1977), 170–87.
Horsey, M. *Tenements and Towers: Glasgow Working-Class Housing, 1890–1990*, Edinburgh, 1990.
Ingold, T. Building, dwelling, living. How animals and people make themselves at home in the world. In Strathern, M, ed, *Shifting Contexts. Transformations in Anthropological*

Knowledge ('The Uses of Knowledge: Global and Local Relations', Association of Social Anthropologists Decennial Conference), London, 1995, 57–80.

Jackson, A, ed. *Way of Life and Identity* (North Sea Oil Occasional Paper, no 4), London, c 1981.

Jackson, A, ed. *Way of Life and Integration and Immigration* (North Sea Oil Panel Occasional Paper, no 12), London, 1982.

MacSween, A. The brochs, duns and enclosures of Skye, *Northern Archaeology*, 5/6 (1985), passim.

Madigan, R, Munro, M, Smith, S. Gender and the meaning of home, *International Journal of Urban and Regional Research*, 14 (1990), 625–47.

Miles, I. *Home Informatic: Information Technology of Everyday Life*, London, 1988.

Morrison, I. *Landscape with Lake Dwellings: The Crannogs of Scotland*, Edinburgh, 1985.

Norberg-Schulz, C. *The Concept of Dwelling: On the Way to Figurative Architecture*, New York, 1985.

Rapaport, A. *House Form and Culture*, New Jersey, 1969.

Roberts, B K. *Landscapes of Settlement: Prehistory to the Present*, London, 1996.

Rowlands, M. The politics of identity in Archaeology. In Bond, G C, Gilliam, A, eds, *Social Construction of the Past: Representation as Power*, London, 129–43.

Royal College of Surgeons, *Report of the Convention on Accident Prevention and Life Saving*, Edinburgh, 1963.

Scherer, J. *Contemporary Community: Sociological Illusion or Reality*, London, 1972.

Scottish Development Department. *Housing for the Disabled* (Scottish Housing Handbook, no 6), Edinburgh, 1979.

Stoklund, B. *Folklife Research: Between History and Anthropology. A 75th Anniversary Lecture*, Cardiff, 1983.

Tuan, Y. *Landscapes of Fear*, Oxford, 1979.

Varwell, A. *Way of Life: In Search of Meaning* (North Sea Oil Panel Occasional Paper, no 5), Glasgow, 1981.

3 Privacy

HANNAH AVIS

I mean at mum and dad's house, and in university flats and places you could always shut your bedroom door and it would make a private place, my place, but it's Mike's bedroom too and what right do I have to lock him out of it? Really the only place is the bathroom, the only one with a lock on is the bathroom door and you don't spend that long on the loo![1]

Ideas of home and privacy are powerfully, and continually, linked in popular discourse. The spaces called home are often characterised as being those to which people retreat from the pressures of life lived in public and where they find time to be alone. However, homes have come to be understood as sets of spaces which carry profound symbolic and ideological significance; amongst other things, homes serve to structure people's social interaction both within and beyond them.[2] Further to this, housing studies have increasingly shown that the ideals informing and structuring constructions of home exist in tension with the lived experience that individuals report, in ways that problematise the seemingly neat coupling between home and privacy.[3] This chapter considers the ways in which women living in Scotland's capital city at the end of the twentieth century experienced the relationship between home and privacy. It examines the stories of the women's home lives and shows how they negotiated the spatial restrictions of their tenement flats and the emotional demands of intimacy in their search for privacy.

The material presented is based on doctoral research carried out within a group of home-owning women living in Edinburgh during the late 1990s.[4] The narratives are those told by a group of younger, middle-class people, all of whom, when interviewed, were in the first five years of homeownership and were living in tenements. Focusing upon such a group is useful as the women, and their stories of home, offer a means of examining and representing the day-to-day significance of the social changes which took place in the later part of the twentieth century and continue today. Developments in women's political participation have impacted upon the relationships which they enjoy, modifying the dynamic and complex relations between men and women, work and home, and the ideas of private and public space, for example. The case study presented here is of importance in this regard, as it is amongst members of the middle classes that distinctions between public and private, both people and spaces, have traditionally been most rigidly adhered to and change is potentially the most dynamic.[5] The location of Edinburgh is also of note as the city's service sector has created a sizeable pool of middle-class occupations that are often considered appropriate for women and are indeed often filled by women.

Underpinning the collection and analysis of the women's stories is a feminist understanding of home as a highly-gendered socio-spatial landscape. Feminist researchers have asserted that the imagery of home as a haven is too simplistic and have urged for a discussion of the ways in which gender is key to the structuring and experience of home.[6] In such work it is understood that women and men experience the home in very different ways according to the social, and in particular the patriarchal, expectations which circulate about the differing roles that they have. Hunt, for example, discusses how roles within, and experiences of, home are to a very large extent determined and distinguished by gender.[7] In her examination of people's home lives, Hunt encountered a common understanding that men are primarily involved in domestic life through financial contributions brought into the household as a result of paid work, while women's engagement with home is centred upon the idea and delivery of a domestic haven. This linking of women with the home has many implications for the lives which women lead and the experiences of home that they have; these range from the 'second shift' of domestic work which is undertaken while women are at home to limitations on the non-domestic work that they undertake, both in terms of hours and distance from home.[8] Such understandings and expectations also have implications for the stories of privacy that women relate. Women are often found to experience little or no privacy at home, being either on view all the time or prevented from demarcating private space and time from family space and time and the domestic responsibilities that are assumed to be theirs.[9]

IDEAS OF PRIVACY

> [Mary] Having a bit of time and space to myself, like I do need to a bit of time alone ... yeah, where I won't be bothered and can just think through the day or quietly read or something, like switch off from things and, and be alone.[10]

When speaking about privacy the women in this study most often constructed narratives of times and spaces in which they could be by themselves. As Mary suggests, claiming private time and space is thought of as being about getting away from other people and pressures. The women hope to distance themselves from work – both paid and domestic – and also consider events, work through feelings and so on. The ideas of privacy that emerge echo Goffman's characterisation of private space as the 'off-stage' areas of a person's social world.[11] The privacy sought is space and time in which there is no need to perform. In private the women hope not to have to adhere to public expectations and demands, and privacy is thought of as times and spaces in which they can relax. Indeed, in the narratives the women construct there are many instances of the use of words and phrases such as 'comfy', 'letting it all hang out', 'just being yourself', 'no pretence', and 'having time off'. As Charlotte describes:

> It's like, your home and your private space is like your artist's palette and all the things that you do are the different colours and you can mix it and you can, you know, you can wash it again ... like you go through the day and the palette gets all mixed with colours, browns and yellows and greens and whatever, and then you come home and you know, bit by bit some of the colours come off and then you are clean again ... and you, I think you need some private place to do 'cos it is like your emotional washer or cleanser.[12]

Highly significant in Charlotte's description of her private space is that she firmly locates it in her home. This is a feature of many of the discussions of privacy conducted with the women in this research; their idea of privacy is very strongly linked to their understandings of home. It is to their homes that the women retreat. This linking reflects the construction of home as a haven from work and as a place to which people return in order to rest and restore themselves.[13] For Charlotte it is important to be able to go home in order to sort through the day and to get ready for the following day. She positions her home as the place in which she can engage in those emotional processes which she understands to be private and not for public display.

HOMES AS PRIVATE SPACES

The development of the home as a private space in people's patterns of living has been traced to the development of industrial capitalism in the West.[14] As paid work became concentrated in larger, non-domestic spaces the home came to be understood as a place where workers could seek refuge from such work and be restored in order to return to it. This separation can be seen to impact on home spaces in terms of both urban planning and house design. At the planning level, the separation between public and private resulted in urban landscapes where some areas became exclusively residential. Design created dwellings in which certain rooms were deemed fit for public view while those which dealt with the needs of workers and their households, such as kitchens and bedrooms, were situated in the back or, where they existed, the upstairs regions of a dwelling, out of sight of nearly all except the household members.[15]

Sharply delineated distinctions between private and public spaces at home have, however, always been contested and blurred as individual households go about their lives within the spatial constraints of their dwellings. The limitations of space and the expectations of society have meant that rooms are often used for more than one set of purposes. A living room may be used, for example, for family relaxation and at other times for receiving guests or for working. In many dwellings, therefore, observing the division between public and private areas is less a matter of separation than it is a negotiation of space and time, with rooms being used by both members of the household and those who can be considered outsiders.[16]

In addition, boundaries around private space have been stretched throughout the twentieth century, and into the twenty-first, by the spilling over of domestic concerns into more traditionally public areas such as shopping streets and work places, television programmes and magazines.[17] Homes are being increasingly conceived of, and promoted as, the means through which a person's chosen lifestyle can be attained and displayed; through various media the home is put forward as the space in which individuals aspire to, and possibly create, a way of life.[18] The architecture of homes, and the interpretations which individual households make of their basic house design, can be understood to be bound up with popular ideas

Figure 3.1 Exterior shot of Edinburgh tenement, 2005. This type of flat was built in large numbers in many parts of Edinburgh in the late nineteenth century to house industrial workers and their families. Originally the dwellings consisted of a kitchen-living room with a bed recess, a smaller bedroom (often also used as a 'good room' for more formal entertaining), a water closet and sometimes an internal bed closet. Large families could be accommodated in such dwellings. By the turn of the twenty-first century these flats mostly took the form of a lounge, often with a dining area where the bed recess had been, which could be used for formal entertaining, a bedroom which those from beyond the household were not expected to enter, a kitchen (usually in the former bed closet) and a tiny bathroom. Source: M Mulhern.

PRIVACY • 45

Figure 3.2 The lounge, formerly the kitchen-living room, of a one-bedroom flat off Easter Road, Edinburgh, 2000. The pair of sofas shows that this room is expected to serve part of the time as space for entertaining rather than as a dedicated place of privacy. Such flats are now mainly occupied by single people or couples. The density of occupation might be dramatically lower than in the past but privacy can still be hard to find for couples living in this and other types of tenement accommodation. 'Colony' flats can be seen in the background. Source: S Storrier.

about private life and space and the way in which people might aspire, value and realise privacy. While much modern housing takes into consideration the changing workings of the family and individual, in cities such as Edinburgh many people continue to live in tenement blocks whose built form is in keeping with earlier ideas (Fig. 3.1). It thus might be assumed that the inhabitants of such dwellings experience a sharper difference between public and private space than those living in more modern homes, and that household spaces and relations continue to be affected by earlier ideas of work, home and family which influenced domestic architecture and arrangements. However, as the following analysis of women's narratives of privacy shows, modern household relations do not necessarily mean that privacy is mapped onto the nineteenth-century tenement flats of Edinburgh (Fig. 3.2).

THE POSSIBILITIES AND PROBLEMS OF FINDING PRIVATE SPACES AT HOME

Stories of privacy within homes identify a variety of strategies for claiming space within which to be private. Common throughout the narratives of privacy and home is the view that bedrooms are the most private of spaces within flats. The bedroom is regarded as being the one area of a flat to which visitors are not given access. It constitutes the space in which a sense of escaping from the wider world is felt most and it is the most protected of rooms in the dwelling. Laura explains:

> I definitely have public and private ... but in reality I have a very small really private area which is just the bedroom ... pretty much people can go anywhere really, there's not a lot of private space, yeah, really it boils down to the bedroom, no one goes in there, that is mine and Matthew's space and the door is never open to other people.[19]

Bedrooms are thought of as spaces in which household members can claim privacy from those who visit the home. They are characterised as spaces in which standards of tidiness may be less rigorously maintained and very personal and intimate effects might be kept, and are often thought of as places where the intimacy of relationships is held beyond the gaze of others. Such rooms are characterised in the women's narratives as places of retreat, as spaces that lie beyond the pressures of life outside. Indeed for some of the women interviewed, their bedrooms are decorated and furnished with particular attention paid to comfort so that they might be places of luxury and indulgence, away from the demands of their wider lives.

However, while bedrooms are thought of as being the most private spaces in homes, they are also often shared with a partner. Thus they are understood to be private spaces when considering the household as a whole but do not necessarily offer a space which is exclusive to any individual member. In their narratives a number of women identify the challenges that sharing presents to them in realising a sense of privacy at home. Mary, for example, found that sharing her bedroom was a particular challenge when she began to live with her partner:

> It took me a long time to get used to always dressing in front of him and things like that ... and we would share space and all but it was hard for me to get used to that level of sharing.

> HA: Did you feel like your personal space was being invaded?

> Yes, that's a very good way of putting it, it was like my home is meant to be my personal, private space and here I am having to put my underwear on in front of someone.[20]

While the bedroom is off limits to those beyond the household, it is not necessarily off limits to others within it and this can take getting used to. Grace echoes such sentiments when she contrasts her marital home with her childhood experiences:

> I find it hard not having enough private time and space compared to when I lived at my mum and dad's where I had my own room ... that

was one of the big changes, sharing space in quite the way we would have to.[21]

This experience of bedrooms prompts narratives of aspiration in which the private spaces hoped for are rooms for the exclusive use of the women. Also linked to the understanding of privacy highlighted above – of times and places alone in which to be relaxed and to be 'yourself' – the women desire a home in which there will be enough space for them to have a room set aside for themselves. As Caroline hopes:

> I would just really like somewhere that was mine, where I could put my things and do the things that are important to me, you know, if I wanted to write a letter or ... just do something by myself then I could, just to have a bit of space that was mine would be really nice.[22]

Likewise, Grace wants:

> A music room because that is something that is mine, it is me ... it's not something that I do with Jack and I want it to be completely my own room because at the moment my things just get lost in and amongst stuff and there's no space for me to sing or anything, I mean I can't really be singing an aria in one corner while he's watching TV in the other, can I?[23]

Spaces which are clearly marked out as their own form an important feature of the women's narratives; to be able successfully to claim a sense of privacy they feel that having a space at home which is theirs and theirs alone is a requirement. Creating a spatial boundary within their home, a boundary which serves to delineate their own space from that which other members of the household might enjoy, is understood to be a means of creating privacy. As Ward suggests,[24] the concept of privacy includes the idea of setting up boundaries which separate space, objects and relationships into those that are shared and those which are not and are therefore private. Within the home this can mean setting aside rooms for particular people and for activities which are understood to be private and beyond the broader interests and relationships of the household in general.

Finding space in which to enjoy private activities and time is not easy within the relatively small tenement flats in which the women in this study live, however. There are few rooms in the flats and often the stories of privacy which emerge involve a considerable amount about spatial negotiation and compromise. Space restrictions mean that rooms often have to fulfil more than one function at a time and matters of privacy are often discussed within stories of adaptation and adjustment. Mary, for example, took a corner of the kitchen as her private space:

> It is a bit typical but I have the kitchen, well a wee corner and I've made that into my area where I can go to get away ... so, there is the old *chaise longe* from the junk store and the book shelf with some of my stuff on and I come in here and just have a bit of stuff

> about me and a bit of time to, to, well just to be a bit calm and quiet.[25]

So Mary takes a part of the kitchen and adjusts it so that she can have a space which is clearly marked out as hers and in which she can claim some privacy. It is not separated from the rest of the kitchen by any physical barrier, but through the positioning of the *chaise longe* and the accumulation of personal items such as books, photographs and ornaments Mary creates a sense of a private space into which she can retreat to have time to herself.

Another strategy is the allocation of time alone in rooms which are usually shared spaces. Such a strategy is often identified in women's stories of home as they negotiate the multiple roles required of them at work and home.[26] Amongst the women interviewed, time is often used as a means of claiming privacy within their small living areas, with varying degrees of action taken to ensure time alone. For example, Laura requires some space to herself after work and so agreed with her partner that she would have the living room for her exclusive use while she had a cup of tea and watched some television:

> I come home from work and need just a bit of time to throw off stuff from the day, you know get it out my system and I can't do that if Matthew is around me, 'cos like it is hard to say 'get out of my face' ... I don't want to hurt his feelings, he would do it but I feel mean, so ... anyway, so most days I get in and he will stay in his wee little office-boxroom while I watch *Neighbours*[27] in the living room with a cuppa and get over things and after that I feel ready to share the evening with him.[28]

Laura's story reveals a further strategy for claiming time in which to have some privacy; she marks out time in the living room as hers by watching a particular programme on television. Watching the television is an effective way of shutting down interaction between her and her partner and Laura uses it as a tool to ensure that she has time which she can call her own when she comes in from work. Women in particular often report difficulty in claiming time to themselves because when at home they feel the pressure to do domestic tasks. Making time without housework is often achieved through the use of television, radio, listening to music or reading. Just as Laura watches *Neighbours* on the television as a means of making time and space for herself, Mary retreats to the kitchen to read:

> I suppose that reading is a bit of tactic, I mean, I enjoy reading, yeah I like it a lot and it is a way to make some time for myself ... like now I'm reading and so I'm not up to talking or doing things, yeah, I like to read but I, I do use it as a tactic to get some time for myself away from Ryan and everything.[29]

The use of an activity such as reading or watching television is often cited as a successful way of claiming private time and space within the spatial limitations of small tenement flats. However, it is not always easy to achieve. Caroline, for example, finds claiming time alone in her home space is more difficult as the flat in which she and her partner lives has only one living area and one bedroom. This means that to have the solitude needed to achieve private space she or her partner often have to leave the flat:

> In a flat this small it is really difficult to be on your own unless the other person is out ... like to do this [the interview] Jonathon either has to go sit on the bed in the bedroom or go out 'cos him sitting at the table in here wouldn't give us the space to do this 'cos he'd be there ... and, I mean at times, and I know this sounds drastic, but sometimes in the past we would have a row and one of us would have to leave the flat entirely so that we could each have a bit of space to calm down and get things in perspective.[30]

Caroline's experiences are dramatic, but they are illustrative of a common theme in the testimonies of the women; that they make use of time alone in rooms or indeed the whole flat as a means of claiming and enjoying privacy. The use of time as a means to claim privacy is, however, a problematic strategy. Often in their stories of privacy there is a tension between the idea of intimacy that the women associate with co-habitation and the desire to have time and space to themselves. Nell mentions:

> I found actually living with him quite hard ... once you share a home that is where you go to and sometimes I need a bit of de-stress time and, well it happens because he cares I know, but I was being a grumpy cow and he would try and make me feel better, like 'what can I do? ... make you a cup of tea?' ... and sometimes, oh, I wanted, I was like 'JUST LEAVE ME ALONE', but he was just being kind ... I had to allow him to be kind, didn't I?[31]

For Nell, there is a conflict between the level of sharing and care she understands to accompany an intimate relationship and the level of privacy that she wants to be able to enjoy at home. On coming in from work she would like to have time alone in order to unwind from the day but feels that to make such a claim would be to shut out her partner with whom she shares the private space of her home. She feels guilty at wanting to have time to herself and while she does sometimes push her partner away she is uncomfortable about claiming time and space just for herself. Nell is not alone in this. Laura, for example, also tries to negotiate such feelings:

> I can't lie and say that it doesn't create problems because working out alone time and space can be difficult.
>
> HA: In what way?
>
> Oh, God this makes me sound pathetic but, well it's about hurting his feelings isn't it? Like, it is hard to say 'get out of my face' ... he should be the one who I share it all with, you know what it's like, sharing

things in marriage ... being giving, but I want to keep some things to myself and sometimes I feel bad about that.[32]

The complexities of these narratives are ongoing and the negotiations into which the women enter in order comfortably to claim time and space for privacy are a constant feature of their home lives. As Laura's narrative suggests, her need for privacy does not go away and nor does the difficulty that she has with wanting and claiming it. She is not alone in this regard and there is in the narratives of this research a frequent revisiting of the strategies employed to obtain private time and space. These negotiations are sometimes prompted by the women themselves and involve a reassertion of themselves and their need for privacy. Mary, for example, tells of how she protects her space in the kitchen:

> I do have to remind him, you know sometimes there will be some of his stuff on my shelves and I remind him he has to move it, 'cos it is my space and I want to keep it that way, and, and I know, it makes me sound a bit unreasonable but, well ... it's not a lot of space and it's not a lot to ask and so I remind him and, and he doesn't seem to mind.[33]

However, negotiations are often more discursive as the lines around space in the home are reconsidered and redrawn between the women and their partners. These represent less of a reassertion and more of a series of compromises that are demanded and conceded by the women. When Nell continues to reflect on her need for privacy and how she works that out with her partner she suggests that there is an ongoing system of balancing out of needs which occurs:

> It has been a case of talking about it and explaining what it is that I need, and, and making compromises.
>
> HA: Such as?
>
> Well, compromises like, taking the cup of tea I suppose, yeah now I let him make the tea and then he kind of retreats and I drink it on my own but still, still there are days when I don't even want to do that and so now I find ways of just telling him that and I hope that he doesn't mind.[34]

There is a strong sense in these testimonies that having private time and space is problematic not only because finding the time and space within homes is difficult but also because it tests what the women understand to be intimacy, relationships such as marriage and sharing a home with your partner. Desiring privacy is characterised as cutting across the expectations which the women have of sharing a home and of being in a relationship such as companionate marriage. Laura, for example, understands marriage to be about sharing and she wants privacy so as to keep some of herself to herself. In most stories this remains an unresolved tension and the women continue to claim time alone while remaining uneasy about doing so.

CONCLUSION

The women represented here understand their homes should provide them with private spaces into which they can retreat from their work and the other pressures of their more public lives. They aspire to owning houses which have enough space for a room which is exclusively theirs; the ultimate experience of privacy would be to have space to themselves and time in which to do those things they enjoy such as reading or singing. However, the practicalities of the flats in which they live make realising this difficult as there is little space in which they can construct the necessary boundaries or markers around their privacy. Spatial restrictions mean that they often have to find compromises and may mark out small sections of a room, or they may negotiate time alone in a room in order to claim an experience of privacy. Another commonly used strategy is suspending interaction with others by watching TV, listening to music or reading a book, for example. There is a further tension, however, in that the women feel a conflict between their desire for privacy and their ideals for intimate relationships. Popular discourses of intimacy and relationships encourage them always to share their space and their time with their partners, wanting time and space without such sharing is viewed as somehow failing. The search for a sense of home which encompasses intimacy and individuality creates tensions in which privacy becomes difficult to claim and sustain. Thus, it can be seen that emotional demands and spatial limits of cohabiting in tenement flats result in women telling narratives of home and privacy which are tense and subject to the constant reworking of popular understandings of femininity.

NOTES

1. 'Nell', interview 3:18. The names of respondents, and those to whom they refer, have all been replaced by pseudonyms.
2. Madigan, Munro, Smith, 1990; Saunders, Williams, 1988; Somerville, 1992; see also Chapter 2 Security.
3. See for example Darke, 1994; Madigan, Munro, 1999; McDowell, 1983.
4. The women are all aged under 35 and are in employment which can be considered as middle class, being either professionals or in managerial positions (see Savage, Barlow, Dickens, Fielding, 1992). Perhaps most importantly in the context of this study, they all self-identify as being young and middle class. While not all were born in Edinburgh, or indeed in Scotland, all the women lived in the city for a number of years prior to the study being undertaken, having been drawn there to undertake university study and subsequently staying on, or for employment.
5. Davidoff, Hall, 1987; Domosh, Seager, 2001; Spain, 1992.
6. Bowlby, Gregory, McKie, 1997; Darke, 1994; Mackenzie, Rose, 1982; McDowell, 1999.
7. Hunt, 1989.
8. Hochschild, 1989; Hanson, Pratt, 1995.
9. Hunt, 1989; Madigan, Munro, 1999; Munro, Madigan, 1993.
10. Mary, interview 2:15.
11. Goffman, 1959, suggests that in everyday life individuals have to present

themselves to others in particular ways which display their understanding of their social position. This performance, he claims, is potentially taxing and one from which people at times need to escape. Thus Goffman suggests that people find 'off stage' time and space in which they can rest from public demands and scrutiny. Such times and spaces are often in the 'back regions' of life, such as the home.

12. Charlotte, interview 1:31.
13. Saunders, Williams, 1988.
14. Allan, Crow, 1989; Mackenzie, Rose, 1982; Madigan, Munro, 1991.
15. Matrix Group, 1984.
16. Madigan, Munro, 1991.
17. Bondi, 1998.
18. Chapman, Hockey, 1999; Pratt, 1981.
19. Laura, interview 1:9.
20. Mary, interview 2:21.
21. Grace, interview 2:27.
22. Caroline, interview 1:7.
23. Grace, interview 2:27.
24. Ward, 1999.
25. Mary, interview 2:16.
26. See Hochschild, 1989; Madigan, Munro, 1999.
27. *Neighbours* is a long-running daytime and early evening television soap opera.
28. Laura, interview 3:31.
29. Mary, interview 2:18.
30. Caroline, interview 1:6.
31. Nell, interview 3:14.
32. Laura, interview 3:31.
33. Mary, interview 2:19.
34. Nell, interview 3:16.

BIBLIOGRAPHY

Allan, G, Crow, G, eds. *Home and Family: Creating the Domestic Sphere*, Basingstoke, 1989.

Bowlby, S, Gregory, S, Mckie, L. 'Doing Home': Patriarchy, caring and space, *Women's Studies International Forum*, 20 (1997), 434–50.

Bondi, L. Gender, class and urban space: Public and private in contemporary urban landscape, *Urban Geography*, 19 (1998), 160–85.

Chapman, T, Hockey, J. The ideal home as it is imagined and as it is lived. In Chapman, T, Hockey, J, eds, *Ideal Homes? Social change and domestic life*, London, 1999, 1–13.

Darke, J. Women and the meaning of home. In Gilroy, R, Woods, R, eds, *Housing Women*, London, 1994, 11–30.

Davidoff, L, Hall, C. *Family Fortunes: Men and Women of the English Middle Class, 1780–1850*, London, 1987.

Domosh, M, Seager, J. *Putting Women in Place: Feminist Geographers Make Sense of the World*, New York, 2001.

Goffman, E. *The Presentation of Self in Everyday Life*, London, 1959.

Hanson, S, Pratt, G. *Gender, Work and Space*, London, 1995.

Hochschild, A. *The Second Shift: Working Patterns and the Revolution at Home*, New York, 1989.

Hunt, P. Gender and the construction of home life. In Allan, G, Crow, G, eds, *Home and Family: Creating the Domestic Sphere*, Basingstoke, 1989, 66–81.

Matrix Group. *Making Space: Women and the Manmade Environment*, London, 1984.

Mackenzie, S, Rose, D. Industrial change, the domestic economy and home life. In

Anderson, J, Duncan, S, Hudson, R, eds, *Redundant Spaces in Cities and Regions*, London, 1982, 159–94.

Madigan, R, Munro, M. Gender, house and 'home': Social meanings and domestic architecture in Britain, *Journal of Architecture and Planning Research*, 8 (1991), 116–31.

Madigan, R, Munro, M. 'The more we are together': Domestic space, gender and privacy. In Chapman, T, Hockey, J, eds, *Ideal Homes? Social Change and Domestic Life*, London, 1999, 61–72.

Madigan, R, Munro, M, Smith, S. Gender and the meaning of home, *International Journal of Architecture and Planning Research*, 14/4 (1990), 625–47.

Munro, M, Madigan, R. Privacy and the private sphere, *Housing Studies*, 8/1 (1993), 29–45.

McDowell, L. City and home: Urban housing and the sexual division of space. In Evans, M, Ungerson, C, eds, *Sexual Divisions: Patterns and Processes*, London, 1983, 142–63.

McDowell, L. *Gender, Place and Identity: Understanding Feminist Geographies*, Cambridge, 1999.

Pratt, G. The house as an expression of social worlds. In Duncan, J, ed, *Housing and Identity*, London, 1981, 135–80.

Saunders, P, Williams, P. The construction of the home: Towards a research agenda, *Housing Studies*, 3/2 (1988), 81–93.

Savage, M, Barlow, J, Dickens, P, Fielding, T. *Property, Bureaucracy and Culture: Middle-Class Formation in Contemporary Britain*, London, 1992.

Somerville, P. Homelessness and the meaning of home: Rooflessness or rootlessness?, *International Journal of Urban and Regional Research*, 16/4 (1992), 529–39.

Spain, D. *Gendered Spaces*, Chapel Hill, 1992.

Ward, P. *A History of Domestic Space*, Vancouver, 1999.

4 Storage

DAVID JONES

STORAGE FURNITURE

Storage furniture has been found in a diversity of forms in all types of housing because it is essential for keeping food and personal belongings secure, dry and reasonably safe from animal pests. The category embraces items that are purely functional and therefore relatively unnoticed, or those which are highly prestigious and prominently displayed in the home. Individual pieces can be either severely plain or attractively decorated. In Scotland, decorated items are most likely to be found in small dwellings rather than large houses.

Kists and girnals
Plain storage boxes represent this furniture sub-group at its simplest level. The earliest surviving domestic examples, made of stone slabs bedded in an earthen floor, can be seen in the Neolithic village houses at Skara Brae and Knap of Howar, Orkney. Their domestic descendants, in the form of portable rectangular wooden 'kists', endured as essential items in one- or two-roomed dwellings until the middle of the twentieth century and could be found also in the servants' quarters of large country houses.

Kists performed a multitude of storage uses. It would be inaccurate to state that they had an absolutely fixed role in any household, but perhaps their most characteristic use in this country was as a receptacle for personal belongings taken to seasonal or residential employment.[1] Every young woman or man leaving home to enter domestic service or work on the land from the early 1800s until the 1940s was provided with a strongly made travelling kist with carrying handles. Until replaced by tin trunks, they were a ubiquitous feature of the nineteenth-century farm bothy or servants' garret. Margaret Bochel in *Dear Gremista* describes and illustrates the travelling kists used by Nairn fisher girls.[2] They contained working and Sunday clothes, linen items, a coat, knitting wool, assorted crockery and cutlery. The 'shottle' was an inner fitted box, almost always at the left hand side, which held knitting needles, bandages, a gutting knife and a 'wisker' (the waist-tied knitting sheath designed to take one end of a needle, thus freeing one arm).

Kists were usually made of deal (pine or fir) and commonly painted to disguise this relatively inexpensive timber. Aberdeenshire kists were often

Figure 4.1 Walter Geikie: Farmhouse kitchen interior, Cousland, Midlothian, c 1830. Source: Reproduced by kind permission of the Trustees of the National Galleries of Scotland.

painted dark brown or a distinctive maroon colour, in imitation of mahogany,[3] while further south, in Forfarshire, now Angus, and Fife, for example, they were grained, as were children's cradles, in sophisticated mimicry of oak and other cabinet woods. Their construction, too, could differ according to regional tradition. Kists from the Northern Isles, particularly Shetland, regularly adopted a sturdy wedge shape, being tapered towards the top on all sides. Their lids often had a lip which was of a slightly curved profile at either end and which fitted closely over the kist walls. These features may have been taken from Scandinavian tradition[4] but also the kists were designed to be particularly robust and stable because they were taken to sea by fishermen. Sea-going kists sometimes had rope handles.

In Aberdeenshire and Banff, kists had lids that were 'dyked' or fitted around a raised inner lip. This made an especially tight fit and was known as a 'tea-caddy top'. Further south, in Fife and the Lothians, for example, kists often had flat or overhanging lids.[5] In the home, including the farm servants' chaumer or bothy, as well as on board ship, the kist could double as a seat. Walter Geikie's sketches of small farmhouse interiors in the Edinburgh area c 1830 illustrate how kists could be used as bench seating at a table or placed in rows immediately in front of box-bed openings (Fig. 4.1). In this guise, they could be known as 'bunks' or 'bunkers'. Kists could also be used outside; they can be found, for example in field corners, having been used at the end of their days as feed containers for stock animals. In the late twentieth century, kists, once part of the repertoire of pine storage furniture found only in the working-class dwelling and servants' quarters, became fashionable in the middle-class home. Their decorative painted finishes were generally removed using the acid stripping technique.

A specialised type of kist, and one which was not usually portable, was the meal kist or 'girnal'. These were also known as 'girnels', 'girnalls', 'gernels', 'meal kists' or 'meal arks' according to use and region. In Scotland, they were found in greatest numbers in the agricultural districts of the east coast, but very similar high-sided wooden bins with sloping tops were found in other countries, such as Ireland, where oatmeal or wheaten flour was the staple diet. Ordinary flat-topped kists were regularly used for storing meal, but the specially made girnal had a sloping fold top that lifted on hinges to reveal one, two or, occasionally, three compartments which held different kinds of grain product, usually oatmeal and wheaten flour. There are interesting exceptions, such as the flat-topped girnal made in the form of a 'Scotch' chest (illustrated below), in the collections of the NMS[6] (Fig. 4.2).

Whatever their exterior shape, many examples had an inner shelf beneath the lid at the top where a flour scoop or cup and sieves were kept. Because the kitchen was so frequently the living room of the small house, girnals were often decorative, with shaped back boards and small shelves for ornaments or drawers for cutlery or other kitchen things. Very showy nineteenth-century examples might be faced with sham drawers to simulate

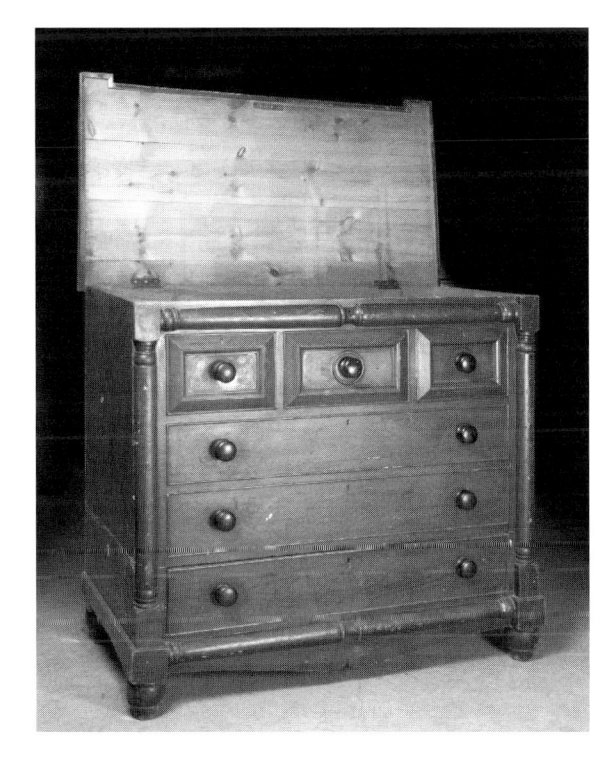

Figure 4.2 Meal girnal in the form of a Scotch chest. Source: Reproduced by kind permission of the Trustees of NMS, SH.2002.4223.

a bureau, or given a glamorous grained finish in imitation of various different cabinet woods. This last decoration was particularly prevalent in coastal districts of the north east such as Kincardineshire.

To keep the contents fresh over the space of a year (from harvest to harvest), the domestic girnal was usually positioned in a dry place beside the hearth, where the meal was also conveniently at hand for cooking purposes. Great attention was paid to the packing because the exclusion of air prevented fouling, and to this end I F Grant relates that the contents of each compartment were 'generally tramped with the bare feet',[7] possibly a task performed by children. Many girnals were raised on either bracket or turned bun feet, for which there are several likely reasons. The first, and most practical, was to avoid the entry of vermin, as well as contact with a damp floor, but the addition of fashionable feet could also complete a girnal's disguise as a bureau and would certainly provide valuable foot space for a person reaching inside. The scale of a girnal or meal kist might be relative to the size of the landholding which it served, and it thus follows that particularly monumental examples could be found in prosperous farmhouses in counties such as Ayrshire and Aberdeenshire. Bulk supplies of other foods tended to be kept in barrels and were not always stored in the kitchen. Potatoes, for example, might be kept outside in a pit or, if inside, in a barrel or sack which might be housed in the 'other' room.

The indoor storage of meal in girnals had been part of farming life for centuries, but similar items were used in towns and cities, particularly in institutions where many stomachs had to be fed. For example, the fashionable wright Alexander Peter was engaged by the managers of the Orphan Hospital in Edinburgh to make meal arks for their kitchens in 1741.[8] Arks or girnals were also found, until very recently, in shops. William Rodger & Son of Cupar and St Andrews, Fife, for instance, had large metal slope-topped meal bins opposite their counters until the end of the twentieth century. These were removed as a consequence of a combination of hygiene regulations and the advance of pre-packaging which made the sale of loose goods redundant.

Presses and dressers
Perhaps the most useful piece of storage furniture in the eighteenth-century farmhouse kitchen was the press, or shelved cupboard, used for the safekeeping of both food and kitchen utensils. A simple twentieth-century example, belonging to farmer William Munro of Clashnabuil, Easter Ross, is illustrated in Fig. 4.3. Perthshire kitchen presses display the most distinctive regional tradition and many impressive early examples survive from that county.[9] The upper stage of these cupboards has deep-panelled doors concealing a shelved interior, raised upon and rebated into a wider lower stage comprising one row of drawers over a two-door cupboard. They are commonly mistaken for linen presses. Food could be kept in the upper cupboard and pots and dishes in the lower, but the dark and deep recesses of such storage cannot have been conducive to tidiness. Later examples, from

Figure 4.3 Provision cupboard belonging to William Munro, farmer, Clashnabuil, Easter Ross. Source: William Munro, Clashnabuil, Easter Ross.

the nineteenth century, have glazed upper stages and lower parts in the form of chests of drawers; they appear to have been used more for display and the storage of a greater number of small objects.

R Ross Noble has demonstrated that the dresser was introduced considerably later than the kitchen press and that it took over some of the functions of the food/utensil cupboard.[10] The dresser had much more display space for china and ornaments, indicating, wherever it appeared, either an increase in disposable income or a change in the priority of family spending. The open shelves of Scottish dressers were usually filled with rows of earthenware plates and ashets leaning forwards against guard rails. J C Loudon noted that the dresser was 'the cottager's sideboard' (see Fig. 13.6) and that where the owner's proud array of plates leaned forwards this was to protect their faces from the dust.[11] On the very bottom shelf, bowls might be 'whammelled', that is turned upside down, or piled foot to foot, also to avoid dusting. This was particularly a tradition of western districts such as Argyllshire. Where a dresser had drawers in the lower stage,

STORAGE • 59

as most did, these were used to keep tablecloths, towels, cutlery and cleaning stuff, in addition to any papers and the assorted miscellanea which might gather in a kitchen. Latterly, transistor radios and alarm clocks were items commonly kept on dresser tops. Because the top surface of a Scottish dresser was usually much deeper than that of its English or Welsh counterpart, it was also a convenient place to put a television set.

The lower part of the dresser, or 'pot board', could be either open or enclosed and was used to store larger cooking pots and dishes, or perhaps an iron girdle if this was not hung on the wall. Enclosed dresser bases could also be used to hold groceries and bottles. Some dressers had central cupboards, at drawer level, which were ventilated with a row of battens. It has been suggested that these were used as winter nesting boxes for laying chickens. An example in the collection of Angus Folk Museum, Glamis, has a ventilated door which gives access to a dark compartment at the left, fronted by a sham drawer. This arrangement does indeed appear to have been designed as a poultry nesting box and not simply as a cupboard for meat. However, other examples have at their base level a continuous row of battens across the front of a space which seems to have been created as racked storage for large plates.

The Scottish dresser tradition can be broadly divided into two types: Highland and Lowland forms. The Highland type had a plate rack and lower stage with pot board, but these components were usually made separately. The top surface was always designed for working, with a generous lip at either end to stop baking ingredients from spilling over the edge. Some West Highland and Island dressers had sloping cowl tops which fitted the roofline of the low thatched houses they occupied. The Lowland dresser, on the other hand, was distinctive in that it usually possessed a lower section only, with no multi-shelved plate rack above. This type was also provided with a deep, functional top and anti-spill 'baking-board' sides. The work surface of a Lowland dresser was sometimes made from sycamore, a close-grained timber which does not harbour bacteria and can be repeatedly scrubbed. Although these dressers are relatively simple pieces of furniture, they are always characterised by their harmony of proportion; they are generally higher than three feet (0.91 m) and relatively deep from front to back. In this way they are similar to the lower stage of a kitchen press, which they resemble also in their common arrangement of three drawers over paired cupboard doors. Large versions of the practical Lowland dresser with plain worktops were standard in country-house kitchens all over Scotland; there are very good examples, for instance, at Monzie House by Crieff, Perthshire. In the smaller domestic setting they were rarely more than five feet long (1.51 m) and usually had some type of raised decorative feature on the top, either in the form of a stage, a shelf, a row of small drawers or a shaped back board. Because Lowland dressers were mass produced in the nineteenth century they were bought by customers from all parts of Scotland, including the Highlands and Islands.

Within the broad Highland/Lowland design distinction several local variants have been identified. For example, dressers of South Uist have characteristic scalloped edge decoration (see Fig. 12.8), Banffshire dressers are recognisable by their low proportions and tall shaped backboards with fancy moulded edges, while perhaps the most distinctive type within the Lowland tradition is the 'Carluke' dresser from Lanarkshire, which features the use of local fruitwood inlay. Again, the scale of a dresser reflected the size and prosperity of its domestic location and this explains the appearance of particularly handsome examples from large farms in Aberdeenshire and Perthshire.

Cupboards and shelves
Where the dresser was without cupboards it was frequently complemented by a corner cupboard for keeping cups, saucers, best teapots, glasses and possibly valuable beverages such as tea or bottles of spirits. For those who had money to spend, corner cupboards could be made to match as part of a dresser 'set', but very few of these survive together. Cheaper corner cupboards could be made 'fixed', that is let into the plaster or fitted on slips nailed to the wall. Walter Geikie shows even simpler versions which are just three triangular shelves let into the corner of a room.[12] Other typical cupboard types include a very tall, narrow, freestanding form, rather like a gun cupboard, which was found in small dwellings in both Highlands and Lowlands. They do not appear to have been made for a specific storage purpose but were a versatile and very space-saving shape which could fit easily into an awkward spot. Holdings which illustrate the range of these cupboards can be seen within the collections of the NMS[13] and at Bod of Gremista Museum, Shetland. The Shetland examples have characteristic wooden pin hinges that derive from Norse tradition, with upright pieces called 'charl pins' and 'harr' cross pieces.

Hanging shelves for china plates and bowls appear to have developed separately from dresser 'bases'. They were frequently used either by themselves or in combination with dressers or kitchen presses. A common form would comprise four or five shelves with guard rails and a deeper, boxed-in shelf at the bottom for bowls or perhaps spoons and other eating implements. They were usually painted and went by different dialect names in various regions; for example, in Shetland these shelves were known as a 'lem rack'.[14] There is an excellent collection of hanging shelves from several Angus Glens farmhouses and cottages in The Retreat folk museum in Glen Esk.

Other shelved storage in the small dwelling took advantage of whatever useful corner or space presented itself. The interiors of box beds, for example, were often fitted with a shelf at one end or even around all three sides. Rectangular kitchen tables were relatively uncommon until the later nineteenth century, but where they did occur, they had a close framework of stretchers forming a lower shelf, like a pot board, upon which large bowls or perhaps a bath tub could be usefully supported (see Fig. 23.4).

Although box beds were designed for sleeping, their storage function in the small house cannot be overlooked. Their flat tops, when not used as extra sleeping accommodation, could be used as shelf space for any large objects while the void between mattress and floor could be used to keep shoes, chamber pots, working equipment or, in some cases, 'hurly' beds on wheels. The inside of the bed also gave useful space to keep bulky items and provided hanging space for clothes.[15]

Storage furniture in larger homes
Larger town and country houses had a different repertoire of specialised storage furniture suited to a range of public rooms (see Fig. 29.4). In Edinburgh, where a considerable proportion of the New Town housing stock was rented in the late eighteenth and early nineteenth centuries, house furnishing displayed a startling uniformity.[16] This was partly because the houses themselves, along with complete *équipages* of furniture, were often rented out by entrepreneurial cabinet makers. The principal firms were Dalgliesh and Forrest of Potterrow, off West Nicolson Street, and William Trotter of Princes Street. A New Town dining room for instance, would have contained a sideboard with stage top, usually situated in its own wall niche (Fig. 4.4). Drinking glasses were most commonly kept inside the baize-lined stage, although deeper bottle wells within were not unknown. Bottles of wines and spirits were usually kept in a cellaret cupboard to one side. Freestanding octagonal cellarets, for use beneath a plain sideboard table, followed a distinctive Edinburgh pattern with curved brackets at the top of the legs. There is a good provenanced example by Young and Trotter of Edinburgh at Newhailes House, Musselburgh, Midlothian.[17] An item for storing alcohol outside the dining room was the liquor case, or 'guarduvine' as it was known in the West of Scotland and also in Northern Ireland, where Scottish cabinet-making traditions had taken root. The liquor case was a square, lidded box raised on four legs with a lock and usually six fitted bottle wells. It was distinct from the dining room cellaret, with which it is frequently confused, and could be used in the drawing room.[18]

Other supplies which might have been regularly stored in a fashionable drawing room between c1780 and 1850 were yarns, embroidery materials and tea. Edinburgh tea tables were specially made with a lockable frieze drawer divided into three 'canisters' for two types of tea plus sugar.[19] They were combination pieces which could be used as a tea or card table. Larger quantities of tea were kept in a 'store tea chest' which contained one large and two small canisters with lifting handles.[20] These could be decorated with stringing detail in contrasting woods or perhaps were japanned.

Amongst the repertoire of bedroom furniture in the fashionable country house of the early eighteenth century there is one piece of storage furniture which is worthy of mention. The 'lady's closet', a tall, narrow cabinet with mirrored upper stage and chest of drawers base, is unique to Scotland and seems to correspond with the room of the same name which occurred in seventeenth-century houses. Like the room, the piece of furniture was designed to store precious items such as jewellery and expensive tea things. The interior was fully fitted with small drawers and looking glass.

Figure 4.4 Stage-top sideboard, made by James Japp, Montrose, Angus, 1828.
Source: Mr and Mrs Stansfield, Dunninald, Angus.

The small fold-down flap beneath the door was probably not for writing but for placing cosmetics on. There are good examples of lady's closets at Dumfries House, Ayrshire, and Lennoxlove, East Lothian.

A sample survey of Edinburgh inventories and newspaper advertisements reveals that the chest of drawers became a relatively common household item only in the early nineteenth century, as a piece of furniture for use in the dressing room rather than in public rooms. William Trotter, for example, recommended three-quarter-sized chests of drawers for the dressing rooms in his estimate for the furnishing of Sir Duncan Campbell's town house at number 3 Moray Place in c1831.[21] Scottish chests of drawers are characterised by the appearance of three, rather than two top drawers. The most celebrated form is the so-called 'lum' chest which has a deep square drawer for storing hats, flanked by pairs of smaller drawers, one of which usually has a sham double front disguising a single deep drawer. The earliest labelled example of this pattern to have been traced is a chest of drawers by James Mein the Elder which is signed and dated 1825. Much earlier examples of the design appear in the English region of East Anglia and it is possible that it was transmitted to eastern Scotland through maritime links such as the herring fishery. During the nineteenth century, the lum or 'Scotch' chest became an increasingly common item in the small house where it had been a rarity before c 1830. It became ubiquitous in tenement dwellings in the west of Scotland, where mass-produced examples assumed quite gigantic proportions; but they were always relatively space-saving pieces of furniture because they were tall rather than deep and thus did not occupy too much floor space. A lum chest in a 'single-end' flat could be very versatile, storing clothes and bedding with space on the top for a display of china and ornaments.

OTHER FORMS OF STORAGE

Storage is made possible by more means than the use of furniture. Many homes have been constructed incorporating storage space within their structure (see Fig. 44.6). Larger homes may possess cellars, attics and box rooms, while built-in cupboards and shelves have been found in a variety of Scottish dwellings. In cottage homes use could be made of the open roof space. Rope or twine slung from roof couples or beams provided useful storage, particularly for food items such as dried and smoked fish or meat. Edible provisions were kept away from domestic animals and vermin by suspending them from the ceiling, usually in cloth bundles, but the most common item to be stored in this way was meat in the process of curing. Since it was rare for small cottage kitchens to have ceilings until the second half of the nineteenth century, meat hooks had to be fixed high up in the rafters, but wall hooks of all sizes were used to hold objects in daily use, such as cups, jugs, sieves and flat irons.

The storage of fuel was an essential consideration in the small house but in many cases a container was not used. Peats, for instance, could be conveniently and relatively cleanly piled up in the corner of a room. Kindling too, could be stacked next to the hearth for easy use. Keeping coal in a built-in bunker was principally a habit of urban tenement dwellers but was subject to regional variation. In Edinburgh and Dundee the bunker was customarily built into the window recess next to the sink while in Dumbarton, Glasgow and Paisley it could be found in the lobby. Bunkers usually had a flat top with hinged lid and were designed to hold two one-hundredweight bags of coal.

Inventory records of deceased householders give a good idea of the range of storage receptacles, other than furniture, to be found in the small rural house. Perhaps of prime importance was the water barrel, usually kept in a corner and raised off the floor in the same manner as a washing 'boyne'.[22] Walter Geikie illustrates these in his sketches of farmhouse interiors in the Lothians made between 1820 and 1830. Other goods such as butter, vinegar or fish might also have been kept in wooden staved casks or kegs, but smaller quantities were kept in earthenware jars or pots.

Ceramic storage vessels

The history of ceramic vessels in Scotland is a long and involved one and only a small number of salient points can be discussed here. In some areas of the country there are strands of remarkable continuity in the form and type of small containers from the Iron Age to the early twentieth century. In relation to ceramics, an example of this continuity is the Hebridean 'crogan' or hand-thrown, low-fired earthenware pot. The 'crogain' (pl) of the Western Isles were rudely shaped by hand from naturally occurring red or grey clay, dried and sealed with a milk glaze over an open fire. They were closed for storage with dressed animal skins tied with a thong or a cord. The contents were usually fluids such as milk, fish oil or medicinal liquids, but they are associated also with the storage of butter and salt. Very comparable items have been found in Brittany, Scandinavia and Eastern Europe. Hugh Cheape has made a detailed study of one Scottish 'crogan' tradition; that of Barvas on the west coast of Lewis.[23] He has shown that this pottery, known as 'Barvas ware', in some respects echoed Neolithic storage forms but also copied far newer types such as tea sets and tea pots. It was last made in 1935, by which time it was viewed by tourists as a picturesque anomaly with extraordinary links to the distant past.

One common type of Scottish ceramic storage vessel was the pig. This was an open-topped earthenware container available in a variety of sizes and usually employed for holding liquids. Pigs could be fitted with a lid and were closely related to another type of vessel known as the mug.[24]

Change from a clan-based subsistence economy to a capital economy, as well as improved distribution to remoter parts of the country, increased the nineteenth-century market for mass-produced earthenware and stoneware. By 1880, it had become common to find Leith or Glasgow-bought storage jars in places such as Shetland. Some of these containers, produced by several potteries, were made in deliberate imitation of the barrels which were their immediate functional predecessors. For example, the stoneware casks by the Barrowfield pottery of Glasgow made during the 1880s featured mock wooden staves, imitation iron bands and even a grained finish. Some were specifically for storing whisky, but others were used for a multiplicity of domestic supplies. The form endured in all parts of the country, but some potteries produced decorative finishes which were more striking than others. Aberdeen's Seaton Pottery, for example, made storage barrels with agate and splash-glaze patterns during the 1880s and 1890s.

Ordinary serviceable earthenware and stoneware vessels, made chiefly in Glasgow and Portobello potteries, seem to have been the main products of the Scottish ceramic industry during the nineteenth and early twentieth centuries. Butter dishes, bread safes, meal jars, preserve, sugar and extract containers provided a staple repertoire, but some newer lines, such as tobacco jars and bleach bottles, were successively introduced.

A significant new appearance in the household was the brand name, of either manufacturer or retailer, which was written prominently and permanently on the outside of storage containers. 'Possilware', 'Parazone', or the name of the local grocer became part of a new language of advertising which was able to penetrate inside the home. Almost certainly related to this was the custom of putting personal names and personalised messages on decorated jars. This was particularly prevalent in late nineteenth-century Scotland and it reflected the widespread practice of giving attractive storage containers as presents, particularly to newly-wed couples (Fig. 4.5). Personalised storage was not likely to be received as a present in the upper-income household, but otherwise there was little difference in the appearance of ceramic storage articles at different social levels.

The advent of tinned goods in the early twentieth century, the increased number of shops – which assumed some of the responsibility for the storage of foodstuffs – and the gradual demise of seasonal tasks such as pickling and bottling, dispensed with the need for many of the containers so far discussed. So too did the introduction of refrigeration from the mid twentieth century onwards and later the use of home freezers. There were some exceptions however. Jars for home-baking ingredients were still a prominent kitchen item in the middle of the twentieth century, but by the millennium, only tea, coffee and sugar containers commonly remained from an older repertoire. By the 1960s these were likely to be made from metal or plastic rather then wood, earthenware or stoneware. In this decade also, small glass jars of cooking herbs and spices made a widespread appearance. They were sold pre-packed by supermarkets and used for making versions of Mediterranean food, or for decoration, grouped together on a wall-mounted rack. Storage jars for dried macaroni have been used in Scottish kitchens since the early twentieth century,[25] but by 2000 they were much more common as different kinds of wheat pasta have become a staple

Figure 4.5 Meal barrels made at the Seaton pottery in Aberdeen. A wedding gift to Mrs Munro in 1893 kept at Tarrel, near Portmahomack, Ross-shire. Source: Crissie White.

food, often replacing bread or potatoes as the favoured carbohydrate (although many people keep dried pasta, along with other dry goods such as rice and pulses, simply in the packet in which it was purchased). In the following century it has been become preferable for some to buy fresh pasta in disposable packaging, so that even these storage items might be becoming redundant. In general, improvements in disposable packaging have had considerable impact on the number of food and some other smaller storage items – such as those used to hold cleaning agents – kept in the home. At the same time, a wide new range of glass and especially plastic containers, including disposable bags, exists for use in fridge and freezer.

STORAGE IN THE HOME TODAY

Today, storage of non-food items is more of an issue than ever as; despite the decreasing size of many households, the number of possessions working- and middle-class families are likely to have is greatly increased. The ownership of collections of objects by each member of the household – audio or video recordings, books and clothing, for example – is the expected norm and some individuals may collect on an impressive scale. The quantity of tools and materials considered necessary for the maintenance of the home, indoors and out, may also be substantial. To compound the problem of where to keep things, modern privately built housing tends to be small, especially bedrooms where many personal items are kept, and in addition is often poorly supplied with built-in cupboard and other storage space. Twentieth-century local-authority housing, despite usually being built with a regulation minimum area of storage space, is also frequently cramped.

A wide range of storage objects and furniture has been created in recent decades to hold collections in as space-saving and as organised a way as possible, for instance CD racks, which complement re-vamped versions of traditional items such as freestanding and wall-mounted shelves. Fitted storage furniture, can be found in many bedrooms, some bathrooms and most kitchens. Following a post-war period in which pieces of furniture that performed a single function had become popular, multi-purpose items are now back in vogue. For example, sofa beds in which the bedding can be kept, and 'cabin' beds with wardrobe and drawer space underneath the sleeping area are quite common.

NOTES

1. See Chapter 33 'Time Out' Work.
2. Bochel, 1979.
3. A widely used dark maroon paint was made by the Aberdeen firm Isaac Spencer & Co. It was used also in Orkney.
4. See the Viking kists from Maestmyr, Sweden.
5. A good example of a Fife-made kist with overhanging lid is that made for the session records of Carnbee Kirk, by Pittenweem, by James Mackie in 1847.

6. Collections of the NMS, Port Edgar store (2001), W 3156.
7. Grant, 1995, 173.
8. Records of Lands and Buildings in Edinburgh, Inveresk and Berwickshire Pertaining to the Orphan Hospital, SRO, GD 417/214/22.
9. There are two very good eighteenth-century Perthshire kitchen presses in the collection of the Highland Folk Museum at Kingussie, Inverness-shire. For further discussion of food storage see Chapter 23 Food.
10. Noble, 1992.
11. Loudon, 1839, 294.
12. Walter Geikie Drawings collection, Print Room, NGS.
13. Cupboard from Fala Dam, Peebles, collections of the NMS, Port Edgar store, W 3342.
14. Old Scots 'lame', earthenware, c1500.
15. See also Chapter 6 Heating and Sleeping; Chapter 7 Rest.
16. For detailed information on the repertoire of furniture found in the Edinburgh New Town house, 1780–1830, see Jones, 2000.
17. Owned by the NTS.
18. Jones, 2000.
19. Jones, 2000.
20. Jones, 2000.
21. See Jones, 1988.
22. A wash tub constructed of wooden staves.
23. Cheape, 1988.
24. Fenton, 1988.
25. Dried macaroni was first mass produced in Scotland by James Marshall & Co of Glasgow in 1935.

BIBLIOGRAPHY

Bochel, M. *Dear Gremista*, Nairn, 1979.
Carruthers, A, ed. *The Scottish Home*, Edinburgh, 1996.
Cheape, H. Food and liquid containers in the Hebrides: A window on the Iron Age. In Fenton, A, Myrdal, J, eds, *Food and Drink and Travelling Accessories: Essays in honour of Gosta Berg*, Edinburgh, 1988, 6–27.
Cruickshank, G. *Scottish Pottery*, Princes Risborough, 1987.
Grant, I F. *Highland Folkways*, Edinburgh, 1995.
Fenton, A. Pigs and mugs. In Fenton, A, Myrdal, J, eds, *Food and Drink and Travelling Accessories: Essays in honour of Gosta Berg*, Edinburgh, 1988, 38–49.
Fleming, J A. *Scottish Pottery*, Wakefield, 1973.
Jones, D. *Looking at Scottish Furniture: A Documented Anthology, 1570–1900*, St Andrews and Glasgow, 1987.
Jones, D. Scotch chests, *Regional Furniture*, ii (1988), 38–47.
Jones, D. Everyman's furniture in Lowland Scotland. In Fladmark, J M, ed, *Heritage, Conservation, Interpretation and Enterprise*, London, 1993, 325–35.
Jones, D. *The Edinburgh Cabinet and Chair Makers' Books of Prices, 1805–25*, Cupar, 2000
Loudon, John Claudius. *Encyclopaedia of Cottage, Farm and Villa Architecture and Furniture*, London, 1839.
Noble, R R. Highland dressers and the process of innovation, *Regional Furniture*, VI (1992), 36–46.

5 Lighting

ALEXANDER FENTON

One of the difficulties in thinking about past days lies in the fact that at different periods people were used to different sets of circumstances. Their daily actions, times of working, sleeping, eating, etc were conditioned accordingly. On the Aberdeenshire (Buchan) croft where I was brought up, there was no electricity and only the kitchen, which was also the living room, had a regular light in it in the evenings. The rest of the house was mostly in darkness. But familiarity with the place allowed us to run up the stairs without putting a foot astray, and to lay hands on books or other objects that were wanted for some immediate purpose. This kind of ability must have been even more marked in earlier days, when people were yet more accustomed to the dark. Even the available lights were dimmer. When I moved to Edinburgh, I took with me the paraffin lamp that had sat by me on the kitchen table when I was doing my evening homework on the croft. As a demonstration of past days to my two small daughters, I closed the curtains of the dining room, lit the lamp after having polished the glass and trimmed the double wick, and switched off the electric light. The contrast in light intensity brought surprised exclamations, and then they began to realise that the room appeared brighter as their eyes became more accustomed to the lower light source. They were also aware of the impression of a pool of light with a bright centre that diminished in strength towards the edges. The paraffin lamp was for sitting beside or clustering around, creating a degree of intimacy quite different from that of the more even spread of the stronger light sources of later times. In past days the lack of light may well have contributed to a sense of privacy in crowded conditions, when, for example, a husband and wife might have been in a box bed in the kitchen, and a passing traveller accommodated on a temporary bed of straw on the floor.

It would be interesting to know if the lack of light by night and sometimes by day in dark houses had an effect on people's eyesight. Eyes are to be used, and poor light may have in some cases made them sharper, though peering at papers or books, or trying to do fine stitching might not have been entirely favourable. Outside, there was no bright street lighting either. When I used to go to a farm in north-east Scotland, leaving Edinburgh behind, the first thing I did on the night of arrival was to walk down the unlit road, however dark, seeing only a farm kitchen light here and there, and at other times the stars, and if the moon was full – and nearly blinding from its lack of competition – there would be delicate moon shadows, those of fencing posts reaching out onto the road, and, at the back of stone dykes, an increased darkness. Such conditions stirred the imagination. So did the poorly lit kitchen in earlier days, with the glow from the fire and a constant slight movement and crackling of embers, and curls of smoke rising or eddying as a stray breeze penetrated door or window edges. Such circumstances could indeed encourage superstitious speculation.

Where the countryside remained in darkness, there were lights in the streets of at least the major towns from medieval times. In Edinburgh, there was a town regulation of 1519 that no one should be found on the high road after 9 o'clock at night without a lantern or 'bowatt' in his hand. This injunction was repeated in 1530, when the operative time was 8 o'clock. And so it was in Glasgow, Stirling, Ayr, Aberdeen, etc. Lights were not only to be carried, save on clear moonlit nights, but they had to be maintained on the walls of churches and public buildings, and all neighbours who had their dwelling on the forestairs on the high streets had to ensure that lanterns were hung up. In the fifteenth and sixteenth centuries public lighting, at least in towns, was an obligation upon the inhabitants. The 'bouet', as it was known within my memory in the north east, was a rectangular frame of wood, with a hinged side door, with horn or eventually glass panels to let the light pass through from a candle inside.[1] It was the equivalent of the later paraffin-burning stable or byre lantern, with its metal frame and glass sides (Fig. 5.1). But the contrast with today's well-lit streets, where even the shop windows often remain lit all night, can hardly be greater, and this outside brightness must inevitably lead to similar expectations in the home, even if the television screen's attraction may for a time lead to dimming of the main lights.

We had made progress on the Aberdeenshire croft, however, since my schooldays. The lamp with the wick was replaced by a Tilley lamp, hung from the ceiling, in which a light with a mantle was fed by pressure. The lamp had to be pumped from time to time to maintain the flow of paraffin. We never achieved the glory of electric light. Such change to an improved

Figure 5.1 A bouet, in this case three-sided. Source: From Ross, A. *Scottish Home Industries* (1895), Glasgow, reprint 1976, 33.

form of lighting, though still paraffin fuelled, spanned the middle of the twentieth century, on the croft as elsewhere in rural districts. In towns, however, lighting by gas and electricity had become common by the end of the nineteenth century.

In earlier days, before industrialisation had generalised the use of fuels such as paraffin and gas, or a light source like electricity, it was necessary to be self-sufficient. The hearth itself, whether placed in the middle of the floor or against the gable wall, was good enough for many purposes, though firelight is fitful and scarcely adequate for activities like reading for longish periods. For these, supplementary sources were needed, as also to aid movement within the house.

TEMPORARY LIGHTS

If a light was needed elsewhere, a temporary strategy was to take a burning peat in the tongs and carry it.[2] This could be used outside also. Anyone wanting to go 'through the hoose', according to an Aberdeenshire report, would take a bit of peeled and dried rush as a wick, dab one end in dogfish oil and the other in peat ash to stop the oil dripping, and light it at the fire.[3] A light made of hemp fibres was used in Renfrewshire to provide light for doing small jobs about the house,[4] and early nineteenth-century sources from Ayrshire and Angus speak of the pith of hemp, or of a wild carrot or parsnip, dried and used as a small candle.[5]

A brand of bog fir or a torch made of cotton rags dipped in tallow was known as a 'ruffie'. This would serve for showing a visiting stranger to his bed, or for going out to fetch peats from the stackyard.[6] The same name was used in the south west of Scotland, for example in the parish of Tongland in Kirkcudbright. There, in the 1790s, when the goodman of the house conducted family worship in the dark evenings of winter, a ruffie was lit to let him read the psalm and a portion of Scripture, before he prayed.[7] Such a practice was widespread in Britain. It is recorded of Suffolk that on Sunday evening the Bible would be brought out and all in the assembled company would read a bit, 'by the aid of a rushlight candle held in an ancient stick ... This was handed round to each reader in turn, since the smelly light gave only sufficient illumination to see in the darkness, but it sometimes meant singed eyelashes if they were not careful.'[8]

There was, of course, always a fire risk associated with naked lights. A burning peat, no matter how firmly gripped in the tongs, could throw sparks, especially if encouraged by an outside wind, and a fire could start, just as it might do with the stub of a discarded cigarette nowadays. Roofs were often of thatch, and the inner structures of homes contained much dry timber. Timbers near the hearth could be well charred, but the charring itself provided a form of protection from the fire, though accidents could still happen. When I visited a croft with a wooden-canopy ('hingin-lum')[9] type of fireplace at Ardoyne, Insch, Aberdeenshire in 1961, the upright planks at the sides of the hearth had caught alight the day before when for some reason there had been a big fire – perhaps to boil the 'hens' pot' – and had to be torn off. No one seemed to be particularly alarmed. So risks there were, no doubt greater overall than those of the present day, though a faulty electrical circuit or, even more, a gas leak can do much damage, however well regulated the installation.

THE FIR CANDLE

A common form of lighting, which could be for a short-term purpose or a longer one, was the fir candle. Bog fir was in wide use in areas of peat mosses, from which supplies were regularly dug. Four people were reprimanded by the Kirk Session in Keith, Banffshire on 26 June 1715 for 'holling [digging] fir on the Sabbath morning'.[10] Fir candles made from bog fir were also used by the Highlanders in Atholl, Perthshire, and by the folk in the higher parts of Angus.[11] The resinous fir was split into suitable lengths with a stout knife, and dried. This might be done by sticking the splinters into the links of the crook, above the fire, or they might be laid on a brander,[12] or on a square, ribbed wooden frame called a 'calchen', which was hung in the chimney.[13] The splinters were about one-and-a-half to three feet (46–91 cm) long. In north-east Scotland, on the occasion of the reading of a newspaper on winter evenings, an unusually large peat fire was prepared, which 'cast a warm ruddy light around', and one of the family sat next to the reader with a sheaf of fir candles to supply light.[14] As they burned, from time to time they had to be 'snited', ie the 'aizle' or charred tip had to be knocked off so that they continued to give a clear light.[15] The fire was obviously expected to contribute to the lighting here, and this likelihood is emphasised by information dating from the 1790s in relation to the parish of Kilmarnock in Ayrshire. There a particularly bright burning type of coal was called 'fire coal' and had also the name 'seeing coal', so emphasising the amount of light it could emit.[16]

Fir candles were held in various ways. Single ones could be held aloft in the hand, and an account of around 1730 tells that for want of candles, when a better light was needed than that of the fire alone, the Highlanders:

> Provide themselves with a Quantity of Sticks of Fir, the most resinous that can be procured: some of these are lighted and laid upon a Stone; and as the Light decays they revive it with fresh Fuel.[17]

This sounds very much like a cross between a fire and a lamp.

Alternatively, there were blacksmith-made iron clips with sharp ends that could be stuck into a wall. A jointed bar, similarly fixed, might be used and it allowed for some adjustment of the position of the light. But the most prominent form was the 'peerman' (poor man), also variously called the 'peer page',[18] 'clivvie' (piece of wood or iron with a cleft), and in Gaelic 'bodach' (old man),[19] which could stand on floor or table. It consisted of a stick about three feet (91 cm) long placed in a hole in a stone used as a base, with an iron holding device at the top end. This could be positioned

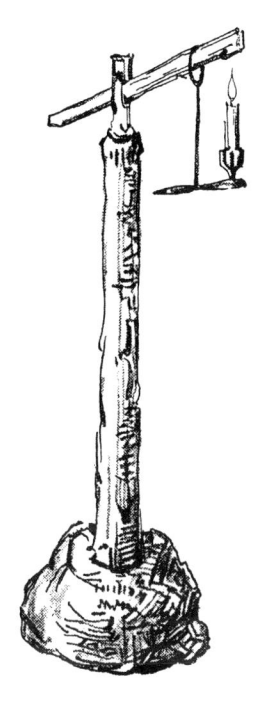

Figure 5.2 A peerman. Source: Alexander Fenton.

alongside a reader, and 'one to light the candles, fix them in the homely candlestick, and keep them trimmed, was close at hand'[20] (Fig. 5.2). There were, of course, individual variations. For example, a peerman from Banffshire had fixings for two fir candles.[21] By the end of the nineteenth century they were becoming museum-worthy curiosities, and the bundle of fir candles on the mantelpiece in the farm kitchen at White Stripe, Enzie, Banffshire, then in constant use, and the peerman fixed in a block of wood at the nearby farm of Hill Park, had become 'neo-archaic' objects which marked the end of a long road.[22]

THE CRUISIE LAMP

Of the devices which required fuel, the cruisie lamp is certainly the best known, and has a form which is very widely distributed in Europe, although most commonly in the form of a single shell, usually of earthenware, which acts as a reservoir for the oil, with a spout on which lies a wick. The cruisie had other localised names in Scotland – 'collie' in the Northern Isles, 'guse neb' in Caithness, from the shape of the spout, and 'eelie dolly' in north-east Scotland. In the same area the two parts of the cruisie were called 'shalls' (shells). The Gaelic name is 'crùisgean'.

The standard Scottish form is of iron, and is blacksmith made. Many smithies had anvil blocks with moulds hollowed out in them for shaping cruisies. Typically, the Scottish cruisie had two boat-shaped shells, one fitting inside the other. The lower shell caught any oil drips from the upper one. At the back of the lower shell was an upright piece which ended in a swivel with an iron spike and a hook, for sticking in the wall or hanging up, and from the front of this upright there stood a notched tongue of metal that held the upper shell and allowed it to be tilted forward increasingly as the level of oil dropped.[23] It has been suggested that the notched bar is unique to Scotland, but it is known from Ireland also[24] (Fig. 5.3).

The fuel was fish-liver oil. The livers of saithe, pollack and cod were left for some time in a jar, then boiled.[25] An account from Skye says that 'the oil was dark, like port wine, but thin and good.' In north-east Scotland it was called 'black oil'. It was made there from the livers of gutted fish, left in an earthenware pot or 'craggan' until partially liquid. Then the decayed livers were set over a slow fire to dissolve them completely. The pure oil was poured off into a craggan, and the refuse was thrown away.[26] The practice in Ireland was for the livers of hake or cod to be put in a big vat or tub. Salt was shaken over them. They were left for a time in the open for the sun to draw out the oil, then this was skimmed off with a large shell, and put in a well-corked jar or bottle. But tallow, lard and even butter could be used at times. If the woman of the house wanted to go to some dark corner, she would dip a rush in the oil, light it and carry it in her hand.[27]

Clearly there were different techniques for processing the oil, and the Irish evidence is for open-air work, but when the simmering was done in the house, the smell of livers combined with that of peat smoke and other odours would have created a very strong atmosphere which was part of the everyday business of living, and part of the work of preparing light.

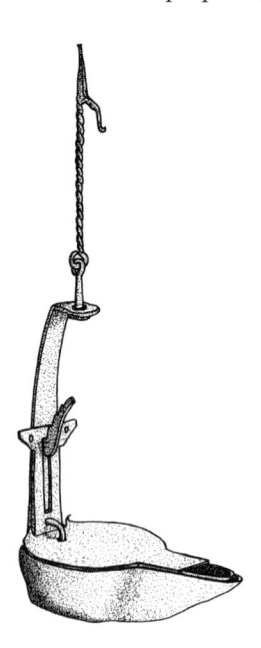

Figure 5.3 A cruisie, possibly from Angus. Drawn by Colin Hendry in 1978. Source: SLA128F C.6781.

The wick of the cruisie was usually the dried pith of rushes, although cotton thread or a piece of worsted could also be used. Evidence from Ireland tells that the rushes were picked in late summer, peeled, apart from one or two strips to keep them intact, dried in the sun, and then dipped in melted fat.[28] This would have made them into small tapers, but for use as cruisie wicks, such dipping was not necessary.

With their low level of illumination (although this could be enhanced by using more than one strip of pith together), their smells (in Banffshire and around the cruisie was also known as 'Reekie Peter'),[29] and the need to have them in a fixed place – moving around with them would have risked spilling the oil – cruisies were quick to disappear, along with fir candles, in the latter years of the nineteenth century as better forms of lighting were introduced on a commercial scale. They had been part of the household scene for centuries, however, and even formed part of superstitious beliefs in the north east of Scotland. 'If ... "the shalls" (the cruisie) are turned towards the door when a new female servant makes her arrival, she will in no long time leave the service.'[30] However, the death knell of the cruisie was well and truly marked by a story told of my grandfather, who passed his apprenticeship as a blacksmith at Kinmundy in Aberdeenshire in 1879. He had a cruisie-making mould and an antique dealer asked him to make a supply of cruisies for sale in his shop. But he also asked my grandfather to leave them in the rain for a time, to add the rusty authenticity of age. Whether or not these could be regarded as genuine, the cruisie, like the peerman, had become a museum object or an archaic specimen by the end of the nineteenth century. The change must have come quickly. In the 1870s, it was worth the while of a small trader with a boat to render saithe livers and sell the oil in jars as cruisie oil, at 1/6 a gallon, in Tayvallich and North Knapdale in Argyll. But later in that century, paraffin lamps with wicks had largely replaced the cruisies.[31] However, cruisies sometimes continued to be used in outbuildings or in the mill, as was reported in the 1890s from Shetland, after simple paraffin lamps had replaced them in homes.[32]

THE CANDLE

Candles made of wax or tallow are of ancient vintage. Candlemakers were prominent amongst medieval burgh craftsmen, and the quality and price of candles was controlled by town statutes. They were for people of all social levels, although the lower levels would scarcely have made as much use of them as their better-off contemporaries. Those who were rich enough to have their goods listed in a testament could possess candle vats, no doubt for holding the wax or fat in which the wicks were dipped, and wicks themselves are also specified (Edinburgh, 1647). An Aberdeen source refers to five pairs of riven sheets for making candle wicks (1519). The prepared candles were kept in creels or chests. Candlesticks could be ornate, prestigious items of silver or brass.[33]

A shopkeeper in Perth, William Arnot, had a candle works which operated into the early nineteenth century. He sent large quantities of rendered tallow to the London market, and he and others did a lot of business in candles, although by the 1830s the trade had fallen off. This was partly due to the introduction of gas, and partly to the considerable supply of candles from Aberdeen, 'and many genteel families, where gas is not yet introduced, use wax and sperm candles'.[34] So at this date there was still an upper-class emphasis on the use of candles. A Shetland estate can be used as a further example. There, in late October in the nineteenth century, animals were slaughtered and salted and candles, in their hundreds, were dipped by the servants. 'The process was a tedious, untidy, sickening one, for it was nothing but dipping wicks into hot tallow all day long, until the candles were considered thick enough.'[35] Candles were also made at lower social levels, however. A Lanarkshire shepherd remembered their making when he was a small boy around 1880. Braxy grease, the fat from a diseased sheep, was used. A wooden stick with wicks twisted and hanging from it was held over a tub of melted tallow. The wicks were dipped and re-dipped until the required thickness of candle was reached and then they were laid away to harden.[36] A Galloway account of 1877 relates that the tallow was melted in the kail pot, with warm water put in the bottom of the pot to keep the fat fluid. One stickful of candles was being dipped, while another was drying, supported on an upturned chair.[37] No doubt candles, when stored, could provide an occasional tasty nibble for mice and rats, which well know how to make the best of any resources available.

The equivalent of the peerman in the south west of Scotland was the 'carle' or 'kerl'. This was for holding candles, in a bracket fixed to a vertical rod in such a way that it could be moved up and down. It was meant to stand on the floor, and had a loop at the top for moving and carrying:

> The end of the loop was usually turned up to form a small hook upon which the candle snuffer was hung. The candle bracket could be moved down the vertical rod until the flame of the candle was opposite the eye of the spinning wheel as would be required, or upwards for reading, sewing, etc.[38]

These candlesticks, therefore, were very much intended to serve working purposes, providing evening light for spinning and sewing activities.[39]

The paraffin-wax candle had its period between 1820 and 1850, when it spread as a cheaper light source than beeswax or tallow candles, and it was replaced by paraffin lamps and gas and electric light later in the nineteenth century.[40] The town of Tain, for example, had been lit by gas shortly before 1841, and gas 'had been introduced into almost all the respectable houses' (Fig. 5.4).

With such improved, and for the most part dependable, light sources there was probably an explosion of home crafts and hobbies in the nineteenth century. This concept remains to be fully explored and tested, but it is likely that the sewing of samplers, for example, with their varieties of stitching, received a strong stimulus. Board games would have been easier

Figure 5.4 Gas-light fitting in a house in Edinburgh *c* 1900. Source: SLA 128F 56/38/6.

to play in evening leisure time.[41] It may be also that movements which fostered household industries, such as the Scottish Home Industries Association (founded by the Countess of Rosebery in 1889), were greatly helped by the extension of the daylight hours into better-lit evenings.[42] The standard textile activities of earlier days – carding, spinning, weaving, etc – on the other hand, could be carried on quite well under gloomier conditions, although in villages and towns where weaving was undertaken to any considerable extent, additional windows were inserted into homes to make it easier to see. Most weaving took place at ground level, partly on earthen floors to ease the noise, but in some places, such as Errol, Perthshire, and Kirriemuir, Angus, there were weaving 'galleries'. A large building in The Square, Errol, had such a gallery. To light it, there were small windows under the eaves. The sills were level with the floor, and the upper parts of the looms projected into the large roof space.[43]

NATURAL LIGHT

There is a painting of the kitchen cum living room of a home in Perthshire, dated to 1850, in the collections of the NMS (Fig. 5.5). The room is lit by a fixed window with six panes of glass, the wall at each side being splayed to increase the flow of light into the room as much as possible. There are no curtains, which may indicate that people were less concerned with privacy than they are nowadays, or perhaps the prestige of having curtains at all was first shown by hanging them in best room windows, as was happening in Tain, Ross-shire, by 1841 (see Note 47 below). The items on the windowsill and on the floor near the window are there because of the proximity of natural light. On the sill there is a mirror, and some small items which go with the making up of the hair of the lady of the house. The table is to the left of the window, against the wall. Immediately beneath the window is a hard-backed wooden chair with what looks like the folds of a newspaper on the seat, and to the right is a horizontal spinning wheel. There is a glowing fire in the room, but there is no indication of any source of artificial light. In urban tenement blocks, such as those at Dalry Meadows, Tynecastle, Edinburgh in 1896, the sink was placed immediately in front of the kitchen window or the only window if the home consisted of one room.[44] In such ways, the functional use of natural light can be seen (see Fig. 15.4), as well as the role of the kitchen or living-room window as a focal point of daytime activities, jointly with the hearth where women cooked and baked by day and by which both sexes sat in the evening, talking or carrying out various auxiliary tasks.[45]

Natural light was once – broadly speaking up till the 1760s and much later in the more peripheral areas – a relatively scarce commodity in everyday dwellings. The pages of the *OSA* of the 1790s and of the *General Views of the Agriculture* of the various counties of Scotland from the late 1790s into the early 1800s provide a good deal of information about farm buildings, including the nature of their windows. Windows, like doors, were not

Figure 5.5 'A Woman's Place'. Perthshire interior of about 1850. Source: SLA C.8339.

regarded as structural, and were inserted after the walls had been built (see Fig. 28.3). They might even be removed, as in the case of farm-workers' housing in Berwickshire in the 1840s, when the occupants were going to a new dwelling. Window openings were not necessarily glazed. In the parish of Whithorn, Wigtownshire, for example, 'about 60 years ago [ie in the 1730s] ... In the farmers' houses there were no windows of glass. The light was admitted through openings on each side of the house, and that in the windward side was filled with straw in blowing weather.'[46] Somewhat more sophisticated were windows infilled with wooden boards, in which there could be small panes. In Clackmannanshire, window openings could be fitted with two boards which opened in the middle (and were presumably hinged at the sides), with a small pane of glass in the upper part of each board. Such window openings or light slots, as they are sometimes called, boarded or unboarded, were about 12 to 15 inches (30–38 cm) square, according to a report from East Lothian. This size lent itself to what seems to have been a common infill of four small glass panes, which might be regarded as full glazing.[47] These were being increasingly replaced by larger, glazed windows from the 1760s. Wooden boards are also mentioned in Perthshire, Angus and Banffshire, and were evidently quite common, here and elsewhere, until the third quarter of the eighteenth century, with a few survivors at later dates.[48] In remoter areas, such as in parts of Shetland, there were thatched houses without windows, but with two to six openings in the roof which let out smoke and let in light. They were too small to provide any very strong light, however. Where there were window openings, these were sometimes filled by a semi-transparent bladder, or an untanned sheepskin cleaned of wool and stretched on a frame.[49]

Glazed windows were, however, to be found in churches, manses, castles and in more elegant houses from medieval times. The craft of glazier, 'glasinwricht', 'glaswricht' or 'glasser', was well established by the early 1400s. A Window Tax was applied in England under William III, in 1695, but in Scotland not until the 1740s. It was repealed in 1851. However, its impact on the fenestration of housing in Scotland is hard to assess. It has been held to be responsible for blocked up window spaces, though there is little evidence to prove that this was so, and in any case, the tax does not appear to have been collected effectively in Scotland for the housing of the lower classes.

The questions of light and ventilation hang together. In 1841, a surgeon in the town of Tain, Easter Ross, wrote a report on its sanitary conditions, and his comments undoubtedly have a much wider application in Scotland:

> With respect to their houses the chief defect seems to be want of ventilation. The windows, which are often composed wholly of wood, are far too small, and are too seldom left open; and I need hardly add that the want of a free circulation of air, besides predisposing to disease, is most prejudicial to invalids, and especially to the bedridden ...

> The houses are not well constructed with respect to ventilation. In instances not a few, the chimneys being ill framed, the houses are subject to smoke, and are consequently filthy. In order to obtain a desirable degree of *warmth*, every opening by which the external air might be admitted is carefully blocked up.

As far as recently built farm-servant housing was concerned, floors were of earth and clay, there were well-constructed fireplaces and chimneys, walls were whitewashed with lime, all the rooms were as a rule used for sleeping, and in the best rooms, curtains were often seen. But in general, the windows were 'closely shut up with boards'.[50]

Even in rural houses built in the nineteenth century, there were windows not made to open. This was as much due to questions of economy as to questions of heat conservation. Sash-and-case windows became regular features of buildings in both town and country, in England as in Scotland, and their existence was a matter of surprise to foreign visitors, who did not understand why they should be so prevalent when they were so difficult to make draughtproof.[51] But at least windows which could be opened allowed for ventilation, and being larger than their predecessors, allowed more light to enter. This applied also to attic skylights. Street doors with glass panels are nowadays not uncommon for lighting a lobby. Fanlights above the door are more ornamental than functional, and as a rule too small and too high to provide much light. The internal lobby door was and is very often glazed.

There is here a subject for further detailed investigation. Some houses, such as the blackhouses of Lewis, might have no windows at all and the light would come from the open door and from the smoke hole, as well as one or more light openings, not always glazed, in the roof. Single-width houses might have a door and two windows at the front, small gable windows and a solid rear wall. During the mainly nineteenth-century spread of but-and-ben type homes, a small window was added centrally at the back, to light the closet or milk house which lay between the two rooms. Urban dwellings built in large housing blocks were probably worse off than rural homes for natural light sources. It took the coming of gas and electricity to improve lighting conditions there.

Nowadays light is part of the consumer society. It is freely used in ornamental fashion, for example illuminating small Christmas trees in the home and whole streets and large buildings and other objects on festival occasions in towns and cities. Many homes keep candlesticks in some forgotten corner of a glory hole for emergencies like power failure, and battery-operated hand torches still play an important if occasional role. But the desire for more intimate space remains strong and even in open-plan housing the inhabitants will have separate islands of light from side lamps for the relaxation of reading or for carrying out small domestic jobs. In this they may resemble their forefathers, but the source of the power that they enjoy is supplied from a central external source, the light fittings are mass produced, and there is no longer the self-sufficiency of earlier times which are, in fact, not as remote as might be thought.

NOTES

1. DOST, s v Bowat; SND, s v Bouet.
2. Martin, 1987, 164.
3. Information from Mrs Johanna Murray, aged 87, Post Office, Blackhills, Aberdeenshire, sent to the *Aberdeen Press and Journal*, April 1963.
4. SND, s v Keppock (1837).
5. SND, s v Kempit.
6. Horne, 1907, 123.
7. OSA, IX (1792), 328: Tongland, Kirkcudbrightshire.
8. Jobson, 1959, 33.
9. See Chapter 6 Heating and Sleeping.
10. Gordon, 1880, 88.
11. Steele, 1826, 17.
12. Grant, 1961, 184.
13. SND, s v Calchen (recorded 1808); an example was put into the National Museum of Antiquities of Scotland in the late nineteenth century. See Mitchell (1897–8), 184.
14. Gregor, 1874, 20–1, 33; see also Chapter 21 Reading and Study.
15. Linn, 1883, 113.
16. OSA, II (1795), 93: Kilmarnock, Ayrshire.
17. Burt, 1974, II, 136.
18. Gregor, 1866, s v.
19. Linn, 1883, 111.
20. Gregor, 1874, 33; Grant, 1961, 184–5.
21. Mitchell, 1897–8, 183.
22. Wallace, 1898–9, 55–6.
23. See Goudie, 1882–3; Mitchell, 1896–7.
24. Goudie, 1882–3; O'Neill, 1977, 30–1.
25. Martin, 1987, 162.
26. McGregor, 1879–80, 145; Gregor, 1874, 22.
27. O'Neill, 1977, 30–1.
28. O'Neill, 1977, 30–1.
29. Linn, 1883, 110.
30. Gregor, 1881, 30.
31. Fraser, 1962, 25, 32.
32. Smith, 1892–3, 295.
33. DOST, s v Candil(l), Candill-kist, Candilmakar, Candil-making, Candilsteke.
34. Penny, 1836.
35. Edmondston, Saxby, 1888, 97.
36. SLA, NMS, information collected by J L Mackenzie in August 1959.
37. Trotter, 1877, 163.
38. *Galloway Gazette*, 8 December 1951.
39. Maxwell, 1887–8.
40. For details, see O'Dea, 1958, 49–65; also, for example, Briggs, 1988, 399, 401–2.
41. See Chapter 9 Needlework; Chapter 11 Games and Sport; Chapter 25 Childcare and Children's Leisure.
42. Ross, 1976, 62.
43. Fenton, Walker, 1981, 172; see also Whittington, 1967.
44. Reproduced in Carruthers, 1996, 45, 63.

45. See Chapter 18 Auxiliary Work in the Rural Home.
46. *OSA*, V (1983), 552: Whithorn, Wigtownshire.
47. Fenton, Walker, 1981, 113, 127, 129, 144.
48. Fenton, Walker, 1981, 121, 123, 126, 136.
49. *NSA*, XV (1845), 97: Sandsting and Aithsting, Shetland.
50. Cameron, 1841, 3, 6, 17.
51. See, for example, Barley, 1971, 69–70.

BIBLIOGRAPHY

Barley, M W. *The House and Home: A Review of 900 Years of House Planning and Furnishing in Britain* (1963), reprint London, 1971.
Briggs, A. *Victorian Things*, London, 1988.
Burt, E. *Letters from a Gentleman in the North of Scotland to his Friend in London* (1754), Edinburgh, 2 vols, 1974.
Cameron, J. *Report on the Sanitary Condition and General Economy of the Town of Tain and the District of Easter Ross, Made to the Poor Law Commissioners*, npp, 1841.
Carruthers, A, ed. *The Scottish Home*, Edinburgh, 1996.
Edmondston, B, Saxby, J M E. *The Home of a Naturalist*, London, 1888.
Fenton, A, Walker, B. *The Rural Architecture of Scotland*, Edinburgh, 1981.
Fraser, A. *Tayvallich and North Knapdale*, Glasgow, 1962.
Gordon, J F S. *The Book of the Chronicles of Keith*, Glasgow, 1880.
Goudie, G. The crusie, or ancient oil lamp of Scotland, *PSAS*, XXII (1882–3), 70–8.
Grant, I F. *Highland Folk Ways*, London, 1961.
Gregor, W. *The Dialect of Banffshire*, London, 1866.
Gregor, W. *An Echo of Olden Time from the North of Scotland*, Edinburgh and Glasgow, 1874.
Gregor, W. *Notes on the Folklore of the North East of Scotland*, London, 1881.
Horne, J, ed. *The County of Caithness*, Wick, 1907.
Jobson, A. *An Hour Glass on the Run*, London, 1959.
Linn, J. Peer-men, *Transactions of the Banffshire Field Club* (1883), 110–3.
Martin, A. *Kintyre Country Life*, Edinburgh, 1987.
Maxwell, Sir H E. Notes on the 'carles' or wooden candlesticks of Wigtownshire, *PSAS*, XXII (1887–8), 113–8.
McGregor, Rev A. Notes on some old customs in the island of Skye, *PSAS*, XIV (1879–80), 143–7.
Mitchell, A. Some notes on Scottish crusies, *PSAS*, XXXI (1896–7), 121–46.
Mitchell, A. A description of some neo-archaic objects, from various parts of Scotland, recently added to the Museum, *PSAS*, XXXII (1897–8), 181–194.
O'Dea, W. *The Social History of Lighting*, London, 1958.
O'Neill, T. *Life and Tradition in Rural Ireland*, London, 1977.
Penny, G. *Traditions of Perth*, Perth, 1836.
Ross, A. *Scottish Home Industries* (1895), reprint Glasgow, 1976.
Smith, J A. Notes on some stone implements, etc, from Shetland now presented to the Museum, *PSAS*, XVII (1892–3), 291–9.
Steele, A. *The Natural and Agricultural History of Peat Moss or Turf Bog*, Edinburgh, 1826.
Trotter, R de Bruce (ed Saxon [pseud]). *Galloway Gossip Sixty Years Ago: Articles Illustrative of Manners, Customs and Peculiarities of the Aboriginal Picts of Galloway*, Choppington, 1877.
Wallace, T. Notes on some antiquities in Enzie, Banffshire, *PSAS*, XXXIII (1898–9), 55–6.
Whittington, G. The imprint of former occupations and the Improvement Movement on house types in Fife, *Folk Life*, 5 (1967), 52–7.

6 Heating and Sleeping

ALEXANDER FENTON

The concept of good heating – as for good lighting – has varied through time. On my parents' Aberdeenshire croft in the 1940s to 1950s, the main rooms had hearths, but I was never aware that fires were lit except for in the kitchen. The beds, however, would be warmed up by an earthenware 'pig' or later a rubber hot-water bottle before we went to bed, and in the winter mornings we would wait till the rattle of the poker on the kitchen grate told the world that Mother had the fire going. Then we dressed as fast as we could and headed for the kitchen, and breakfast. But as a matter of fact, no one would have thought of saying the bedroom was cold. Cold was just a fact of life, countered by an extra cover on the bed, an extra jersey, physical activity or sheer stoicism. The contrast with modern times, when more and more houses are being built with, or upgraded to include, central heating and double glazing (replacing the draughty sash windows of former days), could hardly be greater. The expectation now is that even in winter, tee-shirts and bare arms are appropriate to the room temperatures inside. It may be that people have become less resistant to illnesses caused by draughts, on the one hand, and on the other they may have become more tolerant of dust mites and other creatures.

THE OPEN FIRE

The kitchen fire at the home croft had an open grate. It is a characteristic of the great majority of Scottish and indeed British homes at all social levels through the centuries – as compared with other parts of Europe – that open hearths with chimneys or smoke vents prevail. Few homes had the ovens or stoves which have marked other countries of Europe in more recent centuries. This is not to say that open fires were unknown in these countries in earlier times and are to be found in survivals in the economically less advanced areas and in seasonally occupied small dwellings such as those sited in the hill pastures for the summer grazing of stock.[1]

In most Scottish homes, only one fire was normally in actual daily use (see Fig. 28.3). This served for heating, cooking and baking, and disposing of rubbish. Everything which was combustible, including bones and other food refuse which could not be consumed by the livestock or domestic creatures, went into it and eventually found its way to the arable fields in the form of ashes, which were tipped first onto the ash middens. Ash pits sunk into the floor near fireplaces were widespread. I photographed examples in north-east Scotland, in a farm kitchen near Keith, Banffshire, and in a croft in the heart of Buchan, in the 1980s, when they were still in use. An eighteenth-century chapbook tells of the use of such a pit as a place of toilet relief for the folk sleeping overnight in the kitchen of a farmhouse near Dalkeith, Midlothian – a much warmer solution than going outside.[2] Clearly, sanitation and the need to apply all possible manurial substances to the crops on the land formed part of the mental construct of the rural populace of the time. In the dark evenings, the hearth was the source of light as well as heat, enabling a variety of activities to take place, such as, for the women, spinning, carding or knitting; and for the men, the making of small items of wood, bent grass or straw. In the words of a Banffshire observer in 1874:

> All were busy. One of the women might be knitting, another making, and another mending, some article of dress. Of the men, one might be making candles from bog fir – 'cleavin can'les' – another manufacturing harrow tynes of wood, a third sewing brogues, and a fourth weaving with a *cleek* a pair of mittens.

And if there were children of school age, homework would be done by the fireside also.[3]

The main period of agricultural improvement got into its stride from about the 1760s, and for the wider social levels was accompanied by housing changes which included gable chimneys for smoke extraction. As part of the process, in the late eighteenth and nineteenth centuries coal-burning grates came into increased use, at least in the best rooms.

Coal is referred to in twelfth-century documents, and was being systematically worked by the end of that century and the beginning of the next.[4] It was soon, of course, in regular use in higher-class homes. For example, in the *Accounts of the Lord High Treasurer of Scotland*, coals were got for 'Lady Annis chalmer' in 1551, by which time its use was well established. Illustrations in books show that in the grander Scottish houses fireplaces could be very ornate, works of art in which the hearth and its surrounds were designed as a unit, with mirrors or paintings or classical motifs in stucco. Examples can be seen in Kinross House, Kinross-shire, in the ballroom, drawing room and present-day study, with mirrors or paintings in the panels above.[5] At 5 Charlotte Square, Edinburgh, can be seen in the drawing room a brazier or basket type of grate, which was there during the occupancy of the 4th Marquess of Bute.[6] The fireplace with an ornamental grate or fire basket is the focal point in the lavishly ornamented entrance hall at Arniston House, Midlothian, designed by William Adam in 1726. It is surrounded by mock-medieval woodwork in Sir Walter Scott's nineteenth-century 'great hall' at Abbotsford, Roxburghshire. In the great dining room in the House of Dun, Angus, also designed by William Adam, it is the centre point of an exuberance of stucco decorative motifs dating to 1742.[7] But in spite of all this glory, the essential element for producing heat was an iron grate for burning coal, and until the coming of cheap power – gas or electricity – for heating, servants had to continue to carry coal and keep the hearths clean.

Coal was brought by sea from the coal-bearing districts of Central Scotland; from 1612 onwards there are numerous reports in the *Aberdeen Shore Work Accounts* of boats from Leith, and Borrowstounness (Bo'ness) and Queensferry, West Lothian, with coal as part of their cargoes.[8] Shallow-draught boats could land coal at suitable beaches, so that the possibility of getting coal was not limited to towns with harbours. The practice of landing coal on beaches was still going on a few years before World War I. For example, a photograph dating to a little before 1907 shows four coal carts waiting in shallow water by the Cromarty Firth, till the tide had gone far enough out for them to reach the boat.[9]

A major period of change relates to the late eighteenth century, when the situation was still mixed in some places, with brushwood, furze, broom, etc being used in summer and coal in winter. It was said in 1790–1 of the parish of Stranraer, Wigtownshire, that the chief fuels were peat and turf, usually transported 3 or 4 miles (4.8–6.4 km) – which, when added to the value of time spent in casting and in the drying processes, would amount to no little expense – so that:

> Many of the inhabitants burn coals in their rooms. These are brought from Air [*sic*] or Irvine by sea, and cost the purchaser about fourteen pence the herring-barrel, including the price of leading from the shore. A family, keeping only a regular fire in the kitchen, and another in a parlour [these would be the better-off], must expend six or seven guineas a year for firing alone.[10]

There were quite frequent imports to Aberdeen of 'bolls' (6 bushels) of coal from Leith and from certain Fife towns in the early seventeenth century. On the other hand, there are also records of the arrival of boatloads of peat, one of which was specified, in 1669–70, as being for 'the provost's chalmer'.[11] Where peat was plentiful, it continued long in service, in some areas up to the present day.

FIRESIDE ACCESSORIES, HEATING AND COOKING AIDS,[12] AND ORNAMENTS

The central hearth in the middle of the floor offered no encouragement for decorating the fire area. The addition of a stone back (see below) to a central hearth meant that there was a raised surface on which to set objects such as the tea caddie (see Fig. 21.5), but the real breakthrough came with the mantelpiece which usually accompanied the move of the fireplace to a gable wall. This could then accommodate the tea caddie, a clock, ornamental tins full of oddments, family photographs, and sometimes jars or vases of flowers. A brass rod could be slung below the mantelpiece, with hooks at either end, on which the bonnet of the man of the house might be hung, as well as drying socks and other items of clothing. A fender served to keep burning embers from escaping onto the floorboards, or the carpet or linoleum which covered them, so reducing the fire risk, and the tongs and poker usually leaned against one side of it. A pair of bellows was often hung nearby, and on the floor was the coal scuttle and shovel, with the more elegant examples, such as those in brass, being reserved for the best room. An armchair usually sat at each side of the fireplace, one of these being regarded as the man's seat. In earlier days, up until the early nineteenth century at least, there appear to have been quite specific seating formalities, at least in homes of higher status.[13] A traveller who visited a small landed proprietor at Burrafirth, Aithsting, Shetland in 1824 found that the master had his 'high seat' at the end of the room, with two rows of servants along each side of the room, between which he had to pass before paying his respects to the laird. Later he was given a bed in the ben end, which doubled up as a granary. The bed was of straw spread on the floor, with blankets and clean sheets over it. Later, when he visited a farmer's house at the Hill of Aithsness, Vementry, he entered, through the byre, 'a spacious room with a central fire, clay floor, and soot covered walls, with two long forms for the servants of each sex, and a high and separate chair for the mistress of the house.'[14]

In the houses of the better off, such as those of professional people, there might be a round or rectangular footstool, covered with a piece of embroidery, and above the fireplace there could be a more or less ornate mirror (see Fig. 12.7). In wealthy households, fire screens of a wide variety of styles were to be found, in acknowledgement, perhaps, of the strong heat from a good coal fire.[15] In humbler homes, there were often devices on which to dry washed clothes, such as the 'winter dykes', in two folding sections.

For the floor-level fire, among the devices used to raise the cooking pots, etc, above the embers was an iron tripod or a four-legged, square equivalent with cross bars, or the vessels themselves might have legs. Conveniently placed stones would also serve the purpose. When the coal-burning raised grate came into fashion, and fireplaces had migrated from the centre of the floor to a wall, there was a considerable increase in the range of smith-made ancillary devices for similar purposes. Amongst the simplest was an iron frame with its front bars, sometimes in the form of claw supports, attached to or lying on the ribs of the grate, and the back resting on the rear part of the fire. These were not mass produced and distributed through ironmongers' shops, with a specific name, but were of local manufacture, and had a variety of names (Fig. 6.1).

In the case of the central hearth, the crook and the links for supporting utensils over the heat hung from a beam lying athwart the room, from eave to eave. Once the hearth migrated to a wall, the replacement for this was the 'rantle tree', a cross piece of wood or iron fixed in the chimney for hanging the chain and hooks (see Fig. 5.5). Where the wall was of stone, or where the cheeks of the chimney were of slabs of stone, a 'swey' could be attached. This was normally of iron (though rare wooden examples have been noted), and consisted of a horizontal bar that stretched across the fire

Figure 6.1 The kitchen range at Smithy Croft, Pitglassie, Auchterless, Aberdeenshire, early 1960s. There is an iron swey from which dangle the crook and the links. A pair of tongs and a small fire shovel hang from a hook. There is a hob at each side of the open range; the left-hand side contains an oven (not visible). An iron pitowerlie can be seen behind the kettle. The hooks on the metal struts supporting the mantelpiece are used here for hanging a man's bonnet and a towel. The mantelpiece holds a clock, two earthenware piggie banks, a tea caddie and a can for nick-nacks. Source: SLA W012423.

and could be swung in and out on a vertical pivot. The chains and pothooks hung from it.

THE FUEL OF THE POOR

For those who were extremely poor, at the lowest social levels, regular heating and even the preparation of a warm meal were matters of difficulty, perhaps even more extreme in the towns than in the country, where there was more scope for foraging. In the 1790s in the parish of Kirkinner, Wigtownshire, the 'elding' or fuel which could be picked up consisted of small sticks, brushwood, whins or dried broom stalks, gathered by women and children during the daylight hours in winter, and in the evenings they warmed 'their shivering limbs before the scanty fire which this produces'.[16] In the parish of Kells, Kirkcudbrightshire, peat was the main fuel, but 'the poorer sort' had recourse to 'broom, furze, and other brushwood, to supply the deficiency'.[17] In the parish of Glenorchy and Innishail, Argyll, in 1792–3, 'the sufferings of the lower classes, with respect to firing, in wet years, during the rigour of winter, can only be conceived by such as have felt them.'

A few years before, it was noted, the poor people in the West Highlands had been obliged to burn much of their household furniture to repel the cold and prepare food, and old people and young children mostly kept to their beds.[18] Amongst a list of 'abuses' in the Forest of Atholl, Perthshire, in 1708, was the allegation that 'children poor people & cottars are ordinarily allowed to carry burdens of withered sticks out of the wood; they seldom miss to bring hooks & axes to cutt green timber that it may wither against the nixt tyme they come.' On the one hand, this was an abuse, and on the other it was long-term planning. However, the tenantry in general do not appear to have had such a privilege, so the estate was, in a small way, looking after its poor.[19]

In general practice, the type of fuel could vary throughout the year. In 1792 at Abernyte, Perthshire, the peasantry burned only broom and furze, which they frequently had for the cutting, or for a small price, for the light lands were overrun with broom which had been sown for fuel about 40 years earlier. But in winter, turf and coal were used.[20] In Forgandenny, Perthshire, furze, broom, bits of wood, and tree branches were the common fuel stuffs, with coal from Fife in the winter.[21] The same story was repeated elsewhere, with the variation also that where peat or turf was normally used in the country districts, in towns, such as Campbeltown in Argyll, it was mainly the poorer folk who used it.[22]

In a house in the coastal town of Stonehaven, Kincardineshire, I have seen, in the 1970s, a kettle being boiled – though it took a long time – by burning the pages of a newspaper, rolled up into a tube, then tied in a knot, so saving firewood and coal that had to be paid for. This was the kind of shift that the urban poor had to learn. But it frequently happened in both country and town that the only heating was combined with the process of cooking, after which the fire was allowed to die out.

In areas where peat and brushwood were not available, dried animal dung, sometimes mixed with sawdust or coal dross, was used as fuel. The practice was known from the early sixteenth century in Orkney, and was still within the memory of older people in the 1960s. The dung was gathered in summer, dried, and used as winter fuel. Not only the poor used it. By all accounts it gave quite a good fire, although it 'emitted an unsavoury reek that tickles the inexperienced nostril. The islanders eat with relish fish that have been smoked with these dried dung cakes.'[23] Animal manure as fuel is recorded in the Western Isles in the seventeenth to nineteenth centuries, and Gaelic dictionaries have a name for it, 'buacharan'. In east central Scotland, it formed part of fuel mixtures: in Angus, a 'dall' was a large cake made of sawdust mixed with cow dung, and in Fife a 'daw' was 'a cake of cow's dung, baked with coal dross, and, when dried in the sun, used by the poor for fuel'.[24] In Perthshire, some of the poor also mixed the dross of coal with cow dung as fuel.[25] In south-west Scotland, 'bachruns' were cakes of ox dung, used in place of peat by the poor, and gathered off the fields in autumn for winter fuel.[26] It is not clear if it was the owners of the animals who used such fuel themselves, or if they allowed poorer folk to use it, but there must inevitably have been at times a conflict between dung as fuel and dung as manure. Evidence also comes from northern England, Ireland, and a number of other European countries, so the practice was widespread. In Scotland it was probably in occasional use, mainly but not exclusively amongst the poor, in many more places than are mentioned above.[27] However the major type of fuel was peat.

PEAT

The casting and drying of peat requires planning and organisation, usually involving several members of a community, and its use was often a right which went with the tenancy of land. As a fuel, peat has the ability to burn on a flat surface and there is no need for an under draught as there is with coal. Peat-fired hearths were usually at floor level, on flat slabs or cobbling. Archaeological excavations, for example of Norse houses, have shown that a living room might have temporary hearths in different parts of the floor at different times, and with peat as a fuel, such movement would be readily possible.[28]

It is estimated that ten per cent of the surface of Scotland is still covered by peat, in spite of the centuries of use to which it has been subjected. It was burned in open fires and in the cast-iron stoves which spread, especially in island regions, in the Victorian period. The peat was carried home from the peat banks, and built into neat stacks alongside the houses. A day's supply was taken into the kitchen as required and sometimes, as in Glenesk in Angus, there was a specially built, wood-framed 'peat neuk' just alongside the fire. This arrangement was the equivalent of the later coal bunker, and the coal scuttle which was filled from it.

PEAT-BURNING HEARTHS

The earliest types of hearth were placed centrally on the kitchen floor. Examples have survived into the twenty-first century, even if only in buildings now preserved as museums.[29] A disused millstone has been known to serve as a base for a central hearth, as in a house at Calbost in Lewis. It was in daily use when photographed in 1964[30] (Fig. 6.2). Central, round fires also survived in Shetland and Orkney into the twentieth century.[31] The next stage was to use a back stone against which a fire of peat could be carefully set up to last as long as possible and still emanate warmth. A large, sandy turf, which did not burn readily, was set against the stone, and the more combustible peats were laid on end in front of it, angled so that their heads came together over the glowing core of peat ash.[32] A good depth of ash would also keep the warmth, and was used to 'smoor' or smother the fire and still allow the heart of it to stay alive through the night, ready to be brought into life on the following morning, rather more speedily than could be done with a coal fire, which needed sticks and paper to get it going anew. It was, however, possible to keep a coal fire in all night, by covering it first with clinkers, and then when they were glowing, putting a crust of damp

Figure 6.2 The central hearth in a blackhouse at Calbost, Isle of Lewis, 1964, set on the top stone of a horizontal meal mill. Fresh, slightly damp peat has just been put on the fire. The crook and the links hang from a rope tied to a crossbeam. There are two cast-iron kettles and a pair of tongs. Source: A Fenton.

dross on top – or even potato peelings. Nowadays it is possible to light a fire instantly by means of a match, but in the days when a strike-a-light had to be used with a flint and tinder, it might take much more time and patience. It was, therefore, sensible to smoor the fire, so that in the morning it was self-propagating.

There is no doubt that the people who inhabited the houses in the days before chimney refinements, and indeed before any chimneys at all, were inured to the smoke and probably hardly noticed it. According to Captain Burt, writing in the first half of the eighteenth century, in winter the Highlanders:

> ... have no Diversions to amuse them, but sit brooding in the Smoke over the Fire till their Legs and Thighs are scorched to an extraordinary Degree, and many have sore Eyes, and some are quite blind. This long Continuance in the Smoke makes them almost as black as Chimney-Sweepers.[33]

The practice of sitting before the fire with the legs apart and letting the heat play on the thighs and crotch, especially by women who would have to pull up their long skirts in earlier days, was widespread, and in north-east Scotland was known as getting an 'up-throwe heat'.[34] Over-frequent indulgence could result in a mottling of the skin of the legs, called 'tinker's tartan', or in Gaelic 'breacan Màiri Uisdean'. This, as much as anything, throws some light on the nature of heating as a source of pleasure, for repeated indulgence.

A work-related problem caused by having a hearth in a fixed place was that for a woman spinning at the fireside the wheel was constantly in one position, so only one side of the worker was heated. This point was noted by the minister of the parish of Kirkintilloch, Dunbartonshire, in 1790.[35]

THE HINGIN LUM: AN INNOVATION

Another major hearth form, mainly built against a gable, was characterised by a framed canopy of wood or of laths plastered over with clay or cow dung, and was commonly known as a 'hingin lum' (Fig. 6.3). It seems to have spread from an innovation in the Lothians from about the 1770s and had reached Shetland by the early nineteenth century.

Also in southern Scotland, there were other contemporary methods of smoke extraction. One was the 'round-about fireside', said in 1802 to be universal in the county of Peebles. It was circular, more or less in the centre of the floor, and had a frame of lath and plaster, or spars and mats, suspended over it from the roof like an inverted funnel. It was of such a size that the whole family could sit round the fire within its circumference.[36] In Roxburgh there was a similar type, consisting of a funnel of wattle covered with a mixture of straw, mud or clay.[37] Around 1832, the kitchen of a Selkirkshire farm had a similar fireplace nearly in the centre of the floor, with a wooden settle on one side and various seats on the other. This seems to have been large enough to form a kind of unit, so to speak a room within a room, where the family could sit in the light and warmth from the fire.[38]

In the 1840s, there is a Lanarkshire record of a large 'hingin lum' at the gable of a farm kitchen. It came so far out into the room that there was a 6-foot (1.8 m) space between the fire and the wall, and here a wooden 'lang settle' was placed for the young farm lads to sit. The hood was described as a square-mouthed box, about 5 to 6 feet (1.5–1.8 m) wide, placed about six feet (1.8 m) above the fire and contracting as it rose to two feet (61 cm) square, at the point where it was taken up through the roof ridge against the gable.[39] This seems to be a larger version of the Lothian brace, and it was, like the inverted funnel type, capable of providing a demarcated area within the bigger entity of the kitchen.

Also in Lanarkshire, later in the nineteenth century, the era of photography has provided evidence for yet another form of heating, without a canopy or funnel. A photograph of the kitchen at Eastfield Farm shows a wooden swey, seemingly attached to the back wall. From this dangles a pot, and below the pot is an iron brazier on four legs. These stand on the stone flags which cover the floor, but underneath the brazier is a square hole in the floor to receive the ash and debris from the fire. The smoke appears to rise into the room, and there is no sign of a chimney. A woman sits in front of the fire but a little way from it, carding wool for use on the spinning wheel beside her. A leg of pork or mutton, no doubt wrapped in a cloth, hangs from the ceiling joists (Fig. 6.4).

This photograph brings with it the message that there could be great variety in forms of heating, much of which may have gone unrecorded. However, the use of a brazier for heating is sometimes to be seen in sketches of eighteenth and nineteenth-century interiors. For example, David Allan (1744–96) has a drawing of a freestanding iron grate or brazier under a chimney canopy, with the crook and links, and a three-legged pot with two lugs and a one-piece bow handle.[40] Probably investigation of such sources, and their localisation, would show a fairly wide use of such braziers, which are to some extent like open grates, especially in or near urban areas. They are likely to have been used also on the paved floors of the great chimneys

Figure 6.3 Hanging lum, Ardoyne, Insch, Aberdeenshire, 1961. The hanging lum or chimney projects over the open hearth of the fire. An iron swey is fixed in the left-hand side of the hearth and the crook and links suspend a large iron lidded cooking pot over the embers. Source: SLA SLAW0100334.

Figure 6.4 A photograph, probably of the late nineteenth century, showing the kitchen at Eastfield Farm, Lanarkshire. The fire, in an iron brazier, sits over an ash pit. Above it is a wooden swey, supporting a pot. Dried meat hangs from the ceiling beams. Source: SLA 3/26/21.

of big houses and castles, but this is an area of research which remains to be undertaken.

THE HEARTH OUTSHOT

In considering the origins of phenomena such as the inverted funnel or the wide canopy chimney, the concept of the 'trickle-down' effect, according to which aspects of higher-class culture come to be more widely adopted at lower levels, may be relevant. The great fireplaces of medieval castles had, up to the fifteenth century, projecting hoods of stone. Some of these fireplaces contained recesses that acted as buffets or dressers, where displays of plate became common in the fifteenth century.[41] This suggests that they had a social role to play which went beyond the mere functional purpose of providing warmth. Like the hingin lums discussed above, they provided more intimate areas within larger spaces, and may have inspired the hingin lums themselves.

They may also have inspired the hearth outshots, which are associated with a living space which has its own small window. Their main concentration is in south-east Scotland, with outliers north of Edinburgh and along the coast, and others to the west, north and south of Glasgow. They are built as outshots, accessible from the main kitchen area through a wide, arched entrance, and it is not without significance that such a stone arch, whether for an outshot or any other kind of fireplace, was originally called the 'brace', a word which later developed differentiated senses, one of which applied to the hingin lum.[42]

THE IRON FIRE GRATE, THE KITCHEN RANGE AND THE ENCLOSED STOVE

The widespread adoption of coal as a leading fuel, for heating as well as for cooking, was a major step in modernisation, bringing the industrial era firmly into the Scottish home. The sequence of events was neatly summed up by a sharp observer in 1824:

> The fire in the middle of the floor is first transferred to the gable; a canopy with a chimney is next placed over it; those who formerly sat near the fire, then sit within the fireplace; in progress of time, this is contracted so as to exclude them; and lastly, this eventful history ends in Carron grates and Bath stoves and registers, in bright brass and brighter steel, the pride of housewives, the dread of chilly guests, and the torment of housemaids.[43]

Though this was written in 1824, the elements mentioned survived well into the twentieth century and can still be found. Fire grates were first made by fixing iron bars between two hobs, with an ash pan below. As already suggested, the general concept may be linked with braziers, but an important factor was the provision of flat-topped hobs on which kitchen equipment could be laid. From this the next logical step was the kitchen range. The first one, said to have been designed in 1780 in England by Thomas Robinson, had an open fire with bars between metal hobs, an oven at one side and a tank for hot water at the other.[44] In Scotland, the founding of the Carron Iron Company at Falkirk, Stirlingshire, in 1759 led eventually to the production of domestic stoves and grates. These could have a high-status market. The designs and decorative features were by the Adam family, Robert, James and John, who together stimulated an architectural fashion based on classical features. James even became a partner in the Carron Iron Company. The elegance of the exceptionally large variety of designs and sizes produced by Carron to suit the needs and pockets of different classes of customer was truly noteworthy.[45]

By the 1820s and 1830s, kitchen ranges in all sizes, shapes and forms had become dominant in domestic kitchens. They still provided heating as well as cooking facilities, since in mainstream homes the kitchen remained the living area (see Fig. 4.1). In the other rooms, however, there were simply barred grates for coal fires, rarely lit except when visitors came in cold times or in the case of illness. Ranges were no doubt at first seen as status symbols, but soon came to be made by small county-town foundries, and specimens became available to suit all pockets. They could be supplied

Figure 6.5 A cast-iron stove, built into a formerly open hearth, at Estabin, Orkney, 1968. Source: A Fenton, SLA 4/10/9.

as adaptations of existing hob grates, which must have facilitated their spread, and local ironmongers were major agents in the distributional system. One example examined on a Caithness croft in the 1970s, when it was still in use, had an oval plaque on the oven door bearing the legend 'D R Simpson & Sons, Ironmonger, Wick'. This did not indicate the maker, but the source of local distribution. It had an oven or stove on the right and a grate on the left, and that was all. The kitchen range in my parents' croft, used at least until the 1980s, likewise had an oven on the left, an open grate, and on the right a 'blank' hob which served only to give symmetry and to support pots, pans and kettles. In bigger kitchens, this side would contain a water boiler with a tap at the front, and this is the basic style which became general. As long as the fire was in, hot water could be available for washing.

The design of kitchen ranges reached a peak by about 1890, with all kinds of improvements, and some, as in examples in Glasgow, even had tiled surrounds in the *art-nouveau* style. In urban situations especially, the 'grate or range ... was included as a rule in the moveable furniture of

HEATING AND SLEEPING • 95

the house and was taken by the tenants when they moved.'[46] This points very clearly to the uniformity in the style of building of blocks of single-end or room-and-kitchen accommodation.

An alternative to the kitchen range, which may have produced a better form of heating, was the enclosed cast-iron stove which could equally well consume coal, peat or wood – or a mixture of these – and was being inserted, sometimes into the place of the older floor-level hearth (Fig. 6.5). Amongst the favoured types of stove were those produced by the firm of Smith & Wellstood, in particular the 'Enchantress' and the 'Victoress'.

A manuscript diary kept by Mrs Ann Still of Smoogrow, Orkney, in 1878 to 1882 tells a success story. On 17 October 1878 she bought a stove at a sale. Her open fire smoked mercilessly when the wind was in the wrong direction, but on 10 January 1879 she got the stove fitted, '& room much warmer'.

SEPARATION OF HEATING AND COOKING

The gradual reorganisation of space within the home to allow for separation of heating and cooking facilities has been long going on. The kitchens of castles, monasteries and big houses generally were always separate, but at lesser social levels one fire normally served both purposes and the kitchen was also the living room and frequently provided sleeping space as well. In the cottages of but-and-ben type, which spread from around 1800, the first signs of differentiation can be noted. The kitchen fire might continue to be of an old type, for example at floor level, and peat-burning, with a smoke hole or wooden 'hingin lum', and eventually it could be a kitchen range, whilst the best-room fire could be a grate for coal, with a chimney in the thickness of the gable wall. The best-room fire would be lit only rarely, however, although it presaged more modern times, with the almost complete victory of coal over peat.

Other fuels also came along. On my parents' croft, a Rippingille paraffin-fired oven came to be used in the kitchen for baking and cooking, and the range became primarily a source of heat only. The production of paraffin from the shales of West Lothian and Midlothian began in 1865, but it was in America, where shale oil was also found, that paraffin-burning stoves and cookers developed. Heating with gas and electricity required vents and connections of various kinds, and so fires were often conveniently sited where the fixed hearth had been. But with night-storage heating, gas central heating, or under-floor heating, the hearth as a focal point has disappeared, and many new houses do not even have chimneys, especially where they are built in smoke-free zones, for which there was legislation in the 1940s to 1950s. The whole of a modern living room can be used in a way which was never possible in the past, in equal comfort at most points, and an even spread of light from the main points on the ceiling helps to make such generalised use possible.[47] Yet the instinct for a focal point remains strong, even if it is now the television set rather than the hearth, and as often as not the main lights are turned off and the cosier pools of light from side lamps are preferred.[48]

A BBC publication of 1992 on the British home makes the following statement:

> By 1956 houses were getting adventurous. These houses longed to be free, to do away with dividing walls ... The architects of the Modern Movement wanted to abolish rooms – unhealthy and rigid boxes – altogether. They wanted free-flowing, omni-directional space within which there would be zones of activity. Walls and doors would be dissolved to produce a continuum of space and function – what came to be called open plan ... The new houses of the Fifties and Sixties were made possible by the availability of economic central heating. One reason why the old front rooms were seldom used was simply because it required too much labour to tend two fireplaces. Central heating had two main consequences: rooms could be opened up and there was no further need of a fireplace.[49]

There is some irony in this, for houses of the old byre-dwelling or blackhouse type, with people and animals under one roof, and a central fire in the dwelling area (and probably heat coming also from the animals in the byre end) may be regarded as the original open plan, although it took the mediation of central heating to make the return.

In spite of advances in heating, the poor or under-privileged still suffer cold. Huddling round a fireplace or a single-bar electric heater and confronting glacial bedrooms remains a reality for many in private rented accommodation or those who lack sufficient capital to improve their bought homes. Local authorities and housing associations, however, are increasingly installing central heating and double-glazed windows in social houses, as, for example, at Wester Hailes in Edinburgh. But at all social levels old instincts are strong, and though central heating can provide background warmth there is often a desire also for a good blast of heat from a more conventional source.

Besides new ways of using the living room, it is worth noting that other rooms are nowadays often used as living and working space. Bedrooms, dining rooms and family rooms may all be pressed into service, and may be the scene of office-type, computer-based activities.[50] Here is another story that remains to be told.

SLEEPING AT HOME[51]

In terms of comfort and warmth, there was a close association in earlier times between the fire and sleeping arrangements (see Fig. 26.1). The bed itself and its coverings were a source of warmth, as was sharing a bed, which was not unusual. In 1797, Edward Clarke visited a house on the island of Mull in which dry bracken and heather on the earth floor was the bed for a family of nine children, a grandmother, and an adult female idiot 'harboured in charity'. There were also two dogs, a cat, three kittens and a pig.[52] In some crowded situations, people would lie 'heids an thraws', with feet and heads alternating:

> A laigh [low] hut, where sax thegither
> Ly Heads and Thraws on Craps [heads] of Heather[53]

This practice was known throughout Scotland.

In stone-rich areas such as St Kilda and parts of the Hebrides, there were houses partitioned into two spaces, one serving as a byre for the cattle in winter and one as a multifunctional kitchen, sitting room and bedroom for the people. The bed spaces, said to be capable of holding at least three persons, were within the thickness of the wall. The entrances were low and narrow.[54] An illustration of a St Kilda house published in 1870 shows a bed, there called a 'crûb', within a thickened gable wall, and with its entrance just opposite the central hearth[55] (Fig. 6.6).

Edward Clarke described St Kilda examples as follows in 1797:

> Round the walls of their huts, are one or more small arched apertures, according to the number of the family, leading to a vault like an oven, arched with stone, and defended strongly from the inclemency of the weather, in which they sleep. I crawled on all fours, with a lamp, into one of these, and found the bottom covered with heath; in this, I was informed, four persons slept. There is not sufficient space in them for a tall man to sit upright, though the dimensions of these vaulted

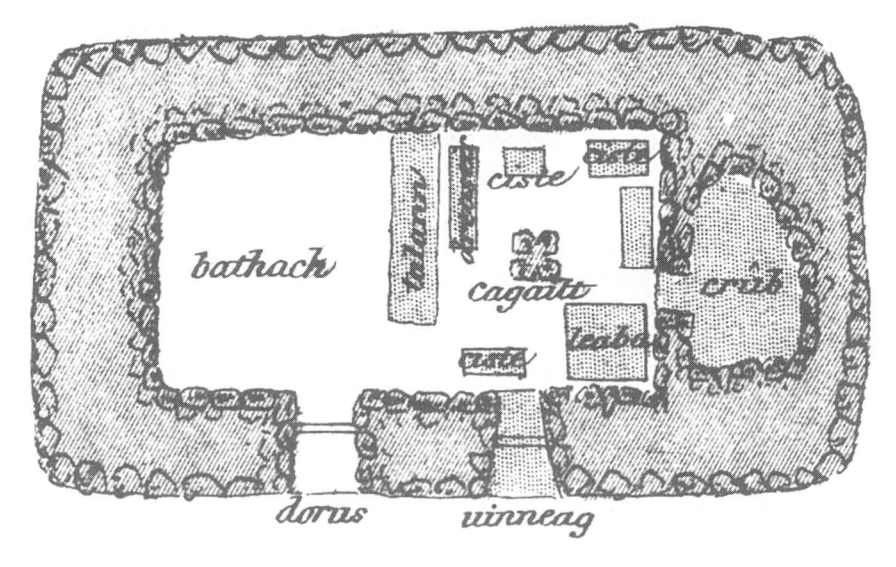

Figure 6.6 A blackhouse or 'byre dwelling' in St Kilda, with a byre at one end and living space at the other. A bed, 'crûb', lies within the thickness of the stone gable wall. Crown Copyright: RCAHMS.

dormitories varied with every hut, according to the number it was required to contain, or the industry of its owners.[56]

The Rev Neil MacKenzie arrived in St Kilda in 1830. He described the older houses he saw as roughly circular, with walls 6 or 7 feet [1.8–2.1 m] thick, in which were bed spaces, roofed with long slabs, and with an entrance from the interior of the house of about 3 by 2 feet [91 cm by 61 cm]. But the houses occupied at the time of his arrival were larger, more oval-shaped, and had double-skin walls – that is, there was an inner wall and an outer wall, with a 4-foot [1.2 m] space between filled up with earth, except for the spaces formed by the beds.[57]

The obvious similarity between such wall beds and recesses within the thickness of the walls of prehistoric burial chambers is noteworthy, but does not fall to be discussed here. The same kinds of building materials might have led to similar structural solutions within comparable culture zones. But as times change, techniques can change. For example, in Orkney, there are 'neuk (corner) beds' in the kitchens of older farmhouses. In this island group, excellent flagstone for building was to be found, the walls are single, and the beds are formed by protruding outshots. Examples of full outshots have been recorded at Nether Benzieclett, Langalour and Mossetter (where there is an outshot row of four), of a half neuk, with only a shallow outshot, at Winksetter, and at Kirbister, where the bed is actually in the thickness of the wall.[58]

These beds separate the sleepers from the kitchen itself. However, at least from Viking times, there was sleeping accommodation within the space of the kitchen or 'hall'. At either side of the long, rectangular hearth, the Vikings had a raised platform, a *pallr*, which served for sitting on by day and for sleeping on by night. This must have been the nature of the 'broad benches' on which people lay around the central fire of a Shetland house in 1614 to 1618.[59] This type of bed is continued by the long wooden settle with back and arms, found generally in Scotland and widespread in Europe, known as the 'restin chair', 'muckle chair', 'lang seat' or 'deece' (dais) in Lowland Scotland and as the 'béing' or 'beinge', or 'séise', 'séiseach' in Gaelic-speaking areas.[60]

A parallel to these, in seasonally occupied buildings, is the raised division of stone and turf, rarely of wood, which marked the end of the bed space (this taking up a good part of the floor room) and served at the same time as seats before the fire. This arrangement is found in the beehive shielings of the Western Isles and elsewhere, in the fishing lodges of Shetland fishermen, and in the 'peat hooses' which sheltered those taking home peat from a remote part of the island of Fetlar in Shetland. In these, a plank-covered bed space occupied at least half the available space. When work started, straw was carried in bags and spread on the planks, or heather could be used. This layer was covered with sacking, and over that were laid sheets, blankets and patchwork quilts each made of two sewn together to get the needed width for all the family workers, male or female, to sleep.[61]

There were other types of beds with potentially early antecedents. In a report of the 1880s, a bed in a cottar house in Orkney was described as follows:

> Parallel to, and about 2 feet [61cm] from the horribly damp north wall, a row of flags was set upon edge, fixed in the earthen floor. The trough formed by the damp wall, for back, the damp earth covered with a little straw or heather for bottom, and having the cold flagstones for front, was the bed of the Orkney peasant during the greater part of last century.

The bed in this Orkney house, at floor level and with stone surrounds, is undoubtedly reminiscent of the presumed bed spaces near the hearth in the prehistoric village of Skara Brae in Orkney. It may also more probably be seen, not as a survival, but as a floor-space fixed-feature variant of the outshot neuk bed, at a lower economic level than that of the farm. Outshots, which included bed spaces, were part of the architectural development of Faroese houses,[62] and were to be found in one form or another not only in the western half of Scotland, but also in Ireland, Northern England, the Scandinavian countries, Brittany, Belgium, the Netherlands and north-west Germany.[63]

The enclosed nature of such alcove beds, which in effect, like some of the forms of hearth, created a private space within a larger entity, is similar to that of box beds. It was suggested that the introduction of box beds towards the end of the eighteenth century was amongst the greatest improvements which took place in the domestic economy.[64] They were of wood, closed on three sides and roofed. The front had hinged doors, sliding doors or curtains. According to an Orkney writer, they resembled 'a little wooden house with sliding doors, and when they are shut it is said to be very warm'.[65] They were used, often in pairs, along with presses and doors, to create space divisions within houses, which might lead to a kitchen area and a back room area, as in a Berwickshire farm-servant's house of 1809,[66] or they could be so placed that the house had an entrance lobby with three doors, one leading straight ahead into a closet, one into the kitchen and one into the 'room'. The resulting separation of the kitchen from the immediacy of the outside door certainly improved its heating. Two of the possible ways in which box beds and presses could be sited to divide spaces can be seen on the plans of two crofts in Aberdeenshire (Estate of Hatton), Oldwood Croft and Ordley Croft (Fig. 6.7).[67] In combination with a press or cupboard, the unit was called in north-east Scotland a 'bun breist' (bound breast). Box beds remained in sporadic use well up to the second half of the twentieth century. When I visited the so-called 'blackhouse' at 42 Arnol, Lewis, in the early 1970s, when it was in active occupation and not yet preserved as a museum, there was an old lady lying in the box bed which was visible at the far side of the kitchen, through the slight haze of peat smoke from the central fire.[68] I photographed a box bed at Bewan in North Ronaldsay, Orkney, in 1964,[69] and noted one in use on a farm in Glenesk, Angus, in the 1970s. All were cosy-looking places of comfort on stormy nights, but possibly too cosy when fully occupied, as in an example seen in a single-roomed house in 1847 in Orkney. The visitor asked where the parents and six children slept at night:

> Weel ... The wife and I lie wi' oor heeds at the head o' the bed, the two eldest lie with their heads at the foot o' the bed, the peerie [little] t'ing [the baby] lies i' his mither's bosom, the ain next the peerie t'ing lies i' mine, and the middle two lie on a shelf at the foot o' the bed, over the heads o' the eldest twa. An' truth I can tell you we are no' cauld gin I close the bed doors.[70]

Box beds, however, were difficult to clean. A Perth historian commented in 1836 that they were very unhealthy, 'and soon became the stronghold of such numerous colonies of intruders, that the only effectual expedient to get rid of these nocturnal visitors was to burn them out, by throwing the wood work to the street and making a bon-fire of it'. The other common form of bed, with four short posts and a wooden base, was regarded as much healthier.[71]

Kirbister in Orkney, which was occupied until 1962, before it became a museum, shows a fairly standard succession of bed types. The wall bed in the kitchen was replaced by a box bed, and in the other rooms there were beds of the normal freestanding rectangular form with four legs. In some cases, mattresses were stuffed with straw, sometimes in coastal districts with a kind of seaweed, *Zostera marina*, which was proof against fleas, and often with chaff. The 'caff bed' was a widespread phenomenon. Those of higher rank had feather beds, of goose or eider-duck feathers.

It should not be imagined that crowded sleeping spaces were only to be found in rural areas. The situation, in fact, could be much worse in the industrial towns. There, in line however with the old rural tradition, there was often a recessed bed in the kitchen. According to the Report of the Royal Commission on the Housing of the Working Class, 1885:

> In Scotland they have a bed recess in the kitchen, they cannot do without that, a Scotchman always likes to lie in the kitchen, and of course that saves fires. They are near a fire in the winter time. It is the favourite bed in the house.

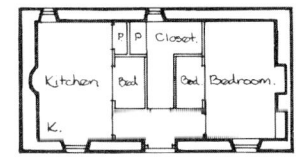

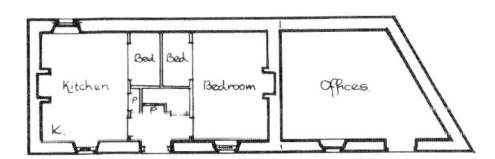

Figure 6.7 The positions of the box beds in two small Aberdeenshire crofts: left, Oldwood croft, estate of Hatton; right, Ordley croft, Auchterless. Source: Fenton, A, Walker, B. *The Rural Architecture of Scotland*, Edinburgh, 1981, 191.

In the one-roomed dwellings, the 'single ends', conditions could be even more crowded than in the room-and-kitchen homes. In one example, noted in 1931, there were two beds, with the end of the children's bed pushed partly under the end of that of the parents, to save space. There were eight in the family.[72]

SLEEPING AWAY FROM HOME

Some indication of the nature of sleeping accommodation in inns in the seventeenth to nineteenth centuries can be got from the accounts of travellers. Neither comfort nor warmth was always guaranteed. Bed bugs were widespread and the traveller had to tolerate them. John Taylor, the 'Water Poet', experienced them when he stayed in Glenesk, Angus, in 1618.[73] Comments in later times show their prevalence. James Hogg, the Ettrick Shepherd, found his bed full of insects in an inn at Invershiel, Wester Ross, where there was one room for the guests.[74] In the cities, bugs, fleas, scabies and lice were often-present dangers. Oral information gathered in recent times by the People's Story Oral History Archive of Edinburgh City Museums and Art Galleries tells of gallant attempts to clear infestations with hot water and carbolic acid, and of the need, sometimes, to call in the Sanitary Inspector.[75]

In the 1720s to 1730s, Captain Burt got a 'lodging room' in Kelso in the Scottish Borders, with a bed in which the curtains were dirty, but otherwise there was 'exceedingly good Linen, white, well aired and hardened, and I think as good as in our best Inns in England'. But in the Highland areas, people's beds in ordinary houses were in a corner on the floor. Burt noted a bed with a small quantity of straw and two blankets, for a woman and two children, and another for two young fellows in the opposite corner. At the end of Loch Ness, in a large room, there were nine folk in similar beds, including children and a young woman who was naked under the blankets; this was not a poor house. Burt's comments could be quite scathing. In some of these remote inns:

> … your Chamber, to which you sometimes enter from without-Doors, by Stairs as dirty as the Street, is so far from having been washed, it has hardly ever been scraped, and it would be no Wonder if you stumbled over Clods of dried Dirt in going from the Fire-Side to the Bed, under which there often is lumber and Dust that almost fill up the Space between the Floor and the Bedstead. But it is nauseous to see the Walls and Inside of the Curtains spotted, as if every one that had lain there had spit straight forward in whatever Position they lay.

At one place, he found that the chamber pot was a hole in the ground by the bedside.[76]

James Boswell, during his tour with Samuel Johnson in 1773, makes frequent mention of sleeping arrangements. In Aberdeen, he slept in a little box bed which was wheeled into the dining room. This does not appear to have been the same as the 'hurly bed', a low bed or truckle bed which, in the housing of the working classes, stood below another during the day and was pulled or 'hurled' out for use at night.[77] In a turf hut at the end of Loch Ness, they saw a bed of wood and heath behind a wicker partition, with the blankets rolled up into a heap. There were two adults and five children there.

It seems that Boswell and Johnson had brought sheets with them, which suggests that they were scarce commodities in the Highlands. They sometimes slept on fresh hay, with the sheets spread over it, as at Glenelg Inn, Inverness-shire, and would strip for sleeping or leave their clothes on. Johnson's version of the stay at Glenelg pokes fun at Boswell, saying that, 'Mr Boswell being more delicate, laid himself sheets with hay over and under him, and lay in linen like a gentleman.' In Skye, at the farmhouse of Coirechatachan, the beds had curtains, and Boswell shared a room with two others:

> After I was in bed, the minister came up to go to his. The maid stood by and took his clothes and laid them carefully on a chair piece by piece, not excepting his breeches, before throwing off which he made water, while she was just at his back.[78]

THE OLD WORLD AND THE NEW

There is a great contrast between the old world and the new. What remains of the old world has to be interpreted with some care, for ways of thinking and attitudes are conditioned by time and circumstance. Even a hundred years ago – indeed as little as 20 years ago – life was very different. It is one of the tasks of ethnology to try to understand and explain how people thought about family and community, work and play, seasonal and personal occasions, and so on, in the past, as well as at the present, which has the advantage of personal observation as a research tool. The individual worlds of everyman are constantly expanding. Mobile phones and Internet computer facilities give access virtually everywhere, and radio and television bring instant news about anything, anywhere in the world. The slower pace of earlier existence, and the more limited spheres within which people operated, are now almost unimaginable.

Heating, which implies shelter, and food are amongst the subject areas of most importance in everyday life. Accordingly, some of the major aspects of warmth and heating, and of the comforts of bed, have been assembled here. They show that the quest for ever-improved forms of shelter, and for better facilities within that shelter, led to long-term, slow forms of endogenous development, as in the case of the central hearth, which acquired a back, then moved to a wall, each stage leading to corresponding developments in means of smoke extraction. The coming of the kitchen range and iron fire grates, from the mid eighteenth century, marked a new era, involving the distribution from fixed centres of manufactured items. Indigenous development was replaced by consumer products, as the industrial era set in.

But the tradition of an open fire survived in town as well as in country, almost as a marker of national identity, and can be remarked on by foreign visitors at the present day. The open hearth and chimney is still a source of draughts, but conditions are changing. Central and local legislation and regulations which implement conservation policies have created smokeless zones in cities, and this has helped to speed up the general adoption of certain types of central heating, including, on occasion, solar-powered heating, which are not dependent on solid fuel. Cavity walls, often insulated with foam, and double-glazed windows accompany these, and chimneys are becoming obsolete. The world of today is being totally renewed, though it is fortunate that enough of the old has survived until relatively recently for a number of the steps in the changes to be observed in reality. Now they are largely documented by museum and archival collections.

NOTES

1. For the situation in Denmark, see Michelsen, 1968, which includes a chapter on 'the open hearth'.
2. For further details, see Fenton, 1997, 195–7; Fenton, 2001, 9–10.
3. Gregor, 1874, 22–3.
4. National Coal Board, Scottish Division, 1958, 34.
5. Walker, 1990, 33–5.
6. Mays, 1997, 100.
7. Carruthers, 1996, 108, 117, 131: numerous examples are given in this volume.
8. Taylor, 1972.
9. Gordon, npd, Fig. 33.
10. *OSA*, V (1790–1) (1983), 527: Stranraer, Wigtownshire.
11. Taylor, 1972, 74, 79, 112, 269, 303–4, etc (coal); 183, 391, 429, 588 (peat).
12. See also Chapter 23 Food.
13. See Chapter 7 Rest.
14. Hibbert, 1822, 538, 545.
15. Carruthers, 1996, 10, the morning room or parlour of a lawyer's home at Balerno, Midlothian, in 1897–8 and a Shetland croft kitchen in 1910.
16. *OSA*, IV (1792), 172: Kirkinner, Wigtownshire; *OSA*, XI (1794), 241: Eccles, Berwickshire; *OSA*, XI (1794), 73: Ure, Berwickshire; *OSA*, XI (1813), 507: Headrick, Berwickshire.
17. *OSA*, V (1791) (1983), 153: Kells, Kirkcudbrightshire.
18. *OSA*, VIII (1792–3) (1983), 132: Glenorchy and Innishail, Argyll.
19. Leneman, 1986, 186.
20. *OSA*, XI (1792) (1976), 25: Abernyte, Perthshire.
21. *OSA*, XI (1792) (1976), 186: Forgandenny, Perthshire.
22. *OSA*, XI (1791–2) (1983), 55: Campbeltown, Argyll.
23. Gorrie, 1869, 368–9.
24. *SND*, s v Dall, Daw (1825).
25. *OSA*, XVIII (1796), 562.
26. MacTaggart, 1824, s v.
27. For fuller details, see Fenton, 1972.
28. For example, Small, 1968.

29. Fenton, 1989a.
30. Illustrated in Fenton, 1989c, 30; see also Sage, 1889, 10–11.
31. Fenton, 1997, 196–7.
32. Fenton, 1997, 198, Fig 74.
33. Burt, 1974, II, 135–6.
34. *SND*, s v Upthrou, Tinker.
35. *OSA* (1790) IX (1978), 80: Kirkintilloch, Dunbartonshire.
36. Findlater, 1802, 39–40.
37. Douglas, 1813, 29.
38. Bathgate, 1901, 27.
39. *NSA*, VI (1845), 474, 831–2.
40. Allan, D. *Sketches of Scottish Life* (in the NMS collections).
41. Mackenzie, 1927, 117–8.
42. See Marshall, 1983–4.
43. MacCulloch, 1824, III, 14.
44. Yarwood, 1981, 84.
45. Campbell, 1961, 1, 77.
46. Worsdall, 1979, 44.
47. See Chapter 5 Lighting.
48. Some of the contents of this chapter are digested from Fenton, 1989a, 65–75; Fenton 1998; Fenton, 2001.
49. Rivers, et al, 1992, 29–3.
50. See Chapter 10 Television, Video and Computing.
51. See also Chapter 7 Rest.
52. Quoted in Harman, 1997, 145.
53. Ramsay, 1953.
54. MacAulay, 1764, 43–4. Cited in Stell, Harman, 1988, 3–4.
55. Stell, Harman, 1988, 4.
56. Clarke, 1824, 231, 270. Cited in Harman, 1997, 145.
57. MacKenzie, 1911, 18–20. Cited in Harman, 1997, 146.
58. Fenton, 1997, 116–32.
59. MacGillivray, 1953, 50.
60. Sinclair, 1953, 72: Grant, 1975, 171.
61. Fenton, 1997, 235, 572, 578.
62. Stoklund, 1996, 62–3.
63. McCourt, 1956; Walton, 1961
64. Dennison, 1884.
65. Omond, 1911, 9.
66. Reproduced in diagram form in Fenton, 1999, 195.
67. See Fenton, Walker, 1981, 191.
68. See Fenton, 1989a, 13.
69. See Fenton, 1997, 193.
70. Dennison, 1884, 275.
71. Penny, 1986, 21.
72. Clark, 1996, 68–71.
73. Cited in Hume Brown, 1978, 119.
74. Hogg, 1981, 81.
75. Clark, 1996, 71–2.
76. Burt, 1974, I, 17–18, 116–7, 152; II, 65.
77. For an illustration, see Carruthers, 1996, 68.
78. Boswell, 1979, 100, 105, 111, 124, 176; Johnson, 1979, 43.

BIBLIOGRAPHY

Bathgate, J. *Aunt Janet's Legacy to her Nieces* (c1832), Selkirk, 1901.

Boswell, J. *The Journal of a Tour to the Hebrides, with Samuel Johnson, LLD* (1786), Oxford, 1979.

Burt, E. *Letters from a Gentleman in the North of Scotland to his Friend in London* (1754), Edinburgh, 2 vols, 1974.

Campbell, R. *Carron Company*, Edinburgh and London, 1961.

Carruthers, A, ed. *The Scottish Home*, Edinburgh, 1996.

Clarke, E D (ed W Otter). *The Life and Remains of Edward Daniel Clarke*, London, 1824.

Clark, H. Living in one or two rooms in the city. In Carruthers, A, ed, *The Scottish Home*, Edinburgh, 1996, 59–82.

Dennison, W T. Remarks on the agricultural classes in the North Isles of Orkney. In *Report of Her Majesty's Commissioners of Enquiry into the Conditions of the Crofters and Cottars in the Highlands and Islands of Scotland*, Edinburgh, 1884, 270–80.

Douglas, R. *General View of the Agriculture of the County of Roxburgh*, Edinburgh, 1813.

Fenton, A. A fuel of necessity: Animal manure. In Ennen, E, Wiegelmann, G, eds, *Festschrift Matthias Zender. Studien zu Volkskultur, Sprache und Landesgeschichte (Memorial Volume for Matthias Zender: Studies on Folk Culture, Language and Rural History)*, Bonn, 1972, 2 vols, II, 722–32.

Fenton, A. *The Island Blackhouse* (1978), Edinburgh, 1989a.

Fenton, A. The hearth as a marker of social change: The Scottish example. In Gunda, B, Lukács, L, Paládi-Kovács, A, eds, *Ideen, Objecte und Lebensformen. Gedenkschrift für Zsigmond Bátky (Ideas, Objects and Life Forms: Memorial Volume for Zsigmond Bátky)*, Székesfehérvár, 1989b, 65–75.

Fenton, A. *Scottish Country Life*, Edinburgh, 1989c.

Fenton, A. *The Northern Isles: Orkney and Shetland* (1978), East Linton, 1997.

Fenton, A. Hearth and kitchen: The Scottish example. In Schärer, M R, Fenton, A, eds, *Food and Material Culture* (Proceedings of the Fourth Symposium of the International Commission for Research into European Food History), East Linton, 1998, 29–47.

Fenton, A. *Scottish Country Life* (1976), East Linton, 1999.

Fenton, A. Fires and firing: An overview. In Wood, M. *The Hearth in Scotland* (SVBWG, Regional and Thematic Studies, no 7), Edinburgh, 2001, 5–24.

Fenton, A, Walker, B. *The Rural Architecture of Scotland*, Edinburgh, 1981.

Findlater, C. *General View of the Agriculture of the County of Peebles*, Edinburgh, 1802.

Gordon, A, ed. *Views of Old Ross-shire*, Tain, 1976.

Gorrie, D. *Summers and Winters in the Orkneys*, London, 1869.

Grant, I F. *Highland Folk Ways* (1961), London, 1975.

Gregor, W. *An Echo of Olden Time from the North of Scotland*, Peterhead, 1874.

Harman, M. *An Isle Called Hirte: A History and Culture of St Kilda to 1930*, Isle of Skye, 1997.

Headrick, J. *General View of the Agriculture of Angus or Forfarshire*, Edinburgh, 1813.

Hibbert, S. *A Description of the Shetland Islands*, Edinburgh, 1822.

Hogg, J (ed W F Laughlan). *James Hogg, Highland Tours: The Ettrick Shepherd's Travels in the Scottish Highlands and Western Isles in 1802, 1804 and 1804*, Hawick, 1981.

Horne, J, ed. *The County of Caithness*, Wick, 1907.

Hume Brown, P. *Early Travellers in Scotland* (1891), Edinburgh, 1978.

Johnson, S. *A Journey to the Western Isles of Scotland* (1775), Oxford, 1979.

Leneman, L. *Living in Atholl: A Social History of the Estates, 1685–1785*, Edinburgh, 1986.

MacAulay, K. *The History of St Kilda* (1758), London, 1764.

MacCulloch, J. *The Highlands and Western Isles of Scotland, Containing Descriptions of their Scenery and Antiquities ... in Letters to Sir Walter Scott*, London, 4 vols, 1824.

MacGillivray, E. Richard James, 1592–1638: Description of Shetland, Orkney, etc. In Orkney Record and Antiquarian Society, *Orkney Miscellany, vol I*, Kirkwall, 1953, 50.
MacKenzie, N. *Episode in the Life of the Rev Neil MacKenzie: At St Kilda from 1829 to 1843*, privately printed, 1911
Mackenzie, W M. *The Mediaeval Castle in Scotland*, London, 1927.
MacTaggart, J. *The Scottish Gallovidian Encyclopedia*, London, 1824.
Marshall, R. The ingleneuk hearth in Scottish buildings: A preliminary survey. *Vernacular Building*, 8 (1983–4), 28–48.
Mays, D, ed. *The Architecture of Scottish Cities*, East Linton, 1997.
McCourt, D. The outshot house type and its distribution in County Londonderry, *Ulster Folklife*, 2 (1956), 27–34.
Michelsen, P. *Ildsteder og Opvarmning på Frilandsmuseet (Hearths and Heating in the Open Air Museum)*, Nationalmuseet, Copenhagen, 1968.
National Coal Board, Scottish Division. *A Short History of the Scottish Coal-Mining Industry*, Edinburgh and Glasgow, 1958.
Omond, J. *Orkney 80 Years Ago*, Kirkwall, 1911.
Penny, G. *Traditions of Perth* (Perth, 1836), facsimile reprint, Coupar Angus, 1986.
Ramsay, A (eds M Burns, J W Oliver). *The Works of Allan Ramsay, vol 2: Poems, 1728*, Edinburgh, 1953.
Rivers, T, Cruickshank, D, Darley, G, Pawley, M. *The Name of the Room: A History of the British House and Home*, London, 1992.
Sage, D. *Memorabilia Domestica*, Wick, 1889.
Sinclair, C. *The Thatched Houses of the Old Highlands*, Edinburgh, 1953.
Small, A. A viking longhouse in Unst, Shetland. In Niclasen, B ed, *The Fifth Viking Congress*, Torshavn, 1968, 62–70.
Stell, G P, Harman, M. *Buildings of St Kilda* (RCAHMS), Edinburgh, 1988.
Stewart, J H. Insights into the building industry in eighteenth-century Perthshire, *Vernacular Building*, 8 (1983–4), 49–61.
Stoklund, B. *Det Færøske Hus I Kulturhistorisk Belysning (The Faroese House in a Cultural Historical Perspective)*, Copenhagen, 1996.
Taylor, L B, ed. *Aberdeen Shore Work Accounts, 1596–1670*, Aberdeen, 1972.
Ure, D. *General View of the Agriculture of Roxburgh*, London, 1794.
Walker, N H, ed. *Kinross House and its Associations*, published privately, 1990.
Walton, J. The built-in bed tradition in North Yorkshire, *Gwerin*, 1 (1961), 161–70.
Wood, M. *The Hearth in Scotland* (SVBWG, Regional and Thematic Studies, no 7), Edinburgh, 2001.
Worsdall, F. *The Tenement, A Way of Life: A Social Historical and Architectural Study of Housing in Glasgow*, Edinburgh, 1979.
Yarwood, D. *The British Kitchen: Housewifery since Roman Times*, London, 1981.

7 Rest

NAOMI E A TARRANT

DAYS OF REST AND HOLIDAYS

The Bible gives rest a prominent part in people's lives. The Sabbath was kept by all major Christian denominations as a day when no work should be done.[1] Whether it was restful to go to church and attend long services was not necessarily the point, hard manual labour was not permitted. The day of rest did however allow people to enjoy a walk in the fresh air, even if they could not run or play games, and there might be the chance to catch up on sleep once religious obligations had been met. For some, though, there was little rest on the Sabbath; farmers with stock, for example, or the ministers themselves.

Others who lacked a day of rest included many married women, as Colin MacDonald notes, speaking of life in the nineteenth-century crofting communities of Cromarty but equally applicable to most married women in Scotland at the time:

> For most of the women in the Glen, from the day they were married till the day they were laid to rest in the old churchyard, anything in the nature of real respite was practically unknown. It must not be inferred from this that on the whole they lived unhappy lives; far from it; for theirs was a life of service to others in which the greatest happiness is found. But oh! How they must have longed for an occasional week off the chain, those brave self-sacrificing women.[2]

Their husbands, for whom a day of rest was more likely to be a reality, could also enjoy a small number of holidays in the year, as MacDonald again records:

> For the men, apart from New Year's Day, the two great holidays of the year were the day of the games and rent-day – 'Lath a' Mhàil'.
>
> The games afforded to the men the opportunity of putting on their Sunday suits. A real enjoyable day witnessing, or taking part in, the various sports, and a trouble-banishing spree at the finish.[3]
>
> On 'Lath a' Mhàil', every man donned his Sunday suit, paid his due, took off his dram from the factor and looked the world in the face.[4]

Without the Bible's instruction it is doubtful if people would have been granted any regular day without working. Scrooge, the caricature of the grasping, unfeeling and miserly employer portrayed by Charles Dickens in *A Christmas Carol*, who resents giving his assistant Bob Cratchett Christmas Day off, was probably not very different from some employers in the nineteenth century. It was, however, much more difficult to deny workers their Sabbath rest as this was prescribed by the Bible.

For those able to enjoy it, the day of rest was not closely associated with leisure activities until the mid twentieth century. As late as the 1960s many employers required their employees to work on Saturday mornings leaving Sunday the only day completely free. As long as the churches enforced the attendance of services on Sunday, the scope of the other activities which could be undertaken on that day was limited. Typically people rose late, went to church, had a substantial lunch and followed that either by an afternoon sleep, or if younger, by a long walk in the countryside. The weekend only came into being in the present-day sense after Saturday working ceased and church attendance relaxed.

Whilst Roman Catholic countries enjoyed holidays for various religious festivals, the number of such days in Protestant countries was far smaller. The fight to get regular paid holidays was one which was taken up in the nineteenth century as part of the greater struggle for better working conditions. One of the consequences of winning the battle was that a whole new industry grew up serving these holidays and this demanded that some people work on Sunday. This, coupled with the gradual secularisation of Britain during the twentieth century, especially the second half, led to Sunday becoming virtually indistinguishable from any other day of the week for a considerable proportion of the population.

Employers of all kinds were fearful that their employees would slack. Unless workers required to be seated to undertake their occupation, such as clerks in offices (and even then not always), seats would not be provided. This often had an injurious effect on the health of people, especially after the establishment of large factories with very long hours of work and the development of department stores where shop assistants could be on their feet for ten or more hours a day. Comfortable seating at home therefore became much more of a necessity.

HOUSES

Most people in Scotland until the second half of the twentieth century lived in one or two rooms and this is true of both the countryside and urban areas.[5] Even the homes of the wealthy in earlier times did not have many rooms, those which existed usually being multi-functional. In the great majority of Scottish homes, therefore, the same spaces were used for sleeping and waking activities, although the situation was ameliorated by many of the latter being conducted outside the dwelling.

Various types of single-roomed dwellings existed including shieling huts and 'benders' or 'gellies'.[6] Shieling huts were constructed for the summer migration to the hills with livestock, whilst benders, tent-like structures, were used by the Travelling People during the summer months. In the nineteenth and early twentieth centuries, unmarried male farm workers, and other groups of workers, were often housed in bothies, usually single-room structures.[7] Sometimes these were simply parts of the outbuildings divided off, but they could also be separate buildings. They were plainly furnished, usually with built-in beds and perhaps a chair or two, farm labourers storing their possessions in their own kists – also used as seating – which moved with them from job to job[8] (Fig. 7.1). In the Highlands and Islands and elsewhere the most common type of house was a single-storeyed dwelling which accommodated livestock at one end and humans at the other[9] (see Fig. 6.6). The larger examples had one or more rooms primarily for human use[10] (see Fig. 6.7).

Improvements in country dwellings for workers in the late nineteenth and early twentieth centuries gave an increasing number of people a greater number of rooms to dwell in, usually one for living and one for sleeping, with grander houses having a parlour and perhaps an additional space for coal as well as a pantry. However all the living areas would probably have beds of some kind so that the maximum number of people could be accommodated.

Dwellings in towns and cities were also small. In Edinburgh buildings rose upwards for several storeys, the various floors being occupied by

Figure 7.1 The Bothy at Gagie, Angus, late nineteenth-early twentieth century. On the left is a form or settle and on the right a kist or chest. The men wear practical garments for farm work. Source: SLA 121B C.9197.

families of different social status. The pressure for housing in the city often resulted in poorer families living in a single room – in the worst instances more than one family shared a room – so that the buildings became overpopulated and prone to the accumulation of dirt and the occurrence and spread of disease.

The smallness of dwellings ensured that furniture was kept to a minimum. The cost also helped to control the number of possessions a household owned, as did the fact that most people rented property and needed to have belongings which could be easily transported when they moved to new accommodation, which could be fairly often.

SLEEPING[11]

Where other cultural areas have seen the ground or floor as adequate for resting upon, whether for sitting or for lying, Europeans have tended to favour the use of something that raises them off floor level. The laying of heather or brushwood as a base was enough to offer some protection, an ancient tradition and one that was still in use by the poorest people in highland areas in the eighteenth century and beyond.[12]

Beds as such were usually the most substantial item of furniture in a room, and in the medieval period they were regarded by royalty and the nobility as more than places to sleep. Bedrooms were often used for receiving important visitors and the costly furnishings of the bed would indicate the owner's status. In the home of a less wealthy and influential household the bed was still the most important piece of furniture in a private chamber, but the room might also hold a chest containing precious items such as silver vessels or money, whilst other chests held clothes. The only chair might be one for the head of the household; otherwise small stools and the chests or seats in the deep window embrasures were used as seating.

Over time, as the richness of the room furnishings increased, so did the elaboration of the bed. By the seventeenth century there were several different types of bed available. Richly carved wooden beds with solid roofs were a particular feature of the late sixteenth century and appear to be associated with a school of carvers in Aberdeen. Other roofed beds were covered with cloth, the wooden components being unseen. Curtains on a bed helped to create a feeling of a room within a room, giving a sense of privacy and intimacy, and also served to protect the occupants from draughts and the cold.

Less elaborate beds consisted of wooden bases, and were considered suitable in particular for servants' rooms. Small moveable beds on wheels, known as 'hurly' beds, which could be pushed under a larger bed during the day, were also provided for servants. In families with more numerous offspring these might be the beds of the younger children. Box beds, freestanding beds enclosed by wooden panelling, were known in Scotland from at least the sixteenth century. They could be found in a range of homes, from those of the well-to-do, in highland areas especially, to the dwellings of some farm servants, and were frequently used as room dividers, being built out from a wall. Similar beds were built into a wall recess and were closed off with panelling during the day, giving the appearance of a continuation of the wall. Late nineteenth- and early twentieth-century flats commonly had alcoves built into the kitchen or living room, just large enough for a bed. These were a more sophisticated example of the 'crûb', the bed in a wall alcove found in the houses on St Kilda and in other parts of the Western Isles[13] (Fig. 7.2). Other beds folded up into the wall, usually into a cupboard, and could be opened out at night. In the nineteenth century further types of folding bed were often disguised as pieces of furniture such as freestanding cupboards, whereas today they are more likely to be sofas or chairs.

By the mid nineteenth century beds with roofs were being replaced by a type of bed with a wooden headboard and foot board, often elaborately carved in mahogany, and with matching wardrobe, chest, dressing table, washstand and chair, forming a suite of furniture (Fig. 7.3). Suites, containing varying numbers of items, remained very popular until the 1960s, when the vogue for built-in furniture started.

Figure 7.2 'The Wake', by Alexander Carse, early nineteenth century, showing the interior of a house with the corpse on an alcove bed, forms for seating at the table and a chair on the right for the seated mourner. The women wear sober dresses with white aprons, white caps and probably black ribbons, with a large white triangular collar at the neck, whilst the men are dressed in suits with knee breeches and hats or bonnets. Source: Reproduced by kind permission of the Trustees of the National Galleries of Scotland.

Figure 7.3 Blue Room, Montgomerie House, Ayrshire, 1948, showing the range of furniture to be found by the late nineteenth century in the bedroom, when most houses had separated the bedroom from the living room. There was still the need to have seats in the bedroom, for example a stool for the dressing table, a chair at the writing desk and a comfortable armchair for reading. Crown Copyright: RCAHMS. Reproduced courtesy of Mrs Robert Garnett and the late Colonel and Mrs Evelyn Garnett.

Iron bedsteads with coiled metal springs or a mesh base became popular in the second half of the nineteenth century. These were considered less likely to harbour vermin and as they could be mass produced they were used wherever large numbers of beds were needed, such as residential schools and hospitals. Fancy brass headboards and foot boards were also sold for more expensive beds.

In the second half of the twentieth century the interior sprung mattress was introduced. This placed springs within a deep padded cover, complemented by further springs in the base. Usually sold as a base and mattress together, with head or foot boards chosen separately, such a bed is much more comfortable than those with coiled spring bases. Various kinds of bed have been introduced in the recent past including orthopaedic beds to help those with bad backs. Some beds have adjustable head and foot ends that can help ease other types of medical problem. Waterbeds are used to help people with poor circulation or by those who for whatever reason cannot lie in a firmer bed.

In the past, cradles constituted the normal provision for babies. Some examples were of wood, whilst others were made, far more cheaply, of plaited straw or rushes. Most had a canopy over the baby's head to protect him or her from the draughts, and the drips of water common in many dwellings. In the nineteenth century cradles for the babies of the wealthy were often heavily draped with fabric, and might even have curtains. They were frequently on stands so that they could swing and the baby was at a higher level, meaning less bending for the mother or nurse.

Small beds do not appear to have been made for children until the nineteenth century when cots were developed. Most children before this date had to share standard beds, usually with their siblings. The twentieth century saw the development of separate beds for each child and placed more emphasis on special beds for children, such as bunk beds, one above another, and beds with side rails to stop small children falling out during the night.

Bedding
Helping to make for a better night's sleep are the various articles of bedding. The simplest was the plaid wrapped around to keep a person warm. Plaids were extremely useful for travellers who might have to camp out at night, especially likely in highland areas where there were few inns and the distances between houses where travellers might lodge too great. For those fortunate enough to have an actual bed to sleep on there would be some kind of sack loosely filled with chaff or straw. Colin MacDonald, living on a late nineteenth-century croft in the area between Dingwall and Strathpeffer, Easter Ross, wrote of the joys of a new chaff bed:

> The real good chaff was much in demand in those days of broad-bottomed beds. In the best circles the huge sack that we knew as the 'chaff bed' was filled with fresh chaff once every year. As the year went on the bed grew flatter and harder, and by the end of the twelve-month it made but a middling couch. But oh, the ecstasy of sinking into the luxurious depths of a newly filled chaff bed![14]

On top of the sack there might be one or more mattresses tightly filled with materials such as horsehair. Both sack and mattresses were replaced by the sprung mattress. Other materials used for mattresses include rubber and synthetic foam. These, unlike traditional mattress materials, are suitable for those who suffer from asthma and certain allergies. More recently there has been a vogue for futons, a Japanese form of padded bedding usually placed on a low wooden base. This has the advantage of being able to be hidden away in a cupboard during the day, so allowing the room to be used for daytime functions.

The mattress was complemented by pillows and bolster. Over these went the washable linen, later cotton, sheets and pillowcases. Blankets or plaids were used in differing quantities to suit the temperature, with a decorative cover on top, which might or might not match the curtains. In the twentieth century the padded eiderdown on top of the bed became popular and remained so until the 1960s when the duvet or 'downie' became almost universal as the preferred form of bed covering. The duvet and cover replaces sheets, blankets and the decorative top cover, and the fitted sheet and pillowcase are usually co-ordinating. In some cases curtains and even wallpaper also co-ordinate with the duvet cover.

It appears that people in the past were generally less concerned about quiet and darkness for sleeping in than we often are today. Long hours of hard physical work probably meant that they slept whenever their heads touched the pillow. However, cold and bed bugs were guaranteed to disturb anyone's rest. Bed bugs appear to have entered the British Isles in the sixteenth century and spread rapidly in the overcrowded towns and cities.[15] No level of society was free from their attacks and great efforts were made to eradicate them, sometimes with noxious potions. They were a considerable hazard for travellers staying at inns as hostelries were not always scrupulous in their cleaning. The introduction of metal bedsteads and the use of fewer draperies, together with a better understanding of how to combat the bugs, meant that by the late nineteenth and early twentieth centuries the problem was largely confined to the overcrowded slums of the cities.

SEATING

Beds were often used for seating in the past as they were the most comfortable items of furniture in the home. However, separate types of seating developed to cater for different occasions or user. Most seating took the form of wooden stools, with chairs originally being reserved for the wealthy or at least the head of household (see Fig. 4.1). Both chairs and stools were fairly solid, made of wood and with no padding to the seats, cushions sometimes being supplied instead. The desire for greater comfort and the acquisition of more possessions in the sixteenth and seventeenth centuries led to an increase in the number of chairs, although those examples with arms were still seen as the prerogative of the privileged. Stools and chairs also became more comfortable with padded seats and the craft of the upholsterer developed to accommodate this new feature. Developments continued throughout the eighteenth and nineteenth centuries, as seating became more varied and comfortable. By 1900 most homes would have had at least one padded armchair (see Fig. 21.1).

The industrial manufacture of furniture in the second half of the nineteenth century lowered the cost of seating, allowing the types of wooden chair often found in kitchens to become the standard type of seat. But not everyone was able to or needed to buy factory-made chairs. In the islands, where wood was in short supply and there was little cash with which to buy items, chairs would often be made from any pieces of driftwood which happened to be washed ashore, leading to some interesting variations in types.[16] Also in the islands, and in highland areas, were chairs made from plaited grass and wickerwork using basketry techniques.[17]

Stools remained very common items for seating and several would have been in use in poorer homes or in kitchens. Some were low, often used by children and called 'creepies' in some parts of Scotland. Other stools were higher; for example those used in a kitchen where the sitter needed to be able to work at a table. In grander houses stools were often provided as part of a suite of furniture and would have been used by those considered of least importance in the domestic hierarchy. Specialised stools were required for some activities, for example playing the piano. In areas where wood was expensive or rare, stools could be made from turf, or coils of rope,[18] and driftwood could be fashioned not only into chairs but also stools of unusual shapes, depending on the pieces salvaged. Anything, in fact, which could be used to form a stool was utilised.

Kists or other types of chest were also used as seating in poorer homes. Kists were usually fairly plain, large wooden boxes, used to store dry food (such as oatmeal), clothes, tools or personal items. Where possible there would be several kists in a dwelling, each one kept for a particular use. With a cover or a cushion placed on them they could form auxiliary seating, otherwise they could be used as work surfaces. Another multi-purpose piece of furniture regularly found in Scots homes was a settle or bench. This would have a back and arms and was long enough for a person to lie on, but would be used primarily as seating (Fig. 7.4). Sometimes these items of furniture were similar to long kists in form, with box-like storage under the seat, while some had a folding table down the centre.

In the mid and late twentieth century the basis of a sitting room would be the three-piece suite, consisting of settee and two armchairs, all matching. Another popular seating form was the rocking chair, originally developed to help mothers rock babies to sleep. At various times poufs – low, soft, leather or fabric-covered, backless and armless seats – have replaced stools. One of the most recent forms is the bean bag, a large sack loosely filled with polystyrene 'beans' which moulds itself to the form of whoever sits in it, therefore making it a particular favourite of those who prefer to lounge rather than to sit. Lounging was not approved of until fairly recently. Chairs were seen as something to take the weight off the feet but not necessarily to relax in and had straight backs as people were expected to sit upright in them. Many women did not even touch the back of a chair when they sat. The more relaxed attitude to seating probably started with the development of the club chair for men, where the deep upholstered seats and arms helped to create a comfortable chair to sink into (see Fig. 10.3). By the 1920s comfortable seating for everyone was available.

Seat types, sitting and hierarchy
For most of history sitting has been intimately bound up with social distinctions between groups and individuals.[19] In order not to offend when sitting

Figure 7.4 Shetland chairs, with and without arms, one latterly used as a commode; and a settle. Source: Fenton, A. *The Northern Isles: Orkney and Shetland*, (1978), East Linton, 1997, 166.

in company, one has to be very sure that the right to do so exists and one will be expected to use the correct form of seating.

At all levels of society, and until recent decades, the most important male in the family usually had the best seat. This might be a chair with arms or a high back, echoing somewhat the thrones used by monarchs. The wife might also sit on a chair, usually a simpler one, but the children would be on stools. In Caithness, for example, seats often existed in pairs, the chair with arms being for the man and that without for the woman, in particular to facilitate spinning. In the homes of the wealthy seating could also reflect other aspects of hierarchy. The backless stools, found in sixteenth- to eighteenth-century homes, were used not only by women but also by men of the lowest standing within the household. In the case of the monarch, even members of his or her government might be relegated to stools.

Hierarchy was shown not only in the type of seating used but also when one was able to sit. In the medieval and early modern periods seating tended to be the preserve of the most important members of a household. The head of a house would sit for activities such as eating, reading, writing and generally whenever in the house. Men of consequence would never sit on the floor or lounge. Women might sit to do needlework or dressmaking, to draw, read or write letters. They would also sit to breast-feed infants but for most other activities they would probably stand. In general, until the second half of the twentieth century, one did not sit in the presence of an older person or someone of a higher social standing unless asked to do so, and such formality would have extended to close members of the family (see Fig. 13.6). Servants would certainly not be allowed to sit in the presence of their superiors, except on those rare occasions when they dined or feasted at the same time. That such ideas are not entirely extinct is seen in many parts of present-day Europe where all must wait until the monarch is seated before seating themselves; if the monarch stands then so must the rest of the company, the only exceptions being those who are physically infirm. Somewhat counter to the general trend was the nineteenth-century etiquette in which in mixed company seats would be given up to women, and men would wait for women to be seated before they sat down themselves. Usually, however, these would be gatherings of social equals and the men would not necessarily consider giving up seats to women known to be their social inferiors.

NOTES

1. Observance of the Sabbath is also of great importance in Judaism. Jewish communities have existed in Scotland since the early nineteenth century.
2. MacDonald, 1936, 74.
3. MacDonald, 1936, 74.
4. MacDonald 1936, 76.
5. Carruthers, 1996.
6. See Chapter 33 'Time Out' Work; Chapter 44 Communities on the Move.
7. See Adams, 1991; also Chapter 33 'Time Out' Work.
8. See Chapter 4 Storage; Chapter 33 'Time Out' Work.
9. See, for example, Fenton, 1989, 12–13, 20–3.
10. Even in these spaces poultry, for example, could be housed.
11. See also Chapter 6 Heating and Sleeping.
12. Grant, 1961, 168; see also Burt, 1974, I, 116. Ordinary people's beds (around and in Inverness) were on the floor, in a corner, with a small quantity of straw and two blankets.
13. See the plan of a St Kilda house around 1870 in Stell, Harman, 1988, 4; See also Chapter 6 Heating and Sleeping.
14. MacDonald, 1947, 40.
15. Tarrant, 1998, 69–71.
16. Grant, 1961, 172.
17. Grant, 1961, 172.
18. See Chapter 18 Auxiliary Work in the Rural Home.
19. See also Chapter 29 Status.

BIBLIOGRAPHY

Adams, D G. *Bothy Nichts and Days: Farm Bothy Life in Angus and the Mearns*, Edinburgh, 1991.

Burt, E. *Letters from a Gentleman in the North of Scotland to His Friend in London* (1754), Edinburgh, 2 vols, 1974.

Carruthers, A, ed. *The Scottish Home*, Edinburgh, 1996.

Davidson, I, ed. *Caring for the Scottish Home*, Edinburgh, 2001.

Fenton, A. *The Island Blackhouse* (1978), Edinburgh, 1989.

Grant, I F. *Highland Folkways*, London, 1961.

MacDonald, C. *Echoes of the Glen* (1936), reprinted Edinburgh, 1997, as *Life in the Highlands and Islands of Scotland, Part One*.

MacDonald, C. *Croft and Ceilidh* (1947), reprinted Edinburgh, 1997, as *Highland Life and Lore, Part One*.

Stell, G P, Harman, M. *Buildings of St Kilda* (RCAHMS), Edinburgh, 1988.

Tarrant, N E A. *Going to Bed*, Edinburgh, 1998.

8 Music

GARY J WEST

INTRODUCTION

There have been very few households within Scotland, whatever their positioning in respect of time, place or social hierarchy, which have not been enriched with music. Those householders who could afford to, often paid for its provision, while the majority made do with their own efforts, or indeed those of their more talented neighbours and visitors. In pre-industrial times, Highland chieftains supported their pipers, bards and harpers, the daughters of Edinburgh socialites learned spinet, virginal and lute from French and Italian masters, while the 69 individual musicians being paid by the court of King James IV on New Year's Day in 1506 testify to the importance placed on the spectacular household consumption of the musical arts by the ruling élite.[1] And if the folklore surrounding that particular sovereign's son and heir is to be taken literally, James V's clandestine 'gaberlunzie' wanderings around the Scottish countryside would have opened his ears to the diverse musical riches to be found within the homes of his subjects across the length and breadth of the nation.[2] This diversity remained a constant in the centuries to come: infants being dandled to the gentle sounds (but sometimes brutal lyrics) of a Gaelic lullaby;[3] wool being spun to a rhythmic tale of heroic deed or bloody treachery; pious offerings of metrical psalms or *Solitary Thoughts* in praise of The Almighty;[4] a ribald parlour gathering, delighting in the bawdy muses of the 'secret songs of silence'.[5]

The twin processes of industrialisation and urbanisation may have reshaped the domestic settings within which such music was played and listened to,[6] but the rural exodus towards townhouse, tenement and miners' row did nothing to silence the music of the people. By the mid nineteenth century, accordion, melodeon and mouth organ had joined the fiddle and trump (jew's or jaw's harp) on the list of the most popular working-class instruments, while the new concentration of such large numbers of children within the limited urban space encouraged a massive creative surge of rhymes, clapping songs and other ditties which enlivened the kitchen as well as the stairwell, street and playground (see Fig. 34.6).

Yet it was not nineteenth-century urbanisation, but rather twentieth-century technology which was to revolutionise the music of the Scottish home, especially as regards consumption rather than production. Radio, gramophone, television and hi-fi (and its successors) have all brought a huge range of hitherto little-known musical genres into the domestic setting, while the widespread adoption of personal playback devices, especially by the young, has served radically to reduce the communal nature of music appreciation within the home. Yet the longer-term impact of these technological developments in terms of the ongoing relationship between people, music and the home has yet to be uncovered.

ÉLITE HOUSEHOLDS

Professional musicians in Scotland – as elsewhere – have often relied upon the patronage of the social élite in order to earn a livelihood, and so the households of the aristocracy have long acted as the setting of much music making and appreciation. The cultural life of the Highland chief's household, for instance, certainly up until the eighteenth century, was rich indeed:

> ... with filidh, bard, clarsair, piper, fool, all part of a considerable retinue of intensively trained and highly skilled people ... an evening's entertainment recounting the deeds of the chieftain and his ancestors could have been of epic proportions.[7]

By the dawn of the nineteenth century, such luxury was unjustifiable in the harsh social and economic climate of the post-Culloden era, and so clan chiefs could no longer bask in the panegyric eulogising of their personal song poets. But the romantic aesthetic of the new age at least encouraged many lairds, both above and below the Highland divide, to continue (or begin) to sponsor a household piper. In 1842, Lord and Lady Breadalbane were able to call on any combination of nine pipers at Taymouth, Perthshire, to entertain their visiting sovereign, Victoria being quite taken with the 'princely and romantic' scene their presence helped to create,[8] and while this was certainly exceptional, many early nineteenth-century lairds remained keen to support at least one piper.

Those who did not offer permanent employment to musicians were content to pay for their services as and when required. The Mar account book for the late seventeenth century records one-off payments to 'a blind singer at dinner', 'a Highland singing woman', 'Blind Wat the Piper' and 'ane woman harper'.[9] During the same period, John Foulis of Ravelston made regular payments to a range of musicians, especially fiddlers, pipers and drummers, while his daughters were taught to play both the virginal and the viol by one Msr De Voe.[10] The household *Compte Book* of Lady Grisell Hume, wife of a seventeenth-century Borders laird of middle rank, reveals that money was spent on domestic music in three main ways: to pay music teachers for her children; to buy, repair and maintain instruments; and to pay performers (in this case, harpers, fiddlers and singers) for entertaining the household.[11]

Such élite households usually possessed homes large enough to allow for the allocation of a space for music performance and appreciation: the grandest dwellings had ballrooms or great halls, while in more modest abodes the withdrawing room or parlour might be used. Both stood in stark contrast to the standard 'room-and-kitchen' set up in most rural cottages and urban tenements, where the limited space resulted in a fairly flexible attitude to the boundary between the domestic and public sphere, with music, song and dance frequently spilling out of the home and onto the street.[12]

The drawing rooms and parlours of the upper and middle ranks of Scottish society were not only the setting for the playing of much music and song, but also a catalyst for their creation. John Clerk of Penicuik (1676–1755), for one, made a significant, if under-celebrated, contribution to the European classical tradition of his day in the form of a series of five cantatas of outstanding quality.[13] Clerk's contemporaries, William McGibbon (c1690–1756) and James Oswald (1710–69), both products of an east-coast middle-class upbringing, also made a lasting impression on the musical output of the period. In the Victorian era, the drawing rooms of the wealthy played host to a new thirst for romantic Scots song, much of it composed or arranged by some of the women who frequented them. Lady Carolina Nairne (1766–1845), Joanna Baillie (1764–1851), Lady Alicia Scott (1810–1900) and Lady Anne Barnard (1750–1825) were all finally revealed as songsmiths of note in the middle decades of the nineteenth century, earlier sensibilities having demanded that as women, they publish their work anonymously.[14] The themes of their songs – many of which achieved great popularity both north and south of the border – pre-empted those of the kailyard school of novelists of the following generation:

> They are 'morally healthy' in advocating contentment with one's lot; they abound in pictures of domestic peace and comfort; and they show joy in the beauties of nature.[15]

This musical view from the drawing-room window was of a romantic, heroic and egalitarian nation, proud of its recent Jacobite past, and yet safe in the stability of the unionist present. A distorted view it may have been, but it was durable and influential, eventually breaking free of the private home and making its way straight to the very public platform of the music-hall stage.

FUNCTIONAL MUSIC

> Songs of occupation are amongst the most primal things in the history of mankind, and in their simple rhythms and intervals, first evolved by workers for their needs, we find the germs of all Music and Verse.[16]

Music was not always performed within the home for entertainment alone, for it was often employed in a more overtly functional role. Whether used to accompany dancing, or to pass the time while carrying out some laborious task, to set up a working rhythm or to soothe or rock an infant, the voice in particular could be pressed into action for any number of practical uses. These uses were widespread throughout Scotland, but it is within the

Gaelic-speaking areas of the country that the strongest survivals of functional songs are to be found.[17]

In the Preface to his *Highland Vocal Airs* collection of 1781, Patrick MacDonald testifies to the frequent use of songs, particularly by women, when going about their daily and seasonal tasks in and around the home:

> ... and they are sung by the women, not only at their diversions but also during almost every kind of work, where more than one person is employed, as milking the cows, and watching the folds, fulling of cloth, grinding of grain with the quern or hand-mill, hay-making, and cutting down corn.[18]

Of all the labour tasks associated with song accompaniment, the fulling of cloth is perhaps the one which has achieved almost symbolic status in relation to the collective memory of the musical and social past of the Gàidhealteachd (Gaelic-speaking areas). Within the Lowlands, the fulling process was mechanised and centralised in mills by the eighteenth century, and if there ever was a lowland parallel to the Gaelic waulking songs it does not appear to have survived. But in the Outer Hebrides in particular, this form of work song still existed, attached to its original function, well into the twentieth century, for waulkings continued to be practised as a communal task by women in their homes until the 1940s (Fig. 8.1).

The waulking songs were sung at a tempo and rhythm which were closely integrated with each stage of the fulling process. To begin with, when the cloth was wet and heavy, both work and songs were slow, but as the cloth began to lose its moisture, the pace quickened and the song tempo increased accordingly.[19] As the evening wore on, the waulking tended to develop into a highly-charged social gathering, very much the domain of the women folk, and at which stories, gossip and leg pulling would be the order of the day. Indeed, much of this was integrated into the improvised lyrics of the rhythmical songs themselves. As such, the waulking tradition provides an excellent example of the manner in which music, material culture and social organisation can combine to create a symbiosis which was a central element of the culture of many Scottish homes.

The functional uses of music and song within the home stretched well beyond the rhythmical accompaniment of physical labour. The soporific effects of the lyrical human voice, for example, have probably been recognised since the earliest development of civilisation, and both the Gaelic and Scots traditions abound with cradle songs and lullabies. Indeed, it was not only the young whose slumber could benefit from 'sleep music': in the late nineteenth century, the widow of the Reverend George Munro of South Uist employed a local woman primarily to sing her to sleep at night.[20]

Music was also a common element in the domestic celebration of ritual practices associated with both religious and secular custom and belief.[21] Pious musical offerings have rarely been confined to formal buildings of worship, for instance, as gloriously revealed to one visitor who heard 'a mile of singing' as he walked the length of a Lewis crofting township one

Figure 8.1 Waulking Tweed, Eriskay, 1934. Source: SSS BU III f6 5248 Photo: Werner Kisling.

Sabbath evening in the 1880s. 'The singing at family worship,' he noted, 'was so general and continuous that it seemed unbroken from one end of the township to the other.'[22] Rites of passage within the human life cycle, too, provided plenty of opportunities for musical celebration or commemoration within the household. In Shetland, a fiddler was a central figure within several stages of a wedding celebration: the official contract was sealed in the bride's home several days before the main event by dancing to the sound of the fiddle; the fiddler would play at the bachelor's party the night before the wedding; and when the bride was ceremonially bedded, he would provide a musical backdrop there too.[23]

The importance of music in the construction and celebration of cultural or ethnic identities is widely recognised, and it is of little surprise that functional household music in a multitude of forms is a basic ingredient in the complex web of ethnicity which stretches across modern Scotland.[24] Perhaps one brief example can make the point. For David Daiches, growing up in a respected Jewish family in early twentieth- century Edinburgh, music helped to define his 'two worlds':

> ... and round the candle-lit table, Lionel and I wearing little green skullcaps and my father his accustomed black one, we talked in a politer accent than the one we employed in the streets and discussed quite other matters. And as we sang the familiar *shir ha-ma'aloth* at the beginning of the post-prandial grace, to the plangent minor melody

that my father had learned from his father, the Meadows and Marchmont and Arthur's Seat and Blackford Hill were forgotten and we were back in a world which stretched unbroken back through the Middle Ages to ancient Palestine.[25]

COMMUNITY

While music in some form or other must surely have been a feature of virtually every Scottish home, most communities, past and present, would have a selection of houses which might act as the focal point for musical gatherings. Willie Scott recalls the frequency of communal gatherings of this kind in the Scottish Borders in the early twentieth century:

> There was a party in one house one night and a party in another house the next night. An they sort o'gather together. There wasn't a hill cottage but there was a fiddle or two hangin in, ye know. There was always a song or a fiddle to play a tune and there was dancing, a sing-song, just a sort o ceilidh, as they were called in the north o Scotland. We had some great nights.[26]

The role of music as a catalyst in the processes of social integration amongst members of dispersed communities should not be underestimated and has long been a common feature of life throughout rural Scotland (Fig. 8.2). In Shetland, talk was of going 'in aboot da nyht' to share stories and fiddle tunes, while in Gaelic-speaking areas it was the ceilidh house which performed this vital role. Music and song were also ubiquitous in the bothies and chaulmers which, from the early nineteenth until the mid twentieth centuries, acted as home to the unmarried male farm-servant population of lowland Scotland.[27] Physically spartan they may have been, but these communal homes nurtured one of the most dynamic and expressive folk cultures of recent times. The so-called bothy ballads, composed in an earthy Scots vernacular, deal with the issues that were of most interest to those whose world was the 'fermtoun'. Love and sex are themes which lie at the heart of many of them, some leaving little to the imagination, while others take the form of good-humoured but heartfelt protest songs which highlight dissatisfaction with working conditions or tyrannical employers.

While these farm servants' dwellings became most closely associated with the singing of the bothy ballads, musical instruments were by no means in short supply there too. Fiddles seem to have been the favoured instrument, although by the second half of the nineteenth century, the melodeon began to be widely adopted, arriving in great numbers from Germany from the 1860s.[28] The marketing of these instruments was aimed at the domestic, amateur player, with advertisements in the popular press suggesting that no household should be without one as a means of encouraging 'innocent and mirthful recreation in the family'.[29] The message must have worked, for by the turn of the twentieth century the melodeon could be heard in homes throughout Lowland Scotland, although it was by no means uncommon north of the Highland divide too. Its compact form, versatility and relatively low volume also made it a particularly popular choice of instrument amongst the tenement dwellers of Scotland's towns and cities.

While music has certainly served to bring members of local communities together, an unfortunate tendency in the modern world, of course, is for it to drive neighbours apart. Along with the barking of dogs, amplified music from neighbouring homes is the biggest cause of domestic noise complaint, and is a common problem in the tenements and flats of the urban sphere in particular.[30] Although organisations such as the Noise Abatement Society advocate a preventative approach by encouraging the educating of residents in noise awareness, this has failed to obviate the need for the introduction of legislation regarding the issue. The Civic Government (Scotland) Act of 1982, for example, gives the police powers of confiscation over musical instruments, radios, televisions, record players or any other source of noise which has been deemed to be causing annoyance to others. Detailed empirical research would have to be undertaken in order to determine to what extent this situation has influenced patterns of musical production and consumption within the Scottish home, but it is certainly worth noting the increasing popularity of musical 'sessions' and karaoke evenings within the pubs of urban Scotland, which may be acting as a

Figure 8.2 Informal music session, Forrest Glen, Dalry, 1985. Source: SSS NII 8a 8744 Photo: Ian Mackenzie.

surrogate for activities which at one time would have taken place within the private home (Fig. 15.1).

PRACTICE, PERFORMANCE AND TRANSMISSION

Until the second half of the twentieth century, a long-standing scholarly preoccupation with text rather than context – with songs rather than singing, tunes rather than playing – resulted in the development of a wealth of theories on structure, form and variation within the creative arts, but a paucity of ideas concerning practice or performance.[31] Recent studies of a more ethnographic nature, however, have gone some way towards redressing this historiographical imbalance and serve to underline the importance of the domestic home as the hub of the onward transmission, through both time and space, of received traditions of whatever form.[32] 'Strong traditionbearers', the driving force of any artistic tradition, according to John Niles,[33] learn, perfect and transmit their styles and repertoires within a variety of contexts, but the home remains central to these processes, especially as it often becomes the preferred setting for the kind of communal gatherings outlined above. For a twentieth-century 'bard baile' (township song poet), like Iain MacNeacail of Skye, the muse may have come while 'walking the road', but it was the ceilidh house which provided the setting for the performance and transmission of his creations, as his biographer recognises:

> The role of the 'taighean céilidh' in transmitting rural Highland culture cannot be overemphasised ... [I]t was by far the most usual (and natural) environment for oral entertainment traditions, providing endless information about the community, whether through songs, sayings, stories and legends of the past, or news and gossip of the present. From his earliest days, Iain was a keen participant and it was there that he first began to take an interest in songs.[34]

Many of the 'strongest' tradition bearers of all were to be found amongst the Travellers or cairds of central and north-east Scotland, for whom home was frequently a tent or caravan. The folk revival of the 1950s elevated several members of this marginalised community into national stars, but for generations before this belated fame took hold, their songs and stories were taught and learned within 'egalitarian, non-competitive gatherings of kin and acquaintances in domestic surroundings'.[35] For some, such intimate gatherings provided a relaxed, informal environment within which to teach and learn, but others viewed the passing on of the tradition as worthy of a more formalised approach, although still situated within the domestic sphere. The 'acknowledged queen of traditional singers',[36] Jeannie Robertson, belonged to the latter category, as her nephew, Stanley Robertson, recalls:

> If I was asking her for a ballad – asking her to teach me – she was very, very strict, very very hard. 'All right, laddie, I'll learn you this song, but I want you to sing it right, sing it proper, an' sing it real.'[37]

Today, a good deal of formalised teaching and learning of music has been taken out of the private house and transferred to the classroom, village hall or community centre. Yet the home remains the primary venue for the form of performance we call 'practice'. Indeed, it is probably the case that practice serves as the principal form of live music performance in contemporary Scottish households. In musicological terms, however, it is a theme which requires much more attention than it has hitherto received.[38]

MUSIC APPRECIATION IN THE HOME

As theorists such as Nicholas Cook point out, music cannot be defined in terms of the production of the sound alone, but must take note of the role of the listener and indeed the listening environment.[39] James Parakilis illustrates the point thus:

> Classical music is no longer itself when it is used as a background music. It becomes like 'easy-listening' popular music, valued more for its geniality than for its genius. But the change that comes over it is a change in the listening, not in the notes.[40]

Undoubtedly, the twentieth-century revolution in audio technology has initiated a 'change in the listening', greatly altering the dominant patterns of musical consumption and appreciation within most Scottish homes. While live music *making* carries on regardless, the ubiquity of recorded music within most households today – often different forms playing within several rooms simultaneously – has served to redefine significantly the role of the listener. Music has thus become less of an *event*, and more part of the *space* – aural wallpaper perhaps:

> ... Music becomes a form of architecture ... the sound becomes a presence, and as that presence it becomes an essential part of the building's infrastructure.[41]

Jonathan Sterne is here referring to the use of background music within the very public space of American shopping malls, but his findings might usefully be applied to the changing soundscape of domestic space also. And yet, the same technology which may encourage an architectural role for music, can also facilitate a very direct and highly personalised relationship between music and the individual. Headphones and portable playback devices enable the listener to engage repeatedly and in minute detail with a musical passage in a way that would be virtually impossible in a 'live' or even communal setting. Any suggestion that widespread adoption of advanced audio technologies within the home has necessarily 'dumbed down' music appreciation must therefore be treated very cautiously.

The birth of the British Broadcasting Company in 1922 (from 1927 the British Broadcasting Corporation), introduced a revolution which was to bring music into the homes of the British public in a whole new way, and open up a brand-new chapter in the story of music appreciation. From the outset, it was music rather than speech which dominated the radio airwaves: in 1926, for instance, music made up four hours and 40 minutes of the seven hours of daily output. Opera, classical symphonies, chamber music and 'light' or 'dance' music were the most common broadcast genres, most of this music being played 'live' within the studio: there was very little reliance on gramophone records in pre-War British broadcasting.[42]

Indigenous Scottish music was not well represented on the airwaves in the early years of the BBC, but it was by no means absent altogether. The first bagpipe to be heard on the radio, for instance, was that of Willie Ross, who played 'The Lament for James MacDonald of the Isles' in March 1927 as the finale to a programme consisting largely of a lecture on the pibroch tradition presented by Seton Gordon.[43] Through the 1940s and 1950s, series such as *Country Magazine*, *As I Roved Out* and Ewan MacColl's *Radio Ballads* provided a broadcast platform for the kind of traditional music and song that was still being played 'live' within many Scottish homes. But it was the 'foreign' sounds that were coming into these homes that were having the greatest impact on listening tastes. Blues, jazz, country and western and rock and roll provided a very different aural backdrop to the post-War domestic life of the Scottish people from that which had gone before. In time, of course, this directly influenced the patterns of music *making* as well as appreciation within subsequent generations of the Scottish population.[44]

While the broadcasting age introduced a new chapter in the history of music appreciation within the home, the gramophone and its subsequent derivatives constituted an entire new book (see Fig. 10.2). The final quarter of the twentieth century saw a massive growth in the consumption of home-entertainment products within the UK, with music provision being a major factor in this trend. Between 1973 and 1998, UK sales of music albums (LPs, cassettes, CDs and singles) virtually doubled, rising from 146.4 million to 289.5 million.[45] By 1996–7, 78 per cent of adult men and 76 per cent of adult women cited listening to recordings (including tapes and CDs) amongst their home-based leisure activities, although this figure varied between social groups: only 67 per cent of unskilled men and 64 per cent of unskilled women made this claim.[46] It is also clear that gender and socio-economic status are significant contributing factors to variations in audience figures for radio listening, but not for watching television: in the UK in 2001, 93 per cent of professional males but only 83 per cent of unskilled males listened to the radio, while 99 per cent of both groups watched television. Within the female population, 96 per cent of professional women but only 78 per cent of unskilled women listened to the radio, while 98 per cent of both groups watched television.[47] Perhaps more telling is the fact that the average weekly spending on items related to music listening per UK household in 1999–2000 was £2.20: the corresponding figure for musical instruments was 20 pence![48]

Is it the case, then, that the music culture of the home at the dawn of the twenty-first century is one of passive appreciation rather than active music making? This would be a rather dangerous conclusion to draw. Statistics provide part of the story only: CDs and stereos are not lovingly passed down through the generations as 'family heirlooms' in the same way as pianos, violins and bagpipes, and there is no need to spend money in order to sing! It should not be assumed, either, that listening to recorded music is a largely passive activity. Recent research by John Sloboda and others suggests that individuals' emotional response to music can still be powerful despite the fact that they find themselves engaging with it while carrying out some other form of activity, such as washing up, socialising, exercising or working. Music, even when consumed in such an apparently passive manner, 'increased emotional state towards greater positivity', especially when individuals were able to exercise choice over the music they were listening to.[49]

Perhaps, then, the role of 'background' music in the home is less passive than has hitherto been recognised. Not only is it likely to act in a positive sense psychologically, it could also be argued that it fulfils at least some of the same roles as functional music in traditional society. This may well go some way towards appeasing those incredulous parents whose children study to the accompaniment of a booming stereo on the basis that it helps them to think!

CONCLUSION

The story of music making and appreciation within the Scottish home is a very long and heterogeneous one, and in this chapter it has been possible to provide only a brief overview of some of the main developments and issues, and indeed merely to hint at the level of complexity involved. As is undoubtedly the case with other cultural themes, tracing the musical goings-on within the private household is much more problematic than conducting a parallel investigation within the public sphere. While some excellent work is being done by the EERC and others to promote the dissemination and analysis of a greater range of 'ego documents' – diaries, household account books, private correspondence, etc – access to such material remains frustratingly limited. Archive-based research within such sources as probate inventories, manufacturers' catalogues and press advertisements can certainly shed useful light on the availability and social distribution of musical instruments, but reaching solid conclusions on who played them, and why, when and where, is much more difficult. Many of the answers to such questions in relation to the past may well remain elusive, but there is much scope for working towards a more solid understanding of the relationship between people, music and the home within contemporary Scottish society. An ethnographic approach is the most promising – observation, participation, surveying, recording – and focused investigation of this kind will surely bear fruit. But there remains much work to be done.

NOTES

1. Purser, 1992, 82.
2. Popular tradition suggests that King James V was given to touring his kingdom

dressed as a beggar. The ballad entitled 'The Gaberlunzie Man' is reputed, by some of its singers, to have been inspired by his exploits.
3. One of the most popular lullabies, 'Griogal Cridhe [Beloved Gregor]', for instance, tells of the heartache of a widow for her slain husband and does not spare the listener the details of his fate: '... they spilled your blood and put your head upon an oaken stob, separate from your body'.
4. 'Smuaintean Aonranach [Solitary Thoughts]' is the title of one of the many evangelical hymns which have been popular in Gaelic-speaking Presbyterian areas since the eighteenth century. This song can be heard on the SSS recording, *Music From the Western Isles* (Scottish Tradition Two), University of Edinburgh and Tangent, 1971, reissued University of Edinburgh and Greentrax Recordings, 1992.
5. 'The Secret Songs of Silence' is the title of an unpublished manuscript housed at Harvard University, containing a collection of 'blue' songs gathered by the northeast collector, Peter Buchan (1790–1854).
6. The urbanisation process was extremely rapid within Scotland: according to a survey by Jan de Vries, 1984, 39–48, in 1700 Scotland was the tenth most urbanised nation in Europe, but by 1850 it was second only to England (as defined by percentage of total population living in settlements with a population of at least 10,000).
7. Purser, 1992, 129.
8. Queen Victoria, 1972, 21–3.
9. Colville, 1932, 32.
10. Hallen, 1894, xxxix.
11. Kelsall, Kelsall, 1993, 126–7. The authors cite evidence from other household account books to show that such employment of musicians was standard practice within lowland landed families during this period.
12. Tommy Blackhall of Falkirk, for instance, recalls his father and uncle regularly playing their whistles and melodeons outside their homes: 'When the blackoot was on, they had tae play in the hooses, but on the simmer nichts they played ootside in the streets and there were great street activity in thae days.' Douglas, 1992, 60.
13. Purser, 1992, 164–72.
14. Scott, 1989, 92–9.
15. Scott, 1989, 93–4.
16. Tolmie, 1997, v.
17. These too, eventually found their way into the drawing rooms and concert halls of the twentieth century through the collecting and arranging work of Marjorie Kennedy-Fraser (1857–1930).
18. Quoted in Tolmie, 1997, vi.
19. Margaret Fay Shaw provides a succinct account of the process: 'The ends of a length of newly woven cloth are sewn together to make it a circle, and the cloth is then placed on a long trestle table and soaked with hot urine. An even number of women sit at the table, say twelve with six a side, and the cloth is passed around sunwise, to the left, with a kneading motion. They reach to the right and clutch the cloth, draw in, pass to the left, push out and free the hands to grasp again to the right. One, two, three, four, slowly the rhythm emerges.' Shaw, 1955, 6.
20. Purser, 1992, 127. According to tradition, sleep music is one of the three basic strains of Celtic music, the others being the music of laughter and the music of sorrow.
21. See Chapter 30 Custom.
22. As reported by the Rev Malcolm MacPhail, Kilmartin, Argyll, in a letter to the *Oban Times*, October, 1898. Quoted in the notes to SSS, 1994.
23. Cooke, 1986, 11.
24. According to Norwegian anthropologist Fredrik Barth, 1969, one of the defining criteria of an ethnic group is that its members share fundamental cultural values

realised in overt unity in cultural forms. These cultural forms, in the majority of cases, include music, song and dance traditions.
25. Daiches, 1987, 5.
26. Douglas, 1992, 8.
27. See Chapter 33 'Time Out' Work.
28. Eydmann, 1999.
29. Eydmann, 1999, 599.
30. www.noiseabatementsociety.com.
31. For a brief discussion of performance theory, see Porter, 1995.
32. See, for example, Campbell, 1999; Douglas, 1992; McKean, 1996; Porter, Gower, 1995; Williamson, 1981.
33. Niles, 1995, 234, 'strong' tradition bearers are distinguished from their more passive counterparts by displaying heightened levels of five traits: engagement, retentiveness, acquisitiveness, critical consciousness and showmanship.
34. McKean, 1996, 111.
35. Porter, Gower, 1995, xxxix.
36. The phrase is that of Fred Woods, 1979, 16.
37. Porter, Gower, 1995, 87.
38. Empirical research by Campbell, 1999, involving an in-depth investigation of adult learners of the fiddle in Edinburgh, suggests that home-based practice is a highly complex field of enquiry, owing to the variety of strategies and methods employed by learners as well as notable variations in these individuals' interpretation of what 'practice' actually is.
39. Cook, 1992, 11.
40. Parakilas, 1984, 15.
41. Sterne, 1977, 23.
42. Briggs, 1961, 275.
43. Donaldson, 2000, 424.
44. As Cruickshank, 1993, 192, asserts, it was the rural centres of Scotland, rather than the larger towns and cities, which witnessed the earliest popularity of rock music. This lends support to the claim that broadcasting and the gramophone were highly influential in the dissemination of such new musical styles, for the opportunities to hear this music played live were limited within the rural sphere. Nonetheless, some local entrepreneurs, such as Albert Bonici, were indeed very successful in their attempts to lure the top acts to the Scottish rural periphery. For example, the Beatles played Bonici's club in Elgin in 1963. Cruickshank, 1993, 192.
45. Office for National Statistics, 2000, 213.
46. Office for National Statistics, 2001, 226–7.
47. Office for National Statistics, 2001, 226–7; the influence of visual media on music consumption and appreciation within the home is undoubtedly a highly-significant issue, but more research needs to be carried out before solid conclusions can be drawn on this. A useful starting point for a scholarly discussion of the influence of television on popular culture in general is provided by Storey, 1996; See also Chapter 10 Television, Video and Computing.
48. Office for National Statistics, *Family Expenditure Survey, 1999–2000*, Social Trends Dataset, ST 301307, <www.statistics.gov.uk/statbase>.
49. Sloboda, 1999, 450–5. This research runs counter to earlier claims by Adorno, 1994, that 'popular music', consumed in any form, promotes passive listening.

BIBLIOGRAPHY

Adorno, T. On popular music. In Storey, J, ed, *Cultural Theory and Popular Culture: A Reader*, Hemel Hempstead, 1994, 197–209.

Barth, F. *Ethnic Groups and Boundaries: The Social Organisation of Culture Difference*, Bergen, 1969.
Briggs, A. *The Birth of Broadcasting*, vol 1, Oxford, 1961.
Buchan, P. The Secret Songs of Silence, unpublished MS, 25241.9* Houghton Library, Harvard University.
Campbell, K. 'Learning to play Scots fiddle: An adult learning perspective', PhD thesis, University of Edinburgh, 1999.
Colville, K. 'Scottish Culture in the Seventeenth Century (1603–1660)', PhD thesis, University of Edinburgh, 1932.
Cook, N. *Music, Imagination and Culture*, Oxford, 1992.
Cooke, P. *The Fiddle Tradition of the Shetland Isles*, Cambridge, 1986.
Cruickshank, S. Scottish rock music, 1955–93. In Scott, P H, ed, *Scotland: A Concise Cultural History*, Edinburgh, 1993, 191–4.
Daiches, D. *Two Worlds*, Edinburgh, 1987.
Donaldson, W. *The Highland Pipe and Scottish Society, 1750–1950*, East Linton, 2000.
Douglas, S. *The Sang's the Thing*, Edinburgh, 1992.
Eydmann, S. As common as blackberries: The first hundred years of the accordion in Scotland, 1830–1930, *Folk Music Journal*, 7/5 (1999), 595–608.
Hallen, A, ed. *Foulis of Ravelston's Account Book* (Scottish History Society, vol XVI), Edinburgh, 1894.
Kelsall, H, Kelsall, K. *Scottish Lifestyle 300 Years Ago* (1986), Aberdeen, 1993.
McKean, T. *Hebridean Song-Maker: Iain MacNeacail of the Isle of Skye*, Edinburgh, 1996.
Niles, J. The role of the strong tradition bearer in the making of an oral culture. In Porter, J, ed, *Ballads and Boundaries: Narrative Singing in an Intercultural Context*, California, 1995, 231–40.
Office for National Statistics, *Social Trends*, No 30, London, 2000.
Office for National Statistics, *Social Trends*, No 31, London, 2001.
Parakilas, J. Classical music as popular music, *Journal of Musicology*, 3 (1984), 1–18.
Porter, J. Toward a theory and method of ballad performance. In Porter, J, ed, *Ballads and Boundaries: Narrative Singing in an Intercultural Context*, University of California, 1995, 225–30.
Porter, J, Gower, H. *Jeannie Robertson: Emergent Singer, Transformative Voice*, East Linton, 1995.
Purser, J. *Scotland's Music*, Edinburgh, 1992.
Queen Victoria. *Our Life in the Highlands*, Newton Abbot, 1972.
Scott, D. *The Singing Bourgeois: Songs of the Victorian Drawing room and Parlour*, Oxford, 1989.
SSS. *Gaelic Psalms from Lewis* (Scottish Tradition Six), University of Edinburgh and Tangent (1975), University of Edinburgh and Greentrax Recordings, 1994.
Shaw, M F. *Folksongs and Folklore of South Uist*, London, 1955.
Sloboda, J. Music – Where cognition and emotion meet, *The Psychologist*, 12/9 (1999), 450–5.
Sterne, J. Sounds like the Mall of America: Programmed music and the architectonics of commercial space, *Ethnomusicology*, 41/1 (1977), 22–50.
Storey, J. *Cultural Studies and the Study of Popular Culture: Theories and Methods*, Edinburgh, 1996.
Tolmie, F. *One Hundred and Five Songs of Occupation from the Western Isles of Scotland* (1911), Llanerch, 1997.
de Vries, J. *European Urbanisation, 1500–1800*, London, 1984.
Williamson, L. What storytelling means to a Traveller, *ARV: Scandinavian Yearbook of Folklore*, 37 (1981), 69–76.
Woods, F. *Folk Revival: The Rediscovery of a National Music*, Poole, 1979.

9 Needlework

NAOMI E A TARRANT

Needlework describes the making by hand of garments and unadorned household textiles such as sheets (together termed 'plain sewing'), or more decorative items, such as embroidered tablecloths. Both plain sewing and embroidery have been undertaken to meet household needs or to provide a source of income, while embroidery has also been deemed a leisure activity. Needlework can in addition cover allied and primarily recreational forms of handiwork, such as knitting, crochet, *macramé* and some types of lace making. Whatever form it has taken, needlework in Scotland has been a largely home-based matter, and one overwhelmingly in the care of women and girls.

NEEDLEWORK AS DOMESTIC TASK

Needlework by the women of a household was a necessity in the past. For many centuries it was the main means whereby the dwelling could be adequately supplied with household textiles and the family decently clothed, although people from outside might be employed to undertake some very specialised work, such as tailors making formal outer garments. When readymade goods started to appear in Scotland is not known, but they were increasingly available by the later nineteenth century. However women with the correct skills would usually prefer not to waste money on such less well made items.

Nearly all commentators on female education from the seventeenth century onwards saw needlework as the backbone of what a girl should be taught, no matter how wealthy or poor were her family circumstances. Before schooling became general for girls they were taught at home by their mothers or in wealthier homes by a governess. By the early nineteenth century the teaching of needlework in schools had become formalised. Plain-sewing exercise books with graded steps were printed to help prepare girls to sew and darn anything required of them.[1] Knitting was often included in the curriculum and also some forms of embroidery, and the training would be finished by executing a sampler of alphabets and numerals (Fig. 9.1).

It was common for the women in a family, once they could sew adequately, to make all the members' underwear, including the men's shirts, which could also involve initialling, numbering and dating each piece so that it could be identified at the laundry. It was only the very wealthy who would have their underwear made for them outside the home, at least until the

Figure 9.1 Smith family sampler, Lintmill, near Fraserburgh, Aberdeenshire, late nineteenth century. Source: SLA 69M.

latter part of the nineteenth century when such items started to become available as readymade garments.

Hand knitting has existed as an industry for many centuries[2] but for long it was also a very useful accomplishment for the housewife, as all kinds of items could be created for the home and family. Very few hand-knitted garments survive from the past, however, as the pieces normally made were mostly intended to be hidden and thus had a low status. They included stockings, gloves and scarves, as well as some forms of intermediate garments, worn over undergarments but not intended to be seen, such as thin woollen spencers (a form of close fitting cardigan), under-waistcoats, petticoats and comforters for the chest. Given the lack of warmth in most homes, these knitted garments were essential to help combat the cold. One item which does tend to survive is a type of large bedspread, often made in white cotton, and examples of this are a testament to the skill of the nineteenth-century knitter.

Until World War I, national schools, the first state schools created after the 1872 Education Act, laid emphasis on plain sewing by hand and the making of useful garments such as night dresses and chemises. In the inter-War period girls were sometimes able to go on to domestic science colleges to further these skills. Another development of the period was the introduction of sewing machines into some establishments, including a number of state schools, but not all, as the machines were expensive to buy and maintain. After 1944 and the introduction of a new state-school system for all, responsibility for education was divided between primary and secondary schools, and the latter were seen as the place for teaching needlework to girls. Simple embroidery might still be taught in primary schools but it became much more a home-taught technique, as did knitting. However, after 1945 the need for women with hand-sewing skills in the garment industry was greatly reduced, while more women preferred to buy readymade clothes or to make their own clothes at home, rather than use a dressmaker. This led to the decline of teaching of all kinds of needlecrafts in schools by the third quarter of the twentieth century.

NEEDLEWORK AS A SOURCE OF INCOME

Plain sewing and embroidery were skills with which a woman could earn her living. School needlework training in the nineteenth century was the basis for many skilled and semi-skilled jobs, such as lady's maid, milliner, dressmaker or, more humbly, sewing woman. The use of women as sweated

Figure 9.2 Mrs Mabel Pritchard folding and packing hand-woven and hand-knitted garments in polythene bags prior to their dispatch to customers. Dunkeld, Perthshire, 1963. Source: SLA 69M 57/13/24.

labour in the domestic manufacture of readymade clothing is well documented from the nineteenth century,[3] although the industry probably existed from at least the seventeenth century with the rise of institutional and military requirements.[4] Women were also involved in professional embroidery work. Ayrshire needlework – a white embroidery on fine white muslin, used on baby robes and caps and women's dresses, collars and capes – was a mid nineteenth century industry carried out in the home.[5] While the clothes-making industry and its labour were found predominantly in urban homes in the nineteenth century, Ayrshire needlework in Scotland, and lace making in certain areas of England, were rural industries employing the wives of agricultural labourers. Their earnings could substantially affect the living conditions of households at a time of low rural wages.

Other opportunities existed for earning an income from needlework carried out in the home. Some families employed a sewing woman who came in on a regular basis to mend household linen or a dressmaker who came in to make clothes for children or servants or the everyday clothes of the women of the family (Fig. 20.5). Knitting could also be commissioned from a local woman. This only became less common by the mid twentieth century (Fig. 9.2).

NEEDLEWORK AS LEISURE

By the nineteenth century, in middle-class homes plain sewing was considered work and would be done in the morning when no visitors were expected. In the afternoons, during visiting hours and in the evening, recreational activities took over and embroidery would be considered suitable. The horror of idle hands was probably to account for the emphasis on women always having something to hand in the way of needlework.

The luxury of doing needlework as a leisure activity – one of the relatively few forms of female indoor recreation available in the past – was originally confined to those women who did not need to earn their own living. The numerous small embroidered pictures which survive from the mid seventeenth century onwards are most likely to have been made by the unmarried girls of wealthy families, although it is difficult to verify this as very few of the pictures are signed and dated. In the late eighteenth century there was a vogue for satin-ground pictures worked in black thread or human hair, these sometimes being memorials showing a weeping woman at a grave. Other popular topics were reproductions of engravings from plays.

The greater quantity of needlework of all types that survives from the mid nineteenth century onwards has led to an erroneous view that Victorian women generally indulged in more hours of leisure than those of previous periods. In fact the population increased dramatically after 1800, and more hands made for greater output. At the same time there was a rise in the number of women who did not have to earn their own living (although the proportion of women in this category fluctuated in response to general economic conditions). For these individuals, the fear of being left destitute by the death of a husband or father meant that needlework skills had to be kept in good order. During most of the nineteenth century needlework remained the best hope of securing employment for women without other forms of training, even if it was only work as an unpaid sewing woman in a relative's house.

The increase in needlework in the mid to late nineteenth century can also be seen as part of the rise in the number of material possessions which families owned. Houses had more rooms containing more furniture, ornaments, books and household linen, while people acquired more clothes. The nineteenth century was also a period of enormous technological change and new inventions were continually placed before the public. Amongst these were new forms of lighting. Beeswax candles had been very much for the upper classes. In more modest homes, malodorous and not particularly effective tallow candles had been used more often. Lighting had usually been provided on an everyday basis in humble homes by rush lights, splinters of resinous wood (known in places as 'fir candles')[6] or simple oil lamps such as the 'cruisie', none of which gave a very strong or a very persistent light. For the poorest members of Scottish society darkness might normally have been dispelled from their homes only by the light of a modest and sometimes ephemeral fire. All of these methods gave way to various types of more sophisticated oil lamp and later to gas and electric lighting. Good electric light is a mid to late twentieth- century feature as before the 1950s it was regarded as expensive and a single 25-watt bulb in a room or staircase commonly had to suffice. However, by directing the available light downwards even a low light level can be effective. Middle-class Victorians often gathered round a large centre table so that all members of the family could take advantage of the available light. The use of globes filled with water to magnify the light from a single candle or rush light was used by groups of women in the nineteenth century who embroidered or made lace for a living at home, not by other types of domestic needle worker.

During the nineteenth century there was an increasing number of books published on needlework and various new techniques for domestic handiwork were introduced. In the second half of the century and in the early part of the twentieth century, dozens of new magazines for women of all income levels were produced, nearly all of them including dress patterns and directions for knitting or other techniques.[7] The popularity of each technique can be charted through these magazines, with knitting remaining a staple throughout the later nineteenth century and most of the twentieth century.

Knitting
Knitting as a home leisure pursuit became very popular after about 1840 when the wife of a fancy goods merchant in Edinburgh, Jane Gaugain, produced one of the first books published in English for the domestic knitter.[8] Her work went to several editions.

Many of the items produced, partly out of necessity, partly by way of leisure as mentioned above, were examples of underclothing. Hand-knitted outerwear was not seen as fashionable until the twentieth century. The machine knitting industry, however, had grown apace during the nineteenth century, greatly helped by the fashion for knitted woollen underwear introduced by Dr Jaeger in the 1870s.[9] The greater use of specific clothing for various leisure pursuits helped to make knitted garments more popular, and the machine-knitwear manufacturers began to produce well-designed and fitted items of outerwear which were to be copied in many homes. By World War I golf and country outfits regularly included knitted jackets and sweaters. The rise of a more sporting lifestyle in the 1920s helped to popularise further knitted garments and women's magazines were not slow to include patterns for their readers. From the 1920s to the 1960s just about anything was knitted, and jumpers, cardigans, berets, gloves and scarves for men and women were popular attempts (see Fig. 31.5). More ambitious were women's suits and dresses. Patterns for baby and children's wear were also given and the new mother could expect gifts of all kinds of knitted garments for her baby (Fig. 9.3).

Figure 9.3 86-year-old retired fishwife Mrs Belle Ann McAngus of Hilton, Fearn, Ross-shire, knitting in 1975. Source: SLA 69M 24/12/5/647.

Knitting in recent years, however, has suffered a decline. It is cheaper to buy both machine and hand-knitted items than it is to make them and even knitted baby clothes are not made in the same quantity as newer fabrics are increasingly important.

Crochet and tape lace
Crochet was another form of recreational needlework which rose to popularity in the nineteenth century. It is most often found in the form of small mats for tables; in the 1920s and 1930s whole table settings of crocheted place mats were fashionable. Crochet was also used for trimming chemises and for the collars and cuffs of dresses. Bobbin lace was too difficult for most women to undertake as a hobby but tape laces were popular in the early 1900s. These were created from machine-made narrow tape which was laid on a design, for a collar or other item, stamped on green or blue cotton. An open-work pattern was created by joining the meanders of the tape together by narrow bars of thread and when this was complete the cotton backing was removed, leaving the article in the form required.

Canvas work and other handcraft embroidery
Canvas work became very popular in the early nineteenth century with the introduction of soft wool threads dyed in what seemed to contemporary eyes an amazing range of colours. Known as Berlin woolwork, this technique remained in vogue for nearly the whole century and was used to illustrate Bible stories, historical incidents and create floral patterns.[10] Items were worked in either cross stitch or *petit point* (half cross stitch), and were sometimes embellished with coloured beads. The patterns were usually produced on squared paper, and coloured by hand in water-colour paints, another home industry for women and children. The patterns were then sold in fancy goods stores, or else issued in women's magazines, such as the *English Women's Domestic Magazine*, edited by Samuel Beeton. An additional amount of money had to be paid to secure both the patterns and the hand-coloured fashion plates in these magazines. As well as pictures for the wall, various garments and sundries were made including men's waistcoats, smoking caps, braces, slippers, purses and tobacco pouches. Women's items usually consisted of slippers and bags, but by far the largest number of articles were for the home and included footstools, chair and sofa seats, bell pulls, fire screens and valances for windows.

Small embroideries were made for sale at bazaars, which were often held to help raise money for some charity.[11] Other handcraft items might include the netted purses which until the later nineteenth century were used by both men and women. Pen wipes, usually consisting of several small pieces of felt within a decorative cover, were also popular, and were necessary for cleaning the dip pens used in the nineteenth century. After the introduction of the Penny Post in 1840, small holders for postage stamps were added to the list of bazaar goods. Like book markers, the stamp holders were often made of a punched card called Bristol board, easy for cutting and for embroidering letters using the regular grid of holes. Small items made for the bazaars were also suitable to make and give as presents at birthdays and Christmas to family and friends, and they could be given to members of the opposite sex without causing offence by being too intimate. Children in particular were encouraged to make their own gifts for the family and these small items of handiwork were very suitable for them to do.

Patchwork
Another technique which enjoyed a certain amount of popularity in the nineteenth and early twentieth centuries was patchwork, often allied with quilting, to produce bed coverings. There are references to patchwork quilts in house inventories, but there does not appear to have been a particularly Scottish quilt style, unlike County Durham and Wales.[12] Today the making of patchwork quilts is immensely popular and there are many quilters' groups producing work of both a functional and a decorative nature.

MEN AND NEEDLEWORK

Needlework in the home has not been confined to women and girls. There are references in letters, reminiscences and journals to men working cross stitch and *petit point*. Queen Mary (1868–1953), the wife of King George V, and a keen needlewoman, taught all her children, including the boys, how to do *petit point*. When in her seventies, she helped raise dollars for Britain after World War II by making a needlework carpet, which was sold in the United States of America.

Another type of needlework which men are known to have undertaken consists of elaborate patchwork covers made using scraps of army uniforms and usually ascribed to army tailors.[13] Occupational therapy for wounded service men started at least with the Crimean War in the 1850s. By World War I embroidery and other forms of needlework were being undertaken in this context in hospitals.[14] Also during the War, sailors and soldiers were expected to be able to mend their clothes in whatever temporary accommodation they possessed, and 'housewifes' or 'huswifes', small sewing kits, were issued to them.

NEEDLEWORK IN THE HOME TODAY

Today needlework and other related crafts are still popular hobbies. Counted thread on canvas, erroneously described as tapestry,[15] is probably the most popular, with special promotions being held in department stores every year. Mostly the work is sold as kits where the pattern is printed onto the canvas, or onto squared paper, and the correct amount and number of coloured wools, silks or cottons are supplied to complete the piece. The technique is particularly popular with actresses and others whose jobs involve waiting around for long periods because it is easy to carry in the hand and does not demand total concentration.

Very few people now make their own clothes, but there has been a resurgence of interest in obtaining dressmaker-made garments for special occasions such as weddings or eveningwear, or for those with unusual figures. Those (usually women) who undertake this work have often trained as dress designers at college but work from home. In addition, there are now textile artists, again working largely from home, but often using a dedicated studio space, who earn their living entirely by their work. They create a wide range of objects, from wall hangings to three-dimensional pieces designed to stand on tables.

Examples of all the various items mentioned above can be found in the collections of the NMS and in other museums throughout Scotland.

NOTES

1. Various books were published from the early nineteenth century onwards with graded instructions for each step in plain and fancy sewing. These were often illustrated with small, actual examples of work. Several were published in Dublin for use in the National Female Schools of Ireland, but in fact they were used far more widely.
2. Bennett, 1986; Barnes, 1977; Fryer, 1995.
3. Walkley, 1981.
4. Lemire, 1997. Lemire indicates that there is some evidence to suggest the situation in Scotland was very similar to that in England during this period; see also Chapter 19 Working from Home.
5. Swain, 1955.
6. Grant, 1995; See also Chapter 5 Lighting.
7. *The Girl's Own Paper* started 1880; *The Englishwomen's Domestic Magazine* started 1854; de Dillmot, Therese. *Encyclopedia of Needlework*; Miss Lambert. *The Handbook of Needlework*, London, 1842; Caulfield, S F A, Saward, B C. *Dictionary of Needlework*, London, 1882; etc.
8. Gaugain, 1840. The name of this book is not consistent in the various editions.
9. Jaeger, 1884.
10. Proctor, 1972.
11. The kind of items sold can be seen in James Collinson's painting 'The Empty Purse', 1857, in the Tate Gallery, London.
12. Rae, 1987.
13. One in the Town House, Biggar, Lanarkshire, the Royal Crimean Hero Tablecloth, was made by a local tailor, Menzies Moffat (1829–1907); for army quilts see Rae, Tucker, 1995.
14. A soldier in the Liverpool Scottish Regiment, who was badly gassed during World War I, made black satin cushion covers embroidered with red roses in spider's web stitch for all his female relations. Personal communication.
15. Tapestry is woven on a loom, it is not an embroidery technique.

BIBLIOGRAPHY

Barnes, I. The Aberdeen stocking trade, *Textile History*, 8 (1977), 77–98.
Bennett, H. *Scottish Knitting* (Shire Series), Princes Risborough, 1986.
Fryer, L G. *Knitting by the Fireside and the Hillside: A History of the Shetland Hand-Knitting Industry, 1600–1950*, Lerwick, 1995.

Gaugain, J. *The Lady's Assistant for Executing Useful and Fancy Designs in Knitting, Netting and Crochet*, Edinburgh and London, 1840.

Grant, I F. *Highland Folk Ways* (1961), Edinburgh, 1995.

Jaeger, Dr G (trans anon). *Dr G Jaeger's Clothing Reform for Men, Women and Children*, London, 1884.

Lemire, B. *Dress, Culture and Commerce: The English Clothing Trade before the Factory, 1660–1800*, London, 1997.

Proctor, M. *Victorian Canvas Work: Berlin wool work*, London, 1972.

Rae, J. *The Quilts of the British Isles*, London, 1987.

Rae, J, Tucker, M. Quilts with special associations. In *Quilt Treasures: The Quilters' Guild Heritage Search*, London, 1995, 170–7.

Swain, M H. *The Flowerers: The Origins and History of Ayrshire Needlework*, London and Edinburgh, 1955.

Swain, M H. *Historical Needlework: A study of influences in Scotland and Northern England*, London, 1970.

Swain, M H. *The Needlework of Mary Queen of Scots*, London, 1973.

Swain, M H. *Scottish Embroidery: Medieval to modern*, London, 1986.

Tarrant, N E A. *Samplers in the Royal Scottish Museum*, Edinburgh, 1978.

Walkley, C. *The Ghost in the Looking Glass: The Victorian Seamstress*, London, 1981.

10 Television, Video and Computing

JAMES STEWART AND KATHY BUCKNER

A REVIEW OF QUALITATIVE RESEARCH IN SCOTLAND

Relatively little qualitative research has been undertaken on the use of screen technologies – games, television or computers – in domestic environments in Scotland, unlike in many other European countries.[1] Research on media has concentrated on aspects of production, while that on computers has focused on their business and educational use. There is some work on public access although this often concentrates on training as much as recreation. Media research in Scotland is concerned with national culture, the production of Scottish content and the organisation and financing of the Scottish media industry.[2] However, the users and viewers of media content have been rather ignored. This chapter reviews the findings of a series of research projects (see Appendix) undertaken in Scotland between 1995–2003, which, amongst other matters, deal with the place of media and information and communication technologies (ICTs) in home life. In particular it gives examples of how people live with, enjoy, cope with and reject computers and television devices, and how computers are used to link the home to work and the outside world in general.

The review is based on the work of researchers from a number of Scottish Universities. At Queen Margaret University College (QMUC) work was initiated by Davenport and colleagues in 1995/6 and focused on appropriation patterns and the use and integration of technology within the domestic environment.[3] In parallel with this, Gillham and Buckner investigated the evaluation and use of interactive multimedia CD ROMs (primarily encyclopaedias) in the home.[4] From 1999 to 2002 Buckner and Gillham undertook an in-depth study involving repeated interviews with 12 households, investigating their use of email for social communication. Significant Other grams (SO-grams), were used in conjunction with structured interviews to investigate social networks. Patterns of interaction and perceptions of the use of technology in supporting interpersonal communication and friendship development were investigated.[5]

The EU-funded Living Memory Project[6] provided insights into the use of technology by older people in the home. More recently (1999–2001), the UTOPIA project (funded by the Scottish Higher Education Funding Council and involving a consortium of four Scottish Universities: Dundee, Abertay,

Glasgow and Napier)[7] has investigated the use of technology by older people living in Scotland through questionnaires, focus groups, individual interviews and workshops.

The data on television and game use, as well as the oral testimonies relating to the place of technologies in the home, are drawn from a PhD research project at the University of Edinburgh.[8] The study involved in-depth interviews and observations among the social networks of 25 people, undertaken over three years in the late 1990s. The conclusions of the study focus on the social nature of ICT consumption, the many problems of adopting and using ICTs, resistance to adoption, and the ways barriers between home and work are challenged by new technology.

Work focusing on the future use of ICTs in the home has been undertaken at Napier University (1999–2001) as part of interaction design research.[9] Some of the issues highlighted during 'investigative technology tours' of five households in central Scotland included: the location of technology and its affect on use; reasons for its use or disuse; matters affecting technology lifecycles, such as losing remote controls and battery replacement; learning how to use technology from family members and others and from manuals; and, privacy and security.[10]

Many research fields examine the place of ICTs in the home, in everyday life and in relation to the individual. These include sociology of consumption, design, gender studies, family sociology, telework research, technology studies, media and cultural studies,[11] and consumer research[12] in various forms. Work in these areas has a concern for both the practical and symbolic aspects of technology. One important theoretical contribution is 'domestication'.[13] This examines how people bring technologies into the home, how these are integrated into the physical space, everyday routines and social relationships of the dwelling, and how technology in the home reflects the household's relationship with the outside world. This chapter touches on all these aspects plus additional matters addressed by researchers in other countries – from the physical location of the television and computer, use habits, disputes and control, the symbolic meanings of these technologies, and the way they link individuals in the home to the larger community and education and work. There are new issues relating to ICTs in the home today, specifically concerning the home as a workplace, with the gradual rise of teleworking and IT-enabled small businesses in which the home is a digital hub for work, learning and social, economic and political participation.[14] Access to technology at home has become a political issue with concerns arising over the 'digital divide' and the potential for social exclusion which ownership or lack of ownership creates.

OWNERSHIP AND UPTAKE

Television, video and video-game ownership in the UK is very high, and in many households, especially those with children, there are multiple televison sets and often multiple video recorders. Computers and Internet access are relative newcomers. Access to computers is on the increase across the UK, but overall Scotland has a lower adoption rate than wealthier parts of the UK (Fig.10.1). However the rate of increase in access to the Internet between 2000/1 and 2001/2 was greater in Scotland (12 per cent) than in England (7 per cent) and Wales (9 per cent). Adoption of computers and Internet in the home is correlated with urban living, younger age, higher educational level, the presence of children, and higher income.[15] Gender has significance in this context also. Women are less likely to have access to the Internet, and households with a female head of household are less likely to have computers.[16] There may be other influencing factors, such as use of Internet by local employers, the development of local Internet-based services and local government policy.[17]

Region	1998–9	1999–2000	2000–1	2001–2
London	16 per cent	25 per cent	40 per cent	48 per cent
England	11 per cent	20 per cent	34 per cent	41 per cent
Wales	7 per cent	15 per cent	22 per cent	31 per cent
Scotland	8 per cent	14 per cent	24 per cent	36 per cent
UK	10 per cent	19 per cent	32 per cent	

Figure 10.1 Households with home access to the Internet by Government Office Region and UK country 2001/2, Family Expenditure Survey (April 1998 to March 2001); Expenditure and Food Survey (April 2001 onwards).[18]

TELEVISION, VIDEO AND VIDEO GAMES

The television (TV) is undoubtedly the principal leisure-time technology in the home. There was a time when watching television together was a mainstay of household life, with living-room television sets replacing the radio in the 1950s and '60s as the focus of family or communal household entertainment (Figs 10.2, 10.3). A limited number of television channels and no extra gadgets meant that the whole viewing population watched more or less the same programmes, so that from the 1960s to '80s television became a central part of national culture. As the radio had done in earlier times, the television shaped household life, with domestic activities being arranged around favourite programmes, and programmes being broadcast with a specific routine in mind; women's programmes in the day time, children's broadcasts until 6pm, then the news and programming for men coming home from work and family entertainment.

Over the last 20 years much has changed. From one box in the living room with a few broadcast channels, the TV has proliferated to the extent that most homes have two sets, and many have three or more, in kitchens, bedrooms, offices and playrooms. Television, as a technology, has also been extended with video recorders, home computers, games consoles and multichannel TV via cable or satellite. In many cases the TV is moving towards becoming a computer, with teletext, onscreen menus, interactive features,

Figure 10.2 Mr and Mrs Brimfield, Arbroath, Angus, relaxing at home by their radiogram. Source: SLA 59/57/11.

Figure 10.3 George Hastie in his sitting room, Inverkeithing, Fife, 1960s. The television dominates the room. Source: SLA 128A(2) CP63.D 626632.

games and email. These changes have had a considerable effect on the place of TV in the home. Household members can watch alone or in different rooms; there is negotiation not only over the channel but whether to watch TV or play games. Multi-channel TV has also changed the programming available, and means that more money can be spent on having television through subscriptions and 'pay per view'. People use TV for a variety of reasons which extend the idea of 'leisure' use; these include relaxation and 'unwinding', as a background presence while doing other things, and for education and childminding. Television can be watched in a group, as the image of early radio and television days suggests so strongly. However social watching is not only about sharing and interaction around the television, but also disputes and negotiation over programmes and watching times. Video gaming has now become one of the most popular entertainment forms and a firm part of many households' activities, raising many of the same social issues as television. This section illustrates these aspects of television and video games with direct quotations from transcripts.[19]

AMOUNT AND CONTENT WATCHED

The television can be used as a background presence, to provide a noise, or be dipped in and out of while doing other things, 'Sometimes it's on just in the background. If it's something interesting I stop and watch it ... Sometimes just monitoring news and stuff like that.' This is particularly true for those living alone, or for those who spend much time by themselves in their bedrooms. The number of hours that the TV is on can be substantial. Many people feel that they watch too much television and that access to more channels would increase their viewing and reduce the time spent on other activities:

> JS: Do you ever think you watch too much TV?
>
> [laughs] Possibly. Square eyes.
>
> JS: How many hours of TV you watch a day?
>
> Em, too many ... Ken, probably about seven or eight hours a day mostly on TV. (teenager)
>
> I don't think I'll even bother about cable television because I would become a total couch potato; I'm bad enough as it is. I would start actually watching the programmes instead of just having it as a background noise 'cos you'd be going oh I want to watch that, I want to watch that you know, like old sci-fi, the sci-fi channel and I would never have that off you know, no matter what shite's on, it'd be like I want to watch that laugh. Junkie. (artist)

While young people have the choice of watching or doing something else, some older people with limited energy or mobility find themselves 'stuck in front of the box':

I am coming to the conclusion that I am watching it too much.

JS: Why is that?

Because I am sitting here and ... well ... you feel in the evening you have to sit, I would rather be out in the workshop doing something, but of course certain programmes that I like to watch ... but the trouble is you watch everything in between the good programmes as well, that is the problem. (retired person)

As these quotations indicate, there are some items considered worth watching but many shown in between which are not deemed worthwhile. Some people are very choosy about what they watch; others do not care as the content is not the motivation behind viewing, while some are frustrated by what is available.

MOTIVATIONS FOR WATCHING TV

Television is used by many people as a form of relaxation, whilst others perceive it as being not only entertaining but also educational:

> Well, it's highly educational, highly entertaining. I'm very careful about what I choose to watch, there's a multitude of game shows which I wouldn't touch with a barge pole, even the intellectual quiz shows I won't watch because they give me an inferiority complex.

Watching TV in order to spend time with friends or family members is also important. However, there is often a difference of opinion over what to watch. There are a number of solutions to this. One way is to let others decide, 'I tend to leave the choosing to L or I, 'cos I am not really too choosy about TV a lot of the time.' Another option is to watch television in a different room, 'There is another box through there, I will go through or she will go through if we have conflicting interests.' (retired person) A further strategy is to stay in the same room, ignore the TV and do something else:

> JS: Do you watch things you don't want to watch sometimes?
>
> No. If I remain seated in the room I'll be doing a crossword puzzle. (retired person)

Ambivalence towards television is commonly expressed through a desire to control how much one watches. For parents this may extend to limiting their children's viewing:

> We've got a few children's CD ROMs which she uses quite happily. But I don't ... television, don't let her sort of use that too much, it's the same, wouldn't like to do anything too often. Just sort of keep the variety. But I let her use it for an hour or so at a time. Maybe a couple of times a week. Or if things are pretty hectic and she demands it to get some peace and quiet she can use that. (mother)

Since it is difficult to control what is shown on TV, some people prefer to use videos:

> Well we've been very conservative about what they view and they don't actually watch television, they just see videos. Sounds awful but it's all, it's very controlled in that respect. And that's chiefly because we remain in a state of tension where M would get rid of the TV altogether and I won't have it because it's my best childminder and I think therefore we tend to have what we would consider sort of good-quality story videos, rather than the junky cartoons that are zappy and fast in America. Having said that of course the greatest thing is to collect all those Disney classics as they come along and the kids have a fair number of them. (mother)

These quotations reveal tension in relation to control and household relationships. Parents find that the TV is a wonderful way to keep children quiet when they have other things to do or want to keep them out the way. However, there is sometimes disagreement associated with the responsibility of looking after children, and this is seen in parents' beliefs about whether children should be allowed to watch TV:

> R has a routine on a week day evening and they get it nearly every evening she's in charge. When I'm in charge they never put the video on at all, I mean they just don't expect it so I mean that's it. They get really excited and upset if they don't get a video when she's there. When I'm there it doesn't even occur to them to ask because they know it's a loser. Completely different. (father)

VIDEO

Until 2002/3 the video recorder was the main adjunct to the television. This performs several functions, including time shifting, recording while the television is being watched, playing pre-recorded tapes, saving information from the television for leisure or professional use, and recording for a programme library. Today many households will have more video tapes on their shelves than books. As with TV, the use of the video is a fairly social process, with household members often recording programmes for each other:

> D watches much more TV than I do, in as much as she will watch programmes between 7pm and 9pm. If there's something which I specifically want to see, I will record it so that we can both watch it later at night. (retired man)

At the same time there exists the perennial problem of knowing how to work the machine, or being frustrated with the interface and the generally increasing complexity of many devices:

> Yes, I don't even know how to use it. My wife puts a tape on for me,

whenever, I don't bother about it. I should, she says, if ever she, if she goes I'll have to learn how to use it, but until then, I'm not too bothered about it. (retired person)

Videos and computers can be daunting devices to bring into the homes of those who have little technical confidence:

I know quite a few people who are quite intimidated by them. Like my sister and like A's ex-girlfriend, she doesn't go anywhere near computers. Or videos. Ken, she doesnae quite ken how to work them so she's quite scared of them. Quite powerful things if you think about all the information they've got in them. (teenager)

However those who do know how to use a video can do so in what might seem surprising ways:

But he does experiment a lot and he watches all the cookery programmes on the TV and if he likes a recipe he videos, watches it back and then cooks that meal which is lovely. (wife)

In general, recording from the TV is a decreasing practice. For those who are not particularly interested in specific programmes on the television, there is little use for videos:

JS: Do you record anything?

Occasionally, I have a video but I've had that video player for about 15 years. It's still going it's so rarely used. But it is used for the odd programme. Again I don't feel, I don't feel strongly about TV. I watch it when I feel like watching it but I don't watch it a lot. (artist)

And some of those who do like to record never get round to watching the tapes:

I often video films but it's mainly films that I video and I never watch them. So I've got this unbelievably huge collection of unlabelled tapes with classic films on, usually missing the first ten minutes. I just don't know, repeatedly I say oh that's going to be really good, I'll definitely, I'll watch this one and I tape it and we buy new tapes ever so often. I've hundreds of them. And when will I ever see any of them? (husband)

From 2002, new video technologies such as DVD have been rapidly replacing older tape technologies. Current DVD players are used for rental or purchased disks. DVD recorders and 'personal video recorders' based on computer hard disks are going to introduce new ways of sifting and storing television, and the development of 'on-demand' television services offer users increasing flexibility of use, and temptations to spend money.

COMPUTER AND VIDEO GAMES

Computer and video games played on the TV have become one of the main forms of media and home entertainment in the last ten years. Two salient features are the social nature of gaming and the problems caused by its 'addictiveness'. Whilst for girls playing with the computer is often a solitary experience, for boys, playing games together is intensely social, sometimes competitive and sometimes collaborative:

> If it's competitive, like the fighting games, we usually don't tell each other the moves and that ken, because you want to kick ass, ken, but if it's like Tomb Raider or that we help each other because that's a one player game. One person plays, the other one helps out and suggests things. (teenager)

Once all or most of the boys in an area have computers a kind of 'local economy' develops in which games are swapped and traded:

> So the games just get passed about between us. R has got Soul Blade just now. That's the most recent game I've bought. He's got it. (teenager)

Computer games often take up huge amounts of time:

> My dad's got a Nintendo 64. He got that just before Christmas. Because we used to have ... Griselda 2 Link and my dad, that was his favourite game back then so the Nintendo 64 came in, a big thing about Zelda Links, 'Oh I'll have to get it.' So he's never off it. He's still playing it, he's been playing it since Christmas time. (teenage son)

This might be acceptable for unemployed or underemployed teenagers, but for couples it can cause trouble:

> Hours and hours and hours disappear down the drain so I took it off and I've still got them in my room but I won't put it back on again because it was just, I was, it caused too much domestic friction. (husband)

As children's entertainment, computer games are often compared favourably to television:

> Parents appeared wary of their children watching TV too much ... but see use of the computer, even the playing of games, as a more active and educationally-beneficial activity. 'It's taken us away from the television which is maybe a good thing ... with this (the computer) it gets you up off your backside ... you're using your brain' ... 'their time watching television has decreased which is good.'[20]

However, while games do require skill and co-ordination parents still wish their children to do other things:

A lot of what R does, I wish he could do other things. He's a bright laddie but doesn't have any qualifications. He gets these Playstation games and breaks them apart, the only way you can do that is if you know what you're doing. (father)

COMPUTER ADOPTION AND THE PLACE OF COMPUTERS IN THE HOME

Recent studies[21] of Scottish households suggest that influence and purchasing power in relation to screen technologies lies predominantly in the hands of the male members of the household. Men are more likely to suggest purchasing a computer and will usually play a more dominant role than women in seeking information about hardware and software, in deciding which product to purchase and where to buy it, and in determining how much money to spend on it. Buckner and Gillham[22] found that 80 per cent of such decision making was undertaken by males.

The main reasons for purchasing a computer are to support the work of adults or the learning of children, for example in doing homework, to become more IT literate, or to become generally more informed. Buckner and Gillham[23] found that most households with a computer only owned one machine, differing quite significantly from TV ownership. Where there was more than one computer, the second was most likely to be a laptop owned by the employer of one member of the household and used primarily for work. One informant stated that he bought his computer because of his, 'interest in technology, for educational use by the children, and because of a manly requirement for toys!'[24]

There appear to be increasingly fuzzy boundaries between work and home life. One man notes that it is with some reluctance that he brings work home:

> I work at home 'of sorts'. I do very, very, very little work from here because my attitude is that all my resources are at work and I don't want to bring work home, but occasionally, very, very occasionally, if there's a particular email memo or something that I want to send around I might go upstairs, type it out on the PC and see what it looks like, make a copy on a disk and take it into work and send it then.

Others will cross the boundaries in the opposite direction with social activities such as accessing personal email being undertaken at work and then brought back to the home. An informant uses different types of media to satisfy his requirements: he reads emails off the screen at work, 'and then [prints them off to] take home to show my partner'.[25]

Excessive use of a particular technology in the workplace can act as an inhibitor to its use at home. This affects the use of email:

> My sister ... she prefers it, she's not at work so she's happy to use email socially. For me ... it seems too much like work ... [I'd] rather pick up the phone.[26]

and the telephone:

> I think nowadays I talk far less at home on the phone because I'm spending far more time at work phoning and I'm so pissed off with phoning that I regard it not as the pleasure that I once did but sort of something that has to be done.

and computers in general:

> after having spent eight hours a day banging away in front of a screen I think I'm pretty loathe at some points to go home and do it again.

More than half of the households observed by Buckner and Gillham[27] located their computers away from shared spaces and where they could be used without interfering with communal activities such as eating or socialising. Preferred locations included home offices, workrooms, studies or spare bedrooms:

> ... having a home office which is where I do the sort of home finance and stuff ... for example I do the banking on the computer and do the household accounts on a spreadsheet.

Some households keep the computer in a communal room but appear to be uncomfortable with its presence and attempt to conceal it from general view. For example, one elderly couple made a fabric cover (see the bulky covered object to the left of the curtain in Fig. 10.4) with which they covered the computer when it was not in use, rather like a budgie's cage is covered over at night.[28]

It is difficult to measure the extent to which owning a computer affects

Figure 10.4 Concealing the PC. Source: J Stewart.

other home-life activities. Self-reporting is often unreliable, offering only incomplete or unsubstantiated evidence. Davenport, et al observed that in their study:

> ... most respondents ... claimed that the computer does not have an effect on leisure activities but one 21-year-old participant noted, 'I see it affecting younger people, friends of the family that I know ... cousins have computers and they tend to ... they go and they play with the computers instead of going out and playing in the garden and that sort of thing, like I used to.'[29]

ICTS: LINKING THE HOME TO THE OUTSIDE WORLD

Whilst face-to-face communication is almost always seen to be the most effective medium for communication because of its 'richness',[30] there are many instances in which telephones and computers, particularly through the use of email, are deemed to be extremely important in maintaining social networks. Social networks include family, friends, acquaintances and sometimes work colleagues. They may include people who are geographically close as well as those who are geographically distant. They include people who may or may not have ready access to telephone and computer technologies. In the UK, email use for social communication is now more common than traditional letter writing.[31] Because of its uptake there is now the critical mass to support a variety of activities such as making arrangements:

> My mother has her 80th birthday coming up, we have to make arrangements for that, and phone calls are not practical, so it's much easier for me to do it by email.

the maintenance of long-distance relationships:

> ... it's become a regular communication [medium] with my daughter in Hong Kong. When she was tuned in, she had a hotmail address and that's great.

and communicating without talking and taking control of the communication process:

> ... a bit like telephoning but without having to talk [or listen, the informant later admitted] ... The control is in your own hands.[32]

There is a perception amongst some people that email can be a solitary activity:

> You used to run up stairs and say something to somebody, and now you just email them, it's isolating people more and more.[33]

Although people often use email, because of its attributes of speed, brevity and convenience they do not necessarily prefer it to other forms of media such as telephone and letter.[34] Most people prefer receiving letters because of their physical attributes such as how they feel and smell and the character of the handwriting, and some individuals consider that they support the development of deeper relationships better than email. One person interviewed felt that it was, 'less intimate than conventional mail, and less likely to cement a relationship ... emails are less sociable despite their chatty nature.'[35]

In an attempt to personalise email people will often annotate their messages with 'smilies' or send attachments of photographs of friends and family members. Although lacking some of the qualities of a letter, emails are thought to be beneficial in the *formation* of some social contacts. Carol felt that, 'email encourages me to communicate with those who may not be closest friends' whilst Lauren suggests that it, 'generates communication which may not otherwise occur'.[36]

When conventional mail arrives through a letterbox there may be a rush, by more than one member of the household, to pick it up. Similarly, email is often opened with a feeling of excitement, 'I check for new mail first thing after switching on – [it is] exciting.'[37] Some people find that the computer, with its capability for organizing information into folders helps them manage their email correspondence more efficiently than conventional mail. Email may be sorted on arrival into categories such as junk, work or social; it may be discarded without reading; it may be forwarded to another family member; it may be responded to immediately; it may be stored in a folder for re-reading or action in the future.

Managing personal details often involves a complex mix of documents and lists which may be stored on paper or electronically. Address lists and contact details may be held within email software, in a physical address book, in a mobile phone or on notes around the desk, on walls or notice boards. These multiple storage devices are frequently linked in complex ways, with for example, the PC being used to print address labels for Christmas cards or it being used to send text messages to another person's mobile phone. Other important information may be attached in clearly visible locations around the computer. In Fig. 10.5 there is a white strip on a small desk drawer visible behind the laptop screen; this is a hand-written note of this retired person's email address and Internet service provider (ISP) password.[38] Computers are also often surrounded by a mix of filed paper documents, and photographs in albums or frames, often with a pen and paper nearby for jotting down important information or supplementing what can be done with the computer.

Computers are not usually used in isolation from all other types of media and neither are the skills acquired on them. Transferability of skills acquired with other technologies is very apparent when dealing with computer technologies, for example:

> The audio clip was played by all subjects ... they recognised the icon bar under the picture, as one subject said 'because there's a bar at the bottom ... like your stereo'.[39]

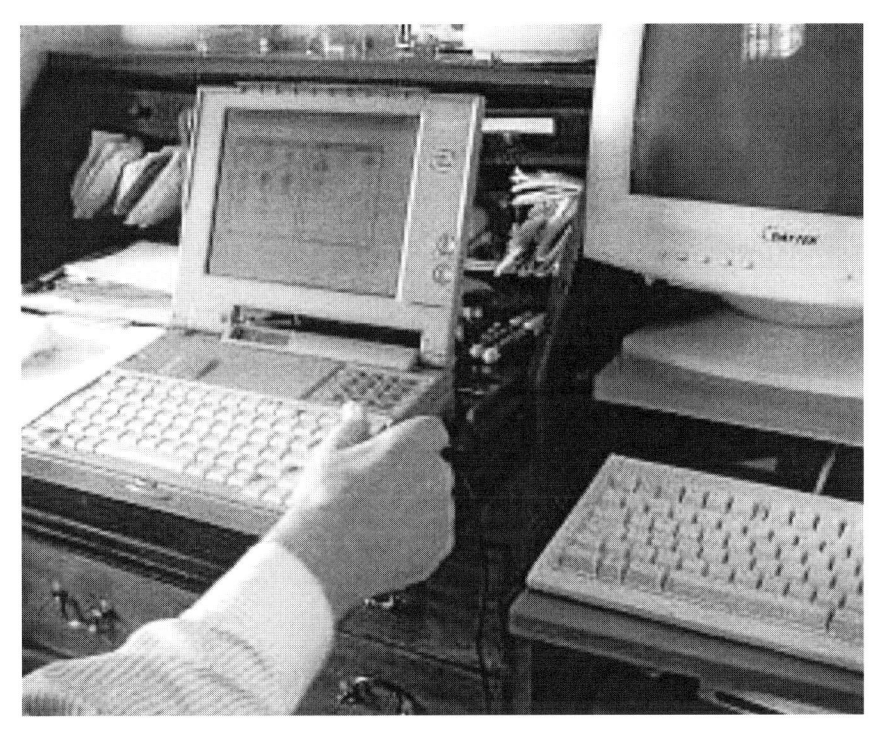

Figure 10.5 PC surrounded by a mix of documents and writing tools. Source: J Stewart.

Skills are not universally acquired by everyone and there may be gender differences in the way in which techniques are adopted by different household members. Gillham and Buckner[40] found that some participants (not necessarily male) were reluctant to use the keyboard at all. They preferred using the mouse to navigate through search categories rather than typing a keyword as a search term. Such matters may well have a significant influence on the extent to which ICTs are adopted in the home. However, of greater significance to adoption levels is the increasing accessibility of fast Internet access through broadband connections provided by cable and telephone-service providers. Although cost is an inhibitor to uptake for some households, competition between providers is likely to bring this down to an affordable level.

SOCIAL CONSUMPTION

The adoption and use of technologies, as with most activities in the home, is fundamentally social. The purchase and employment of equipment is negotiated by household members, often with agreed limits or purposes. There is frequently a division of labour in the use of technology, corresponding to the tasks it is used for, cultural norms and the inclination and skills of the household members. These divisions and agreements tend to follow patterns – particularly between parents and children, and between men and women. However people also set their own personal rules, whether they are living alone or not. But, as with all limits, they are there to be tested and broken, causing tensions in relationships, but also in individuals, such as those who feel they watch too much TV. These technologies also modify the relationship between those within a household and the outside world, and blur the boundary between home-based activities and outside activities, especially work.

The research so far conducted all shows that technology provokes problems between household members. This includes arguments over what to watch on TV, or just uneasiness over knowing that others do not want to watch something ('we both find we can't enjoy a programme if we know that the other half isn't really interested in it'). Other difficulties include stress over the costs of pay-TV subscriptions, the time spent playing video games, or restrictions on TV viewing. The presence of a computer in the house is recorded as implicated in the break up of one couple.[41] There are also issues over spending money on new technology. One man wanted to get an Internet connection, but knew his wife would not allow him to spend the money. Another got in trouble when he spent the holiday money on a new television.[42]

The increased use of and interest in technologies among men is well recognised. Stewart[43] found that women were often excluded from conversations about computer technology which men were having:

> I suppose it would be with my brother, and my husband. I usually switch off, I'm just not into computers. I know that everybody else seems to be worried about this bug that's going to bite in 2000 and everything will go whoosh and anarchy will prevail. (wife)

> But the machine that we've got upstairs won't take CDs. It will be a topic of conversation but I can't imagine it will be the main topic of conversation, M will kill it dead. (husband)

> It'll kill me dead. (wife)

However, in the case of particular technologies there are some interesting examples of reversal of the norm. Two couples[44] reported that the husbands did the typing for their wives. These are professional women who have never learned keyboard skills, relying on computer-using men to type for them:

> But I just write it down, and M [husband] does it. Sometimes it's just a question and he types so quickly, and he's always busy and using his machine. Sometimes I'll just speak and he types it in.

> I spend all day writing it and scribbling. You know I work so hard that it's probably 11 o'clock at night and it's to be sent off the next morning and so T [husband] types it because he's quicker cos I always leave them till the last minute, don't I?

The usual division of acceptance and use of technology is between the young and the old. But of course statistics obscure the fact that many older people are proficient IT users and many younger people are not. Stewart[45] reports teenagers who are not interested in the Internet and one 93-year-old who is as proficient and engaged with the technology as anyone. Nevertheless, the UTOPIA project found that the use of many technologies decreases with age, the exceptions being those that are either more established or 'easier to use', such as telephones and televisions.[46] Older people may feel that they are too old to use newer technologies as two very common expressions reveal, '[I am] too old to bother with more modern things' and 'I am too old to learn new tricks'.[47] However, technology is changing very quickly and much is being done to make it simpler and cheaper and to create new forms more relevant to the needs of a wider range of people, as well as extending existing uses – from pay TV to video games and Internet access – to a much greater proportion of the population.

CONCLUSION

These studies throw some light on issues surrounding technology in the home. Gender and age differences, for instance, are typical, but at the same time atypical examples abound. The social aspects of technology use are clearly visible within Scottish homes, with technology entering into the range of normal household relationships. Whether these studies show any particular 'Scottish' aspects of technology use is debatable, although the experiences in some more or less isolated rural areas show which ICT can make possible home businesses that would otherwise be very difficult to sustain.[48] There is room for a great deal more research in the field, both quantitative and qualitative, looking at the experience of home technology in different Scottish regions. Work in progress on the design and use of ICTs in many fields, including community and economic development, education, telework and matters relating to older and disabled people, now takes much more account of the place of such technologies in the home, and could be brought together in the future.

NOTES

1. Especially that funded by the UK Programme for Information and Communication Technologies (PICT) programme in the 1990s, and European research under the EC 4th and 5th Framework Programme (2000–3), COST actions and Information Society Technologies (IST) Programmes. For some of the leading current work the European Media, Technology and Everyday-Life Network (EMTEL) is the main source.
2. Hetherington, 1989; Meech, Kilborn, 1992; MacInnes, 1993.
3. Davenport, Higgins, Gillham, 1996.
4. Gillham, Buckner, 1997.
5. Buckner, Gillham, 2001; Buckner, Gillham, 2003.
6. LiMe, 1997–2001.
7. UTOPIA, 2002–3.

8. Stewart, 2002.
9. FLEX, 1999–2001.
10. Baillie, 2002.
11. Morley, 1986; Lull, 1990; Moores, 1993.
12. Dholakia, et al, 1996.
13. Silverstone, et al, 1992; Lie, Sorensen, 1997.
14. Stewart, 2003.
15. Foley, 2000.
16. Scottish Executive, 2001.
17. For a review of IT uptake and the issues of the digital divide in the UK at the time of this research see Stewart, 2000.
18. http://www.e-envoy.gov.uk/oee/oee.nsf/sections/about-epolicy-analysis/$file/uk_stats-tables.htm#hhabyregion
19. Quotations throughout this chapter, unless otherwise stated, have been taken from Stewart, 2002.
20. Davenport, Higgins, Gillham, 1996.
21. Buckner, Gillham, 1998; Stewart, 2002.
22. Buckner, Gillham, 1998.
23. Buckner, Gillham, 1998.
24. Buckner, Gillham, 1998.
25. Buckner, Gillham, 1998.
26. Buckner, Gillham, 2003.
27. Buckner, Gillham, 1999.
28. LiMe, 1997–2001.
29. Davenport, et al, 1996.
30. Daft, Lengel, 1984; 1986.
31. van Raalte, 2002.
32. Buckner, Gillham, 2003.
33. LiMe, 1997–2001.
34. See Buckner, Gillham, 2001.
35. Buckner, Gillham, 2003.
36. Buckner, Gillham, 2003.
37. Buckner, Gillham, 2003.
38. LiMe, 1997–2001.
39. From Gillham, Buckner, 1996.
40. Gillham, Buckner, 1996.
41. Stewart, 2002.
42. Stewart, 2002.
43. Stewart, 2002.
44. Quoted in Stewart, 2002.
45. Stewart, 2002.
46. Eisma, et al, 2003.
47. Eisma, et al, 2003.
48. Stewart, 2002.

BIBLIOGRAPHY

Buckner, K, Gillham, M. Using email for social and domestic purposes: Processes, practices and attitudes. In Sloane, A, van Rijn, F, eds, *Home Informatics and Telematics: Information, technology and society* (IFIP WG 9.3 International Conference on Home Oriented Informatics and Telematics [HOIT 2000], 'IT at Home: Virtual influences on everyday life', June 28–30 2000, Wolverhampton), Boston, 2000, 87–98.

Buckner, K, Gillham, M. Email in a social setting: Towards an understanding of interpersonal communication motivational attributes (HOIT 2003), Irvine (California), 2003.

Daft, R L, Lengel, R H. Information richness: A new approach to managerial behaviour and organization design, *Research in Organizational Behaviour*, 6 (1984), 191–233.

Daft, R L, Lengel, R H. Organisational information requirements: Media richness and structural design, *Management Sciences*, 32 (1986), 554–71.

Davenport, E, Higgins, M, Gillham, M. Designing a probe to explore home information systems in the UK, *Online and CD Rom Review*, 20/2 (1996), 75–9.

Dholakia, R R, Mundorf, N, et al. Bringing Infotainment Home: Challenges and choices. In Dholakia, R R, Mundorf, N, Dholakia, N, eds, *New Infotainment Technologies in the Home: Demand-side perspectives*, New Jersey (1996), 1–18.

Eisma, R, Dickinson, A, Goodman, J, Syme, A, Tiwari, L, Newell, A F. *Early User Involvement in the Development of Information Technology Related Products for Older People* (submitted for peer review, 2003).

Foley, P. *Whose Net? Characteristics of Internet users in the UK* (report submitted to the Information Age and Policy Development Communications and Information Industries Directorate of the Department of Trade and Industry), Leicester, 2000.

Gillham, M, Buckner, K. User evaluation of hypermedia encyclopedias, *Journal of Educational Multimedia and Hypermedia*, 6/1 (1997), 77–90.

Hetherington, A. *News in the Regions: Plymouth Sound to Moray Firth*, Basingstoke, 1989.

Hüttenmoser, M. Kindheit und Schule – Kontinuität und Wandel [Childhood and school, community and knowledge]. In Hugger, P, ed, *Handbuch der Schweizerischen Volkskultur* [Handbook of Swiss Folkculture], Basel, 1992, 3 vols, Vol I, 92–4.

Lie, M, Sorensen, K H. Making technology our own? Domesticating technology into everyday life. In Lie, M, et al, eds, *Making Technology our Own? Into everyday life*, Oslo, 1997, 1–23.

Lull, J. *Inside Family Viewing: Ethnographic research on television audiences*, London, 1990.

MacInnes, J. The broadcast media in Scotland, *Scottish Affairs*, 2 (1993), 85–98.

Meech, P, Kilborn, R. Media, culture and society, *Media, Culture and Society*, 14 (1992), 249–59.

Moores, S. *Interpreting Audiences: The ethnography of media consumption*, London, 1993.

Morley, D. *Family Television: Cultural power and domestic leisure*, London, 1986.

van Raalte, J. Record email usage signals last post for letters: Email overtake letters sent and received from UK homes, <http://www.netvalue.com/corp/presse/cp0046.htm (accessed 3 September 2002).>

Scottish Executive, System Three, MORI Scotland. *Consumer Durables* (Scottish Household Survey Bulletin, no 5), Edinburgh, 2001.

Silverstone, R, Hirsch, E, Morley, D. Information and communications technologies and the moral economy of the household. In Silverstone, R, Hirsch, E, eds, *Consuming Technologies: Media and information in domestic spaces*, London, 1992, 15–31.

Stewart, J. The Digital Divide in the UK: A review of quantitative indicators and Public Policies (paper for the conference 'The Digital Divide'), Bremen, Germany, 2000, <http://www.digital divide.de/pdf/UK.pdf.>

Stewart, J. The Social Consumption of ICTs: Insights from research on the appropriation and consumption of new ICTs in the domestic environment, *Cognition, Technology and Work*, Forthcoming (2003), <http://link.springer.de/link/service/journals/10111/contents/02/00111/.>

APPENDIX: STUDIES AND PROJECTS

Baillie L. 'The Home Workshop: A method for investigating the home', PhD thesis, School of Computing, Napier University, 2002.

Buckner, K, Gillham, M. Household Information Systems: An evaluation of local and Internet information in the home, QMUC funded project, 1998.

Gillham, M, Buckner, K. Access Strategies of Home Users of Interactive Multimedia Information Systems, Elsevier/LIRG Research Award for 1996.

Flexible Knowledge-Based Information Access and Navigation using Multimodal Input/Output (FLEX), EU 4th Framework Esprit IV 29158, 1999–2001.

Living Memory Project (LiMe), EU 4th Framework Esprit IV 25621, 1997–2001.

Usable Technology for Older People: Inclusive and appropriate (UTOPIA), Universities of Dundee, Abertay, Glasgow and Napier, funded by Scottish Higher Education Funding Council, 2002–3.

Stewart, J. 'Encounters with the Information Society: Personal and social issues in the appropriation of new media products in everyday life: Adoption, non-adoption, and the role of the informal economy and local experts', PhD thesis, Research Centre for Social Sciences, University of Edinburgh, 2002.

11 Games and Sport

JOHN BURNETT

INTRODUCTION

This chapter looks at two important aspects of play in the context of the dwelling. In order to understand its place in the home, we must take a broad definition of play; one which goes further than the idea that it is a pleasant but pointless way of spending time. Play is not pointless. It is full of intentions and meanings and is a way of interacting with the social and physical worlds around us.

The play of infancy involves the use of hands and eyes, the building of an understanding of the outside world and of relationships with parents and siblings.[1] It exercises the mind, including the memory (Figs 28.5, 42.2). The play of childhood teaches strategy, fairness, the need to follow rules, and the use of the imagination. The child can relish the pleasure of play, and learn to compete – and to cheat. Adults and adolescents play too. Their play has similar function to that of children: it reinforces elementary principles of social interaction and develops, or at least maintains, intellectual, creative, imaginative or physical skills. Play allows the growth and reinforcement of aspects of the self, in a context of little risk. Play for adults can also constitute a pleasurable but ephemeral return to childhood for its own sake.

Much play is verbal. Maclagan's classic of contemporary recording included word games, riddles and tongue twisters among *The Games and Diversions of Argyleshire* (1901). In Peebles around 1830:

> After tea there were songs, with perhaps a round of Scottish proverbs – a class of sayings which, from their agreeable tartness, found scope for exercise in ordinary transactions, and were more especially useful in snubbing children, and keeping them in remembrance of their duty.[2]

Here the functions of the play are socialisation (the songs), and the establishing of norms of behaviour through the proverbs. Other forms of play encourage dexterity and creativity. When the child imagines playing a social role, she or he is thinking about being an adult, and adults can use role playing for fun, to show their individuality, or to raise questions about themselves. When there are imagined roles, there may be different rules.

Play in the home includes many activities which are treated elsewhere in the Compendium series; singing games, storytelling, toys, guising, dancing and ludic aspects of calendar customs (such as 'hunt the gowk'). Music and song are not necessarily playful in themselves, but in a social situation they can be, and the play may be purposeful, as when the herd laddie was playing the whistle:

> He tried a spring for wooers, though he wistna what it meant,
> But the kitchen-lass was lauchin' an' he thought she maybe kent.[3]

He did not understand courting, but she did.

Play takes place in the home, and around it. The nature of the spaces available controls the possibilities. The rich have always lived in buildings with ample furniture and the houses of the middling classes usually have had space within and often gardens outside. The urban poor have frequently lived in overcrowded flats; the close, the tenement backyard, and the street were parts of home life if not of the dwelling itself.[4] Patterns of play were different in the country where children were expected to do some work, but this often included the possibility of play. A girl might learn work songs such as waulking songs, and herds were notorious for their tendency to diversion.

To one kind of mind, play, laughter and the imagination are threats to social and religious discipline. The Reformers put down much overt play, either in the floodtide of the late sixteenth century, or in the more intolerant periods of the seventeenth, especially around 1640. In the Highlands and Islands the evangelical revival at the beginning of the nineteenth century crushed public play such as horse races at Michaelmas, and shinty, and also domestic fun such as songs, music and stories. 'Better is the small fire that warms on the little day of peace than the big fire that burns on the great day of wrath.'[5]

In view of the immense variety of play in and around the home, and the minute survival of evidence, it is not possible to treat the subject systematically. The remainder of this chapter therefore outlines six areas of games and sports.

BOARD GAMES

Board games have been played by adults since the Middle Ages. In the *Táin Bó Cuailnge* that great hero Conchobor spent a third of his day playing *fidchell*, whose nature is unknown. Codified games were played first in the houses of noblemen and royalty, and later in public places and in ordinary houses. These games are important because they have endured for many centuries, and have been played all over Europe. The first games of which we have certain Scottish evidence are board games: chess, draughts, backgammon ('tables') and *hnefetafl*.[6] They were played by people of high status, those who had a table on which to place a board. Chess is first recorded in a Sanskrit text of *c* AD 620, and it was widely known in northern Europe by 1000. We know the night James I was murdered (in 1437) he had been playing chess and the tables as well as listening to singing, piping and harping. The modern rules date from *c* 1475 (Fig. 11.1).

'Tables' was the medieval name for backgammon (c 1650). William Cunningham, a Renfrewshire laird, was staying in Edinburgh in December 1676 when he bought 'a suit of chesse men'; the following year he was back in the east and lost money 'at tables'.[7] The game persisted. In Perth at the beginning of the nineteenth century:

> Backgammon was much played by private parties; and in the Coffee-room, which was then in a wooden house ... So much was this game the rage with the merchants in their back shops, and even on the counter, that many of them hurt their business by it; – customers declining to enter those shops which were continually full of loungers engaged in this game.[8]

What we now call draughts is first noted in Scotland in 1685 as 'the dams' and the board became the 'dam brod'. The word is from the French *jeu de dames*, which passed to most European languages (German *Damespiel*, Spanish *damas*, Czech *dáma*). The game was more common in Scotland than in England, with the Scots playing on the dark squares, the English on the light, until in the nineteenth century the Scots method was adopted south of the Border.[9] There are names for standard openings, such as 'Laird and Lady', the favourite of a laird and his wife from Cambusnethan, Lanarkshire; 'Maid o' the Mill', after a Lanarkshire miller's daughter; and 'Ayrshire Lassie'.

Figure 11.1 The Stuart family, probably father and daughters, playing chess at their house at Roslin, Midlothian, c 1900. Source: SLA 107H 56/45/6A.

Tod and Lambs, once a common board game in Scotland, was first recorded in thirteenth-century Spain. The object was for fifteen slow-moving pieces to hem in one powerful one. As Fox and Geese it was played in England in 1633, but Tod and Lambs is not recorded until about 1800.[10]

Race games reached Europe from the Middle East in the sixteenth century. The first famous one was the *gioco dell' oca* (game of the goose) given by Francesco de Medici to Philip II of Spain c1580 and reproduced in many countries.[11] London map and print engravers and publishers developed race games in the eighteenth century including, by the 1780s, geographical ones based on England and Europe. One of the first to include Scotland was John Wallis's Tour of Europe (1794) which visited Glasgow, Edinburgh, Aberdeen and 'Johnny Groats'. These games stand at the beginning of the invasion of Scottish childhood play by its commercialised English counterpart. In the twentieth century, the commercial development of board games produced many more complex types, most of which failed to ignite popular interest although a few became very popular indeed. Monopoly was devised in Philadelphia in 1934–6, and a London version was produced in 1936. In the 1990s it was joined by sets based on the cities of Edinburgh and Glasgow, and on Glasgow Rangers and Celtic football teams. In 2000 the British Association of Toy Retailers voted Monopoly the 'game of the century'. Almost all board games, except the most abstract, reflect something of the society in which they were devised: chess is feudal, and Monopoly capitalistic. Trivial Pursuit, fashionable in the 1980s, was a structured general knowledge quiz which had great appeal as after-dinner entertainment for young professionals; some companies used it as a way of measuring the competitiveness of new recruits.

PLAYING CARDS

Playing cards and card games were probably invented in the Middle East and arrived in Europe in the 1370s. They spread rapidly across the Continent, although they have not been detected in Scotland until 1556.[12] James VI was an enthusiast for Five Finger or Maw.[13] After the restoration of the monarchy in 1660 gambling was fashionable, and it is likely that it was in this period that two French terms, *monsieur* for the knave and *piquet* for clubs, were transformed into the Scots 'munsie' and 'picks', though they are not recorded until the early nineteenth century. The engraving of plates for printing sets of good-quality cards was a costly business. The Edinburgh goldsmith Walter Scott engraved by far the earliest-known Scottish pack in 1691. Each card shows the arms of one or more landed family.[14] Since the early eighteenth century 'the curse of Scotland' has been the nine of diamonds, because it resembles the nine lozenges in the Earl of Stair's arms.[15]

Until at least the end of the nineteenth century card playing was the subject of moral disapproval. Robert Baillie, Principal of Glasgow University, accused the Bishop of Ross in 1638 of being, 'ane admitter of fornicators, a companier with papists, and usuall carder on Sonday'.[16] At Banff in 1705 a prospective dominie, 'could keep no order in the school ... and was a habitual drunkard and cairder', playing from seven in the evening to seven in the morning at Baillie Watson's house.[17] Robert Burns said his friend Gavin Hamilton 'drinks, an swears, an plays at cartes' in the 1780s.[18] Yet the habit was spreading, and Alexander Carlyle of Inveresk, Midlothian, thought himself the first minister to play cards 'behind unlocked doors'.[19] At Perth about 1810:

> *Cards* were played by parties of tradesmen at home, and in public houses over a bottle of ale; as the game was uniformly catch the ten, and played merely for pastime, the thing was allowed by respectable houses ... By times the drinking of gills superseded the good old custom of a bottle of ale, and persons of very loose morals took the opportunity of introducing various games at cards.[20]

A pack of playing cards was still called 'the deil's testament' in Shetland in 1949.[21]

By the 1820s card playing had become common among all kinds of people across the Lowlands. In Galloway the popular card games were 'Catch the Ten', 'Brag' and 'Beggour my Neebour'.[22] Card playing was sufficiently widespread for there to be terms, depending on locality, for the kind of naïve play in which one plays all one's winners first – Paisley play, Jeddart (Jedburgh, Roxburghshire) play, or among the sophisticates of Anstruther and Pittenweem in Fife, Crail play.[23] After the Napoleonic Wars, patience spread from France to Britain. Patiences (for there are many kinds) are unusual in that they are complex games for one person.[24] This category was to grow rapidly after computer games became available (Fig. 11.2).

Cards, like dominoes, have been a major part of public-house life, which is to say the home life of many urban Scots males. One can only guess that working-class gambling was rare in 1800 but more common by 1900. Although there is a vivid description of a Sunday-afternoon domestic gambling den on the south side of Edinburgh in the 1920s,[25] tenements often lacked sufficient space for a man to relax with his peers, while the pub on the corner of the street was available for the price of a nip.

GAMES AND SPORTS IN AND WITH HOUSES AND GARDENS

The first structures to include provision for sports were religious houses. The young men had time to spare, quiet courtyards with masonry walls, and either flagstone or beaten earth in front of them. About 1275 the game of *cache* was invented in Picardy. It spread over Europe and 'caich' reached Scotland by 1475.[26] The murder of James I in 1437 took place in the tennis court at the Franciscan Friary at Perth. At this date the racket was yet to be invented and the ball was struck with the gloved hand. This game developed into 'real' (or royal) tennis, for which James V had a court built at

Figure 11.2 Playing cards on Lewis, probably c1970. The man on the right seems to be playing patience. Source: SLA 107H 39/5/17.

Falkland in 1541.[27] Caich continued to be played by ordinary people in public spaces. As 'handball' it was played by boys at Portpatrick, Wigtownshire, in the 1840s, and at Galston, Ayrshire, it survived until 1939.[28]

There is a group of games in which the house itself is a component. In 'chicky-mellie' a long piece of thread was attached to a window frame. A large button was tied on, six or eight inches (15 or 20cm) from the end. The other end, much more distant, was in the hands of boys who pulled it to make the button rattle on the glass.[29] In the north east, on Auld Yule, young people spent the night visiting friends and drinking sowans, which they spattered over the doors and windows of houses belonging to the less sociable or less liked members of the community. In the parish of Savoch in Aberdeenshire in 1860 a mixture of turpentine and tar was used.[30] In this, there is an element of charivari, but this kind of play can become openly threatening. The Glasgow poet Stephen Mulrine imagined a juvenile gang frightening a woman in a tenement. What will she feel:

> If they play keepie-uppie on the clean close wall
> An blooter your windaes in wi the ball?[31]

Girls' ball games developed rapidly after the invention of the rubber ball about 1830, and most of them require a flat wall off which to bounce the ball – not something which could be taken for granted before the spread of brick. The Opies described ball bouncing as 'less a competitive game than a pursuit in which skill is required for personal satisfaction'.[32] Boys' ball games were generally more competitive, though the use of a small ball enabled boys to develop ball control without having to take account of other players; the characteristic close-heid tanner-ba merchant. In back courts overlooked by tenements, and elsewhere, girls played 'peevers', also known by several other names including 'peeverie beds', 'the beds', in north-east Scotland 'hoppin beddies', and in Dalkieth, Midlothian, 'pickies'.[33] But in 2002 the English name, hopscotch, is almost universally used and the game is not as common as it was.

In the second half of the twentieth century changes in working-class housing altered the game-playing pattern. The tenement back court and the urban street largely disappeared as places to play. More houses had gardens and most blocks of flats had green space around them. Suburban housing was laid out without large hedges between properties (fewer possibilities for hide and seek), and with accessible patches of green near houses where parents could keep half an eye on their free-range children. As always, the games played depended on the balance of age and gender in the group.

THE HOME AND OUTDOOR SPORTS

The rapid growth of organised sport in the late nineteenth century is one of the largest changes in the history of leisure in Scotland. The garden of the suburban villa had, in the eighteenth century, a space for archery or bowling. At the end of the nineteenth century there were public parks with space for most sports although the need to purchase equipment limited the range of people who could play. In all cases, players had clothes to be laundered, and supporters had a need to talk about their new enthusiasm.

Perhaps some mothers and wives took on the work of laundering, as an increasing number of flats had kitchen ranges, but often the work was done in a back green washhouse by women who were paid for it. As James Curran's song, probably dating from the 1890s, said:

> It wad tak a dozen skivvies
> His claes tae wash an scrub
> Since oor Jock became a member
> O that terrible fitba club.[34]

Football was a game to be watched as well as played. Men discussed football in pubs, and at home they deaved their wives and families with it. The evening paper, such as Glasgow's *Evening Times* (1876), and specialist weeklies of which the *Scottish Athletic News* (1882) was the first, were read at the kitchen table. After World War I a further cultural form was added – the football pool. Here is Danny Shields, riveter, in his flat in Partick, Glasgow, doing the football coupon on Thursday night:

> All his force was concentrated on the pinkish sheet. The coupon was a document that might have bewildered even an intelligent layman, but to Danny it presented no difficulties save those of decision ... After an hour of it he pushed the paper away with a sigh.[35]

The point of mass sport was escape from the mundane, including domesticity, to tribal sectarianism and collective passion, and the ecstasy of being lost in a crowd with only one thought. The role of the home was to be utterly different from Hampden or Pittodrie, at least until the 1990s when the largest football clubs started to sell domestic items bearing team colours or symbols. The 2001 Glasgow Rangers catalogue offered mugs, glassware, nightwear, a choice of four duvet covers, pillow cases and wallpaper. There was also an advent calendar ('every day contains a solid milk chocolate'), which more traditional supporters would have seen as being outside the club's Protestant tradition.[36]

TECHNOLOGY AND GAMES IN THE HOME

Indoors, more, and more different, games were available for those who could afford them from the eighteenth century onwards. Chess, and other board games of the kind already described became more widespread. They were joined by various handcraft hobbies which aimed to make something ostensibly useful, such as fretwork. Among boys the most popular handcraft toy was Meccano, invented by Frank Hornby in 1901, an ingenious kit from which engineering models could be made.

Poorer families had fewer possessions and their children fewer toys. Yet there is no reason to suppose that they had less pleasure from play, or that their imaginations ranged less widely. A simple toy of carved wood may leave more scope for creativity than a box of painted soldiers. There were certain occasions when a child might gain access to unfamiliar forms of play, for example, during a long-term illness. Before the widespread use of antibiotics in the late 1940s, a sick child might have to spend weeks or months in hospital or being nursed at home, and have the pleasures of a rocking horse or colouring books.

The first games which used computer graphics were developed in the 1970s, descendants of arcade pinball machines, which themselves had as ancestors the aristocratic *trou-madame*, a game played on a specially made board in which the object is to roll balls into holes which give different scores.[37] The first widely used computer for game playing was the Atari Video Computer System of 1977, made in America. In 1985 the hand-held Nintendo player swept the market, but from 1989 it suffered severe competition from the Sega Genesis. Since then, these two Japanese firms have remained dominant. Most of the games available on these platforms were either race games or 'shoot 'em up'. The IBM PC-XT, released in 1981 as a business personal computer, was the starting point for a rapid development in power. The games available on PCs and their clones (and on Macintosh computers) were much more varied, including the building and management of pets, cities and railway networks, and other fantasies. In a game that simulates the growth of a family, neglect of a baby leads to the intervention of the Social Services: again we see play as a preparation for adult life. The presence of a specialist software house in Dundee, DMA Design, led to Abertay University starting a master's degree in computer-games software in 1997.[38]

Television started to spread quickly after the coronation of Queen Elizabeth in 1953, and by the early 1960s was in most homes. It now dominates the leisure time of children and adults, and so there is less active play. Handcraft recreations declined in popularity; the *Meccano Magazine* was last published in 1981.

ROLE-PLAYING GAMES

Play requires the players to manipulate representations of themselves, even if the reproduction is no more than a piece on a board. Far more of the personality is involved in role-playing games. Boys have played for centuries at being soldiers. The son of a shepherd, raised at Linton in Roxburghshire in the 1920s, remembered:

> Sometimes too we took part in the hoosies game, and acted as husbands or sons. It always intrigued me how the girls 'put on the English' when they played 'hoosies'. We would join in with the girls at playing 'schools' as well, when the main activity of the 'teacher' was wielding the tawse![39]

Status and gender differentiation are obvious in this quotation. We all have similar memories, and they are significant because they recall childhood images of adulthood and the world in which we were about to live.

CONCLUSION

The difficulty of discussing the history of play in the home is illustrated by the unsystematic treatment it has been given here. The existing literature on games and sports concerns the ways in which they have been played, but not their social roles, or their distribution in different social situations. There is an urgent need for oral history and contemporary recording. The Opies have shown the richness of life in the school playground; there is far more to say about the home, where play helps each of us to become ourselves.[40]

ACKNOWLEDGEMENTS

I am grateful to several colleagues for helpful observations: Heather Brunton of the Scottish Life Section of the NMS; and Susan Storrier and Professor Alexander Fenton of the EERC.

NOTES

1. Winnicott, 1971, 53–85.
2. Chambers, 1872, 17.
3. Murray, Charles. *Hamewith*, 1911. Reprinted in MacQueen, Scott, 1966, 463.
4. There are a number of accounts of tenement childhoods of which one of the most interesting, because it also describes life in a council house, is Hanley, 1958; A more analytical approach is suggested by Hand, 1993–8.
5. Cited by Lochhead, 1948, 177.
6. For *hnefetafl*, see Murray, 1952, 55–64.
7. Cunningham, 1887, 84.
8. Penny, 1836, 116.
9. Parlett, 1999, 242–70.
10. *SND*, s v Tod; Penny, 1836, 117 gives a good description.
11. Goodfellow, 1991, 14, 17, illustrates two early commercial examples, one Italian and one English.
12. Lauder, 1864, 329; *DOST*, s v Card.
13. Parlett, 1991, 189; some idea of the nature of the game is given by Parlett, 1979, 148–52.
14. NMS H.NS 211; heraldic cards are rare and the Scottish ones appear to be based on a set issued by Claude Oronce Finé at Lyons in 1659 – see Hofman, 1973, 87.
15. *SND* gives a discussion of the other possible sources of the name.
16. Baillie, 1841, I, 161.
17. Cramond, 1891–93, II, 179.
18. Burns, R. 'Holy Willie's Prayer'. In Kinsley, 1968, I, 77.
19. Cited by Lochhead, 1948, 88.
20. Penny, 1836, 118.
21. *SND*, s v Deil VII.C23.
22. MacTaggart, 1824, 458; For 'Catch the Ten', see Parlett, 1979, 438–9; for 'Brag', Parlett, 1979, 395–8, Parlett, 1991, 101–4; and for 'Beggar my Neighbour', Parlett, 1991, 74–5.
23. *SND*, s v Crail.
24. Parlett, 1979, 423–4. In general very little evidence exists for solo play, patiences and certain girls' ball games being two important exceptions.
25. Denton, Wilson, 1994, 43–5.
26. Gillmeister, 1981.
27. Puttfarken, Crichton-Stewart, 1989, 18–25.
28. *NSA*, IV, 146; See also Burnett, 2000, 45–50.
29. *SND*, s v Chicky-mellie.
30. *Peterhead Sentinel* (13 January) and (3 February 1860); Fenton, 1974.
31. Mulrine, 1971, 5.
32. Opie, Opie, 1997, 128.
33. For the last two, I am grateful to Professor Alexander Fenton and Mrs Marguerita Burnett.
34. As sung by Adam MacNaughtan.

35. Blake, 1935, 35–6.
36. Glasgow Rangers FC. *Official Merchandise*, Winter 2001.
37. There is an example of *trou-madame* at Castle Fraser, Aberdeenshire.
38. See Chapter 10 Television, Video and Computing.
39. Purves, 2001, 14.
40. Opie, Opie, 1979.

BIBLIOGRAPHY

Baillie, R. *Letters and Journals*, 2 vols, Edinburgh, 1841.
Blake, G. *The Shipbuilders*, London, 1935.
Burnett, J. *Riot, Revelry and Rout: Sport in Lowland Scotland before 1860*, East Linton, 2000.
Chambers, W. *Memoir of Robert Chambers*, London, 1872.
Cramond, W, ed. *Annals of Banff*, 2 vols, Aberdeen, 1891–93.
Cunningham, W. *The Diary of William Cunningham of Craigends*, Edinburgh, 1887.
Denton, H, Wilson, J C. *The Happy Land*, Edinburgh, 1994.
Fenton, A. Sowans in Scotland, *Folk Life*, 12 (1974), 41–7.
Gillmeister, H. The origin of European ball games: A re-evaluation and linguistic analysis, *Stadion*, 7 (1981), 19–51.
Gomme, A B. *The Traditional Games of England, Scotland and Ireland*, 2 vols, London, 1894–8.
Goodfellow, C. *A Collector's Guide to Games and Puzzles*, London, 1991.
Hand, O. Society's bairns, *SS*, 32 (1993–8), 126–30.
Hanley, C. *Dancing in the Streets*, London, 1958.
Hofman, D. *The Playing Card: An Illustrated History*, Leipzig, 1973.
Kinsley, J, ed. *The Poems and Songs of Robert Burns*, Oxford, 3 vols, 1968.
Lauder, W. *Ane Compendious and Breue Tractate Concernyng ye Office and Dewtie of Kyngis* (1556) (Early English Text Society), London, 1864.
Lochhead, M. *The Scots Household in the Eighteenth Century*, Edinburgh, 1948.
MacilleDhuibh, R. Wood-sense and raven-black ('The Quern-Dust Calendar'), *West Highland Free Press* (17 March, 1995).
Maclagan, R C. *The Games and Diversions of Argyleshire*, London, 1901.
MacQueen, J, Scott, T, eds. *The Oxford Book of Scottish Verse*, Oxford, 1966.
MacTaggart, J. *The Scottish Gallovidian Encyclopaedia*, London, 1824.
Mulrine, S. *Poems*, Preston, 1971.
Murray, H J R. *A History of Chess*, Oxford, 1913.
Murray, H J R. *A History of Board Games other than Chess*, Oxford, 1952.
Opie, I, Opie, P. *The Lore and Language of Schoolchildren*, Oxford, 1979.
Opie, I, Opie, P. *Children's Games with Things*, Oxford, 1997.
Parlett, D. *The Penguin Book of Card Games*, London, 1979.
Parlett, D. *A History of Card Games*, Oxford, 1991.
Parlett, D. *The Oxford History of Board Games*, Oxford, 1999.
Penny, G. *Traditions of Perth*, Edinburgh, 1836.
Purves, A. *A Shepherd Remembers*, East Linton, 2001.
Puttfarken T, Crichton-Stewart, M. The royal tennis court at Falkland. In Butler, L St J, Wordie, P J, eds, *The Royal Game*, Kippen, 1989, 26–35.
Winnicott, D W. *Playing and Reality*, London, 1971.

12 Artistic Expression and Interiors

ELIZABETH CUMMING

INTRODUCTION

>Sunday clothes, lozenge pattern, discomfort.
>Drugget on floor for parties.
>Decorative efforts. Fairy lights.
>Crimson carpet in drawing room. Salmon-coloured watered silk effect panelled in white and gold on walls.
>Father's foreign travels. Chinese sword and helmet.
>Maltese lace collar.
>Grey fur-edged boots. [1]

As Catherine Carswell's (nee Macfarlane) Proustian recollections of her late nineteenth-century parental home and its middle-class domestic life indicate, there are two principal aspects to the alterable appearance of a home's interior, namely decoration and, within house furnishing, ornamentation. These stand apart from the more basic house crafts of structural alterations and the installation of services and fulfil several quite specific roles within the human psyche. They may demonstrate the status (either achieved or to which aspired) of the household, or at least of its prominent members. Such status may be social, economic, intellectual or political. Allied to this are an awareness of fashion and a capacity to follow it. In the Macfarlane home, the types of paper on the walls spoke of the financial standing of a shipping merchant and the colour of a room an understanding of fashion, while the carpeted, papered and panelled drawing room as a unit revealed the 'artistic' nature of the dwelling.

Identity is also powerfully expressed in interiors. The particular objects in a home, along with the textures and colours of walls and floors, can be the most personalised indicators of an individual's nature, tastes and hopes. Regional or national identity can also be readily asserted. In addition, decoration and ornamentation may help demarcate and grade areas within the home and do the same for those household members particularly associated with them. But alongside these more functional motivations – and sometimes separated by the finest line – comes the opportunity for artistic expression, for the delight in practising creative judgement and skills and the enjoyment of the resultant aesthetic values. All these aspects help fabricate what the French social historian Michelle Perrot has labelled the 'private kingdom', a 'place of pleasures and servitudes, conflicts and dreams'.[2]

Today forms of decoration and furnishing, and the use of ornament, witness a degree of artistic choice and arrangement in almost all Scottish homes. Whether inherited, purchased or otherwise obtained, the selection and combination of specific items, plus the choice of colour and texture of walls, floors and ceilings is an art in itself. As few householders have had the will or means to commission a totally integrated interior masterminded by architect-designers, they have sought, like the Macfarlanes, to create a home of value themselves. Sometimes this has necessitated or allowed design to be combined with their own labour.

The types of products used to decorate or ornament the home often reflect both Scottish and British values. This chapter draws on those aspects of British commercial production (such as textiles and wallpapers) which inform the choice of goods available to Scottish householders. Yet, while manufacture in Scotland has sometimes been identified as predominantly commercial or industrial, the Scottish home has also long contained hand-made decorative, and sometimes functional objects – most often in wood, horn, metal, ceramics and woven or stitched textiles – whose making and display are peculiarly localised forms of artistic expression. Added to this, the revitalisation of the domestic crafts within the British Arts and Crafts Movement in the late nineteenth century found expression across Scottish society in a remarkable double expansion of professional training and amateur practice which persisted throughout the first half of the twentieth century. This helped to energise interior-design aesthetics, although predominantly for the middle classes. Its modern legacy lies in the values we now place on skills (some aspired to today by do-it-yourself enthusiasts) and on the many values of tradition.

A lack of implementation skills among those who cannot afford to employ others places constraints on artistic expression, as can limitations on funds or the availability of materials and objects, community norms, the position of the individual within the household and the wishes of other household members. Yet at times these hindrances are perceived as challenges which may actually make the creative process more satisfying.

There are constraints, too, on our accurate understanding of decoration and ornamentation in the past. However hard we study the fragmented past we can never fully appreciate the original meanings and attitudes of distant cultures. However, it is possible at least to explore certain aspects, particularly within the documented recent past, a time of accelerated and self-conscious social change and a period which has yielded a spectacular mass of artefacts for the consumer. The first section of this chapter examines some meanings of room decoration, and the second the values of ornament within interior space. With the growth of a range of home-improvement agencies in the twentieth century, the third section looks at the home as site of modern artistic enterprise.

DECORATION OF THE HOME

External walls of homes do not usually require a surface application, other than harling, perhaps, to serve their function as weatherproof boundaries of shelter. Internal walls and ceilings, however, may be painted or otherwise clothed in a visual display. This application, which may often make reference to poetic language as well as the social position and aspirations of the inhabitants, enhances the physical ambience of a room and simultaneously declares its purpose and status within the dwelling as a whole. However, until the industrial expansion of the chemical industries in the mid nineteenth century in which Glasgow had a vital role, colours and decorative materials were important only in the upper-class home. It was in the period 1895 to 1914 that the commercial availability of paints, papers and textiles, allied to a corresponding development of the interior decorating trade, first peaked. In Scotland, as throughout Britain, only the urban middle classes benefited to any visible extent.

In the last decade of the twentieth century, and now at the beginning of the twenty-first, a dramatic swell in the number and scale of direct suppliers of house furnishing and decorating materials has offered the vast majority of Scottish householders an almost limitless choice of means of self-expression. With both property ownership expansion and a marked rise in property prices throughout the 1980s, the desire to maximise dwelling values has motivated modernisation and decoration. Within the last decade, mail order and especially Internet shopping have reduced differences between rural and urban areas. Yet still restricted in personalised domestic expression are the very poor and those members of Scottish society of more comfortable but still modest means who, for various reasons, are unable to undertake decorating themselves. Then there are those who live in residences, hospitals and hostels, for example – people with virtually no control over the visual character of their daily habitat.

Surface and space
The language and patterning of a room's internal boundaries, its walls, ceilings and floors, set the stage for its complementary, often deeper, personalisation through furnishing and, more specifically, ornamentation. These 'stage-set' surfaces of a room contextualise and augment the artefact contents of the space and contain elements of both tradition and innovation (see Fig. 21.1). In the Scottish home they also frequently blend the expressiveness of the vernacular with the control of philosophical judgement. The spatial and elemental values which emerge from this localised synthesis might be said to correspond to the mental landscape of such indigenous arts as poetry and storytelling, music and sculpture. In instances where there is particularly cohesive design integration, a room's expression may present an extraordinary dialogue of form and imagination. The applied materials that commonly validate the surfaces of these interiors include paint and papers, wood panelling and carving, and textiles.

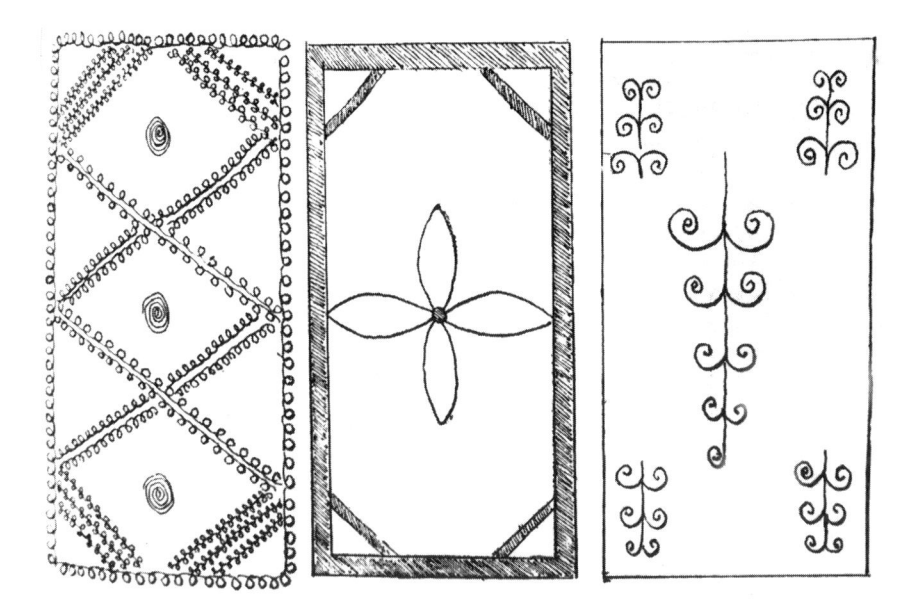

Figure 12.1 Doorstep designs, Dunsyre, Lanarkshire. From Munro-Denholm, R. Flagstone designs: Some details of the collection being made by Scottish Home and County, *Transactions of the Glasgow Archaeological Society*, new series, IX/II (1938), 95–100. Source: SLA 128E C.11006.

Paint

As elsewhere in Europe, the domestic folk art of Scotland has been rooted in two-dimensional decorative devices, with many of their graphics descended from the images of Celtic and Pictish cultures. Two-dimensional decoration deserves recognition across the country and at a range of social levels. In small rural homes, decoration of the threshold, the floor of the entrance passage and hearthstones with traditional abstract patterns of spirals, waves, the 'bull's-eye' or 'tangled-threid' patterns (more recently, icons of thistles or horseshoes) served to mark out points of social exchange[3] (Fig. 12.1). Such pipeclay decoration of flagstones also marked out regional status, being widespread in the more affluent agricultural areas of central and lowland Scotland, from Argyll to the Borders, Fife and Angus. The practice died out only recently.

For the wealthy Scottish homeowner, the decoration of the dwelling with paint and representational imagery provided applied symbols of power, and denoted both a material and ideological transference from foreign cultures. In the later Middle Ages wall painting had been used as an iconographical device in ecclesiastical buildings from Glasgow Cathedral, Dryburgh Abbey, Berwickshire, and the church at Fowlis Easter, Angus, to St Magnus Cathedral, Orkney. Although there was thus a precedent for the decorated interior, the popularity of ceiling painting within the domestic

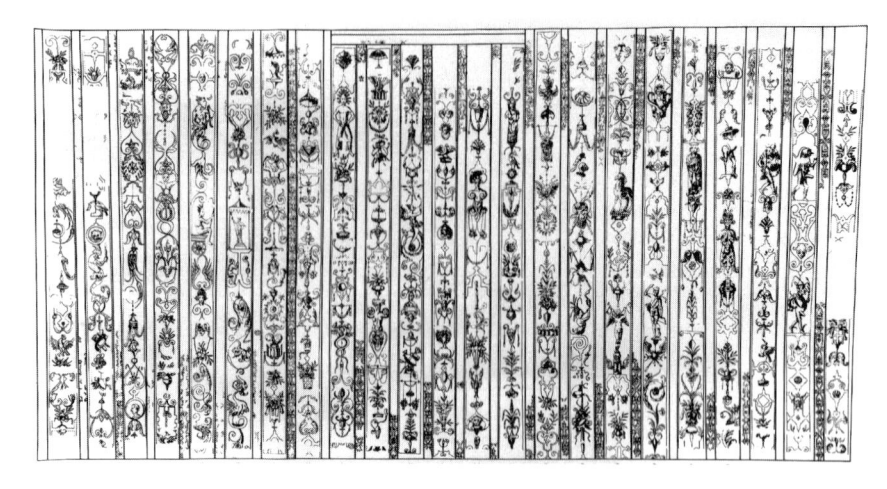

Figure 12.2 Late sixteenth-century ceiling decoration at Prestongrange, Prestonpans, East Lothian, dated 1581. Crown Copyright: RCAHMS.

dwelling in the late sixteenth and early seventeenth centuries was principally the result of direct influence from Italy and the Low Countries. Images mostly came from emblem books, with non-figurative work generally allowing more scope for individual expression, in occasional upper-wall decoration, as well as ceilings. Figurative scenes were often mythological or illustrated incidents from the lives of Biblical characters such as Abraham, Cain and Abel, or David. Exposed beamed ceilings (showing the underside of floor planks) of the late sixteenth century were made of imported softwoods and were enhanced and animated by colour and form to provide fashionable icons of status for the upper and merchant classes. For merchant-class townhouses in Glasgow and Edinburgh, wall painting had become an accepted part of interior decoration by the start of the seventeenth century[4] (Fig. 12.2).

Other forms of decoration
There have been other forms of decoration in wealthy Scottish homes beyond sculpture and painting. Many of the interiors to have survived from the seventeenth century possessed wall panelling, which used wood imported from the Baltic and could be plain or painted. There were also embroidered hangings and woven tapestries. The latter, frequently depicting classical subjects, were expensive and imported from Flanders, or from England after the foundation of the Mortlake workshops in the early part of the century. Royal household furnishings included linen and damask hangings embroidered by women at the start of the sixteenth century.[5] For the bedrooms of the élite, a matching set of tapestries and painted cloths or hangings might be commissioned. For other rooms, wall tapestries might be partnered by locally embroidered or tapestry cushion textiles, and table and

sideboard carpets. This Low Countries taste was revived in the 1890s and 1900s by architect Sir Robert Lorimer (1864–1929) when he was involved in restoring sixteenth- and seventeenth-century tower houses and their gardens, partly in response to the English Arts and Crafts re-evaluation of tradition. Clients of Lorimer, whose dining room in Edinburgh was equipped with a seventeenth-century tapestry as well as his own furniture, included Sir William Burrell.

Lorimer also revived hand-wrought ironwork and plasterwork. The hall at Craigievar, Aberdeenshire, had been furnished with a massive plasterwork crest overmantel in the first quarter of the seventeenth century. The moulds used here had also been employed in England and elsewhere in Scotland (for example, at Glamis, Angus and Muchalls, Kincardineshire).[6] Such plasterwork was executed by itinerant workmen who worked to their own designs and also from pattern books. Stucco was recommended in the eighteenth century by William Adam (1689–1748) on the grounds that it could be worked in any colour, or marble veined, and also hardened through washing. In Joseph Enzer's work on the walls and coved ceilings at the House of Dun, Angus, Jacobite symbolism is encoded into the classicised details (Fig. 12.3).

Figure 12.3 Saloon, the House of Dun, Angus, designed by William Adam with stucco work by Joseph Enzer, 1742. Source: Reproduced by kind permission of the Trustees of the National Trust for Scotland.

Ceilings
In general, the ceiling as a dominant visual statement of power had lost much of its importance and meaning by the twentieth century, but was revived spectacularly in one modern example, the nine panels commissioned, with companion window blind designs, from sculptor Eduardo Paolozzi for Cleish Castle, near Kinross (1972–3). W H Playfair and William Burn were two architects who maximised the ceiling's potential in early Victorian Scotland, and the interiors illustrated in R W Billings' popular *Baronial and Ecclesiastical Antiquities of Scotland* (1845–52) concentrated on the strengths of space, walls, and especially ceilings as the enduring and vital features of a room. Initially published three years after Queen Victoria's first visit to Scotland, his book coincided with the popularity of Scott's home, Abbotsford, Roxburghshire, and new work at Balmoral, Aberdeenshire, and Holyrood, Edinburgh, in the 1840s and 1850s.

Status and intellectualism
The emphatic embellishment of ceilings (perhaps using heraldic devices) was just one personalised form of the expression of upper-class status. Desired status may be political, economic, social or intellectual, or a combination of these four. Intellectualism was particularly displayed through classical imagery. Eighteenth-century Italianate antiquarianism led to many different forms of this – some were lightly stylistic, others more rigorously iconic – which were set to dominate interior work for Scotland's richer patrons for three centuries. The landscape painted panels of James Norie (1684–1757) and his son in Edinburgh William Delacour (*fl* 1740–67) (for example, at Yester, East Lothian) provided domestic echoes of the Virgilian pastoral. Early 'flock' wallpapers, dusted with patterns made by powdered wool ends and introduced from the Low Countries in the seventeenth century, had also depicted classical or floral subjects. These were revived over a century ago in the embossed, flax-based canvas frieze and wall coverings (echoing a vogue for gilded leather stamped in the Italian style) offered by William Scott Morton's Tynecastle Company, Edinburgh, for upper middle-class houses in Britain and America.[7] The Edinburgh home of spirit merchant Arthur Sanderson was fitted out thus in the 1890s and with furnishings reflecting a wide range of western decorative art.

The houses by Alexander 'Greek' Thomson (1817–75) for Glasgow's new mercantile bourgeoisie also enabled clients to display educated taste. His exoticised Classicism was inspired by a range of sources including, from England, the philosophical writings of Edmund Burke about the sublime and beautiful, and the dramatic, apocalyptic paintings of John Martin. By this date the dining room was important as the domestic hub of middle-class intellectual discourse. The decoration of Holmwood House (1857–8) in Cathcart, south Glasgow, reproduced scenes from John Flaxman's published *Illustrations of the Iliad*, where figures in brown were set on a blue ground. The drawing room upstairs – the usual arrangement – had scenes from Tennyson's newly published 'Idylls of the King' painted by Hugh Cameron c 1859–60 (a date that matches the Pre-Raphaelite mural work and fittings of Philip Webb's Red House at Bexleyheath, Kent, for William Morris). Ceiling decorations followed the progression of an evening, with the sun of the dining room yielding to night stars in the upstairs drawing room. This mix of published image and new artistic expression was suited to the social position of Thomson's patron, paper manufacturer James Couper, but may also be seen as a Grecian design counterpart to the Italianate Hall of Ossian painted in 1772 by Alexander Runciman at baronet Sir James Clerk's Penicuik House, Midlothian. At Penicuik the Scottish subject had been painted in a style studied at first hand in Rome. Flaxman's images at Cathcart applied second-hand Greek culture to a smaller, semi-rural residence. Thomson's latter-day Enlightenment balance of mind and emotion certainly found expression in spatial form, but equally in the decorative details of panelled doors, doorframes and fretwork ornamentation. The solo acanthus palmette, a decorative feature much used in Thomson's domestic and ecclesiastical buildings, provided punctuation on stonework and window frames (Fig. 12.4). The home at Cathcart observed the concept of a house as a total work of art, laden with literary and intellectual reference.

Working alongside a growing number of commercial decorators, Regency and Victorian professional artists, often based in Edinburgh, were employed to equip town and country houses with Italianate or French-style decorative panels which were normally worked in the studio on canvas and

Figure 12.4 The drawing room, Holmwood House, Cathcart, Glasgow, designed by Alexander Thomson, 1857. Source: Reproduced by kind permission of the Trustees of the National Trust for Scotland.

let into ceilings or onto walls. These included John Zephaniah Bell (at Muirhouse, Edinburgh), Phoebe Traquair (Aberlour House, Banffshire), William S Black (8 Charlotte Square, Edinburgh) and Mary Hill Burton, John Duncan and Charles Mackie (Ramsay Garden, Edinburgh). Some individual houses, however, were used by British artists as scenic props for mural work. William Bell Scott's 1860s decoration of the staircase of Alice and Spencer Boyd's Penkill Castle, Ayrshire, illustrated a Scots poem, James I's rediscovered 'The King's Quair', in an essentially British Pre-Raphaelite Romantic style. This use of a Scots building to create a visual poem continued at Robert Rowand Anderson's Mount Stuart, Bute, for the Marquess of Bute. Fitted out over the course of a century with diverse rich materials, from marbles and painted ceilings to coloured glass and tapestries woven in Edinburgh at the Butes' own Arts and Crafts workshop, Mount Stuart maintained a Renaissance tradition of patron and artist collaborating on the creation of the house as a work of art.

Where the Scottish dwelling, as site of artistic expression, has contributed to European culture is in the depth of its intellectual synthesis of space and surface. In the finest examples, the home is presented as more than decorated structure but as a balance between culture and nature. The designs of Robert Adam (1728–92) for Culzean Castle, Ayrshire, or of his brother James for tenements in College Street, Glasgow, symbolised the harmony and modernity of the Scottish Enlightenment. Charles Rennie Mackintosh (1868–1928), like Thomson, also may be viewed as a latter-day Enlightenment designer who worked with light and form to thread together imagination and the empirical truth of the natural world.

Thomson and Mackintosh both produced intellectually synthesised design in the Victorian and Edwardian periods. Mackintosh's client for his most celebrated house, Hill House (1902–4), at Helensburgh, Dunbartonshire, was the publisher Walter Blackie whose father and uncle had occupied Thomson houses. What gives Hill House its character is its imaginative dialogue of intimate and open forms and spaces, in part managed through wall stencilling and carpet design and in part through the subtlest workings and textures of dark and light furniture and walls. By the mid nineteenth century, with both leisure and work in modern society securely located outside the home for urban upper middle-class men, such as Blackie, the dwelling was regarded as the antithesis of the male workplace and thus 'the province of women and children'.[8] While it is true that British, including Scottish, middle-class homes were generally gender coded, with lighter colours used for areas such as drawing rooms and bedrooms associated with the female members of the family, and darker tones for those used by males, such as libraries and dining rooms, the contributing elements were complex.[9] For a family home, such as Hill House, Mackintosh in fact blended gendered details and colours within individual rooms and in specific items of furniture.[10] Creamy, 'pure' lightness might have echoed the current fashion for *dixhuitième*-style interiors, but equally reflected publicised links between cleanliness and health (Fig. 12.5).[11]

Figure 12.5 Mrs Blackie's bedroom, Hill House, Helensburgh, Dunbartonshire, designed and furnished by Charles Rennie Mackintosh, 1902–4. Source: Reproduced by kind permission of the Trustees of the National Trust for Scotland.

If Mackintosh's creams orchestrated his interiors, calling into play pinks, greens, blues and silvers as well as the dark stains of some furniture, then elsewhere so also did wall, window (curtaining or blinds, plus garden), floor and ceiling tones shape a room. 'Scotland's first interior decorator', David Ramsay Hay, was initially employed by Sir Walter Scott at Abbotsford in the early 1820s.[12] Hay rejected the late eighteenth-century neo-classical fashion for light, sunny or muted plain colour washed walls and white paintwork in favour of decorative painting imitating wood or even leather. This type of work developed into the popular painted 'wood graining' with which the doors, dados, even white marble mantelpieces, of late nineteenth- and early twentieth-century homes were decorated. Hay went on to produce new colour theories for interior work. His contrasting room colours worked alongside the assorted arrangements of furniture and objects that typified the mid Victorian middle-class room. The creation of the bourgeois home was encouraged by the publication not only of books but also of 'taste' journals, the increasing availability of urban professional decorators now with their own showrooms, and the provision of exhibitions featuring Scottish, Empire and foreign design (see Fig. 21.1).

The 1901 International Exhibition held in Glasgow showcased the new Glasgow interior. The pavilion presented by local store Wylie & Lochhead was selected for the British Arts and Crafts Exhibition in Budapest the following year.[13] The Glasgow 'notion', to quote the contemporary German commentator Hermann Muthesius, was 'a unified organic whole embracing colour, form and atmosphere'.[14] The extent to which the public adopted the 'new-art' style was varied. A single object might suffice, but many Glaswegians lived with *art-nouveau* fixtures. As well as having stair and room windows fitted with painted glass, communal entrances to many tenements were lined with ceramic tile dados. 'Wally closes' featured tiles bordered in pink, yellow or green *art-nouveau* flowers.[15] Most middle-class tenement housing in other cities also had closes hygienically supplied with coloured tiles.

Materials, manufacture and retail
By the end of the Victorian period the choice of commercial decorative items available to the middle-class householder in Scotland was vast (see Fig. 29.4). It included wallpapers supplied by British manufacturers such as Jeffrey & Co in Manchester or Hayward & Son and Morris & Co in London. Some papers were designed by Scots. William Black, for example, worked for Jeffreys in the 1880s. Heavy embossed friezes, apart from Tynecastle Tapestry, Edinburgh, were supplied from England by Anaglypta and Lincrusta Walton. In printed textiles, G P Baker, Morris & Co and Alexander Morton & Co of Darvel, Ayrshire, all supplied upmarket cottons and linens, with Morton and Templeton of Glasgow producing quality rugs and carpets to designs by local or English designers. Ceramic tiles for fireplaces, kitchen ranges and ceramic bedroom door furniture (fingerplates and handles) were provided by a number of Midlands suppliers. Metalwork, including lamp fittings, door furniture and fireside fenders, were mostly of Birmingham or London manufacture, with some decorative cast ironwork umbrella stands, bootscrapers and other moveables made in Scotland by the Falkirk Iron Company or Walter Macfarlane & Co of Glasgow.

The term 'conspicuous consumption' was first used in 1899.[16] Until this date the amount of house refurbishment which regularly took place was strictly limited, with curtains, for instance, remaining in use for as long as 30 years. While the surge in decoration in the last quarter of the nineteenth century was certainly a matter of taste, it resulted from a growth in the number and type of urban outlets selling household goods. Some older medium-sized drapers, such as J & W Campbell (founded 1817) of Glasgow, had extended their range of goods to become stores serving the wider home needs of an increasingly urbanised population.[17] The development also of multiple shop retailers catering for both trade and domestic customers set them alongside the growing number of stores such as Anderson's Royal Polytechnic (1845) in Glasgow which gradually increased its departments over the next 40 years. The Scottish Co-operative Wholesale Society (1868), like its English predecessor (1863), enabled the new local Retail Societies to acquire and offer foods, clothing and household goods, including furnishings, at modest prices with loyalty dividend repayments. In addition, the sewing-machine trade, started in 1856 with the opening of the first British shop in Glasgow by the Singer Manufacturing Company,[18] allowed dress and furnishing textiles to be made up cheaply and thus changed more frequently. All these shops served different areas of the market, but their success depended on the expansion of public transport, middle-class aspirations and the new 'leisure' time afforded the 'lady of the house'.

In the twentieth century, shop types again changed, particularly over the final 30 years. Further urbanisation and mass building in the 1920s and 1930s led to an increase in both specialist shops (including ironmongers, and fabric, paint and tool retailers) and the variety retail store. The enthusiasm for do-it-yourself house improvements in the post-War years was encouraged by new popular magazines, 'ideal home' exhibitions – part-funded by the press – and television and radio programmes on home and garden improvement. It also overlapped with the mid 1960s expansion of interior-design courses at British art and technical colleges.

With more women working outside the home and increased disposable family income, the interior of many dwellings could be made more comfortable through the installation of fitted kitchens, carpets and central heating. Home 'improvement' reflected changing family living patterns and opened up opportunities for the younger generation. Where a family had used only public rooms by day, the fitting of coal, oil or gas-fired central heating or electric storage heaters meant that 'teenagers' could spend more time in their own bedrooms, which they often individualised with wall posters of 'pop' musicians or other figures. The recent rise of out-of-town shopping has provided Scotland with 'green-belt' superstores catering for the mass consumer, who may subscribe to home magazines, watch television programmes, and surf the Internet for advice.

ORNAMENT AND FURNISHING

'Artistic expression' does not involve only the semi-permanent decoration of a domestic space. A temporary festive display within the home to commemorate a religious festival, a personal anniversary or a national celebration such as the end of war are manifestations of the same phenomenon. Display tokens of celebration, be they Christmas-tree decorations, Christmas and birthday cards, balloons or banners, have been purchased from stores and specialist retailers since the late nineteenth century. Their domestic arrangement is still one of the most common visible signs of artistic expression.

Personalising space
There are many other ways to personalise space. Creating a sanctum is normally a material activity, although a recent study of households occupying rented and owner-occupied post-War flats and houses in the Glasgow area has concluded that where lack of space in a home does not allow this, women especially create personalised, mental space through physical 'busyness'.[19] But for most people in the western world, Scots included, the conscious arrangement of objects is a personalisation of space which turns a house into a 'home', and this most basic domestic instinct has many forms of expression. The artistic free development of space became visible in Scotland in the first quarter of the nineteenth century (and a little earlier

in continental Europe) with the loosening of furniture display, releasing items permanently from their previous formal placing against room walls. The introduction of such informality coincided with the parlour giving way to the drawing room and its complement, the dining room, among the middle classes as prime space for social intercourse, although the parlour in fact lost little of its popularity as the 'best room' for both lower middle-class and well-off working-class tenement occupants throughout the century.

Objects and identity
Time may colour our understanding of an interior in different ways. The sharp distinction between a house filled over a period and one speedily furnished, particularly with modern pieces, has been viewed in class mobility terms since the rise of the industrial bourgeoisie. Since the nineteenth century, novelists, whose writings in fact may often reveal (at times indirectly) far more than a surviving object or illustration about social mores and class moralities, have also echoed and encouraged these distinctions. In the 1820s Susan Ferrier was the first female Scottish novelist to comment on society attitudes. A century later 'O Douglas' (Anna Buchan) entertained middle-class female readers with caricatures of interiors and class in a gentle narrative of manners, property and possessions. In *The Proper Place*, Rutherfurd, an ancestral Borders country mansion, is sold by its rural *nouveau pauvre* gentry to an upwardly-mobile Pollokshields, Glasgow family who view it and its historic contents as a museum of a house, not a comfortable home.

How a room is filled with artefacts has always been a matter of taste. In the most celebrated fictional antiquary Jonathan Oldbuck's private 'sanctum sactorum':

> One end was entirely occupied by bookshelves, greatly too limited in space for the number of volumes placed upon them, which were, therefore, drawn up in ranks of two or three files deep, while numberless others littered the floor and the tables, amid a chaos of maps, engravings, scraps of parchment, bundles of papers, pieces of old armour, swords, dirks, helmets, and Highland targes ... The floor, as well as the table and chairs, was overflowed by the same mare magnum of miscellaneous trumpery, where it would have been as impossible to find any individual article wanted, as to put it to any use when discovered ... [20]

But Oldbuck's clutter of relics of historical, political or academic achievement, the end result of many pleasurable quests for individual artefacts, was not so unusual among professional artists and antiquarians in the nineteenth century. Artist Waller Hugh Paton recorded his father at home near Dunfermline, Fife, in 1848, in an interior which if more organised was equally object-centric and in some ways a condensed version of Abbotsford. Such a real composite interior may be viewed quite differently by its creator and by others not witness to its accumulation. Objects change their meanings through time and according to shifts in knowledge and taste. Heirlooms, for instance, eventually lose their contexts. Obviously an interior as an entity also evolves its own character as old objects are discarded and new objects are acquired.

Some historic interiors were recorded by watercolourists and by social and documentary photographers. One such room was in the lodge Corriemulzie Cottage on Deeside, refitted for Victorian visitors and described in a gazetteer as a cross 'between a Swiss chalet and an Indian Bungalow'.[21] An 1863 photograph of the drawing room shows a plain country upper-class room furnished entirely with modern materials and pieces. Side tables in mahogany had Jacobean twist legs and buttoned upholstered chairs were arranged informally. From a contemporary watercolour, the curtains are known to have been yellow patterned in green with undercurtains of Swiss net.[22] The *moiré* wallpaper had a broad botanical frieze border which also framed an overmantel mirror. The marbleised mantelpiece was topped by a fashionable velvet cover on which several figures, probably ceramic, were arranged. But what located the room nationally was the arrangement of stuffed stags' heads on each wall – not the massed antler trophies papering many a Baronial entrance hall, but ornaments of power over nature fitted around paintings and prints: a further herd lined the walls of the dining room. In this aspect of the decoration, the room was far from unique (Fig. 12.6). It echoed a number of other Highland sitting rooms, including one at nearby Balmoral, recorded by artist James Giles prior to modernisation for Queen Victoria.

Figure 12.6 Drawing room, Corriemulzie Cottage, Deeside, in 1863. Crown Copyright: RCAHMS. Source: Gow, I. *The Scottish Interior, Georgian and Victorian Décor, etc*, Edinburgh, 1992, 83.

The upholstered chairs at Corriemulzie Cottage, ostensibly created for comfort, expressed the confidence of British high-Victorian society, in their individual form and weighty occupation of the total room space. Similar displays may be seen in other rooms photographed in the 1860–90 period. Their typically informal arrangement was replaced in the twentieth century by the invention of the readymade suite of a pair of sprung chairs and matching sofa, advertised in department stores in tight square room settings. Intended equally for the small-roomed bungalow, modern cottage or council house and the older residence, they maintained their popularity throughout the century with traditionalists of all classes. But from the later 1960s, with the increased employment of women and greater disposable income, central heating could abolish the real need for such formal arrangements.

Regardless of room layout and usage, the domestic 'altar', where family objects and icons may be openly arranged and displayed, has remained at the core of the Scottish dwelling. The dressed mantelshelf, emphasising the hearth as the heart of the home, remained the focus of displays of assorted ornaments in the century between 1850 and 1960 (Fig. 12.7). In more recent homes, the top of a television set, sideboards and

Figure 12.7 Parlour mantelpiece display at 111 New Street, Fisherrow, Midlothian. Source: SLA 59/42/4.

shelves above heaters have provided additional or alternative surfaces for family celebratory and portrait photographs and artefacts. Where space and funds allowed, from the 1870s credenza and wall and floor china-display cabinets were added. Glass, serving as dust protector, also applied museum (and thereby communal and educative) values to the Victorian public room. Glass vitrines or domed 'shrines' containing stuffed birds, clocks or china figures occupied many a surface from the 1860s to the 1880s, including the mantelshelf. Frequently this shelf, perhaps dressed in velvet or brocade edged in crocheted lace or tassels, was accompanied below by a dressing of fire irons and a smart cast-iron or brass fender. In this age of photography, it was surmounted by a mirror, sufficiently large and wide in more affluent homes to reflect a living image of the family and their amassed possessions. The direct identification of artefact arrangements with their owners was repeated through visual links between fashion and interior design, and in the growth of portrait photography.[23]

Such an ideal of home and hearth as focus and symbol of the family was repeated across the land and throughout society. A crofter's daughter recalled her family's 1900s Lewis croft-house parlour in which the fireplace had 'tile irons and the mantelpiece had a fringe, drapery, with little bobbles. Well, we had that and we had a lovely mirror, ornaments with jingle bells, you know, tassels, glass things ...'.[24] The description of this interior also demonstrates the availability of furniture and fittings and the pride the family took in them. The main bedroom had a 'lovely grate', and a fine circular mahogany table, while the parlour downstairs echoed earlier upmarket homes. It was furnished with 'a carpet on the floor and it had a sofa! A sofa! Padded chairs, you know. None of your wooden chairs ... it also had a rocking chair, the style of it, you have no idea!' A second bedroom had another table and 'a lovely clock'. The parlour however had 'a very special French clock, had a figure stuck on it ... you must not forget that my grandfather could tell the time from the sun, when the sun hit a certain point every Saturday, and the clocks were wound and adjusted'.[25]

The readymade and the homemade
While the display of possessions as part of 'conspicuous consumption' might indicate either inherited or accumulated wealth, this description demonstrates the personal satisfaction which an object may endow on a room's user, quite regardless of monetary value. For all classes of Scottish society the moveable non-functional artistic object intended to ornament a dwelling, that is, the 'ornament' – bought new or second hand – was an essential part of the proper fitting of a public room by 1880. Ornaments fell into three main categories: ceramic wares (from the 1880s to World War I blue and white ware was popular among all classes); embroidered textiles; and painted or, more commonly, engraved wall pictures. In the 1900s Neil Munro's character Erchie could define them as 'wally dugs on the mantelpiece, worsted things on the chairbacks, a picture o' John Knox ower the kist o' drawers'.[26] The popular Staffordshire (or Scottish) 'wally dog' was a moulded and decorated

ceramic ornament sold as pairs and in different sizes through local purveyors of household goods from Stranraer to Lerwick. Their popularity encouraged the creation of 'Wemyss-ware' cats and pigs. Steel engravings and chromolithographs were bought through both market traders and specialist art dealers. But textile decorations, long since a female discipline, always remained homemade, with matching embroidered tablecloths and serviettes representing popular subjects of middle-class skills from the mid nineteenth century. Small wall textiles included the stitched linen sampler of the eighteenth and early nineteenth centuries, an artwork in which the maker both demonstrated her needle proficiency and directly iconicised herself in place and time. By the 1880s this had been supplanted by Berlin 'art' woolwork or useful tea stands and cosies fashioned in beadwork, craft forerunners of the twentieth-century interclass homecraft movement which crossed the boundaries between art and function.

A sense of evolution within domestic space has been presented by creative handcraft over the last century and a half. Nature and its values, patterns and colours have assumed a place within most home environments. The popularity of the public park and garden forged a wide interest in plants and in their diversity, colours, growth and change. Within the late Victorian drawing room or parlour, prominent space was occupied by pots in ceramic or in metal, particularly brass and copper, placed on legged or tabletop wooden stands. Worked in beaten soft metals these stands, together with decorative window boxes, were popular domestic crafts learnt at urban Arts and Crafts classes in the 1890s and 1900s. Containing houseplants such as ferns, aspidistras and geraniums they created a personalised and satisfying garden inside the ordinary home. In addition, the now inexpensive flowering spring bulb both celebrated the seasons and lent colour and perfume to a room. The growth, care and positioning of old and new species of household plants has remained a classless area of personalised expression in the home, with each wave of handcraft fashion incorporating appropriate new accessories. Thus in the late 1960s and 1970s, white cotton *macramé* holders, made in the home to patterns published in craft magazines, were often suspended from ceilings or from painted steel or plastic wall brackets. Maintaining a low-key position in the subsequent 20 years, in the last two decades plants again have occupied centre stage in many middle-class homes. Encouraged by the growth of the domestic conservatory room in the mid to late 1990s, the 'home improvement' do-it-yourself store located in today's out-of-town urban 'green belt' nearly always now includes a garden centre selling bulbs and plants, tools and home display utilities. As is the case with other domestic artefacts, the supply of these goods is a matter of both creation and satisfaction of consumer demand.

ARTISTIC ENTERPRISE AND HOME IMPROVEMENT

Almost every Scottish home was supplied with 'special' artefacts in the course of the last century, and Victorian display principles have survived to the present day. Some objects were collected, others were made for family use, all were arranged, and not necessarily by women. The 'best' room of a croft house in Shetland, for example, might display souvenirs brought home by the fisher head of the house, including Russian wooden bowls from the Baltic, 'Riga bowls', as well as a pair of 'wally dogs'. Model ships made on voyages might also be on show in this part of the dwelling which also housed the sleeping quarters.

Other, more functional handmade items were used in the 'best', or 'ben', room including the handtufted woven rug, or 'taat', in bright colours and geometric patterns which formed the bed coverlet, accompanied sometimes by a woollen quilt recycling worn garments.[27] 'Clouty' rugs for floor or door, and 'clouty coverins' for beds made from rags furnished many Scottish homes, using up woollen or tweed clothes fragments, stockings, even hats, all of which were frequently redyed in the brightest colours possible from natural tree bark and plants.[28] These rugs were looped or embroidered, often by men, and lined before use. Uncut looping gave a particularly firm surface for a floor clouty. The Shetland taat, while part of this tradition, was also related to Scandinavian rugs, notably the Finnish 'ryijy', but unlike the latter did not also serve as a floor cover. In the everyday 'but' end, however, was one form of artistic arrangement ubiquitous across Scotland. The dish rack, developed also into the display dresser in many homes, stored and displayed everyday plates and other items placed decoratively and functionally in a forward position to keep them clean and to prevent slippage (Fig. 12.8).

In the Shetlander's croft house, household goods were made for both use and profit. Horn spoons were carved, garments including shawls were knitted, and wool carded and spun. These were all artistic but also quotidian, communal activities. At the opposite end of the creative spectrum, the craft of making beautiful, but not necessarily functional display objects within the home was part of the middle-class Arts and Crafts ethic in the first four decades of the twentieth century. Glasgow School of Art taught modern crafts in its Technical Art Studios. Women who had trained there taught and also produced beaten *repoussé* metalwork boxes and panels, wood staining and chipcarving, pokerwork, textile embroidery, and glass and china painting. Their goods were sold through charity and professional exhibitions. Across Britain, home crafts were further encouraged from the 1920s by the Leicester company Dryad through its new materials and tools mail-order service. In addition, a range of British magazines such as *Bestway* and *Weldon's* promoted home crafts with issues focusing on popular crafts such as tooled and embossed leatherwork (magazine covers, calendar backs, and blotters as well as bags and purses), rugmaking, basketry, shell work, enamelling on glass, lacquer work, and, not least, raffia, a particularly versatile material which could be turned into embroidered cushions, chairbacks, wall pictures, tray cloths and dinner mats. These objects, for some of which patterns were supplied, were recommended for the modern, smaller British home of 'bungalows, country

Figure 12.8 The 'but' end of a South Uist croft with dresser made by Archibald MacPherson and display of household ceramics. Sources: South Uist Museum. Photograph: SLA 9/19/8.

and seaside cottages', and as such, they fitted many of the dwellings of rural and urban Scotland.

Scotland's textile industries responded to the home crafts movement, particularly the fashion for making clothes and household embroidery including table runners, cushions, chair backs and arm rests. The amalgamated Paisley company of Coats and Clark published pattern brochures and advertised that they could supply every kind of thread for 'every kind of stitch'. Patons and Baldwins of Alloa, Clackmannanshire, published brochures for 'tapestries old and new' which included patterns for embroidered panels to fit stools, chair seats and pole screens in the 'Louis XVI' revival style. In Dundee, Donald Brothers offered canvas for home and shopping bags as well as pattern leaflets. Their 1930s 'Old Glamis' range of decorative fabrics, needlework and 'Old Glamis Helvelyn' rug canvas promoted a sense of tradition. A typical 'Old Glamis' brochure was illustrated with detailed demonstration photographs, arranged by Ann Macbeth, doyenne of the Glasgow Style art-embroidery movement. As well as using a developed postal delivery service, materials and brochures were sold through haberdashers and department stores, with multiple shop city outlets such as John Smith & Co (Wools) Ltd of Edinburgh and Glasgow expanding to meet demand. The Smith company was one which issued the hugely popular 'Fairistytch art embroidery' floral and nursery printed domestic linens throughout the 1920s and 1930s.

These promotions of decorative crafts matched the expansion of suburban housing before and after World War II. The modernist aesthetics promoted by the 1951 Festival of Britain were those of harmony, with a range of new colours (cherry red, mint green, sunny yellow, sky blue, with black and cream) used by many manufacturers to create a sense of well-being and aid post-War recovery. Again firms like Dryad and magazines reached out across the nation, with the Fred Aldous Co's 'Atlas' leaflets promoting home-delivered goods such as 'lampshade craft' using natural parchment or synthetic 'crinothene' materials. Areas of the country served by television transmitters received craft programmes led by F J Christopher, editor of the popular *Craftworker* magazine for the amateur and the new professional. Magazines and programmes both stressed family values. Wood veneer pictures, carved meerschaum figures and wooden musical boxes were particularly popular. The name of the paper *Popular Handicrafts and Practical Homecrafts* epitomised the invisible line between creative crafts and home improvements, and as a fund of new or revitalised companies, mostly based in England, offered goods from wall tiles to rug-making kits. New materials to improve the home included Thomas De La Rue & Co's 'Formica', a kitchen surface laminate fitted in 'half an hour's pleasant work'. While such activities were more British than Scottish, the philosophical, social and economic values of making artefacts in and for the home had been localised and Edinburgh's major exhibition for the Festival of Britain, *Living Traditions of Scotland* presented modernity in terms of heritage and modern craft.

This personalisation of domestic space in the post-War period was advertised as an ideological and, not least, financial investment. This was not so very new. The slogan of 'Fairistytch art embroidery' had been 'buy Fairistytch and prosper'. By the 1960s 'craft' improvements to the home included indoor clad panelling for which kits of pine or cedar were available. Weatherproof mock shutters on the outside of a window added 'dignity and charm' but also added '£s to the value' of a home as well as expressing a degree of urban nonconformity. 'Larch Lap' fencing and concrete 'Portland' screenwalling blocks created privacy in the garden and granted variation to a street of private housing. The choice of a fireplace by 1970 stated background, income and aspiration. Adam-style mantelpieces contrasted with the 'hand-finished' 'Minster' reconstructed-stone fire surrounds and kits which could extend a fireplace across an entire wall. Marble with onyx panel inlay and a steel surround for an electric fire was also popular in suburban Scotland, but few suppliers were Scottish. Within the 1960s and 1970s 'living' room, heterogeneous assemblies of furniture and colours, with adjacent walls dramatically coloured and textured, opened up interior design and artistic expression. While householders bought furniture as either reproduction or modern design, the growth of the conservation movement and of folk culture during the 1965–75 decade – in the aftermath of urban redevelopment across the country, particularly in Glasgow and Dundee – helped produce a taste for historic authenticity among some younger middle-class professionals. Related to the folk-culture movement, the anti-consumerist 'good life' ideals of the later 1970s and 1980s also expanded the rural craft movement, assisted financially by the Highlands and Islands Development Board and the Scottish Development Agency.

With revitalised urban and, to a lesser extent, rural employment from the 1980s, and greater personal wealth, including increased home ownership among the working classes, the last twenty years have witnessed a relatively dramatic increase in structural and other major alterations to the fabric of homes. This has facilitated an expansion in domestic artistic expression. The early 2000s growth in home magazines, allied to the perennial Scottish Ideal Homes Exhibition and Royal Highland Show, multiple DIY-shop expansion and various new television programmes, has encouraged increasingly frequent redecoration and refurbishment of property. National awareness and confidence has reinvigorated the conservation movement with its validation of the past. With allied advances in museum presentation of 'material culture', the intellectual values we again place on historic objects may also now influence the way we seek to organise and augment our surroundings. But the modern Scottish home still expresses and nurtures our wellbeing, a double – even reciprocal – process which Kant recognised as creating an essential 'internal order' and a 'civility' imbued with 'a passion of its own'.[29] And if a dwelling essentially is the 'social organisation of space',[30] then decoration and ornamentation will continue to personalise this space to reward our individual and collective imaginations.

NOTES

1. Carswell, 1997, 30.
2. Perrot, 1990, 344, 1.
3. Carruthers, 1996, 22–3, designs were recorded by the Scottish Rural Women's Institute in the 1930s.
4. Apted, 1966, 69.
5. Marshall, 1983, 52.
6. Mackay, 1995, 39.
7. Bowe, Cumming, 1998, 60.
8. Kinchin, 1996a.
9. See for instance Kinchin, 1996b.
10. Kirkham, 1996a.
11. Laermans, Meulders, 1999, 120.
12. Gow, 1992, title page, 40.
13. Kinchin, Kinchin, 1988, 68.
14. Muthesius, 1987, 51.
15. Worsdall, 1991, 32.
16. The publication date of social economist Thorstein Veblen's *The Theory of the Leisure Class: An economic study of institutions*. See Cieraad, 1999, 120.
17. Jeffreys, 1954, 19.
18. Jeffreys, 1954, 401.
19. Madigan, Munro, 1999, 115.

20. Scott, c1995, 21–2.
21. Gow, 1992, 83.
22. Thornton, 1984, 292.
23. Kinchin, 1996a, 170–1.
24. Jamieson, Toynbee, 1992, 37–8.
25. Jamieson, Toynbee, 1992, 38.
26. Munro, 1993, 102; See Chapter 9 Needlework for further information on homemade ornamental textiles.
27. Tait, 2000, 53–4.
28. Murray, 1960, 27.
29. Edelman, 1984.
30. Cieraad, 1999, introduction, x, John Rennie short.

BIBLIOGRAPHY

Apted, M R. *The Painted Ceilings of Scotland, 1550–1650*, Edinburgh, 1966.
Bath, M. *Renaissance Decorative Painting in Scotland*, Edinburgh, 2003.
Bourdieu, P. *Distinction: A Social Critique of the Judgement of Taste*, London, 1984.
Bowe, N G, Cumming, E. *The Arts and Crafts Movements in Dublin and Edinburgh, 1885–1925*, Dublin, 1998.
Busch, A. *Geography of Home: Writings on Where We Live*, New York, 1999.
Carruthers, A, ed. *The Scottish Home*, Edinburgh, 1996.
Carswell, C. *Open the Door!*, London, 1920.
Carswell, C (ed J Carswell). *Lying Awake: An Unfinished Autobiography and Other Posthumous Papers* (London, 1950), Edinburgh, 1997.
Cieraad, I, ed. *At Home: An Anthropology of Domestic Space*, New York, 1999.
Donnelly, M. *Scotland's Stained Glass: Making the Colours Sing*, Edinburgh, 1997.
Douglas, O. *The Proper Place*, London, 1926.
Edelman, B. *La Maison de Kant*, Paris, 1984, 25–6. Cited in Perrot, M, ed, *A History of Private Life, vol IV: From the Fires of Revolution to the Great War*, Cambridge (Mass) and London, 1990, 342.
Fenton, A. *Country Life in Scotland: Our Rural Past*, Edinburgh, 1987.
Gow, I. The Northern Athenian Room, *ROSC*, 5 (1989), 1–12.
Gow, I. *The Scottish Interior: Georgian and Victorian Décor*, Edinburgh, 1992.
Gow, I, Rowan, A. *Scottish Country Houses, 1600–1914*, Edinburgh, 1995.
Hay, D R. *The Laws of Harmonious Colouring: Adapted to Interior Decorations, Manufactures, and Other Useful Purposes*, 4th edn, Edinburgh, 1838.
Jamieson, L, Toynbee, C. *Country Bairns: Growing Up, 1900–30*, Edinburgh, 1992.
Jeffreys, J B. *Retail Trading in Britain, 1850–1950*, Cambridge, 1954.
Kaplan, W, ed. *Charles Rennie Mackintosh*, Glasgow and New York, 1996.
Kinchin, J. The drawing room. In Carruthers, A, ed, *The Scottish Home*, Edinburgh, 1996a, 161.
Kinchin, J. The 'masculine' and the 'feminine' room. In Kirkham, P, ed, *The Gendered Object*, Manchester, 1996b, 12–29.
Kinchin, P, Kinchin, J. *Glasgow's Great Exhibitions: 1888, 1901, 1911, 1938, 1988*, Wendlebury, 1988.
Kirkham, P. 'Living Fancy': Mackintosh furniture and interiors. In Kaplan, W, ed, *Charles Rennie Mackintosh*, Glasgow and New York, 1996a, 227–61.
Kirkham, P, ed. *The Gendered Object*, Manchester, 1996b.
Laermans, R, Meulders, C. The domestication of laundering. In Cieraad, I, ed, *At Home: An Anthropology of Domestic Space*, New York, 1999, 118–29.
Logan, T. *The Victorian Parlour: A Cultural Study*, Cambridge, 2001.

Mackay, S. *Behind the Façade: Four Centuries of Scottish Interiors*, Edinburgh, 1995.
Macmillan, D. *Scottish Art, 1460–2000*, Edinburgh, 2000.
Madigan, R, Munro, M. Gender, house and 'home', *Journal of Architecture and Planning Research*, 8/2 (1999), 116–32.
Marshall, R K. *Virgins and Viragos: A History of Women in Scotland from 1080 to 1980*, London, 1983.
Munro, N (intro B D Osborne, R Armstrong). *Erchie and Jimmy Swan*, Edinburgh, 1993.
Murray, J. Clouty rugs and coverings, *Scotland's Magazine* (June 1960), 27–9.
Muthesius, H (ed D Sharp). *The English House (Das englische Haus)*, London, 1987.
Parker, W M. *Dobie and Son Ltd, 1849–1949*, Edinburgh, 1949.
Perrot, M, ed. *A History of Private Life, vol IV: From the Fires of Revolution to the Great War*, Cambridge (Mass) and London, 1990.
Putnam, T, Newton, C, eds. *Household Choices*, London, 1990.
Robinson, P. Tenements: A pre-industrial urban tradition, *ROSC*, I (1984), 52–64.
Rybczynski, W. *Home: A Short History of an Idea*, Harmondsworth, 1987.
Savage, P. *Lorimer and the Edinburgh Craft Designers*, Edinburgh, 1980.
Scott, W (ed D Hewitt). *The Antiquary* (1816), Edinburgh and Columbia, c 1995.
Snodin, M, Howard, M. *Ornament: A Social History since 1450*, London, 1996.
Stamp, G, McKinstry, S, eds. *'Greek' Thomson*, Edinburgh, 1994.
Stevenson, S, et al. *Light from the Dark Room: A Celebration of Scottish Photography*, Edinburgh, 1995.
Swain, M. *Scottish Embroidery Medieval to Modern*, London, 1986.
Tait, I. *Rural Life in Shetland and Guidebook to the Croft-House Museum*, Lerwick, 2000.
Tarrant, N. *Textile Treasures: An Introduction to European Decorative Textiles for Home and Church in the NMS*, Edinburgh, 2001.
Thornton, P. *Authentic Décor: The Domestic Interior, 1620–1920*, London, 1984.
Wainwright, C. *The Romantic Interior: The British Collector at Home, 1750–1850*, London and New Haven, 1989.
Worsdall, F. *The Glasgow Tenement: A Way of Life: A Social, Historical and Architectural Study* (1979), Edinburgh, 1991.

13 Spirituality

DAVID REID

INTRODUCTION

This chapter does not attempt to be an exhaustive account of spirituality in the Scottish dwelling nor does it claim to be an in-depth exposition. Its intention is to present some of the manifestations of domestic spirituality and piety which have been practised in the homes of the people of Scotland from early times until the present day and which have been recorded in written form. It deals with the expressions of domestic spirituality of the Christian mainstream and cannot examine all the multifarious expressions of spiritual observance extant in Scotland.

> Our spiritual life is the only precious thing we possess. It is that without which all else is dust and ashes. It is that without which, in spite of all our academic knowledge, we shall become, in St Paul's own words, as sounding brass or as a tinkling cymbal.[1]

DEFINITION

The word 'spirituality' is in frequent use nowadays. People may aver that while they may not be religious they are certainly spiritual. However the word does not have a long history. The *Catholic Encyclopaedia*, published in 1907, makes no mention of it but the *New Catholic Encyclopaedia* of the 1970s has eight articles which use that word. In one of the most up-to-date dictionaries spirituality is defined as follows: 'Spiritual quality: the quality or condition of being spiritual'.[2] It is listed by Collins under 'spiritual' as a noun, with no definition.[3] It is also defined as, 'Of the soul, relating to the soul or spirit, usually in contrast to material things'.[4]

When 'faith' is mentioned in the Gospels it refers to spirituality; it means trust in God, belief in and devotion to God.[5] Christians are instructed to let their light (their faith) shine before men, they are to set it on a candlestick and not to hide it. In the Epistles they are told that they must be able to show concrete evidence of their faith. This being so then spirituality must be capable of demonstration. The method by which faith and spirituality are expressed by those whose lives have not been specially set apart (as the lives of monks, for example, are) is through 'piety', defined as: *'religious devotion, strong respectful belief in a deity or deities and strict observance of religious* principles in everyday life; or *devout act*, an action inspired by devout religious principles'.[6]

One of the main problems of recovering the evidence of the practice of spirituality in the dwelling is that accounts have concentrated on the development of relatively sophisticated ideas about Christian life, 'a spirituality of ideas is inherently élitist and so the religious world of the ordinary Christian scarcely found a place in such accounts'.[7]

ANCIENT SPIRITUALITY

We have little or no evidence of spirituality or piety in the homes of our remote ancestors. There is no doubt that they were religious or had religious beliefs, as is evidenced by prehistoric temples, sacred circles, ritual burials, etc, but what went on in the houses of villages such as Skara Brae or in the dwellings in and around the brochs we cannot say.

In Biblical times, we know that the Hebrews in captivity in Babylon and with no access to their place of worship in Jerusalem succeeded in preserving the continuity of their worship through the new medium of the synagogue.[8] This change was probably facilitated by the already dwelling-based expression of their beliefs and customs acquired in Egypt.[9] In ancient Rome, people worshipped the 'household gods', Janus the god of the door (there was also a god of the hinges), Vesta the goddess of the hearth, the Lares and Penates, who were the gods of the store room, and the Lemures who were the shades of the dead. All of these and more had to be worshipped or placated by each household. Before the introduction of Christianity into Britain, Romanised Britons had adopted the lifestyle of Rome and it is possible that the beliefs of Rome also influenced the Celtic native population. It may even be that in what we now call Scotland, the translocation of Roman household gods had taken place.

Roman influence in Scotland began with the invasions of Agricola in the first century AD. By the fourth century, Christianity, by this time a Roman religion, had certainly established itself around Whithorn. In 1891 a stone pillar was discovered there with the inscription:

> TE DOMINU[S] LAUDAMUS LATINUS ANNORU[M] XXXV ET FILIA SUA ANN[ORUM] IV [H]IC SI[G]NUM FECERU[N]T NEPUS BARROVADI (We praise Thee, O Lord, Latinus aged thirty-five and his daughter aged four [rest here], the descendant of Barrovados made this monument).[10]

This early Christian inscription cannot tell us anything about the spirituality of those to whom it refers or who raised the monument – although the mingling of Roman and Celtic influences can be seen in the names (Latinus being that of a Romanised Briton and Barrovados being British) – but it is an expression of Christian piety and shows that such a spiritual awareness existed among the dwellers in this part of Scotland at that time. The stone is dated as being of the fourth or fifth century.

CELTIC SPIRITUALITY AND ITS SURVIVAL IN GAELIC-SPEAKING AND OTHER PARTS OF SCOTLAND

Around the fourth century there was a tightening of church discipline. In AD 305 the Council of Illiberis in the Iberian Peninsula made observance of the Lord's Day compulsory. There exists a Gaelic manuscript titled 'Càin Domnaig' (Law of The Lord's Day) which dates back to the beginning of the ninth century but was compiled from earlier sources. The Rev Dr Donald MacLean considers the Càin to be older than the Synod of Whitby which took place in AD 664.[11]

The Lord's Day begins at evening prayer on Saturday and lasts until morning prayer on Monday. Remnants of this long Sabbath were still evident after World War II when the swings and roundabouts in children's play parks all over Scotland were immobilised on Saturday evenings and not unlocked until Monday morning. Even in the cities, Saturday-night dances had to end well before the onset of Sunday. In the Càin, work, travel, commerce, shaving, haircutting and other forms of personal grooming were all forbidden. Walking to a place of worship, fetching medical help (including the midwife), tending to the needs of guests, catching criminals and similar works of necessity were permitted. Sabbatarianism, therefore, did not begin in Scotland with John Knox. The Calvinists were merely continuing in the ways of the ancient Celtic Church.

Many of the beliefs and domestic practices of Celtic Christians[12] in Scotland were rescued from imminent oblivion by Alexander Carmichael during the latter part of the nineteenth century. In his monumental collection of folklore, *Carmina Gadelica*, we have a record of the daily spirituality of the laity preserved in prayers, poems and songs.[13] That these survived for so long is surprising, although the scarcity of ministers and priests in the remoter parts of the land may have been influential in that the hierarchy of the Church could not dictate much of daily life as was the case elsewhere.

From Janet Currie of South Uist Carmichael recovered a version of the Càin:

> Duan an Domnuich
> The Lord's Day, the seventh day,
> God ordained to take rest,
> To keep the life everlasting,
> Without taking use of ox or man,
> Or of creature as Mary desired,
> Without spinning the thread of silk or satin,
> Without sewing, without embroidering either,
> Without sowing, without harrowing, without reaping,
> Without rowing, without games, without fishing,
> Without going out to the hunting hill,
> Without trimming arrows on the Lord's Day,
> Without cleaning byre, without threshing corn,

> Without kiln, without mill on the Lord's Day.
> Whosoever would keep the Lord's Day,
> Even would it be to him and lasting
> From setting of sun on Saturday
> Till rising of sun on Monday.[14]

Whether Carmichael's claim that 'poems similar to this can be traced back to the eighth century' is true or not, it demonstrates that Scots have been quite consistent in their exercise of spiritual piety on the Sabbath over many centuries.

Carmichael's collection also reveals that prayer occurred in the dwelling at various times during the day. It began when eyes opened in the morning[15] and at the reviving of the fire on the hearth.[16] Morning ablutions were an opportunity to remember the Virgin and Christ.[17] When darkness fell, the fire on the hearth in the heart of the house was smoored, or covered over with ash, to keep it smouldering until revived the following morning, and this act was also accompanied by ceremony and prayer.[18] Having gone through the day's darg, prayer also accompanied preparations for sleep.[19]

Indeed most aspects of everyday life were covered by prayer. In the byre end of the dwelling, Saint Brendan was invoked to increase the yield of the family cow. The quern was blessed as corn was ground before the fire. The baking and eating of bread was a spiritual act:

> The *bannag* is symbolic of Christ, and is broken and eaten by the family with becoming reverence and solemnity. After the Bannock has been cooked the mother takes up the 'clach bhannag', bannock stone, against which the cake was supported before the fire, and tenderly hands it to her daughters, in emblem of Christ.[20]

Prayer and ritual had particular importance at turning points in the individual's life, the home being the setting for most of these rites of passage. When the woman of the house was in labour with child, Carmichael tells us that the midwife would stand in the doorway of the house and invite Saint Bride (the midwife of Christ) to enter the house:

> Bride! Bride! Come in,
> Thy welcome is truly made.
> Give thou relief to the woman,
> And give the conception to the Trinity.[21]

When the time for death came there were particular rituals and prayers also:

> The soul-peace is intoned, not necessarily by a cleric, over the dying, and the man or woman who says it is called 'anama-chara', soul-friend. The soul-peace is slowly sung – all present earnestly … beseeching the Three Persons of the Godhead and all the saints of heaven to receive the departing soul of earth. During the prayer the soul-friend makes the sign of the cross with the right thumb over the lips of the dying.[22]

Although most of Carmichael's research was conducted among the people of the Western Isles, such expressions of spirituality in the dwelling were not confined to that area. Thomas Pennant, writing in 1769 of customs in the Highlands in general, records that when a corpse was laid out in the house, a platter was placed upon its chest, containing two separate piles, one of earth and the other of salt, the earth representing the corruptible body and the salt the immortal soul. We also learn from Pennant that the lyke wake (corpse watch) was an essential part of the obsequies. This was not a simple vigil of prayer beside the corpse but an elaborate ritual in the dwelling of the deceased and was a ceremony used at all funerals. On the evening of the day of the death, the family and friends of the dead person gathered in his or her house. Some brought their musical instruments with them, usually bagpipes and fiddles. The next of kin led the company in a melancholy and tearful dance around the corpse. This continued until dawn. But, like many of today's after-funeral parties, the wake could take on the character of a 'céilidh'. Pennant states, 'Thus, Scythian-like, they rejoice at the deliverance of their friends out of this life of misery'.[23]

In the highlands of west Aberdeenshire domestic spirituality of a 'Celtic' type was still alive in the nineteenth century:

> [Margaret MacGregor of the Laggan] ... went to Mass every day that it was celebrated and had a seat close to the altar. She was well educated and had many books. Her piety was the admiration of the countryside. All day long she worked and prayed at intervals. She had an hourglass that told her the time for prayer and the time for labour and she passed from her knitting to her prayers as methodically as possible. She composed and repeated constantly Gaelic prayers. I sometimes brought her meal or other food and learnt these prayers from her own lips.
>
> > When you extinguish the light below the ashes,
> > Pray that God who made the world may be near you;
> > Bow your knee to the ground and bethink you
> > That judgement may be closer than the dawn.[24]

An observation by H G Graham shows that the custom of a prayer for every action of the day also survived, partially, into the eighteenth century in the cities of Glasgow and Edinburgh:

> Religious observances attended the simplest acts of social life – not probably with much meaning, but as traditional customs from more fervent ages.
>
> English travellers at the early part of the century were much amused at the frankly pious practices of the people. No refreshment, however slight, could be partaken without a formal blessing being asked. Drinking a glass of ale was preluded by a grace; the progress of a dram to the lips was suspended by the utterance of a prayer –

sometimes of no mean dimension ... 'if you crack a nut with them, there is a grace for that'.[25]

Perhaps the best-known grace in the English-speaking world, and beyond, is the *Selkirk Grace*, written by Robert Burns in the later 1700s. Its fame is due to its being recited at every celebration of his birth:

> Some hae meat, and canna eat
> And some wad eat that want it
> But we hae meat and we can eat
> And sae the Lord be thankit.

Graces remain the norm at wedding receptions, funeral meals, formal dinners, etc, even if used only to give the signal to begin eating. They are also still used in many Christian households, despite the growing infrequency of commensality.

PRE-REFORMATION SPIRITUALITY

Records of the spirituality of ordinary people in the now English-speaking areas of Scotland are sparse. Perhaps, as the Gaelic tongue slowly receded, the memory of the spiritual experience of its speakers went with it. The perceived inferiority of the people, their culture and their language would encourage a forgetting of the past among many who wished to assimilate into the new majority culture.[26]

We know, however, something about the upper classes in these areas. In the dwellings of the nobility there was often a domestic chaplain who celebrated the sacraments of the Church. In less grand dwellings there was usually a place set aside for worship, 'every country house had its small room, or closet, used as an oratory ... '[27] These were often beautified by paintings of religious scenes on the ceilings and walls. The houses of the richer merchants, too, were sometimes similarly decorated. Although the Church had to all intents and purposes kept study of the Bible out of the homes of the people, it had allowed them other aids to spirituality. There was no discouragement of religious pictures or of crosses and crucifixes (Fig. 13.1). These fashions did not die with the establishment of Protestantism. Fifty-two years after the Scottish Reformation, Glasgow Presbytery dealt with three painters who were accused of painting the crucifix in many houses,[28] and Thomas Boston, in his *Memoirs*, speaks disparagingly of 'the Presbyterian silly woman' who had 'a picture of Christ on the cross, hanging on the wall'. This was 200 years after the Scottish Reformation.[29]

It is thought that the use of rosary beads as an aid to private and domestic prayer was encouraged by the mission of Saint Dominic (1170–1221), so they may have been introduced into Scotland by the first Dominican monks to settle here. The beads act as a simple reckoner for formulaic, contemplative prayer (Fig. 13.2). In Provost Skene's House in

Figure 13.1 Crucifix on the wall above the mantelpiece at Sanna, Ardnamurchan, Argyll, 1930s. Source: SLA 128A 51/60/5 D.

Aberdeen there are paintings which illustrate the Mysteries of the Rosary:

> The devotion to the Fifteen Mysteries in which fifteen decades of 'Ave Marias', are recited, each decade being preceded by the 'Paternoster' and followed by the 'Gloria Patri'. Ordinarily only a third part of the Rosary, a so-called chaplet, is said on one occasion. To assist the memory the prayers are commonly counted on a string of beads.[30]

The whole in order is a summary of the lives of Jesus and the Virgin Mary.

Rosary beads would be of little use either publicly or domestically if some instruction in their use was not available. McRoberts assesses that rosary beads were well enough known in seventeenth-century Aberdeen and that books dealing with the devotion would have been circulating among Catholics in the city.[31] A century earlier Catholic tractates could say, 'Being busy with their Rosaries is the proper occupation for right-thinking women.'[32] The importance of rosary beads is shown in the lament of Aithbhreac Inghean Corcadail (c 1460) for her husband, the MacNeil of MacNeil:

> O rosary that recalled my tear,
> Dear was the finger in my sight,
> That touched you once, beloved the heart
> Of him who owned you till tonight.

SPIRITUALITY • 203

Figure 13.2 Selection of personal items, including a wooden rosary, recovered from the sixteenth-century English vessel, the *Mary Rose*. Source: SLA 122B/NMS 29/9/86.

> I grieve the death of him whose hand
> you did entwine each hour of prayer,
> My grief that it is lifeless now
> And I no longer see it there.[33]

The literacy of the upper classes was minimal in the centuries preceding the Reformation[34] and the Church had contributed to this by forbidding the laity to read the Bible, which was practically the only source of reading material. Priests used the confessional as a way of discovering what was going on in the privacy of the home and who was surreptitiously reading, not only the Bible but other banned material. Sir David Lindsay in one of his poems shows a priest questioning a peasant girl:

> Quod he, ken ye na heresie?
> I wyt not qhat that is, quod she.
> Quod he, hard ye no Inglis bukis?
> Quod she, my maister on them lukis.
> Quod he, the bishop that shall knaw
> For I am sworn that for to schaw.[35]

In 1526 Tyndale's translation into English of the New Testament was published in Antwerp. Copies were soon imported into Scotland and:

> consigned to persons of tried principles and prudence who circulated

them in private with great industry. One copy of the Scriptures supplied several families. At the dead hour of night when others were asleep they assembled in a private house, the sacred volume was brought from its concealment and while one read the others listened.[36]

In 1578 there was printed or, more probably, reprinted, the *Gude and Godlie Ballates*. It contains much that is obviously didactic, some of it dating from before the Scottish Reformation.[37] This material, in the hands of the people, would have been of considerable assistance in the practice of their spirituality in the home. In the original editions there are instructions about Baptism and Communion. The Lord's Prayer, the Creed, the Ten Commandments, graces for before and after meals and devotional songs are all included. There is evidence that the pre-Reformation Church was concerned about the people's possession of such literature.:

> ... the Provincial Council, a purely ecclesiastical body, which met in 1549 decreed that each bishop and the head of each monastery should appoint, 'inquisitors of heretical pravity, men of piety, probity, learning, good fame and great circumspection, who should make the most diligent search after heresies, foreign opinions, condemned books ... [38]

The Church was trying to ensure that religious expression and manifestations of private faith continued to be under its own control. Being a priest was no protection. Thomas Forret, parish priest of Dollar, Clackmannanshire, was tried for heresy and denounced for reading the Bible.[39] The reasoning behind the Church hierarchy's discountenancing of the reading of the Bible by the people is summed up thus:

> No doubt it is a laudable thing that a man should aspire to study for himself the oracles of God in Scripture. But the task is so difficult, the possibilities of error so great, and the consequences of error so terrible, that no man should embark on such study unless he has prepared himself for it by a thorough training in theology.

The Reformation was to overthrow such thinking and release the Bible for the use of the people. John Knox wrote a letter in 1556 in which he said that all may sit down with the Bible and that they did not need the guidance of a theologically trained minister to learn what was God's Word.[40]

POST-REFORMATION SPIRITUALITY

John Knox showed by his actions of 1556 that celebration of the Sacraments need not be tied to church buildings and consecrated altars:

> ... his practice was at every possible opportunity to gather the faithful into the drawing room of some gentleman's house, or some other convenient place, and there to administer the Sacrament of the Lord's

Supper ... Wherever two or three could be gathered together, there Knox administered Communion ... [41]

The reformed view of such matters shared in the concept of the sacredness of the common dwelling, as shown by those from whom Alexander Carmichael collected his oral histories in the nineteenth century. Schmidt claims that:

> The reformers had dismantled the rituals of community of the mediaeval church without creating replacements ... As sacramental festivity diminished, family devotions grew in importance; religion became more of a domestic affair than it had been before, more private in its emphases.[42]

However, some of the disciplines of the Roman Church lent themselves to retention and adaptation by the Reformed Church:

> Sacramental fasts in various forms among the Scottish Protestants – on the day of the communion itself, in private, or even during the whole week before – were prevalent as early as the 1560s, evidence that the penitential and Lenten practices of their Catholic forebears were never wholly set aside.[43]

Nineteen years after the Scottish Reformation of 1560, the Scottish Parliament ordered that all householders with at least 300 merks of rent annually should supply themselves with a Bible and a psalm book. Some Kirk Sessions appointed people to ascertain that this had been done, and to prevent books being passed from house to house they were required to have the person's name marked on them. It was also required that the head of every household catechise all members of his family and the servants.[44]

Gilbert Burnet, a Scottish minister who became Bishop of Salisbury and who was no lover of the Reformed Church, paints a picture of the domestic visits of the parish minister (Fig. 13.3) and how the influence of this recurrent visitation influenced the spirituality of the common people:

> They [the ministers] used to visit their parishes much, and were so full of the Scriptures, and so ready at extempore prayer, that from that they grew to practise extempore sermons: for the custom in Scotland was after dinner or supper to read a chapter in the Scriptures: and where they happened to come, if it was acceptable, they of the sudden expounded the chapter. They had brought the people to such a degree of knowledge that cottagers and servants could have prayed extempore. I have often overheard them at it ... I was astonished to see how copious and ready they were in it. By these means they had a comprehension of matters of religion, greater than I have seen among people of that sort anywhere.[45]

The practice of spirituality in the dwelling was not always confined to the public rooms of the house but sometimes took place in the bedroom and

Figure 13.3 'Presbyterian Catechising' by John Philip, RA (1817–67). Source: Dickson, Nicholas (ed T W Foulis). *The Kirk and its Worthies*, London and Edinburgh, 1914, 168.

even in the bed itself. Archibald Warriston, a well-known lawyer, catechised his wife in bed on the morning after their wedding in St Giles, Edinburgh, and 'was ravisched with her ansuears and blessed god for hir'.[46] For some, however, the bed was not such a happy place. In evangelical Christianity, Christ sometimes became an instrument of separation between husband and wife. This was evident in the times of evangelical religious revival such as that experienced in Cambuslang, Lanarkshire, during the first half of the eighteenth century:

> Christ's love lifted the saint ... into another realm where worldly affections were put aside. One woman noted that she was filled 'with such a love to Christ, that I could have been content never to have seen my husband ... ' Similarly, another woman ... disgusted by her own carnality ... was repulsed by thoughts that 'my husband should come near to me,' but at the same time found her heart 'leaping out of love to Christ'. Her Saviour's love was cleansing, that of her husband polluting. Agape took the place of eros.
>
> ... Catherine Cameron, in her passionate search for the Bridegroom, related that 'I had great delight in prayer, when I would sometimes have gone to Bed, I thought I would have Christ between my Arms: He was a Bundle of Myrrh to me & sweet to my soul.'[17]

Catherine Cameron made much of her preparation for Communion. This was the centrality of evangelical faith at this time:

SPIRITUALITY • 207

She spent a week in preparation, devoting much time in particular to secret prayer, 'out in the fields' ... 'I was so ravished with the love of Christ that night ... that I could sleep little, and all the next morning and day, I was in the same frame: and saying as the Spouse of Christ, My beloved is mine & I am his'.[48]

Men generally found that they could not identify with Christ in exactly the same way as women often did. They found difficulty in seeing themselves as the Bride of Christ. It was far manlier to spit on the hand, as it were, and strike a bargain with him. One of the ways was to draw up a personal covenant:

On bended knees in a posture of reverence and humility, the saints took up their pens in the sight of God and bound themselves (and often their families as well) to the Lord. Far more tangible and durable than spoken words or silent thoughts, these documents helped assure the saints of their worthiness and helped them to give focus to the difficult, and sometimes nebulous task of self examination.[49]

There is little information about how Catholics exercised their domestic spirituality in the years after the Reformation. The typical home would not have possessed a Bible, but it might have had a New Testament. There were aids to worship such as the *New Manual* about which not very much is known, although Fr Robert Francis Strachan is recorded as recommending it to his parishioners on Deeside in 1691. There was, however, a striking difference between the idealised form of family spirituality epitomised in 'The Cotter's Saturday Night' of Robert Burns and that of the Catholic cotters. William Anderson says:

To my mind the most striking contrast between Mr Strachan's ideal family worship and the Ayrshire picture is the total absence of singing. Burns mentions Dundee, Martyrs and Elgin, and unquestionably psalm singing is a feature of Lutheran and Huguenot worship, whether public or private, and so it was and is in Protestant Scotland ... Post-Reformation Scotland also produced no popular Catholic hymn writers. This seems odd, and surely is a grave defect. Here we are dealing only with family worship though families are unlikely to praise God by singing if singing is discouraged as it was by Bishop Hay.[50]

Fr Strachan, in an effort to guide his parishioners in their family worship, provided a list of psalms in sixteen categories that were to be *said*, not sung.

The formal teaching of the Roman Church concerning family worship, however, was almost identical to that of the Reformed:

The advantages of public prayer in a family are so many and so great that it were too long to recount them and therefore it is much to be recommended to all masters and mistresses of families that every morning and evening all the house may meet together in some set place to make this prayer. If this were done, children and servants would be sweetly accustomed to prayer and would thereby receive some profitable instructions of which they have but too great need many times.[51]

Detailed orders for Morning and Evening Prayers were given. The services ended with the Angelic Salutation (Hail Mary) and the Blessing.

The practice of family prayer in devout Presbyterian homes was somewhat more demanding than the structured form of the Catholic. Families would begin their prayers fasting and they might not rise, on some occasions, from their knees until well into the afternoon. Even where this extreme form of prayer was not practised it was expected that the head of the household would withdraw at certain periods of the day:

> ... in high flats of Edinburgh there was one tiny closet built off the dining room, to which the head of the household withdrew for his devotions. Even the flat occupied in Riddel's Close by David Hume had one of these tiny praying apartments; which in his case was a sad superfluity.[52]

There were voices which criticised the style and form of worship and spirituality both in the church and in the home. One in particular was offended by the 'miserable mixture of nonsense, errour and blasphemy' presented as family worship. He recommended that liturgical prayer be used as a means of raising the standards of Scottish worship in general, 'What rude and shocking expressions! What blasphemous petitions have I heard! How often have I trembled, when the ignorant and proud *enthusiast* kneeled down with his family to his extemporary worship.'[53]

But not all family worship was chaos. John Mitchell, whose boyhood was spent in the parish of Beith in Ayrshire in the 1780s, looks back to an ordered exercise of domestic spirituality:

> ... parents, spent much time explaining it [The Shorter Catechism] to their children ... In households, according to the manners of the good old times, especially those professing godliness ... family worship was observed everyday together with Catechetical and Biblical instruction, particularly on the Lord's Day, and the omission of these duties, which in later times there is reason to fear has existed to a mournful extent were [*sic*] then comparatively rare, and considered as hardly compatible with the existence of personal piety (Fig. 13.4).[54]

Mitchell's comments show that change was being acknowledged. In the eighteenth and nineteenth centuries cities were growing rapidly, populations were on the move, not only within the country but exiting it. Industrialisation was affecting every facet of life, communications were improving all the time, ordinary people were demanding representation in government and science was opening the eyes and minds of thinking men

Figure 13.4 'Family Devotions'. A man leads his household in prayer. From a magazine of the 1870s (*Sunday at Home: A Family Magazine for Sabbath Reading*, London, 1861). Source: D Reid.

and women. The old answers and the old certainties were now challenged. During Victoria's reign and before 'there was a general failure to acknowledge the mass paganism of the depressed classes'.[55] Some ministers did recognise it and knew that the old type of Christian spirituality was no longer customary in hundreds of thousands of Scottish homes and that they could do little about it. In 1849 The Rev James Begg, minister of Newington Free Church, paints a vivid picture of conditions in Edinburgh:

> We plunged into a black opening more like the mouth of a coalpit than the entrance to human habitations; we were almost knocked down by the horrid vapour by which we were assailed; the population of the ruinous tenement was greater than that of a considerable country village, and these human beings were in far more uncomfortable circumstances than any sensible farmer's cattle.[56]

The slum tenements of Glasgow were no better. James Begg, however, showed more spiritual courage and evangelical stamina than at least one minister in Glasgow who obviously had given up any attempt to be the parish minister in the areas of his parish where the Church no longer had any meaningful Christian ministry:

> Once a year he [the parish minister] approached the mouth of each several 'close' in his district – down whose dark vista of sin and

misery, however, he never penetrated – and there, uncovering his head with due solemnity, and lifting his gloved right hand, he sought the Divine blessing to rest on 'all inhabitants, young and old of this close.' The annual visitation thus ended, he went on his way.[57]

In 1824 the United Secession Church built the first church in Partick at Dowanhill (the parish church lay across the Clyde in Govan). In 1831 the Presbytery required the session to provide details on the 'state of personal and family religion in the congregation'. With regard to domestic worship the session stated that:

> ... in several families it is kept up morning and evening. The neglect or irregular performance must in many instances be traced to spiritual indifference, in a few to a false and criminal delicacy, but in others to real want of time from the long hours demanded at public works.[58]

A comment from a non-Presbyterian viewpoint on the subject of family worship is given in Dean Ramsay's *Reminiscences of Scottish Life and Character*. He wrote from a minority position, being a minister in the Scottish Episcopal Church at a time when it looked to the High Church tendency in the Church of England for direction. He says:

> Take as an example, the practise of family prayer, many excellent and pious households of the former generation [of Episcopalians?] would not venture upon the observance, I am afraid, as they were in dread of the sneer. There was a foolish application of the terms 'Methodist' [not a Scottish application], 'saints,' 'over-righteous' where the practice was observed. It was to take up a rather decided position in the neighbourhood; and I can testify that, less than fifty years ago, a family would have been marked and talked of for a usage of which now throughout the country the exception is rather the unusual circumstance ... These remarks apply perhaps more especially to the state of religious feeling amongst the upper classes of society.[59]

Ramsay's words regarding the prevalence of domestic worship in his day in certain sections of society are borne out by the following account:

> In a well-to-do and pious Episcopalian home in Edinburgh in the 1880s, family prayers were the custom ... The signs of Sunday are all about me as I run downstairs dressed in my Sunday frock ... The round table in the dining room is laid for breakfast and the big brass kettle is beginning to sing on its gas ring attached by rubber piping to one of the jets of the chandelier. My father has cleared a space for the enormous fat family Bible and found his place. The first bell is rung, for prayers. The second bell will mean breakfast. On weekdays there is a third bell, for porridge, between prayers and breakfast proper.[60]

The Church was no longer able to exercise its former disciplines in the modern world of industry and urbanisation. Recognising this, the General

Assembly of the Church of Scotland in 1836 issued a letter to be read by all its ministers to their congregations.

> In compliance with the solicitations of many who watch for your souls ... we have resolved to issue this brotherly exhortation on the sacred and indispensable duty of Family Worship – not as if we had any recent ground for apprehending that it is likely to fall into more extensive neglect, but because we know too well that it is by no means universally practised ... It is not for us to unfold the laws of the spiritual world, or to demonstrate why and how it is that the communications of heavenly influence and favour are in any degree suspended on the frequency and fervency of our supplications. But this we know, that, as in old time the father of the faithful commanded his children, and his household after him, to unite with him in the exercises of a holy life ... [61]

The General Assembly was on the defensive and was trying to lock the stable door despite the evidence that many of the horses had bolted some time before, and it showed little concern for the spiritual or physical plight of those whose new stables rivalled the Augean and in whose reeking stalls the practice of a spiritual life was well nigh impossible:

> The coming of the Industrial Revolution and the growth of illiteracy and mass paganism in concentrated areas must have made the Shorter Catechism and family worship totally unknown in a large section of the population. The Assembly's Pastoral Letter on Family Worship is not so much evidence for its presence as for its absence.[62]

Among the humbler classes in rural areas spirituality in the home still survived. David Marshall supervised the fishing on Loch Leven and lived in Kinross. When he died in 1902 his minister said of him:

> His own house was to him the brightest spot on earth – a place made sacred by meditation and prayer. The evening before he died he conducted family worship which included a very comprehensive prayer as if he was committing all whom he loved into God's safe keeping.[63]

A NEW APPROACH

Alistair Phillips was born in 1909. His memories of family worship in the earlier part of the twentieth century remained clear; the old way of doing things still prevailed in some households:

> ... while praying at family worship he [Uncle George] referred at least once a day to 'the well that has been opened for sin and for uncleanness', and of this I had the clearest architectural picture, even down to the bucket and the rope and the winding gear ... My

complete familiarity with the prayers was only to be expected for we took the Books twice a day, immediately after breakfast and before bedtime ... I just needed to hear one familiar word to be able to fill in for myself the dozen which would follow it ... The order of service of worship was unchangeable.[64]

The unchangeable nature of Uncle George's prayer was probably a consequence of using the orders of service for family prayers issued by the General Assembly of the Church of Scotland. There are prayers for every day, morning and evening. Updated versions were still being produced well into the twentieth century and there are modern publications from various religious viewpoints which fulfil the same purpose.

In the twentieth century the churches in Scotland were existing in a changed world yet, in the main, they were still trying to operate as if things were as they had always been, and this was truer of the Church of Scotland with its weight of history and establishment. It is now difficult to find any clear evidence of domestic spirituality being practised as it once was.

The influx of Catholic Irish, in the second half of the nineteenth century and the first half of the twentieth, altered the Christian face of Scotland. Some parish ministers in the cities and towns of west central Scotland were finding that there were now more Catholics in their parishes than Protestants. It was easy to tell the religious allegiance of a family on entering a home. Roman Catholics would usually have, and sometimes still have, pictures of the Sacred Heart of Jesus on the wall and sometimes a small shrine in a corner of the room, with a statue of the Virgin and of the saints, where prayers could be said. The crucifix and rosary beads also identified the Catholic home. Protestants had no equivalent but many dwellings had huge leather-bound family Bibles with the dates of marriages, births and deaths of members of the family. It was not unusual among Protestants for individuals to read the Bible from beginning to end many times, and frequently these marathons were recorded, dated and initialled on the flyleaf of the Bible (Fig. 13.5). Latterly in the twentieth century, the large family Bibles ceased to be used but families wished to give them a decent burial, as it were. Many parish ministers found themselves the unwilling recipients of those monuments to a past domestic piety.

There were numerous revivals among the fishing communities of the east coast in the twentieth century. The effect was to enliven spirituality in homes and their influence is still apparent in the denominational constitution of many of the former fishing villages. Among certain sections of the population this sometimes meant separating themselves from all who did not believe as they did, and such individuals even refused to sit at the same meal table as members of their families who did not share their views, or to go to sea with unbelievers. Domestic and public spirituality was based firmly on a literal interpretation of the Bible.

The Influence of the Iona Community, founded by George MacLeod in 1938, has also played a part in the spirituality of the homes of those

Figure 13.5 'He Walks a Portion wi' Judicious Care' by G Paul Chalmers, RSA. Source: Dickson, Nicholas (ed T W Foulis). *The Kirk and its Worthies*, London and Edinburgh, 1914, 200.

affected by its ethos. Its membership is not confined to Scotland alone but its strength lies there. It is its Rule which gives it its cohesiveness, 'It is ... clear from the correspondence and discussion that take place when people seek to join the Community that one of the strong attractions is the Rule, the possibility of discipline within highly pressurised lives.'[65] There are five parts to the Rule. Firstly, all Members and Associate Members undertake a daily devotional discipline of prayer and Bible reading. Secondly, there is an economic discipline and members are obliged to account to one another for their use of money. They are asked to give away ten per cent of their disposable income. Thirdly, there is the obligation to use their time wisely and well. Fourthly, members are committed to work for justice and peace in society. Fifthly, there is the duty to meet periodically with the other members and to give an account of their diligence and to share news and concerns. Though comparatively small in numbers the spirituality of the Community has affected the lives of many in many churches worldwide.

The radio had been used, almost from its beginning, by the Church as a means of entry into the homes of people who would never set foot in a parish church. Speakers of the quality of George MacLeod elicited a strong response from listeners. Here was a way back into the homes of the people. After World War II the idea of radio mission was born. Local churches of all denominations, apart from Catholics and a few of the lesser denominations, were asked to inform the people in their areas of responsibility of the broadcasts and to mount a door-to-door visitation afterwards to assess the response. However, there was little evidence of any great and lasting spiritual revival in the dwellings of Scotland as a consequence of the radio mission. Perhaps this was because of the inertia of most of the ministers and ordinary members of the congregations:

> Ninety out of every hundred sat back and did nothing about the Mission except to listen to some broadcast which attracted them. In proportion to their activity, so were the results in their parishes ... they sat back and waited for radio miracles to happen. There was no such revival ... [66]

But, there were some results. In one parish in Aberdeen which had embraced the idea of the radio mission there was a Confirmation class of eighty. There were other congregations which reported similar local responses. Radio had proved that it could help to revive spirituality in the homes of some people and although it could not take on the duties of the local congregations, it could be a valuable aid to evangelism on the ground.

Another result was that the churches were now much more receptive to the idea of radio mission. From this came the Tell Scotland movement of evangelism. The main idea was that there should be in every congregation a group of devout people who would gather round the Bible in their own homes, see what it was saying to the world, and then carry their findings and convictions out into everyday life:

> It was the application of the Parable of the Leaven; the grain of mustard seed; the creative minority; the dynamic core at the heart of the Church, Christ's Body on earth ... the place of the layman was crucial.[67]

The spirituality of their homes was to be shared with the community. At one point there were around 800 dedicated groups of various denominations who were committed to this form of evangelistic outreach, and even if their efforts had no effect on the public in general, their own domestic spirituality was certainly enhanced. It is worth noting that at this period (1950s) formal church membership in Scotland was at an all-time high.

In 1955 it was mooted that the American Evangelist, Dr Billy Graham, should be invited to Scotland. He had been in England a year earlier and the conservative evangelicals had been impressed by his methods. The organisers of Tell Scotland were not of one mind on the subject. It was not what was originally envisaged by them:

> Then we were persuaded to change our strategy [against the advice of those such as George MacLeod], thereby sowing the seeds which were to choke a promising Movement. Like a bombshell, into the middle of our small groups technique was thrown the explosion of

Mass Evangelism through the person of Billy Graham and his All Scotland Crusade.[68]

There is no doubt that the All Scotland Crusade by means of public gatherings, radio and television changed the spiritual life of some Scots, but it also coincided with the beginning of a spiral of decline in church membership in Scotland from which the churches have not yet recovered. The Tell Scotland movement and its influence in the homes of ordinary people slowly died, 'The content of Billy's message was not really in touch with our modern society. It would have been superb in 1855 instead of 1955 ... '[69] Graham, who had filled Hampden Park in 1955, came back just over six years later and Ibrox Park was only one third full. The tide was on the ebb.

Religious programmes on radio and television still succeed in attracting a large audience and their style is sometimes reminiscent of the mass evangelism of the past. Is this the modern expression of spirituality in the electronic Scottish dwelling?

> Throughout Christian history spirituality has changed shape, often subtly but sometimes substantially ... Our understanding of what the word broadly seeks to express (that is, the theory and practice of the Christian life) has evolved as individuals and historical or cultural environments change.[70]

But some ancient expressions of spirituality are still with us today. Cardinal Thomas Winning, Archbishop of Glasgow, died in the year 2001. In *Scotland on Sunday*, an article reports that some Catholics are now praying to him:

> We have had dozens of letters from people saying they were actually praying to the Cardinal to ask him to do good on their behalf and answer their prayers. This is more than just remembering him in their prayers or praying for him. People are saying that they expect now that he's in Heaven he will continue to be able to do good for them.[71]

VISITING THE DWELLING PLACE IN THE TWENTY-FIRST CENTURY

Home visits by the clergy or elders in former times were directed chiefly at ensuring members of the flock were properly instructed in the tenets of the faith and in prayer, or for the administration of sacraments such as baptism and marriage (Fig. 13.6). Today's visits are much more socially orientated with the visitors in a far less inquisitorial role. Mostly the visits are welcomed, but occasionally the visitor may have to contend with a rival in the corner of the room which competes for attention, the television set. To call at a time when a favourite programme is being shown creates a situation in which it can be difficult to speak of 'eternal things'. Sometimes the sound is turned down but the picture left flickering away as a reminder that the visit should not be too long.

Figure 13.6 'A Scottish Christening' by John Philip, RA (1817–67).

On serious occasions, such as visiting the sick or the dying, the attitude is appreciative. At such times people are seeking assurance from the Church that there is a spiritual dimension to their lives. When there is a death in the home they expect to be comforted by the sacraments, prayers and Scriptures of the Church and any perfunctory approach is resented. The Church of Scotland minister officiates at the funerals of his or her own congregation and at most funerals of those people who have no live church connection. The same happens for weddings and baptisms. Like the NHS, it is free at the point of need. The Roman Catholic priest and the ministers of the smaller denominations attend, mostly, to the spiritual needs of members of their own congregations. Nowadays the Humanist Society also offers its expertise for funerals and non-religious counselling.

Many older people now spend some of their later years in residential homes. They are visited regularly by their own church representatives and there are visits by groups of the laity, who may also visit people in need in other domestic settings. Short services of worship are held in the communal areas of the home. There are in addition voluntary groups who provide programmes of entertainment. Most homes have a chaplain appointed. Sheltered-housing complexes are included in the responsibilities of the local parish and non-parish clergy.

Hospitals and hospices are served by appointed chaplains, some full-time and some part-time. They attend to the spiritual needs of both the staff and the patients. Many hospitals have chapels or meditation areas. The chaplains offer counselling to the relatives of patients if it is wished. Chaplains of the main faiths are appointed to all prisons. They celebrate the appropriate sacraments and lead worship for those who wish to participate. Counselling is an important part of their work.

BEYOND THE MAINSTREAM OF WESTERN CHRISTIANITY

Owing to the constraints of chapter size, this contribution has had to be selective. In dealing with the mainstream of western Christianity it is recognised that it has been necessary to exclude many expressions of spirituality in the modern Scottish home. There are strong Orthodox Christian communities of several varieties contributing to Christian life in Scotland. There is also an active Church of the Latter Day Saints, which is well known for its family-based spirituality. There are Seventh Day Adventists, Jehovah's Witnesses, Plymouth Brethren and others too numerous to list here.

Observations of the domestic expression of other world religions, such as Islam, Judaism, Buddhism, Hinduism and Sikhism, which are now Scottish religions, are also worthy of discussion. Among a small number of immigrant groups, local religions may sometimes be practised in exile. 'New Age' spiritualities abound and each has its place in the history of spirituality in the Scottish dwelling.

NOTES

1. Baillie, 1962, 17.
2. Rooney, 1999.
3. McLeod, 1989.
4. Rooney, 1999.
5. Rooney, 1999.
6. Rooney, 1999.
7. Sheldrake, 1995, 19.
8. Davis, 1944.
9. *Holy Bible*, Exodus, 6: 6–9.
10. Mackenzie, 1930, 110.
11. MacLean, 1930.
12. There is, in some of Carmichael's material, obvious reference to the beliefs of pre-Christian times, as seen, for example, in the invocation of the sun and moon, the archangels Michael, Ariel, Uriel, et al (thinly disguised gods of the ancient times), and Saint Bride, whose attributes merge seamlessly with the pre-Christian goddess of the same name.
13. Carmichael has been accused of polishing the rough gems which he found and presenting them as fine jewels. Some have gone as far as to compare his work to that of MacPherson's *Ossian*. Be that as it may and be they polished or unpolished, they are widely regarded as sparkling.
14. Carmichael, 1900, 219.
15. Carmichael, 1900, 69.
16. Carmichael, 1900, 233:
    I will kindle my fire this morning
    In presence of the holy angels of heaven,
    In the presence of Ariel of the loveliest form,

      In the presence of Uriel of the myriad charms.
      Without malice, without jealousy, without envy,
      Without fear, without terror of anyone under the sun,
      But the Holy Son of God to shield me ...
17. Carmichael, 1900, 57:
      I bathe my face in the nine rays of the sun,
      As Mary bathed her son in the rich fermented milk.
18. Carmichael, 1900, 237.
19. Carmichael, 1900, 85.
20. Carmichael, 1900, 225.
21. Carmichael, 1900, 166; For further information on lifecycle rites of passage see Chapter 30 Custom.
22. Carmichael, 1900, 116.
23. Pennant, 1979, 98–9; See also Burt, 1974, I, 185–6; Another description of the lyke wake is given in Sanderson, 1904, 158.
24. Dilworth, 1956.
25. Graham, 1969, 334–5; in Carmichael, 1940, 315 the following is recorded:
      Be with me, O God, at the breaking of bread,
      Be with me, O God, at the close of my meal;
      Let no whit adown my body
      That may hurt my sorrowing soul.
      O no whit adown my body
      That may hurt my sorrowing soul.
26. The perception of the superiority of the 'Teuton' over the 'Celt' is of long standing. There is an old poem which claims that God created the first Highlandman out of a horse turd: Watson, 1995, 143. This sentiment is echoed in Victorian times, 'Ethnologically the Celtic race is an inferior one, and attempt to disguise it as we may, there is naturally and rationally no getting rid of the great cosmical fact that it is destined to give way – slowly and painfully it may be, but still most certainly – before the higher capabilities of the Anglo-Saxon' *Fifeshire Journal* (11 September 1851). Cited in Bradley, 1999, 119.
27. Graham, 1969, 337.
28. Strathclyde Regional Archives, the Mitchell Library, Glasgow, Glasgow Presbytery Records, 35: 248–9, 251, 317. Cited in Schmidt, 1989, 253–4.
29. Boston, 1776, 275.
30. Cross, 1958.
31. McRoberts, 1954.
32. Law, 1901, 171.
33. Watson, 1995, 75.
34. Thomson, 1901, 44. 'An impression exists that the nobles, whose sole profession almost was arms, despised letters as belonging properly to priests and traders. It is this impression that Sir Walter Scott embodies in the words he puts into the mouth of old Bell-the-Cat, father of the poet-Bishop of Dunkeld:
      'Thanks to St Bothan, son of mine
      Save Gawain ne'er could pen a line.'
35. Thomson, 1901, 47.
36. Thomson, 1901, 50; See also Chapter 21 Reading and Study.
37. Ross, 1957, 5. 'The earliest known edition of the *Gude and Godlie Ballatis* belongs to the year 1567. There is strong reason to suppose earlier editions existed, perhaps about 1540 or 1546, and even before then many of the poems in the book may have been circulated as broadsheets. Several of them are translations from German hymns of Luther and his followers.'
38. Thomson, 1901, 53.

39. Calderwood, 1842, 127.
40. Knox, J, *Works*, vol IV, 137ff. In McEwen, 1961, 31.
41. McEwen, 1961, 56.
42. Schmidt, 1989, 18.
43. Schmidt, 1989, 239.
44. Gordon, 1992, 578.
45. Pagan, 1988, 123.
46. Pagan, 1988, 130.
47. Schmidt, 1989, 163.
48. Schmidt, 1989, 120.
49. Schmidt, 1989, 136.
50. Anderson, 1967.
51. Anderson, 1967.
52. Graham, 1969, 337.
53. Schmidt, 1989, 182.
54. Mitchell, 1939, 75, 285–6, 301–05. Cited in Drummond, Bulloch, 1975, 102–3.
55. Drummond, Bulloch, 1975, 215.
56. Blakey, 1978, 33.
57. Blakey, 1978.
58. Dickie, 1926, 37.
59. Ramsay, 1911, 24.
60. Smout, Wood, 1990, 128–9.
61. The General Assembly of the Church of Scotland, 1865, 1, 2, 7.
62. Drummond, Bulloch, 1975, 103.
63. Marshall, 1986, 116.
64. Phillips, 1984. In Kamm, Lean, 1985, 28.
65. Shanks, 1999, 65.
66. Falconer, 1978, 72–3.
67. Falconer, 1978, 77.
68. Falconer, 1978, 77.
69. Falconer, 1978, 81.
70. Sheldrake, 1995, 40.
71. Sister Roseann Ready in *Scotland on Sunday* (13 Jan 2002), 7.

BIBLIOGRAPHY

Anderson, W. Catholic family worship on Deeside, 1691, *The Innes Review*, 18 (1967), 151–6.
Baillie, J. *Christian Devotion*, London, 1962.
Bennett, M. *Scottish Customs from the Cradle to the Grave*, Edinburgh, 1992.
Blair, A. *Croft and Creel*, London, 1987.
Blakey, R. *The Man in the Manse*, Edinburgh, 1978.
Boston, T (eds A Murray, J Cochran). *Memoirs*, Edinburgh, 1776.
Bradley, I. *Celtic Christianity: Making Myths and Chasing Dreams*, Edinburgh, 1999.
Brown, T. *Annals of the Disruption: With Extracts from the Narratives of Ministers who Left the Scottish Establishment in 1843*, Edinburgh, 1893.
Bruce, W. *The Nor' East*, London, 1915.
Burleigh, J. *A Church History of Scotland*, London, 1960.
Burt, E. *Letters from a Gentleman in the North of Scotland to his Friend in London* (1754), Edinburgh, 2 vols, 1974.
Calderwood, D (ed T Thomson). *History of the Kirk of Scotland* (Woodrow Society, vol 1), Edinburgh, 1842.

Cameron, D. *Willie Gavin, Crofter Man: A Portrait of a Vanished Lifestyle*, Edinburgh, 1995.
Carmichael, A. *Carmina Gadelica*, vols 1–2, Edinburgh, 1900, vol 3, Edinburgh, 1940.
Cross, F L, ed. *The Oxford Dictionary of the Christian Church*, London, 1958.
Daiches, D, ed. *The Selected Poems of Robert Burns*, London, 1980.
Davis, J D (revised H Gehman). *The Westminster Dictionary of the Bible*, London and New York, 1944.
Dickie, W. *History of Dowanhill Church, 1823–1923*, Glasgow, 1926.
Dickson, N (ed D Malloch). *The Kirk and its Worthies*, Edinburgh, 1914.
Dickson, N. *The Kirk Beadle: A Collection of Anecdotes and Incidents Relating to the Minister's Man*, Glasgow, 1891.
Dilworth, M. Catholic Glengairn in the nineteenth century, *The Innes Review*, 7/1 (1956), 11–23.
Donaldson, G. *The Faith of the Scots*, London, 1990.
Drummond, A, Bulloch, J. *The Church in Victorian Scotland, 1843–74*, Edinburgh, 1975.
Duncan, A. *Scotland: The Making of the Kingdom*, Edinburgh, 1978.
Dwelly, E. *The Illustrated Gaelic–English Dictionary* (1901–11), Edinburgh, 1993.
Edgar, A. *Old Church Life in Scotland: Lectures on Kirk-Session and Presbytery Records*, second series, London, 1886.
Falconer, R. *The Kilt Beneath my Cassock*, Edinburgh, 1978.
Fleg, E. *Why I am a Jew*, London, 1943.
Forrester, D. *On Human Worth*, London, 2001.
General Assembly of the Church of Scotland, Committee on Aids to Devotion. *Family Prayers*, Edinburgh and London, 1865.
Gordon, A. *Candie for the Foundling*, Edinburgh, 1992.
Graham, H. *The Social Life of Scotland in the Eighteenth Century*, London, 1969.
Hardinge, L. *The Celtic Church in Britain*, London, 1972.
Henderson, I. *Scotland: Kirk and People*, Edinburgh and London, 1969.
Herberman, C G, et al, eds. *Catholic Encyclopaedia*, London, 1907.
James, W. *The Varieties of Religious Experience*, London, 1902.
Kamm, A, Lean A, eds. *A Scottish Childhood*, Glasgow, 1985.
Kellett, E. *A Short History of Religions*, London, 1954.
Law, T G, ed. *Catholic Tractates of the Sixteenth Century* (Scottish History Society), London, 1901, 171. Cited in McRoberts, 1972.
Lees, J. *Life and Conduct*, Edinburgh, 1922.
Mackenzie, D A. *Scotland: The Ancient Kingdom*, Glasgow, 1930.
MacLean, D. The Law of the Lord's Day in the Celtic Church. In Mackenzie, 1930, 151.
MacLean, C. *Going to Church*, Edinburgh, 1997.
MacLeod, F. *The Winged Destiny: Studies in the Spiritual History of the Gael*, London, 1910.
MacLeod, N. *Reminiscences of a Highland Parish*, London, 1891.
Marshall, D. *A Victorian Countryman's Diary*, Kinross, 1986 (original diary from 1847).
McEwen, J. *The Faith of John Knox*, London, 1961.
McGraw-Hill. *New Catholic Encyclopaedia*, New York and London, 1967–79.
McLeod, W T, ed. *The New Collins Dictionary and Thesaurus*, Glasgow, 1989.
McRoberts, D. Provost Skene's house in Aberdeen and its Catholic chapel, *Innes Review*, 5/2 (1954), 119–24.
McRoberts, D. The Rosary in Scotland, *The Innes Review*, 23 (1972), 81–6.
Meek, D. *The Scottish Highlands: The Churches and Gaelic Culture*, Geneva, 1996.
Mitchell, J. *Memories of Ayrshire* (Scottish History Society, miscellany VI), Edinburgh, 1939.
Murray, J. *Life in Scotland a Hundred Years Ago: As Reflected in the OSA of Scotland, 1791–9*, Paisley, 1907.
Pagan, A. *God's Scotland? The Story of Scottish Christian Religion*, Edinburgh, 1988.
Pennant, T A. *A Tour in Scotland, 1769* (Warrington, 1774), Perth, 1979.
Phillips, A. *My Uncle George*, Glasgow, 1984.

Presbyterian Church in the United States. *The Book of Church Order*, Richmond, Virginia, 1949.
Ramsay, Dean. *Reminiscences of Scottish Life and Character*, Edinburgh, 1911.
Robertson, A. *Lifestyle Survey: Church of Scotland Board of Social Responsibility*, Edinburgh, 1987.
Rooney, K, ed. *Encarta Dictionary of World English*, London, 1999.
Ross, I, ed. *The Gude and Godlie Ballatis*, Edinburgh, 1957.
Sanderson, W. *Scottish Life and Character*, London, 1904.
Schmidt, L E. *Holy Fairs*, Princeton, New Jersey, 1989.
Shanks, N. *Iona–God's Energy: The Vision and Spirituality of the Iona Community*, London, 1999.
Sheldrake, P. *Spirituality and History*, London, 1995.
Smout, T. *A Century of the Scottish People, 1830–1950*, London, 1986.
Smout, T, Wood, S. *Scottish Voices, 1745–1960*, Glasgow, 1990.
Thomson, G W. Literature in the Scottish Reformation, *Transactions of the Aberdeen Ecclesiological Society*, XII (1901), 44.
Watson, R, ed. *The Poetry of Scotland: Gaelic, Scots and English, 1380–1980*, Edinburgh, 1995.
Wilkinson, R. *Memories of Maryhill*, Edinburgh, 1993.

14 Clothing

NAOMI E A TARRANT AND ALEXANDER FENTON

Clothing has a number of functions. It can provide warmth and it can be decorative. It can act as an identifier with an important symbolic function, marking status differences on the one hand and on the other, common identity, as for school pupils, occupational groups, religious communities, members of the military, football fans or those in youth cults – the list is endless. This chapter, however, is not so much concerned with the public face of clothing as with its nature, making and maintenance in association with the home and the immediate surroundings of the local community.

THE NATURE OF DRESS IN SCOTLAND

There is virtually no evidence for, or information about, clothing in the prehistoric or early historic periods, although a few articles of dress have been found preserved in peat bogs. An example is the hood of woollen fabric, with a long, knotted fringe, from Orkney,[1] in the collections of the NMS.

For the medieval period there is a lack of firm evidence for the nature of clothing in Scotland, especially in Highland areas. There are very few illustrative sources and the written evidence is sparse. Definitions of terminology are not always clear. There has also been a tendency to see later developments as having an earlier history than the evidence warrants. The fact that very little survives in the way of actual garments before the early eighteenth century also hinders understanding of how people dressed, again especially in the remoter regions of the country. A certain amount of insight, however, can be gained from funeral monuments and gravestones.

From what evidence is available it seems that there were national fashions to be observed in the medieval period, but there was no folk dress as we would understand it today. Those at the bottom, the poor, had to dress in what they could find, afford or were given. Servants, and even courtiers at times, wore a uniform. This might consist of wearing particular colours and badges, or it could be more elaborate. Monarchs gave sets of clothing to their courtiers at various times in the year, such as Easter or Christmas, and to celebrate royal weddings or christenings, each person receiving a set in cloth carefully graded to their position. These gifts were important not just for their practical value, but also because they underlined the receiver's position within the strict social hierarchy of medieval Scotland.

For later periods, clothing in Scotland has, on the whole, followed the general trends of European dress. Significantly, the minister of Barvas, Lewis, said in the 1830s that the, 'formation of the female habits, with their whole appearance, closely resembles that of the "wandering Bavarians", or the Swiss "buy-a-broom" singers, who itinerate through this country'.[2] Those with money and at the top of the social scale would not have wanted to appear outlandish to any visiting dignitaries from abroad. Such Scots travelled outside Scotland and observed different ways of dressing, noting what was fashionable and suitable for each grade in society. Although there was growing academic interest in folk dress in the nineteenth century which led to searches for such a thing in Britain, all that could be found was occupational dress.

Today, clothing worn in Scotland almost entirely follows trends set outside the country and reflects the involvement of most of its inhabitants in a global culture. There is no longer even a standard fashionable dress which, up to the 1980s, influenced most manufacturers of clothing as they followed the cat-walk styles put out by the designers of Paris, London, Milan or New York. Clothing styles of the mass market change less from year to year than they did, although this trend may be less marked for adolescents, and it is increasingly up to each individual to put together a style based on his or her particular interests and requirements. Clothing is more casual on the whole but it is possible for those who wish to dress more formally to do so without being seen as out of step and on certain occasions formal wear is still expected. However, what we wear remains not wholly governed by personal taste, occasion, or by practicalities such as the climate, the job or the pocket, despite what dress commentators imply. There are other factors which influence people's decisions on what they should or want to wear. For example, certain religious dress codes require general European-style dress to be either ignored or adapted, such requirements taking precedence whether or not the style of dress fits the climatic circumstances of living in a northern European country. As has often been demonstrated, humans can learn to live with many types of inconvenience in clothing if it fulfils other needs.

A TIME OF CHANGE: THE LATE EIGHTEENTH TO MID NINETEENTH CENTURIES

The period from the late eighteenth century to the mid nineteenth century was one in which the domestic wellbeing of the 'working classes' visibly improved, first in the lowland areas and in and around towns, and then creeping on to the more highland and remoter-lying parts of the country. In many districts the old ways of hand spinning and weaving were being superseded, and the heavy woollen items of homemade clothing were being replaced by commercially produced cottons and muslins. A review of references to clothing in the *NSA*, published in 1845 but mostly written in the 1830s, pinpoints the main aspects of change over the preceding 30 or more years.

Figure 14.1 'The Arrival of the Country Cousins', by Alexander Carse, 1812. The fashionable women wear light cotton dresses, the low necks filled in with false shirts for daytime wear and none is wearing indoor caps. Their young country cousin wears a similar dress but has a neck handkerchief which is crossed and tied at the back. The older woman has a large shawl, probably printed cotton and from one of the Scottish printfields, a dark dress and white apron, with a close fitting cap on her head indicating that she is a married woman. The red hooded cloak over her arm was a popular outdoor garment, though now rather old fashioned for town wear. This in its turn had replaced the plaid for most women by the end of the eighteenth century. All three men wear similar outfits, probably only distinguishable from each other by the quality of the woollen cloth. Source: By kind permission of His Grace The Duke of Buccleuch and Queensbury, KT.

In the south-eastern parish of Hounam, Roxburghshire, for example, the minister noted that the style of living and of dress had improved very much over the preceding 30 years, especially amongst servants.[3] This kind of change went hand in hand with economic and social developments such as an increasing range of shops selling food, furniture, furnishings and clothing, and, in homes, the more or less regular drinking of tea and the consumption of white bread and an increased amount of butcher meat. Buildings were being greatly improved also, with better forms of hearth for heating and cooking. There was an increasing move away from the making or processing of everyday items of use in the home to simply purchasing what was needed in a shop.

Modernising tendencies could only rarely be avoided by the 1830s to 1840s. In Tweedsmuir, Peeblesshire, the shepherd's plaid was still universally worn,[4] although in Kilbride, Bute, older people wore the old-style dress and younger people, especially 'the softer sex', followed mainland fashions.[5] In

Urquhart and Logie Wester, Ross-shire, the spinning wheel and loom had almost gone, the making of linen at home – at one time a main employment of the women – had been nearly forgotten and woollen stuffs had been exchanged by young people for, 'the more gaudy but less substantial fabrics of Glasgow or Manchester'.[6] In Tillicoultry, Clackmannanshire, shawls and tartans made by skilled merchants were as much in request as the serges of former times.[7] The adoption of lighter fabrics must have led to an increased use of equipment such as flatirons and goffering irons, the latter to be found as rusting and unused curios in homes throughout Scotland within the last couple of decades.

MOVEMENT OF PEOPLE, AND EMULATION, AS TRIGGERS OF CHANGE

Apart from general economic trends, there were several influences which accelerated change. For example, in Dornoch, Sutherland, the inward movement of people from southern areas brought about innovations in dress, especially that of the young of both sexes.[8] Conversely, the seasonal outward movement of Highlanders to the southern parts of Scotland to shear the harvest undoubtedly led to the acquiring and bringing back of items of dress of lighter and more colourful materials; or if people did not spend money on such matters, at least they gained knowledge of them. Those from Sutherland who went south and to Caithness for the herring fishing adopted Lowland dress and habits to a considerable extent.[9] Proximity to towns with dressmakers' shops also played a role, and southern forms of dress were being widely imitated in the north and west.[10]

Another widespread instrument of change was emulation by female servants of the fashions of higher classes. The ministers who wrote the parish accounts in 1845 were often censorious. The dress of females in Kelso, Roxburghshire, is neat, said one, 'that of female servants (in particular) showy to an extent that might be advantageously abated'. In Bedrule, in the same county, the dress of women and girls was often too gaudy and expensive for their station. In Ochiltree, Ayrshire, young females, especially servant girls, were fond of dressing finely, and it was not unusual to see them arrayed in gowns, with other parts of their dress corresponding. In Lesmahagow, Lanarkshire, the style and manner of female dress were somewhat expensive, 'the servant girl dressing as gaily as the squires' daughters did thirty years ago'.[11] This widespread tendency in the more lowland areas was further stimulated by the practice in big houses of passing on worn clothes to the servants. An account from Culzean Castle in Ayrshire records that in 1784 a servant called Pegie refused a gown worth 10/-, and in 1794 another person named Babie would not take a cloak of similar value. New clothing, however, might be provided for special occasions such as funerals. When the father of a dairymaid at Culzean died in 1795, she was given a 'cheap' gown worth 7/-.[12]

Church attendance on Sunday also had a strong effect on dress. A special effort was made to be neat on the Lord's Day. The 'Sunday's best' worn by the common people in Melrose, Roxburghshire, was said to be very little different from that of the upper classes, and in Dalry, Ayrshire, the change in dress over the 15 years before 1836 meant that it was little, if anything, inferior to the Sunday clothes of the inhabitants of Edinburgh and Glasgow in the same walk of life.[13] In Garvock, Kincardineshire, clothing at church was similar to that of the minister's or laird's family of 40 years before.[14] Of course, the prestige of being well kitted out at church had on occasion an opposite effect, for lack of suitable clothing could be made an excuse for not attending. Children of the lower classes were especially affected by this. It was said that their parents '*cannot afford* to give their children *two suits of clothes at once*, and are *ashamed* to send or take them to the kirk in rags'.[15]

THE OLD AND THE NEW

The examples given above indicate some of the major factors which influenced change: economic improvement, movement of people from core areas to peripheries, seasonal migration for work purposes from peripheral to core areas, emulation of higher classes, the pressures of looking decent at church. To assess the extent and nature of change, it is necessary to set it against some examples of the older or more conservative features of clothing in the domestic situation.

Old people, said the minister of Alvie, Inverness-shire, in 1835, are rather slovenly in their dress.[16] This was no doubt true, since their homespun woollen clothes would be worn for a long time and would rarely, if ever, have seen an iron, with baggy knees, for example, the order of the day. Furthermore, as was noted in a Berwickshire parish, smoky fires gave 'a smell to the clothes which is strong and offensive to such as are not accustomed to it'.[17]

In the old days, clothes were for the most part homemade, perhaps being sewn from cloth obtained from the local weaver. Wool was the main material, spun by women and girls, often at the fireside, on the distaff and spindle or later on the spinning wheel. Although changes were well advanced by the 1830s, relics of older forms of clothing remained.

The plaid
The plaid, seen as particularly Scottish according to travellers' accounts of the seventeenth and eighteenth centuries, was still common in the 1840s. Plaids were universal outer garments for both sexes – simple long, wide lengths of woollen cloth which could be wrapped around the body in various ways. They could be slept in, and could serve as bed covers. Because wide fabric could not be woven easily on handlooms, most plaids were made up of two widths sewn together. The length varied, but was often around 10 feet (3–4 m). Manipulating this amount of woollen fabric required skill, and certainly in the eighteenth century the fabric itself, made from tightly twisted

yarn woven to give a material that could shed water, had a very hard feel and did not drape particularly well.[18]

In the south of Scotland, shepherds continued to wear the plaid or 'maud' in the mid nineteenth century, and for them it was an item of work clothing, this no doubt ensuring its retention almost until the present day. Amongst its uses, one end could be made into a pouch for carrying a lamb. It could be of a natural grey colour, made from the native wool, but the style of the Borders was to have black and white or blue and white checks.[19] It was in widespread use in Dumfriesshire, where the 'people ... seem never to forget that a black and white homemade plaid (be it in the month of June or January, and the weather what it may) is essential to complete their dress when they go abroad, to church or market'.[20] The plaid was highly praised in Old Cumnock, Ayrshire. It was said to be worn there by all, summer and winter and, 'when tastefully put on, it contributes very much to give elegance of form and appearance, more especially to the female figure; which, perhaps, is one reason why Ayrshire women are so much noted for their fine form'.[21]

As well as being a substitute for a cloak, the plaid could be draped around the lower half of men's bodies and belted at the waist to form a type of skirt, with the rest carried over the shoulders. Underneath, a linen or woollen tunic might be worn, and a short-waisted doublet. Wealthier men, including clan chiefs, might also wear trews below, which made a more comfortable outfit for wear on a horse. Trews were cut on the cross of the fabric in order to fit the leg and lower body closely. A relic of fifteenth-century fashionable footed hose, they were developed to cover the lower part of the body when tunics rose to waist level.[22]

The origin of the separate kilt is not clear. It has been attributed to a wish to create a better garment for working in than the cumbersome belted plaid, and certainly it would be less of a hindrance when doing manual work.[23] Cattle, however, were the main source of wealth for Highlanders and for droving the plaid was very suitable. Whatever the origin of 'the kilt' as it became known, it was in use by the mid eighteenth century.

By the time of the *NSA*, Highland forms of dress were dying out, to be replaced by types popular in the south of the country.[24] This was sometimes deplored, for it was claimed that as a result people were less hardy and more subject to disease.[25] But the 'Highland garb' was kept for young children, for whom it may have been thought healthier,[26] and for 'extraordinary occasions'. It was also used as a kind of status symbol, for young gentlemen and those who wanted to emulate them would wear the philabeg and other paraphernalia of the 'garb of old Gaul'.[27] The present-day concept of Highland dress as a symbol of middle-class aspirations, something to be worn by men at weddings and other festive occasions, was in the course of being born.

Probably it was in the Hebrides that the home making of woollen cloth continued most.[28] Here the checked patterns were more colourful than elsewhere but without the range of specific attachment to clans of today's often commercially promoted tartans,[29] although as early as 1845, the local minister at Ardersier, Inverness-shire, was observing an 'endless' variety of clan and fancy tartans displayed in the cloaks and shawls of women.[30] 'Plaid' and 'tartan' appear to have described types of fabric, both being a twill weave of some kind, but their exact early appearance is uncertain.

Head coverings

Women's head coverings were sensitive indicators of change (see Figs 7.2, 13.6). Married and elderly women often continued to wear the 'toy' or the 'mutch' in the mid eighteenth century. The toy was a cap of linen or wool, flat-crowned, with a back flap reaching to the shoulders. Known from the seventeenth century, the word is of Dutch origin. Its close neighbour, the mutch, was a close-fitting cap of white linen or muslin, with a goffered, gathered or trimmed border. It too was used especially by married women. Both types were tied under the chin. The mutch is referred to from the fifteenth century, and is also Dutch in origin. In Kilmory, Bute, many of the married women continued to wear the 'high cap' or mutch, instead of the bonnet.[31] In Banffshire, it was recorded that some years earlier women had gone to church bareheaded, but that they now put on a white muslin cap, presumably a form of mutch, or a straw bonnet.[32] In Clyne, Sutherland, no young damsel was to be seen without a neatly-made cap, with tastefully braided hair and dress according to the imported fashion[33] (Fig. 14.2). In Ardclach, Nairn, younger women formerly wore a snood, now being replaced by a cap or straw bonnet, and matrons who had been accustomed to wearing the toy were changing to the mutch, and avoiding the bonnet, 'which would be considered by most of them unbecoming their station'.[34] In Durness, Sutherland, the heads of women were often covered with gauze or muslin

Figure 14.2 'Highland Dance', by David Allan, about 1780. The men wearing kilts have shorter jackets than those wearing breeches, whilst the seated man on the right has trews and a long coat. The women wear short jacket-like bodices with striped petticoats, probably of a locally made woollen fabric, with white aprons. Married women wear a white cap and white neckerchief. Unmarried women wore their hair loose, bound with a ribbon. All are wearing buckled shoes. Source: Reproduced by kind permission of the Trustees of the National Galleries of Scotland.

caps, and now and then a straw bonnet, and umbrellas had begun to be as numerous as greatcoats or mantles.[35] In North Uist straw bonnets had largely replaced the old, neat head dress, where the hair was kept together with a single comb and held in place with 'a slight kerchief'.[36] Caps would only be worn by married women, although at home, types of mutches were worn in bed as night caps, by boys as well as adults.[37]

From the fourteenth century, there are references to bonnets. Bonnets had numerous variations: when worn by royalty they could be of a rich crimson and made of cloth such as satin; they could be round or four cornered, and some of them were double, probably to provide insulation in cold weather; there were 'night bonnets' for wearing in bed; and a range of colours was possible – red or crimson, black or white, or blue. Blue was particularly frequent, being noted from *c* 1568, and after around 1700 'blue bonnet' was used specifically of a flat-topped, round cap, without a snout. By a process of metonymy, and with varying degrees of opprobrium, it could be applied to wearers of such caps, namely Scotsmen, then to religious groups, Covenanters and Presbyterians, while in Scott's *Rob Roy* a 'band o' blue bonnets' was synonymous with marauders.[38]

Another name for the same item was 'braid bonnet', with reference to its often considerable width, about 11 inches (28cm) in diameter, and 'Kilmarnock bonnet', since the bonnets were often made in that Ayrshire town. A 'Kilmarnock' could also be 'a knitted woollen conical skull cap, worn as a night cap, or by indoor workers such as weavers and shoemakers'. Similar items of indoor head gear were the 'cowl' and 'pirnie' or 'pirnie cap'. These could be striped in a variety of colours, such as red, blue and white.[39]

Hand-knitted and felted, bonnets were made by specialists, the craft being established in Dundee by 1496, although by the 1840s the making of broad bonnets in the town had almost died out.[40] Fynes Moryson comments on the blue bonnets worn by the Lowland men he saw on his travels around Scotland, published in 1617. John Taylor similarly notes their wear in the Highlands the following year. The earliest surviving bonnet is likely to be that found on a body on Dava Moor, Moray, which probably dates to the late sixteenth century.[41]

From blue bonnets developed the other styles of headwear found in the nineteenth century such as the feathered bonnets of the military and the glengarry.[42] In many places there was an age-specific relationship; for example, in Moray in the mid nineteenth century farmers might still wear the old-style, broad, flat bonnets, but the glengarry was adopted by young men and boys,[43] while in the parish of Carmylie, Angus, only 'a few blue bonnets of goodly extent' were to be seen amongst the old men.[44]

Shoes

Shoes could be homemade. 'Rullions' or 'rivlins' were shoes or sandals of untanned hide, with the hair side outwards (see Fig. 18.1). They were moulded to the shape of the foot while still pliable, and were in wide use from at least the fifteenth century. In the 1830s it was said of the men in

Ardclach, Nairn, that they were their own tanners and shoemakers, and in Orkney that many young men made shoes for themselves and others in the winter evenings.[45] This activity was no doubt widespread.

Readymade shoes have a long history. Sometimes they formed part of the wages in kind given to farm workers. An East Lothian farmer's account book for 1729–70 records the mending and soling of shoes in 1751 and 1765, the putting on of a pair of 'beend soles', and 'solling & hell topping' in 1768. Two pairs of shoes costing 5/- were part of a wage in 1763.[46] Figures on gravestones can provide good illustrations of footwear and dress. For example, an Angus stone of 1759 shows a man with a stout pair of heeled boots or shoes.[47]

Body clothes
In terms of body clothes, at the beginning of the nineteenth century in Barvas, Lewis, and in North Uist, blue kelt was reported as almost the only dress worn by men, whilst the women wore variously striped clothing materials with under dresses of plaiding, all homemade. Kelt is a word of Gaelic origin. In earlier times it was a coarse cloth, of grey, black or black and white wool mixed, used for outer garments such as large and short coats and gowns. It was the equivalent of frieze.[48]

However, in many cases, garments made from new types of cloth, 'the cottons and calicoes of Glasgow', were beginning to replace some of these items. In Glensheil, Ross-shire, the ordinary male clothing was a short coat or jacket and trousers of kelt, blue or chequered of different colours, homemade, with a cotton waistcoat, shirt and neck cloth, a flat or cocked blue felt bonnet, stockings and shoes, the latter made by each individual for himself of home-tanned leather. Stockings, however, were spun and 'wrought' by the women. The women themselves, until forty years before, had worn worsted materials spun and dyed in the home, usually blue with narrow stripes of red, and woven by country weavers. Now such clothes were worn in the field or when at other forms of labour, but at church and social gatherings cotton garments were donned. The shift, a chemise-like undergarment, was always of cotton. In Durness, Sutherland, the previous 20 or 30 years had seen the disappearance of tartan or kelt coat and trousers, spun and dyed at home. Young men wore mostly fustian jackets and trousers or the lighter tartan of shops, and here and there the blue and fancy cloths of Leeds. The blue mantle and well-spun blue gown of the women had been replaced by prints and Merinos.[49]

Cloth
Cloth, until the mid nineteenth century, was all handmade. For most Scots the fibres available until the nineteenth century consisted of wool and linen. Both are hard wearing. Wool is best for creating warm clothing, necessary given the country's climate, but is less easily washed than linen, which was used for making clothing worn next to the skin. Here it could absorb body fluids and keep them from soiling woollen outer garments and causing them

to smell. Underwear also helped to protect the body against the roughness of some outer clothing or from dye rubbing off on the skin.

Linen does not take dyes easily but it can be bleached to a good white. Wool on the other hand takes dyes well and a wide range of colours could be created using the available plant and mineral resources. There was much dyeing of wool at home. Natural substances which could produce different colours were well known to the women, especially, it seems, in highland areas. Part of the preparation was the keeping of a tub of domestic urine near the fire.[50] The combination of wool and linen therefore gave Scots the basics for making warm, colourful clothing. The only stumbling block was lack of wealth, for the more processes required to turn sheep's wool and flax into clothing the more expensive the garment.

All the processes from spinning the raw fibre to making the finished garment were hand operations, many undertaken in the home. Gradually, during the eighteenth and early nineteenth centuries, machines were developed which could imitate hand operations and do the work more quickly, although spinning wheels are still in use for handcraft purposes and were to be found in many houses in the Highlands and Islands within the last few decades. It was said in 1834 of the parish of Westruther in Berwickshire, for example, that 50 years before all clothes had been homemade. 'Women's gowns and petticoats were woollen stuffs of various colours, and men's clothes were spun by the women, and sent to the weaver or dyer.' But by the 1830s, nearly all the weavers had gone, there was little home manufacture of clothing items and it was, in fact, cheaper to buy clothes readymade.[51] Similarly, in Traquair, Peeblesshire, the spinning wheel had been almost laid aside, 'nearly every article of raiment being now purchased instead of being made at home'.[52] Eventually the massive factories built in the nineteenth century could produce cloth in a fraction of the time and at a fraction of the cost of hand-produced fabric.

Nineteenth-century mechanisation of every aspect of making cloth and clothes gave people far greater options (see Figs 19.2, 19.3). Cotton virtually replaced linen as the fabric of choice for underclothes and bedclothes. Once the problem of machine spinning had been overcome, cotton could be produced more cheaply than linen. It was also easier to launder and iron, and took dyes very well. Silk too became cheaper and there was a good deal of experimentation with combining fibres, such as silk and cotton to create cheaper silks. Fibre combination had been used earlier but it became more common in the later nineteenth century. Synthetic dyes replaced the old natural dyes. There were also other treatments which fabrics were subjected to, the most notorious being the weighting of silks with mineral salts. When used on dress linings or petticoats it created a lovely frou-frou noise as women walked. Today it poses a conservation problem for museums because the salts have eaten into the silk fibres, causing them to split and disintegrate.

ACQUIRING CLOTHES

Clothes were acquired through various means. Many garments were made at home by female members of the household.[53] Those with money had items made by tailors and dressmakers of repute based in leading burghs. Others used local tailors and dressmakers who may not always have had much expertise in their craft. A widespread custom was to employ the services of an itinerant tailor who would travel from place to place on request, receiving board and lodging in return for making clothing which was expected to be very durable. Such travelling craftsmen were not necessarily held in high regard, as is made clear by the contemptuous or jocular names they were given in Scots; 'pricker', 'prick the clout' or 'prick the louse'. Every parish had its quota of tailors of one sort or another, and most graveyards contain stones in their memory. On these can sometimes be seen the tools of the tailoring trade: scissors or shears, smoothing and pressing irons, a pressing board, a bobbin, a pin or bodkin, or a thimble.[54]

The skills of the tailor and seamstress had developed alongside the new fourteenth-century fashion in Europe for tightly fitted garments.[55] The tailor was responsible for outer garments whilst the seamstress mainly sewed underwear, although the latter was usually made at home by the women in the household. As in most medieval trades and crafts, tailors were a closed society regulated by a guild or incorporation, with tight rules to discourage non-members from practising the craft. Although incorporated quite early they did not usually constitute one of the wealthier crafts.

By the mid seventeenth century tailors were under threat from ready-made clothes and from the new fashion for women of a loose garment known as a mantua.[56] This was in Scots a 'mantea', 'manta', 'manto' or 'mantie', recorded from the second half of the seventeenth century. It was an unshaped robe based on a simple T-shape like a kimono. Mantuas were popular with women as undress, that is wear for non-formal occasions, but they could be made in sumptuous fabrics. Because they were unshaped, mantuas required no skill in cutting out, which was the tailor's real expertise, a good tailor being able to cut out a garment without marking out the pattern on the cloth. Tailors eventually lost the battle to make mantuas and by the nineteenth century most women's clothes were made by women, except for riding habits. In fact, female dressmakers were commonly called 'mantie makers'.

A major period of change occurred in the first half of the nineteenth century, when economic circumstances were improving generally. By the mid nineteenth century, the fashion for buying readymade clothes was well established. The manufacture of readymade clothing is an industry whose early development is only now beginning to be understood.[57] The rise of standing armies and the wish to clothe them in uniform was clearly a driving force behind this development. Some items of costume had been readymade well before this. The simple basic garments, such as linen or woollen under and outer tunics, would have been easy to make in a general, 'one size fits all',

Figure 14.3 A German Gypsy family, about 1906. Most of the items worn by this group are the same as those worn by Scottish agricultural labourers. The exotic look is created by the way the clothes are worn and a certain dandyism in the men's dress, which would have been out of place in the everyday wear of a farm labourer. We do not know if these garments were acquired first or second-hand. Source: SLA 119(1) 56/28/27.

which could then be altered if necessary. However, readymade clothing needs an assured market and in an essentially rural economy where barter or self-sufficiency rather than money were the main means of procuring food and clothing, readymade clothes were not profitable, although matters were changing by the 1840s. In towns or seaports where there was a population divorced from its rural roots, readymade clothes had a strong appeal, but the evidence from the *NSA* shows that by the mid nineteenth century, they were appreciated throughout most of Scotland, especially amongst women. It was an option particularly relevant for those away from home.

Within households, the passing on of clothes and altering or remaking them was very common. There was also a very thriving second and even third and fourth-hand market, of a kind which still exists today in the form of charity shops, such as those run by Oxfam; jumble sales and retro clothing shops. Indeed, until the mid twentieth century, clothes were worn until they were mere rags. There was always someone lower down the social scale for whom the cast-offs would be suitable. Even today some clothes which are well past being respectable wear are still put to use, such as shirts bought second hand to wear for dirty jobs.

MENDING

With cloth and clothes being either laborious to make or expensive to buy, mending them when they were worn was a necessity for all but the very wealthy. Apart from replacing lost buttons and other fasteners, tears in garments had to be mended as invisibly as possible. Girls might learn to do this by making darning samplers, where they imitated in thread the weave of various fabrics. Darning stockings was a particular chore, and if they got too worn new feet might have to be knitted and attached. Men's shirts regularly had the collars and cuffs turned when they were worn. Women's skirts in the later nineteenth century had brush braid sewn inside the hem so that it was protected from dirt and wear. One of the best methods of preserving clothing from undue wear and dirt was to employ protective clothing. False sleeve ends and aprons were the most obvious extra garments worn for protection. A coarse type of apron for rough indoor or byre work by women was a 'brat', so called from the fifteenth century, and a widespread protective garment in the twentieth century was the 'peenie' or pinafore, often of flower-patterned cloth, homemade or bought from a shop, travelling van, or through one of the clothes catalogues which were widely browsed over.[58]

CLEANING CLOTHES[59]

People in the past, especially in the pre- and early industrial periods, did not have the washing and cleaning facilities – including the convenient and ample water supply – of the present day. Most individuals undertook some form of heavy manual work and so sweated regularly, and the open fires of many homes were fuelled by peat or wood which left their smoky aromas in the clothes of the dwellers.[60] People were, however, used to stronger smells than we would tolerate today when it is possible for clothes and bodies to be fragrant or at least odourless at all times.

Washing was a major operation until the widespread adoption of the home washing machine in the mid to late twentieth century (see Figs 17.7, 17.8, 20.2). The washing of clothes and household linen was labour intensive and arduous work, depending on the number of people and possessions in the dwelling. It is said that some people believed in being sewn into their underwear for the whole winter. But those who could afford it changed their linen regularly so that at least the clothes which came into contact with their bodies were reasonably fresh. It is not always easy to find out how often this happened or how frequently wash days occurred. Many writers have quoted accounts which suggest that washing be done only once a year. However, whilst living in Inverness in the 1720s, the Englishman Edward Burt had his linen washed once a week. He was living away from home and perhaps did not have too many changes of clothing with him.

In the 1770s Susannah Whatman, living in Kent, wrote one of the few surviving household instruction manuals to have been created before the mid nineteenth century.[61] In it are clear instructions for the laundry maid. She was to start at midday on Monday to wash the small and fine things of the female members of the family as well as the gentlemen's shirts and neckcloths. She had to rise early on Tuesday morning to do the main wash. The clothes were to be dried outside if the weather permitted. On Wednesday she folded the clothes, on Thursday or Friday she mangled them. Great detail is gone into on mangling which Susannah obviously regarded as very important. Mangling was a means of giving a glaze to dry heavy linens, and was rarely used for garments. In contrast ironing is not mentioned by Susannah.

It is clear that Susannah Whatman regarded washing as a weekly event. It was also the custom in the royal household, for example, but by the 1830s it appears to have been accepted that a yearly or quarterly wash was preferable.[62] Others washed at monthly or six-weekly intervals. This assumed that the household would have enough underwear and bed and table linen to last for that length of time. It did not mean that members only changed those items on a yearly or monthly basis. Such practices would have resulted in a vast quantity of dirty clothing and linen requiring storage somewhere and must have exacerbated problems of stain removal. The reason for an infrequent wash was the sheer upheaval washing caused and the effort involved. Many households, however, sent their laundry out because they did not have the facilities or space to do the work at home. This was true of those who lived in cities and towns, including students, or those travelling away from home. The very wealthy, who had homes in the country, when in town would send their dirty washing back so that it could be dealt with in more spacious surroundings.

Infrequent washing of fabric items indicated that one was sufficiently wealthy to afford enough sets of clothes and linen to see one through. But gradually during the nineteenth century the weekly wash came to be considered the normal course for those in the middle and lower-income brackets. The classic weekly wash was immortalised in the song, 'Twas on a Monday Morning When I Beheld My Darling ... ' Margaret Paxton, born in 1905, remembered washdays at Gattonside, near Melrose:

> Ma mother did the washin' wi'a scrubbing board and a wooden tub. What a job! And filling pails of hot water intae it. Monday wis her washing day – all done by hand.[63]

However, Nellie Traill, born in 1916, remembers her mother washing more than once a week:

> She sometimes wid have a washin' this day and then maybe half wey through the week she'd some mair tae dae, because there wis a lot o' us and we had tae dae a lot o' washins. Ye didnae have many spare claes. Ye were lucky if ye had twae chinges and that wis it. Ma mother

had to keep washin' a' the time. There was washin' oot every day near enough. And when there was a new bairn on the go and nappies, ye'll ken what that wis.[64]

Washing day had become a late nineteenth- and twentieth-century ritual, only declining when the electric-powered washing machine allowed working women to wash whenever it suited them. By the late nineteenth century manuals and women's magazines detailed every process so that the new housewife of good means understood what was required and could make sure her instructions were carried out by whoever was doing the laundry. Laundry work was also added to some school curricula so that pupils would know how to do it when they had their own homes or when they went to work, perhaps as laundry maids.

Before chemical bleaching in the nineteenth century, white linen was unlikely to have been the brilliant white that we associate today with white fabrics. Most linen was bleached outside by the sun on bleachfields, areas of a town, usually fields on the outskirts, where women could take the washing and lay it on the ground. In countries with indifferent sunshine for most of the year this might be difficult. It is therefore possible that the linen was bleached once a year outside to restore its whiteness but for the rest of the year the washers concentrated on getting the cloth washed, dried and pressed, a not inconsiderable job. The development of chemical bleaches enabled bleaching every time the garment or piece was washed.

Washhouses with boilers became a feature of new houses in the nineteenth century, which helped to keep the whole messy business away from the normal life of the home. Scottish tenements were often provided with a washhouse which had to be shared between eight, ten or more households depending on the number of flats on the stair. Each household would be allowed the use of the washhouse for a day turn about. Molly Weir's family in Glasgow lived in a tenement of twelve houses so that wash day came round only every 12 days. She comments:

> With the meagre wardrobes we all possessed it must have been a nightmare trying to keep families clean and in dry clothes for twelve days between washing days ... and a wet day a tragedy.[65]

For others there was the 'steamie', the public washhouse, where women could take their washing, rather like the more modern launderette. Launderettes were self-service establishments which replaced the laundries where many middle-class families had sent their washing, particularly household items. As more and more homes acquired washing machines so the old laundries declined but did not disappear completely. However, for those without a washing machine, including those renting rooms, the launderette offers a cheaper alternative to the laundry.

Dry cleaning of clothes which could not be washed was attempted with varying degrees of success.[66] It would appear that certainly by the eighteenth century dyers were also doing some cleaning of clothes, using various well-tried methods to remove stains. Clothes were often sent to them to be re-dyed, particularly black if there was a death in the family, but also to refresh any older garments. If an item was a light colour it could be re-dyed a darker shade. Silk and heavier woollen garments were the most usual items sent to be cleaned but all kinds of items were re-dyed.

STORING CLOTHES[67]

Clothes which were removed before going to bed had to be stored overnight. By the mid nineteenth century airing clothes at night was seen to be important because it helped to stop the accumulation of fusty odours within them. Therefore underclothes might be draped on a chair. For those who had very few clothes storage was not a problem. Many poor people, for example, would be wearing all the clothes they owned on a cold night. During a cholera scare, it was found that at Bracadale in Skye, 140 families had no change of day or night clothes.[68]

Only the very wealthy in the past would own enough clothes to have a choice of what to wear every day. For most people the clothes worn one day would be worn again the next, and so on, except on special occasions and for going to church, when new or best outfits might be put on. However, if a person possessed more than one outfit these needed to be stored and kept clean, dry and away from any vermin. The simplest storage solution was a nail driven into the wall, found in bothies as well as in servants' bedrooms, even in grand houses.

Clothes and other possessions were mostly kept in kists or chests, of various kinds and sizes, until well into this century. Servants, for example, would go from place to place with a kist which held all their possessions (see Fig. 7.1). The main disadvantage of a kist was the difficulty of getting at items at the bottom. Another problem was keeping clothes from getting too creased, although people in the past were not quite so concerned about creased clothes as their descendants are. However, for some fabrics, velvet for example, creasing was a significant problem and other delicate materials, or trimmings, could easily be crushed and lose their freshness.

The freestanding press was another piece of furniture which could be used for clothes storage. This possessed shelves, making it easier to access items than in a kist. Presses could also take the form of built-in wall cupboards and it is not always clear in inventories which type is meant. Some built-in cupboards in houses had wooden pegs onto which certain garments could be hung by the armhole. Coat hangers only came into use in the late nineteenth century, although dress skirts had been provided with small hanging loops sewn into the waistband for many years previously. Mid to late nineteenth-century wardrobes often had small metal hooks inside onto which the skirt loops could be hung.

As more people became wealthier in the eighteenth and especially the nineteenth century they acquired more clothes and needed additional places to store them. Different types of clothing demanded new forms of storage.

Many more small items were required to complete a toilette, such as separate white collars and undersleeves, and gloves were used in greater quantity than in previous periods. Chests of drawers and dressing tables were items of furniture developed to deal with the greater quantity and variety of clothes and accessories (see Fig. 15.4). Chests of drawers, of various designs, became popular during the eighteenth century. These had three or four long drawers with possibly two smaller drawers at the top. The Scottish or lum chest, a nineteenth-century development, had a larger central drawer for storing a man's hat or a woman's bonnet. Chests became taller in the mid nineteenth century with more drawers which could store both clothes and household linen. In the nineteenth century the dressing table developed into a relatively specialised piece of furniture with several small drawers for jewellery, gloves, handkerchiefs and other small items.[69]

Stand-alone wardrobes are found in eighteenth-century furniture-makers' catalogues. At this period they usually possessed shelves at the top and drawers below. In the twentieth century wardrobes were developed which were especially suitable for men's clothes in which trousers were usually hung over the rail of a coat hanger. Such wardrobes were often not as tall as those wardrobes for women's clothes which could still include full-length dresses. Another innovation was to have separate drawers in the wardrobe for shirts, pyjamas, collars and gloves, with spaces for hats and shoes and a rail to hang ties over. There might even be small drawers to take collar studs and other small items. Suites of bedroom furniture therefore often had separate wardrobes for the husband's and the wife's clothes.

Today, built-in wardrobes, including drawer storage, are often provided within dwellings. New, imaginative and cheaper ways of storing clothes have also been developed for those who cannot afford or do not like built-in furniture. These include dress rails for hanging clothes, fabric hanging shelves and boxes and baskets of various sizes.[70]

NOTES

1. Illustrated in Catalogue of the National Museum of Antiquities, Edinburgh, 1892, 352.
2. *NSA*, XIV (1845), 647: Barvas, Lewis.
3. *NSA*, III (1845), 199: Hounam, Roxburgh.
4. *NSA*, III (1845), 66: Tweedsmuir, Peeblesshire.
5. *NSA*, V (1845), 27: Kilbride, Bute.
6. *NSA*, XIV (1845), 377: Urquhart and Logie Wester, Ross-shire.
7. *NSA*, VIII (1845), 72: Tillicoultry, Clackmannanshire.
8. *NSA*, XV (1845), 7–8: Dornoch, Sutherland.
9. *NSA*, XV (1845), 128: Edderachillis, Sutherland.
10. *NSA*, XIV (1845), 73: Kingussie, Inverness-shire; *NSA*, XIV (1845),123: Kilmallie, Inverness-shire.
11. *NSA*, III (1845), 326: Kelso, Roxburghshire; *NSA*, III (1845), 292: Bedrule, Roxburghshire; *NSA*, V (1845), 110: Ochiltree, Ayrshire; *NSA*, VI (1845), 34: Lesmahagow, Lanarkshire.
12. Aitchison, 2001, 60, 64.

13. *NSA*, III (1845), 64: Melrose, Roxburghshire; *NSA*, IV (1845), 60: Kirkmahoe, Dumfriesshire; *NSA*, V, 223: Dalry, Ayrshire.
14. *NSA*, XI (1845), 42.
15. *NSA*, IV (1845), 181: Penningham, Wigtownshire; *NSA*, VI (1845), 824: Wandell and Lamington, Lanarkshire.
16. *NSA*, XIV (1845), 90: Alvie, Inverness-shire.
17. *NSA*, II (1845), 76–7: Westruther, Berwickshire.
18. We do not know what seventeenth-century tartan felt like or how it was woven. The 'hard' tartan, so called because of its feel, is similar to the fabric used to make some women's riding habits in the eighteenth century. It was made of finely combed yarn but in the 1830s it began to be replaced by the softer Saxony wools. This new type of fabric had better draping qualities, and allowed the greater amount of fabric needed for the deeply pleated kilts which became a feature of the nineteenth and twentieth century. Cheape, 1991, 52.
19. *SND*, s v Maud, Plaid; *NSA*, III (1845), 292: Bedrule, Roxburghshire.
20. *NSA*, IV (1845), 159: Johnstone, Dumfriesshire; see also *NSA*, IV (1845), 115: Moffat, Dumfriesshire; *NSA*, IV (1845), 281: Kirkpatrick Fleming, Dumfriesshire; *NSA*, IV (1845), 433: Westerkirk, Dumfriesshire.
21. *NSA*, V (1845), 483,
22. Bennett, 1980, describes, with pattern, the earliest-known surviving trews outfit dated to 1744.
23. Dunbar, 1979, 12–14.
24. *NSA*, XIV (1845), 107: Moy and Dalarossie, Inverness-shire; *NSA*, XIV (1845), 323: Kiltearn, Ross-shire.
25. *NSA*, VII (1845), 63: Kilninver and Kilmelfort, Argyll.
26. *NSA*, XIV (1845), 73, 79: Kingussie, Inverness-shire; *NSA*, XIII (1845), 73: Knockando, Moray.
27. *NSA*, XIV (1845), 428: Laggan, Inverness-shire.
28. *NSA*, XIV (1845), 154: Uig, Ross-shire; *NSA*, XIV (1845), 157: Harris, Inverness-shire; *NSA*, XIV (1845), 319: Sleat, Inverness-shire.
29. Orr, 1920–1.
30. *NSA*, XIV (1845), 474: Ardersier, Inverness-shire.
31. *NSA*, V (1845), 59: Kilmory, Bute.
32. *NSA*, XIII (1845), 257: Rathven, Banffshire.
33. *NSA*, XV (1845), 156: Clyne, Sutherland.
34. *NSA*, XIII (1845), 34: Ardclach, Nairn.
35. *NSA*, XV (1845), 97: Durness, Sutherland.
36. *NSA*, XIV (1845), 147: Barvas, Lewis, Ross-shire; *NSA*, XIV (1845), 172–3: North Uist, Inverness-shire.
37. *SND*, s v Mutch.
38. *DOST*, s v Bonet; *SND*, s v Blue Bonnet.
39. *SND*, s v Cowl, Kilmarnock, Pirnie.
40. *NSA*, XI (1845), 21: Dundee.
41. Bennett, 1983, 553.
42. Bennett, 1983, 546–65.
43. *NSA*, XIV (1845), 474: Ardersier, Inverness-shire.
44. *NSA*, XI (1845), 301: Carmyllie, Angus.
45. *NSA*, XIII (1845), 33: Ardclach, Nairn; *NSA*, XV (1845), 19: Orphir, Orkney; see also Chapter 18 Auxiliary Work in the Rural Home.
46. Fenton, 1966, 26, 42–3, 47.
47. Willsher, Hunter, 1978, 16.
48. *DOST*, s v Kelt; *SND*, s v Kelt.
49. *NSA*, XV (1845), 96–7.

50. See, for example, Shaw, 1999, 53–5.
51. *NSA*, II (1845), 76–7: Westruther, Berwickshire.
52. *NSA*, III (1845), 46: Traquair, Peeblesshire.
53. See Chapter 9 Needlework.
54. Willsher, Hunter, 1978, 57 (Fife, 1705), 106 (Angus, 1780); Willsher, 1985, 34.
55. Tarrant, 1998.
56. Sanderson, 2001.
57. Lemire, 1997; See also Sanderson, 2001.
58. *DOST*, s v Brat; *SND*, s v Brat, Peenie.
59. See also Chapter 17 Housework.
60. See Chapter 6 Heating and Sleeping.
61. Whatman, 1987.
62. A Lady, 1975.
63. See MacDougall, 2000, 41.
64. MacDougall, 2000, 185.
65. Weir, 1988, 110.
66. Mansfield, 1968.
67. See also Chapter 4 Storage.
68. *NSA*, XIV (1845), 297: Bracadale, Skye.
69. See also Chapter 15 Toilette.
70. For example the catalogue of The Holding Company, 2001, shows leather boxes that can double as coffee tables, various small plastic boxes for jewellery and makeup, hanging shelves of canvas for shoes, underwear and shirts and under-bed bags for larger items.

BIBLIOGRAPHY

Aitchison, J. *Servants in Ayrshire 1750–1914* (Ayrshire Archaeological and Natural History Society, Ayrshire Monographs, 26), Ayr, 2001.
Bennett, H. A murder victim discovered: Clothing and other finds from an early eighteenth-century grave on Arnish Moor, Lewis, *PSAS*, 106 (1974–5), 172–82.
Bennett, H. Sir John Hynde Cotton's Highland suit, *Costume*, 14 (1980), 95–09.
Bennett, H. The Scots bonnet. In O'Connor, A, Clarke, D V, eds, *From the Stone Age to the 'Forty-Five: Studies Presented to RBK Stevenson,* Edinburgh, 1983, 546–66.
Bennett, H. *Scottish Knitting*, Princes Risborough, 1986.
Cheape, H. *Tartan: The Highland Habit*, Edinburgh, 1991.
Dunbar, J T. *History of Highland Dress* (1962), London, 1979.
Dunbar, J T. *The Costume of Scotland*, London, 1981.
Fenton, A. The Begbie farm account book, 1729–70, *Transactions of the East Lothian Antiquarian and Field Naturalists' Society*, X (1966), 26–47.
Henshall, A, Maxwell, S. Clothing and other articles from a late seventeenth century grave at Gunnister, Shetland, *PSAS*, LXXXVI (1951–2), 30–42.
Henshall, A. Clothing found at Hunsgarth, Harray, Orkney, *PSAS*, 101 (1968–9), 150–9.
A Lady. *The Workwoman's Guide* (1838), 2nd edn, 1840, reprinted 1975.
Lemire, B. *Dress, Culture and Commerce: The English Clothing Trade before the Factory, 1660–1800*, London, 1997.
MacDougall, I. *Bondagers: Eight Scots Women Farm Workers*, Edinburgh, 2000.
Mansfield, A. Dyeing and cleaning clothes in the late eighteenth and early nineteenth centuries, *Costume*, 2 (1968), 24–9.
Maxwell, S, Hutchison, R. *Scottish Costume, 1550–1850*, London, 1958.
Maxwell, S. Two eighteenth-century tailors, *Transactions of the Hawick Archaeological Society* (1972), 3–29.

Orr, S. Clothing found on a skeleton discovered at Quintfall Hill, Barrock Estate, near Wick, *PSAS*, LV (1920–1), 213–21.

Sanderson, E. 'The New Dresses': A look at how mantua making became established in Scotland, *Costume*, 35 (2001), 14–23.

Shaw, M F. *Folksongs and Folklore of South Uist* (1986), Edinburgh, 1999.

Tarrant, N. *Going to Bed*, Edinburgh, 1996.

Tarrant, N. *The Development of Costume* (1994), London, 1998.

Walkley, C, Foster, V. *Crinolines and Crimping Irons: Victorian Clothes, how they were Cleaned and Cared For*, London, 1978.

Weir, M. *Molly Weir's Trilogy of Scottish Childhood*, London, 1988.

Willsher, B. *Understanding Scottish Graveyards*, Edinburgh, 1985.

Willsher, B, Hunter, D. *Stones: A Guide to Some Remarkable Eighteenth-Century Gravestones*, Edinburgh and Vancouver, 1978.

Whatman, S. *The Housekeeping Book of Susannah Whatman, 1776–1810*, London, 1987.

15 Toilette

SUSAN STORRIER

Let there be no doubt about it – how we look counts for a great deal. Ebin is not alone in recognising that:

> the body is the physical link between ourselves, our souls, and the outside world ... the medium through which we most directly present ourselves in social life; our use and presentation of it say precise things about the society in which we live, the degree of our integration within that society, and the controls which society exerts over the inner man.[1]

The three most important elements in the physical presentation of the individual are costume (both clothing and accessories), personal hygiene, and what might be called grooming or 'toilette'; that is, the additional care, enhancement and adornment of the body itself. This chapter offers a framed exploration of some aspects of toilette in Scotland, as expressed primarily within the domestic context.

ATTITUDES IN SCOTLAND TOWARDS APPEARANCE AND TOILETTE

As far as it is possible to discern, Scottish attitudes towards matters of toilette and the body's appearance in general seem, to a substantial degree, to be no different from those found in many other parts of the world. Yet it is not to say that there have not been some particularly Scottish, or at least northern European, emphases. This section looks at some ideas, of varied provenance, which have found prominence in Scotland.

Some Scottish Ideals and their Converse
Each of us has our own particular set of notions concerning appearances, yet most people's views usually have much in common with biological imperatives and shared cultural values playing remarkably large roles. The arts often dwell on our hopes and aspirations and ideals of beauty are thoroughly explored there. Scots song and poetry, for example, make clear what is most desirable in the appearance of women:

> For Nancy's hair is yella like gowd.
> And her e'en are like the lift [sky], sae blue.[2]

As this example indicates, one prominent ideal of feminine beauty focuses on golden or brown hair (which should also be long and curly) and blue eyes, along with rosy cheeks in an otherwise white – and faultless – complexion, even, white teeth and a slender but rounded figure. The image is one of health and youth and by implication fertility. The arts of Gaelic Scotland have at times elaborated on the male appearance in at least as much detail as womanly comeliness. Here ideals of beauty tend to be more closely entwined with matters of character or position[3] and describe in physical terms not only health and relative youthfulness but also strength. Within the Gaelic panegyric code, for example, esteemed physical features for men include soft white skin, blue eyes and long dark curly hair, as well as tallness and a commanding stature.[4]

Commentaries on remarkable beauty, real or imagined, abound, but there are fewer references to what makes for a less valued appearance. The following proverb however makes clear the relative ranking for colouring:

> Is fhearr a bhi dubh na bhi bàn; is fhearr a bhi bàn na bhi ruadh; is fhearr a bhi ruadh na bhi carrach; is fhearr a bhi carrach na bhi gun cheann
>
> [Better be black than fair; better be fair than red; better be red than scabby; better scabby than no head].[5]

The view that red hair is not only inferior but somehow dangerous is found within traditional belief:

> Bean ruadh dhubh-shuileach, cù lachdunn las-shuileach, fear an fhuilt dhuibh 's na fiasaige ruaidhe – na trì comhlaichean a 's mios' air bith
>
> [A red-haired, black-eyed woman; a dun fiery-eyed dog; a black-haired, red-bearded man – the three unluckiest to meet].[6]

At the more specific level, extreme criticism can be found in Elizabeth Grant's description of one early nineteenth-century gentleman whose looks were felt to be very lacking indeed:

> ... his personal appearance ... was truly hideous. He was very lame, one leg being many inches shorter than the other, and his countenance, harsh and heavy when composed, became demoniack when illumined by the mocking smile that sometimes relaxed it. I always thought him the personification of the devil on two sticks, a living, actual Mephistopheles.[7]

Although there appears to have been more liberty to call a man ugly (a man's appearance and his feelings about this seemingly sometimes regarded as of less consequence than those of a woman), women considered not sufficiently attractive have not uncommonly found themselves censured, in this example through the diffuse lens of folktale. In 'The Ugly Queen' a young mother, transforming herself into an old but rich woman in order to rescue her abducted child, is described as '... the ugliest old crone, warts on her face, nose as long as anything, chin hanging down ... '.[8] For both sexes deformity, no matter how minor, contributes to the effect of a poor appearance. For women it is otherwise enough to be old. For men symbols of weakness in body and character count most against them.

Looking Good: 'Beauty', 'Attractiveness' and 'Decency'
BEAUTY
It could be argued that 'beauty', although an often-used word, in bodily terms essentially describes something with considerably more impact than simply looking good – an appearance with the ability to transcend the ordinary, perhaps to the extent of imparting a sense of the immortal. Jessie Kesson, for example, in her autobiographical novel, *A White Bird Passes*, recalls her mother in her youth, saying, 'her beauty made the bright world dim'.[9] Although such beauty may be the ideal, the distant goal, and the facet of physical appearance most celebrated in the arts and media, the more quotidian concepts of either attractiveness or decency are far more likely to make their presence felt in real life.

ATTRACTIVENESS
Looking attractive brings benefits in many walks of life. Pretty children, for example, may find themselves favoured at school and among family and friends and in adult life popularity, wealth and fame often seem to adhere more to the good looking than the plain. Some employment opportunities are only open to those deemed suitably good looking. While in other workplaces looking attractive is not obligatory but may go some way towards helping to obtain and maintain employment and assist advancement. Physical characteristics which make for attractiveness draw from beauty ideals but also emphasise normality: youthfulness, health, regularity of facial features and deemed suitable height and weight are usually key elements, lacking which a person may be labelled unattractive or plain.

Attractiveness is usually most potent in sexual terms. Although forming a partnership within which to create offspring may not always be prominent in the mind of individuals sexually attracted to one another, it seems that, subconsciously at least, we usually find others who look capable of producing and adequately supporting healthy children the most desirable. Numerous elements, however, are involved in sexual attraction, and only some of these concern the body. Confidence, charm, sense of humour, intelligence, modesty, assertiveness, good manners, tact and eloquence, for example, may be seen greatly to shape a person's degree of desirability. Other matters, such as wealth, education and correct social status can sometimes be even more significant.[10]

Of course, being sexually attractive is by no means a universal concern. Some adults reject the whole business for reasons including personal preference and religious observance. Pre-pubescent children will not be concerned and traditionally the old and very ill and sometimes the disabled have often been regarded, no matter how incorrectly, as also beyond the mating game.

DECENCY

Looking acceptable in terms of social correctness – looking 'respectable' or 'decent' – is a far wider matter than attractiveness and one long taken very seriously throughout most of Scottish society. The individual's appearance is used here to state to the world an assessment of his or her social position, either within the general social hierarchy, or within a specific hierarchy (such as that of the workplace), or membership of a sub-group (for example a youth cult). This is no easy matter. A person's position changes repeatedly through life and at any one time he or she may have more than one social existence with ideas of what is correct within a level or group changing in response to fashion.

The blurring of ideas of sexual attractiveness and decency sometimes occurs. Young women, for example, may be viewed as socially correct by being sexually attractive, even in contexts where sexuality is ostensibly not an issue.

Enhancing and Altering the Appearance

While decency can, in general, be achieved equally by those naturally favoured or disfavoured in appearance terms, attractiveness is much more dependent on the blessings of nature. However, when the requisite physical attributes are lacking in some way it may be possible to employ artificial assistance, although it has not always been practicable or acceptable for people to do this – especially openly – depending on factors such as when and where they have lived and their age, gender, marital status, wealth and social class.

Culture and Religion, Vanity and Deception

Maintaining a decent appearance seems to have been an obligation long laid on Scots. Undue concern with attractiveness, in contrast, has been at times discouraged or even condemned by religious and civil authorities, although usually without much sustained success. This is because attractiveness can be regarded as containing an element of narcissism, beyond the bounds of self-respect.[11] Religious condemnation of such 'vanity' seems to have occurred as early as the medieval period and what has been regarded by some as the beginnings of individualism in appearance.[12] Despite the ineffectualness of such church disapproval:

> ... the legal approach of the Middle Ages became a tradition which was given new strength by the rise of the Protestant Church and its strong belief in modesty and simplicity in dress. Worldly show was considered sinful and its wearer an empty vessel not fit for society![13]

Leneman and Mitchison claim that 'by 1660, for most of Scotland, the dominant features of Calvinism were well established',[14] and these might be assumed to have commonly brought in their wake an emphasis on an unadorned, plain appearance (maybe accompanied by an additional stress on 'decency'). Certainly adoration of good fortune in terms of physical appearance was being frowned upon over a century-and-a-half later[15] and extensive or overt attempts to enhance or create artificially an attractive appearance were still unlikely to have been tolerated then amongst large parts of the rural population. Yet the impression is that authoritative strictures at no time fully convinced the Scottish population that vanity was completely undesirable and attractiveness altogether unnecessary. Those in more marginal positions within society seem to have had more liberty in some of aspects of life[16] and it may well have been that slightly greater licence in terms of appearance – very much tempered by poverty and other constraints – was sometimes allowed them.[17] There were also occasions – festivals, fairs, christenings, weddings and, ironically, church attendance (an opportunity especially seized upon by those of courting age) – when an inversion of the normal order allowed for more emphasis on looking good[18] (see Fig. 13.6). That said, common sense dictated that on all occasions attractiveness had to be approached with caution so as to avoid not only the moral damage of vanity but also the practical matter of deception.[19] However, by the eighteenth century, ideas suppressing attractiveness, derived from the Church or from everyday level-headedness, were being challenged by much of wealthy society, while in urban areas, with relative prosperity and an influx of population, there was more dissent and individualism at a general level. By c 1770, 'the Church's capacity to repress had clearly become reduced in all classes of society. Signs of an overt and non-intellectual interest in sexuality, expressed in ways which at the beginning of the century would certainly have been denounced, can be seen.'[20] Chapbooks were widely read, often containing bawdy material, and an ostentatious sexuality developed among the upper classes.

Although the nineteenth century saw a great number of people on the move and the previous moral order slackening further as a consequence,[21] it also witnessed the middle classes and aspirants to that status return somewhat to notions of modesty and naturalness in everyday appearance. It was only from the 1930s or so that all classes were granted a more or less similar degree of leave, although not the same means, openly to enjoy and enhance their degree of physical attractiveness. This is not to say that all of Scotland's present-day population is able or wishes to take such liberties. Many people during working hours or when at school are expected to adhere to a temporary suspension of sexual attractiveness. Members of certain religious groups[22] may decline to be involved in such matters. And others too may feel obliged to accept or may prefer a certain degree of disengagement in this area of life.

Gender, Age, Marital Status and Wealth
GENDER
Women and girls have frequently been regarded as being more concerned with making themselves look attractive, and there is no doubt that more

emphasis has been placed on females' appearance than that of males. Yet at times – such as in much of the eighteenth century – men have also made prominent efforts and at other periods of history they have had to make their toilette covert or subtle but often little or no less demanding than that of women.

AGE AND MARITAL STATUS

We are expected to pay greater or lesser attention to our appearance, especially in terms of attractiveness, at different times in our lives (Fig. 14.2). In part this is a reflection of our differing engagement in life's principal roles and events. Generally speaking, children, and especially boys, may be expected at the everyday level to be less involved (Figs 25.2, 28.7). Whether this is a blessing or a bane depends on the individual but certainly many children appear to have dreaded being dressed up by adults for special occasions. Schools have tended to play down the idea of attractiveness, although at the same time often adhering strictly to notions of decency through the use of uniforms. Eugenie Fraser, of Scottish-Russian parentage and growing up in northern Russia, writes of her school there that:

> Discipline was strict – the rules laid down had to be observed. Hair was not to be worn loose hanging around the shoulders. It had to be braided and if wished the braids could be placed on the crown of the head and tied together. Uniform had to be tidy and spotless. No jewellery of any kind, with the exception of a watch, was tolerated.[23]

Similar rules operated in many Scottish schools.

The onset of adolescence sees most boys and girls embark upon a period of intense concern with looking attractive. Molly Weir writes of herself and a teenage friend in inter-War Glasgow that, 'we were becoming very absorbed with our grooming now, and stayed in at least two evenings a week for the sole purpose of washing our hair, doing our nails, going over our clothes and mending any straps or stockings which needed attention.'[24] As people seek for sexual experience, efforts to enhance the appearance become paramount and are directed towards the everyday as well as special occasions, although employers as well as schools may try to thoroughly trammel matters. Anna Blair writes, for example, that 'juniors' in Glasgow offices at the beginning of the twentieth century were forced to keep their hair in 'pleats' (braids) until they were 18 years old.[25]

Adulthood might be thought to offer more freedom with appearance, but it seems that this is not often the case. In part this has to do with the fact that for many Scots adulthood has been equated with married status and an associated suppression of sexuality, including sexual attractiveness, beyond marriage. The effects of this seem to lie more heavily on women than men.[26] Certainly Scottish autobiographies and biographies speak repeatedly about the obligation (and often the associated demoralisation or frustration) placed on married and other mature women to take on a degree of calmness or staidness, sometimes dowdiness or even drabness, in stark contrast to the show and vivacity expected in their courting years. Kesson, for example, describes her middle-aged grandmother as a solid respectable dame in the house (where she was in fact in public) but outside, in the privacy of the garden with only the company of her granddaughter, she became girl-like, singing ballads and love songs, and dancing. Once inside again her grandmother, 'huge and hurried in her white baking apron, had become old and wise again'.[27]

For married women on low incomes the exigencies of poverty may have had the greater importance. Kesson refers to the married women of 'The Lane' in Inverness in the early twentieth century as 'grey ... prisoners clamped firmly into the dour pattern of its walls and cobblestones'.[28] Patrick McVeigh, speaking of East Lothian in the early twentieth century, has this to say:

> Of course many of the people we thought of as very old were, by today's standards, not even middle aged ... but poverty and defeat, and often booze as well, had aged them all ... Even 'respectable' people, however, if they were working class, aged quickly. This was especially true of women. A diet high in carbohydrates and low in protein quickly turned most women to fat but, in any event, no wife of a man with a family could afford to dress decently or to have her hair dressed, or even to look after her teeth. At forty, then, wives were already tired and old and drab, and they accepted this as their fate. They even managed to have some measure of contempt for the only women who contrived to stay smart in their appearance, almost invariably the spinsters who were presumed incapable of 'getting a man'.[29]

However, many well-to-do women (in contrast to very rich women who were given more liberty) seem to have also been subject to some of the same processes and the need to display premature, exaggerated, or at least unconstrained ageing.[30]

Hair has a very important place in this discussion. It was, and remains, an indicator of a woman's age and marital status and, as a woman's 'glory', a potent symbol of attractiveness, especially sexual attractiveness.[31] Until the twentieth century, across the country and at all social levels, women's hair was almost always worn long.[32] It was only allowed to fall loose, however, while girls were young and unmarried[33] (Fig. 15.1). Married women and unmarried women of more mature status usually put their hair up or covered it in some way, although this might sometimes be delayed for the former until maternity.[34] Molly Weir describes her favourite race at the school sports day, that in which the mothers ran:

> I loved the sight of their rippling hair flying unbound in the breeze and was delighted at how young they looked when they were pink-cheeked and dishevelled like this ... I had a notion they quite enjoyed the free feeling of their hair tumbling down their backs, and the

Figure 15.1 Evelyn and Alexa Thompson at home in Academy Street, Ayr, c 1900. Evelyn is at the piano. Flowing long hair was acceptable for girls (in this case aged around 16 and 13 respectively) but not women. Source: SLA 128A(1) C.23830 510593.

admiring comments made by the men on the 'fine heids o' hair' they had.[35]

In the twentieth century it became the custom for married women to cut their hair short, sometimes for their wedding day,[36] and there is still some expectation that women, especially of the middle classes, and now of any marital status, will, at the onset of middle age desist from having very long or otherwise showy hair.

What then of the appearance of people beyond middle age? There is remarkably little evidence available. For men, at the broadest level, life after adolescence seems to have been more or less constantly marked by an understated appearance which requires care but does not attract excessive attention. Many older women appear to have continued, perhaps in a more intensified form and sometimes maybe gladly, with the self-negating customs formed in maturity, but there are indications of others, notably those of moderate to great wealth, returning somewhat to the liberties of adolescence, either becoming more carefree about others' opinions and perhaps moving towards eccentricity or else returning to the notion of show and glorying in their good looks.[37]

WEALTH: MATERIAL AND TIME

Perhaps the most important factor in the individual's involvement with toilette, after desire or a perceived need to be involved, is time. A time-rich person with little material wealth can do a fair amount in regard to their appearance but the time-poor cannot. Thus people raising children, especially mothers, have often been unable to undertake much in the way of toilette, reinforcing their disengagement for other reasons.

Of course, the materially poor are often also time poor and in any case poor people have often suffered in terms of appearance by not being able to afford attractive and suitable costume, accessories and cosmetics, by having less access to adequate sanitation and comfortable health and rest-promoting housing, a less nutritious diet, a higher risk of illness and accident and more stress in their lives. And demoralisation of any sort often strips people of pride in their appearance.

As a consequence, looking good has sometimes come to be seen as the right of the wealthy and almost impertinence on the part of the poor. Many servants and other manual workers, for example, have not been expected to look better than their employers deem a necessary minimum,[38] with, at one time, intemperance in domestic servants' dress considered 'an evil to be carefully guarded against'; in men it could be affectation but in women it might be worse, leading to temptation and extravagance.[39] Something similar could happen, but more profoundly, in educational or charitable institutions where the residents might on occasion have their heads shaved and wear clothing reminiscent of a prison environment, although the stated justification behind such measures was usually cleanliness or for commercial reasons.[40]

Not surprisingly, the poor have often been unimpressed with the obligation to look plain. The fact that individuals of great physical beauty are born into all classes challenges the idea nicely and the satisfaction in this is echoed in many types of folk literature, and more recently in written and visual works of fiction, where the poor but good-looking character makes good over the rich, powerful and artificially attractive. Yet the notion that good looks are for the privileged could, in the days before cosmetic surgery could radically alter the body, bring unhappiness for the better off also. This was especially the case for wealthy women, particularly when their social roles were essentially limited to those defined primarily by reproduction – that is as wives or mothers. The expectation to look good would be less concerning to a man of similar social rank – looking extremely respectable would probably be enough – and in any case a powerful and wealthy man could often acquire looks vicariously by his choice of marriage or other sexual partner. Rich women, who for all their efforts could not look attractive, did not have this opportunity and such individuals were often ridiculed and otherwise effectively punished by their society.[41] Also to be considered is the way in which privilege can put good looks at risk; riotous, or at least generous, living (with ample alcohol and other drugs and rich food, inadequate sleep and restricted opportunities for healthy exercise), the stress of gaining and maintaining position and residence in inner urban environments – which although prestigious may have a poorer quality of environment[42] – can all take their toll and offset superior access to beauty resources.

Fashion
Some people care for following fashion – be it in clothing, manners, grooming or anything else – and others do not.[43] But there is more to the story than personality alone. Of those who wish to be fashionable, some are able to fulfil their desires while others cannot.

Class is a very significant factor. The upper classes are more likely both to set and be able to keep up with the latest fashions and 'the middle classes have usually attempted to imitate their extravagances'.[44] Those in the lower classes have sometimes had to make do with whatever they could find or afford but at other times have managed to not disregard fashion, either creating their own trends or selecting elements from the styles of the upper classes.[45] For long it was members of the Court and the social élite who set the tone, but from the late nineteenth century theatre, music-hall and cinema stars became the leaders of fashion.[46] Today fashion experts such as celebrity hairdressers, along with a range of media and sporting stars, all supported by commercial advertising, create new trends.[47] One can argue that people of all classes are as likely to be influenced by these.

Nationality and geographical location also have an effect. The residents of some countries are thought to be more fashion orientated than others and keeping up with the latest trends may be more of a concern in a city than in a remote rural area.[48] Also affecting interest in following fashion is again the age of the individual.[49]

TOILETTE IN THE HOME

Toilette, along with clothing and hygiene, has a close relationship with the home on a number of levels. The dwelling can constitute the scene of preparation, or a place of privacy and of storage of materials and equipment, and sometimes a manufactory of these. It can also have the role of haven, a place to escape from the pressures of looking acceptable. Conversely, our abodes can act as an important platform on which the self, physical and otherwise, can be displayed to others, either from within the household or from beyond. Finally, the dwelling can constitute an arena in which the values and prohibitions surrounding the physical presentation of individuals can be discussed and reinforced in everyday conversation.

Studying Toilette in the Context of Scottish Homes
Little has been written about toilette in comparison to, say, costume. It is not always viewed as a suitable topic for serious academic research and in any case only very limited evidence is available. To a fair degree this is because people have often been reluctant to record many practices and notions

related to grooming as these are considered private, sometimes shameful, or simply too everyday to be very interesting. The amount of information available also varies according to region, the period of study, whether the area of investigation is urban or rural, and the social class, sex and age of the individual. For example, we know remarkably little about the appearance of ordinary people, especially women, in pre-modern Gaelic-speaking areas.[50]

Modern academic investigations of toilette are rare, with research funding given to what seem more pertinent matters. Historical studies focus primarily on the upper classes about which there is greater evidence available and say relatively little about people in humbler positions. Children and older people are often excluded both from source material and seemingly from the consciousness of writers on the subject. Various other groups of people in more 'marginal' social locations can also be overlooked in this context. Women's toilette may receive substantially more attention than that of men. In addition, studies do not focus exclusively on the home; indeed the larger part of discussions often deals with aspects of grooming in places such as hairdressers and barbers. Of that which has been written about toilette, most relates, in the context of the British Isles, to England. European-wide studies tend almost completely to overlook Scotland, although it is possible that much of the general discussion is applicable to this country.

Aside from their brevity, the fragments of evidence available are rarely entirely impartial or without other limitations. They come from various sources: a range of objects from prehistory to the present day and visual representations of these (until the nineteenth century only those belonging to people of higher status survive. Conversely, in modern times the widespread availability of such common items can result in a lack of interest in preserving artefacts); paintings and other art works depicting individuals and groups (painters of the well-off have endeavoured to make their patrons look as good as possible, but those few who have depicted poorer people – for example eighteenth-century *genre* artists such as David Wilkie and Walter Geikie – have often created 'warts an a'', if not semi-grotesque, depictions of their subjects); snippets of information in travellers' accounts (each to some extent with their own axe to grind); fairly stark and fleeting entries in diaries, recipe books,[51] almanacs and herbals; and – perhaps the most telling but often obliquely placed and extremely fragmentary evidence – glimpses through biography, song, tale, poetry, fictional writing and especially oral accounts and autobiography. Photography, TV, film, advertisements, journals and magazines give more substantial, although generally extremely slanted, information. Here it is unusually attractive people who are depicted and various commercially derived concerns limit what is presented. Any investigation of toilette in the Scottish home has thus to be incomplete and can in some areas draw only the most tentative of conclusions.

Places Beyond the Home
In Scotland various places beyond dwellings, have offered a helping and often costly hand towards looking good and many effectively work in

tandem with the home. Amongst these are barbers and hairdressers, important parts of the urban scene from the eighteenth century, but with earlier antecedents.[52]

Beauticians and beauty parlours are more recent phenomena. It was not until the mid nineteenth century that the 'rich and daring' began secretly visiting prototype beauty parlours in England.[55] When these establishments first became popular in Scotland we do not exactly know, but there is now at least one example to be found in most towns and even substantial villages. They remain, in social terms, essentially female equivalents of the old barbers' shops in that they serve also as places of meeting and exchange of information. They join the ranks of not only hairdressers but dentists, gyms and sports clubs, fitness and dieting groups (The Women's League of Health and Beauty, or Weight Watchers, for example), tanning centres (Fig. 15.2) and nail bars, all concerned with maintaining or improving our appearance, and in big cities might be allied to the services of alternative therapists and cosmetic surgeons working out of commercial clinics or private hospitals. Yet despite the almost bewildering choice of places to go, staying in and dealing with one's appearance at home remains the most popular choice for many, whether for reasons of privacy or comfort, or because remoteness, cost or incapacity impede access to professional services.

Making possible many beauty practices in the home are a range of substances, in raw or prepared form, obtained commercially from several places. The more overtly cosmetic might be purchased from a perfumer's

Figure 15.2 Tanning centre, Edinburgh, 2005. Source: M Mulhern.

shop, common by the eighteenth century in most of the principal towns.[53] Dealing with health and other concerns as well were apothecaries, druggists and chemists.[54] Today, large 'chemist' chain stores, which house within their vastness something of the character of all these shops, as well as smaller chemists or herbalists emphasising primarily health and medicine, make a very wide range of 'beauty' substances within easy reach of most of Scotland's population. Department stores and some fashion stores, and especially large supermarkets, play increasingly important roles.

Tools and Accessories in the Home
A remarkable range of beauty equipment has been found in Scottish homes over the centuries, most being of simple technology. Until the nineteenth century and the advent of mass production, the majority of items were made either at home or in the local area, with only the most privileged members of society relying on objects 'imported' into a region. As might be expected, the wealthy possessed a greater number of toilette items and those of superior quality. Yet much of the same basic equipment was probably to be found at almost all social levels. The frequent representations of combs and mirrors on symbol stones, such as that at Fairygreen, Collace, Perthshire,[56] clearly relate to the upper levels of Pictish society, with mirrors the preserve of those of high rank. Combs however could be made painstakingly but otherwise without difficulty from local resources.[57] Other basic items we might suppose to have been found in the majority of homes include bowls for mixing preparations and rags used for a plethora of toilette purposes, from applying substances to the skin and hair, to perhaps very occasionally carrying a cleaning agent such as soot to the teeth. Ties for hair and blades[58] to cut hair of different sorts, and nails, might well have been found in most homes too since pre-history. With these few items it is quite possible to maintain a good appearance.

Specialised tools and equipment, usually made at a distance from the place of use by people separate from the household, were increasingly common amongst the wealthy from the Middle Ages. They not only made particular tasks easier to carry out but, especially when ornamented, served to display their owners' status. They could make fine gifts and valuable heirlooms too and being used for private purposes were usually in individual ownership. As some of the few personal objects to be found in homes in the past it may be guessed that much meaning was often attached to them. This eventually became the case among the humbler classes as well, who might further use items as a gesture towards aspired social status. By the nineteenth century a wide range of objects was being marketed towards the lower middle classes, although we are not sure how widely bought and used these actually were. A century later cheaper materials, such as plastics, and more-efficient manufacturing processes had lowered prices to the extent that many items were within the range of just about everyone's pocket. By the end of the twentieth century relatively low prices increasingly typified many electrical and battery-powered items also. These and the widespread use of disposable items have done away with a good deal of the centuries'-old meaning of toilette equipment. Nevertheless, many Scottish dwellings still house a forebear's set of treasured ornate hair brushes or razors while 'beauty accessories' remain popular as gifts.

Furnishings and Spaces
Private space is important for most, but not all, toilette practices. Some, such as tying up the hair, shaving or applying lipstick, normally cause no scandal when undertaken in front of others but there are those which require at least a degree of privacy. Timetabling has always been important in small Scottish homes in this regard. In the past, the effect of substantial quantities of smoke drifting around inside and poor lighting[59] may have helped individuals to achieve a sense of privacy in crowded homes of one or two rooms where the setting aside of space primarily for toilette purposes and the accommodation of substantial numbers of toilette items would have been difficult if not impossible. This would have been enhanced by a tacit understanding on behalf of all reasonably mature-minded household members that turning a blind eye was sometimes necessary. And perhaps it is easier for one's activities to go unnoticed in a busy room than in a quiet one. Towards the other end of the social scale, the amount of privacy which those with servants in their home could expect is debatable. The wealthy could enjoy more space and more locks, plus the understanding that servants were not to report what they had seen after serving at their masters' and mistresses' toilette. The Adams advised domestic servants to 'always remember to hold the secrets of the family sacred, as none, not even the least of these, may be divulged with impunity',[60] yet clearly they sometimes were.[61]

Bedrooms were multi-purpose rooms even in large dwellings until at least the eighteenth century[62] and in those homes in which they existed were the site of most toilette and personal-hygiene activity (see Fig. 7.3). Even the extremely well-off often washed in their bedrooms, perhaps at a washstand with a bowl and ewer (jug) of water (Fig. 15.3), maybe occasionally making use of a portable bath in front of the bedroom fire.[63] If room was limited, grooming could be carried out using general items of furniture, such as a chest on which equipment could be laid out.[64] Where there was more space, bedrooms might contain, in addition to the washstand, various items of furniture devoted to grooming, including an object called the 'lady's closet'.[65] By the late eighteenth century they might house the newly fashionable draped dressing table, laid out with a wide range of equipment and a looking glass, the latter ornate and perhaps with a matching toilet set. Less elaborate mirrors were placed in secondary bedrooms in great houses and were in general use by those of more moderate means. By the mid nineteenth century bedroom furniture was available (see Fig. 29.4), to more of society, in suites including a full-length mirror and a dressing table, which might have a range of drawers for storing toilette equipment[66] (Fig. 15.4). Dressing tables, either as part of suites or separate items of furniture (such as the commonly found kidney-shaped item with adjustable triple mirror) became

Figure 15.3 Tillycorthie House, Udny, Aberdeenshire, 1920s. Bedroom with washstand. Courtesy of Aberdeen Library and Information Services. Source: SLA 128A(1) K9/22.

still more widespread in the twentieth century when, thanks largely to council housing, poorer Scots usually had considerably more space to live in.[67] At the same time toilette products and accessories to be used and stored at home became far more readily available to those on lower incomes. Dressing tables continued to enjoy great popularity until the last decades of the twentieth century and the revived fashion for modular, multi-purpose items of bedroom furniture.

'Still rooms', for the preparation of lotions and remedies, for example, were long found in some grand dwellings.[68] They were not the only rooms closely tied to matters of toilette. By the latter half of the eighteenth century there had been in England a 'change from Bog House to bedroom and dressing room', the latter containing a toilet table and commode and items such as basins, stands, stools, looking glasses and boxes.[69] Dressing rooms, lady's closets and occasionally powder closets (for dressing wigs) began to be found about the same time in some larger Scottish homes, although the inclusion of the first of these increased in the nineteenth century to the extent that even relatively modest houses possessed them. Dressing rooms were usually positioned next to a bedroom and were '... essentially the husband's room, where his clothes were kept and there was usually a single bed ... Dressing rooms would have had wardrobes and chests for clothes and a

TOILETTE • 257

Figure 15.4 Dressing table in a house in Ayr, 1979. The table is next to the window so that maximum light falls on the person sitting before the mirror. Brushes, cosmetic and trinket boxes and perfume bottles find space on top of the dressing table with drawers for further storage below. Source: SLA 128A(1) S.10477.

washstand with matching china.' They might also be decorated in a sporting theme or otherwise given a masculine air.[70]

Although dressing rooms were sleeping as well as toilette space, the latter function defined them. Bathrooms lay more emphasis on personal hygiene, but grooming is an important secondary use[71] (see Figs 16.3, 16.4). By the eighteenth century they were to be found in an increasing number of large homes but were still rare. At the end of the following century bathrooms had become fairly common in urban areas at least, even in ordinary homes, although they often contained only a bath and no wash-hand basin. However, many of the smallest homes remained without a bathroom.[72] It was only in the late twentieth century that virtually all homes could boast such a room and the advent of adequate means of heating this not only encouraged more detailed washing and bathing but made the space more comfortable for lingering in over an elaborate toilette, leading at times to urgent conflict between household members' demands!

People: The Makers of Toilette and the Recipients
Toilette may be an often private matter but it is not always a solitary one. Marjorie Plant describes how most women in eighteenth-century rural Scotland had to adopt simple hairstyles because there was no professional skill available. This meant setting their hair in ringlets, which were kept off the face by a snood and sometimes a comb. Such plain hairdressing continued throughout town and country for those with little interest in fashion.[73] The women in a household might also trim their hair from time to time without the services of a hairdresser. Remoteness, poverty, infirmity and lack of interest could all keep people from leaving their homes and accessing professional services. Nevertheless it did not necessarily mean that they always had to undertake their own toilette.

Members of religious communities cut and shaved each other's hair[74] and in many other homes the household has proved an important source of grooming assistants. These might be family members, for example children assisting parents[75] and vice-versa, and wives helping husbands (Fig. 15.5). Household servants were often required to attend to at least part of their

Figure 15.5 'The Golden Fleece' by Keith Henderson. This 1930s painting of the Bruce family outside their home in Glen Nevis, Inverness-shire, might exaggerate the rough-and-ready nature of home hairdressing, but until the later twentieth century most homes of ordinary means in rural areas usually had to be self-reliant in such matters. Source: SSS Vg 5 7053.

masters' and mistresses' toilette.[76] In very large households the gentry might even have their own private hairdressers, for both men and women.[77] The tradition of having a social inferior attend to one's toilette is an ancient one. A fashionable Roman woman, for example, often had a slave, an *ornatrix*, who acted as a skilled cosmetician.[78] Indeed the origins of the hairdressing profession are to be found in those same Roman slaves.[79]

The local community can provide a source of friends to assist girls and young women with their grooming, at least on an occasional basis and usually in preparation for special events. When people died, their last toilette or 'laying out' might be performed by an experienced neighbour in the days before commercial undertakers.[80] Professionals might be employed to work within the dwelling. In the eighteenth century it became extremely fashionable to have one's appearance attended to at home by such people[81] and today it is still possible to have hair or complexion dealt with there by a paid specialist.

For whom is all of this? Anyone we seek to impress or we fear will censure our appearance. This might be members of the opposite sex or one's peer group, or society in general. Perhaps the greatest critic of an individual's appearance, however, is that person him- or herself.

Learning about Toilette Matters
Beauty as well as culinary preparations and medicinal remedies were made from recipes kept and handed down from generation to generation[82] but, 'many herbals and other printed materials were available to those who could read them and had money to buy them. From these they could learn about the plants themselves and how to grow them, which parts should be used, when to gather and their virtues.'[83] Instruction later became available via a plethora of ladies' magazines and journals and, in the 1990s, equivalents for men. These encourage people to spend a long time on personal grooming and adornment and aim at persuading them to buy numerous products and accessories. Printed sources have been joined by an array of advice to be found on the Internet. However, the passing on of information by word of mouth, at work or school, during recreation[84] or gleaned from professionals whilst visiting their premises and then brought home, still plays a very important – perhaps the most important – role.

When? Daily Routines and the Special and the Everyday
There is a distinct rhythm to toilette, as defined over the day, week, scholastic term, or year as well as at special points in the individual's life. Upon rising and in preparation for going to bed are usually the most important times of the day, but some Scots have been required to attend to their appearance at the end of the working day or before main meals ('changing for dinner' for example). Before and during the Sabbath[85] more attention has often been paid to appearances and dances and their modern equivalent of 'nights out' continue to fill bathrooms and bedrooms with earnest groomers on Friday and Saturday nights. School children might be expected to dress up for

Figure 15.6 Young people on an outing, Ness, Lewis, probably the 1950s. Much effort has gone into the costume and grooming of both men and women. Source: SLA 99B(4) 45/29/27.

end-of-term and annual events, and Christmas and New Year and other holidays, as well as excursions, still see many people making an extra effort with their appearance, as also birthdays, graduations, anniversaries, weddings and funerals, for example (Fig. 15.6).

Substances
The substances used in Scottish homes to maintain or enhance appearance have been legion. Many were created in the dwelling from locally derived ingredients and some were the result of involved processing. The materials included animal fats such as lard, other foodstuffs and common wild plants such as cowslip, elderflower and violet,[86] as well as more localised species, the latter lending a good measure of regional variety to cosmetics. Cultivated plants were also made use of in traditional recipes.[87] Except in those houses in possession of a still room, these materials were usually prepared, often by the housewife, in the kitchen.[88]

When efficient markets and then shops came to prominence, people were able to make use of the apothecary's skill or employ common bought household substances, such as fullers' earth,[89] or imported plant and animal derivatives such as whale oil, almond oil and wine.[90] A number of these substances still have some importance, for example lemon juice used to remove uneven skin pigmentation, although for well over a century there has been a vast industry which creates innumerable specialised readymade products which are now widely available from pharmacies, cosmetic shops and supermarkets or by mail order and sometimes commercial 'house parties' in almost all of the country. Curiously, while many are products of the international pharmaceutical industry, others are labelled 'traditional', 'natural', 'handmade' or 'fresh' preparations, imitating, echoing or copying locally made cosmetic substances from the British Isles and far beyond.

Practices
The remainder of this chapter looks at a range of practices relating to the care, adornment and enhancement of the body, focusing on two topics; skincare and shaving (references are included which point the reader towards information on practices relating to other aspects of hair care,[91] and to those dealing with the facial features,[92] nails[93] and the figure[94]). The aim is to present a small sample of the astonishingly varied body of toilette practices which, as a whole, reflect the range of human ingenuity and at times the extent of human fortitude. Although innovations have occurred throughout history, with an acceleration in the rate of new 'discoveries' from the nineteenth century and especially the early decades of the twentieth century, there remains a sense of timelessness about many other practices. Technological change in grooming has not always resulted in the displacement or replacement of phenomena with earlier origins. Indeed the old and new may exist rather comfortably alongside each other.

SKIN: MAINTAINING THE SKIN IN GOOD CONDITION
The early history of soap manufacture and use is discussed elsewhere in this volume.[95] In the eighteenth century, those Scots who could afford soap often preferred to apply it to the skin as a component of 'wash balls'.[96] By the end of the nineteenth century soap was being widely used on its own for toilette purposes in Scotland, in a plethora of aromas and colours, its popularity in part a response to advertising campaigns, most notably by Pears.[97] Recently there has been a move towards more use of liquid soaps with moisturising properties, including hand soaps and shower gels.

Soap, however, is just one of the substances which has been used to clean skin. A wide range of creams, lotions, washes and cleansing bath additives have been used in Britain since Roman times, despite full bathing generally being unusual until the nineteenth century.[98] At times these preparations were greatly preferred to soap as they could condition the skin while cleansing it.[99] Some ingredients[100] appear to have required a good measure of stoicism – dung, urine, seaweed, pine needles, peat or mustard, for example. Others seem much more attractive – ingredients such as rose petals, camomile, thyme, balsam, myrrh, fir-wood oil, and foodstuffs such as milk (from various species), wine (including champagne), flours and meals of various sorts, starch or bran. But further examples are clearly dangerous. These include mercury, oil of vitriol (a highly corrosive acid made from sulphur dioxide) and other acids, white lead (a mixture of lead carbonate and lead hydroxide) and albeit weak solutions of bleaches. Water, either warm or cold or occasionally in the form of steam or dew (in traditional belief at its most potent when collected on any 'quarter day'), was sometimes regarded as enough (rainwater was especially favoured), but it could also be considered harmful. Some very careful people refrained from cleaning their faces in any way so as to preserve their complexions. The contrast with today could not be greater when skin is often pampered with numerous commercial preparations, carefully made from petrochemical and plant-derived compounds and subject by law to rigorous chemical testing.

Creams and tonics have also been employed with the prime function not of cleansing but of maintaining the appearance, softness and suppleness of skin and many women, and no small number of men, have had their own favourite mixtures or ingredients. Again they have been homemade or commercially and then industrially produced, although the homemade is by no means entirely out of the picture today, as those who have made up a cucumber and yoghurt face pack will testify. As with cleansers, the substances used as conditioners have been more or less appealing and sometimes harmful. They include such varied ingredients as elderflower, almond oil and fir-club moss, as well as cucumber, cream, lemon juice, ammonia, and borax.[101] Before the pharmaceutical era got under way in earnest, the most widely used skin conditioners were a range of herbal preparations and 'cold cream' made from scented lard.[102] Bran and oatmeal were also popular and effective skin conditioners, available to virtually everyone.

There are other means of helping keep the skin in good condition. Scrubbing and massaging (or 'shampooing') with towels, brushes, flannels, sponges, loofahs or special gloves[103] or abrasive creams (and with pumice stone on the soles of the feet and the elbows), can stimulate the blood supply to the skin, helping in the renewal of cells. The technique has varied little since early times, although beauty parlours can now offer the employment of sophisticated of skin-abrasion machinery and electric massagers are available for home use. A further, and important, element in keeping skin in good condition is the maintenance of a health-promoting lifestyle.

When the skin possesses minor abnormalities or is damaged through environmental factors such as sunlight, by disease, accident and, more debatably, ageing, various strategies have been used, by females and males, to reverse, offset or conceal the effects. Freckles, perhaps provoked by exposure to the sun's rays and often charming to others but deplored by the bearer, could be tackled with an infusion of honeysuckle, buttermilk (both of which also served to soothe or prevent sunburn), or diluted lemon juice. Rough skin could be made smoother by glycerine soap.[104] 'Folk' cures were widespread in Scotland for the removal of birthmarks and warts. David Rorie, for example, reports from Fife on attempts to remove birthmarks by rubbing the affected skin with the hand of a dead blood relation.[105] Small blemishes on the skin could be removed by the use of soliman (a sublimate of mercury), while the scars of smallpox might be hidden with a mixture of white lead, vinegar and sulphur,[106] or, in the seventeenth and eighteenth centuries, covered, by young and old alike, with patches; tiny, black fabric shapes.[107] In the nineteenth century grease paint became available and this, used with a little powder, could also disguise problems.[108] The twentieth century not only witnessed the creation of a very wide range of commercial products, such as after-sun and anti-wrinkle creams and concealer sticks, serving similar purposes, but also saw the development of surgical and related techniques such as facelifts, peels and dermabrasion. The home now has a new role in toilette; that of a place of recovery from surgical intervention.

Sometimes it is not enough simply to possess healthy skin and there is a desire to ornament the complexion. One form of decoration is the acquisition of a tan, or, in the past, the maintenance and enhancement of paleness. Both have sometimes been dangerous pursuits. Whiteness, associated with a lack of outdoor work and thus for long a symbol of status,[109] could be achieved by keeping the skin away from sunlight. Outdoor female workers, such as the bondagers of south-east Scotland, wore protective headgear to this end, in this case the well-known 'ugly' or 'crazy' sun bonnet.[110] Members of the gentry too, both male and female, might don a hat in bright weather to protect their complexions. Less kindly was the application of skin whiteners such as lemon water mixed with poisonous sublimates, white lead, mercury water and ceruse (a mixture of white lead and vinegar which could lead to gastric troubles and shaking).[111] Tanning came into vogue with foreign holidays 'to the sun'. Scots are now aware of the risk of skin damage, including skin cancer, which can result when tanning is carried out, even with care, yet we seem strangely reluctant to dispense with the practice of exposing ourselves to the sun, or more commonly, a tanning machine. Safer by far, but not always very convincing, is 'fake' tan out of a bottle.

Commentators mention the early use of woad to colour the skin in ritual and battle contexts.[112] Permanent colouring might be achieved through tattooing, another practice apparently with a venerable history.[113] In terms of cosmetics as we would recognise them, the skin of the face, as well as the neck, chest and arms, might be tinged. The Romans used powder to whiten the complexion and prevent shine and a pink colouring to return blush to the cheeks.[114] In England, Elizabethan women sometimes painted their faces and pencilled blue veins on their bosoms.[115] By the late seventeenth century white lead was at times being replaced by talc and a light rouging, perhaps with red leather or woods named as hailing from Brazil, Spain or China. Rouges could also be homemade of harmless organic dyes – red sandalwood or cochineal for example, mixed with vinegar or oils – but might make use of poisonous metallic paints. At times the combination of powder, cold cream (a foundation to which the powder would stick) and rouge might be glazed with egg white to give a marble-like finish. Other substances used to colour the skin included ceruse, vermilion (mercury sulphide or 'cinnabar'), yellow ochre and textile dyestuffs.[116] Colours could be applied with the tip of a finger, in pencils, on brushes or flannels or a hare's paw.[117]

Most of this activity was for members of the upper classes, both male and female, and perhaps some middle-class women. However by Victorian times the fashion for a natural, freshly washed look, extended to virtually all of the middle classes.[118] As Huggett writes, 'demure respectability was imposed on the wives of newly rich husbands',[119] but it also was laid on the husbands themselves. Such decency extended to the 'respectable working' classes, although to either side of these groups, women of the highest classes continued to paint, while those at the lower end began for the first time to do so, thanks to the increased availability and lower cost of cosmetics. The extent to which colouring practices were used in Scotland before the 1920s and especially the 1930s and beyond the highest echelons, accustomed to residing in the south, is highly debatable. Respectable, but not the richest, Scots women were for long renowned for their lack of powder or colour application[120] (commonly thought to be cosmetics in general, but as we have seen this certainly was not so) and during the early twentieth century, when worthy English women now began to apply colour, those north of the border often still regarded the practice as rather unsavoury.[121] Amateur 'theatrics', sometimes conducted in those homes of sufficient proportions to admit a gathering, might have been an exception to this Scots reserve with colour cosmetics.[122]

By the 1930s the application of colour, thanks to the influence of music hall, cinema and magazines, was more acceptable in Scotland (indeed it was now almost demanded of young urban women), ready-prepared products could be far more easily obtained and Scots women quickly caught up with their English and other counterparts. Today, aside from various forms of colour applied to the features, foundation, coloured moisturisers, face powders and blusher, sometimes even glittered gel, adorn the faces of many women in Scotland, the beauty industry advertising these products especially through television. Scots males are also now more likely to use cosmetics, at least of the moisturiser or fake-tan variety.[123]

Immigrants from parts of Africa and Asia have brought new skin-colouring techniques to Scotland in recent decades, of which the use of henna dyes on the skin of the hands or feet is a prominent example. Members of immigrant communities are joined by a few more innovative members of the wider Scottish population in making use of this practice.

Shaving

Although men have been clean-shaven since earliest times, the wearing of facial hair has also been valued, for ease and as a sign of masculinity and maturity.[124] The impression is that most of Scotland's mature male population before the eighteenth century sported beards of some sort and only the habit of wearing wigs changed this (facial hair was not thought to 'go' with wigs), and then only for those of reasonable means in more-accessible parts of the country.[125] After wigs fell out of fashion there was a gradual return to facial hair for all men, beginning with sideburns, so that by the 1880s it was considered very commendable for a man to wear a full, indeed luxurious, beard. Men of fashion were concerned that these should both smell and look good[126] and it was not unknown for surreptitious dyeing to occur.[127] Hair care extended to 'moustache' cups, developed around 1850 and popular for about 50 years,[128] which had a perforated tray inside to keep lip hair out of the beverage. However, by the beginning of the twentieth century beards and sideburns, were going out of favour except for the very elderly, artists, intellectuals and eccentrics, although moustaches gained further prominence for a time (Fig. 15.7),[129] their sometimes remarkable qualities often the result of professional care. By the 1950s completely clean-shaven faces were in vogue and it was only the young, anti-establishment types of the 1960s and 1970s who changed this. Men now have the liberty to grow facial hair, in a variety of forms, or be clean-shaven, although the latter is by far the commonest response.

In the past, many men seem to have been rather reluctant to shave themselves and shaving was thus for long more a matter of town, where there were plenty of often inexpensive barbers to be found, than country, at least for those not of aristocratic description. In eighteenth-century Galloway, for example, razors were not in general use and beards were simply clipped with scissors on Saturday night. Even the very wealthy living in the countryside could sometimes be put to it to find someone to shave them.[130] Samuel Pepys, to take an English example, had to use pumice to smooth his beard before he developed the skill to shave himself with an open razor.[131] When shaving became more widespread and a little more home-orientated in the early nineteenth century, it still might not take place every day (which became more common in the following century). Indeed shaving could be carried out far less frequently, either at home or at the barbers, once or twice a week being common.[132] Morning was not always the commonest hour for this task, especially for men with early starts to the working day. One Angus bothy man stated that 'Ye didna hae muckle time tae wash and shave in the morning so ye mostly had a guid wash and shave at nicht.'[133]

Shaving has significance as an act of maturity but also expresses other ideas. From the detailed descriptions of their fathers' shaving habits recounted in a number of Scottish men's autobiographies, it seems that the act could highlight the bond between father and son.[134] Shaving another has often possessed a tone of servitude. A Scots soldier in India during World War II recalled his feelings when he saw other soldiers being shaved by an elderly Indian attendant, '... but you lay in your bed and the Indian lad came round and shaved you while you were lying in bed. A soldier doesnae need tae be shaved. He shaves himself ... And these old men, you know, shavin' lads. To me it was all wrong.'[135]

Various items were needed to facilitate shaving with an open razor: a mirror if one was shaving oneself, sometimes in the form of shaving stands

Figure 15.7 By 1900 beards were generally no longer favoured. With one exception these fashionable younger men in Campbeltown, Argyll, sport moustaches instead. Source: SLA 123A(1) 61/3/26/McGrory Collection 6435/1992 Argyll and Bute Library Services.

(mirrors with bowls and brushes); a basin or shaving jug for water; soap (in bar form or as 'shaving balls', now shaving 'foam'); a shaving brush (before the introduction of these from France in the 1750s a sponge or simply the fingertips were used to extend the lather) and cup or mug; a kettle for heating water (and of course an adequate water supply); a leather strop and a hone or razor stone (which might need a few drops of oil);[136] a towel; and the razor itself,[137] in poorer households often an object of considerable status and personal meaning. Ralph Glasser, in the early twentieth-century Gorbals, Glasgow, describes his father's routine after stropping this razor:

> Then he held it up to the light and twisted it back and forth, inspecting it for specks of dust on the fine edge, and the blade glittered with a vibrant mirror-like sheen, as if it answered him in private communion, as the magic sword Durendal might silently have spoken to its heroic master. He then folded the blade into its ivory guard and placed it gently in the shaped blue velvet cushion in its slender leather case and closed it.[138]

To complete the shave, some type of aftershave (which might simply be a splash of brandy),[139] lotion, cream or pomade, might be applied to close the pores and leave the skin smooth and fresh-smelling.

As long as razors remained open they acted somewhat as a disincentive to frequent shaving at home, which only became more popular with the development of the first real safety razor in 1874. However it was the invention of the disposable razor by King Camp Gillette in 1895 in the USA which saw those in urban areas visiting barbers far less frequently. The stainless-steel blade was patented by Wilkinson Sword in 1956, with bonded systems introduced in the 1970s. In 1951 Philishave brought out the first battery-operated razor,[140] which made shaving still easier, although many men still prefer a wet shave.

Shaving, of course, is not confined to facial hair or to men for that matter. Women also may wish to remove or reduce the visibility of facial hair or body hair (as have men sometimes) and techniques have included not only shaving but plucking, waxing, bleaching and, occasionally, 'sugaring' (a technique in which a ball of caramelised sugar is rolled over the skin, removing hairs by the root as is goes), traditional in, for example, parts of the Middle East and recently made known in northern Europe by the presence of immigrants and travel abroad to that and other areas. Electrolysis and laser treatments are available in beauty parlours and specialist clinics for those wishing a more permanent hair-removal solution. No information is available about body-hair removal in Scotland until the twentieth century and it may be possible to assume that as bodies were covered up a lot more in the past so little went on in this regard. The growth of the mass media, foreign holidaying and the introduction of efficient forms of heating into homes and other buildings has ensured that Scots can now witness and copy the fashions of warmer climates and wear fewer clothes than ever before, even in the depths of winter. Exposing flesh has meant the need in many cases to remove body hair.

The scalp has also been shaved on occasion. Shaved heads have become fashionable in the last 30 years or so, among young men and, to a much lesser extent, young women keen on achieving a trendy 'hard-man' (or 'hard-woman') look.[141] This imitates the inmates of prisons or other institutions in the past. Other reasons for shaving the head in the past included the elimination of lice, offsetting the effects of fever, as a way of removing the fine hair at the nape of the neck to set off a haircut, and to create tonsures for members of male religious communities.[142]

CONCLUSION

Whether regarded as shameful or a source of pride, challenging or tedious, delightful or dreadful, matters of appearance have had a very close relation with Scottish homes and Scottish domestic life. Our evidence for the past is limited but there is still much to be investigated and presented and the present day offers vast scope for academic exploration, for example, by surveys of current practices within households, or a consideration of the impact of modern mass media on matters of domestic toilette.

ACKNOWLEDGEMENTS

The author is most grateful to John Burnett, Mrs Marguerita Burnett and Professor Alexander Fenton for reading and commenting on a draft of this chapter.

NOTES

1. Ebin, 1979, 5.
2. 'Oh, Nancy's Hair is Yellow Like Gold'; for equivalents in Gaelic song see, for example, Shaw, 1986, 'Cha Labhair mi 'n t-Òran [I will not Utter the Song]'; Caldair, 1912, 'Oran Gaoil le Donnchadh Macantsaoir [Love Song by Duncan MacIntyre]'.
3. Newton, 2000, 93.
4. Newton, 2000; see also, for example, Campbell, 1969, 17.
5. Nicolson, MacInnes, 1951, 182, 238–9; contrast this with McClintock, 1943, 97, quoting de Boullayne le Gouze, a French commentator who, in 1644, stated that 'people with red hair are considered the handsomest in Ireland. Women who are freckled like a trout are esteemed the most beautiful.'
6. Nicolson, MacInnes, 1951, 52.
7. Grant, 1988b, 68; Elizabeth also vilifies a certain Young Charles Grant, 'I thought him hideous, tall, thin, yellow, grave, with sandy hair and small light eyes, and a shy awkward manner ... ' Grant, 1988a, 341.
8. Williamson, Williamson, 1987, 206.
9. Kesson, 1992, 125.
10. For commentary on the importance of modesty in women see, for example, Dunbar, 1962, 95; for the effect of good manners Grant, 1988b, 75; Bennett, 1992, 81, quotes the proverbs, Wha may woo without cost? and, A fair maid tocherless will get mair woers than husbands; see also Grant, 1988a, 150.
11. Woodforde, 1995, xi.
12. Huggett, 1982, 15.
13. de Marly, 1986, 12; see also Dunbar, 1962, 91.
14. Leneman, Mitchison, 1998a, 4.
15. Grant, 1988b, 9, relates that although much admired elsewhere for her looks this did not happen in her early nineteenth-century home, 'personal beauty being little spoken of in the family'.
16. Leneman, Mitchison, 1998b, 13.
17. For example, see Kesson, 1992, 48, for a description of a Traveller woman's adornment.
18. Leneman, Mitchison, 1998b, 13.
19. Many sources indicate how well recognised as a form of deception physical appearance is. Proverb, for example, offers, Is minig a bha dreach breagh air maide mosgain [A rotten stick is often nice to look at], Is mór thugam, 's is beag agam [Great appearance and little value], Na toir breith a réir coltais; faodaidh cridhe beairteach 'bhi fo chòta bochd [Judge not by appearance: a rich heart may be under a poor coat], Nicolson, MacInnes, 1951, 276, 282, 331; see also Ó Baoill, 1972, 6–11, for 'Comhairle air na Nigheanan Òga [Advice to Young Girls]' by seventeenth- and eighteenth-century poet Sìleas na Ceapaich [Sìleas MacDonald] for an example of verse dealing with this theme.
20. Leneman, Mitchison, 1998b, 34.
21. Leneman, Mitchison, 1998b, 17, 30, 34, 35.
22. See McVeigh, 1999, 63.

23. Fraser, 1991, 176.
24. Weir, 1998, 323.
25. Blair, 1998, 51.
26. See Höskuldsson, 1975, 4, 6, for a discussion of sexuality in women's youth, maturity and old age as expressed in northern European literature.
27. Kesson, 1992, 73–4.
28. Kesson, 1992, 125.
29. McVeigh, 1999, 74.
30. See, for example, Grant, 1988b, 98, for a description of the ageing of one Lady Napier.
31. See Huggett, 1982, 11–12; also Bryde, 1979, 159, for a general discussion of the meaning attached to both long and short hair.
32. For example, MacDonald, 2003, 37, says of early twentieth-century Lewis, 'hair styles did not change. The hair was parted in the middle and gathered in a cue or bun at the back of the head. The "crowning glory" was never cut, and anyone doing so was looked upon as being common'; in some parts of society long hair remained obligatory for women in positions of responsibility. Blair, 1998, 51, 55, indicates that nurses in Glasgow were for long into the twentieth century not allowed to cut their hair. In at least one case a girl had to wear a false bun at work after getting her hair cut fashionably short.
33. Nicolson, MacInnes, 1951, 102, include the proverb, Cha ghruagaichean gu léir air am bi am falt fhéin [All are maidens that wear their own hair]; the significance of long hair for single girls as expressed in traditional belief in the Northern Isles is shown by Marwick, 1986, 90–1.
34. See Dunbar, 1962, 100, for headdress customs in Gaelic Scotland; Nicolson, MacInnes, 1951, 102; McClintock, 1943, 65; Grant, 1988b, 270.
35. Weir, 1998, 218; see also Blair, 1998, 184.
36. See Blair, 1998, 103, for example.
37. Grant, 1988a, 343, for example describes the Leddy Borlam who was said to be not far from 90 years of age, upright, active, slender, richly dressed for her station and with a pleasant countenance.
38. See for example de Marly, 1986, 46.
39. Adams, 1989, 6.
40. Huggett, 1982, 48; See also Kesson, 1992, 110. Here the author describes having her head shaved before going to an orphanage.
41. The desperation which some women in this category felt is revealed by Burnett, 1999–2000, describing the case of 'Madam Rachel' (Sarah Rachel Leverson) who was found guilty in London in 1868 of obtaining money by false pretences by offering to make women look young at her New Bond Street shop. The case focused on Mrs Mary Borrowdail who had spent over a thousand pounds in a futile attempt to enhance her looks and who was ridiculed for this by her own counsel during the trial.
42. Bynum, Porter, 1993, 1287, 1292.
43. The place of preference in the context of appearance is examined by Corson who, in discussion of hair styles, outlines four non-discrete categories. He lists the fashionable (conspicuous and conventional), the conservative (inconspicuous and conventional), the individual (inconspicuous and non-conforming) and the eccentric (conspicuous and non-conforming). Corson, 1980, 21.
44. Dunbar, 1962, 91.
45. Corson, 1980, 19.
46. Boucher, 1967, 413; Huggett, 1982, 50.
47. Corson, 1980, 19.
48. For example de Marly, 1986, 51 says that the Swiss considered the English 'to not

be servile about fashion'; Dunbar, 1962, 92, goes as far as to say that 'in the remote country districts there was of course "no style" or "fashion"'.
49. For example, Plant, 1952, 182, claims that fashions were followed by many in eighteenth-century Scotland, 'but there were exceptions, particularly among the old ladies, who liked to assert their independence'.
50. McClintock, 1943, 77, 140.
51. For example, Recipes and Remedies: The first Scottish recipe book, MS.
52. For the history and nature of these in Scotland see Allen, 1977, 43; Bryde, 1979, 160; Corson, 1980, 19; Durban, 1984, 28; Glasser, 1987, 30–1; Kidd, 1992, 41; Huggett, 1982, 55, 63; Leneman, Mitchison, 1998a, 6; Plant, 1952, 206; Worsdall, 1981, 190.
53. Plant, 1952, 210, writes that by 1790s these existed in every main street, with Edinburgh's being the best.
54. Macnaughton, 1987, 67; see also Kesson, 1992, 12–13.
55. Woodforde, 1995, 56. The infamous Madame Rachel had such an establishment in New Bond Street, London.
56. See Ralston, Inglis, 1984, 46–7.
57. See Durban, 1984, title page.
58. Bryde, 1979, 160. Bronze and iron knives were in common use as razors, for example among the Romans of the first century AD.
59. See Chapter 5 Lighting.
60. Adams, Adams, 1989, 21.
61. Gibbon, 1982, 17, 18, gives a fictional account of servants gossiping about their masters' domestic goings-on.
62. See Chapter 7 Rest.
63. Allen, 1977, 27.
64. Tarrant, 1996, 197.
65. See Chapter 4 Storage.
66. Tarrant, 1996, 198–9.
67. See Blair, 1998, 9.
68. Woodforde, 1995, 51.
69. Wright, 1960, 112.
70. Tarrant, 1996, 196–7.
71. See Chapter 16 Personal Hygiene.
72. Allen, 1977, 34.
73. Plant, 1952, 205.
74. Bryde, 1979, 160.
75. See for example Grant, 1988a, 52; Grant, 1988b, 7.
76. See Bayne-Powell, 1956, 135, for information on waiting gentlewomen serving unmarried men or widows and the figure facing 161, 'Progress of the Toilet' (a lady being dressed by her maid); also Maxwell, Hutchinson, 1958, 90.
77. Plant, 1952, 207.
78. Woodforde, 1995, 47.
79. Huggett, 1982, 10.
80. See Chapter 26 Infirmity.
81. Plant, 1952, 206–7. A hairdresser might call each day or a household member might do the work (which included shaving gentlemen and carrying out pedicures). Wig makers also frequently visited; Douglas, 1992, 134, presents 'The Jolly Barber', a bawdy tale of adultery which shows how well such a person could get to know the household!
82. Woodforde, 1995, 51.
83. Roberston, 1997, 156.
84. See Weir, 1998, 323.
85. Grant, 1988a, 215, Miss Elphick (Elizabeth's tutor), threw her clothes on in

haste and completed her toilette by wiping her face and hands and quickly smoothing down her hair. She claimed that she never spent more than ten minutes dressing in the morning and thought people could clean themselves properly on Sundays.

86. See Beith, 1995, 213, 215–7, 24–5–0.
87. Robertson, 1997, 156.
88. Allen, 1977, 25.
89. Spalding, 1895, 36.
90. Bayne-Powell, 1956, 176, for example, mentions a sunburn cream made from nettle seed, parsley seed with kernel of pear, six eggs, distilled lemon, bean flour and white wine.
91. For *hair maintenance* (washing, cleaning and drying) see Allen, 1977, 40, 43; Durban, 1984, 29, 31; Grant, 1988a, 251; Huggett, 1982, 55, 68: (conditioning) Allen, 1977, 43; Corson, 1980, 467; Durban, 1984, 23; Huggett, 1982, 46, 55; MacKenzie, 1996, 55, 58; Recipes and Remedies: The first Scottish recipe book, MS; Wright, 1960, 242.

    For *hair styling* (brushing and combing) see Allen, 1977, 43; Durban, 1984, 23, 27; Huggett, 1982, 37: (cutting hair) Allen, 1977; Blair, 1998, 15, 17, 18; Bryde, 1979; Corson, 1980, 95; Durban, 1984; Fraser, 1991, 213; Huggett, 1982; MacKenzie, 1996, 55–8; Maxwell, Hutchinson, 1958; McClintock, 1943, 7, 31, 40, 63, 89, 123: (colouring) Allen, 1977, 41, 43; Beith, 1995, 65; Durban, 1984, 23; Huggett, 1982, 12, 21, 46, 55; Plant, 1952, 209; Woodforde, 1995, 55: (plaiting, tying back and putting up hair) Durban, 1984, 29; McClintock, 1943, 97: (curling and straightening) Allen, 1977, 41, 43; Blair, 1998, 17, 18; Durban, 1984, 28, 31; de Garsault, 1961, 6, 7; Huggett, 1982, 21, 26, 29, 30, 35, 36, 45–6, 50; Plant, 1952, 207: (dressings) Allen, 1977, 42; Durban, 1984, 23; Huggett, 1982, 35, 46, 55; Plant, 1952, 209.

    For *adorning hair* see Huggett, 1982, 32; Kesson, 1992, 12, 13, 21, 47; Leneman, 1993, 51; MacDonald, 2003, 45; Maxwell, Hutchinson, 1958, 162.

    For *artificial hair* see Allen, 1977, 40; Beith, 1995, 65–6; Bryde, 1979, 158, 163–6; Durban, 1984, 6, 7; de Garsault, 1961, 10; Huggett, 1982, 14, 15, 24, 32, 37, 48, 55; Plant, 1952, 207; Woodforde, 1995, especially 93, 110.
92. For the facial features *in general* see Woodforde, 1995, 54.

    For the *eyes* see Beith, 1995, 215; Smout, Wood, 1990, 31; Spalding, 1895, 40; Woodforde, 1995, 51, 55.

    For the *eyebrows* see Corson, 1980, 572; Huggett, 1982, 18; Spalding, 1895, 40; Woodforde, 1995, 54.

    For the *nose* see Woodforde, 1995, 23–4.

    For the *lips* see Spalding, 1895, 40; Woodforde, 1995, 51, 54.

    For the *ears* see Kesson, 1992, 48; MacDonald, 2003, 37; Plant, 1952, 182; Weir, 1998, 363–6.

    For the *teeth* see Allen, 1977, 44–5, 47–9; Denton, Wilson, 1991, 137–8; Kerr, 1986, 189, 197; MacKenzie, 1996, 54–5; Plant, 1952, 203–5; 'Recipes and Remedies: The first Scottish recipe book', MS (recipe for dentifrice); Spalding, 1895, 41; Woodforde, 1995, xviii, xxi, 29, 59, 65, 67, 69, 70–2; Wright, 1960, 245–6.
93. For *nails* see Spalding, 1895, 44–5; Rorie, 1994, 237.
94. For *garments* designed to alter *the figure* see Ewing, 1977, 51; de Marly, 1986, 57; Smout, Wood, 1990, 95; Spalding, 1895, 44; Woodforde, 1995, xviii, 1–16.
95. See Chapter 16 Personal Hygiene; Chapter 17 Housework.
96. Plant, 1952, 203. They might also contain rice, flour, starch, white lead and arris root.
97. Allen, 1977, 24; Wright, 1960, 242.
98. See Chapter 16 Personal Hygiene.
99. Allen, 1977, 35.

100. For those mentioned see Allen, 1977, 24, 35; Plant, 1952, 203, 210; Recipes and Remedies: The first Scottish recipe book, MS; Woodforde, 1995, 49, 50; Wright, 1960, 240, 246.
101. Beith, 1995, 215; Spalding, 1895, 37.
102. Spalding, in 1895, offers a recipe for cold cream said to be, 'both cheaper and nicer than that purchased at the chemist', 39.
103. Wright, 1960, 246; Spalding, 1895, 38.
104. Beith, 1995, 233; Spalding, 1895, 37. Spalding also mentions that Americans had preventative measures against sunburn in cold cream and pure almond oil.
105. Rorie, 1994, 243–4.
106. Woodforde, 1995, 49, 50.
107. Allen, 1977, 37; Woodforde, 1995, 54; Plant, 1952, 210.
108. Spalding, 1895, 39.
109. de Marly, 1986, 51, writes, 'doubtless all those straw hats protected complexions from the sun, for to be tanned like an Ethiope was considered decidedly inferior!'
110. Cheape, 1983, 101.
111. Allen, 1977, 37; Woodforde, 1995, 49, 50.
112. See Beith, 2000a, 2000b, 2001.
113. For the debate surrounding the Picts and tattooing see Ritchie, 1994.
114. Woodforde, 1995, 47.
115. Woodforde, 1995, 51; yet de Marly, 1986, 12, records that one German visitor to England, merchant Samuel Kiechel, in 1585 said that, '... the women there are charming and by nature so mighty pretty as I have scarcely ever beheld for they do not falsify, paint, or bedaub themselves as in Italy or other places.' England might be an important source of cosmetic innovation for the upper classes in Scotland but seems itself to have been relatively conservative.
116. For these substances and procedures see Allen, 1977, 37–8; Corson, 1980, 467; Plant, 1952, 209; Spalding, 1895, 38–9; Woodforde, 1995, 53–4.
117. Corson, 1980, 470; Spalding, 1895, 40.
118. Allen, 1977, 37.
119. Huggett, 1982, 44.
120. Plant, 1952, 209.
121. The wearing of makeup was often associated with prostitutes. For example, de Marly, 1986, 14, claims that prostitutes in England were known for makeup, hoisted-up skirts and tawdry finery indoors where the law could not see; see McVeigh, 1999, 122 for a description of an English woman wearing makeup in a small Scottish community.
122. Spalding, 1895, 40.
123. Woodforde, 1995, 57.
124. Bryde, 1979, 159.
125. For example, see McClintock, 1943, 27, 30, 31, for descriptions of beards among the men of the Gaelic-speaking areas of Ireland and Scotland before the eighteenth century.
126. Woodforde, 1995, 51.
127. Allen, 1977, 43.
128. Durban, 1984, 21.
129. A 'beaver' was a derogatory term for a bearded man. Moustaches gained popularity immediately before World War I and, in slighter form, during World War II. See Mansfield, Cunnington, 1973, 289, 335; Corson, 1980, 572.
130. Plant, 1952, 203, writes that in the eighteenth century Sir John Foulis tried to train a servant to shave him and later obtained a boy to do it.
131. Durban, 1984, 11–13, 18.
132. See MacKenzie, 1996, 5; Durban, 1984, 12.

133. Adams, 1991, 31.
134. See for example MacKenzie, 1996, 5.
135. Eddie Mathieson. Quoted in MacDougall, 1995, 236.
136. Durban, 1984, 18.
137. For the history of open razors see Wright, 1960, 233-4.
138. Glasser, 1987, 32.
139. de Garsault, 1961, 5.
140. Durban, 1984, 15, 18.
141. The 1970s and early 1980s saw punks partially shave the head (the 'mohican' was the most famous of such styles). By the 1990s a completely shaved head (perhaps with a few millimetres of growth) had become a more widespread youth fashion, one which has continued to be popular with that and the following generation.
142. Corson, 1980, 95; Fraser, 1991, 213. Hair can be shed in substantial quantities after serious illness. It was thought that shaving the head would allow the hair to grow back to full thickness; see also MacKenzie, 1996, 58.

BIBLIOGRAPHY

Adams, D G. *Bothy Nichts and Days: Farm Bothy Life in Angus and the Mearns*, Edinburgh, 1991.
Adams, S, Adams, S (ed A Haly, introduction by P Horn). *The Complete Servant: Regency Life Below Stairs*, Southover, 1989.
Allen, E. *Wash and Brush Up* (1976), London, 1977.
Bayne-Powell, R. *Housekeeping in the Eighteenth Century*, London, 1956.
Beith, M. *Healing Threads: Traditional Medicines of the Highlands and Islands*, Edinburgh, 1995.
Beith, M. Deànamh an leighis [Making the cure], *West Highland Free Press*, no 1493 (8 December 2000a).
Beith, M. Deànamh an leighis [Making the cure], *West Highland Free Press*, no 1496 (29 December 2000b).
Beith, M. Deànamh an leighis [Making the cure], *West Highland Free Press*, no 1497 (5 January 2001).
Bennett, M. *Scottish Customs from the Cradle to the Grave*, Edinburgh, 1992.
Blair, A. *Miss Cranston's Omnibus: Recollections of Glasgow Life* (1985, 1991), Edinburgh, 1998.
Boucher, F. *A History of Costume in the West*, London, 1967.
Bryde, P. *The Male Image: Men's Fashion in Britain, 1300-1970*, London, 1979.
Burnett, J. Madame Rachel and the meat trade, *ROSC*, 12 (1999-2000), 123.
Bynum, W F, Porter, R. *Companion Encyclopedia of the History of Medicine, vol 2*, London and New York, 1993.
Caldair, D [Calder, G], ed. *Orain Ghàidhealach le Donnchadh Macantsaoir* [*Gaelic Songs of Duncan MacIntyre*], Edinburgh, 1912.
Campbell, J L, ed. *Hebridean Folksongs: A Collection of Waulking Songs by Donald MacCormick in Kilphedir in South Uist in the Year 1893*, Oxford, 1969.
Carruthers, A, ed. *The Scottish Home*, Edinburgh, 1996.
Cheape, H. The Lothian farm servant, *Northern Studies*, 20 (1983), 79-104.
Corson, R. *Fashions in Hair: The First 5,000 Years* (1965), London, 1980.
Denton, H, Wilson, J C. *The Happy Land*, Edinburgh, 1991.
Douglas, S, ed. *'The Sang's the Thing', Scottish Folk, Scottish History: Voices from Lowland Scotland*, Edinburgh, 1992.
Dunbar, J T. *History of Highland Dress*, Edinburgh and London, 1962.
Durban, G. *Wig, Hairdressing and Shaving Bygones* (Shire Album, 117), Princes Risborough, 1984.

Ebin, V. *The Body Decorated*, London, 1979.
Ewing, E. *History of Children's Costume*, London, 1977.
Fraser, E. *The House by the Dvina: A Russian Childhood* (1984), London, 1991.
de Garsault, M (ed J Stevens Cox). *The Art of the Wigmaker* (1767), London, 1961.
Gibbon, Grassic L. *Cloud Howe* (1933), London, 1982.
Glasser, R. *Growing Up in the Gorbals*, London, 1987.
Grant, E. *Memoirs of a Highland Lady*, vol 1 (1898), Edinburgh, 1988a.
Grant, E. *Memoirs of a Highland Lady*, vol 2 (1898), Edinburgh, 1988b.
Höskuldsson, S S. Women and love in the novels of Laxness, *Northern Studies*, 6 (1975), 3–19.
Huggett, R. *Hair Styles and Headdresses* (History in Focus Series), London, 1982.
Kerr, N W. Dental examination of the Aberdeen Carmelite collection, late mediaevel, 1300–1600. In Cruwys, E, Foley, R A, eds, *Teeth and Anthropology*, Oxford, 1986, 189–97.
Kesson, J. *The White Bird Passes* (1958), London, 1992.
Kidd, D I. *To See Ourselves: Rural Scotland in Old Photographs*, Glasgow and Edinburgh, 1992.
Leneman, L. *Into the Foreground: A Century of Scottish Women in Photographs*, Edinburgh, 1993.
Leneman, L, Mitchison, R. *Sin in the City: Sexuality and Social Control in Urban Scotland, 1660–1780*, Edinburgh, 1998a.
Leneman, L, Mitchison, R. *Girls in Trouble: Sexuality and Social Control in Rural Scotland, 1660–1780*, Edinburgh, 1998b.
Livingstone, S. *Scottish Festivals*, Edinburgh, 1997.
MacDonald, K. *Peat Fire Memories: Life in Lewis in the Early Twentieth Century* (Flashbacks, no 16), East Linton and Edinburgh, 2003.
MacDougall, I. *Voices from War and Some Labour Struggles: Personal Recollections of War in our Century by Scottish Men and Women*, Edinburgh, 1995.
MacKenzie, J A. *A Mallaig Boyhood* (Flashbacks, no 5), East Linton, 1996.
Macnaughton, Edwin G. A Highland pharmacy, 1882–1972, *ROSC*, 3 (1987), 67–76.
Mansfield, A, Cunnington, P. *Handbook of English Costume in the Twentieth Century, 1900–50*, London, 1973.
de Marly, D. *Working Dress: A History of Occupational Clothing*, London, 1986.
Marwick, E. *The Folklore of Orkney and Shetland* (1975), London, 1986.
Maxwell, S, Hutchinson, R. *Scottish Costume, 1550–1850*, London, 1958.
McClintock, H F. *Old Irish and Highland Dress*, Dundalk, 1943.
McVeigh, P. *'Look After the Bairns': A Childhood in East Lothian* (Flashbacks, no 8), East Linton, 1999.
Newton, M. *A Handbook of the Scottish Gaelic World*, Dublin, 2000.
Nicolson, A, MacInnes, M. *Nicolson's Gaelic Proverbs*, Glasgow, 1951.
Ó Baoill, C, ed. *Bàrdachd Shìlis na Ceapaich* [Poems and Songs by Sileas Macdonald], c.1660–c.1729, Edinburgh, 1972.
Plant, M. *The Domestic Life of Scotland in the Eighteenth Century*, Edinburgh, 1952.
Ralston, I, Inglis, J. *Foul Hordes: The Picts in the North East and their Background*, Aberdeen, 1984.
Recipes and Remedies: The first Scottish recipe book, MS.
Ritchie, A. *Perceptions of the Picts: From Eumenius to John Buchan*, Rosemarkie, 1994.
Robertson, U. *The Illustrated History of the Housewife, 1650–1950*, Stroud, 1997.
Rorie, D (ed D Buchan). *Folk Tradition and Folk Medicine in Scotland: The Writings of David Rorie*, Edinburgh, 1994.
Shaw, M F. *Folksongs and Folklore of South Uist* (Oxford, 1977), Edinburgh, 1986.
Smout, T C, Wood, Sydney. *Scottish Voices, 1745–1960*, London, 1990.
Spalding, E, ed. *Every Woman's Book of the Toilet, Self-Culture, Etiquette, Home Occupations, Employments, Amusements, Correspondence, etc*, Port Sunlight, 1895.

Tarrant, N. The bedroom. In Carruthers, A, ed, *The Scottish Home*, Edinburgh, 1996, 181–202.
Weir, M. *Trilogy of Scottish Childhood: Shoes were for Sunday; Best Foot Forward; A Toe on the Ladder* (1970, 1972, 1973), Edinburgh, 1998.
Williamson, D, Williamson, L. *A Thorn in the King's Foot: Stories of the Travelling People*, Harmondsworth, 1987.
Woodforde, J. *The History of Vanity*, Stroud, 1995.
Worsdall, F. *A Glasgow Keek Show: Glimpses of City Life*, Glasgow, 1981.
Wright, L. *Clean and Decent: The Fascinating History of the Bathroom and the Water Closet and of Sundry Habits, Fashions and Accessories of the Toilet, principally in Great Britain, France and America*, London, 1960.

16 Personal Hygiene

MILES OGLETHORPE WITH A SECTION BY ALEXANDER FENTON

In twenty-first-century Scotland, a high level of personal hygiene is considered the norm and the presence within the home of paraphernalia associated with achieving cleanliness is taken for granted. It is only those in exceptional circumstances who are without a water supply in their dwellings. Moreover, a considerable proportion of modern consumer culture is devoted to the manufacture and marketing of products designed to assist the processes involved in personal washing and cleaning, in bathing and in the acceptable management of human waste. Yet even at the beginning of the twentieth century primitive sanitary arrangements were still usual in both urban and rural dwellings, and Scottish society was emerging from a century dominated by public health crises and, by today's standards, horrifying domestic squalor.

It is extremely difficult to find references to toilet habits either within or outside the home for the period before the eighteenth century, and examples of rooms designated for such activities within buildings appear to be very unusual. The first known instances in Scotland relate to baths and bathing in the Roman period, in the first and second centuries AD. Traces of baths were, for example, found during the archaeological excavations of forts on the Antonine Wall.[1] Thereafter, for many hundreds of years toilet activities are described very infrequently as they were rarely considered worthy of mention,[2] although archaeological evidence has offered some insights.[3] Still more difficult to obtain than facts are attitudes towards toilet matters, especially before the modern period. However, monastic evidence suggests the routine imposition of personal hygiene, and monasteries and abbeys were usually located near to reliable supplies of running water which were diverted to run through troughs used as *lavatoria* (for washing hands) and separate rows of *necessaria* (latrines).[4]

In the nineteenth century sanitation became a public issue, driven by the consequences of rapid urbanisation and public health crises. The century was one of great change, witnessing the creation of increasingly sophisticated municipal water-supply and sewerage schemes, and subsequently the evolution of the bathroom and WC as a common feature of twentieth-century dwellings.

BEFORE THE BATHROOM: WASHING AND CLEANING

Washing requires water, perhaps a cleaning agent such as soap to make the process more effective, a means of drying and a perceived need for cleanliness on behalf of the individual. Warrack, when commenting on living conditions in the fifteenth century, observed, '... but to tell the truth, neither in Scotland nor elsewhere had habits of personal cleanliness yet come into fashion. Washing, so far from being a habit, was only occasional ... '[5] Yet this may be only partially true of Scotland's past. Before the rise of the bathroom, bathing or washing of the whole body might in many cases have been infrequent, in the worst examples limited to the individual's entry into and exit from the world, but hands were regularly washed before and sometimes after eating in medieval religious communities[6] and this must also have been the case in other homes. Even before the discovery of the principles of transmission of disease, washing of hands to prevent disgust was normally required before the preparation of food. Likewise, laving after particularly messy chores and before handling delicate or light-coloured fabrics was necessary. Other parts of the body might also have seen water. Dirty faces would not be tolerated for long, in some circles at least; elsewhere it might be grooming enough to pass the hands over the face. For those without footwear, but with ready access to a water supply, foot washing may have taken place. Special occasions would usually have demanded extra effort. Before weddings and similar events, and the Sabbath, it was necessary to pay more attention to cleanliness and appearance in general.[7]

The frequency and manner of washing on an everyday basis, however, depended on many factors: societal notions about acceptable levels of cleanliness during particular historical periods, the social, economic and geographical circumstances of the household, and the individual's position within it. Gender and age may also have played a role. One ever-present factor is personal inclination.

Washing is more likely to be a frequent occurrence when a suitable water supply is nearby. In addition to water drawn from wells, washing and cleaning in the past would have utilised, where available, natural running water supplies. If individuals could not go to the water to wash, then more able-bodied family members or servants would be required to fetch it, for use inside and outside the dwelling (see Fig. 17.6). Washing outside by the children of a well-to-do Scottish household resident in eighteenth-century London is mentioned by Elizabeth Grant:

> In town, there was a large, long tub stood in the kitchen court, the ice on the top of which had often been broken before our horrid plunge into it. We were brought down from the very top of the house, four pairs of stairs, with only a cotton cloak over our night gowns, just to chill us completely before the dreadful shock ... Nearly senseless I have been taken to be dried in the housekeeper's room ... [8]

The outdoors was the place not only for the early morning washing of small children's faces and necks but also for the cleansing of hands between rougher tasks,[9] the facilities anticipating some of the roles of the modern utility room. Washing could take place inside the dwelling with the aid of portable water-carrying devices and furniture such as buckets, tubs, jugs, basins and washstands (see Figs 15.3, 29.4). In this case the water might be warmed over the fire or heated by other means,[10] but this was usually done – and then only sometimes – for small infants or very senior members of the family, or when household members were ill. In modest households of more recent times, where it existed, the washstand, complete with towel rail and matching ceramic ewer (jug), basin and sometimes slop basin, was a substantial purchase and a source of considerable domestic pride.[11] Such stands and sets continued to be found in early twentieth-century homes without bathrooms or in guest bedrooms in larger dwellings.

In many Scottish homes there occurred a period in which piped water was available, sometimes as a communal tap serving all the households in a building, but no bathroom existed. Where each home was provided with its own water supply, washing took place at the sink, no doubt more frequently than before, and perhaps more extensively, although clean hands and faces remained enough in some homes. Many Scots can recall being fully bathed in the sink as small children.[12]

Tooth and hair care, so dear to most modern hearts, were generally less of an issue in the past, at least until the necessary products (sometimes backed by health campaigns) became available and popular.[13] Before then, in the case of the latter, it was more a question of adornment than hygiene.

BEFORE THE BATHROOM: BATHING

Bathing in the home
As with washing and cleaning of parts of the body, full bathing could take place outdoors at the water source or make use of a range of portable objects, used inside or out. These included large wooden tubs of the kind used for washing clothes[14] and latterly tin baths, allowing an all-over wash, executed in stages, rather than bathing as such to be carried out. Such devices were effective in keeping people clean, but in the era before piped water, boilers and adequate drainage, the difficulties involved in carrying sufficient water into the home, heating it, and then getting rid of the waste water, often meant they were not very regularly used.[15] However a certain permissiveness in terms of the freshness of the bathing water could make the process more practicable, with the same bath water serving several bathers, the order of use being according to household status or state of dirtiness. Also inhibiting frequent bathing was the multi-purpose nature of many devices within small homes. John Alexander MacKenzie describes a fixed but unplumbed bath in his Mallaig, Inverness-shire, home in the 1930s:

> In addition to running hot and cold water and a flush toilet we had a bath. This was in the scullery and had a wooden cover. I am not certain that the hot water was supplied by a separate boiler and

fireplace sited adjacent to the bath because this equipment was very rarely used. In order to use it the cover on top had to be cleared of domestic essentials – jam jars, cups and saucers, tea pot, and the paraffin stove with the kettle on top, with the tea caddy close by.

Even after all that was cleared, the wooden cover had to be lifted off and a temporary home found for it. Is it any wonder that the bath was so little used?[16]

Privacy is clearly much more of an issue for bathing than washing. Generally it appears that bathing with others, or witnessing the process of bathing, was acceptable, but only if those involved were of the same sex, although this might be relaxed for small children.[17] When Miss Baillie burst in on a visiting Mr Macklin, bathing in a tub in his room, in Elizabeth Grant's Inverness-shire home she:

> ... would not appear for dinner. Mr Macklin, who was full of fun, would stay upstairs if she did; she insisted on immediate departure ... Such a hubbub was never in a house before ... 'If she had had common sense,' said Miss Ramsay, 'she would have held her tongue; shut the door and held her tongue, then no one would have been the wiser.'[18]

The upper classes could enjoy the precursors of modern baths and bathrooms. Sources of the sixteenth and seventeenth centuries associate baths with stoves for heating the water. 'Bath fats' or 'bathing fats', wooden tubs, were used in the royal household in the 1470s, as also in higher-class town and country houses. A 'bathing house' is mentioned in 1695 by Grizell Baillie in East Lothian, and in the seventeenth century the term 'bathstoll', incorporating the Norse word for a room, occurs sporadically; it probably means a room containing a tub for bathing in.[19]

Public baths

In some towns and cities, public baths compensated for the limitations of many dwellings and also provided an opportunity for social interaction. Warrack noted that public baths, which had begun to play a significant part in urban life in European cities as early as the thirteenth century, and which had acquired a dubious reputation in some places, did not seem to feature prominently in Scotland until the second half of the seventeenth century with the appearance of 'bath stoves' and 'sweating balnes', early hot-air or steam baths. These had their origins in the Near and Middle East, where hot-air and sweating baths on the Roman model had survived and prospered through the Dark Ages.[20] The Arabs re-introduced them to the Iberian peninsula, and returning crusaders spread them into continental Europe where the Italian name *bagnio* was adopted, later being replaced by the term 'Turkish bath'. Similarly, the use of hot-vapour 'Russian' or 'Finnish' baths had spread through Europe from the Baltic, and these are considered to be the antecedents of the modern sauna.[21]

The earliest bagnios to be introduced to the British Isles were created in London during the sixteenth century, but it was not until 1654 that the first was established in Edinburgh.[22] Thereafter, several others were set up in the city, and one in Perth, but there is little evidence of others being established elsewhere in Scotland. Several of the British bagnios are known to have had bedrooms for clients, as was the model elsewhere. It was this facility which tended sometimes to cast doubt upon the morality of particular institutions, especially in Europe, but it appears not to have been reported widely in Edinburgh, even when the services of masseurs, known as 'rubbers and rubberesses', were added to the repertoire in the eighteenth century. The major source of dissatisfaction with these establishments was much more likely to be the dubious health cures with which some were associated.

It was, nevertheless, the enthusiasm of the medical profession in its search for remedies for a variety of ailments which promoted the concept of the bagnio in Scotland. The Incorporation of Surgeons and Barbers opened a bagnio in Edinburgh in 1704, designed explicitly for 'noblemen, gentlemen, ladies and others,'[23] who were invited to purchase season tickets. The Royal College of Physicians constructed its own bagnio in the Pleasance in 1712, which was rebuilt and renewed in 1728 and 1742. In 1731 the Edinburgh Royal Infirmary installed its first bathing establishment, but proper facilities were not added until a public appeal yielded funds in 1752, and it was only in 1769 that warm baths were installed for the patients in the hospital itself, rather than for paying visitors alone. This reflected a belief amongst doctors of the time in the medical benefits of cold baths, which were widely considered to cure a number of common complaints. It seems, however, that the paying clientele had less faith in the medicinal qualities of the bagnios, and attendances declined towards the end of the eighteenth century, partly because of their expense, and partly because of the problems of acquiring adequate supplies of water. In contrast, commercial bath houses were beginning to prosper, especially in Leith, where four establishments were in operation by the 1840s and were successfully using seawater instead of precious fresh water supplies.[24] Such places, and those like them, catered mostly for the needs of the relatively wealthy. It was not until the late nineteenth century that reliable city water supplies permitted the establishment of large-scale municipal baths and washhouses for the masses in Scotland.

There was also substantial investment in public baths, encouraged initially by the passing of the Baths and Washhouses Act in 1864.[25] Glasgow Corporation had originally considered establishing public baths as early as 1816, following the building of philanthropist William Harley's baths in Bath Street in 1800. A project at Rutherglen, to be modelled on the floating baths of the River Seine, was considered, but no scheme resulted, possibly because the state of the Clyde at the time would have added to, rather than removed, dirt from any person or item being washed. Despite the establishment of a washhouse at Glasgow Green, little else was possible until the arrival of a water supply from Loch Katrine in the 1860s. Thereafter, municipal baths were built throughout the city. New baths were opened at Greenhead, for example, on the site of the earlier washhouse, and contained two swimming ponds (one for men and one for women), 27 private baths for men and seven for ladies. There was also provision for 40 washers at any one time within the washhouse. By 1913, the Corporation reported that there had been 821,329 attendances of washhouses during the year.[26]

Glasgow's provision of public baths and washhouses was repeated in other Scottish cities, and throughout the United Kingdom as a whole. In addition to swimming baths, some included Turkish and Russian baths, and these were particularly popular in large industrial towns. Many of the baths also possessed washhouses or 'steamies', the sophistication and scale of which increased in the early twentieth century. As well as providing compartments with washing tables, troughs and an allocated drying horse, washing machines were gradually introduced, with hydro-extractors (spin driers), and adjoining mangling rooms with rows of electric irons.[27] Public washhouses and baths therefore became essential parts of life in the twentieth century, and their importance only faded after World War II with the introduction of launderettes, with the availability of increasingly cheap and reliable domestic washing machines, and with a rapid rise in the quality of housing, including the appearance of the bathroom in most dwellings.

The most significant exception to this pattern was provided by the coal

Figure 16.1 The old way in which miners took their bath. Source: *Miners' Welfare Fund Annual Report*, 1937, 13.

Figure 16.2 The new way in which miners took their bath. Source: *Miners' Welfare Fund Annual Report*, 1937, 13.

industry, the majority of whose employees lived in villages away from the main conurbations and who became unavoidably and extremely dirty whilst working their shifts both underground and at the surface (in processes such as coal picking). Miners routinely returned to homes with no bathroom or plumbing and did not have the option of using municipal public baths (Fig. 16.1). Although the homes of miners were to be improved substantially in the second half of the twentieth century, the provision of pit-head baths became a major political issue, resulting in the establishment of the Miners' Welfare Fund (MWF) in 1920. This obliged all coal owners to pay a fixed levy on every ton of coal produced by their mines, and in the 1920s and 1930s permitted for the first time the construction of baths at many collieries in the UK. Many of the finest examples in Scotland, such as Polkemmet, West Lothian, Arniston, Midlothian and Michael, Fife, were designed by the MWF's architect John Austin Dempster in the 1930s, and their importance is reflected in their iconic architectural quality.[28] There is no doubt that pit-head baths transformed the lives of many miners, providing an effective means of taking a regular shower and compensating for the shortcomings of their homes (Fig. 16.2).

For some communities, another route to cleanliness was occasionally provided by Lever Brothers, who introduced portable bathing facilities to schools in an attempt to promote their soap products. These temporary facilities were erected in school playgrounds such as in Kelvinhaugh, Glasgow,

PERSONAL HYGIENE • 283

and took the form of a canvas tent. The provision of this facility reflected increasing concern at the potential for the spread of disease amongst children when they were congregated in large numbers at school.[29]

BEFORE THE BATHROOM: EXCRETORY MATTERS

For most people excretory matters appear to have routinely involved the use of a commode, close stool, chamber pot or privy, the last possibly through an appropriate orifice in a building. One of the better known manifestations of the latter was the 'garderobe', which is described as a wardrobe extension to a bed chamber, being a euphemism for a privy.[30] Garderobes were in fact small rooms often built into the thickness of the walls or, in medieval buildings, formed by a projection, and usually had a small window and a wooden seat set over a vertical chute which directed the human waste either onto the ground, or in the case of some castles, into a moat.

Some people had the choice of using facilities situated in outhouses, such as the medieval example recently excavated in Kirk Close in Perth.[31] By the late nineteenth century, toilets were situated on landings in the common stairs of some tenements. In these circumstances, such were the large numbers of people sharing the one facility that most used it merely to receive the contents of chamber pots and other vessels used within the adjacent dwellings.[32] However, just in case, cut sheets of newspaper were often provided in the toilet. For those without running water, the traditional way of neutralising the odours emanating from human waste accumulating in or near to a dwelling was to cover it with ashes, and this tended to combine to produce a valuable fertilizer.[33] Indeed, ash closets were being installed well into the twentieth century, both in homes, and as public toilet facilities.

WATER SUPPLY AND WASTE REMOVAL

Water was originally acquired by both urban and rural dwellers on an ad hoc basis, taking advantage of natural sources such as springs, streams, lakes and rivers, and by installing rainwater cisterns. With rising population densities and increasing cultivation and industrial activity, came escalating demand for water, and more sophisticated means of supply were devised such as the sinking of wells and the construction of reservoirs, dams, watercourses and pipelines. For the most part, however, water supply was irregular, and there were no regulated or controlled means of disposing of the increasingly large quantities of sewage. Writing in 1873, William McDowall described the grim situation prevailing in Dumfries, where the River Nith was the principal supply of water and conduit for sewage:

> With the exception of what water was furnished by a few wells and private pumps, all the water used for domestic purposes was carried by hand or carted in barrels from the Nith by four old men, who doled it out in tin pitchers or cans, from door to door, at the rate of five

canfuls a penny. The river, when swelled by heavy rains, which was often the case, became thick with mud; and it was constantly exposed to a more noxious pollution, caused by refuse poured into it from the town. The quality of the water did not improve by being borne about in barrels of suspicious aspect; and often, indeed, the liquid drawn from them during the summer acquired a taste-me-not repulsiveness by the presence of innumerable little objects, pleasant to no one save an enthusiast for entomology.[34]

The situation was typical of most towns and cities, with such 'water caddies' delivering river or spring water to households, particularly where there were no local wells or pumps.

The first formal water supply schemes in Scottish cities were in Edinburgh (1674), Stirling (1774), Greenock (1796), Glasgow (1806), and Paisley (1834).[35] These were, however, usually limited to the wealthier classes, and the general living conditions for most people in the towns and cities continued to deteriorate with rapid urbanisation.[36] The extent of the problem is well illustrated by the fact that Glasgow's population increased by 33,000 between 1831 and 1841, but there was no corresponding increase in housing stock.[37] By 1861, the census confirmed that 226,000 families (half the Scottish population) lived in one-roomed dwellings.[38] Nowhere were the problems of poor sanitation better documented than in the capital city itself. Brotherstone, when writing about sanitary reform, noted that:

> The chronicles of foreign travellers during the eighteenth century are full of pithy descriptions of the more unpleasing habits of the people; it was a set piece for every foreign visitor to Edinburgh to describe the remarkable nightly shower of slops which rained from the High Street onto the causeway to the accompaniment of warning cries 'gardyloo'.[39]

He also observed that attempts at reforming the sanitary customs of the people were remarkable for their lack of effect, and that in 1608 King James had complained that the throwing down of middens or 'fulzie' was '... nocht only uncumlie and uncivill, but lykwayis verie dangerous in tyme of plaig and pestilence, and verie infective of itself'.[40] By the mid nineteenth century, urban squalor in Scotland had, if anything worsened rather than improved.

The state of the cities has prompted unflattering comment from sanitary historians. Again, Brotherston comments that 'Scotland had a time-honoured tradition of dirt, both inside and outside the house; filth and squalor prevailed everywhere, but they reached their extreme in the towns.'[41] Much of the population had migrated from rural squalor, and had been augmented in the towns by large numbers of Irish immigrants who had often escaped even harsher conditions at home. Sanitary reports of the period repeatedly describe groups of overcrowded dwellings, at the heart of which were always communal dunghills. In a normal tenement there was no water supply, no sink, no WC, and no connection to a common sewer. Water and waste had to be carried up and down stairs, and there was much spillage in the process, prompting the comment that, 'Common stairs and passages are often the receptacles for the most disgusting nuisances ... There are no domestic conveniences, even in the loftiest tenements, where they are most needed.'[42]

A typical description of the time detailed the following scene in Glasgow:

> There are large midden-steads, some of them actually under houses, and all of them in the immediate vicinity of windows and doors of human dwellings. These receptacles hold the entire filth and offal of large masses of people and households, until country farmers can be bargained with for their removal.[43]

One of the principal problems was that the content of the midden was a source of profit for the landlords, and any attempts by the authorities to remove it were greatly resented.[44] The middens tended to include stable and byre manure, and were traditionally left to accumulate until there was sufficient to cover the rent.[45] In the absence of sewers, most tenement houses had been equipped with privies or trough closets, sometimes a whole stair housing dozens of families sharing a single privy, which was usually located in the back green, adjacent to ash bins. After mixing the excrement with ash, the resulting mixture was taken away by the 'night soil' man in dung carts during the night. Although these descriptions relate to Glasgow, the same situation was repeated in some form or another in every Scottish town and village. The situation was exacerbated by the new water companies' lack of interest in supplying water to those who could not pay for it. Indeed, in many cases, the arrival of the new water enterprises reduced the amount of water available to most people living in a town or city.

Inevitably, it was disease that proved to be the major force behind sanitary reform, especially following Snow's discovery in 1849 that cholera was a water-borne disease.[46] Even though the middle and upper classes had in most cities begun to escape to leafier new suburbs, major outbreaks of typhoid and cholera in 1831, 1847, 1848 and 1853 did not strike only the poor. In response to the growing crisis, The Nuisances Removal (Scotland) Act of 1856 led to the formation of a Committee of Nuisances aimed at introducing sanitary reform to Glasgow, which, in combination with the Police Act of 1862, resulted in the formation of the city's Sanitary Department in 1862. To many, this date marks the beginning of the public health movement in Scotland, and was followed in subsequent years by other cities' efforts, including those of Edinburgh in 1867.[47]

Much of the impetus behind the improvements in public health in the second half of the nineteenth century emanated from the increased powers of the local authorities, but although a Baths and Washhouses Act had been passed as early as 1846 in England, it was not until 1892 that similar legislation was passed for Scotland.[48] The Scottish local authorities therefore initially focused on the issue of water supply. The total inadequacy, unreliability and poor quality of private water supplies led to the incorporation into municipal control of most cities' supplies in the mid nineteenth century. Dundee, which was one of the first cities to consider opening municipal baths in 1844, was the first to end private ownership and control of water supply in 1840.[49] Glasgow followed in 1855, the ensuing Loch Katrine scheme justly receiving international attention as it approached completion in the 1860s.[50] Other cities to follow suit included Greenock in 1867, Edinburgh and Dundee in 1869, and Inverness in 1875.

The provision of large-scale water supply was made possible by the development of new materials available for the manufacture of pipes and mains. In 1674, Edinburgh's first piped water supply had been provided in three-inch (7.6cm) diameter lead pipes, which carried water from Comiston to a reservoir at Castle Hill. The pipe was increased to four-and-a-half inches (11.5cm) in diameter in 1722, and augmented by supply from Swanston in wooden pipes in 1760. However, it was the advent of cast iron which, after 1787, transformed Edinburgh's water supplies.[51] In the early nineteenth century, Scotland had emerged as one of the world's most important manufacturers of pig iron, and had developed a great reputation for high-quality iron castings.[52] The best known Scottish iron company was Carron (founded in 1759), but several others grew to specialise in the manufacture of cast-iron water mains and associated products, one of the most important producers being Thomas Edington of Glasgow, whose Phoenix Foundry supplied mains pipes for towns and cities throughout the world.[53] Other important companies included Walter Macfarlane & Co, who produced a wide range of architectural cast-iron products, and also manufactured an assortment of sanitary appliances, including cast-iron baths, water closets, wash-hand basins and urinals.[54] Furthermore, as Scotland's engineering industries grew and prospered, companies such as Glenfield Kennedy of Kilmarnock, Ayrshire, came to specialise very successfully in products associated with water supply, such as pipes, valves and hydraulic engineering fixtures particularly relating to reservoirs. Equally significant was the developing Scottish expertise in the production of 'pressure pipes', capable of containing high-pressure supplies of water, steam and gas. This required vertical as opposed to horizontal casting, and was developed by D Y Stewart at the Links Foundry in Montrose, Angus, in 1846.[55]

Supplying urban populations with water was, of course, only part of the problem. Just as significant was the task of finding a means of taking away the waste, the volume of which was greatly enlarged by the increasingly plentiful supplies of water. In the case of Glasgow, the need for more sewers was recognised comparatively early, the first significant developments occurring in 1790. The sequences of serious epidemics accelerated municipal action from the mid nineteenth century onwards, and by the 1890s, sewage treatment had begun to cope with the huge quantities of effluent generated.[56] Nevertheless, the practice of selling 'night soil' to farms and market gardens, sometimes as part payment of rent,[57] survived for much of the nineteenth century, as did the sale of urine, the properties of which were of great value to the textile industries.

The extent of the problem is well illustrated by a report prepared in 1891, which noted that:

> The provision made for the disposal of the excrement of the inhabitants of these tenements demands immediate attention. Several places are noted where there is no provision whatsoever, but in our opinion, the privy (a public facility in the back yard) is in no case a sufficient provision for flatted tenements. It is never used, and cannot in the nature of the case be used by females, and seldom by children. The result is that every sink is practically a water closet, and the stairs and courts and roofs of the outhouses are littered with deposits of filth cast from windows.[58]

In the last decades of the nineteenth century and in the early twentieth century, therefore, the development of sewerage infrastructure occurred rapidly, and its high quality was such that it was only in the late twentieth century that major renewal programmes became necessary. These were in part necessitated by increased standards imposed by the European Union.

BATHROOMS

Roles

The popularity of bathrooms increased in the latter half of the nineteenth century in response to the enhanced provision of reliable water supply and drainage in towns and cities, where house building for the middle and upper classes accelerated rapidly, particularly within new suburbs. Older houses might witness the conversion of a bedroom into an oftentimes gaunt and draughty bathroom. By the late twentieth century, after several decades of slum clearance and social-housing programmes, bathrooms or shower rooms came to be considered as essential components of almost all dwellings, and a considerable number of homes now possess more than one. Despite being sometimes supplemented by a WC or cloakroom and a utility room, there is often conflict between the roles carried out in these confined spaces, as dealing with human waste is slightly at odds with the washing, cleaning, bathing and beautification of people and the conflict is felt more keenly the more people share the facilities.

Bathrooms can play other roles. Amongst these is the display of household status. Bathrooms and shower rooms are expensive to equip and decorate, yet having a fashionable bathroom has often been seen as a priority in home decoration terms, especially in recent decades with the rising popularity of 'do-it-yourself'. But creating a lavish bathroom, or bathrooms, is not just part of keeping up with the Joneses, it is also a way of increasing the market value of properties. Bathrooms can in addition reveal the relative status of people within a household. In wealthy nineteenth-century homes, bathrooms were usually for the use of the family members

Figure 16.3 & 16.4 A typical modern bathroom in working-class housing. Two views taken in 2002 of the bathroom in a former local authority flat in Edinburgh. The wash-hand basin is the original installed in 1959, while the enamelled bath (now with instant electric shower 'over') and the WC are 1980s replacements bought from a do-it-yourself chain store and fitted by the tenant/owner himself. Timber panelling has been popular in bathrooms since the nineteenth century. Here its installation was another 1980s DIY project. The double glazing (and central heating – both fitted in the 1990s) made it practicable to spend time on grooming and to relax in the bathroom. The need for increased storage is one result. Shelves hold an ample selection of toiletries and cosmetics, ornaments and even a pair of candlesticks, while a basket under the wash-hand basin contains toilet paper (and, in a lower compartment, medicines, with a set of scales tucked underneath). Source: S Storrier.

and their guests, not the servants. En-suite bathrooms are in some respects a return to the old idea of having toilet facilities close to one's sleeping space for reasons of convenience and privacy, but they are also beset by the old problem of proximity of indelicate odours and sounds. Despite their drawbacks, en suites possess a high status and in recently-constructed homes they are often devoted exclusively to parental use. Very grand modern dwellings might have en-suite facilities for important guests, but rarely are the household's children considered in this context, being left to

PERSONAL HYGIENE • 289

share the main or 'family bathroom' with each other and, where there is no cloakroom, with casually visiting guests.

Bathrooms can also be an important site of artistic expression as their walls and floors are frequently intensely decorated and they are often bedecked with semi-ornamental objects. With their locked door they offer moments of respite for household members otherwise denied privacy. Their healthcare overtones are revealed by objects such as weighing scales and medicine cabinets (Figs 16.3, 16.4).

Products
Just as the Industrial Revolution accelerated urbanisation and created the terrible sanitation crises of the nineteenth century, it was the same revolution which delivered the technologies necessary to connect homes to water-supply and sewerage systems. It was also responsible for supplying a wide variety of products without which the modern bathroom would not be complete. The more obvious examples include cheap, effective soap, absorbent textiles for towels and flannels, and latterly, 'toilet tissue' (preceded by various materials, including dried moss). Of these products, soap was one of the most important and had been produced in crude form for many centuries by boiling fat or oil in caustic soda in an emulsifying process known as 'saponification'. This was not a particularly pleasant process because animal fats (especially tallow) were usually the principal source of fat. Much of the soap production was designed for use in the textile industries and normally comprised either 'hard soap' in bars or tablets, or soft soap (made using potash instead of caustic soda). Apart from the tax on soap levied by Oliver Cromwell and not repealed until 1853, the main obstacle to the wider availability of soap was the shortage of raw materials, particularly alkalis. The difficulty was alleviated by the invention of the Leblanc process in 1791 which solved the problem by deriving caustic soda from common salt (sodium chloride). The first chemical factory in Britain to utilise the process was at St Helens in 1814, but Tennant's St Rollox Works in Glasgow soon commenced production, discovering that the immense quantity of toxic fumes emitted required the building of a chimney 420 feet (128 m) tall.[59] Tennants themselves became one of the more important Scottish soap manufacturers.

The cheap supply of alkali brought down costs and permitted higher-quality oils to be used, thereby enabling the creation of a wider range of products, including scented toilet soaps and powders for personal and domestic use. With the establishment by William Hesketh Lever (later Viscount Leverhulme) of his Port Sunlight factory in Cheshire in 1885, the market was transformed, and by 1900 an aggressive marketing campaign had put Lever into a dominant market position, the main resistance coming from the Co-operative Wholesale Society. Throughout the remainder of the twentieth century, soap products were consistently among the most heavily advertised consumer goods in the world, and were constantly the subject of monopoly investigations.[60] The consumer was suddenly confronted with advertising campaigns which heightened awareness of personal hygiene and which also grew to focus on new ranges of products such as toothpastes, household cleaning fluids, and soapless detergents designed for the domestic laundry market. The latter, and intense competition from the emerging market leaders Unilever and Proctor & Gamble, resulted in the closure of the Scottish Co-operative Wholesale Society's soap factory at Grangemouth in 1965.[61]

Plumbing
Despite prevailing medical opinion that bathing in cold water was beneficial, the populace at large greatly preferred hot baths, and the performance of the soaps which they purchased benefited significantly from the use of hot water. This was made possible at first simply by the heating of water over a fire or on a range, but was subsequently assisted by the acquisition of purpose-built ranges such as those manufactured by Smith & Wellstood of Bonnybridge, Stirlingshire, in the 1820s.[62] Such fixtures permitted the hot water to be drained and taken to a basin or bath for use, and it was not until the end of the nineteenth century that plumbing systems were routinely installed in houses, and in the case of hot water, attached to boilers. The latter became much more widely available and affordable, at first in urban areas with the increased availability of town gas and, in the mid twentieth century, of electricity. The installation of heating in many bathrooms was also a further incentive to frequent bathing.

Floors and walls
Meanwhile, the arrival in the home of plumbing and the presence of comparatively large quantities of water required the use of sturdy, water-resistant materials, and promoted the use in bathrooms of waterproof flooring such as floorcloth and linoleum (much of which was produced in Fife, especially in Kirkcaldy).[63] Other favoured waterproof materials included ceramic bricks and tiles, which protected otherwise porous walls and floors vulnerable to rot.[64]

Sanitary ware
At the heart of the bathroom is its sanitary ware. In the early twenty-first century a bathroom is expected to contain a bath or shower (or both), a lavatory (wash-hand basin), and usually a water closet, although the latter may be housed separately in an adjacent, smaller room. More luxurious bathrooms may contain a bidet, or even an extra lavatory basin. In newer housing, the bathroom frequently does not have a window and is required by law to have ventilation ducts and fans. There are, in addition, many fittings and accessories which may be added to a bathroom, including mirrors, cupboards and towel rails, as well as personal touches such as chairs, magazine racks and bookshelves. The focus of any bathroom, nevertheless, remains the sanitary ware itself.

BATHS

The dominant item in the room is likely to be the bath. In 1929, A B Searle observed that earlier types of mass-produced bath were often made from sheet metal, generally zinc or copper, and usually were coffin shaped.[65] Wright (1960) illustrates a variety of other types and shapes, including 'Sitz' and 'Hip' models.[66] These baths were superseded by vitrified-enamel cast-iron and fireclay baths, changing shape to achieve the customary near-parallel sides, although corner and recess baths had also been developed. The fireclay baths were comparatively heavy and cumbersome to install, but their advantages over metal baths included great resilience, their heat-retention qualities and the fact that if chipped, iron baths tended to rust. Noteworthy is the range of auxiliary uses for baths in some Scottish homes, from serving as wash tubs to coal stores and repositories for live shellfish.[67] Showers, meanwhile, appeared in the earliest bathrooms in a variety of forms, ranging from detached compartments to simple fittings above a bath. Most elaborate of all were showers integrated into the bath itself, as illustrated by the *Baignoire Shanks brevetée avec Douche de Tête et en Cercle* in the Shanks French catalogue of 1912[68] (Fig. 16.5).

Figure 16.5 The *Baignoire Shanks brevetée avec douche de tête et en cercle*, an example of a bath-shower combination, appearing in the 1912 Shanks catalogue, *Appareils Sanitaire* of 1912, produced for the French and Belgian markets.
Source: Shanks catalogue *Appareils Sanitaire*, 1912.

Figure 16.6
Example of a pedestal lavatory basin from the J & M Craig catalogue of 1914. Source: J & M Craig catalogue of 1914, 6.

LAVATORIES

The second most prominent fixture in a bathroom is usually the lavatory, which can be described as a form of washing bowl, sometimes on a pedestal, sometimes wall-mounted, or in a cabinet, to which is attached a permanent supply of hot and cold water, and a waste pipe connecting it to the drains. The earliest lavatory basins evoked the form of a traditional washstand but with an outlet at the bottom of the basin to which a waste pipe could be attached. Modern lavatories range from plain simple designs to extravagant models comprising part of extensive, often highly decorative, suites[69] (Fig. 16.6).

WATER CLOSETS

In sanitary terms, perhaps the most important of the fixtures is the 'water closet' or WC,[70] which can be described as a form of basin, many different patterns being available,[71] which should be so shaped as to be self cleaning with a specified quantity of water (originally two gallons [9 litres]).[72] Early models were, however, not necessarily very effective, usually being built from cast-iron, lead, leather and wooden components (Figs 16.7, 16.8). One of three types of pedestal closet, the 'wash out' abandoned the concept of a valve to close the bottom of the pan, favouring instead a small amount of water sitting in a 'U' or 'S' bend, acting as a seal after each flush. Wash-out closets became popular in the last decades of the nineteenth century and by 1900 were being produced in great numbers by Scottish manufacturers.

Figure 16.7 An example of an old, soiled pan closet, illustrated by Hellyer, 1886, in order to demonstrate the inadequacy of this type of WC which comprised an upper earthenware basin with a shallow copper pan containing three to four inches of water as a seal at its base. The pan could be tipped away to discharge the contents into a lower cast-iron receptacle connected to the drainage system. The principal disadvantage was that it was immensely difficult to maintain both the pan and the space beneath in a clean condition. Source: Hellyer, 1886, 191.

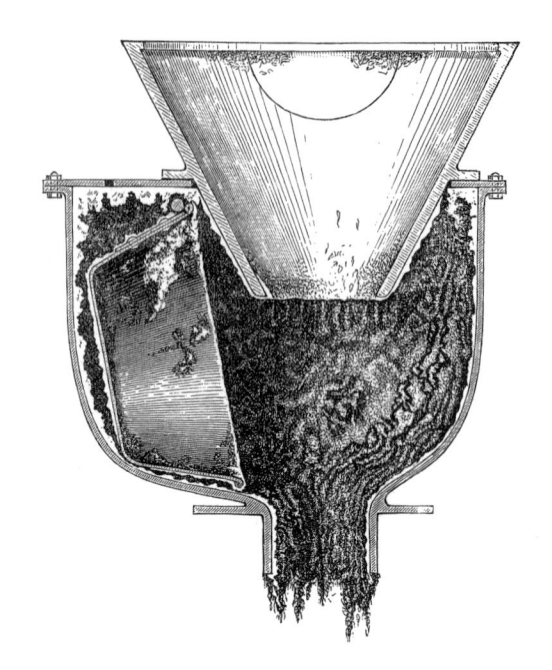

Syphonic closets were being manufactured in Scotland by 1900. In these the water and contents in the basin are carried away by means of a syphonic action. Wash-down closets grew to become the most common type of WC in use in the United Kingdom. The flush was exerted directly on the water in the basin, driving the waste away more efficiently and improving the self-cleaning properties of the closet.

The introduction of WCs to the home was made possible by the extension of sewerage systems and the availability of cesspools into which they could discharge. This was itself dependent upon the development of affordable non-porous (usually internally-glazed) pipes to carry away the waste without leakages. Although many early WCs were installed in outhouses, replacing privies, in time the WC was incorporated within the home and subsequently often within bathrooms. This change was welcome not only because of the enhanced hygiene, comfort and convenience of such arrangements, but also because it had the advantage of alleviating the problems caused by severe winter weather and damage caused by frozen pipes.[73]

BIDETS

The bidet originated in France and is said to have been first created by Parisian furniture makers in the seventeenth century, when it was known as 'the ladies' confidant'.[74] There is little evidence that it ever gained widespread popularity in Scotland or the United Kingdom as a whole, although the Scottish sanitary-ware industry was producing bidets for export in substantial numbers by 1900. Statistics provided by plumbing companies

Figure 16.8 An example of a valve WC at Bielside House in East Lothian. Valve closets were the first mass-produced WCs and operated on a system by which a pan with an opening at the bottom was sealed with a leather-face valve. The action of flushing involved an often complex arrangement of handle, lever and counterweight. The wooden cabinet work conceals a very complicated mechanism. Crown Copyright RCAHMS: C36047.

in Britain suggest that, in the early twenty-first century, over 90 per cent of households in France, Italy and Portugal have bidets, as compared with less than five per cent in Britain.

THE MANUFACTURE OF SANITARY WARE

By 1900, trade catalogues reveal that much of the demand for bathroom fixtures was for ceramic appliances. In Scotland a well-established heavy-ceramics industry based on high-quality fireclays was able to exploit the rapidly growing new markets.[75] The most significant areas specialising in fireclay and earthenware sanitary ware were to be found in Ayrshire (notably Kilmarnock and the Irvine area), and Renfrewshire (centred on Paisley and Barrhead). Smaller companies also prospered elsewhere, with numerous examples in Glasgow and other cities. Significant other manufacturers included Steel's of Edinburgh (Niddrie Fireclay Works) and Buick's of Alloa, Clackmannanshire (Hilton Fireclay Works).

The bigger Scottish companies included J & M Craig and Shanks, whose products ranged from the basic to the exotic (Fig. 16.9) and who were

Figure 16.9 Shanks of Barrhead patent water closet, with blue floral pattern, and matching cistern. The view also shows the mosaic-pattern floor, ornate wall tiles, and a later (c1930s) pedestal lavatory basin (Twyford). 'Belmont', Dalkeith, Midlothian. Crown Copyright: RCAHMS C37108/CN.

mass producing bidets by 1900. In addition to Craig's Maelstrom and other pedestal water closets, the company was well known for its decorated range of appliances, such as the 'Chrysanthemum'. Other Ayrshire companies included Howie (who was responsible for equipping many pithead baths at coal mines), and the Bourtreehill Coal Co, the latter's 1902 catalogue containing a wide range of sanitary products. These included selections of white-glazed bricks which were widely used not only in the walls of bathrooms and washhouses, but also in prisons, hospitals and other institutions. Their advantage was the ease with which their glazed, non-porous surfaces could be cleaned, this quality resulting in their use throughout industry in electrical switch and generating houses and engine houses of many types. In addition, a common architectural use for these bricks was to maximise light in the light wells of large, multi-storeyed buildings, and in poorly lit rooms.[76]

As the twentieth century progressed so domestic plumbing expanded into existing buildings in many series of conversions, extensions and adaptations, and the demand for Scottish sanitary ware continued to increase. Although private house-building schemes were an important additional source of demand, the largest market was that created by huge public-sector house-building programmes, which, because of comparatively high standards and the imposition of building regulations, required well-appointed, indoor bathroom and WC facilities in every new dwelling.[77] Following the success of the Miners' Welfare Fund, demand was further bolstered by rules and regulations enforcing the provision of washroom facilities in factories, offices and other institutions, including schools. One of the more unusual markets to be exploited by Scottish manufacturers was for London Board Schools, for whom the Bourtreehill Coal Co made a range of fixtures in the early 1900s, including the near indestructible 'Infantile' closet.

From the 1950s, the Scottish sanitary-ware industry began to decline for a variety of reasons but particularly because of reductions in house building. Contracting market conditions permitted Shanks, perhaps the most famous name in the industry, to assume a dominant position by taking over most of its surviving competitors, hence WCs earning the nickname 'the Shankie'. Shank's Barrhead factories were unrivalled in technical terms, pioneering fully vitrified or vitreous china, a much lighter, stronger and potentially cleaner ceramic material from which most sanitary ware is now made.[78] Vitreous ware rapidly replaced heavier fireclay and earthenware products, one of its many advantages being that its glaze was less prone to craze, the entire body of the clay being fully fired, as opposed to only the external glaze and engobe. Barrhead also developed battery-moulding techniques which revolutionised production at the Victorian Pottery. However, Shanks themselves were merged with the English firm, Armitage, in the 1980s, later becoming part of the Blue Circle Group. The new parent company finally closed down the remains of the Scottish industry by shutting the Barrhead factory in the early 1990s, but the sanitary-ware production was revived by the workforce at Shanks, who formed 'Barrhead Sanitary Ware', setting up a new factory in Hillington, Glasgow.

Sources of evidence for bathrooms and toilet facilities in the modern period
Despite the immense importance of hygiene in the home, the issues do not seem to have troubled historians or other writers, with the exception of the sanitary pioneers and revolutionaries of the Victorian era, and certain autobiographical authors such as Finlay J MacDonald.[79] Even architects and draughtsmen appear to have been reluctant to include details of bathing and toilet facilities within their designs and reports. In the cases where they have been included, few details and specifications of the fixtures are provided. However, some examples do begin to creep into the designs for Glasgow tenements in the early twentieth century,[80] for instance, and some fine designs for large houses in the same period contain significant detail. Notable was Robert Lorimer, whose enthusiasm and professional interest in plumbing led him to design a water closet of his own called the 'Remirol' (his own name spelt backwards), in association with the British Medical Association.[81]

Unfortunately, these and other examples appear to reveal no distinctively Scottish pattern. Both large and small dwellings have WCs separate from bathrooms, and some do not. Others have separate WCs as well as a WC in the bathroom. Some have a downstairs WC and cloakroom, usually also with a lavatory basin (sometimes matching). Very few of the architects' drawings and specifications provide details of the models of appliances and accessories installed. It is, for example, interesting to note that bidets feature extensively in trade catalogues of the period, but rarely seem to appear in building plans.

The low profile of the bathroom, lavatory and WC in Scottish architectural records is disappointing and may be attributed to the socially difficult nature of the subject of personal hygiene, and to the fact that very quickly, in the course of only a few decades, bathrooms and toilets became a central and standard part of daily life. This is a shame because Scotland has played an important role in the sanitary revolution through the large-scale design and manufacture of sanitary ware and accessories, and more significantly, through its pioneering, municipally inspired water-supply and sewerage schemes. The achievements of the past have only recently been highlighted within the water industry itself, where renewal work by the new Scottish water authorities has confirmed the extraordinary engineering work carried out in the late nineteenth century, much of which was working satisfactorily over 100 years later.

In 1902, H J Jennings observed that in Britain '... there is a room which no self-respecting householder can do without, and that is the bathroom. One can only marvel at the astonishing fact that prior to 20 or 30 years back, the majority of small and medium-sized houses, and perhaps 50 per cent of the larger ones, were built without bathrooms.'[82] In 1914, although Glasgow had managed to replace many of its privies with WCs, over half were still shared with other dwellings.[83] The extent of progress since then is well illustrated by the fact that, by the early 1990s, 97 per cent of Scottish households were linked to public water authority supplies.[84]

RURAL TOILET MATTERS: ALEXANDER FENTON

On the eve of the French Revolution, there was much discussion in France about the effects of accumulations of human and other forms of waste on the air and on the people. It was said, however, that 'Excrement was much more of a danger in the town than in the country. Louis-Sébastien Mercier envied the peasants who relieved themselves in the fields while town dwellers risked putrid fever by sitting on deadly commodes.'[85]

What was the situation in rural Scotland? It was certainly better than in the towns, but relatively little has been written about the subject. Relief outdoors was no special problem. Excrement at the back of a dyke was soon cleared by the action of rain, birds and insects, and in any case it helped to add fertility to the soil. The subject can be considered under three main daily occasions: at night before going to bed, overnight and in the morning. Variations in procedure between men and women, the latter seeking greater privacy, must also be borne in mind.

From my own knowledge of a rural parish in the first half of the twentieth century, I can say that in preparation for the night men would go outside even in rain or snow and urinate somewhere in the close (yard), in the kailyard[86] or at any convenient spot, not necessarily far from the door. By night, all the family, old or young, male or female, would at nature's call make use of the chamber pot, variously known as the 'po', 'chanty' or 'dirler',[87] under each occupied bed. An older name is 'wash mug', recorded from 1735.[88] These were not necessarily covered by any lid. They could be quite full by morning, when they were emptied into a slop pail that sat under a shelf in the scullery before being carried out for the contents to be deposited on the midden. Morning relief in homes where sanitation was lacking was in the wooden outside lavatory, with a bucket (which was eventually also emptied onto the midden) under the wooden seat, and sheets of newspaper torn into neat rectangles and hung on a nail, if the woman of the house was tidy-minded, or otherwise simply left lying as loose sheets on the side of the seat. Such wooden toilets were as a rule supplied by the estates to which the farmers and crofters paid their rents, but many places, still in the twentieth century, were not as far advanced. In that case – and this was the advantage a farm had over a cottar house – recourse was had to the warm byre, where there was straw for wiping and where the refuse could be readily disposed of with the rest of the muck. There was, in good weather, an alternative. The wired-off hens' run would often encircle half the house, from back to front, and the secluded area at the back, shaded further perhaps by a bourtree (elder) bush, made an occasional toilet area. Waste disposal was no problem, for the hens cleared everything up. Alternatively, a friendly plantation of trees would provide the necessary seclusion. A bunch of grass then served for wiping. That outside relief was customary amongst all classes is suggested by a passage in the thirteenth-century poem, 'The Bruce'. When staying in Carrick, King Robert was in the habit:

> For to rys arly ilk day
> And pas weill fer fra his menye
> Quhen he wad pas to the prevé
> And sek a covert him allane
> Or at the maist with him ane

But when enemies plotted to attack him, three of them hid:

> In a covert that was prevé
> Quhar the king oft wes wont to ga
> His preve nedys for to ma.[89]

So King Robert behaved like a normal man, seeking privacy for personal needs, though in the open air, whereas the eighteenth-century approach at royal level is marked by the installation at the palace of Versailles of English-type water closets, reserved for the king and Marie Antoinette, who thus 'were among the first individuals in France to experience a new privacy'.[90]

Many rural houses in Scotland had no water or electricity. In the 1920s, when there was a rising tide of complaints about the poor quality of farm-servants' housing, there are frequent references, in for example the journal of the Scottish Farm-Servants' Union, to housing conditions. Typical examples are:

> We ... have asked that existing cottages should be made wind and watertight and that water should be brought to the houses, and the decencies observed by providing each house with a water closet or earth closet.[91]

A meeting with the Permanent Secretary of the Department of Health for Scotland on 13 March 1929 was not very fruitful. Joseph F Duncan, the General Secretary of the Union, pointed out that statutory inspections by local authorities of dwelling houses in their areas had not been systematically carried out, and it was the opinion of the Union that:

> ... at least 50 per cent of the farm cottages in Scotland were without proper water supplies or sanitary conveniences, that in some districts the proportion rose as high as 90 per cent of the farm cottages which were without these necessities, and that 75 per cent of the farm cottages were not 'in all respects reasonably fit for human habitation', although that was an implied condition under Section 2 of the 1925 Act.[92]

However, even if attention was drawn to farm-servant housing in this kind of way, there were plenty of crofts and small farms which were no better off. The farther back in time, the less likely that any special provision would be made for sanitary purposes. Exactly how matters were organised, where even chamber pots were lacking, can occasionally be glimpsed from the 'ruder' forms of literature. For example, in an eighteenth-century chap-book, John Cheap the Chapman stopped at a farm near Dalkeith, Midlothian, and got on so well with the farmer that he was offered a shake-down bed beyond the fire in the kitchen. There were three women in a bed in the same kitchen, and:

> ... they not minding that I was there, first one of them rose and let her water go in below the chimney-grate ... then another rose and did the same, last of all got up the old matron ... and as she let her dam go, she also with full force, when done, let a fart like the blast of a trumpet, which made the dust on the hearth-stone to fly up like mist about her arse ... [93]

Such a practice was unlikely to be rare, especially where there were central hearths or hearths with convenient ash pits. The ash absorbed the urine, and when the pit was emptied, it went to fertilise the fields.

Urine was seen as a valuable substance. Not only could it be used as a form of manure, but it also served in remedies. In the late nineteenth century in Fife, urine was applied as a cure for 'rose' (erysipelas) and a child's wet diaper was used to wipe out the mouth and throat passage of a sufferer of thrush.[94] A major use for urine, however, was as a detergent, a source of ammonia, in the processing of woollen textiles. Accordingly it was stored up in homes in tubs or large jars and left to stand for some time for use as a lye in bleaching, for example. It also served as a cost-free form of soap in blanket washing, which according to one writer took place once a year, using urine that had been stored for six months.[95] It had a number of names, such as 'maister', 'stale', 'strang', 'wash'[96] and in Gaelic 'maistir'. The word appears to be related to German *Mist*, Middle Dutch *mest*, dung. Urine was also used as a mordant in dyeing. A recipe collected from Peigi MacRae of South Uist by Margaret Fay Shaw gave the instructions for dying indigo blue:

> Save the household urine until sufficient in a big clean tub beside the fire. The temperature must be about 85° to 90° and must not vary. Put the indigo in a little bag of muslin and steep it in the urine. Every three days squeeze it, rubbing it with the hands to take the colour out. An ounce [28g] of indigo to a pound [454g] of wool, but as the strength varies, take what you think will do.[97]

Urine, therefore, in times not too remote from the present, was a valuable commodity, to be collected carefully and not squandered lightly. Perhaps this is the reason for the apparently casual behaviour observed by James Boswell in Skye in 1773:

> After I was in bed, the minister came up to go to his. The maid stood by and took his clothes and laid them carefully on a chair piece by piece, not excepting his breeches, before throwing off which he made water, while she was just at his back.[98]

NOTES

1. A good example is the bathhouse and latrine found at the Roman fort at Bearsden on the Antonine Wall, for details of which see Breeze, 1984.
2. Beith, 1995, cites references to the early medicated baths and sweating cures, also observing that the Vikings were much cleaner than their popular image suggests.
3. A rare example of an archaeological discovery related to personal hygiene is that of a medieval latrine with surviving wooden seat discovered at Kirk Close in Perth. Yeoman, 1995, 59.
4. Robertson, 1997, 41; Cowan, 1986, 278; and Bulloch, 1958, 44.
5. Warrack, 1920, 157.
6. Bulloch, 1958, 44.
7. Grant, 1988, 216; see also Chapter 15 Toilette.
8. Grant, 1988, 63–4.
9. Grant, 1988, 94.
10. See Chapter 6 Heating and Sleeping.
11. Blair, 1991, 27.
12. Clark, 1996, 78.
13. John Alexander MacKenzie recalls being given a lecture, a toothbrush and a tin of Gibbs Dentrifice by the dentist visiting his Inverness-shire school in the 1930s, but to no avail. 'I never used my gift, but my mother found the brush and the moistened contents of the tin very good for getting into the cracks and corners of the brass candlesticks which adorned the mantelpiece'. MacKenzie, 1996, 55; See also Chapter 15 Toilette.
14. See Grant, 1988, 111.

15. Not all households were so easily put off regular bathing. See Blair, 1991, 26.
16. Mackenzie, 1996, 35.
17. Clark, 1996, 78.
18. Grant, 1988, 111.
19. *DOST*, s v Bath, Bath-fat, Bathing, Bathstoll.
20. Some archaeologists have suggested that a number of the many 'burnt-mound' sites found in Scotland may be the remains of old sweathouses dating potentially from the Bronze Age to medieval times, and operated by Gaels in a similar fashion to those of the Scandinavians and native-American groups. Beith, 1995, 136–7; see also Danaher, Lucas, 1952.
21. Boog Watson, 1979, 57.
22. Boog Watson, 1979, 57.
23. Boog Watson, 1979, 59.
24. Boog Watson, 1979, 65.
25. Cross, Cross, 1930, 5.
26. Glasgow Corporation, 1914, 98.
27. Cross, Cross, 1930, 29.
28. Miners' Welfare Fund, 1921–2.
29. Blair, 1991, 28.
30. Pride, 1975; Robertson, 1997, 41.
31. Yeoman, 1995, 61
32. Denton, Wilson, 1991, 9.
33. Harrison, 1998–9, 71.
34. McDowall, 1873, 797.
35. For details of the development of water supplies in Scottish cities, one of the best sources is Groom, *c*1892.
36. Gauldie, 1974, 75; Gibb, 1983, 144.
37. Brotherston, 1952, 48.
38. Brotherston, 1952, 49.
39. Brotherston, 1952, 79.
40. Brotherston, 1952, 79.
41. Brotherston, 1952, 79.
42. Brotherston, 1952, 81.
43. Chalmers, 1930, 5.
44. Dunghills were described as probably the most lucrative part of an estate. Symons, 1839, 51; see Chapter 32 Paying for Dwellings.
45. Several towns and cities, including Stirling (in 1529), tried to control the growing problem of filth in the streets by imposing laws forbidding the laying of middens in the street for more than 24 hours. Harrison, 1998–9, 69.
46. Gauldie, 1974, 78.
47. Gauldie, 1974, 87.
48. Parker, 2000, 24.
49. Information on the development of Scottish water supply and sewerage infrastructure was provided by the Scottish and Northern Ireland Plumbing Employers' Federation.
50. A description of the Loch Katrine project, and earlier schemes, can be found in Burnet, 1869.
51. Colston, 1890.
52. Gloag, Bridgewater, 1948, 78.
53. Hume, 1974, 63.
54. Weekly issues of *The Builder* in the 1860s regularly advertised the wares of Walter Macfarlane & Co (for example, 7 July 1860, xv). It is also worth noting that in the same issues (in the 1860s) *The Builder* described itself as a journal for the architect,

engineer, operative and artist. By 1881 it claimed to be catering for the architect, engineer, archaeologist, constructor, sanitary reformer and art lover.
55. Gloag, Bridgewater, 1948, 80.
56. Hume, 1974, 136–7, describes the development of sewers in Glasgow. In addition, information provided by Jim Robinson of the Scottish and Northern Ireland Plumbing Employers' Federation reveals that the first slow sand filter was developed in Paisley by John Gibb in 1804, the second being established by Robert Thom in Greenock in 1827.
57. Gauldie, 1974, 75.
58. Presbytery of Glasgow, 1891. Cited in Gibb, 1983, 142.
59. Cossons, 1987, 206.
60. Edwards, 1962, 143.
61. Kinloch, Butt 1980, 348.
62. Roberts, Yunnie, 2001, 37.
63. The town of Kirkcaldy grew to dominate the linoleum industry, Nairn (now Forbo Nairn) being the most famous company. The industry has since contracted and diversified, linoleum being mostly superseded by vinyl-based floor coverings.
64. Craig, 1914, 46.
65. Searle, 1929.
66. Wright, 1960, 168, 170.
67. Reid, 2000, 99.
68. Shanks & Co Ltd, 1912, 4.
69. Typical examples of this type of fixture include the 'Chrysanthemum' range depicted in Craig, 1909.
70. Water closets have inspired a number of dialect and slang terms. The *SND* furnishes 'jack', 'little house' or 'wee house' ('tigh beag' being an exact equivalent in Gaelic), 'offie' and 'shankie', while 'nestreis' is to be found in the *DOST*, all meaning privy. Present-day inhabitants of Scotland may be more familiar with 'loo', 'lav' or 'lavvie', 'bog', 'cludgie' or 'cludge', and 'thunderbo'.
71. The best published accounts of the history of the WC can be found in Wright, 1960; Palmer, 1973. Both point to the inventor of the modern water closet as being Sir John Harrington, whose invention is said to have been for the benefit of Queen Elizabeth I of England. Thereafter, not a single patent was taken out in the first 150 years of British patents, the main technical improvements occurring after 1870, particularly with the introduction of heavy ceramic materials.
72. Searle, 1929.
73. Frozen pipes have become less of a problem in modern homes since the widespread introduction of central heating and the occurrence of milder cycles of weather in Scotland.
74. Beaupré, Guerrand, 1997.
75. Douglas, Oglethorpe, 1993, 19; Sanderson, 1990. The latter gives a particularly detailed account of specialised Scottish fireclays.
76. Bourtreehill Coal Co, 1902, 74.
77. Department of Health for Scotland, 1950, 4.
78. By the late twentieth century a growing proportion of the sanitary-ware and plumbing fixtures market was being satisfied by a new range of durable plastic materials.
79. MacDonald, 1983, 20–6; MacDonald, 1985, 112–13.
80. Shaw-Sparrow, c1906, 67, 69–70.
81. Savage, 1980, 127; Shanks & Co Ltd, 1930, 231.
82. Burnett, 1978, 211.
83. Daunton, 1983, 261.

84. Information provided by the Scottish and Northern Ireland Plumbing Employers' Federation.
85. Corbain, 1996, 28.
86. In the chapbook, the 'Ancient and Modern History of Buckhaven', Wise Willie, being unwell, 'drank sea water till his guts was like to rive, and out he goes to ease himself among the kail'. See MacGregor, 1883, 224.
87. 'Po' appears to be an abbreviation of 'chamber pot', using a French pronunciation; the other two names are onomatopoeic.
88. *SND*, s v Wash.
89. Barbour, 1997, 219.
90. Corbain, 1996, 83–4.
91. *The Scottish Farm Servant* (April 1929), 3.
92. *The Scottish Farm Servant* (April 1929), 7.
93. MacGregor, 1883, II, 100.
94. Quoted from the *British Medical Journal*, I (1936), 142. In Buchan, 1994, 19.
95. Johnston, 1923, 353; see also Chapter 17 Housework – For the use of urine in the cleaning of textiles.
96. *SND*, s v Maister, Stale, Strang, Wash.
97. Shaw, 1999, 54–5.
98. Boswell, 1979, 124.

BIBLIOGRAPHY

Barbour, John (trans A A M Duncan). *The Bruce*, Edinburgh, 1997.
Beaupré, F, Guerrand, R. *Le confident des dames. Le bidet du XVIIe au XXe siècle: Histoire d'une intimité (History of the Bidet)*, Paris, 1997.
Beith, M. *Healing Threads: Traditional Medicines of the Highlands and Islands*, Edinburgh, 1995.
Blair, A. *More Tea at Miss Cranston's: Further Recollections of Glasgow Life*, London, 1991.
Boog Watson, W N. Early baths and bagnios in Edinburgh, *The Book of the Old Edinburgh Club*, XXXIV Part 2 (1979), 57–67.
Boswell, J. *The Journal of a Tour to the Hebrides, with Samuel Johnson, LLD* (1786), Oxford, 1979.
Bourtreehill Coal Co. *Catalogue of Fire Clay Goods and Sanitary Specialities*, Dreghorn (Ayrshire), 1902.
Breeze, D J. The Roman fort on the Antonine Wall at Bearsden. In Breeze, D J, ed, *Studies in Antiquity*, Edinburgh, 1984, 48–58.
Brotherston, J H F. *Observations on the Early Public Health Movement in Scotland*, London, 1952.
Buchan, D, ed. *Folk Tradition and Folk Medicine in Scotland*, Edinburgh, 1994.
Bulloch, J. *Adam of Dryburgh*, London, 1958.
Burnet, J. *History of the Water Supply to Glasgow from the Commencement of the Present Century with Descriptions of the Water Works Projected, Executed and from Time to Time in Operation*, Glasgow, 1869.
Burnett, J A. *Social History of Housing, 1815–1970*, Newton Abbott, 1978.
Chalmers, A K. *The Health of Glasgow, 1818–1925*, Glasgow, 1930.
Clark, H. Living in one or two rooms in the city. In Carruthers, A, ed, *The Scottish Home*, Edinburgh, 1996.
Colston, J. *The Edinburgh and District Water Supply: A Historical Sketch*, Edinburgh, 1890.
Corbain, A. *The Foul and the Fragrant*, London, 1996.
Cossons, N. *The BP Book of Industrial Archaeology*, 2nd edn, Newton Abbot, 1987.
Cowan, I B. Ayrshire Abbeys: Crossraguel and Kilwinning, *Ayrshire Collections*, 14/7 (1986), 278.

Craig, J & M Ltd. *Illustrated Catalogue of Sanitary Earthenware Manufactured by J & M Craig (Kilmarnock), Ltd*, 7th edn, Kilmarnock, 1900.
Craig, J & M Ltd. *Illustrated Catalogue of Sanitary Earthenware Manufactured by J & M Craig (Kilmarnock), Ltd*, 9th edn, Kilmarnock, 1909.
Craig, J & M Ltd. *Catalogue of Modern Sanitary Ware Manufactured by J & M Craig (Kilmarnock) Ltd, Kilmarnock, NB*, 11th edn, Kilmarnock, 1914.
Cross, A W S, Cross K M B. *Modern Public Baths and Washhouses*, London, 1930.
Danaher, K, Lucas, A T. Sweathouse, Co Tyrone, *Journal of the Royal Society of Antiquaries of Ireland*, LXXXII/2 (1952), 179–80.
Daunton, M J. *House and Home in the Victorian City: Working-Class Housing, 1850–1914*, London, 1983.
Denton, H, Wilson, J C. *The Happy Land*, Edinburgh, 1991.
Department of Health for Scotland. *Scottish Housing Handbook III: House Design*, Edinburgh, 1950.
Douglas G J, Oglethorpe, M K. *Brick, Tile and Fireclay Industries in Scotland*, Edinburgh, 1993.
Edwards, H R. *Competition and Monopoly in the British Soap Industry*, Oxford, 1962.
Gauldie, E. *Cruel Habitations: A History of Working-Class Housing, 1780–1918*, London, 1974.
Gibb, A. *Glasgow: The Making of a City*, London, 1983.
Glasgow Corporation. *Municipal Glasgow: Its Evolution and Enterprises*, Glasgow, 1914.
Gloag, G, Bridgewater, D. *A History of Cast Iron in Architecture*, London, 1948.
Grant, E. *Memoirs of a Highland Lady*, vol 1 (1898), Edinburgh, 1988.
Groom, F H, ed. *Ordnance Gazetteer of Scotland: A Survey of Scottish Topography*, new edn, London, 4 vols, c 1892.
Harrison, J G. Public hygiene and drainage in Stirling and other early modern Scottish towns, *ROSC*, 11 (1998–9), 67–77.
Hellyer, S S. *The Plumber and Sanitary Houses: A practical treatise on the principles of internal plumbing work, or the best means for effectively excluding noxious gases from our houses*, 3rd edn, London, c 1886.
Hume, J R. *Industrial Archaeology of Glasgow*, Glasgow, 1974.
Johnston, T. *The History of the Working Classes in Scotland* (1860), Glasgow, 1923.
Kinloch, J, Butt, J. *History of the Scottish Co-operative Wholesale Society*, Glasgow, 1980.
MacDonald, F J. *Crowdie and Cream: Memoirs of a Hebridean Childhood*, London, 1983.
MacDonald, F J. *The Corncrake and the Lysander*, London, 1985.
MacGregor, G. *The Collected Writings of Dougal Graham, vol II*, Glasgow, 1883.
MacKenzie, J A. *A Mallaig Boyhood* (Flashbacks, no 5), East Linton, 1996.
McDowall, W. *A History of the Burgh of Dumfries, with Notices of Nithsdale, Annandale, and the Western Border*, Edinburgh, 1873.
Miners' Welfare Fund, *Annual Report, 1921–2*, London, 1921–2.
Nicoll, J. *Domestic Architecture in Scotland*, Aberdeen, 1908.
Oglethorpe, M. The bathroom and water closet. In Carruthers, A, ed, *The Scottish Home*, Edinburgh, 1996, 203–21.
Palmer, R. *The Water Closet: A New History*, Newton Abbot, 1973.
Parker, C. Improving the 'condition' of the people: The health of Britain and the provision of public baths, 1840–70, *The Sports Historian*, 20 (2000), 24–42.
Presbytery of Glasgow. *Report of Commission on Housing of the Poor in Relation to their Social Condition*, Glasgow, 1891.
Pride, G. *Glossary of Scottish Building*, Glasgow, 1975.
Reid, L. *Scottish Midwives. Twentieth-Century Voices* (Flashbacks, no 12), East Linton, 2000.
Roberts, B, Yunnie, P. *The Magic of Hot Water*, Wednesbury, 2001.
Robertson, U A. *The Illustrated History of the Housewife, 1650–1950*, Stroud, 1997.
Sanderson, K W. *The Scottish Refractory Industry, 1830–1980*, Edinburgh, 1990.
Savage, P. *Lorimer and the Edinburgh Craft Designers*, Edinburgh, 1980.
Save Britain's Heritage. *Taking the Plunge: The Architecture of Bathing*, London, npd (c 1970s).

Searle, A B. *An Encyclopaedia of the Ceramics Industries*, London, 3 vols, 1929.
Shanks & Co Ltd. *Appareils Sanitaires, Tubal Works et Victorian Poteries (Sanitary Appliances)*, Barrhead, 1912.
Shanks & Co Ltd. *Bathroom Accessories: Section 14 from 'H' catalogue, Tubal Works*, Barrhead, 1922.
Shanks & Co Ltd. *'M' Catalogue*, Barrhead, 1930.
Shaw, M F. *Folksongs and Folklore of South Uist*, reprinted Edinburgh, 1999.
Shaw-Sparrow, W, ed. *The Modern Home: A Book of British Domestic Architecture for Moderate Incomes*, London, c 1904.
Shaw-Sparrow, W, ed. *Flats, Urban Houses and Cottage Homes*, London, c 1906.
Symons, J. Report from the Assistant Handloom Weavers' Commission (159), BPP (1839), XL11, 51.
Teale, T P. *Dangers to Health: A Pictorial Guide to Domestic Sanitary Defects*, 4th edn, London, 1883.
Warrack, J. *Domestic Life in Scotland, 1488–1688*, London, 1920.
Whitelegge, B A. *Hygiene and Public Health*, London, 1901.
Wright, L. *Clean and Decent: The Fascinating History of the Bathroom and the Water Closet*, London, 1960.
Worsdall, F. *The Glasgow Tenement: A Way of Life: A Social, Historical and Architectural Study* (1979), Edinburgh, 1991.
Yeoman, P. *Medieval Scotland: An Archaeological Perspective*, London and Edinburgh, 1995.

17 Housework

UNA A ROBERTSON

Among a housewife's myriad domestic duties and obligations was ensuring the cleanliness of both her dwelling and its occupants. She might be wife, mother, grandmother, daughter, sister, aunt, niece, widow, orphan, servant or occupy a step or foster or other position within the household, but in all cases, and no matter what the standing of the household, the day-to-day care of the house interior and the household was ultimately the housewife's responsibility, whether she did the chores with her own hands, had assistance from a servant or two, or the work was done on her behalf by a full complement of staff. Despite increasing concern with equality between the sexes in the closing years of the twentieth century and at the beginning of the twenty-first, this obligation on women persists, and recent surveys show that the amount of housework they do still far outweighs that undertaken by men.[1]

Over the centuries the tasks involved in keeping the home clean, for example sweeping floors, polishing furniture and washing windows,[2] and the equipment employed to do these with, such as brooms, dusters and pails, have in some ways altered very little. In other respects there has been considerable change. This includes the proliferation of domestic servants, which reached its apogee late in the nineteenth century. The introduction of a piped supply of water and means of drainage into the great majority of homes was also important as was the manufacture of proprietary products to replace the homemade, while the application of electricity to tools and equipment helped greatly in the battle against dirt. Change did not, though, come at a uniform pace. While some households, especially in the remoter areas, continued using traditional methods, others adapted more readily to newer ways.

For such an ephemeral occupation, no sooner done than requiring to be undertaken again, there is a surprising amount of evidence from the seventeenth century onwards. Visitors to Scotland, both male and female, often remarked on the state of the dwellings they happened upon or the inns at which they stayed, on the manner in which cleaning was carried out and the tools used, and, sometimes, on the women's attitudes towards such matters. Of similar value are published memoirs, whether journals or reminiscences, which shed light on aspects of domestic life. Evidence of domestic, as opposed to culinary, practices can sometimes be found in Scottish cookbooks, although Scotland appears to have lacked the domestic compendiums produced by English authors until relatively recently. A considerable body of evidence lies in often unpublished material relating primarily to more substantial households. Because work was delegated in these homes there was much more record keeping, whether of servants' wages, or in the form of inventories or accounts.

PERCEPTIONS OF SCOTTISH CLEANLINESS

Sir William Brereton, who visited Scotland in 1636, considered Edinburgh's inhabitants as a 'most sluttish, nasty, and slothful people', adjectives he repeated on several occasions. He also accused Scottish housewives of being fearful of wearing out their pewter dishes by cleaning them too often,[3] as did Edward Burt almost a century later,[4] while Dr Johnson thought the silverware would last for ever as it was never cleaned.[5]

In 1698 Celia Fiennes formed a poor opinion of Scotswomen, 'I took them for people who were sick,' she wrote, 'all day idle doing nothing.'[6] The remark was echoed by Isabella Spence in 1811 who wrote that 'they are not particularly industrious'. She also reported, 'The lower orders of the Scotch have an idea that the English spend too much of their time in washing and cleaning their houses, and adorning their persons ... '[7] However, Englishwomen were in their turn seen by foreign observers as doing little but cook, clean, wash and sew, leaving all other responsibilities to their menfolk, including the milking of cows.[8]

The dirtiness and general inadequacies of Scottish inns attracted many complaints. Boswell, for example, during his famous tour with Dr Johnson, used adjectives such as 'sorry', 'indifferent' or 'wretched', although elsewhere they found 'genteel hospitality' or 'an excellent inn'.[9] Similarly, Dorothy Wordsworth complained about the half-finished appearance of so many inns and the dirtiness of their floors. One she praised by saying, 'The house was clean for a Scotch inn.'[10]

THE WORKFORCE

In medieval times women and girls composed a minority of domestic servants. Men and boys cooked and cleaned, and females were employed only as 'waiting women', 'nurses' or 'washing women'. However, from the seventeenth century women were employed in greater numbers as they were less costly (their wages generally being about half that of their male colleagues') and were considered more amenable to discipline.

Women's work became more specialised through time. Instead of often being general servants, able to turn their hand to assorted tasks, such as those employed by Lady Grisell Baillie of Mellerstain, Berwickshire,[11] women were increasingly employed as chambermaids (latterly called housemaids), laundry maids and dairy maids, and also as cooks and kitchen or scullery-maids; all with distinct spheres of work. Also concerned with domestic matters was the lady's maid, who did much of her mistress's laundry work, and the nursery maids who were responsible for the quarters ruled over by Nurse, latterly known as 'Nanny'. Should the housewife be reluctant or unable, for whatever reason, to take charge personally of her household, then, from the eighteenth century onwards, the fashionable answer lay in the employment of a housekeeper who would act as a deputy housewife and take over managerial and hands-on domestic duties, plus, on occasion, other tasks.

The desire for increased domestic comfort and privacy from the closing years of the sixteenth century led to a diminution of the ceremonial and an increased separation of the component parts of the household, in time giving rise to the 'upstairs, downstairs' syndrome. As an ever-larger percentage of the population came to employ servants, so the gap between the social origins of employer and employee narrowed and increasingly harsh restrictions and greater segregation were imposed on domestics.

Where did households find their servants? Traditional sources included the local fair and the children of tenant farmers or those of relations. In the Highlands especially it was not unknown to find among the servants those of the same name as the estate owner.[12] A popular method of recruitment was to seek recommendations from family and friends while shopkeepers frequently acted as intermediaries. Additional sources were charitable institutions and, from the eighteenth century, Servants' Registry Offices and newspaper advertisements. There was often a rapid turnover of staff, particularly among the lower levels of the domestic hierarchy. Many were tempted to move on to find a 'better situation' either with more pay or in a more favourable locality. Equally, though, there were many examples of women serving one household throughout their working life.[13]

By the 1860s worries about the future of domestic service were surfacing. Concern was expressed about the effects of emigration from the Highlands and, in addition, about the ignorance of domestic matters displayed by both servants and their mistresses. Rarely were schoolgirls taught to cook or clean or wash, it was said.[14] Meanwhile, in England there was a campaign to get the subject of 'domestic economy' onto the school curriculum. The aim was not only to prepare girls for their expected future as wives and mothers, but also to ensure that, in the intervening years, they could make themselves useful as domestic servants.[15]

The increased number of women in domestic service, both in Scotland and England, resulted in menservants and their employers acquiring a measure of prestige. This was all the more marked after 1777 when a tax on male servants was introduced which continued, at fluctuating rates, until repealed in 1937. The hair powder worn by footmen for uniformity also carried a tax from 1795 to 1869. As prestigious assets such employees were kept in the public eye, both by their liveries and their work. They were on hand to care for guests, wait at table and accompany their employers when they went out. They would also be expected to deliver messages. Their standing was observed by one French visitor thus.

> Accordingly, while there are certain English noblemen who have thirty or forty menservants, the cooking and the housework that is not seen

are usually done by women, and menservants are employed only for such duties as are performed in the presence of guests.[16]

Any tasks conducted by men behind the scenes were primarily concerned with the drawing room and dining room, but only after the rooms had been thoroughly cleaned by the maids. Footmen polished up mirrors and tables, cleaned silverware and arranged the dining room table. They might also act as valets to guests travelling without their own servants.

In larger establishments the domestic staff – as with the personal servants and outdoor or estate workers – would be divided into distinct departments, each with its own hierarchy and spheres of work (see Fig. 20.4). The house steward or butler, who had overall responsibility and ran the house on behalf of his employer, headed the menservants; the housekeeper was in charge of the maids; and there were the kitchen staff and those working in the stables. Such households were in the minority. Obviously income and servant numbers were linked.[17] In 1861 Mrs Beeton considered an annual household income of £150-£200 necessary to support a maid-of-all-work while £300 maintained two maids. Sample Census figures from 1851 to 1871 show that 34 per cent of households with servants had one servant while 25 per cent had two.[18] There were also innumerable women who worked single-handedly, assisted only by their children, often from their earliest years.

The custom of employing domestic servants virtually disappeared by the 1920s and 1930s. The status of domestic service had been declining for much of the nineteenth century and the work came to be considered demeaning both to those engaged in it and to their families. During the second half of the nineteenth century numerous employment opportunities opened up for women, and these increased still further during World War I. In comparison to the other jobs on offer, domestic service was perceived as being poorly paid, relatively isolated and restrictive to personal freedom.

This situation must also be coupled to the years of the Depression, which witnessed a general diminution in incomes and the necessity for a reduction in expenditure, and also to the way in which many middle-class housewives enjoyed being free of troublesome live-in servants and, instead, engaged a woman to come in daily to 'do the rough'.

EQUIPMENT

Whatever their individual nature, the households of Scotland had much in common when it came to the methods and mixtures used to keep them in good order. The effort expended in cleaning, however, could vary greatly. It goes without saying that a two-roomed cottage, with an earth floor, a central fire and little furniture would need less time spent on it than would a well-furnished four-storeyed terraced house. And the amount of work expended on the latter would be insignificant compared to that lavished on the mansion surrounded by its parkland and estates.

The amount of cleaning undertaken in any one household would also depend on factors such as inclination, aptitude and the time and energy available. Other elements were, for example, the distance from a source of water and the lighting or fuel in use, some means creating more dirt than others. A fire of peats resulted in more than just ash and smoke. Dorothy Wordsworth wrote poetically of how the smoke from a fire in one cottage had encrusted the beams above until, 'they were as glossy as black rocks on a sunny day cased in ice'.[19] Peat would burn on the floor, as would logs, without the need for a grate, built-in hearth or chimney, all of which required cleaning. By contrast these features were essential for coal and even then the fire often smoked uncontrollably if the draught was incorrectly aligned, laying a grimy deposit over the room.

Domestic inventories and account books indicate the tools needed to undertake housework. Many of them, or their updated descendants, are still with us and so need little in the way of explanation. Brooms, besoms and brushes for sweeping often feature (see Fig. 19.4). The simplest were made from a bundle of twigs or heather roots bound together around a handle. 'Heather ranges' were smaller versions for scrubbing out pots and pans. Brushes made by professional makers had the bristles, whether of animal origin or vegetable fibre, inserted into a wooden head. Brushes were also developed for specific functions; 'hearth bisoms', for example, or 'rubbers' (stiff brushes for scrubbing floors or removing dirt from boots), or 'a Cobbwebb brush'.[20] An alternative to such a soft brush was a trimmed goose wing which was recommended for dusting shelves and awkward corners, or for brushing clothes. Feather dusters, made from a bundle of feathers tied to a stick, were popular for dusting ornaments and picture frames. For ornate plasterwork a pair of bellows was called for and the dust was blown out of the crevices. Dusting and polishing an assortment of surfaces required a range of fabrics – silk, linen and flannel, for example – which were often gleaned from worn-out clothing. Lady Grisell Baillie gave strict instructions to her housekeeper, 'See that all the maids keep their dusters and washing clouts dry and in order, and not let them ly about in hols wet, which soon rots and makes an end of them.'[21]

Other necessities were pails and tubs, used for a variety of purposes such as taking ashes from the fire and laundry to the drying green or holding water for scrubbing floors. The most general form was either round or oval, depending on usage, although some might be square or oblong. They were made of wood, with split-withy banding (latterly replaced by iron hoops) and with two elongated staves opposite each other forming raised hand holds or fixing points for a rope or iron handle. Metal tubs and buckets, at first made of galvanised iron and later enamelled, were a nineteenth-century development.

The mop must certainly have evolved from the idea of pushing a cloth across the floor with the aid of a stick. Scotswomen, it seems, did not favour the scrubbing brush, or indeed the mop. Edward Burt, whose *Letters* were penned in the 1720s, described how girls would spread a wet cloth on the

Figure 17.1 An old woman sweeps with a besom made by binding a bundle of twigs or heather roots around a wooden pole. This photograph comes from Surrey but the tool and technique would have been identical in Scotland. Source: Jekyll, G. *Old West Surrey: Some Notes of Memories*, London, 1904, 167.

floor, shuffle it round with their feet and then dry off the floor in the same way. He wrote, 'When I first saw it, I ordered a mop to be made, and the girls to be shown the use of it. But ... there was no persuading them to change their old method.'[22] He found them using their feet for a number of other domestic chores, such as washing vegetables, grinding off the beards and husks of barley and also for laundry work. Not all Scotswomen were so obdurate, however, since mops and assorted rubbers feature in various accounts prior to Burt's visit.[23]

Labour-saving devices were few and far between. There was, in fact, little need for them as long as servants remained plentiful. It was only when staff became scarcer in the years preceding World War I, and virtually vanished thereafter, that labour-saving became a priority.

One early labour-saving device was the knife-cleaning machine. These items gained in popularity in the 1870s and 1880s as they speeded up the cleaning and polishing of steel-bladed knives, common in the days before stainless steel. While knife-handles could be made from different materials the blades were usually of steel, silver blades being reserved for dessert knives. After a meal the knives were collected into a 'knife box' and washed and dried as quickly as possible to avoid rust marks forming. They were then stropped against a knife board and given a final rubbing over with a cloth[24] (Fig. 17.2). The knife machine, also known as a 'knife cleaner', was a revolving drum with slots which held the blades. To this was added an abrasive powder ('Oakeys Britannia Knife Polish', for example) and, by turning the handle, the blades were sharpened and polished simultaneously by alternating felt pads and bristles – as many as four or six at a time.

The carpet sweeper, which collected up the dust as it swept, was an American invention, although many previous attempts had been made to develop such a useful piece of equipment elsewhere. Mr Bissell, of Grand Rapids, Michigan, owned a china warehouse and being allergic to the packing materials he needed to find a solution to the problem of dust and other dirt accumulating on the floor. His sweeper was successful because the bristles were placed in spiralling rows around the roller to lessen resistance on the carpet.

Cleaning by suction was essentially a twentieth-century innovation. The first device to be developed was used in a commercial setting, providing a professional cleaning service for urban dwellers. Smaller models were subsequently produced for domestic use but were often awkward to operate or required two people to work them, one person providing the motive power. True success came in 1907 when J Murray Spangler of Ohio, USA,

Figure 17.2 A stone sink with enamel pails below. Note how the front edge of the sink has been worn away by knives being sharpened against it – an alternative to using a knife board. Source: Reproduced by kind permission of Beamish, The North of England Open Air Museum.

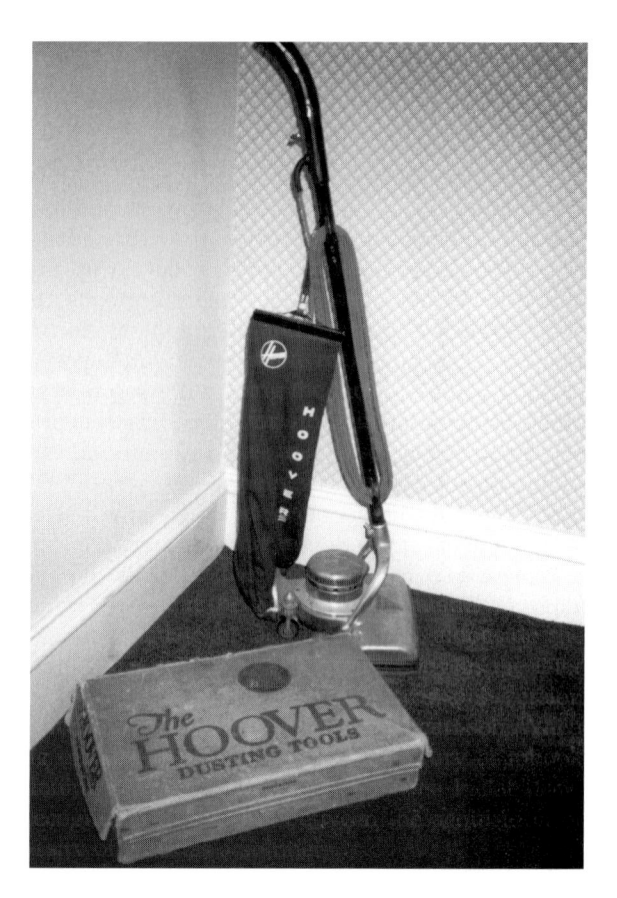

Figure 17.3 The latest in technology from the 1930s, a Hoover vacuum cleaner complete with its box of attachments for specialised cleaning jobs. Source: U A Robertson.

attached a small electric motor to a vacuum cleaner. He then sold the rights to W H Hoover, hence the verb 'to hoover' as against 'to spangler' (Fig. 17.3).

SUBSTANCES

Although much about housework remains familiar, the nature and purpose of numerous cleaning and polishing agents are now largely forgotten, even though many were everyday items until the 1920s. Camstone (caum, cawm or, in England, pipeclay), rotstone (rottenstone), whiting (whitening) and sand were widely used, as were blacklead, sifted wood ash and sweet oil, while Fuller's earth, ox gall, camphor and borax had more specialised applications. Brickdust or 'common brick' was much used in England to scour and to polish metals but did not feature to the same extent in Scotland, although the Duke of Gordon's accounts for 1768 reveal a quantity ordered in London 'for Gordon Castle'.[25]

Sand of differing grades was much in demand. After floors had been scrubbed with a mixture of sand and water and then dried off, handfuls of dry sand were strewn across the floor either as an all-over carpet or in a decorative pattern. This served to absorb any grease or mud. Fine sand scoured cooking pots as well as kitchen tables and shelves, although one authority, Hannah Glasse, warned housewives about its leaving a gritty deposit.[26] It also polished brass and pewter (Fig. 17.4).

Other substances were similarly dual purpose, for both cleaning and polishing. Rotstone was a soft siliceous limestone used in Scotland for scouring although elsewhere it polished metals, glass and stone. Whitening was chalk freed from all sand. Camstone was a comparable product, while the use of a blue caum was peculiar to the Angus area. These substances were employed decoratively on hearths, windowsills, doorsteps and so on but were equally handy for cleaning windows or polishing silver, though Lady Grisell Baillie forbade her butler to use it on her plate, 'A little soap suds to wash it or spirit of wine [ie alcohol] if it has got any spots' was her recommendation.[27]

Storage of both equipment and cleaning materials would depend on the space available. In the larger household, provision was likely to have been in the housekeeper's store room for items bought in bulk. It made sense if those things in use were kept relatively close to where they were most likely to be required. In the case of chambermaids, latterly called housemaids, their cleaning materials were stored in a 'housemaid's cupboard' on each floor. The maids as they worked would carry the materials from room to room in a specially designed 'housemaid's pail', 'box' or 'bucket'[28]

Figure 17.4 Two boys selling sand from a cart. Sand was used for many cleaning jobs around the home. Source: *Book of the Old Edinburgh Club*, vol 2, 1909, 212.

Figure 17.5 A housemaid's box containing the tools and cleaning materials needed for her work around the house. Source: Reproduced by kind permission of Beamish, The North of England Open Air Museum.

(Fig. 17.5). Items used by the footmen in connection with their duties in drawing room and dining room were likely to have been kept in the butler's pantry. Laundry necessities could be found in the laundry.

ROUTINES

Whether the work was being done by servants or by the housewife herself made little difference for the routine stayed the same. When cleaning a room it made sense to start with the fireplace, which required a great deal of attention. After removing the cinders and the ashes, the entire hearth area – including the grate, fire irons, fender and surround – was cleaned, blackleaded, polished and burnished according to composition and the fire was then relaid (footmen, where employed, merely maintained fires). Door handles and other metals in the room were polished and then attention turned to the floor. Smaller carpets were taken out of doors to be shaken or beaten, larger ones were swept in situ. Alternatively, they might be turned over and the dust forced out by being walked over. Not until the nineteenth century was tea cheap enough to be drunk in sufficiently large quantities for the used leaves to be strewn across the carpet prior to brushing. This not only helped to lay the dust but also left a pleasant smell.

Floor cloths began to replace carpets in the nineteenth century. These were later called oilcloths, and were the forerunners of linoleum. They needed a daily dry rubbing and were finished by wiping over with milk, wax polishing leaving them too slippery. Bare boards could be swept, scrubbed with sand and water (which was considered a good way to show the grain of the wood) or dry scrubbed with a mixture of herbs which again stopped the dust flying about and imparted a fragrance to the room.

For cleaning windows two people were recommended, one inside and the other outside. The windows were washed and dried, or merely dusted, then polished with whiting or rotstone which was also used on mirrors.

Wooden furniture was treated with homemade mixtures based on 'sweet oil' (specifically olive oil although the term included any mild-tasting oils), linseed oil or beeswax. For this last, clarified beeswax with enough pure turpentine to moisten it was put into a jar and set on a warm corner of the stove. Once softened and well stirred it was ready for use.[29]

The cleaning of bedrooms involved making the bed to the approved method, which took much time and energy. Another chore pertaining to these rooms was the removal, emptying and cleaning out of the slop pails and chamber pots of the wash stands. It is worth noting that the elegant Georgian townhouses of Scotland were built without a back stair, unlike their counterparts in England, and the maidservants going about their business, their employers and any guests must often have met on the stairs.

Soap and water washed marble floors in halls, though sand and water was considered better. Damp sand also cleaned stone stairs; some was placed on the top step and swept down with the dust. Kitchen and basement floors, generally of stone or slate, required endless scrubbing, as did the tables, shelves and cupboards, while the kitchen grate, latterly the range, demanded a daily cleaning, black-leading and polishing of its brass. Its flues also needed frequent attention.

Some tasks were carried out only occasionally, chimney sweeping for example. In rural parts this was done by hauling a prickly shrub such as gorse or holly or a well-tied bundle of straw up the chimney (with a weight dangling below) to dislodge the soot. Other methods included firing a gun up the flue. In towns there were professional 'sutiemen' who generally worked in pairs because of the large number of flats within tenement buildings, one person going onto the roof to establish which chimney belonged to a particular room. Tenement properties also required their common stairs to be regularly cleaned and a woman would often be paid to do the necessary work, just as at present. Otherwise the residents of the stair would take their turn to clean it each week, according to an established rota. At present, local byelaws can be invoked to enforce such cleaning rotas, as well as the removal of obstructions and litter. These regulations also govern a landlord's obligations with regard to the maintenance of a property.

A particular problem of towns, though not unknown in the country, was bed bugs. Once in a property they were difficult to remove and anywhere advertised as 'free from bugs' would command a premium. In an attempt to rid a dwelling of them, the beds' timbers would be sent away to be boiled, hangings and curtains were boiled or redyed, new feathers were ordered for bolsters and the room itself was fumigated. A similar problem, but one that continues until the present day, especially where numerous properties adjoin, concerns mice, and mousetraps often feature in account books.

The annual spring cleaning, involving a thorough overhaul of the entire dwelling, apparently evolved in response to the introduction of the oil lamp, pioneered by Argand in the 1780s. Periods of intense cleaning had, however, occurred before this. Window hangings went to be callendered (a finishing process for fabrics) and blankets were scoured (given a vigorous washing). Rooms might be painted or cellars cleared out, but none of these activities constituted part of an annual event.

WATER AND DRAINAGE

As long as water was restricted in supply and had to be fetched home, the outcome of the housewife's efforts at cleanliness was limited. A supply brought into or close to the dwelling was of inestimable benefit but not until the nineteenth century was the provision of piped water seriously considered, prompted primarily by matters of public health.

Water supplies in towns were for long enough both inadequate and intermittent. Townsfolk were sometimes able to employ water caddies who would stand in the queue at the pump or the well and carry in a specified number of barrels each day, although they did not work on Sundays[30] (Fig. 17.6). In rural areas, housewives or some other member of the household, including any children who were not employed elsewhere, had to find water as and where they could, sometimes walking several hundred yards, although, in dry seasons, this might be considerably further. Carrying home the necessary supply was not only time consuming, it was also heavy and tiring. Pails or 'stoups' often came in pairs and were a wedding gift from the groom to the bride.[31] Two gallons (9 litres) of water weigh 20 pounds (9 kgs) and that is without the weight of a wooden pail often banded with iron. A single journey's burden of two pails' worth can therefore be compared with the international airline allowance of 44 pounds or 20 kgs.[32]

The obverse of a good water supply is the efficient removal and disposal thereafter of wastewater. Edinburgh was the smelliest of cities, as numerous visitors found to their disgust.[33] For lack of drains, dirty water was emptied into the street or tipped out of the windows. Chamber pots were emptied in the same way. By the 1860s it was claimed, 'For many years past great drainage works have been carried out ... and it may be regarded as satisfactorily drained. In the poorest districts we have large sewers of the best construction.' However, the sewage drained straight into the Water of Leith which in hot weather, 'emits offensive odours. These have of late years been much complained of ... '.[34]

Figure 17.6 Town dwellers could enlist the services of a water caddy. The trade ceased when piped water was introduced and Roderick M'Donald was the last of his kind in Edinburgh. Source: Dunlop, A Hay. *Anent Old Edinburgh*, Edinburgh, 1890, 101.

LAUNDRY WORK

Laundry work consumed much water and here a marked division existed between those with plentiful nearby water and those who had to fetch it from a distance. Wealthier homes, with an ample supply of water at hand, often included laundry premises among their domestic offices, sometimes distinguishing between the 'wet' laundry for the washing processes and the 'dry' laundry for drying, ironing and airing. A drying green with poles and ropes stretched across would be nearby for putting out washing in good weather.

For those without ready access to water it was often easier to take the load to the water source and recognised washing places developed on riverbanks, of a sort still to be seen in rural France and elsewhere. The women could stand in the river itself, pounding the washing upon rocks or beating it with a wooden 'beetle' or mallet. Alternatively, they dealt with their items in tubs on the bank.

Burt, as already noted, had observed Scotswomen using their feet for various domestic tasks (see Fig. 20.2) and found it even stranger that they should, year round, insist on doing their washing in the river. He also observed:

> women with their coats tucked up, stamping, in tubs, upon linen by way of washing; and this not only in summer, but in the hardest frosty weather, when their legs and feet are almost literally as red as blood with the cold; and often two of these wenches stamp in one tub, supporting themselves by their arms thrown over each other's shoulders.

Even more deplorable, he thought, was that they chatted so freely to passers-by while so doing.[35] It does, however, argue a certain determination to achieve cleanliness.

If washing was not done in or beside the river, and the home was small and with no washhouse, items were often washed out of doors in the area immediately surrounding the dwelling (see Fig. 15.5). This was partly a response to the lack of space and light inside and partly to do with the likelihood of steam and splashes damaging interiors.

For those with a built-in washhouse or laundry, along with sufficient water and fuel, washing could be done with soap or with a mixture of certain other substances. Articles for this latter method, called 'bucking', were placed in a deep tub, a perforated tray was placed across the top and lye (water mixed with potash, cow dung or stale urine, sometimes known as 'maister', Gaelic 'maighstir'), was poured in. This soaked through the items and, ideally, was collected into a bowl via a spigot at the base and returned to the tray for a further soaking. The alkali present in the lye reacted with the grease and dirt in the fabrics, thereby reducing the amount of soap needed for the actual washing or obviating it altogether. Lady Grisell Baillie preferred to have her bed and table linen, 'bouckt when the weather will allow [sic] of it'.[36] Her contemporary, the English author Hannah Glasse, found the habit difficult to understand, 'No doubt on it, the Clearer the Water is and the sweeter the Air the better the Clothes will look.'[37] Bucking continued to be used into the nineteenth century when it was noted, 'It rendered the linen beautifully soft and white, with little expense of soap.'[38] This was an important factor since soap was not cheap, carrying fluctuating duties from 1712 onwards. By 1793 the tax was yielding £30,441 in Scotland as against £373,090 in England but it was finally repealed in 1853 in the interests of public health.[39] Notwithstanding, soap requires hot water to be efficacious and this in turn requires a household to possess additional fuel.

Other necessities for efficient laundering included starch for stiffening items such as collars, 'blue' to make linens look whiter, and a variety of substances for finishing individual fabrics. The recommended treatment for chintzes, for example, varied from using bran in the washing water to giving a final rinse in rice water and smoothing the fabric not with an iron, but with a polished stone. Should the colours run into the white ground during washing, then, after rinsing, the fabric was to be soaked in a pail of soft water with the addition of half a pint (0.25 litre) of best white wine vinegar.

To restore the colour of brown Holland chair covers or children's overalls the final rinse was to be in water in which hay had been boiled, although an alternative liquid was water coloured with used tea leaves. Black silk garments would regain their colour if washed in bramble (blackberry) juice boiled up with ivy leaves, while silks in general were to be sponged with potato water. Size and glue were employed on woollens and also on the veils attached to bonnets or hats. The different types of lace required a wide range of substances to restore them to their original condition.

In Lady Grisell Baillie's time at Mellerstain the household routine saw the body linen being washed one week, the next week the bucking, while in the third week the laundry maids had to sit and spin.[40] Set against English examples such frequency was unusual. There, washing was collected for several weeks because to wash more frequently denoted a paucity of supplies. Not until Victorian times did the weekly wash become commonplace.

Households with a full complement of servants would have laundry maids, but others might simply employ one or two extra 'washers' for a day or so to help with the actual washing processes, leaving the starching and ironing to their own servants. For the woman working entirely on her own, washday was very strenuous.

When not done in the river, washing generally required wooden tubs to be filled with hot water taken from an iron pot over the fire. Over time, cauldron and hearth became a single unit which was built into the wash-house as 'the copper'. Articles were boiled in this after washing and prior to rinsing, blueing or starching as appropriate. Gas-fired coppers followed where supplies were available and there were further developments in design and manufacture. The latter-day coppers were often freestanding and set on legs, rather than built in. They might be fitted with a tap for emptying the water even though they still had to be filled manually. The twentieth century witnessed the use of different materials in their manufacture, for example galvanised copper to resist corrosion. Enamelled examples were developed in the 1930s. Where cost was not a serious consideration, electricity replaced gas in some models.

Items were pushed through the water with 'possers', 'dollies' or 'peggy sticks', which somewhat resembled a three-legged stool on a shaft with a cross bar added (Fig. 17.7). The washboard was a nineteenth-century aid. The wooden backing was covered with a ridged zinc or glass surface against which clothes were rubbed to loosen the dirt. Washboards later achieved fame as the 'skiffle boards' of the 1950s music scene.

Numerous 'washing machines' were invented but few proved satisfactory until human energy was replaced by electricity. Many had to be filled and emptied manually, clothes were swirled around by paddles turned by hand and the wet clothes had to be lifted out for further processing and wringing. Electricity transformed the scene but it was only after World War II that washing machines ceased to be considered as luxuries. Since then numerous improvements have been introduced, in the machines, in the washing agents and in the textiles.

Figure 17.7 A girl uses a 'dolly' or 'posser' to push washing through the water. The job is obviously being done out of doors. Source: *Cassell's Book of the Household*, 4 vols, sp edn, London, c 1893, vol 3, 10.

Wealthy establishments had specialised laundry premises which included arrangements for drying. Town dwellers used their own gardens or the communal 'back green', where someone might be paid to 'watch the washing'. In the early days of Edinburgh's New Town, several of the now formal gardens were initially used as drying greens, as were the Meadows, the Queen's Park and the Calton Hill.[41] Tenements often had 'pullishees' (a pulley on a pole with ropes) outside their windows for hanging out the washing. The rural situation was not dissimilar. Washing greens, with a rope or ropes stretched between poles, were possible wherever space permitted, otherwise washing was draped over convenient bushes, hedges or laid out on grassy banks to dry.[42]

In wet weather people managed as best they could, with washing draped around the fire or, if the ceilings were high enough, on racks with pulleys to haul them to the ceiling. Similar provision was made in the laundry complex of the country estate. Here, wheeled racks were sometimes built into a cupboard beside the laundry fireplace and washing draped over the bars dried quickly in the warmth.

Once dry, many items required smoothing, a practice followed for centuries. A simple method involved heating polished stones known as slickstones or slykestones. An alternative was the mangle bat, a flat piece of wood about 30 inches (76cm) long and four inches (10cm) wide, with a carved handle at one end. The damp linen was wrapped around a wooden roller, placed on a flat surface and rolled back and forth with the bat. Dorothy Wordsworth was appalled to watch this being done on the flat-topped tombstones in Melrose churchyard, Roxburghshire, even though she considered the area to be 'exceedingly slovenly and dirty'.[43] Meanwhile, around Dumfries, women laid items on the grass and smoothed out the wrinkles with their bare feet.[44] Smoothing linen by means of a press was a Roman idea. Flat items such as bed and table linen were dried, pulled straight, folded and placed between adjustable boards under a screw which, when turned, increased the pressure.

During the eighteenth century the box mangle appeared in larger households. This was a trough-like box filled with heavy stones, the whole being set within a sturdy wooden frame. The box rested upon several moveable rollers around which items were wrapped and the linens were smoothed as the weighted box moved across them in response to a crank handle being turned. This job was so heavy that a man was often employed to do the turning. By the 1850s the upright cast-iron mangle was becoming commonplace, being more compact and easier to operate. By turning a handle on a large wheel two wooden rollers were worked and a turn screw on the top adjusted the pressure on the items fed through the rollers. The mangle was used both for removing excess water from the washed items and for smoothing them once dried.

Only finer articles were ironed and the irons used came in many shapes, sizes and weights, depending on the intended usage. Some were solid and were heated over the fire or, for cleanliness, a laundry stove; others had a removable inner fitment, the 'shoe', 'slug' or 'filler', which slotted into the outer casing. There were flat irons, variously described as 'box', 'sad' and charcoal (these last had lighted charcoal burning inside them). There were polishing irons and glazing irons (which were similar to a flat iron but with a rounded base). An Italian iron, a 'tally iron', which resembled a poker set at right angles on a metal stick, smoothed ribbons and shallow frills when the items were drawn over it, rather than pressure being applied to the articles from above. The French iron, resembling a mushroom on a long stalk, worked in a similar way but on the uppermost parts of 'puffed' sleeves. For pleats and frills the various forms of goffering irons (from French *gauffre*, a wafer) came into play. Goffering tongs, resembling multi-bladed scissors, were hand held. Another form required the dampened fabric to be slotted in and out between thin wooden slats, pressure was then applied and the contrivance left until the material had dried into the required folds. Yet another form was known as the 'wheeled goffer' with two ridged rollers; the damp fabric was fed between the rollers, the handle was turned and pleats were created (Fig. 17.8).

It was usual to work with a pair of fitments, one heating while the other was in use. Gauging the temperature was a matter of guesswork although spitting on the surface was one test. As different fuels were introduced into the home in general so were innovative methods of heating irons found including the use of charcoal, paraffin, gas and, finally, electricity.

Women, married, single or widowed, could earn money by hiring themselves out as washerwomen to different households. Alternatively they

Figure 17.8 From top left, this selection of implements includes flat, box and charcoal irons; an iron stand and goffering iron; egg, polishing and Italian irons; a laundry stove for heating the irons; and a linen press with drawers. Source: *Cassell's Book of the Household*, 4 vols, sp edn, London, c1893, vol 3, 152–8.

would take in other people's washing to do in their own homes, and they might also advertise a mangling service.[45]

AN EASIER WORKLOAD?

For the poorly-off working woman, in both rural and urban areas, combining a job and household duties could be a struggle indeed. Until quite recently, many girls on getting married were expected to give up paid employment to become full-time housewives and, in the course of time, mothers. If they were ill, their husbands, older children or a kindly neighbour, or a member of the housewife's own family if in the neighbourhood, might help out by doing what they could of the essential chores, the rest being left until the patient had recovered and was able to do them herself.[46]

It was always possible that a housewife would have to go back into the role of breadwinner by force of circumstance while her children were still young enough to need her presence and attention, for example after the death of the husband or if illness, disability or economic decline caused him to lose his employment, or simply his wage was insufficient adequately to support the household. In such cases similar help would hopefully be available from family, friends and neighbours, although payment might have to be made for someone 'to watch the bairns'. Instances are on record of where women chose to go out to work, sometimes leaving their men at home to follow their craft and to care for the children.[47] No doubt these women had to fit other domestic duties around their hours of paid employment.

Cleanliness has long been seen as a domestic virtue, acquiring a moral dimension when linked to the precept of being 'next to Godliness'.[48] Cookery books stressed its importance in the kitchen and the housewife was exhorted to be constantly vigilant in the battle against dirt. Victorian manuals made of housework a never-ending struggle against alien forces. The cleaning workload certainly increased in many respects during the nineteenth century as a result of rising wealth and the greater availability and acquisition of manufactured goods. However, the housewife could offset this by making use of a range of technological advances.

The provision of a piped water supply eliminated one barrier to cleanliness. For townsfolk lacking space within their homes, such as tenement dwellers, communal washhouses and drying facilities were provided, either in the back court or in the basement of the building. Public washhouses or 'steamies', serving a larger number of households, were also created in urban areas. Dorothy Wordsworth noted a public washhouse on Glasgow Green as she passed through in 1803.[49] For some, the steamie was in addition a focal point of neighbourhood life, as was the launderette which became popular in the 1950s. Municipal and commercial laundries developed during the later years of the nineteenth century so that washing could be sent out. Proprietary soap powders replaced the bars of soap bought by housewives when they had ceased making their own washing agents from animal fat, wood ash and other ingredients. Highly-starched clothing went out of fashion and easy-care fabrics were introduced.

Manufacturers provided not only soap powders but also innumerable other products, so the housewife no longer had to buy raw ingredients and make up her own cleaning and polishing substances. Some well-known brand names originated in the Victorian era: Goddard's polishes for silver and other metals and Hedley's 'Fairy' soap are among the earliest, while 'Lux' soap flakes and 'Persil' date from the early twentieth century (Fig. 17.9).

Another significant advance for the housewife was the change from traditional fuels to paraffin and other new forms of oil, and to gas and especially electricity, used in either individual heaters or as part of a central-heating system. Not only was the procurement and use of these fuels easier but they also proved cleaner in use. In the kitchen, the range, with its daily cleaning routine, was superseded by the solid fuel Aga, or the gas or electric cooker. Utensils developed simultaneously, incorporating new materials such as stainless steel or vitreous enamel or they were given special non-stick finishes, all of which contributed to ease of cleaning.

An old proverb claimed that the two greatest plagues in life were a bad servant and a smoky chimney[50] but during the twentieth century the housewife has had less and less to do with either. Piped water, gas and electricity

Figure 17.9 Proprietary cleaning materials and washing powders including, Oxydol, Rinso and several soaps. Betty Brown soap (centre, top shelf) carries two slogans; 'The Ideal Cleanser' and 'British and Best'. Source: Beck Isle Museum, Pickering, Yorkshire.

might have largely taken the place of servants, but they have not done away with housework. The home continues to make demands upon the housewife in matters of cleanliness which would be familiar to women of every era.

NOTES

1. *The Scotsman*, 10 March 2001, 4.
2. Glazed windows were in place in some palaces and churches as early as the fourteenth century but glass was too costly a commodity for the majority of households. By the seventeenth century some Scottish dwelling houses had windows with opening wooden shutters in the lower half and fixed glazing in the upper section. Sash windows, first seen in England in 1680, became fashionable in Scotland during the later years of the seventeenth century. However, poorer dwellings, rural and urban, continued to possess apertures in walls rather than windows as such and these would be covered by rags or hide, or perhaps stuffed with turf, during bad weather. Further information on windows can be found in Chapter 5 Lighting; Chapter 29 Status.
3. Brereton, 1891, 140, 142–3, 'Their chambers, vessel, linen and meat, nothing neat but very slovenly.'
4. Burt, 1815, I, 135.
5. Rogers, 1993, 201.
6. Fiennes, 1888, 171.
7. Spence, 1811, 150.
8. Kalm, 1892, 327.
9. Rogers, 1993, 30, 52, 74, 60, 282; see also Chapter 34 'Time Out' Leisure.
10. Wordsworth, 1894, 52, 184.
11. Baillie, 1911, 278–80; see also Chapter 20 Work in the Homes of Others for information on domestic servants.
12. Grant, 1898, 135, re Annie Grant, 'the accidental daughter'; NAS, Panmure Muniments, 1709–32, for example, 6, 13, James Maule, Servitor, 9, Jean Maule.
13. Baillie, 1911, lii, 'May Menzies'; Grant, 1898, II, 105, 'Mrs Sophy Williams'; Sillar, 1961, 151–62, 'Devoted servants'; for examples in England see Munby, 1891.
14. Colston, 1864, 5–13; Anon., 1862, 418.
15. Buckton, 1885, Introduction.
16. de la Rochefoucauld-Liancourt, 1933, 25.
17. Beeton, npd (but 140th thousand), 8; similar guidelines had been given in Adams, Adams, 1825, 5–6.
18. McBride, 1976, 20.
19. Wordsworth, 1894, 103; for further information on fuels see Chapter 6 Heating and Sleeping.
20. NAS, Seafield Muniments, Account Book 1703–26, 6, 17, 92.
21. Baillie, 1911, 280.
22. Burt, 1974, I, 86–7.
23. NAS, Seafield Muniments, GD248/600, Book of Accounts ... by Alexr Dunbar, 1703–26, 1, 97, 118.
24. Adams, Adams, 1825, 393, the procedure is lengthy and is explained in detail.
25. NAS, Gordon Muniments, 1768, April.
26. Glasse, 1760, 66.
27. Baillie, 1911, 274–5.
28. The housemaid's pail (bucket) was usually square or oblong in shape, made of wood and with an iron handle. It was fitted with a compartmentalised inner tray which could be lifted out. Into the pail the maid would put all the smaller bits and pieces of equipment needed for cleaning and polishing a room and its contents. Not only were items kept together in the pail but they were also easily transported from room to room.
29. Dods, 1828, 491, 'To polish Mahogany'. Alkanet root and rose pink added colour to the polish but were not otherwise necessary.

30. Cockburn, 1856, 353–5.
31. Colston, 1890, 43. In this Edinburgh example the stoups were of a special shape. Colston says, 'The carrier of the stoups, as a rule, made use of a wooden hoop or girr round the person, by means of which he or she was enabled to convey the water more easily without spilling any of the supply.' In rural areas, however, a yoke may have been used if the water had to be carried any distance. In Shetland a much larger container (a 'sae' or 'say') might be slung on a wooden pole and carried between two people. In some areas a pitcher was balanced on the head, which left the hands free for knitting or spinning as the woman walked along.
32. Economy class, 2001.
33. Brereton, 1891, 140; Burt, 1815, I, 19; Ray, 1749, 304; Rogers, 1993, 7.
34. Littlejohn, 1865, 76, 86; for further information on water and drainage see Chapter 16 Personal Hygiene.
35. Burt, 1974, I, 43–4.
36. Baillie, 1911, 279–80.
37. Glasse, 1760, 45–6.
38. Webster, 1844, 1084.
39. Dowell, 1888, 319; see Chapter 14 Clothing for additional information on laundry work.
40. Baillie, 1911, 279–80.
41. Littlejohn, 1865, 113.
42. A legend describes how the Virgin Mary, during the Flight into Egypt, spread out her wet cloak to dry on a rosemary bush. Its white flowers turned to blue in commemoration of the event.
43. Wordsworth, 1894, 265.
44. Spence, 1811, I, 76.
45. In the last few years ironing services have become widely available in many urban areas, echoing the taking in of washing in the past. Often these services are offered by married women with young children who wish to supplement the household income whilst remaining at home. See Chapter 19 Working from Home.
46. When the housewife is prevented from undertaking domestic duties, the household can respond in a number of ways, including sending children to reside temporarily with relatives. Without external support households are forced into self-reliance – a challenge which not all can meet. Some housewives have therefore felt it necessary to keep going at any cost. See for example, Thomson, 2000–1; see also Chapter 24 Childbirth.
47. *OSA*, II (1975), 295. In the parish of Inveresk, a few miles outside Edinburgh, women left their children at home with their menfolk while they themselves transported goods such as vegetables, fruit, salt and sand (for washing floors) into the city and returned home laden with dirty laundry for washing in the purer country water.
48. Attributed to J Wesley (1707–88), although used by others, too. Occurred originally in the writings of Rabbi Phineas ben Yair.
49. Wordsworth, 1894, 53–4.
50. Colston, 1864, 5.

BIBLIOGRAPHY

Adams, S, Adams, S. *The Complete Servant*, London, 1825.
Anon. Modern domestic service, *Edinburgh Review*, 115:235 (Apr. 1862), 409–39.
Baillie, Lady G (ed R Scott Moncrieff). *The Household Book of Lady Grisell Baillie, 1692–1733*, Edinburgh, 1911.

Beeton, Mrs. *Mrs Beeton's Book of Household Management*, London, npd (but 140th thousand).
Brereton, Sir W (ed P Hume Brown). *Early Travellers in Scotland*, Edinburgh, 1891, 132–58.
Buckton, C M. *Our Dwellings: Healthy and Unhealthy*, London, 1885.
Burt, E. *Letters from a Gentleman in the North of Scotland to his Friend in London* (1754), Edinburgh, 2 vols, 1815.
Cockburn, Lord H. *Memorials of his Time*, Edinburgh, 1856.
Colston, J. *The Domestic Servant of the Present Day*, Edinburgh, 1864.
Colston, J. *Edinburgh and District Water Supply*, Edinburgh, 1890.
Davidson, C. *A Woman's Work is Never Done: A History of the Housewife in the British Isles, 1650–1950*, London, 1982.
Dods, M (pseud). *The Cook and Housewife's Manual*, Edinburgh, 1828.
Dowell, S. *History of Taxation and Taxes in England, vol 4*, London, 1888.
Fiennes, C. *Through England on a Side Saddle in the Time of William and Mary*, London, 1888.
Foulis, Sir J (ed Rev A W C Hallam). *The Account Book of Sir John Foulis of Ravelston, 1671–1707* (Scottish History Society, vol XVI), Edinburgh, 1894.
Glasse, H. *The Servant's Directory: Or Housekeeper's Companion, by a Lady*, London, 1760.
Grant, E (ed Lady Strachey). *Memoirs of a Highland Lady*, London, 1898.
Hume Brown, P, ed. *Early Travellers in Scotland*, Edinburgh, 1891.
Kalm, P (tr J Lucas). *Kalm's Account of his Visit to England on his Way to America in 1748*, London, 1892.
Lindsay, M. *The Discovery of Scotland*, London, 1979.
Littlejohn, Dr H. *Report on the Sanitary Condition of the City of Edinburgh*, Edinburgh, 1865.
Macdonald, J. *Memoirs of an Eighteenth-Century Footman*, facsimile edn, London, 1985.
Marshall, R. *The Days of Duchess Anne*, London, 1973.
McBride, T M. *The Domestic Revolution*, London, 1976.
Munby, A J. *Faithful Servants: Being Epitaphs and Obituaries Recording their Names and Services*, London, 1891.
NAS, Gordon Muniments, GD44/52/140, Household Book, 1768.
NAS, Panmure Muniments, GD45/18/770, Servant's Wages &c, 1709–32.
NAS, Seafield Muniments, GD248/600, Book of Accounts ... by Alexr Dunbar, 1703.
Ray, J. *A Complete History of the Rebellion*, York, 1749.
Robertson, U A. *The Illustrated History of the Housewife*, Stroud, 1997.
de la Rochefoucauld-Liancourt, F A F (ed J Marchand, tr S C Roberts). *A Frenchman in England in 1784*, Cambridge, 1933.
Rogers, P, ed. *Johnson and Boswell in Scotland: A Journey to the Hebrides*, New Haven and London, 1993.
Sillar, E. *Edinburgh's Child*, Edinburgh, 1961.
Spence, E I. *Sketches of the Present Manners, Customs, and Scenery of Scotland, With Incidental Remarks on the Scottish Character*, London, 1811.
The modern domestic servant, *Edinburgh Review*, CXV (April 1862).
Warrack. J. *Domestic Life in Scotland, 1488–1688*, London, 1920.
Webster, T. *An Encyclopaedia of Domestic Economy*, London, 1844.
Wordsworth, D. *Recollections of a Tour of Scotland in 1803*, Edinburgh, 1894.

18 Auxiliary Work in the Rural Home

ALEXANDER FENTON

Three of the cornerstones of ethnological theory are the concepts of time, place (geographical location) and social context. The subject of auxiliary work carried out in the rural home, more specifically in the living room or kitchen, lends itself admirably to explanation in these terms. Parts of the steading, notably the barn, may also come into play as wet-weather auxiliary work centres. This was especially true when the home had not yet been changed by the agricultural improvements which spread from the more southerly parts of lowland Scotland from the end of the eighteenth century, and it still had internal communication between living space, byre and barn. Cows and calves were more or less part of the family. Even within the dwelling space, living, working and sleeping areas overlapped.

Auxiliary work relates to a time when, or place where, self-sufficiency was a necessity and all sorts of jobs had to be done by the people themselves, particularly the making of small-scale items of equipment which could not then, or in that place, be bought in shops. It is impossible to give a date which applies to the whole country. The progress of agricultural improvement meant that the more advanced farming areas of south and east Scotland could be up to a couple of centuries ahead of the small farming or crofting areas of the north and west and even of parts of the north-east corner. In such areas older activities, marking forms of self-sufficiency dictated by the nature of the environment and subsistence forms of economy, could persist until well through the twentieth century. The people themselves used raw materials from the local environment to make or repair simple tools and items of equipment which elsewhere had come to be supplied by ironmongers, saddlers and other tradesmen (Fig. 18.1). The social context of auxiliary work was, of course, the family farming or crofting unit, normally without the benefit of help from paid labour. It also had the advantage of giving members of the community with certain types of incapacity 'something to do'.

The concept of 'work' in rural life has given rise to a range of international publications. In 1965 a congress was organised by the *Deutsche Gesellschaft für Volkskunde* [German Society for Folk-Life Study] in Marburg, Germany, on the theme of 'work in its significance for the world of man'. In the published volume of its proceedings it reported the results of research into aspects of house and settlement, tools and equipment, language, tales, song, music and dance, and customs.[1] The Society set up a commission for research into work and implements and the results of a further conference in 1967 were published in 1969. In this case the subject emphasis was the collection of items and their documentation and inventorisation, and a survey of existing questionnaires on the subject was presented.[2] In 1970 there appeared a volume from Sweden on 'work and implements', with contributions on implements, materials and techniques, cultivation, stock raising, trapping and hunting, fishing, food preparation, forestry, transport, buildings, furnishings and clothes.[3] Women's work in the home was amply catered for by a detailed publication from Denmark on the round-the-year tasks of rural women. It examined outside and indoor activities such as the collection of berries, fruit and herbs for later use in the house, gardening, milking and the processing of dairy products, the processing of parts of slaughtered animals and of cereals for food, looking after poultry (and in Denmark, as in Orkney, brood geese sat in stone or wooden recesses in the kitchen), the preparation of daily and festival meals, brewing, textile working and dyeing, and general cleaning.[4]

These and other volumes reviewed the major tasks of the rural year, but nowhere is there mentioned, except perhaps indirectly or implicitly, the range of auxiliary pieces of work, much of it done in men's spare time in the kitchen or living room, or in the barn, in the evenings or sometimes on the worst weather days, which met the small-scale necessities of everyday existence. The communal indoor living space was indeed equivalent to a workshop where a range of jobs was carried out, by the men for mostly outdoor purposes, and by the women as well, who concentrated on domestic needs like the knitting of items of clothing, spinning, woolwinding and many other tasks of a repetitive household nature. This was not a self-conscious form of economy, like the 'cottage economies' which developed on the edges of commons in England (and on the skirts of the Lowlands and later more generally in Scotland), which were proto-industrial in nature and aimed at obtaining income from work such as lace making, spinning, weaving, carrying milk, butter, cheese and poultry to urban markets, making

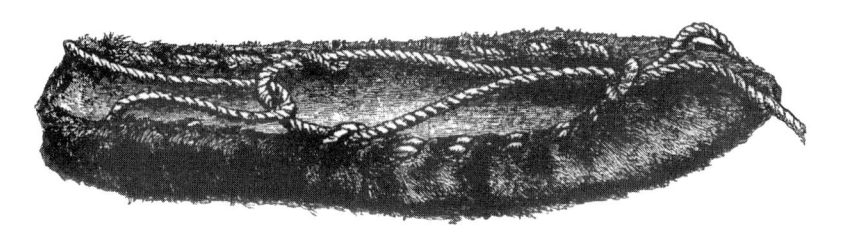

Figure 18.1 Rivlin from Shetland: A shoe of untanned cow hide or sheep skin, worn hair-side out. Formerly widespread in Scotland and known from the early fifteenth century, this is the type of home-made footwear which could be shaped in the kitchen in the evening. Source: Mitchell, A. *The Past in the Present*, Edinburgh, 1880, 93.

shoes, furniture or nails.[5] It was rather a natural, ancient form of social behaviour which was scarcely considered in economic terms.

In seeking sources of possible information, farmers' diaries have been examined. But these are very much concerned with the yearly round of activities in the fields, with the times of putting the cow to the bull and the periods of calving, and they reflect an almost entirely male world. Any mention of women's work is limited to occasions when they also tackled field tasks which were normally those of the men, such as spreading dung, hoeing turnips and the like. It is noticeable, however, that repetitive everyday tasks like the mucking out of animals in byres and stables is never mentioned, nor is the work of the women in milking in the byre and processing the milk in milk house and kitchen. The daily domestic round of the women in the house is also taboo, or at least not considered worth mentioning. 'Mary was down at the Garden, Strichen House, for berries, and I saw she was boiling them at night when I came in to Supper, but I don't question such things as that,' wrote George Gall, an Aberdeenshire farmer, in his diary on 18 July 1868.[6] So, equally, there was little chance that the making or repair of farm or other equipment in the kitchen at night would receive any mention. To a farmer concerned with the seasonal round of work in his fields, domestic activities, and even the inside work of the byre, were not matters of record.[7]

However to anyone concerned with cultural history, it is important to consider the workshop aspects of the domestic living space, for these have centuries of tradition behind them. To some extent, the tasks carried out had a seasonal emphasis. Straw work was a winter occupation, and in harvest there was the winding of ropes for the tying down of thatch on stacks.

Some of the fullest information comes from Orkney, and to a lesser extent from Shetland. It is presented in this short chapter as an example of the kind of activity which must have been carried on throughout much of rural Scotland in earlier days. 'It was out of straw that the farmer produced the greatest variety of articles indispensable for the carrying out of his work,' wrote the Orcadian John Firth in 1920, and this was done indoors 'to occupy his time'. It was unthinkable that time should be wasted, although this concept was not the clock-watching one of the present day. Time was a valuable cultural property under traditional forms of rural existence, in that many activities were carried out almost automatically, by no means according to any clock or timetable, but simply as what was expected and appropriate in the dim kitchen, lit mainly by the glow of the peat fire, as the conversation flowed over the events of the day.[8]

So the men made 'wazzies', which were bands of plaited straw or heather used as horse collars or as seats on which stone-knappers sat when they were breaking stones for road metal, or low seats or stools about 18 inches (46cm) high. In some parts of Orkney these stools were called 'fit wazzies' or 'kringle stuils', the word kringle also referring to a round pad of twisted straw or heather to sit on outside.[9] Quickly made, temporary seats like this must have been very common, and one Orkney source indicates that men would also take them to sea with them to sit on in the boats.[10]

The dialectal evidence shows a wide distribution of the word in Scotland. In the early 1800s in Galloway, at the opposite corner of the country, a 'wassock' was 'a kind of turban', a circular band of straw, worn by milkmaids to relieve pressure when carrying heavy pails on their heads.[11] These are also very likely to have been made indoors.

There were mats called 'flackies' or 'flets' (these were about three feet [91cm] square) to lay on the backs of horses under a wooden saddle, or for sifting or winnowing meal or grain on, or for hanging in a doorway as a screen to keep the draught out. These two names, both of Norse origin, are confined to Shetland, Orkney and Caithness, though the items were certainly also used elsewhere.[12] Then there were 'half lades', woven baskets slung on each side of a saddle for carrying loads of oats or meal (they could hold about three bushels or 24 gallons [109 litres] of grain), and 'maises', of a more open, network construction, similarly arranged on the horse, and used for bulkier items such as peats, and for containing the half lades themselves.[13]

Cup-shaped baskets for carrying on the human back were often named after the substance carried, thus 'seed kubbie', 'peat kubbie', 'bait kubbie', 'fish kubbie', and adapted in size accordingly. They had a band attached for slinging over the back or hanging on a nail in a wall. Another form of basket or creel of heather, straw, coarse grass or reeds was the 'kazy' or 'caisie', which, like the kubbies, were for carrying every sort of substance, from the ash taken from the fire to the midden outside, to butter and eggs destined for the laird's table. Those used for taking meal home from the mill were woven particularly tightly of straw and bands. Seldom would a journey be undertaken without a kubbie or caisie (in Shetland kishie) on the back. The name caisie or cassie extends from Shetland through Orkney and Caithness to Banffshire and Aberdeenshire (Fig. 18.2).

Figure 18.2 Making a caisie in Orkney. The two strands of bent-grass simmon criss cross alternately between each bunch. Source: Fenton, A. *The Northern Isles: Orkney and Shetland*, Edinburgh, 1978, 253.

Much ingenuity could be shown in making baskets, for example to overcome a physical problem. To give an instance, the kishie was built up of successive bands of straw. Binding cords went over and under in pairs which crossed between each band. The body was held with one hand, and the end of one cord was pulled with the other. It was necessary to grip the second cord in the teeth, but if the maker had lost his teeth, he could no longer operate. So as a substitute, he would use a wooden hook fixed to a board strapped across the chest, or an iron hook served, wrapped around with twine and fastened to a belt. The humorous Shetland name for such a device was the 'kishie-makin yackle', or tooth.[14]

Items of straw made for the house were the 'speun cubbie' for holding the spoons in the kitchen, the 'steuls' for sitting on, and straw besoms for sweeping the floors. The farmers also made straw 'beuts' for their own comfort. These were akin to leggings, made in this case of single-stranded straw ropes called 'sookans'. One end was passed under the instep, and the rest was wound several times round the ankle and up to the knee, the upper end being tucked under the topmost round, to give comfort and protection in wet or muddy weather or in snow. These were seldom worn by women in the Northern Isles, however, though in the south of Scotland, the female

Figure 18.3 Winding simmons in Orkney. Ink drawings by W Mason, no date. Source: SLA 72(1) 84.1889 C.165 8/1889.

Figure 18.4 Horsehair rope and winder 'crocan goisid', South Uist, 1965. Source: SLA 72(1) C.10496.

outworkers known as bondagers were accustomed to twisting straw ropes round their ankles when working in muddy conditions.[15]

During the evenings of harvest, ropes known as 'simmons' were wound into great balls a couple of feet (c60cm) in diameter, to be used to rope down the thatch on stacks and on the roofs of houses (Fig. 18.3). These ropes were of straw, grass or a fine kind of heather, not twisted but made into a two-stranded plait. They were used for so many purposes that making them was a regular fall-back job, to ensure that a store would always be available. Simmons of heather were mainly for roping thatched roofs; those of straw were for grain stacks. Sma (small) simmons, made of crushed rushes or of dog's-tail grass, were used for creels, caises and cubbies.[16]

The word 'townit' was usually applied to women's work like carding, spinning and knitting, but it could also be used for men's work in winding ropes in the barn, or sometimes 'abune the fire' to get the benefit of the warmth and light from glowing peats. The performance of this task in the kitchen is emphasised by the story of a man 'who made a bet that if the end of his simmons were passed up through the lum (chimney), he could wind fast enough to supply the men thatching the roof with simmons needed to keep them working'.[17] Auxiliary work by both men and women in the kitchen is referred to in a Shetland verse:

> An windin simmints, spinnin, makkin,
> Dey sat afore da paet fire's licht.[18]

'Townit', with its suggestion of household or communal work in the house, comes from Norwegian dialectal *tonad*, yarn or thread, and the word survived in Shetland and Orkney. The earliest reference, in 1747, even speaks of 'the townet house' in Orkney, meaning a room where the wool was processed by the women.[19] For the sense to have been transferred to men's work indicates that their indoor activity and productivity must have been seen as being in some degree equivalent to that of the women.

Amongst the other items made indoors were animal tethers, using the hair of cows' tails or horses' manes, and even swine's hair. The swine's and cows' hair was mixed together and carded on wool cards, and eased off in the form of 'rowers' or rolls, which were then spun into cords by means of a homemade device called a 'winshoo'. Three strands of the required length were spun separately and then twisted together to make strong tether ropes (Fig. 18.4). These could not be cut with a knife. The wooden swivels which formed parts of the tethers could also be shaped indoors and the hole at each end made by means of a red-hot poker. Another evening task, often done by young folk, was the stripping of the green peel off rushes which had been cut at Lammas and carried home in bunches to make wicks of their white pith for the oil-fuelled 'cruisie' lamps.[20]

These are a few samples of work carried out indoors, often for outdoor purposes. Many others could be added, such as the burning out with a hot wire of holes in calfskin sieves, but in lowland Scotland time has not been kind to the survival of such techniques, although there can be little doubt that they were carried on. The question of auxiliary work, done indoors and often around the fire, is a subject which deserves to be further researched, as an important aspect of the general run of work done under small-scale or subsistence farming conditions. And at the same time, because of the domestic setting, it was part of the environment within which tales and traditions, family history, remarkable events, riddles and proverbs, and to some extent songs, were exchanged between the generations and so for a longer time preserved. Indeed such oral exchanges, within the comfortable and dim atmosphere of the evening kitchen, were greatly enhanced by the presence of visitors, who would take a hand in the work as well as the talk. And so information and lore would disseminate naturally to the wider community.

NOTES

1. Heilfurth, Weber-Kellermann, 1965.
2. Hansen, 1969.
3. Bringéus, 1979.
4. Højrup, 1964.
5. Thomson, 1991, 176.
6. Gall, George. Manuscript Diary, 23 May 1868–29 July 1869 (Atherb, Aberdeenshire).

7. Fenton, 1988.
8. For a discussion of 'time and work discipline' see Thompson, 1991, 352–403.
9. Firth, 1920, 30; Marwick, 1929.
10. Omond, 1911, 10.
11. MacTaggart, 1876, 463; see *SND*, s v Wase.
12. *SND*, s v Flackie and Flet.
13. *SND*, s v Half and Maise.
14. Fenton, 1997, 250.
15. Fenton, 1999, 229.
16. *SND*, s v Simmen.
17. Firth, 1920, 30–2; Marwick, 1929.
18. Vagaland, Fire Lowe, *New Shetlander*, 71 (1964), 21.
19. *SND*, s v Townet. It is to be noted that the use of a room by the women for processing textiles is widespread in Europe – cf the Hungarian term *fonóház*.
20. Firth, 1920, 105, 111–2; see Chapter 5 Lighting.

BIBLIOGRAPHY

Bringéus, N-A, ed. *Arbete och Redskap. Materiell folkkultur på svensk landsbygd före industrialismen [Work and Equipment: Material Folk Culture in a Swedish Rural Parish before the Industrial Period]* (1970), Lund, 1979.

Fenton, A. Farmers' diaries and their interpretation. In Gailey, A, ed, *The Use of Tradition: Essays presented to G B Thompson*, Ulster Folk and Transport Museum, 1988, 123–30.

Fenton, A. *The Northern Isles: Orkney and Shetland* (1978), East Linton, 1997.

Fenton, A. *Scottish Country Life* (1976), East Linton, 1999.

Firth, J. *Reminiscences of an Orkney Parish*, Stromness, 1920.

Hansen, W. *Arbeit und Gerät in volkskundlicher Dokumentation [Work and Equipment in Ethnological Documentation]*, Münster, 1969.

Heilfurth, H, Weber-Kellermann, I. *Arbeit und Volksleben [Work and Folklife]*, Göttingen, 1965.

Højrup, O. *Landbokvinden. Rok og Kærne. Grovbrød og Vadmel [Country Women: Spinning Wheel and Churn, Rough Bread and Frieze]*, Copenhagen, 1964.

MacTaggart, J. *The Scottish Gallovidian Encylopedia* (1824), 2nd edn, London and Glasgow, 1876.

Marwick, H. *The Orkney Norn*, Oxford, 1929.

Omond, J. *Orkney 80 Years Ago*, Kirkwall, 1911.

Thomson, E P. *Customs in Common*, London, 1991.

19 Working from Home

CHRISTOPHER A WHATLEY

From earliest times, the home has been a place of production as well as reproduction and shelter. The division of home and paid work is a comparatively modern development. The home itself or an adjoining workshop was where much if not most paid employment for the production or sale of goods was carried on until the later eighteenth and early nineteenth centuries. This is not to overlook the existence of workplaces beyond the home in medieval and early modern Scotland, such as saltworks, coal mines, bloomeries (for iron production), glassworks and the 'proto-factories' which were scattered through the country. Nor is it to ignore those households where the home continues to double as a place of work. A fairly sharp break, however, occurred with industrialisation and the emergence of large centralised water and steam-powered workplaces, primarily mills, factories, ironworks, shipyards, workshops and, later, bigger retail outlets like department stores and offices. The growth of the non-manual, services sector from the later nineteenth century has been the single most important factor in drawing women into employment outside the home. Developments in transport, notably the railway and later in the nineteenth century the horse-drawn and then the electric tram, hastened the growth of a physical division between residence and place of work.

More and more businesses were concentrated in the urban centres, which were becoming increasingly unpleasant owing to smells, noise, lack of amenity and in some places overcrowding. Accordingly, the middle classes sought to live at some distance from the workplace, housed in detached and semi-detached villas constructed by speculative builders in the suburbs. In due course they were followed out or even leapfrogged by the better-off sections of the working classes, in tenements and terraced housing. For most people the home was to become a refuge from the harshness and impersonality of the workplace, idealised by the Victorians and sought – but not always experienced – as an oasis of domestic calm. The focus of this chapter, however, is the home as workplace; that is those dwellings in which members of the household both resided and engaged in various forms of paid work other than domestic service.

The heyday of paid employment within the domestic environment was almost certainly between the 1730s, when the Scottish economy began its uncertain march towards industrialisation, and the 1840s. It was then that the ideologies of 'separate [gendered] spheres' and the sanctity of the home as a place of male 'peace, seclusion and refuge', as well as the growing importance of the house as status symbol, combined with changing labour demands and the rise of heavy industry to curtail such activity.[1] This was certainly the case where the official figures relating to female occupations are concerned. However census data invariably underestimated the extent to which women, – particularly those who were married, were employed in their own homes. Indeed most of this activity, such as childminding, sewing and knitting, which lay on the fringes of the 'black economy', was unrecorded.[2] The dominant male view in the nineteenth century, underpinned by the drive for a 'family wage', as well as the early and mid Victorian 'cult of the home', was that men's natural place of employment was outside the home while that of women was as devoted mother and wife within it.[3] Paid work from the home, however, continued to be carried out through the nineteenth and twentieth centuries. Throughout all time, and Europe wide, paid (and indeed non-paid) home-based work has always been more likely to be the preserve of the female and the child rather than the adult or adolescent male. It has also had a marked seasonal component. Where home work was carried out in the countryside, the agricultural calendar usually – but not always – played a large part in determining the extent to which home-based female labour was available. Other factors included the women's age and marital status, as well as the numbers and ages of her children. In addition, and as well as being poorly paid, such work was frequently temporary or casual.

THE MEDIEVAL AND EARLY MODERN PERIOD

In both town and country prior to the Industrial Revolution, and wherever it was done, work was invariably conducted on a small scale. Only exceptionally were more than a dozen or so workers congregated in a single location; in fewer than a handful of collieries in the late seventeenth century, for example, or on some construction projects like the royal palace improvements under James IV and V, and the burst of castellated house building by the nobility and lairds between 1480 and 1560. The Newmills woollen manufactory in East Lothian, which in 1683 had around 700 servants, was in terms of its size a highly unusual, factory-like enterprise, although even here some spinners and weavers worked in their homes, using rented looms and wheels.[4]

It should not be supposed that the only alternative to unusual works like this in medieval and early modern Scotland was the family, household or small extended-family workforce. Although the twelfth-century *Burgh Laws* restricted bakers to employing three servants, in the mid sixteenth century six Edinburgh bakehouses were employing between six and nine workers.[5] It has been estimated that some 4,000 water-powered corn mills were in use in Scotland by the early eighteenth century, as well as other modest-sized, non-domestic workplaces such as fulling mills. Salt manufacturing, which flourished in the Forth region from the late sixteenth century, required at least three people to fuel and tend each saltpan. It was a twenty-four hour process, and in a few locations – in the west, at Saltcoats in Ayrshire for instance – saltworkers, usually the wives and children of the male 'master' salters, resided in the panhouse.[6] In this case, rather unusually, the dwelling house was an appendage of the workplace.

Within household units of production, however, whatever their size, all members, old and young, female as well as male, were required to contribute. This was to sustain what in Scotland, prior to the 1760s, was invariably a low or at best a modest standard of living.

In the medieval and early modern period, most Scots lived and worked in the countryside. The bulk of rural dwellers – the sub-tenants and cottars and their families, who were effectively peasants – were largely self-sufficient. Part of the price of their smallholdings was usually several days' labour. Most transactions, primarily the payment of rent, were conducted in kind, even though the commodities often had a money value. Cash was required only for the minimal additional needs of the poorer sorts, to buy salt for example. Even rural craftsmen such as tailors were often paid in kind. Shetland wool knitters continued to be paid in kind, in a variant of the truck system, until the later nineteenth century.[7] For the earlier period, however, the concept of a stable money income is inappropriate; this was a world in which opportunities to garner whatever food was available, from a variety of sources, dominated daily life.[8] Although rural Scotland had become increasingly monetised during the thirteenth and fourteenth centuries, those affected were mainly the aristocracy and gentry.[9] There were rural dwellers, however, who held no land: labourers and skilled agricultural workers who were paid partly in money as well as in kind and attended the burgh fairs mainly to purchase foodstuffs, clothing, pottery and other essentials. They could also offer modest, home-produced goods for sale. Rural demand for urban products does appear to have risen over time, notably in the three quarters of a century after 1450.[10] It was only in the sixteenth and seventeenth centuries, however, that the money economy was driven downwards to include all sub-tenants. By around 1700 only 18 per cent of mainland Scots were more than 12 miles (19 km) from a market centre, numbers of which had mushroomed in the previous four decades.[11]

Nevertheless, in the preceding centuries there were those in the countryside for whom the home was also the principal place of paid work. Primarily these were craftsmen, mentioned above, who supplied many of the needs of the rural community.[12] Some were cottars and most had a small portion of land which they cultivated and used for animal husbandry. Apart from tailors, they included weavers, smiths and shoemakers, some of whom also travelled to provide for their customers by working away in their temporary employers' homes. On the Atholl estate, Perthshire in 1708, for example, it was customary for country dwellers to prepare leather, which was worked up into shoes by a travelling shoemaker in return for an agreed quantity of bere.[13] For most ordinary people, however, historians have pitifully little hard evidence with which to reconstruct lifetime employment patterns, let alone comment authoritatively about the location of work. What is reasonably certain, though, is that prior to the nineteenth century (and sometimes well into it) it is deceptive to associate individuals with a single means of making a livelihood. Multiple occupations were the norm.

It must be assumed that the bulk of paid female work in the countryside in medieval and early modern times was carried on outdoors, assisting during the harvest, for instance, although, as in Shetland well into the nineteenth century, knitting was done at the same time as agricultural tasks.[14] Such contributions to the domestic or household economy were crucial and frequently had a money value. On the smaller family-run farms which typified the south west, dairy farming depended to a substantial degree upon the ability of female members of the household to make butter and cheese and, if there was a town nearby, take it there for sale (see Fig. 23.2), as in other parts of the country where locally available commodities such as coal and fish were transported and sold by women.[15] In the Highlands and Islands too, women working in the domestic environment made butter and cheese. Domestic distilling for local consumption had been carried on in the same region for centuries and from the 1760s until the imposition of the Excise Duty Act of 1823, demographic expansion, urbanisation and industrialisation in the Lowlands generated a massive boom in illicit whisky production which occupied both men and women on a seasonal basis.[16]

Almost all women living on the land spent some time spinning flax by hand for the use of their families, but by the end of the seventeenth century there were large parts of the country in which women were working for the market from their homes for at least some of their time, and particularly during the summer and in the evenings, to supply the growing demand for linen yarn. The rate of expansion of linen exports in the century prior to 1700 was phenomenal.[17] The lowland part of Perthshire, then Scotland's most populous county, was probably the premier flax-spinning district, but many other parts of lowland Scotland were also involved: in 1726 one visitor remarked that 'all the poor people in the west of Scotland spin'.[18] This was prior to the great expansion of the linen trade which occurred from the end of that decade and which resulted in an extension northwards and west of hand spinning, using wheels rather than the less productive distaff. Aberdeenshire, on the other hand, could claim to be the main centre for rural plaid production. One contemporary estimate claimed that as many as 20,000 people were making their living from this trade in the north east in 1624. Most appear to have been sub-tenants whose landholdings were insufficient to maintain their families and who required therefore to turn to part-time household production – carding, weaving and spinning tasks provided by Aberdeen-based merchants – to make ends meet.[19] Woollen-stocking manufacture replaced plaid production, which declined after around 1639. By the 1790s as many as 900,000 pairs of stockings were being made in the best years, most of which were knitted by an estimated 30,000 home workers, mostly females from seven years old, boys and old men. In the north east, landholdings tended to be smaller than in the Lowlands and improvement came late, a situation which generated a large pool of low-cost labour. The lower the real income, the higher was the likelihood that the females of the household would take on additional part-time paid employment. The precise contribution of women and girls to the economy of the early modern rural household, however, is something about which Scottish historians still know too little.[20]

BURGHS

Although the distinction between rural and urban society should not be too sharply drawn – burgh lands, for instance, used both for arable and pastoral farming, were often extensive – it is clear that in terms of both monetisation and commercialisation the 70 or so burghs of late medieval Scotland were considerably in advance of the countryside.[21] It was in and around the burghs, too, that most manufactures were made and where domestic production for the market was concentrated, although, as has been seen, this was not to the exclusion of non-urban locations.

The backlands of the burgage plots provided the main sites for urban industry. Archaeological evidence emphasises the dual role of the household's buildings in the medieval burgh, as a place in which to live and work. Workshops were either located beside or in the same buildings in which family members as well as apprentices and servants slept, ate or socialised.[22] Although the occupational structures of Scotland's burghs differed, a fairly wide range of crafts and occupations was to be found virtually everywhere, serving the needs of both town and country. Typically the list would include baxters (bakers), brewsters (brewers), cordwainers (shoemakers), dyers, fleshers, fullers (who thickened and shrank woollen cloth), hammermen (smiths and metal workers), skinners, spinners, tailors, tanners and weavers. Although much tailoring work was itinerant, tailors also did a great deal of work from home, drawing on the labour of other family members. Unlike some trades, or professions such as lawyers, who in Edinburgh inhabited the central part of the burgh near to the courts and parliament building, tailors, as with bakers, showed relatively little sign of geographical concentration within particular towns but were scattered throughout the burghs.

The home was not the only site of industry in burghs, of course. In textile production some processes could be carried on in the dwelling, mainly spinning and to a lesser extent weaving, but scutching and heckling were more likely to be carried out in workshops close to running water. Neither in brewing was the process entirely home-based. Archaeology confirms the existence of the separate trade of malting, which required space and drying kilns, from at least as early as the fifteenth century. There were also processes for which the home was an inappropriate place of work. Locally made pottery, for example, which grew in popularity during the fourteenth century, was usually made at potteries located on the outskirts of the town, where there was space for hot kilns (a fire risk), and there were sources of water and clay nearby.

Nevertheless much work was undertaken in a domestic setting and some craftsmen, as well as merchants, erected lockable wooden booths (luckenbooths) which abutted onto the fronts of houses lining the principal streets. The booths were used both as workshops as well as retail premises. Over time these structures were incorporated permanently into the buildings against which they were set. The importance of the home as a place for retailing goods was reinforced in Dundee in 1550 when hucksters, small-scale sellers of petty goods, were ordered to operate only from their own doors or windows or at the top of closes where they lived.[23] In Edinburgh, books, pamphlets and broadsheets could be bought from Thomas Keir in 1707, either on the street or 'at his house on the old Posthouse Close'.[24] But through to the end of the eighteenth century and even beyond, for most businessmen, merchants and traders, manufacturers and professionals, the dwelling place doubled as a place of work. Deals were done there, apprentices slept upstairs and goods were stored in lofts and cellars. Lawyers or writers in Edinburgh were not averse to using either their dwelling houses or chambers as auction rooms for property.

As this suggests, in some instances the dwelling housed service industries. Both men and women let rooms or even entire properties, with widows being particularly prominent in this relatively secure type of business. Letting could become more or less full-time, with lodgers being provided by women with other services such as washing or even the making of clothes,[25] and in an indefinable number of cases, the comforts of whoredom, as in the case of Martha Mundell in Dumfries in 1735 who earned the contempt of her neighbours by keeping a 'lewd and disorderly house'.[26]

Women played a major role in urban household income generation, although currently the evidence for this is more visible at the higher levels of society.[27] Lower down the social scale the picture is hazier. In Edinburgh there is evidence that occupational specialisation and professionalisation by males from the later seventeenth century curtailed women's opportunities for paid work within and outside the home. Even so, both beforehand and into the eighteenth century, females 'played a significant part in general household economies', not only as domestic servants and part-time assistants to their husbands (whom married women often replaced on the husband's death, at least until a son became old enough to take over the business – a 'culture of inheritance') but also independently.[28] Clear evidence of an aspiration for independence comes from Dumfries in the early eighteenth century where it was noted (by male officials) with some concern that women were leaving their employment in service and had 'betaken ymselves to chambers' (apparently to work as prostitutes) to 'ye Scandall of the place'.[29]

Again, it is a mistake to assume that *all* women's work was either home based or home related. Many females lived on their own or at least lived outside the family home; from the seventeenth century young single girls became apprentices in sewing schools or were shop workers. In addition, some women managed to retain a degree of independence *within* the home, where economic relations with the husband are more accurately represented as a partnership than as a form of subordination. Women's work of this sort could be concerned with manufacture, retailing or the provision of personal services. It included occupations such as the butterwife, fruitwife, seamstress and lace maker, all involving tasks which could be done partly or largely at home. So, too, could graveclothes making, which was carried on by wives and widows from a broad social spectrum in Edinburgh prior to the end of the eighteenth century when such goods were more likely to have been readymade by specialist (mass) producers. In Edinburgh, too, substantial numbers of women owned property, shops as well as accommodation. Small but increasing numbers of women in other burghs were involved in retailing from shops, not only as extensions of manufacturing premises but also as stand-alone outlets. Some women provided attention during sickness. Wet nursing, which in Scotland as elsewhere in Western Europe flourished during the seventeenth and eighteenth centuries, was initially more usually conducted in the child's parents' home. Although the nobility and upper classes continued to bring a wet nurse in, lower down the social scale – where funds and accommodation were tighter – the tendency was to send the infant out.[30] This was often to the country, where the air was healthier.

Brewing was to become a larger-scale business and increasingly the preserve of males from some time in the sixteenth century. Nevertheless, small-scale brewing and ale selling, usually part-time and informally – thereby evading burgh regulations concerning quality, price and against forestalling – had been carried on from the home for centuries, most often by married women who could call on the assistance of other household members when required.[31] One list of brewers in medieval Aberdeen contains only women's names.[32] Most of the 152 females listed were married, but twenty-three were named in their own right. In addition housewives often sold surplus home-baked bannocks and oatcakes, considered to be a lower order of produce than the bread which was usually baked by males belonging to the baxter craft.[33] This form of domestic production, of cakes, biscuits, jams, jellies and sweets, provided the background in Dundee for the Keillor family's venture into marmalade making at the turn of the nineteenth century.[34] Candle making was another activity which could be carried on, by males and females, sporadically in the home, in accordance with demand and the availability of tallow.

HOME WORK AND THE INDUSTRIALISING ECONOMY

The strengthening of the Scottish economy in the eighteenth century, uncertainly from the 1730s but strongly after *c* 1780, was in large part dependent upon an expansion of home work. This form of economic development, proto-industrialisation, allowed entrepreneurs to expand low-wage production without bearing the heavy capital costs of plant and machinery. Market forces and the efforts of bodies such as the Board of Trustees for Fisheries and Manufactures, established in 1727 partly to encourage loyalty to the British state and the Hanoverians through the provision of employment, brought thousands of previously underemployed and unemployed inhabitants of rural and semi-urban Scotland into the cash economy.[35]

In 1747, for example, the British Linen Company, on learning that the womenfolk belonging to some 400 families in the Leadhills lead-mining district, Lanarkshire, were 'quite idle' and wholly dependent upon the 'labour of their men' for subsistence, gave out spinning wheels and provided training in flax spinning for all women between the ages of ten and sixty. The hope was that each would earn between 16d and 2s 4d per week, although earnings were higher in some regions later in the century.[36] Within two or three decades wheels and flax had been distributed throughout much of Scotland, west as far as Iona in the Hebrides and north to Orkney and even Shetland (Fig. 19.1). Tenants recruited by the Grants of Monymusk in Aberdeenshire were required to bring with them 'at least two spinners', who, when not employed at agricultural work, were expected to spin an agreed quantity of yarn.[37] By mid century perhaps as much as four fifths of the female population of Scotland earned a living primarily by spinning flax, with every household having at least one wheel and some as many as four or five.[38] Even women from the middling and upper ranks spent time on what was a quasi-patriotic venture, the end product of which, in the form of

Figure 19.1 Woman spinning, Harris, 1937. Domestic hand spinning, mainly carried out by females, was the most common form of aid work carried out in the dwelling from the seventeenth century until the early nineteenth century, and carried on after that mainly in the north and west of Scotland. Source: Courtesy of the University of St Andrews Library.

bales of mainly coarse varieties of linen cloth, was despatched south to London and onwards to the slave colonies of the West Indies and the tobacco plantations around the Chesapeake.

The effects of outworking of this kind were mixed. There were the benefits of a cash income and the possibility of making an occasional frivolous purchase such as acquiring a fashionable item of clothing (a practice often condemned by moralising observers). There is evidence, too, that small groups of spinners in Angus (formerly Forfarshire) – a mother and her daughters and servants for example – derived pleasure, self-esteem and a sense of gendered solidarity from work songs and the competition between them to spin the most and best yarn. A certain match-making potential accompanied spinning; those capable of spinning the most were in the greatest demand from males.[39] That opportunities for attracting suitors would be lost was given by one male contemporary as the reason why females in Forfarshire initially resisted employment in the new spinning mills. Against this, however, and the faint traces of a craft pride which are conveyed in some work songs, must be set the harder evidence of increasing effort and longer hours of home work over the course of the eighteenth century, and the expression in poetry and song of regret at the growing tedium and time-dictated rhythms of spinning which had to be accommodated alongside the other fixed demands on a woman's time.[40] A well-known song in the later eighteenth century was 'The Weary Pound of Tow', which in one version includes the couplet 'The spinning, the spinning, it gars my heart sob, When I think upon the beginning o't.' In one early nineteenth-century chapbook variant of the popular spinning song 'The Rock and the Wee Pickle Tow' there is clearly an ironic recognition of the link between the acquisition of better clothes and the necessity of work as a spinner:

> An' we maun hae pearlins [trimmings], an' mabbies [mop caps]
>   an' cocks [headdresses]
> An' some other things that ladies ca' smocks:
> An' how we get that, gin we tak na' our rocks [distaffs],
> An' pow what we can at the spinning o't.[41]

Straw plaiters in nineteenth-century Orkney also frequently worked in company, as did Shetland wool carders in what was called a 'cairdin', a type of communal gathering which carried on into the early twentieth century.[42] It should not be assumed that domestic paid work merely supplemented male income from other activities. This is most striking in parts of Aberdeenshire where at the time of the writing of the *OSA* several parish ministers drew attention to the primacy of wages derived from flax spinning and stocking knitting for the survival of cottar households.[43] Elsewhere, too, female earnings were crucial in raising living standards in the final third of the eighteenth century.[44]

Domestic handloom weaving – at first mainly the province of men and boys – also expanded, as much as four fold between the 1730s and the end of the century. While some weaving was organised in workshops and proto-factories which employed a number of weavers under one roof, most weavers worked in their own houses or adjoining sheds, taking webs distributed by manufacturers and their agents. Frequently the weaver's dwelling was situated above the 'shop' or loomshed at ground level. Weaving was not necessarily carried out in isolation: it was not unusual for a master weaver to employ a journeyman or boy. A father might employ his sons, each with his own loom. George Smith, of Camlachie near Glasgow, had six looms at his dwelling, five of which were worked by his sons.[45] The downside of living and working within a face-to-face community in this instance was that the whereabouts of Smith's looms were well known and accessible to a sizeable armed mob who gathered outside his premises to whip an effigy dressed in clothes identical to those worn by Smith – including his white nightcap – on the grounds that the master weaver and his sons were suspected of working below the prices for webs agreed by the weavers' association.

Even at Carron ironworks, Stirlingshire, considered by some to signal the birth of the modern economy of Scotland, rod iron was distributed to nailers who were employed in their homes in clustered communities across central Scotland, in places such as St Ninians and Kilsyth, Stirlingshire, and Dysart and Pathhead, Fife, although some men and their apprentices worked in nearby 'shops'.[46] In weaving, however, over time and notably from the turn of the nineteenth century, deskilling and an overcrowded labour market sucked women and more children into the trade. By the 1830s around a quarter of the looms in Glasgow and Airdrie, Lanarkshire, were operated by women and child assistants.[47] Elsewhere the proportion may have been as high as one third. Older men were also employed at hand weaving. At its

Figure 19.2 Weaving at Kirriemuir, Angus. Source: SLA 69L(1) C.526(14).

peak in the 1830s handloom weaving in Scotland employed up to 83,000 people, although thereafter the number fell sharply with domestic production surviving longest in fine muslins and shawls and fancy fabrics in the Irvine valley, in Ayrshire, and Paisley respectively (Fig. 19.2).

In the east, too, where linen was the principal cloth, the typical webster as late as 1840 was a domestic outworker, although exceptions were Aberdeen, Dunfermline and Montrose, Angus, where factory working was becoming more common.[48] Virtually everywhere, however, conditions were worsening while the daily hours necessary to eke out a living rose from a typical nine in the 1790s to 13, 14 and even 17 in a place like Kilbarchan, Renfrewshire, in 1834.[49] Wages had begun to fall from the end of the eighteenth century and increasingly destitution visited domestic weavers whose habitations tended to be small, damp and cramped. An exception appears to have been the Borders, where conditions amongst the domestic woollen handloom weavers were judged by government observers to have been relatively good, even though factory working was becoming the norm (Fig. 19.3). In the same region hosiery work, carried out on handframes, was also concentrated in frameshops which could house up to a dozen separate frames, but as the knitters maintained considerable control over the pace of work their work can be considered as 'semi-domestic'. In the outlying villages independent domestic knitters managed to survive on their own until around mid century when growing numbers were called in by masters anxious to centralise and control production.[50]

Figure 19.3 Weaving at Newcastleton, Roxburghshire, 1907. Source: SLA 69L(2) C.5411.

Proto-industrial production, however, had its limitations for the rising class of merchant manufacturers. Domestically produced yarn and cloth were too often of a poor and uneven quality, and long lines of ill-secured credit exposed employers whose materials their workers frequently embezzled. In 1752 Helen Robertson in Orkney was to stand bareheaded at the cross in Kirkwall after being been found guilty of fraud by taking delivery of six pounds of dressed flax for spinning but returning yarn into which had been spun tow and other inferior lint. The 'scandalous workmanship' of the nailers at St Ninians was a cause of deep concern for the Carron partners in 1770, who also disliked the fact that no nailmaking occurred during the harvest period. With the invention of Arkwright's water frame and Crompton's mule, cotton spinning moved into water and steam-powered mills, the numbers of which soared during the last two decades of the eighteenth century. Flax spinning became mill based too, notably in Dundee and outlying towns and villages such as Montrose and Coupar Angus, Perthshire. Fortuitously, by the 1790s, domestic hand spinners in the southwestern Lowlands had begun to find work in muslin flowering, a trade which employed around 25,000 females by the early 1850s.[51] But the transition to mill and factory work was uneven. When, in the early 1790s, David Dale's New Lanark mills, Lanarkshire, were employing 1,157 workers, ten per cent of the females picked cotton in their own homes while the firm also had another 345 people engaged as outworkers, winding and weaving.[52] Over time, however, hand spinning became largely the preserve of the few and the desperate.

UNEVEN DEVELOPMENT

Thus the appearance of the centralised workshop on a wide scale did not lead to the immediate demise of domestic production. Until relatively recent times, many women in the Scottish countryside, wives of the Border 'hinds' (farm workers) for example, continued to manufacture socks, stockings and jerseys for household members, in addition to undertaking an exhausting catalogue of repetitive and dull but essential tasks. There was little or no time for leisure, even for reading.[53] In some trades there was an extension of such activity in the early decades of the nineteenth century – embroidery workers for instance, whose skills could not yet be reproduced by machine. In Shetland, home-based lace knitting began to flourish from the 1830s, and hosiery production, which had been threatened by mechanised competition from the south in the early nineteenth century, survived through the domestic manufacture of speciality products into the twentieth century.[54] Periodic surges in fashion-led demand for Fair Isle knitting – the first was in the 1920s – ensured the survival of outworking (see Fig. 9.2), and although machine-spun yarn and mechanised knitting had become commonplace by the 1960s and 1970s, domestic hand knitting (on a machine) and finishing were still being carried out in Shetland at the end of the twentieth century. While centralised workplaces (for straw plaiting) had appeared in both

Orkney and Shetland in the early 1800s and replaced linen making, by the 1840s most work of this sort was being carried out on a seasonal but more or less full-time basis in the homes mainly of female operatives for steadily reducing wages. By the 1870s, however, the trade had died out in the Northern Isles.

The poor in Victorian Scotland continued to scratch a living together, not infrequently using the dwelling place as place of work. Recently published research has revealed the sheer variety of jobs to which the poor turned their hand, although unfortunately not the numbers of people involved. One Fife woman worked a mangle provided by the parish kirk; others gathered bait and dressed lines or made nets (see Figs 42.2, 42.3), although within the numerous fishing communities on the east and north-eastern coasts these and related tasks were normally undertaken by the fishermen's wives.[55] Near Wick, Caithness, heather was twisted into ropes, while down the east coast of Scotland larger numbers of women wound bobbins for what seemed like endless hours at cruelly low rates of pay. Two shillings a week was the most a bobbin winder could hope to earn.[56] Later in the century, in New Pitsligo, Aberdeenshire, old women, living in the 'oot rooms' attached to the overcrowded single-storeyed family dwellings, made and sold, amongst other things, 'penny wabble', a mild ale[57] (Fig. 19.4).

Figure 19.4 Broom seller, Fife, c1850. The old and the poor often subsisted by making household commodities in their own homes and, as in this case, then selling the products of their labour on the doorstep or in the streets. Source: Courtesy of the University of St Andrews Library.

Figure 19.5 Basket making, Fife, 1950s. In some rural areas small-scale production like this was carried out into the twentieth century by rural craftsmen from a range of other occupations such as tailoring, smith work and shoemaking. Source: Courtesy of the University of St Andrews Library.

On the other hand, while industrialisation *tended* to separate home and paid work (although by no means did it eliminate *unpaid* household labour, demand for which grew, as did numbers of domestic servants), as has just been seen, the home has never entirely lost its place as the locus of some forms of income-generating activity, most of it ill-paid. Low pay and home work, however, were not always linked. Clergymen, for example, conducted much church business from their dwellings. Doctors too saw patients at home; writers and 'men of letters' also worked from home. Nor did the move out of the town centres to the suburbs happen overnight. The first generation of industrialists were reluctant to relocate far from their mills, factories and workshops. In Dundee the homes of the Keillor marmalade dynasty were never more than five minutes' walk from their works premises prior to the 1840s. With mill and factory production beginning later in Dundee than in Glasgow or Paisley, it was only after 1850 that Dundee's leading manufacturers began to move out to the suburbs of Broughty Ferry and Carnoustie.[58]

Evidence from lowland Perthshire reveals the existence of a fairly buoyant rural crafts sector in the nineteenth century. Many of these male tradesmen worked from home; the 1901 census included males in this category ranging from blacksmiths, through sawyers to weavers and wigmakers (Fig. 19.5). While some businesses struggled financially and consequently drew on the 'free' labour of family members, the picture generally is much more robust, with many women playing a significant part in prosperous family firms in the capacity of manager, bookkeeper, shop minder and financial expert. Unpaid work of this kind was crucial to the sustainability of the businesses concerned. Thus although the employment of many such females may often have been part-time and seasonal it was by no means secondary. The concept of partnership is applicable in fishing too, with mussel shelling and baiting of some 1,400 hooks on the line taking as much as nine hours a day of a woman's time, until the decline of line fishing from the 1940s.[59] With household duties in addition, such partnerships were rarely equally balanced. In addition, in many places females were also expected to sell the fish, as they did in Slains parish, Aberdeenshire, in the 1790s by carrying it on their backs into Aberdeen, Old Meldrum and the surrounding country towns. Where females were involved in rural production, they were more likely to be engaged in semi- or unskilled tasks.[60]

THE LAST CENTURY

It was in the southern towns that the worst conditions for domestic workers were to be found. Employers, anxious to keep fixed costs low and whose markets were cyclical or seasonal, had long made use of home workers. In large part this is what had attracted merchant manufacturers in the yarn and cloth trades to the domestic system. But as has been seen already, mechanisation, of spinning from the 1770s, and hand weaving from the early nineteenth century, did not put an end to urban outworking. Often overlooked, and certainly under researched in Scotland, what were frequently described as the 'sweated' trades accounted for over one third of Glasgow's female employment in 1911. The figure understates the real extent of sweating in Glasgow, the Scottish 'centre' of casual, marginal, paid, home employment.[61]

Needlework was the most common form of home-based paid work, followed by laundry and washing, as well as a myriad other activities such as brushmaking, paper-bag or box making and umbrella assembling. Piecework was the norm. Most of the women concerned were married, perhaps with children to support and a husband's wage to supplement, although unlike their first – relatively independent – foray into the labour market, which would have been brought to a close on marriage, the workplace was now the home rather than the mill or factory and the work 'part-time', although often involving long hours at rates of pay which diminished over time. Sack sewers in Dundee in the early twentieth century, for example, were only able to earn one third of what they could obtain in the 1870s. The best a sewer could get was 1s 6d a day, that is if domestic help could be obtained, 1s without. There were women, however, whose husbands were unable to work, and widows too, who wished to remain free of the poorhouse. Single women who worked from home tended to be physically disabled or were judged to be psychologically unfit for regular, supervised employment. Dundee's 250–300 remaining sack sewers around 1903 were described as 'very poor almost destitute women', although before World War I men desperate for work in a city which, unusually, favoured women workers, were offering to collect and sew the 60–68lb (27–31 kg) bundles.[62] There, as in Glasgow, mechanisation, principally in the shape of the sewing machine, was the main cause of long-term deterioration in wages and conditions, with overseas competition adding to the home workers' misery. Poverty experienced in working-class neighbourhoods in Glasgow in the 1930s inspired amongst women a stream of enterprising schemes to supplement the household budget. Making and selling toffee apples and puff candy and collecting and selling jam jars are examples. Collective activity – sharing and 'saving through ménages, baking and lending table covers and good dishes for weddings, birthdays and christenings' – may not have generated additional income, but it reduced household expenditure.[63]

In the last decades of the twentieth century, underpinned by the communications revolution and the appearance and increased use of the home computer, Internet and email (see Fig. 10.5), and assisted by changing attitudes on the part of some employers who no longer conflated attendance at the office or workplace with greatest efficiency, working from home appeared to be on the increase. Men and women can now work at long distances from the head office, and enjoy the benefits of living in the rural north and west for example, provided that their salaries remain near to southern levels. Less salubrious, however, are the conditions of some immigrant workers and others, who in unknown numbers, and under conditions which have thus far eluded the gaze of academic historians, are still employed in twenty-first century equivalents of Dickensian 'sweating'. These include the self-employed as well as 'non-professional' workers – childminders who work from home, women who take in ironing on a casual basis or those who operate as catalogue agents. Overall numbers of home-based workers, however, are almost certainly fewer than they were. While levels of pay, job status and opportunities for both married and unmarried women in Scotland are still lower than for men, a feature of the second half of the twentieth century was the sharp increase in the number and proportion of female workers outside the home, although the bulk of these (90 per cent of 500,000 in the 1990s) were employed part-time. For most Scots home and work are separate. As with earlier periods, however, there is still much investigative work on the internal life of the modern Scottish household and the distribution of labour within the home to be carried out.

NOTES

1. Tosh, 1999, 13, 27–50; see also Chapter 28 Hierarchy and Authority.
2. Gordon, 1991, 18–19.
3. McIvor, 1998, 163–4; Tosh, 1999, 5–6.
4. Hatcher, 1993; Glendinning, et al, 1996, 6–21; Marshall, 1980, 141–4.
5. Lynch, 1988, 278.
6. Whatley, 1987, 103.

7. Fenton, 1997, 463.
8. Mitchison, 2000, 98.
9. Wormald, 1981, 50.
10. Lynch, 1987, 10.
11. Whyte, 1979, 185.
12. Sanderson, 1982, 182–3.
13. University of Edinburgh Library, Special Collections, Athole Papers 1745–6, Dc.1.37.1, Proposal for Improvement of the Regality of Atholl, 1708.
14. Bennett, 1987, 56.
15. Fenton, 1999, 154–62; for the latter topic see also Chapter 33 'Time Out' Work.
16. Devine, 1994, 126–34.
17. Smout, 1963, 233.
18. Paton, 1961, 3.
19. Tyson, 1989, 68–71.
20. Ewan, Meikle, 1999, xxviii.
21. Grant, 1984, 69.
22. Ewan, 1990, 25.
23. Torrie, 1990, 44.
24. *Edinburgh Courant*, 30 May 1707.
25. Sanderson, 1996, 42.
26. Houston, 1994, 80.
27. Dingwall, 1999, 152–60.
28. Dingwall, 1994, 201–8; Mann, 1999, 136.
29. Dumfries Archive Centre, Dumfries Council Minutes, A2/8, 1704–9, 6 March 1704.
30. Marshall, 1984, 47–8; see also Chapter 20 Work in the Homes of Others; Chapter 25 Childcare and Children's Leisure.
31. Ewan, 1999, 127; Mayhew, 1999, 147.
32. Mayhew, 1996–7, 16.
33. Mayhew, 1999, 146.
34. Mathew, 1998, 2–3.
35. Whatley, 2000a, 96–141.
36. Durie, 1996, 30–1; Gibson, Smout, 1995, 354–5.
37. NAS, Grant of Monymusk Papers, GD 345/1025, Agreement, Robert Anderson, Weaver, with Spinners, Spring 1749.
38. Durie, 1979, 159.
39. University of Dundee Archives, MS 11/5/14, History of Flax Spinning from 1806 to 1866 by Charles Mackie, 16–18.
40. Simonton, 1999, 70–1; Whatley, 2000, 152–5.
41. Whatley, 2000a, 223; Whatley, 2000b, 154–5.
42. Fenton, 1997, 272.
43. *OSA*, XV (1973), 28.
44. Gibson, Smout, 1995, 355–6.
45. *Glasgow Courier*, 11 September 1824.
46. NLS, Acc. 5381, Bundle 28 (1), Prices of Nails with Remarks on the Nail Trade, December 1769.
47. Murray, 1978, 28.
48. Murray, 1994, 226.
49. Murray, 1978, 152–3.
50. Gulvin, 1973, 177–8; Gulvin, 1984, 42–3.
51. Collins, 1988, 243–6.
52. *OSA*, VII (1973), 463; see also Chapter 21 Reading and Study.
53. Smout, Wood, 1990, 21.
54. Bennett, 1987, 57–61.

55. Gray, 1978, 13.
56. Mitchison, 2000, 222.
57. Smout, Wood, 1990, 17–18.
58. Tosh, 1999, 16–17; Mathew, 1998, 13–15; Miskell, 2000, 64–5.
59. King, 1993, 58–9.
60. Young, 1991, 54–8.
61. Albert, 1990, 159.
62. Walker, 1979, 91; University of Dundee Archives, MS DA 890, MS, D Lennox, Working Class Life in Dundee for 25 years, from 1878 to 1903, c 1906, 189–91.
63. McGuckin, 1992, 206–14.

BIBLIOGRAPHY

Albert, A J. Fit work for women: Sweated home workers in Glasgow, c.1875–1914. In Gordon, E, Breitenbach, E, eds, *The World is Ill-Divided: Women's Work in Scotland in the Nineteenth and Early Twentieth Centuries*, Edinburgh, 1990, 158–77.
Bennett, H. The Shetland hand-knitting industry. In Butt, J, Ponting, K, eds, *Scottish Textile History*, Aberdeen, 1987, 48–64.
Collins, B. Sewing and social structure: The flowerers of Scotland and Ireland. In Mitchison, R, Roebuck, P, eds, *Economy and Society in Scotland and Ireland, 1500–1939*, Edinburgh, 1988, 242–54.
Devine, T M. *Clanship to Crofters War: The Social Transformation of the Scottish Highlands*, Manchester, 1994.
Dingwall, H. *Late Seventeenth-Century Edinburgh: A Demographic Study*, Aldershot, 1994.
Dingwall, H. The power behind the merchants? Women and the economy of late seventeenth-century Edinburgh. In Ewen, E, Meikle, M M, eds, *Women in Scotland, c.1100–c.1750*, East Linton, 1999, 152–62.
Durie, A J. *The Scottish Linen Industry in the Eighteenth Century*, Edinburgh, 1979.
Durie, A J, ed. *The British Linen Company, 1745–75*, Edinburgh, 1996.
Ewan, E. *Townlife in Fourteenth-Century Scotland*, Edinburgh, 1990.
Ewan, E. 'For whatever ales ye': Women as consumers and producers in late medieval Scottish towns. In Ewan, E, Meikle, M M, eds, *Women in Scotland, c.1100–c.1750*, East Linton, 1999, 125–35.
Ewan, E, Meikle, M M, eds. *Women in Scotland, c.1100–c.1750*, East Linton, 1999.
Fenton, A. *The Northern Isles: Orkney and Shetland*, East Linton, 1997.
Fenton, A. *Scottish Country Life*, East Linton, 1999.
Gibson, A, Smout, T C. *Prices, Food and Wages in Scotland, 1550–1780*, Cambridge, 1995.
Glendinning, M, MacInnes, R, MacKechnie, A, eds. *A History of Scottish Architecture*, Edinburgh, 1996.
Gordon, E. *Women and the Labour Movement in Scotland, 1850–1914*, Oxford, 1991.
Grant, A. *Independence and Nationhood: Scotland, 1306–1469*, London, 1984.
Gray, M. *The Fishing Industries of Scotland, 1790–1914: A Study in Regional Adaptation*, Oxford, 1978.
Gulvin, C. *The Tweedmakers: A History of the Scottish Fancy Woollen Industry, 1600–1914*, Newton Abbot, 1973.
Gulvin, C. *The Scottish Hosiery and Knitwear Industry, 1680–1980*, Edinburgh, 1984.
Hatcher, J. *The History of the British Coal Industry, vol 1: Before 1700*, Oxford, 1993.
Houston, R A. *Social Change in the Age of Enlightenment: Edinburgh, 1660–1760*, Oxford, 1994.
King, M H. Marriage and traditions in fishing communities, *ROSC*, 8 (1993), 58–67.
Lynch, M, ed. *The Early Modern Town in Scotland*, London, 1987.
Lynch, M. The social and economic structures of the larger towns, 1450–1600. In Lynch, M, Spearman, M, Stell, G, eds, *The Scottish Medieval Town*, Edinburgh, 1988, 281–6.

Mann, A J. Embroidery to enterprise: The role of women in the book trade of early modern Scotland. In Ewan, E, Meikle, M M, eds, *Women in Scotland, c.1100–c.1750*, East Linton, 1999, 136–51.
Marshall, G. *Presbyteries and Profits: Calvinism and the Development of Capitalism in Scotland, 1560–1707*, Edinburgh, 1980.
Marshall, R K. Wet nursing in Scotland, 1500–1800, *ROSC*, 1 (1984), 43–51.
Mathew, W M. *Keillor's of Dundee: The Rise of the Marmalade Dynasty, 1800–79*, Dundee, 1998.
Mayhew, N. The status of women and the brewing of ale in medieval Aberdeen, *ROSC*, 10 (1996–7), 16–20.
Mayhew, N. Women in Aberdeen at the end of the Middle Ages. In Brotherstone, T, Simonton, D, Walsh, O, eds, *Gendering Scottish History: An International Approach*, Glasgow, 1999, 142–55.
McGuckin, A. Moving stories: Working-class women. In Breitenbach, E, Gordon, E, eds, *Women in Scottish Society, 1800–1945*, Edinburgh, 1992, 197–220.
McIvor, A. Women and gender relations. In Cooke, A, Donnachie, I, MacSween, A, Whatley, C A, eds, *Modern Scottish History, 1707 to the Present, vol 2: The Modernisation of Scotland, 1850 to the Present*, East Linton, 1998, 161–87.
Miskell, L. Civic leadership and the manufacturing elite: Dundee, 1820–70. In Miskell, L, Whatley, C A, Harris, B, eds, *Victorian Dundee: Image and Realities*, East Linton, 2000, 51–69.
Mitchison, R. *The Old Poor Law in Scotland: The Experience of Poverty, 1574–1845*, Edinburgh, 2000.
Murray, N. *The Scottish Handloom Weavers, 1790–1850: A Social History*, Edinburgh, 1978.
Murray, N. The regional structure of textile employment in Scotland in the nineteenth century: East of Scotland handloom weavers in the 1830s. In Cummings, A J G, Devine, T M, eds, *Industry, Business and Society in Scotland since 1700*, Edinburgh, 1994, 218–33.
Paton, H, ed. *Historical Manuscripts Commission. MS of Lord Polwarth, Vol V, 1725–80*, London, 1961.
Sanderson, E C. *Women and Work in Eighteenth-Century Edinburgh*, London, 1996.
Sanderson, M H B. *Scottish Rural Society in the Sixteenth Century*, Edinburgh, 1982.
Simonton, D. Gendering work: Reflections on home, skill, and status in Europe. In Brotherstone, T, Simonton, D, Walsh, O, eds, *Gendering Scottish History: An International Approach*, Glasgow, 1999, 63–83.
Smout, T C. *Scottish Trade on the Eve of the Union, 1660–1707*, Edinburgh and London, 1963.
Smout, T C, Wood, S. *Scottish Voices, 1745–1960*, London, 1990.
Torrie, E P D. *Medieval Dundee: A Town and Its People*, Dundee, 1990.
Tosh, J. *A Man's Place: Masculinity and the Middle-Class Home in Victorian England*, New Haven and London, 1999.
Tyson, R E. The rise and fall of manufacturing in rural Aberdeenshire. In Smith, J S, Stevenson, D, eds, *Fermfolk and Fisherfolk: Rural Life in Northern Scotland in the Eighteenth and Nineteenth Centuries*, Aberdeen, 1989, 63–82.
Walker, W M. *Juteopolis: Dundee and its Textile Workers, 1885–1923*, Edinburgh, 1979.
Whatley, C A. *The Scottish Salt Industry, 1570–1850: An Economic and Social History*, Aberdeen, 1987.
Whatley, C A. *Scottish Society, 1707–1830: Beyond Jacobitism, towards Industrialisation*, Manchester, 2000a.
Whatley, C A. Sound and song in the ritual of popular protest: Continuity and the Glasgow 'Nob Songs' of 1825. In Cowan, E J, ed, *The Ballad in Scottish History*, East Linton, 2000b, 142–60.
Whyte, I. *Agriculture and Society in Seventeenth-Century Scotland*, Edinburgh, 1979.
Wormald, J. *Court, Kirk, and Community. Scotland, 1470–1625*, London, 1981.
Young, C. Women's work, family and the rural trades in nineteenth-century Scotland, *ROSC*, 7 (1991), 53–8.

20 Work in the Homes of Others

JEAN AITCHISON

As far back as it is possible to know, it has been customary for people of high status in society to have others work for them in their homes. Maintaining a substantial household has always been both a means of enhancing status and a way of allowing masters to devote their time to various pursuits, which in the past could include warfare, hunting, attending the sovereign's court or administering justice. For women belonging to landed families, in both clan and feudal societies, their principal responsibilities were to marry in order to further the economic, political and dynastic aspirations of their fathers and their prospective husbands, to bear and rear the children of the union, and generally to act in their lords' interest. Such a woman's married life 'might be spent in her husband's family home, in subservience to her mother-in-law, but sooner or later she would have her own establishment'.[1] Although she lacked legal status, her husband delegated the running of the household and the rearing of the children to her and she would be served by other women in the household, who attended to her personal needs, delivered and nursed her children, spun, sewed and washed. Those who worked in such a household could expect, amongst other things, to receive their lord's protection in a world where life was often hard, short and brutal.

As their prosperity increased, merchants and craft-guild members of burghs could afford to employ others to serve them either as servants or, where the master craftsman had his home and business in the same building, as craft apprentices. In the countryside, where the majority of the population lived until agricultural improvement took large numbers of people off the land, young people found places of service in the households and on the farms of neighbouring tenant farmers or 'bonnet lairds'. 'Many women, whether or not they were wealthy, had maid servants. These were usually young female relatives willing to learn the domestic skills of the house by sharing in the work.'[2]

There was a further demand for servants with the growth of the professional classes in the seventeenth century, especially ministers as upholders of the established church and lawyers whose 'profession was already a large and important one, for society was conscious of the security implied in a charter or an instrument of sasine safely hidden in a great chest to give hope for the eventual recovery of property in cases of usurpation'.[3] The usefulness of lawyers in bolstering the power of the Crown and the passion of their patrons for litigation meant that men like Lord Auchinleck, a judge at the Court of Session and father of James Boswell, could afford to employ servants in both his Edinburgh town house, and at Auchinleck House on his Ayrshire family estate.[4]

Although Scotland was still not a rich country in the eighteenth century, there were those who became wealthy, often from their activities overseas as tobacco merchants, slave owners or Indian nabobs. Such men were frequently disposed to flaunt their wealth and commissioned splendid houses to be built, which needed a small army of servants to service them. Eighteenth-century aristocrats, with their enlightened ideas and influenced by grand tours and increased contact with England, also began to make great changes to their homes, which meant that their servants had to adapt to the requirements of the new establishments in which they worked.

As industry grew in the nineteenth century, no longer was it a case of the local miller employing a single maid, for the increasingly wealthy steel and shipping magnates, for example, could afford to employ sufficient domestic staff to allow their wives and daughters to live the social life expected of their class. The Victorian period was a time when even members of the lower middle class could aspire to employ a maid, and one can only pity the majority of those girls and women who worked long hours to serve such households (Fig. 20.1).

Figure 20.1 Peeling the tatties. The maid of the Old Kirk Manse, Corstorphine, Edinburgh, c 1880. Source: SLA 5.0714 C.1830.

Figure 20.2 The maid at Sanna, Argyll, trampling the washing, c 1925. Domestic service was increasingly unpopular after World War I, virtually to die out during World War II. Source: SLA 51/13/4 C18225.

World War I caused major disruption in many households as the majority of male servants left to join the armed forces and many of the women and girls went to work in munitions factories or to fill jobs normally undertaken by men. Some former servants returned to domestic service at the end of the War, but many others found work in shops, offices and factories more congenial, although the Depression years of the 1930s forced some back into service (Fig. 20.2). There was official pressure on women to return to or enter domestic service and, although this was not case for men, some male servants also returned to their former employment. At this time it was becoming more common for work in service to be offered to married couples. World War II virtually spelt the death knell of domestic service in the traditional sense.

SERVANTS IN THE SIXTEENTH AND SEVENTEENTH CENTURIES

We can learn a little about those who worked in the homes of others at this period from chance references in estate papers or by consulting testaments left either by themselves or their employers. Although 'many men and women remembered their servants, with small sums of money, grain and clothes',[5] not all servants benefited from the generosity of a deceased employer. One such person was Issobell Broun, a servant girl from St Andrews, who died in 1597 leaving behind no possessions other than her clothing (abulzement) worth 20 shillings (Scots), 'Item the said defunct being

WORK IN THE HOMES OF OTHERS • 359

Figure 20.3 The testament of Issobell Broun, servant girl, 1592. She died intestate, leaving behind no possessions other than her clothing ('abulzement') worth 20 shillings. Reproduced by kind permission of Scottish Archive Network (SCAN). Source: National Archives of Scotland, St Andrews Commissary Court Registers of Testaments (CC20/4/3 p. 216).

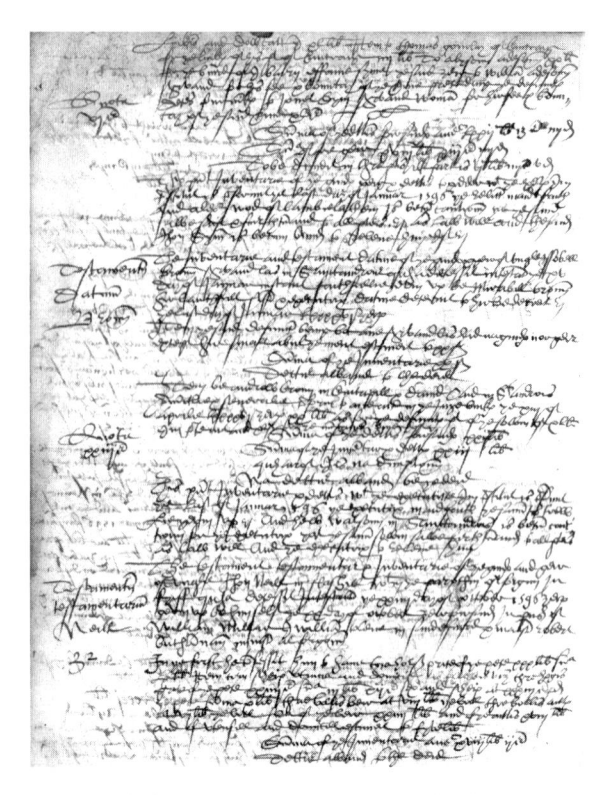

bot ane s[e]rvand las had na guid[is] nor geir except hir small abulzement estimet to xx sh[illings] Su[m]ma of the Inventarie xx sh[illingis].'[6] (Fig. 20.3)

After James VI moved the Royal court south to London, establishments such as that maintained by the Duke and Duchess of Hamilton probably became the most prestigious in lowland Scotland. During Cromwellian times the House of Hamilton suffered the execution of the first Duke in 1649. His daughter, who became Duchess in her own right:

> ... reluctantly paid and dismissed Valentine Beldam (groom), Robert Cole (servant) Robert Naismith (servant) and even Herie the sculleryman. Some of the servants followed her uncle into exile in Holland, others may have stayed on at the palace in the employ of Colonel Ingoldsby, and a few did remain with the Duchess herself, notably her ladies.[7]

The Restoration in 1660 saw a return to having 'always at least thirty servants and sometimes as many as fifty in the Palace'.[8] The majority of these were men, including four professionals: the secretary, the lawyer, the chaplain and the governor.[9] Both the Duke and the Duchess had their own attendants, well-born gentlemen and ladies respectively. The gentlemen might participate in hunting parties and accompany the Duke when he travelled, for they were essentially retainers of a ducal court. Ayr Burgh

Records note that six such male servants of the Duke of Hamilton became burgesses and guild brethren of Ayr on 6 May 1670.[10] The Duchess's ladies, who were addressed as Mistress regardless of their marital status, would act as her companions. Much time would be spent in sewing and embroidery and assisting with the healthcare and education of the children:

> All household linen had to be cut, hemmed and made up into tablecloths, napkins, sheets and pillowberes [pillowslips], as well as shifts and shirts. Textile furnishings such as cushions, bed hangings and chair covers could be ordered, but most were made at home, particularly in remote districts.[11]

The ladies might join in musical activities or games of cards. The pages who served both the Duke and Duchess[12] were the sons of well-born parents and were sent to the Palace to be trained in appropriate social skills and manners. In addition, the Duke had his own personal attendants or *valets de chambre*, and the Duchess her lady's maid. These servants would be responsible for the respective wardrobes, including the dressing of wigs. The running of the household itself was in the charge of a steward. Then there was a master cook, two undercooks, several kitchen boys and a number of pantry boys plus a master baker, a brewer and a butler who supervised the cellars. The nurseries were staffed by Mrs Mary Darlo, in charge overall, with both a wet nurse and an ordinary nurse for each baby, and 'finally one woman who was in charge of two or three of the children'. The Palace staff also included, 'one chambermaid, the washerwoman, the dairymaid and the henwife'. In her latter years, the Duchess 'appointed a female housekeeper'.[13] As befitted their station in life, the Hamiltons employed a 'trumpeter for ceremonial occasions' and musicians to provide entertainment or to give music tuition to the daughters of the family.[14] In relation to other noble houses of the period, the staff employed by the Hamiltons were well paid, with 'the footmen and the lower women servants earning £24 or so each' [Scots annually].[15]

In the seventeenth century, 'women's wages were so low a majority of Aberdeen households (many of them modest indeed) employed one or more female domestics – about half of whom had moved into town from elsewhere'.[16] These servants would be employed on six-month contracts:

> £10 Scots was all the cash, less tips, that senior domestics in Aberdeen earned in a year, and many women earned less. In 1695, the average wage for female domestics (not including wet nurses) was £8.14s, exclusive of room and board.[17]

Most were young (15–24 years of age) and:

> ... living away from parental supervision in crowded quarters amongst mixed company, female domestics were at considerable risk of becoming pregnant: in the third quarter of the seventeenth century, this was the fate of over twenty servants a year in Aberdeen.[18]

Such women could be fined £10 by the Justice of the Peace Court, and be required to appear before the Kirk Session charged with fornication. With no entitlement to poor relief to support them and few, if any, savings, most would have to return to their home parish to give birth to their child, after which they would be recalled to Aberdeen in order to appear before the congregation for three Sundays on the stool of repentance. Much effort was put into seeking out such women and '... imposing financial burdens that made it hard to resist relatively well-paid work as a wet nurse'.[19] In March 1685, a servant named Margaret Rolland was found to be pregnant and had to appear before the Kirk Session. The minister, Dr George Garden, whose wife was also pregnant, hired her as a wet nurse and 'petitioned the elders to suspend their proceedings against Rolland until his child was weaned – a courtesy routinely granted employers, but never unwed mothers nursing their own children'.[20] When the minister's child died, Margaret Rolland made her three appearances before the congregation and was absolved from her sin of fornication.[21] In 1695 there were at least 12, possibly 21, live-in wet nurses in Aberdeen including Christian Thomson who was a wet nurse in the household of a stonemason, offering a service only available to wealthy households in most parts of Europe at that time.[22]

In 1636, Sir William Brereton, a Puritan gentleman from Cheshire, wrote of a village on Lord Roxburgh's estate, about six miles (9.5km) from Dunbar, East Lothian, 'we observed the sluttish women washing their clothes in a great tub with their feet, their coats, smocks and all tucked up to their breech'.[23] The next day he reached Edinburgh, where he had to hold his nose or use wormwood or some other scented plant to counter the smell of his lodgings, 'their houses, and halls, and kitchens, have such a noisome taste, a savour, and that so strong, as it doth offend you as soon as you come within the wall'.[24] He continued, 'Their pewter, I am confident is never scoured; they are afraid it should too much wear and consume thereby; only sometimes, and that but seldom, they do slightly rub them over with a filthy dish-clout dipped in most sluttish greasy water'.[25] He was loath to drink the wine or water from their pots. Again he describes the unsatisfactory method of washing linen which leaves it looking, 'as nastily as ours doth when it is put into and designed to the washing'.[26] It also smelt unpleasant on his bed. Finally, 'To come into their kitchen, and to see them dress the meat, and to behold the sink (which is more offensive than any jakes [privy]) will be a sufficient supper, and will take off the edge of your stomach.'[27] Obviously neither mistress nor servants were too fastidious about their standards of housekeeping.

Sir William was not alone in his poor opinion of Scottish domestic cleanliness. Deprived of his Fellowship at Trinity College, Cambridge, in 1662 by the Act of Uniformity, John Ray and a friend travelled together through the United Kingdom. On 17 August that year, when they reached Dunbar, he wrote:

> The women generally seem to us none of the handsomest. They are not very cleanly in their houses, and but sluttish in dressing their meat. Their way of washing linen is to tuck up their coats, and tread them with their feet in a tub.[28]

In the Highlands and Islands, some clan chiefs still employed traditional officials such as 'poet and bard, piper, councillor and arms bearer'[29] in addition to the men and women who tended to the personal needs of the chief, his wife and family. John Taylor, an Englishman who travelled in Scotland in 1618, records how he visited 'houses like castles' where the master was clad in clothes spun by his wife, daughters and servants, yet 'who kept and maintained 30, 40, 50 servants, or perhaps more'.[30] These numbers would be necessary when the nobleman had to 'give entertainment for foure or five days together to five or six Earles and Lords, besides Knights, Gentlemen and their followers, if they bee three or foure hundred men & horse of them ... '.[31]

When Aphra Behn, the spy, poet, novelist and playwright, visited Colin Du, tacksman to Campbell of Glen Lyon in 1689, she was 'directed to his hut, it standing something apart from the others, and am greeted by a pockmark'd man of middle height, without shoes, stockings or breeches, in a short coat, with a shirt not much longer to hide his Nakedness'.[32] He bowed very civilly to her and invited her 'inside his Dwelling, which is very long but without ceiling or partition, some Cattle at one end and at the other the Familie consisting of his wife, two daughters in their 'teen years, and three servants'.[33] There was a peat fire smouldering in the centre, dividing the beasts from the humans, with the smoke escaping through a hole in the roof. She was served a supper of 'Cow's Blood boil'd to a Cake, with a little milk and a pinch of oatmeal',[34] this being commonly eaten in the Highlands in springtime when stores were running low. At night she was troubled by rainwater coming through the roof 'mingling with the soot on the rafters'[35] and splashing on the floor, with biting vermin, with peat smoke smarting her eyes and the effects of strong punch, 'great Draughts of which I was constrain'd to drink with the rest of the Companie'.[36] From this description one can imagine the living conditions of both master and servants, and draw the inference that the tasks which servants would have to undertake, first and foremost of which would be to attend to the cattle, for therein lay the wealth of the entire household.

ENLIGHTENED TIMES FOR SOME

By the eighteenth century little had changed for some servants. A 1750s description of a blackhouse interior, the typical Highland and Island rural dwelling, goes thus:

> ... The generality of them have no Beds or Bedcloaths. They lie either on the bare floor or on a little straw or ling, with a bit of Blanket or plaid wrap't about them, and as often in their poor rag garment, or stark naked; and man, wife and children, servant man and maids all together on the Ground, as dogs do in a Kennel.[37]

Even in lowland Scotland, where conditions were usually perceived as being better, work in the homes of others frequently involved tasks outside as well as inside. The household accounts for Culzean, Ayrshire, include the following entries:

> May 1766, Langton Lass to get 20 sh & half a crown or 3 sh for shoes or apron if she behave well to have Charge of the Dairy Milk Cows wash an Dress or spin or Do What ever she has time for May 21 1787, Pedry refused to go to the Cows – Constantly for which give her no present £2.[38]

Lady Grisell Baillie of Mellerstain in Berwickshire drew up a set of rules for her servants to follow in 1743. Those for her housekeeper included:

> The dairy carefully lookt after, you to keep the kie of the inner milk house where the butter and milk is, see the butter weighed when churn'd, and salt what is not wanted fresh, to help to make the cheese and every now and then as often as you have time to be at the milking of the cows.[39]

Similar rules had to be observed by the housekeeper at Saltoun Hall, East Lothian, seat of Andrew Fletcher of Saltoun, a member of the pre-Union Scots Parliament.[40]

The eighteenth century saw radical changes in the agricultural system of lowland Scotland which continued well into the following century and changed the pattern of life for many people. Some individuals prospered and were able to take on the tenancy of a substantial farm and employ others. Many lost their existing tenancies, became landless and had to seek an alternative livelihood. Twice a year in most districts, feeing fairs were held where legally enforceable bargains were struck between masters and servants, guaranteeing work for the following six-month term. For unmarried servants this usually meant receiving bed and board in the farmhouse together with cash wages at the end of the term. Men would be employed to undertake agricultural tasks, but women were used for both indoor and outdoor work, often being required to milk cows and help in the fields at harvest time in addition to their normal domestic duties. When the term ended, the contract could be renewed or terminated.

In the 1740s, a female servant in the south Ayrshire parish of Dailly would have received 13s 4d yearly plus an apron or a pair of shoes.[41] Average wages paid for a best dairymaid and servant in Ayrshire rose from 8s 4d per half year in 1720 to £5 per half year in 1809.[42] The wages of an inferior dairymaid and servant went from 6s per half year in 1720 to £3 per half year in 1809.[43] In contrast, a tailor received 2d per day in 1720 and 2s 6d per day in 1809, and in both cases had his food with the family.[44] Such men went from house to house to ply their trade cutting and sewing the homespun cloth woven by a local weaver. There were complaints in Dailly about the rising level of wages for common labourers and servants where male servants were paid £6-£9 and female servants from £2.10s to £4 in 1792.[45]

In areas where industry began to compete for the labour supply, male servants could expect their wages to improve, but female servants remained poorly remunerated. Other forces worked to favour the employment of women and girls. 'In order to raise money for the wars fought during the reign of George III, taxes were levied on masters and mistresses employing servants. The tax applied to the employment of male servants from 1777 to 1792, and to female servants from 1785 to 1792.'[46] Although the tax was paid by the employer, it affected the servants. An entry for the Culzean accounts for Martinmas 1785 records that Betty Trail and Jack were to do all the porter's work because their wages together came 'only to 1/2 a porter's wage besides the Difference between the tax of a man servant & a maid, and the maid spins and does other work'.[47] Although the tax on female servants was not re-imposed, it was for male servants, as a Table of the Assessed Taxes for 1854 shows (see Fig. 20.4).

As the general level of prosperity rose after about 1740 and houses and furnishings improved, servants could expect to encounter better accommodation in the homes of their employers than in the cot houses in which most

A TABLE of the ASSESSED TAXES, chargeable to the 5th April, 1854, and of the NEW DUTIES which take effect from that period.

SERVANTS.	Old Duties: No.	Old Duties: At per Servant (£ s. d.)	Old Duties: Total per Year (£ s. d.)	New Duties (£ s. d.)
MALE SERVANTS—viz.—Maitre d'Hotel, House Steward, Master of the Horse, Groom of the Chambers, Valet de Chambre, Butler, Under Butler, Clerk of the Kitchen, Confectioner, Cook, House Porter, Footman, Running Footman, Coachman, Groom, Postillion, Stable Boy, or Helper in the Stables, Gardener, Park-keeper, Game-keeper, Huntsman, Whipper-in, or by whatever Name or Names Male Servants really acting in any of the said capacities shall be called, or whether such Male Servants shall have been retained or employed in one or more of the said capacities	1	1 6 4	1 6 4	1 1 0 each Servant
	2	1 14 1	3 8 2	
	3	2 1 9	6 5 3	
	4	2 7 10	9 11 4	
	5	2 13 10	13 9 2	
	6	2 16 7	16 19 6	
	7	2 17 8	20 4 3	
	8	3 1 7	24 12 8	
	9	3 7 1	30 3 9	
	10	3 13 1	36 10 10	
	11	4 4 1	46 4 11	
	Above 11	4 4 1 each		
For every Servant kept by Bachelors an Additional Duty of		1 2 0 each		Repealed
WAITERS—Employed by Keepers of Taverns, Coffee-Houses, Inns, Ale-Houses, or other Licensed Houses, or in any Eating and Victualling-Houses, or Hotels, or Lodging-Houses, except occasional Waiters		£ s. d. 1 13 0		1 1 0
MALE PERSONS—Occasionally employed as Servants, where the employer is chargeable for another Servant, or for any Carriage, or for more than one Horse for Riding, &c.		1 6 4		1 1 0
And where not chargeable for a Servant or for any Carriage, or for more than one Horse for Riding, &c.		0 11 0		Exempt
SERVANTS LET TO HIRE—By Postmasters, Inn-keepers, or other Licensed Persons, or Coachmakers, employed as Coachmen, Grooms, Postillions, or Helpers		1 7 6		1 1 0
Under Gardeners		Exempt		0 10 6
Under Gamekeepers		0 11 0		0 10 6
Servants under 18 years of age, and having a legal settlement in the Parish in which their Employer resides		Exempt		0 10 6 Without reference to a Parish Settlement

Figure 20.4 A table of the Assessed Taxes chargeable to employers of male servants, 1854. Source: Courtesy of Ayrshire Archives.

of them had been brought up. Which is not to say that the straw or chaff-filled mattresses, coarse linen sheets, hard, unbrushed 'Scotch' blankets and bare wooden floors in their sleeping quarters did not leave plenty of scope for improvement.

Food for farm families and servants alike in parts of Ayrshire consisted, for breakfast, of pottage made with whey or water taken with plenty of milk, either cheese or herring with oatcakes, or milk with cakes, or frequently both cheese and milk with bread after porridge for breakfast. Three or four times a week there was beef or mutton broth for dinner, thickened with barley, pease, beans or garden stuffs. The meat from the broth was then eaten with potatoes or bread. If there was no meat, cheese or fish was provided, with milk and potatoes or oatcakes. Sometimes in place of the broth, beef or mutton was stewed with potatoes. At other times potatoes were mashed with milk and salt butter and eaten with cheese or fish, or oatcakes were taken with milk. Supper consisted either of pottage and milk or potatoes beaten with butter and milk.[48]

In seventeenth-century Ayr, female domestic servants had been required to wear a particular type of cap to denote their station in life and failing to do so resulted in, 'eight days imprisonment in the blackhole of the Tolbooth'.[49] Men wore less distinctive clothing – breeches, jerkin, shirt, homemade stockings, stout shoes and, when going to church or market, a bonnet. Dress changed little throughout the seventeenth and eighteenth centuries, outer garments being made from a hardwearing type of linen called harn, spun from locally grown flax:

> Younger daughters and maids of a family would wear a jupp or short gown. This was close fitted on the upper part of the body and was well suited for the performance of domestic or farm work. It was made from a locally-woven, stripped, woollen cloth called drugget.[50]

Plaiding, also woven locally, was another common fabric type.

For many, their time in service was a transition between childhood and marriage, after which they could expect to raise the next generation of servants. Some made a long-term career out of service, but for the most part the physical demands made on servants required them to be strong and fit. Hence the preference for employing those who were country bred. Other motivations behind recruiting new staff included the ill-health and death of existing workers. Staff might have to seek fresh positions when an employer was declared bankrupt, as happened in several instances when the Ayr Bank collapsed at the end of the eighteenth century.

An early indication of future changes in the means of obtaining domestic staff occurred in 1701 when John Lawson established Lawson's Intelligence Office in Edinburgh and offered to record 'the names of servants, upon trial and certificate of their manners and qualifications, whereby masters may be provided with honest servants of all sorts, and servants may readily know what masters are unprovided'.[51] He listed a wide range of types of servant and 'offered to supply households in all parts of the country with reliable servants of all kinds'.[52] However, it was not until the second half of the nineteenth century that servant registry offices became a feature in many Scottish towns, although posts in prestigious households were usually found through a superior Edinburgh office. Senior members of staff in large households normally found it necessary to move to a post in a different establishment each time they sought to improve their situation.

Male applicants for a position were interviewed either by the master himself or by the butler, and female servants by the mistress or by the housekeeper. References were absolutely essential. Apart from requiring a written reference from a previous employer, it would help if a verbal recommendation, 'being spoken for', was provided by a relative or friend who was in a position of some influence. Although there were servants who were content, for whatever reason, to remain with a particular household for a long period of time, in general the turnover rate of domestic staff was high. It was not unusual for households to experience staffing difficulties despite legal requirements for servants to be employed. Unless a servant was gainfully employed, or had a licence from the Justices, or was unfit, he or she could be compelled to serve any employer requiring a servant. Failure to comply meant prosecution. In May 1810, for example, a servant girl began service in Edinburgh on the Thursday. While the family was at breakfast on the Monday she left her position without permission. The following day she was apprehended and appeared before the Sheriff-Substitute in Edinburgh who found her guilty, fined her two guineas damages and granted expenses to her mistress. He also granted a warrant to incarcerate her in the Tolbooth of Edinburgh until payment was made. Of her fine money, ten shillings was given to the officer who arrested her, and the remaining one pound twelve shillings was given to the Society for the Relief of the Destitute Sick.[53] No doubt the story was published in the press to act as a deterrent to others who might consider breaking their contract.

Sometimes there were relationship problems between master and servant. In his somewhat racy *Memoir of an Eighteenth-Century Footman*, the author, John MacDonald, is very forthright in recalling just how fraught with difficulties the life of an attractive footman could be. Also likely were clashes of personality between members of staff, and these would not be conducive to the smooth running of the household. Comments in the Culzean accounts provide hints of such tensions:

> June 13th 1793 Jane Anderson, upon Condition her mother never comes to this house, and that Jane never takes an article out of the kitchen. to get 25sh for six months service.

> Nov 16 1784 John Wilson to get £6 & if he behaves well to get something at the genl Assembly or to be made footman from month to month – a month's wage or a month's warning.[54]

Then, as now, people were employed who were not suitable for the tasks they were required to perform. They might have been clumsy or for whatever reason lazy, sullen and unwilling workers. Equally, masters, mistresses and senior members of staff could be demanding and perhaps unfair, so that those who felt themselves to be the victims of an unpleasant regime would hope to find a more congenial situation as soon as the opportunity arose. Female servants would leave to marry, and should any have the misfortune to become pregnant whilst still unmarried, dismissal was inevitable with little support and the prospect of punishment by the Kirk authorities. Kirk Session records provide evidence of a number of cases where the father of the child was either the master of the household, his son, or a male guest or servant. It was not easy for a young maid to reject the advances of men who, in relation to herself, were in positions of power.

When servants became old or infirm, every effort was made to call upon their friends and families to support them. Failing that, parish funds could provide limited assistance, sometimes in the form of the poorhouse. By 1845, this system was under considerable strain and radical changes were made. As the century progressed, and into the early twentieth century, more servants ended their days in the workhouse, stripped of their dignity, with their remaining possessions, including their clothes, becoming the property of the authorities when they died. In Maybole, Ayrshire, at:

> 11pm, 23 Sept. 1858, Elizabeth McMurtrie, 64, a single servant, 'who can sew a white seam'. Except a short time she was chargeable in Barony, Elizabeth has supported herself by service and by taking care of an old couple who are dead. Doctor certifies 'an inward complaint'. Offered relief.[55]

A few, fortunate, long-serving servants would be supported by their former employers. Others in the nineteenth and twentieth centuries might obtain support from a voluntary or friendly society, to whose funds they had contributed during their working life. In August 1834, a distributor of religious tracts in Ayr:

> ... called upon a poor woman, whose sickly appearance and lonely circumstances attracted her attention. It appeared that she had spent the greater part of her life as a servant in the country, and by great carefulness had saved a small sum of money, which she was calculating upon as a support for her, when old age and sickness should arrive. Her hopes, however, were disappointed. She was deprived of her health, being seized with a consumptive disease, and by the misfortunes or dishonesty of a relation whom she trusted, she lost her little stock of money, so that her situation was gloomy enough. The Distributor not only sympathised with her in her affliction, and directed her to the Refuge of the weary and oppressed, but also interested on her behalf some pious and affluent friends, by whom she was supplied with many comforts which she needed.[56]

UPSTAIRS AND DOWNSTAIRS

As wealth increased in the eighteenth century and landowners built new houses, or had their existing houses refashioned, the physical and social division between masters and servants became more pronounced. Perhaps the most extreme evidence of this may be seen at Newhailes, Midlothian, where servants were required both to enter and leave the house via a long tunnel. At Culzean and Bargeny, Ayrshire, the comings and goings of the servants remained hidden behind a wall. No longer were maids and mistress working together for the good of all as had traditionally been the case in much of rural Scotland. One section of the population now assumed control over the other to an increasingly unacceptable degree.

Industrialisation, with its pollution of the physical environment and resulting health problems, was an important factor in encouraging the better off, especially in the nineteenth century, to move from urban areas to the surrounding countryside, their increasing wealth allowing them to commission villas and mansions there. Such buildings provided work for the men needed both to construct and subsequently maintain them; the architect, the masons, builders, plasterers, joiners and painters, for example. On completion, these buildings created a demand not only for furnishings, but also staff to clean and polish, to cook and wait upon the occupants and undertake many more tasks. Mistresses who were unaccustomed to running such a household turned for help to the periodicals of the time such as Mrs Beeton's *The Englishwoman's Domestic Magazine*, which originally appeared as monthly supplements before being published in book form in 1861.[57]

A mistress of the period requiring domestic staff might have an advertisement placed in a newspaper or periodical likely to be seen by potential applicants 'seeking a place'. Equally, she might reply to an advertisement inserted by a servant seeking a new situation. The introduction had to be followed up by an interview, which could be an ordeal for both parties. The applicant had to produce a written reference from her previous employer, which was occasionally forged, although this was illegal. Advice given in 1894 to those conducting the interview suggested that the applicant should be questioned about his or her current situation, length of service there, reason for leaving, age and state of health, the wage received and whether he or she was an early riser.[58] This was to be followed by questions pertaining directly to the situation for which the applicant was applying. In the case of a cook, the interviewer would seek to find out whether the applicant was a 'professional cook', how many family members and how many servants she had cooked for, whether she was used to preparing large dinners and making sweets and entrees. At this stage, agreement had to be reached on whether the cook could select tradespeople supplying the kitchen, place the orders, settle the accounts and be allowed the cook's percentage of these. It was recommended that the interview should conclude with the applicant being made aware of the house rules, and might include such matters as whether the servant would have to pay for breakages and female servants would be allowed 'male followers'.[59] To avoid the necessity of conducting such an interview, a mistress might delegate the responsibility to a senior member of staff, if she employed one, or to a reputable servants' registry.

In larger households, it was, and remained, customary for junior members of staff to be trained in their duties by senior and more experienced servants. This was not usually possible in smaller households where the mistress herself was often unable to provide the necessary training. In such circumstances, a greater or lesser degree of dissatisfaction all round resulted. In an effort to rectify the difficulties, groups of mistresses came together to try to improve matters. One such group subscribed to the Training School for Servant Girls and Temporary Home and Free Registry Office in Glasgow. At their meeting, held on 26 March 1901, Mrs Story,[60] in moving the adoption of the minutes, said that 'in old days mothers sent out girls fairly able to work, but nowadays the girls knew almost nothing, and had everything to learn'. She commented on the excellent work of the institution and how fortunate they were in having people able to train girls thoroughly and take an active interest in their work. The training was, 'such as could be obtained in few cottages and houses of a poorer kind'. The registry and home was at 13 Burnside Gardens, Glasgow, with subscriptions priced at 5s for lady members and 1s for servants annually. The objectives of the establishment were:

> ... the maintenance of a trustworthy register of servants and of situations vacant for the benefit of the members, and the institution of a home where servants can stay between engagements, and receive training in household work, the committee undertaking to protect their interests, and promote their health, comfort, and happiness.[61]

It seems that in 1901 there was a demand for places in the West End of Glasgow, thus making it very difficult for those residing elsewhere to obtain servants. High wages or the attractions offered at the coast or in the country could not compete with the supreme enticements of:

> ... the Exhibition, with its gilded domes, piazza, promenades and bands. No doubt, the vision of a weekly 'night out' with sweetheart and friend listening to the music, enjoying a splash on the water shute and a run on the switchback was all-powerful in its attractiveness.[62]

There was a drive to recruit girls from country areas where there was little work for women, the committee agreeing to care for them and arrange for their training. However, it would appear that at this time the committee hoped 'with increased funds, to extend its usefulness and popularity, and also to introduce training in household work, and perhaps thoroughly practical classes for subjects such as cookery, laundry work, and the table'. This would seem to imply that formerly any training given was of a theoretical nature, probably with some element of moral guidance. Lady members were particularly concerned about girls withdrawing from contracts and unscrupulous individuals 'extorting money from ladies in cases where there was no intention of carrying out the services' and going 'from one lady to another for the sake of arles'.[63]

Institutions which were to have a more lasting influence on training for domestic duties included the Edinburgh School of Cooking, located in Atholl Crescent and opened a few months before the Glasgow School of Cookery which started in 1875 with '70 young ladies in attendance'. An advertisement in a Glasgow periodical, *The Baillie*, proclaimed its class-based curriculum:[64]

Superior Cookery	Tickets 25s/day
Potage Ecossais	
Plain Cookery	Tickets 21s/day
Scotch Soup	
Cookery for the Working Classes	Tickets 3s/day
Broth	

Most of the students would be from the middle classes, learning how to cook or supervise cooking in their own homes, but by 1878 students were being trained to teach cookery not only in Glasgow but at various locations throughout the West of Scotland. By 1908, the school offered a Training for Cooks certificate, Lessons to Cooks in Private Houses, Laundry Classes, as

Figure 20.5 The sewing class at the Glasgow 'Dough' School, c1931. Mending was a necessary skill for many female servants. Source: Courtesy of Glasgow Caledonian University Archives.

well as classes in Dressmaking, Sewing, Knitting, Millinery and certificate courses for both Housewives and Lady Housekeepers. No doubt the work of the school enabled the students to give training to their own domestic staff.

At the end of World War I, the government tried to coerce women back into domestic service. Speaking in the House of Commons in 1919, the Member for Glasgow Springburn said, '... There are hundreds and thousands of positions as domestic servants open to women today, and they will not accept them. They should accept them. It is the most honourable of all occupations for women ... '[65] The Ministry of Labour sent women to be trained at various centres throughout the country. One such was the Glasgow 'Dough' School, which from 1908 had College status.[66] (Figs 20.5, 20.6) On 14 January 1920, *The Bulletin* displayed pictures of former war workers being trained for domestic service, and on 14 December 1927 it featured another group of young women who had just completed their course of training as domestic helps.[67] In 1891, 168,506 people were employed as domestic indoor servants in Scotland,[68] which was the main source of employment for females. In 1911, there were 135,052 female indoor servants[69] and in the Depression year of 1931, the number rose to 138,679,

Figure 20.6 Simple cooking and baking as taught in this class at the Glasgow 'Dough' School, *c* 1931, would be of use to servants working in smaller households. Far greater knowledge and skill would be expected from a professional cook in a large household. Source: Courtesy of Glasgow Caledonian University Archives.

but fell two decades later to 73,278.[70] It would be a valuable contribution to the nation's archives if recordings of the recollections of these now elderly former domestic servants and photographs of them at work could be added to the presently sparse amount of material available.

A CENTURY OF CHANGE

Despite the best efforts of groups of ladies to provide training and their attempts to improve the status of domestic servants, servants themselves were taking action to try to improve their conditions. In addition to voting with their feet when the opportunity arose, reports in *The Times* and the *Dundee Advertiser* feature a short-lived maid-servants' union formed in Dundee in April 1872.[71] Members demanded a weekly half-day holiday, one free Sunday every fortnight, the payment of wages every quarter instead of every six months, and the cessation of the custom requiring them to wear a large 'fern-shaped headdress' known as the 'flag' which 'was regarded as a degrading mark of servitude'.[72] In 1909 the National Domestic Servants' Union was formed.[73] Despite these initiatives, however, many servants chose to emigrate in the expectation of providing a better future for themselves.

The levels of death duties payable after the high casualty figures of World War I forced many owners of large houses to reduce their staff, and in some cases to move to smaller, more convenient properties. This coupled with the increasing development and adoption of domestic appliances – although these had more significance in the later twentieth century – helped to ease the problem of obtaining full-time, live-in staff.

Instead, a plethora of part-time jobs has been created. The charwoman of old has become the cleaner or domestic help, many of whom are supplied by registries. Nevertheless 'the exploitation of women in low-paid jobs in the service sector is still with us'.[74] Au pairs, child minders and baby sitters help to ease the burden for parents unable to employ a nanny and for some, home tutors can supplement children's schooling. Carers and various health professionals visit some older, sick and disabled people in their own homes; likewise hairdressers and social workers. Then there are the tradespeople who undertake work in and around the home: plumbers, electricians, joiners, painters, TV and computer technicians, and so forth. A variety of skilled and unskilled workers also play their part in servicing the home – gardeners, odd-job people, window cleaners, meter readers, posties and those who deliver milk, coal and other items. One has only to look at the classified page of a local newspaper, or through the Yellow Pages directory, to compile a considerable list of people whose job entails work within the homes of others.

NOTES

1. Marshall, 1983, 41.
2. Marshall, 1983, 42.

3. Smout, 1972, 100.
4. Boswell, 1993, 483–7.
5. Sanderson, 1982, 182.
6. NAS, Scottish Documents.com, Testament of Issobell Broun CC20/4/3/198.
7. Marshall, 1973, 62.
8. Marshall, 1973.
9. Marshall, 1973, 63.
10. Lindsay, Kennedy, 2002, 25.
11. Swain, 1986, 53.
12. Swain, 1986, 66–7.
13. Swain, 1986, 69.
14. Swain, 1986, 73.
15. Swain, 1986, 74.
16. Ewan, Meikle, 1999, 212.
17. Ewan, Meikle, 1999.
18. Ewan, Meikle, 1999, 212–3.
19. Ewan, Meikle, 1999, 213.
20. Ewan, Meikle, 1999, 214.
21. Ewan, Meikle, 1999.
22. Ewan, Meikle, 1999, 215.
23. Hume Brown, 1891, 135; see Chapter 17 Housework for further commentaries on standards of cleanliness in Scottish homes.
24. Hume Brown, 1891, 142.
25. Hume Brown, 1891, 142–3.
26. Hume Brown, 1891, 143.
27. Hume Brown, 1891.
28. Hume Brown, 1891, 230–1.
29. Mitchison, 1978, 14.
30. Hume Brown, 1891, 127–8.
31. Hume Brown, 1891.
32. Laidlaw, 1992, 86.
33. Laidlaw, 1992, 86.
34. Laidlaw, 1992, 87.
35. Laidlaw, 1992.
36. Laidlaw, 1992.
37. NAS, Mar & Kellie, GD124/15/1605/1/.
38. NAS, Ailsa Muniments, GD25/9/19.
39. Baillie, 1911, 278.
40. NLS, MS 17080, 134–7.
41. *OSA*, VI, 137, Ayrshire.
42. Aiton, 1811, 517.
43. Aiton, 1811, 517.
44. Aiton, 1811, 517.
45. *OSA*, VI, 137, Ayrshire.
46. Aitchison, 2001, 15.
47. NAS, Ailsa Muniments, GD9/9/19.
48. Aiton, 1811, 654–5.
49. *Ayr Advertiser*, 23 September 1875.
50. Mitchell, 1939, 262–6.
51. Plant, 1952, 159.
52. Plant, 1952, 159.
53. *Ayr Advertiser*, 31 May 1810.
54. NAS, Ailsa Muniments, GD9/9/19.

55. Maybole Record of Applications for Parochial Relief, 1855–65, Ayrshire Archives, A2 075.
56. *The Ayr Monthly Visitor*, first report of the Ayr Religious Tract Society for 1834, Ayrshire Archives, Michael Brown papers.
57. Beeton, 1994.
58. *The Duties of Servants*, 1894, 9–11.
59. Aitchison, 2001, 43.
60. Mrs Story was probably the wife of the then Principal of the University of Glasgow and herself the author of two biographical books.
61. *The Lady's Review of Reviews*, June 1901.
62. *The Lady's Review of Reviews*, June 1901.
63. *The Lady's Review of Reviews*, June 1901.
64. Miller, 1975, 7.
65. Hansard, 5th Series, vol 121, 28 Nov 1919, col 1002–4.
66. Thomson, McCallum, 1998, 15.
67. Source kindly supplied by Carole McCallum.
68. 1891 Census.
69. Horn, 2001, 17.
70. Horn, 2001, 191.
71. Horn, 2001, 130.
72. Horn, 2001, 178–9.
73. *The Woman Worker*, 22 Sep 1909, 275.
74. Sanderson, 1996, 165.

BIBLIOGRAPHY

Adams, S, Adams, S. *The Complete Servant* (London, 1825), reprinted London, 1989.
Aitchison, J. *Servants in Ayrshire, 1750–1914*, Ayr, 2001.
Aiton, W. *Ayrshire: General View of Agriculture in the County of Ayr*, Glasgow, 1811.
Baillie, Lady G (ed R Scott Moncrieff). *The Household Book of Lady Grisell Baillie, 1692–1733*, Edinburgh, 1911.
Beeton, Mrs I. *Beeton's Book of Household Management*, London, 1851–61, reprinted, 1994.
Boswell, J (eds J W Reed, F Pottle). *Boswell, Laird of Auchinleck, 1778–82*, Edinburgh, 1993.
Drummond, J. *Upstairs to Downstairs*, Aberdeen, 1983, reprinted 1991.
The Duties of Servants, East Grinstead, 1894.
Ewan, E, Meikle, M M, eds. *Women in Scotland, c.1100–c.1750*, East Linton, 1999.
Gathorne-Hardy, J. *The Rise and Fall of the British Nanny*, London, 1972.
Horn, P. *The Rise and Fall of the Victorian Servant*, Stroud, 1990.
Horn, P. *Life Below Stairs in the Twentieth Century*, Stroud, 2001.
Hume-Brown, P, ed. *Early Travellers in Scotland*, Edinburgh, 1891.
Jamieson, L. Rural and urban women in domestic service. In Gordon, E, Breitenbach, E, eds, *The World is Ill Divided: Women's Work in Scotland in the Nineteenth and Early Twentieth Centuries*, Edinburgh, 1990, 136–57.
Kelsall, H, Kelsall, K. *Scottish Lifestyle 300 Years Ago*, Aberdeen, 1986, reprinted, 1993.
Laidlaw, R. *Aphra Behn: Dispatch'd from Athole*, Nairn, 1992.
Lindsay, A, Kennedy, J. *The Burgesses and Guild Brethren of Ayr, 1647–1846*, Ayr, 2002.
Lochhead, M. *The Scots Household in the Eighteenth Century*, Edinburgh, 1948.
MacDonald, J. *Memoirs of an Eighteenth-Century Footman*, London, 1790, reprinted 1927.
Marshall, R K. *The Days of Duchess Anne*, London, 1973.
Marshall, R K. *Virgins and Viragos: History of Women in Scotland from 1080 to 1980*, London, 1983.
Miller, E. *Century of Change, 1875–1975*, Glasgow, 1975.

Mitchell, Rev J. *Memories of Ayrshire about 1780* (Miscellany of the Scottish History Society), Edinburgh, 1939.
Mitchison, R. *Life in Scotland*, London, 1978.
Mitchison, R, Leneman, L. *Sexuality and Social Control*, Oxford, 1989.
Mitchison, R. *The Old Poor Law in Scotland*, Edinburgh, 2000.
Moore, L. Educating for the 'woman's sphere'. In Gordon, E, Breitenbach, E, eds, *Out of Bounds: Women in Scottish Society, 1800–1945*, Edinburgh, 1992, 10–41.
Plant, M. *The Domestic Life of Scotland in the Eighteenth Century*, Edinburgh, 1952.
Sanderson, M. *Scottish Rural Society in the Sixteenth Century*, Edinburgh, 1982.
Sanderson, M. *Women and Work in Eighteenth-Century Edinburgh*, Basingstoke, 1996.
Smout, T C. *History of the Scottish People*, Bungay, 1972.
Swain, M. *Scottish Embroidery: Medieval to Modern*, London, 1986.
Thomson, W, McCallum, C. *Glasgow Caledonian University*, East Linton, 1998.

21 Reading and Study

HEATHER HOLMES

This chapter examines recreational reading and study in the Scottish home from the eighteenth century, when books and other printed works were expensive and considered objects of status, to the present day, when they are an integral part of Scottish culture. Because of the complex role which literacy and print culture has had on Scottish society, these activities will be examined across a number of social classes, and in relation to various occupations and geographical areas. The concept of 'home' here refers to a range of dwellings, from places where people live the year round or for a considerable period of time, to temporary and short-term seasonal accommodation.

LITERACY

Central to the activities of recreational reading and study in the home is literacy, which until the early nineteenth century was defined as the ability to read. Before then, writing was considered a separate activity and was taught as a distinct subject in schools.

Within the home, participation in both recreational reading and study was shaped by the great rise in literacy which took place primarily during the period from the eighteenth to twentieth centuries and was stimulated by a number of factors: the increasing importance attached to the written word and a written culture over oral equivalents; the rise of the publishing and printing industries; the upsurge of the popular press (especially after the abolition of the tax on advertisements in 1853, and the repeal of the Stamp Duty in 1855 and the Paper Duty in 1861); the growth of mutual improvement societies[1] and the role of reading in this; greater access to print; and the wider availability of elementary education, especially after 1872. By the 1880s, print culture had become so important that *Mrs Beeton's Housewife's Treasury* declared 'no home was completely furnished where books were absent or only sparsely represented'.[2]

Although significant advances took place in literacy during the eighteenth to twentieth centuries, literacy historians such as R K Webb,[3] Charles Withers,[4] Lawrence Williams,[5] R A Houston[6] and R D Anderson[7] have observed that its spread was uneven throughout Scotland. They have noted regional patterns as well as rural and urban differences (urban areas had higher levels at an earlier date), and variance between social classes, occupational groups and language speakers (Gaelic and Scots).

MATERIALS

Recreational reading and study require space and adequate heating and lighting within the dwelling, plus a range of materials. In the modern period, books and other printed matter (such as that of the periodical press) have been central to both activities, while study has also required writing tools and paper.

Book, library and publishing historians have analysed the development of personal book ownership. For the 1760s, John Crawford observes that, 'book buying was widely but infrequently practised, rather in the manner of house or car purchase today and reflected, both in subject and frequency, the old tradition of intensive book use.'[8] Even though books and examples of other types of printed matter have become accessible to all social classes, some surveys suggest that not all people own large numbers of these. In 1932 Q D Leavis noted that throughout Britain, 'the bulk of the public does not buy many books.'[9] At that and other times, autobiographies, oral testimony and photographic evidence show that in some working-class households few books have been found as possessions. In 1987, the Scottish Arts Council stated that, 'more than half the population buy no books.'[10] Further surveys have demonstrated that book buying is related to social class and education. One survey conducted by Peter H Mann and Jacqueline L Burgoyne in 1969 concluded that, 'the middle and upper middle classes bought 70 per cent more books than the working class.'[11]

Purchase is not the only means of gaining access to reading material. Readers have also borrowed or hired books, often from libraries. From the eighteenth century, public libraries became an increasingly important source of reading matter while the Public Library Act of 1850 allowed for the establishment of free public libraries, leading to a great increase in the book-borrowing public.[12] Material could also be lent by family, friends and neighbours. On farms in the Scottish Borders during the inter-War period Andrew Purves notes that when the farmer's family had read the newspapers they were passed on to the farm servants.[13]

Nevertheless, some homes have possessed sizeable collections of books, and these have fulfilled a number of roles for household members. They can be used as decorative objects, like the bookcases in which they are often housed. Photographic evidence from the 1880s, for example, shows books placed on tables throughout the home[14] (Fig. 21.1). The coffee-table book of the late twentieth century, a highly illustrated large-format publication, is also intended to be exhibited. Books can, in addition, be status symbols, demonstrating the ability of their owners to buy often costly goods. At times of growing literacy, they also indicated the ability to read, although their presence did not always indicate that all members of a household shared the same skill. Depending on the genre to which they belong, books have been associated with learning and scholarly activity. As objects of learning, they have conveyed notions of a civilised life and civilisation. For Marion Lochhead, writing of their role in the home in the eighteenth century,

Figure 21.1 Books used as decorative objects in the drawing room of Thomas Bonnar II, Edinburgh, c1880. Crown Copyright: RCAHMS. Source: Gow, I. *The Scottish Interior, Georgian and Victorian Décor, etc*, Edinburgh, 1992, 102.

they were 'an indication of taste and character'.[15] When handed down from generation to generation books become a symbol of family continuity. Catherine Carswell, who describes her 'Grandpapa's bookcase' in 1950, observes that 'it was the fullest repository of family feeling.'[16]

Within Scottish dwellings, the amount and the location of space devoted to books, reading and scholastic activities has varied greatly. The difference is especially marked in relation to social class and to the size and number of rooms in a house. Until the first half of the twentieth century, the majority of the Scottish population lived in homes consisting of one or two rooms. Household space was at a premium and had to be carefully utilised if a wide range of activities were to be undertaken and all family members provided with sleeping accommodation. Consequently, the number of possessions was kept to a minimum and of those given houseroom each had to have its own, ordered place. In two-roomed homes books were thus usually kept in 'the room', the best room,[17] which was used less intensively than the kitchen. In larger homes, however, more spaces were available in which to keep books. Although the drawing room was 'never a typical space within working-class or smaller rural homes', it was a feature in the houses of the upper and middle ranks, a showplace room where, among other possessions, books could be kept and displayed.[18] They were also to be found in the morning room, the boudoir (originally part of the bedroom suite), and the parlour, a family room where informal meals could be taken.

In these rooms, books and other printed material could form part of the room furnishings in a number of ways. In the mid nineteenth century, Hugh Miller comments that in the north of Scotland the family bookcase could comprise a small movable box; his own personal library was held in a box made of birch bark.[19] Books were also placed on specially constructed shelves, and in a range of cabinets, some of which were glass fronted, and highly ornate[20] (Fig. 21.2). Magazine racks held a range of periodical literature.

Modern furniture manufacturers' catalogues, such as that issued by the Swedish firm Ikea (which has two large stores in Scotland), illustrate a range of bookcases and other furnishings in which to store books. These include wall shelves,[21] wall units,[22] combined wall and bookcase units,[23] modular storage systems,[24] and wall unit combinations which have glass doors and adjustable shelves. Bookcases and storage units combine space for books and magazines, decorative objects (such as glassware) and personal possessions. However, books have had to compete with the ever-increasing amount of storage space required by new types of media. In 1989, one report suggested that 'record cabinets and shelving for cassettes and video tapes are replacing bookshelves,'[25]

Bookcases and storage units are made from a variety of woods, plastics, metals and glass. They are also manufactured in a wide range of sizes. Some are designed to be erected in a number of spaces within the home (Fig. 21.3). Wall shelves, which are considered in one furniture catalogue as

Figure 21.2 Book cases in William Reid's home, Edinburgh, 1894–6. Crown Copyright: RCAHMS. Source: Gow, I. *The Scottish Interior, Georgian and Victorian Décor, etc*, Edinburgh, 1992, 121.

380 • SCOTTISH LIFE AND SOCIETY

Figure 21.3 Flat pack in action. Shelving in a study, Burntisland, Fife, 2003. Source: S Storrier.

being 'vital for organised home life', can be put up in any room.[26] Some bookcases are designed creatively to define room space. The Ikea 'Expedit' bookcase, for example, can be 'secured to a wall, or used as a distinctly trendy room divider'.[27]

Very large collections of books require ample storage space, and usually their own room, the library, to house them. Personal libraries can occupy a converted room within a dwelling, while in some country houses they consist of purpose-built extensions to existing buildings. Libraries are noted in a wide range of dwellings including buildings which provided temporary shelter for those travelling.[28] They were generally found in the homes of the upper social classes and members of the professions. Using inventories, wills, printed catalogues, manuscript lists and sale catalogues throughout Britain before 1855, Robin Alston suggests that some 12.18 per cent of the private libraries established before this date were found in principal country houses;[29] for Scotland he lists 221 such libraries. Among the professional classes, the clergy was the largest owner group, accounting for 16.94 per cent of the total number of libraries recorded. They were followed by doctors (6.80 per cent), lawyers (3.12 per cent) and members of the armed services (1.66 per cent). Fellows of learned societies, such as the Royal Society and Society of Antiquaries had 4.30 per cent.[30]

These libraries were used by a number of people. Although primarily accessed by their owners, some were also consulted by a range of other readers. James Raven states that in England during the late eighteenth and early nineteenth centuries, they were 'designed for use by friends and neighbours',[31] and this was also noted for Scotland. In the early eighteenth century, the Laird of Kilravock, an enthusiastic book collector, 'would lend his books without question to friends far and near'.[32] Marjorie Plant also notes that books from private libraries were circulated to friends who lived at a considerable distance. In Caithness, in the eighteenth century, the Sinclairs of Mey lent a copy of Shakespeare's *Hamlet* from their Barrogill Castle library to a friend who lived twenty miles (34km) away, the return of the book being recorded.[33] In the 1840s the library at Spottiswoode House, Berwickshire, was for the 'general use of inhabitants',[34] while at Monteviot, Roxburghshire, it was utilised by those living on the estate.[35] Until 1826, ministers' libraries, especially those in the Highlands, had a number of purposes. For example, John Mackay of Lairg used his collection of Dutch and theological books to train candidates for the ministry.[36]

The spread of literacy amongst the higher classes and the passion for book collection helped to stimulate the development of the personal library. According to Robin Alston, the domestic library in Britain was largely established in about equal percentages for the periods 1640–1800 and 1801–1850 respectively.[37] By the middle of the nineteenth century, in England, P S Morrish notes that not only was the formal domestic library merging into the main living room but that 'libraries were becoming morning rooms for gentlemen'.[38] In 1934, *The Home of Today* notes that 'if no separate library is provided in a house, accommodation for books may find a place in the dining room, either in the form of bookcases or as separate shelving.'[39] As early as the late seventeenth century, evidence from the House of Binns, West Lothian, shows that General Thomas Dalyell had a personal room which he used as a library. However his Bibles were located in a number of rooms which included the 'dyninge roume' and the 'High Hall'.[40]

SPACES AND PLACES FOR READING AND STUDY

Just as reading matter and writing materials can be stored in a number of areas within the home, so can reading and study be undertaken in various locations. As an activity, study generally requires a more specialised space in the home than recreational reading, for which a seat of sorts and adequate lighting and sometimes heating suffice. Much depends on the size of the dwelling and the space available. Where space is limited, multi-purpose tables and places of rest are used. These can include chairs at the fireside or the reader's bed, and, for study, the kitchen table.

In many homes, participants create their own personal reading and study spaces. From her childhood in a Highland castle in the 1920s, Christian

Miller remembers that she used existing furnishings and fittings to create favourite reading spaces:

> I read anywhere and everywhere; curled up in my father's big armchair whose high back, designed to keep off draughts, formed an effective barrier against interruptions – in the bath, on horseback, in chapel (the book disguised inside the Jesus-decorated dust jacket of my Bible), and up trees, but especially in bed at night.[41]

Migrants aboard ships in the nineteenth century read in their travelling spaces, for example on their bunk.[42] Navvies employed at industrial construction sites in the first decade of the twentieth century, such as the Kinlochleven water works, Argyll, read on top of their beds and around the fireside in their huts.[43]

Spaces in the home can also be devoted primarily or exclusively to reading and study. In some dwellings a space may be created for reading and study within a room of wider purpose, while in larger homes a room or rooms may be given over entirely to the activities. Auxiliary spaces may also be used.

An area for study and reading can be established within a room by furnishing it with suitable items. Desks and chairs for study, for example, can be placed in various locations (see Fig. 7.3). Thus, the inventory of household contents of the Edinburgh lawyer James Drummond included a 'little old writing table' located in his back room which was also used as a bedroom and a store.[44] The 'fore bedroom' of Archibald Campbell had an escritoire, a piece of furniture associated with business activities, which suggested that he undertook such matters in that room.[45] Study could also take place at the bureau, usually located in the principal bedroom, which came into common use from 1760 and remained an important item of furniture throughout the nineteenth century.[46]

In recent years, the increasing use of home computers has led to the development of computer furniture which can provide a self-contained work and study area within a room – most often the living room or bedroom. Furnishings of this sort include computer tables, some having shelves and drawers where files and other paperwork can be stored, and printer tables, which may include space for storing compact discs and floppy discs; sometimes these two items of furniture are combined to form 'workstations' (see Fig. 10.5). These are not infrequently designed as cabinets so that computers and their peripherals can be stored away when not in use. Furthermore, old types of furniture, such as the bureau, have been modified to be used as computer work spaces.[47] The bedrooms of children and teenagers may combine both sleeping and study areas, and be furnished with a bunk bed or loft bed which provides a raised sleeping surface with a study area (with desk and storage) beneath.[48] Purpose-built residential accommodation for students also combines a bedroom with a study area, the latter demarcated by a desk or table, shelving and often a pin board.

Reading and study space can also take the form of designated rooms.

The library is a room immediately associated with both reading and study, although as we have seen other activities sometimes also took place there. In England, from the mid eighteenth century, the library 'had become a focal living and entertaining room for much of the English nobility and upper gentry'.[49] A drawing of the interior of Balmoral Castle, Aberdeenshire, in 1864 shows a billiard table sitting in the library[50] and the library at Manderston, Berwickshire, was turned into a billiards room by 1909.[51]

Other rooms have been set aside for study. Especially before 1872 and the passing of the Education (Scotland) Act, schooling was undertaken in a number of buildings, including houses, which sometimes possessed schoolrooms. There might be such a room within a schoolmaster's house or within the dwelling of upper- and middle-class children taught at home. Christian Miller recalls that her 1920s home had a schoolroom and a room for the governess.[52] Members of the professional classes who studied and worked at home also had their own 'study', which could combine an area for studying and a personal domestic library. Some homes, like that of Sir Walter Scott at Abbotsford, Roxburghshire, had both a library and separate study.[53]

Modern homes rarely contain libraries but sometimes possess studies. As with earlier examples, these usually have secondary functions as 'home offices' (for household administration and sometimes paid work), and they may also serve as sleeping accommodation for guests. In 2002, one furniture manufacturers' catalogue notes that 'these days the home office often needs to double up as a spare bedroom.'[54]

Space devoted to reading and study may also be created out of auxiliary spaces within the modern dwelling, such as that below the stairs or a converted loft. Home-furnishing designers have recognised the potential of using these areas and have designed study furniture to be installed in them.

THE IMPACT OF THE HOME ON RECREATIONAL READING AND STUDY

The experience of reading and study is shaped by the nature of the dwelling in which it occurs. For example, where large families live in small homes with few rooms, readers or students may find it difficult to overcome disturbances and disruptions. But the activities in other instances, especially when space can be devoted exclusively to them, can make for very private and personal experiences. David Daiches emphasises that his father's study in Edinburgh was a personal and private space, 'a sacrosanct room which no maid entered, except when he was out, to dust it'.[55] Writing of her middle-class background in the West of Scotland, Helen Lillie notes that her mother 'shut herself up in her bedroom every afternoon ostensibly to lie down and rest her heart, more likely to enjoy the latest from Boots Library'.[56] According to Peter H Mann and Jacqueline L Burgoyne, reading in bed 'is a well-known but little investigated activity'.[57] In the 1980s, a survey of teenagers' reading habits in Scotland highlighted their preference for reading in this place. For one teenager, bed was a personal space where he or she could read without interruption, 'I mostly read books while I am in bed as I get more peace.'[58] Another combined the pleasure of lying in bed with reading, 'After a long day it's nice to get into bed with a book.'[59]

The appearance, furnishing and layout of rooms can substantially enrich reading and study. Some libraries were highly decorated. A number of country-house libraries like that at Arniston House, Midlothian, were designed by prominent architects who combined exotic marbles, imitation marble and graining, sculptures (including busts of noted scholars) and frieze work.[60] The ways in which the library and study should be decorated and furnished were the subject of intense debate. Dena Attar notes that household books, dictionaries and encyclopaedias gave advice on their layout and furnishings.[61] *Mrs Beeton's Housewife's Treasury*, first published in 1861 and the principal domestic manual of the second half of the nineteenth century, described various aspects of library and study design, such as the best method of heating and furnishing the rooms. The latter included a discussion of fitted bookcases (wood type, size, arrangement of shelves and bays) and the arrangement of other items (reading wheels, reading chairs, stands, ladders, globes and busts).[62] In general, the design of the library embodied and reflected the value of books and their associations of knowledge, scholarship and civilisation. The study has also embodied the values of knowledge and scholarship. The decoration and furnishing of studies very much reflects the individual's interests and tastes. Thus, in 1898, the study of John Miller Gray, first curator of the Scottish National Portrait Gallery, had Chinoiserie and Japanoiserie in woodcuts hung on its walls.[63]

In terms of both methods and use, lighting has a profound impact on reading and study. Lighting has been achieved in Scottish homes by a number of means. According to Ian Gow, 'the skied library was to become a popular feature of eighteenth-century Scottish country houses',[64] in a fashion which is thought to have been established as a result of William Adam's design for the library at Arniston House.[65] Not all homes, however, could incorporate such an architectural arrangement and writers on interior design outlined further ways in which lighting could be arranged for reading and study, often suggesting that the activities should be undertaken in a north-facing room or a room which had a large window. A range of artificial methods were also necessary, and here regional and national patterns can be noted.[66] During the nineteenth century, new, industrial means of lighting, including gaslight, paraffin light and electric light, were introduced, and although limited to only some dwellings, their influence on the design of the home in general and on individual types of room was debated. Thus, in 1835 the leading English gas engineer, J O N Rutter, thought that gas lighting, which was harsh and bright, could be suitably used in the domestic library, but not in the drawing room.[67]

Early forms of artificial lighting, including firelight, had been typified by low levels of brightness, flickering and unpleasant smells, all of which made reading and study more difficult and far less pleasant than after the lighting revolution.[68] Poor lighting also affected people's use of space.

In the nineteenth century households were still being forced to gather around light sources.[69] There are many recorded examples of reading taking place around the fire. Hugh Millar describes reading on winter evenings in the north of Scotland in the mid nineteenth century:

> In the winter evenings his [elder uncle James'] portable bench used to be brought from his shop at the other end of the dwelling, into the family sitting room, and placed beside the circle round the hearth, where his brother Alexander, my younger uncle, whose occupation left his evenings free, would read aloud from some interesting volume for the general benefit – placing him always at the opposite side of the bench, so as to share in the light of the worker.[70]

Matters were similar in urban areas. An advertisement in the *Glasgow Post Office Directory* of 1861 for the Glasgow ironmonger, Finlay's, shows a man and woman sitting beside the fire in their parlour reading newspapers and books[71] (Fig. 21.4).

Today, most homes possess a very high standard of electric lighting in all of their rooms. This usually takes the form of a central light suspended from the ceiling which may be augmented by wall lamps or moveable lamps, such as table lamps and standard lamps, placed in different locations throughout the room. In addition, a range of lights and shades has been specially designed to facilitate reading and studying. Uplighters, which

Figure 21.4 Reading by the fire. Advertisement, 1861.

throw the beam upwards, for example, may incorporate a reading lamp,[72] and desk lamps, including angle-poise lamps, in which the direction of the beam can be adjusted, plus various types of desk-top static light, are common.

THE EXPERIENCE OF READING AND STUDY

The amount of time spent on recreational reading and study in the home has varied throughout time, and between different social classes and occupational groups, owing to a number of factors.[73] Perhaps the most important is the length of work hours. In the nineteenth century, the work hours of the lower classes were long while those of the middle classes were considerably shorter. However, the hours of work of the former group were gradually reduced and by the 1860s the Saturday half holiday was becoming general. For some occupational groups though, such as agricultural workers, the Saturday half holiday was not introduced until the early twentieth century. Long hours meant limited opportunities to read and study. In 1853, Hugh Miller remarked that for his Uncle James, a harness maker, his 'occupation left him little time for reading'.[74] Book historians such as Martyn Lyons have observed that the gradual reduction in the length of the working day gave workers greater opportunities to read.[75] For many housewives and other unpaid domestic workers, a reduction in the range and extent of housework tasks, plus the introduction of labour-saving gadgets around the home had a similar but less profound impact. But for both men and women, whatever enhanced opportunities there were for reading and study had to compete with new forms of entertainment and these will continue to develop in future years.

In the late 1980s, the Scottish Arts Council asserted that 'it would seem unarguable that book reading generally is not today the popular form of entertainment it was in the latter part of the nineteenth century and even in the earlier part of the present century.'[76] Activities which have competed for leisure reading time have not been the same for all individuals, either at one time, throughout time, for every social class or each sex. In the twentieth century and into the twenty-first they have included participation in societies and clubs (such as after-hours school clubs and youth clubs for children and teenagers), listening to the radio, watching television (not only network but also cable and satellite), the reading of specialist magazines (and books of cartoons), computer games, and going out to the cinema, drinking, dancing and partaking in other social activities.[77] Some of these, however, indirectly encourage the reading of books. A number of television programmes and feature films, for example, such as the 2001 *Harry Potter and the Philosopher's Stone* and the 2002 *Lord of the Rings: The Fellowship of the Ring*, are dramatic representations of literary works. The use of Internet has grown in the last ten years and will continue to do so in the foreseeable future. In 2001 the Scottish Executive published its strategy for extending the availability of broadband telecommunications in Scotland. This strategy recognises the importance of broadband telecommunications being available to every community. The strategy also argues for better access to libraries so that they can play their part in the development of e-learning, e-government, e-health and e-business.[78] Web sites can promote reading activities. Sites include those of publishers, libraries and organisations and projects which encourage reading, such as that for the National Year of Reading.[79]

Reading was fitted into a number of slots in the recreational timetable. Although often confined to the weekends, especially Sunday, some reading could take place on weekday evenings as part of the household routine. Historically, reading on Sundays has been an especially widely noted activity. For the last decades of the nineteenth century, W Hamish Fraser states that 'even in poor households ... The Sunday papers were important.'[80] His observation holds true for many households today, of all classes, where the buying and reading of the Sunday paper forms a well-integrated part of the routine activities undertaken on that day. In addition, the increasing importance of Saturday newspapers has had a growing role in weekend reading matters, and for some households, also during the following week. Week-day reading and study patterns are also discussed by Helen Lillie who suggests that, in a middle-class West of Scotland household in the early twentieth century, school homework was undertaken before the evening meal and was followed by reading or other activities such as listening to the radio or playing cards. Further reading could take place at bedtime, before both children and adults went to sleep.[81] There is a report from north-east Scotland during the inter-War years – although no doubt it occurred elsewhere and at other times – that bedtime reading might include the Bible[82] (see Fig. 13.5). For some young children, bedtime reading remains an important element in their daily routine. Reading in bed in the evening is still seen both as a way of relaxing and for preparing for sleep by all ages of people. In the 1980s teenagers spoke about its importance: 'I like reading at night because it makes me tired'; 'after a long day it's nice to get into bed with a book'; 'sometimes I read in bed because it helps my brain to unwind'.[83]

Not all members of a household share the same reading experiences. Preference plays as much of a role as the availability of leisure time and the presence of other forms of entertainment. Physical and mental disability and other forms of infirmity may also have an effect. Significant differences also exist between the reading experiences of men and boys and those of women and girls. In Scotland, as in England,[84] autobiographies of women and girls show the changing role and influence which reading has had at different stages in their lives. Those who mention their early years describe the importance which was attached to the activity then and the part it played in shaping their beliefs and attitudes. Christian Miller, in her reminiscences in the Highlands, states that 'we invented many ways of passing the long evenings, but above all, we read.'[85] As girls grew up, their involvement with books and reading altered. Writing about England for the period 1837 to 1914, Kate Flint notes that 'with the exception of women who read material which had a direct bearing on their professional careers, records of reading after marriage are scarce in autobiographies.'[86] For married women, reading patterns were shaped around domestic activities and routines and notions of what a good housewife was. The experiences of women rural workers in the Scottish Borders during the inter-War years which are mentioned by Andrew Purves are similar to those of women in earlier periods and other countries such as Germany, Australia[87] and England.[88] Purves says that 'Mother, like most of her housewife contemporaries, had little time for reading, although she was very literate and had a good fund of stories.'[89] For Kate Flint, reading was 'a recreation, not an occupation: if they were seen with a book, women were asked if they had no needlework to do'[90] (Fig. 21.5).

However, many women did read as part of their family and domestic duties;[91] to sick children as part of their convalescence (some fathers also undertook this task),[92] and if they had young children, they read to them before putting them to bed. At these times, women encouraged their children to learn and enjoy a well-chosen book. As Christian Miller notes once again, 'Picture books were brought out, and the pages turned. "Look! What's this, darling?"'[93]

Parents have a significant role to play in their children's reading, helping to shape attitudes towards books and literature. Alexander Somerville describes the favourable role his father played in his reading

Figure 21.5 A man reads his newspaper on the left of a central fire with stone-built back at Kirbister, Birsay, Orkney, in the 1960s. Rural men were often photographed reading by their firesides while their wives were usually shown engaged in some type of domestic or auxiliary task next to the warmth, as the spinning wheel on the opposite side of the fire indicates in this photograph. Source: SLA C.801D.

experiences when a young boy, during the first decades of the nineteenth century. When his father saw his enthusiasm for reading poetry he bought him a copy of *The Gospel Sonnets*. It was:

> ... no small thing for a man like him to pay half a week's wages, and send all the way to Edinburgh, for a book of verses for his boy, because he saw that boy eagerly laying hold of every printed poem, song, ballad or verse that could be reached, and in the exuberance of his enthusiasm for making rhymes for others to listen to.[94]

Somerville also indicates that parents assisted in choosing the books or other printed material which was to be read in the home. Other autobiographies mention parental censorship of reading matter, usually where the subject matter was considered inappropriate to a child's age. In describing support for his reading, Alexander Somerville also describes paternal disapproval of the reading of Robert Burns' poetry: 'Still he was not willing for me to become familiar with Burns. He said when I grew older I might read to him to advantage, when I could know what to admire and what to reject; it was hardly fit for me to read poetry while so young.'[95] Christian Miller describes reading in secret after her parents had extinguished her bedroom light. In such a case, she read aided by a torch beneath her bedcovers.[96] This manner of reading was not, however, without its difficulties; she had to keep her 'ears alert for the footfalls of grown ups' and had to deal with dying torch batteries.[97]

Study experiences have also varied from person to person. Children are engaged in full-time education (in the past this could also be part-time), as are many young adults. A working adult may be a member of the professional classes who regularly undertakes work-related study, or be participating in a part-time educational course, including evening classes. Vocational reading can overlap with leisure reading, especially for people who work in the various professions. The examples are numerous and can include solicitors reading professional magazines or textbooks to bring them up to date with developments, doctors reading the *British Medical Journal* or teachers reading the weekly *Times Educational Supplement*. Farmers too may read *The Scottish Farmer* or other examples of the wealth of agricultural literature which has been available. There are also increasing numbers of full-time 'mature' students, while the self-taught reader has become increasingly important from the early nineteenth century.[98]

THE ACT OF READING

In the home, the act of reading can take a number of forms (Figs 21.6, 21.7). It can involve an individual or a group and be a private or a shared experience. It can take place as a single event, or as a number of events, as when a serial publication, published at regular intervals, is read. The reading experience can be passive, for entertainment and enjoyment alone, or it can provoke reaction and discussion among readers or listeners.

Book historians such as Robert Darnton,[99] Alberto Manguel[100] and

silent	aloud
solitary	in company
single event	serial event

Figure 21.6 Forms of Reading.

passive	reactive
single event	serial event

Figure 21.7 Type of Experience.

Jonathan Rose,[101] writing on reading in Europe (including Britain) and other parts of the Western World, have noted that in the eighteenth to twentieth centuries a revolution took place in the way in which books and other printed matter were read, and in the experience of reading in general. In the eighteenth and nineteenth centuries, Scottish culture changed from being largely oral based to being predominantly print based. This transformation was reflected in a move away from the oral reading of texts and the communal and social experience of reading in the home, seen in etchings such as Walter Geikie's 'Reading a Tract' of the 1820s,[102] and paintings such as Sir David Wilkie's, 'The Cottar's Saturday Night', of 1837.[103]

Although the communal reading experience largely gave way to a private one where the individual read alone and silently, it continued in relation to a number of specific groups of readers and specialised reading activities. In the early twentieth century, group reading was recorded amongst immigrant groups such as the Irish navvies working at Kinlochleven, Argyll, and allowed for the dissemination of the written word to those who were illiterate.[104] Communal reading also took place at family gatherings in the Scottish Borders during the inter-War years, for example when the Bible was read on a Sunday.[105] In the latter half of the twentieth century, the oral reading of books became part of new forms of entertainment, such as radio, where programmes can include broadcasts of book readings. Formal reading groups, organised by societies and institutions, have become popular, especially in the late 1990s, and a well-established network has developed throughout Scotland and Britain.[106]

Reading activity in the dwelling is also related to the printed material which can be brought into the home. Periodical publications have shaped daily and weekly reading activities.[107] At weekly or longer intervals, printed matter can arrive from a local library, or other source. Particular publications have been associated with reading on specific days of the week. A number of specialist newspapers and periodicals have been developed for reading on a Sunday, such as Sunday newspapers (The *Sunday Post*, for example), and a range of weekly fiction papers issued on a Saturday. Religious works have had great importance on this day in particular. In the Purves household such material was read on Sunday evening when the family gathered together.[108] Publications of this type included a range of periodicals such as *The Christian Herald*,[109] the Bible, the Catechism and classics of fictional literature. David Daiches records the reading of parallel works on Saturday, the Jewish Sabbath.[110]

The reading of particular genres of literature has been investigated by a number of literary historians and sociologists. They conclude that in Scotland, and throughout Britain, specific types of literature have been associated with particular groups of readers. Adult males were especially associated with the reading of the newspaper press. Andrew Purves writes that in the inter-War years, 'father didn't read many books, but was a keen reader of the newspapers.'[111] For the period 1919 to 1950, when children's reading and publishing rapidly expanded, Joseph McAleer notes that although children now had numerous specific works published for them, they also read material intended for other readership groups; he observes an inclination for girls to read boys' literature, and for children and adolescents to read adult literature. Indeed, the reading of adult titles was especially important among 14-year-olds, as it allowed them to imitate their parents' reading habits and appear 'grown up'.[112] Children's comics were also popular and some of the titles – for boys (girls also had their own) – such as *The Rover* and *The Hotspur*, published by D C Thomson of Dundee, had a large circulation throughout Britain.[113] So important were these magazines that McAleer observes that 'it is difficult to appreciate the hold with which weekly magazines in particular had upon children, let alone their effect on tastes in reading.'[114]

Women read a range of publications. In the same way as children and adult males, a number of magazines was specifically published for women to read. Amongst the most popular titles were again those published by D C Thomson: *The People's Journal* (estimated to have been read by one million people every week in 1989), *My Weekly* (1910–), and *The People's Friend*.[115] Women also regularly read advice texts for routine household and health matters and had cookery books. During the early nineteenth century, especially from the 1840s onwards – when the market for popular literature was developing – novels became an increasingly important part of the new reading public. For that period, Martin Lyons comments that:

> ... novels were held suitable for women, because they were seen as creatures of the imagination, of limited intellectual capacity, both frivolous and emotional. The novel was the antithesis of practical and instructive literature ... Above all, the novel belonged to the domain of the imagination.[116]

For him, 'the nineteenth-century novel was thus associated with the (supposedly) female qualities of irrationality and emotional vulnerability,'[117]

The association of women with romance fiction has continued into the twentieth century. A Scottish study by Bridget Fowler,[118] together with British surveys conducted by Peter H Mann,[119] and research undertaken by Janice Radway in the United States,[120] show the continued importance of romance as a genre for women readers. Although such readers are of a broad range of ages, many are married, work at home and have young children. In a national survey of Mills and Boon romantic novel readers in 1968, Mann concludes that 'Probably the Romances appeal to younger women who are literate enough to cope with a genuine full-length book and who enjoy, and respond to, the ordinary formulae of the romantic story.'[121] Fowler, in her study of women readers in the West of Scotland during the period 1986–8, painted a less-positive picture of the consumers of romantic fiction; they were:

> ... dogged by crisis: the sudden death in their forties and fifties of their husbands, nervous breakdown lasting for years at a time, traumatic marriages in which they were the victims of battering, absentee husbands, long-term unemployment. They were additionally exposed, as women, to charges of intellectual inferiority.[122]

From the late 1960s onwards surveys have highlighted the private nature of this type of reading experience. Reading romantic novels allows women to devote a little time to being by themselves. They seldom discuss the romances they read or their reading experiences, even when titles have been borrowed from friends and relatives. For some women, the novels play an important part in their lives and they read a number of titles each month. Fowler stresses the importance of the 'imaginary world' they create for their readers[123] which provides 'fantasies of power and plenty. Within these fantasies, the wishful images of sexually-attractive heroes are to be unadulterated by alien traces of mundane reality.'[124] The books provide a release and an escape from the everyday world and the daily problems of life.

We know of the personal reading tastes of a number of well-known figures, such as Walter Scott,[125] Henry Sinclair, Bishop of Ross between 1560 and 1565,[126] and the Rev John Beveridge, first President of the Scottish Esperanto Federation, President of the Scottish Beekeepers' Association and a Knight of St Olaf (1857–1943),[127] from the contents of their personal libraries. Information is sometimes available for the less renowned: Ian Mowat notes that a town clerk of Tain in the eighteenth century owned 180 volumes on history, travel and military subjects.[128] The Edinburgh lawyer, Andrew Buckney, who died in 1758, left a small personal library which comprised French language works, including plays and tales, and a range of English fiction, poetry, history, gardening publications, law texts and works relating to general culture.[129]

CONCLUSION AND SUGGESTIONS FOR FURTHER RESEARCH

Recreational reading and study in the Scottish home have undergone many changes in the period between the eighteenth century and the present. Both are activities shaped by changing fashions, in interior decoration, in the use of rooms and in methods of lighting, as well as a range of social and cultural factors relating to work and leisure, and the printed word, education and learning. They have not always been undertaken by all members of a household, at least not to the same degree, nor have reading and study facilities been equally available to all living under one roof.

It is clear that there is much more to be learned about reading and study in the Scottish home, especially the former of these activities. Although there is now a growing interest in reading and reading practices, and much valuable research is currently being undertaken in these areas, relatively little attention has been given to the domestic context. Thus, while a great deal is known about reading activities in circulating libraries, proprietary libraries and mechanics' institutes, and the development of these in Scotland (see, for example, Alistair R Thompson,[130] W R Aitken,[131] Paul Kaufman,[132] John C Crawford[133] and John Minto[134]), little work has been carried out on domestic libraries. Likewise, although study and education in formal educational institutions have received a great deal of attention, the use of the home as a setting for these phenomena has been much neglected. However, some findings by Simon Eliot and Stephen Colclough in their international project, the Reading Experience Database, run jointly by the British Library and the Open University, and the work of Robin Alston, James Raven and Jonathan Rose,[135] have shown the wealth of valuable research which can be undertaken, but at the same time the paucity of knowledge about reading in the home.

Areas for further investigation might include the role of reading and study in the home among particular occupational and social groups (including recent immigrant groups), either within a specific region or nationally, or within a comparative context. Studies of individuals and their reading and study patterns would provide a fuller understanding of the range of reading activities which have been undertaken within households. Research could be concerned with reading at particular times and events, such as festive periods, family celebrations and on a Sunday, and, for certain historical periods, the role of the Bible and religious works in reading experience.

As there is comparatively little known about the interiors of domestic libraries and spaces where reading and study informally took place, further research could be undertaken into the types of furnishing and lighting found there, including the social and cultural factors behind their development and use and their part in shaping reading and study experience. Such research would help to provide a fuller understanding of two related household activities which have played an increasingly significant role within the home over the past three centuries.

NOTES

1. Rose, 2002, ch 2.
2. Morrish, 1994, 33.
3. For example Webb, 1954, 100–14.
4. For example, Withers, 1986.
5. For example, Williams, 2000.
6. For example, Houston, 1985; Houston, 1993; Houston, Tyson, 1991.
7. For example, Anderson, 1995.
8. Crawford, 1992–3, 26.

9. Leavis, 2000, 4.
10. Scottish Arts Council, 1989, 5.
11. Mann, Burgoyne, 1969, 27.
12. See for example Rose, 2002, 59–60, 116–7. Note should be made of the development of mobile libraries. See *Scottish Libraries*, vol 13/5 issue 77 (1999), 8.
13. Purves, 2001, 19.
14. See the illustrations throughout Gow, 1992.
15. Lochhead, 1948, 360.
16. Carsewell, 1997, 32.
17. Clark, 1996, 68.
18. Kinchin, 1996, 155.
19. Miller, 1993, 27.
20. Raven, 1996, 190.
21. Ikea, 2001, 87.
22. Ikea, 2001, 61.
23. Ikea, 2001, 62–3.
24. Ikea, 2001, 64–5.
25. Scottish Arts Council, 1989, 5.
26. Ikea, 2001, 87.
27. Ikea, 2001, 91.
28. Barker, 1971–3.
29. http://www.r-alston.co.uk/privstat.htm (statistics on 06–08–98).
30. http://www.r-alston.co.uk/privstat.htm (statistics on 06–08–98).
31. Raven, 1996, 174.
32. Plant, 1952, 235.
33. Plant, 1952.
34. *NSA*, II (1845), 86.
35. *NSA*, III (1845), 186.
36. Maclean, 1922–4, 95.
37. http://www.r-alston.co.uk/privstat.htm (statistics on 06/08/98).
38. Morrish, 1994, 27.
39. Quoted in Morrish, 1994, 41.
40. Dalyell, Beveridge, 1923–4, 354, 356–8.
41. Miller, 1997, 70.
42. Bell, 1999.
43. Macgill, 1914, 238.
44. Nenadic, 1994, 140–1, 155.
45. Nenadic, 1994, 141.
46. Nenadic, 1994, 144; Gow, 1992, 103.
47. Ikea, 2001, 217.
48. Ikea, 2001, 300–1, 312.
49. See for example Flint, 1995, 103; Raven, 1996, 176, quotation on 188.
50. Gow, 1992, 66.
51. Morrish, 1994, 35.
52. Miller, 1997, 13.
53. Gow, 1992, 42.
54. Ikea, 2001, 46.
55. Daiches, 1997, 25.
56. Lillie, 1999, 50.
57. Mann, Burgoyne, 1969, 19.
58. Scottish Arts Council, 1989, 55.
59. Scottish Arts Council, 1989, 49.
60. Gow, 1992, 15.

61. Attar, 1987, 16.
62. Beeton, 1880, 165, 279, 284.
63. Gow, 1992, 131.
64. Gow, 1992, 15.
65. Morrish, 1994, 33.
66. Bayne-Powell, 1956, 132.
67. Schivelbusch, 1983, 158; see also Chapter 5 Lighting.
68. For a discussion of the impact of lighting on reading in England see Altick, 1957, 92–5.
69. Altick, 1957, 166.
70. Miller, 1993, 33–4.
71. Kinchin, 1996, 160.
72. Ikea, 2001, 322–3.
73. For an overview of recreational reading and study in the home by the British working classes see Rose, 2002, especially chs 1, 2, 3, 4, 5, 11 and 12.
74. Miller, 1993, 33; See also Chapter 7 Rest.
75. Lyons, 1999, 333.
76. Scottish Arts Council, 1989, 5.
77. See Chapter 10 Television, Video and Computing; Chapter 25 Childcare and Children's Leisure.
78. The strategy document can be found at www.scotland.gov.uk/digital scotland/csbc/csbc–00.asp
79. For example www.yearofreading.org.uk and www.scottish-booktrust.com
80. Fraser, 1981, 72.
81. Lillie, 1999, 50–1.
82. McBey, 1993, 4.
83. Scottish Arts Council, 1989, 50.
84. Flint, 1995, 231–2; see also Rose, 2002, who makes extensive use of autobiographies to demonstrate the role of reading in the lives of members of the British working class, especially between the eighteenth and twentieth centuries.
85. Miller, 1997, 70.
86. Flint, 1995, 208.
87. Lyons, 1999, 321.
88. Flint, 1995, 192.
89. Purves, 2001, 19.
90. Flint, 1995, 192.
91. For England see Flint, 1995, 196–9.
92. Miller, 1997, 44.
93. Miller, 1997, 31.
94. Somerville, 1951.
95. Somerville, 1951.
96. Miller, 1997, 71.
97. Miller, 1997, 71.
98. See for example Butt, 2000.
99. Darnton, 1990.
100. Manguel, 1996.
101. Rose, 2002.
102. Macmillan, 1986, plate 97.
103. Macmillan, 1986, plate 45.
104. Macgill, 1914, 283.
105. Purves, 2001, 35.
106. Hartley, 2001, see especially section, 'Tables', 150–69.
107. Purves, 2001, 19.

108. Purves, 2001, 35.
109. Purves, 2001, 35.
110. Daiches, 1997, 82.
111. Purves, 2001, 19.
112. McAleer, 1992, 134.
113. McAleer, 1992, 137.
114. McAleer, 1992, 148.
115. McAleer, 1992, ch 6.
116. Lyons, 1999, 319.
117. Lyons, 1999, 319.
118. Fowler, 1991.
119. Mann, 1969.
120. Radway, 1987.
121. Mann, 1969, 23.
122. Fowler, 1991, 136.
123. Fowler, 1991, 148.
124. Fowler, 1991, 148.
125. Corson, 1963, 44–5.
126. Cherry, 1963, 13–18.
127. Fraser, 1967, 211–2.
128. Mowat, 1979, 9.
129. Lochhead, 1948, 348–9.
130. Thompson, 1963.
131. Aitken, 1971.
132. Kaufman, 1965; Kaufman, 1969, 134–47.
133. Crawford, 1996, 49–61.
134. Minto, 1932.
135. Rose, 2002, see index, 526.

BIBLIOGRAPHY

Aitken, W R. *A History of the Public Library Movement in Scotland to 1955*, Glasgow, 1971.
Altick, R D. *The English Common Reader: A Social History of the Mass Reading Public, 1800–1900*, Chicago, 1957.
Anderson, R D. *Education and the Scottish People, 1750–1918*, Oxford, 1995.
Attar, D. *A Bibliography of Household Books Published in Britain, 1800–1914*, London, 1987.
Barker, J R. Lord Gardenstone's library at Laurencekirk, *The Bibliotheck*, 6 (1971–3), 41–51.
Bayne-Powell, R. *Housekeeping in the Eighteenth Century*, London, 1956.
Beeton, Mrs. *Mrs Beeton's Household Treasury* (1861), London, 1880.
Bell, B. Bound for Australia: Shipboard reading in the nineteenth century. In Myers, R, Harris, M, eds, *Journeys through the Market*, Winchester, 1999, 119–40.
Butt, J. Further education institutions. In Holmes, H, ed, *The Institutions of Scotland: Education, Scottish Life and Society: A Compendium of Scottish Ethnology*, vol 11, East Linton, 2000, 175–97.
Carruthers, A, ed. *The Scottish Home*, Edinburgh, 1996.
Carsewell, C (ed J Carswell). *Lying Awake: An Unfinished Autobiography and Other Posthumous Papers* (1950), Edinburgh, 1997.
Cavallo, G, Chartier, R (trans L G Cochrane). *A History of Reading in the West*, Oxford, 1999.
Celebrating 50 years of mobile libraries in Scotland, *Scottish Libraries*, vol 13/5 issue 77 (1991), 8.
Cherry, T A F. The library of Henry Sinclair, Bishop of Ross, *The Bibliotheck*, 4: 1 (1963), 13–18.

Clark, H. Living in one or two rooms in the city. In Carruthers, A, ed, *The Scottish Home*, Edinburgh, 1996, 60–132.
Corson, J C. Some American books at Abbotsford, *The Bibliotheck*, 4: 2 (1963), 44–65.
Crawford, J. Reading and book use in eighteenth-century Scotland, *The Bibliotheck*, 18 (1992–3), 23–39.
Crawford, J C. The ideology of mutual improvement in Scottish working-class libraries, *Library History*, 12 (1996), 49–61.
Daiches, D. *Two Worlds: An Edinburgh Jewish Childhood* (1956), Edinburgh, 1997.
Dalyell, J, Beveridge, J. Inventory of the plenishing of the House of the Binns at the date of the death of General Thomas Dalyell, 21st August 1685, *PSAS*, 5th series 10 (1923–4), 344–70.
Darnton, R. *The Kiss of Lamourette: Reflections in Cultural History*, London, 1990.
Eliot, S. Some trends in British book production, 1800–1919. In Jordan, J O, Patten, R L, eds, *Literature in the Marketplace: Nineteenth-Century British Publishing and Reading Practices*, Cambridge, 1995, 19–43.
Flint, K. *The Woman Reader, 1837–1914*, Oxford, 1995.
Fowler, B. *The Alienated Reader: Woman and Romantic Literature in the Twentieth Century*, London, 1991.
Fraser, K C. The Beveridge Collection in St Andrew University Library, *The Bibliotheck*, 5 (1967), 211–2.
Fraser, W H. *The Coming of the Mass Market, 1850–1914*, London, 1981.
Gow, I. *The Scottish Interior Georgian and Victorian Decor: A Visual Anthology of the Domestic Room in Scotland Culled Principally from the Collections of the National Monuments Record of Scotland*, Edinburgh, 1992.
Hartley, J (in association with S Turvey). *Reading Groups*, Oxford, 2001.
Houston, R A. *Scottish Literacy and Scottish Identity: Illiteracy and Society in Scotland and Northern England, 1600–1800*, Cambridge, 1985.
Houston, R A. Literacy, education and the culture of print in Enlightenment Edinburgh, *History*, 78 (1993), 373–92.
Houston, R A, Tyson, R E. The geography of literacy in Aberdeenshire in the early eighteenth century, *Journal of Historical Geography*, 17: 2 (1991), 135–45.
Humphreys, K W. The book and the library in society, *Library History*, 7: 4 (1985), 105–13.
Ikea. *2002 Ikea: Prices valid until 20 August 2002*, UK, 2001.
Jones, D. Living in one or two rooms in the country. In Carruthers, A, ed, *The Scottish Home*, Edinburgh, 1996, 37–58.
Kaufman, P. The rise of community libraries in Scotland, *The Papers of the Bibliographical Society of America*, 59: 3 (1965), 233–94.
Kaufman, P. *Libraries and their Users: Collected Papers in Library History*, London, 1969.
Kinchin, J. The drawing room. In Carruthers, A, ed, *The Scottish Home*, Edinburgh, 1996, 155–80.
Leavis, Q D. *Fiction and the Reading Public*, 1932, London, 2000.
Lillie, H. *A New Kind of Life: An Informal Autobiography*, Glendaruel, 1999.
Lochhead, M. *The Scots Household in the Eighteenth Century: A Century of Scottish Domestic and Social Life*, Edinburgh, 1948.
Lyons, M. New readers in the nineteenth century: Women, children, workers. In Cavallo, G, Chartier, R, eds, *A History of Reading in the West*, Oxford, 1999, 313–44.
Macgill, P. *Children of the Dead End: The Autobiography of a Navvy*, London, 1914.
Maclean, Rev Professor D. Highland libraries in the eighteenth century, *TGSI*, 31 (1922–4), 69–97.
Macmillan, D. *Painting in Scotland: The Golden Age*, Oxford, Edinburgh and London, 1986.
McAleer, J. *Popular Reading and Publishing in Britain, 1914–50*, Oxford, 1992.
McBey, J (ed N Barker). *The Early Life of James McBey: An Autobiography* (1977), Edinburgh, 1993.

Manguel, A. *A History of Reading*, New York, 1996.
Mann, P H. *The Romantic Novel: A Survey of Reading Habits*, London, 1969.
Mann, P, Burgoyne, J. *Books and Reading*, London, 1969.
Marshall, R K. The plenishings of Hamilton Palace in the seventeenth century, *ROSC*, 3 (1987), 13–22.
Mays, K J. The disease of reading and Victorian periodicals. In Jordan, J O, Patten, R L, eds, *Literature in the Marketplace: Nineteenth-Century British Publishing and Reading Practices*, Cambridge, 1995, 165–94.
Miller, C. *A Childhood in Scotland*, 1981, Edinburgh, 1997.
Miller, H. *My Schools and Schoolmasters*, 1854, Edinburgh, 1993.
Minto, J. *A History of the Public Library Movement in Great Britain and Ireland*, London, 1932.
Morrish, P S. Domestic libraries: Victorian and Edwardian ideas and practice, *Library History*, 10 (1994), 27–44.
Mowat, I R M. Literacy, libraries and literature in eighteenth and nineteenth-century Easter Ross, *Library History*, 5: 1 (1979), 1–10.
Nenadic, S. Middle-rank consumers and domestic culture in Edinburgh and Glasgow, 1720–1840, *Past and Present*, 145 (1994), 122–56.
O'Dea, W T. *The Social History of Lighting*, London, 1958.
Plant, M. *The Domestic Life of Scotland in the Eighteenth Century*, Edinburgh, 1952.
Purves, A. *A Shepherd Remembers: Reminiscences of a Border Shepherd* (Flashbacks, no 13), East Linton, 2001.
Radway, J A. *Reading the Romance: Women, Patriarchy, and Popular Literature*, Chapel Hill, 1987.
Raven, J. From promotion to proscription: Arrangements for reading and eighteenth-century libraries. In Raven, J, Small, H, Tadmor, N, eds, *The Practice and Representation of Reading in England*, Cambridge, 1996, 175–201.
Raven, J, Small, H, Tadmor, N, eds. *The Practice and Representation of Reading in England*, Cambridge, 1996.
Rose, J. *The Intellectual Life of the British Working Class*, Yale, 2002.
Schivelbusch, W (trans A Davies). *Disenchanted Night: The Industrialisation of Light in the Nineteenth Century*, Berkeley, 1983.
Scottish Arts Council. *Readership Report: A Document prepared by a Working Party of the Literature Committee of the Scottish Arts Council*, npp, March 1989.
Somerville, A (ed J Carsewell). *The Autobiography of a Working Party Man* (1848), London, 1951.
Tadmor, N. 'In the even my wife read to me': Women, reading and household life in the eighteenth century. In Raven, J, Small, H, Tadmor, N, eds, *The Practice and Representation of Reading in England*, Cambridge, 1996, 162–74.
Thompson, A R. The uses of libraries by the working classes in the early nineteenth century, *Scottish Historical Review*, 42 (1963), 21–9.
Warrack, J. *Domestic Life in Scotland, 1488–1688: A Sketch of the Development of Furniture and Household Usage* (Rhind Lectures in Archaeology, 1919–20), London, 1920.
Webb, R K. Literacy among the working classes in nineteenth-century Scotland, *Scottish Historical Review*, 33 (1954), 100–14.
Williams, L. Scottish literacy. In Holmes, H, ed, *The Institutions of Scotland: Education, Scottish Life and Society: A Compendium of Scottish Ethnology, vol 11*, East Linton, 2000, 344–59.
Withers, C W J. 'Moral statistics': A note on the language and literacy in the Scottish Highlands in 1822, *Local Population Studies*, 36 (1986), 36–46.
http://www.r-alston.co.uk/
http://www.scotland.gov.uk/digitalscotland/csbc/csbc–00.asp
http://www.scottish-booktrust.com
http://www.yearofreading.org.uk

22 Gardens

SALLY A BUTLER

DEFINITIONS AND USES

Gardens and yards
One of the difficulties in tracing Scottish garden history is the confusion of terminology. 'Garden' in Scotland often meant something between what we would understand today by an allotment and a smallholding. Yet in some parts of the north east, for example, gardens for growing vegetables, fruits and flowers were frequently called yards.[1] Sometimes references are to the yard and garden (c1586) or to the garden yard, which suggests that these were differing concepts, but elsewhere the terms appear to overlap or even coincide, especially in the last 250 years.

Fourteenth- to seventeenth-century sources indicate that yards grew kale and leeks, herbs, parsley, beets, lettuce and parsnips. They could also contain fruit trees, and a 'grass' yard is mentioned in 1466. A 'sauch yard' (1668) was for willows, from which baskets could be made. In later, post-eighteenth century times, the yard was the usual name for the cottage or kitchen garden. Its practical function is stressed by the term kailyard, dating from the mid sixteenth century. It was dyked, and in any community, rural or urban, the occupants of houses, who paid rent also for their yards, were expected to maintain the dykes.

The earliest gardens in Scotland are those associated with religious communities, later with palaces, colleges, universities and the houses of the nobility. They included fruit trees, particularly apples and pears, and herbs for medicinal and culinary purposes were prominent, especially in monastic and college gardens. There were elaborate flower beds and borders, and parterres.[2] So provisioning, recreation, leisure, pleasure, healing and study (in college and botanical gardens) were primary functions of gardens, and these often continue to be important. More recently, public and private gardens jointly provide for all these functions and are increasingly seen as an important way of enhancing biodiversity within urban areas.

Allotments
Allotments are gardens which are not attached to a dwelling, the standard size of an allotment plot being 300 square yards (251 m²). Originally created to provide land for those displaced by the agricultural improvements of the eighteenth and nineteenth centuries, allotments were intended to offer a

means of self-provisioning and flower growing was not encouraged. This has changed in recent years as allotment gardening has widened its appeal to a greater variety of people. Many now grow flowers along with the traditional vegetables and fruits and allotments are seen as providing recreation, escape and relaxation.

EARLY GARDENS

The earliest Scottish gardens for which we have evidence are those associated with monasteries and abbeys. In dark-age and medieval Scotland all orders appear to have engaged in the important task of gardening. Quoting an earlier source, John Claudius Loudon describes a sixth-century garden attached to the Abbey of Icolmkill (Iona) in the Hebrides:

> On a plain adjoining the gardens of the abbey, and surrounded by small hills, there are vestiges of a large piece of artificial water, which has consisted of several acres, and been contrived both for pleasure and utility. Its banks have been formed by art into walks, and though now a bog, you may perceive the remains of a broad green terrace passing through the middle of it, which has been raised considerably above the water.[3]

In the later Anglo-Norman period, elaborate herb gardens, flower gardens and orchards were developed by heads of the Church, who brought with them knowledge of agriculture, gardening and plants from large monasteries south of Scotland. Although some bishops employed professional gardeners, all individuals resident within monasteries and abbeys worked in the gardens and orchards and for centuries much of the knowledge of and expertise in gardening was held within the walls of these establishments, sited in both town and countryside.

Monastic orchards contained apple, pear and sometimes plum trees. Many of the religious communities grew forest and hedgerow trees as well as garden plants, and encouraged their tenants to plant both. Several had extensive nursery stocks and supplied many lay gardens in their local area. The gardens were not only a source of food and work and a means of recreation for the community's members, but also provided a variety of remedies for the ills of their tenants and of the general population of the surrounding countryside:

> Elecampane for coughs and stomach, Tansy a cure for the gout, Barberry as a febrifuge, St John's Wort for hysteria ... Some herbs were doubtless handed down from the pre Christian era through the monasteries of the Celtic Church, such as Menyanthes trifoliate, the buckbean.[4]

Large lay gardens were quite rare before the mid fifteenth century. This has been attributed to the unsettled and lawless nature of Scotland which demanded fortified dwellings, these having little space for extensive gardens. The royal gardens at Stirling are widely regarded as composing the most important example prior to 1600. Within these gardens were grown herbs, vegetables, flowers and fruit. They included a magnificent knot garden, laid out solely for pleasure, which reached its apogee in the 1530s under James V and his wife Mary of Guise, a keen gardener strongly influenced by French garden design.

GARDENS ON ESTATES

Many wealthy Scots during James V's reign were educated abroad and brought back to Scotland fashionable Renaissance garden ideas, plants and seeds to try out on their lands, marking the beginning of country-estate planning. The gardens at Edzell Castle, Angus, recognised by many to be the earliest (1640) and most perfect of the pleasure gardens in Scotland, are one such example influenced by Continental design styles, '... overlooked by the garden house and sheltered by the most fantastic garden walls, with square recesses for the blue and yellow chequer of Lindsey colours made in flowers and carvings devoted to deities, virtues, sciences and humours'.[5] The walls, essentially all that remains of the original garden, are indeed works of art, yet they also possess practical functions, capturing sunshine and protecting both plants and people from the wind. The surrounding wall has been characteristic of many Scottish gardens, from the high-status pleasure gardens of the wealthy to labourers' humble kailyards. The protection that it, or trees or hedges, offer against wind as well as predators allows greater opportunity for growing a variety of plants.

A contemporary (1654) account of a more modest estate describes a garden which combined utility and ornament. Mr Thomas Stewart, the owner of the Coltness estate in Lanarkshire:

> ... sett himself to planting and inclosing, and so to embellish the place ... The gardens were to be to the south of the house, much improven and inlarged, and the nursery-garden was a small square inclosure to the west of the house. The slope of the grounds to the west made the south garden, next to the house, fall into three cross tarrasses. The tarrass fronting the south of the house was a square parterre, or flour-garden, and the easter and wester, or the higher and lower plots of ground were for cherry and nut gardens, and walnut and chestnut trees planted upon the head of the upper bank, towards the parterre, and the slope bank on the east syde the parterre was a strawberry border ... These three tarrasses had a high stone wall on the south, for ripening and improving finer fruits, and to the south of this wall was a good orchard and kitchen garden, with broad grass walks, all enclosed with a good thorn hedge; and without this a ditch and dry fence, inclosing severall rows of timber trees for shelter; to the west of the house ... was a large square timber tree park, with birches ... rows of ash and plain [probably plane], and in the middle

> a goodly thicket of firs ... north from the house, was a grass inclosure of four akers, with a fish pond in the corner for pikes and perches. All was inclosed with a strong wall and hedge-rowes of trees ... [6]

The terrace is another common feature of the older Scottish country-house garden. Many houses were still being built with defence in mind, on high ground, often with slopes leading up to the house. These slopes needed to be terraced to accommodate fashionable parterres, which require flat ground for best effect.

While landowners could engage themselves in improving their estates, the great majority of Scots were tenants and labourers without the luxury of the time and money necessary to develop a sophisticated garden. Before the early eighteenth century, Scotland was essentially rural, with 90 per cent of Scots living in small, nucleated hamlets. These hamlets, often known as 'fermtouns' in lowland areas, 'had an air of temporariness, reflecting the limited resources available ... A few great aristocratic families owned vast areas and ruled their tenants absolutely.'[7] Very few Scots therefore could be sure of the continued tenancy of the land which they worked and moved frequently. The result was poor horticultural practice, with most gardens being neither drained nor manured.

The gardens and yards of the tenant farmer and his farm workers would have primarily grown food, although there are seventeenth-century accounts of cottage gardens covered with bright orange marigolds. Marigolds seem to have been regarded as a weed in certain areas, but it is possible to speculate that elsewhere they may have functioned as early companion planting to keep crops pest-free. Kail and oatmeal were among the principal foods of the farm labourer in Scotland. As a valuable crop, kail plants were often surrounded by a wall or hedge. One form of such protection is the 'plantiecrue', found in the Northern Isles and parts of Caithness and Lewis. This was a round or four-sided enclosure of turf and stone, which protected seedlings grown in August or September before they were transplanted to gardens.[8] Other vegetables grown in early tenants' and labourers' gardens might include peas, beans and, at a later date, turnips, potatoes and parsnips, but these were rare.[9]

Landowners could make a considerable difference to the quality of their tenant farmers' and farm-workers' living conditions, including their gardens. As early as 1699, Lord Belhaven was urging that the layout of farm buildings should take advantage of the prevailing winds and the sun:

> Yards or gardens should be planted with ash and elm trees, to be used eventually for the upkeep of the house ... The garden plants were to include cabbage and kail, potatoes and turnips, and a few 'turkie beans' or ordinary beans and peas to go with pork.[10]

Curiously he also suggested that the gardens should be placed on the north side of the dwelling house.

The Act of Union early in the eighteenth century brought a renewed

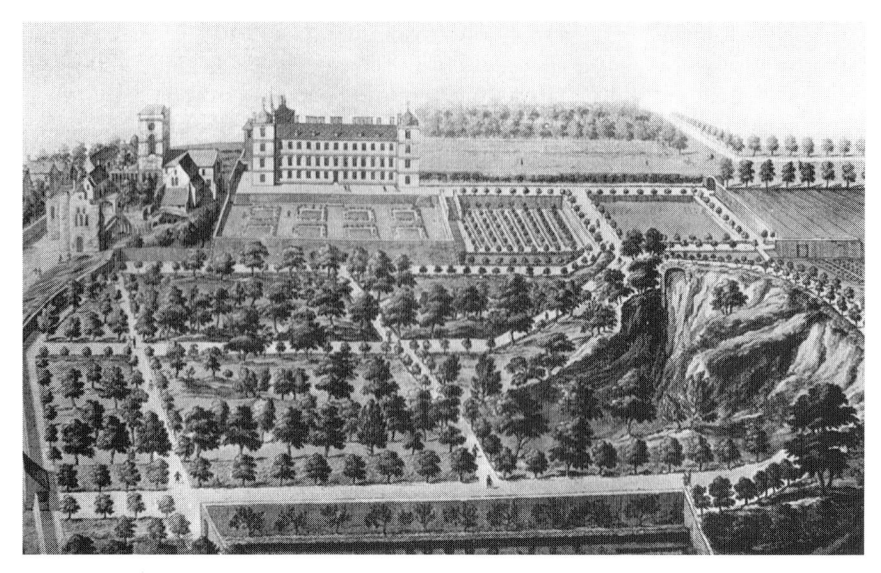

Figure 22.1 The gardens of Culross, Fife, clearly show the formality of the design and the representation of the owners' power over the landscape. Source: Cox, 1935, facing 58.

interest in estate improvement which subsequently transformed the countryside. 'Entire estates were reconstructed by the ruling aristocracy. Whole valleys were affected by the designs, settlements were re-sited; and tens of thousands of trees were planted.'[11] Extensive parklands were designed with straight avenues and rides cutting through plantations of sycamore, alder, ash, birch, beech, elm and Scots pine. The gardens at Culross, Fife, one of the largest estates in Scotland at the time, show very clearly the formality of planting and the clear representation of the landowner's power over the countryside (Fig. 22.1). *The Scots Gard'ner*, written by John Reid in 1683 and the first book aimed specifically at the Scottish gardener, advises a geometric layout for the gardens, 'make all the Buildings and Plantings ly so about the house, as that the House may be the centre all the Walks, Trees and Hedges running to the house.'[12]

While the Scots were creating their great gardens in the ordered, geometric style promoted by Reid, their neighbours south of the border were undergoing a revolution in garden design. The new style, inspired by painters and promoted by poets like Alexander Pope, turned away from the formality, order, and what was thought of as the artificiality of the Baroque style to embrace simplicity and naturalness. The fashionable garden became one of classical buildings and follies set in a landscape of grassy meadows, grazing sheep and clumps of trees. The style took on different, more exaggerated forms as the Picturesque and the Sublime, which sought inspiration in a wilder, untamed nature.

The new style, later referred to as the English Landscape style, was not adopted as widely in Scotland as in England, but it did influence many of the great landowners. The mid eighteenth-century writings of Sir John Dalrymple and Lord Kames established links between garden design, the senses and the emotions. By 1740 Lord Kames was advocating the new style and practising it at Blair Drummond, Perthshire, and others followed. The aesthetic of the Picturesque and the Sublime particularly suited the 'wildness' of the Scottish landscape, especially that of the Highlands. The Hermitage Folly, constructed next to the River Braan by the 3rd Duke of Atholl as part of the Dunkeld House Estate, Perthshire, in 1758, vividly exemplifies this aesthetic (Fig. 22.2). It was later renamed Ossian's Hall and revamped by the 4th Duke, who added mirrored walls and ceilings to reflect the sound and fury of the waterfall which it overlooked. A woodland walk led to the folly and to Ossian's Cave, an eighteenth-century representation of the dwelling of the bard Ossian.

The majority of small country estates in Scotland were laid out between 1735 and 1820. A typical Scottish country house might have a walled garden, orchards, lawns, parks with grass and shelterbelts, and

Figure 22.2 The Hermitage Folly, Dunkeld, Perthshire, built in 1758 by the 2nd Duke of Atholl, is a vivid example of the landscape of the Picturesque and the Sublime. Source: Courtesy of the University of St Andrews Library.

plantations. All of this was often surrounded by a wall.[13] The landed gentry generally were more conservative and slower to change from the formality of the previous generations than the nobility and substantial numbers of Scottish formal designs survived the uprooting which destroyed many earlier English estate gardens.

By the late eighteenth century Scottish kitchen gardens were well known throughout Europe, especially for the growing of peaches and pineapples, which was thought to be a remarkable achievement at such latitudes. By the mid eighteenth century the 'hot wall' was in use on many of the great estates in Scotland. Initially the backs of walls around gardens were warmed using ovens, but this eventually developed into a system of flued walls heated by furnaces. Later refinements included glazed fronts. The Pineapple Folly, designed in the mid eighteenth century on the Dunmore Estate in Stirlingshire, is a fantastical development of the hot-wall type (see Fig. 46.7). Such feats of engineering allowed the forcing (forced ripening) of fruit which would not otherwise have survived the climate. Conservatories fulfilled similar functions. By the mid nineteenth century the Buccleuch Estate at Dalkeith, Midlothian, boasted the grandest conservatories in the country. Melons, cucumbers and cherries were to be had year round, while two vine houses 50-foot (15 m) long produced many varieties of grapes, and pineapples were grown in a house 40 feet (12 m) in length.

In addition to the development of elaborate kitchen gardens and formal and 'natural' landscape gardens, the interests of the aristocratic landowners encompassed reforestation. The 4th Duke of Atholl, the 'Planting Duke', not only created great conifer plantations on his estate at Blair Atholl, Perthshire, but is credited with pioneering commercial reforestation in Scotland. Between 1774 and 1826 he planted fourteen million larch trees and thirteen million trees of other species.[14] Landowners urged their tenants also to plant trees, but this was widely rejected as being injurious to crops in field and garden.

As well as large-scale tree planting, land drainage, pasture improvement and the introduction of modern methods of husbandry, agricultural reforms during this period included changes in tenancy arrangements. Both agricultural improvements and the spread of individual farms (as against farming townships) led to a fundamental change in labour relations. Farm workers were paid, in cash and kind, and had cottar houses, often with kailyards and gardens attached. The standard of gardening improved. Further encouragement came in 1830 when the 'Highland and Agricultural Society of Scotland offered premiums for improved farm-servants' houses, and the better keeping of gardens'.[15]

Unlike the placement of gardens on a large estate where a choice position could be selected, the siting of a cottage garden, especially one belonging to a farm worker, was much more constrained and usually could take little regard of soil type, shelter or aspect. Many gardens were too small, often placed on the north side of the house, and were frequently surrounded by ash trees, which shaded the growing area and depleted soils.[16] Cottagers

Vegetables

Potatoes	several varieties
Cabbages	large early, early York, Scotch cabbage for the late crop, Winter kail, Savoys and German green or red and green airlies
Carrots	Orange
Onions	Strasburgh and Deptford
Turnips	early Dutch or white, Dutch yellow, field yellow, large yellow
Beans	long pod, Lisbon, broad Windsor, Turkey
Pease	the Charlton, blue Prussian, marrowfat
Leeks	Scotch and flag, chives or cives
Parsley	Sallad, common cress, ice and cabbage lettuce

Flowers

Marigold	Polyanthus (spinks)
White narcissus	Pinks
Wallflowers	Carnations
London Pride	Red and cabbage rose

Herbs and Medicines

Thyme, Spearmint, Southernwood	Tansy: for worms in children
Elccampane (a panacea)	Horehound (in the north of Scotland)
Rue	Chamomile

Figure 22.3 Plants found in late eighteenth- and early nineteenth-century cottage gardens. Source: Neill, 1813, 24.

were well aware of the importance of a garden and often their choice of dwelling, when they had the option, was influenced by this, the whole family being involved in looking after the garden. As time and leisure increased the garden became more than a source of food; it became a place of recreation and leisure for the rural lower classes as it had been for the landed gentry for centuries.

Although the great majority of cottage gardens produced solely potatoes and greens, some gardens produced many types of vegetables, flowers and culinary and medicinal herbs (Fig. 22.3). The variety of produce related directly to the size of the garden and, most importantly, its owners' security of tenure. There was little incentive to spend time and energy developing a garden which would be lost the following year. The biggest and best gardens often supplied a full year of produce which could include fruits. It was common to find red and whitecurrant bushes, gooseberries and possibly apple, pear or cherry trees.[17] The reduction of the price of sugar in the early nineteenth century, with the use of sugar beets, encouraged the greater planting and use of fruits and berries for jam and other preserves.

Village gardens, whether owned or rented, were generally better kept than those of tenant farmers and farm workers. The dwellings were normally occupied by artisans, labourers, manufacturers or mechanics who were financially better off than those who worked on the land. Competition between neighbours and the emulation of grander types of garden meant that more time and effort was invested in these establishments. Those village gardens found close to gentlemen's seats were acknowledged to be superior with a greater variety of plants, some villagers working in the big house's garden and having access to additional horticultural knowledge and materials.

EARLY TOWN GARDENS

Gardens were also cultivated in medieval burghs. The burgage, the normal unit of ownership, was a long strip of land with a narrow frontage on the street. The house was built along this street frontage, while the 'backlands' served a variety of purposes, including garden cultivation. Some burghs, under pressure of population growth, built over these areas; others, such as Haddington, East Lothian, and the Canongate in Edinburgh, have retained some of this open space until the present day.

James Gordon's maps of seventeenth-century Aberdeen and Edinburgh show many of the backlands being used for gardens. His 'Bird's Eye View of Edinburgh' of 1647 shows the Canongate backlands filled with parterres, knot gardens and kitchen gardens, which must have been well-established prior to his publication[18] (Fig. 22.4). The mid sixteenth-century garden of the Regent Moray in the Canongate was reported to have been elegant, 'It contains some venerable pear trees, a magnificent weeping

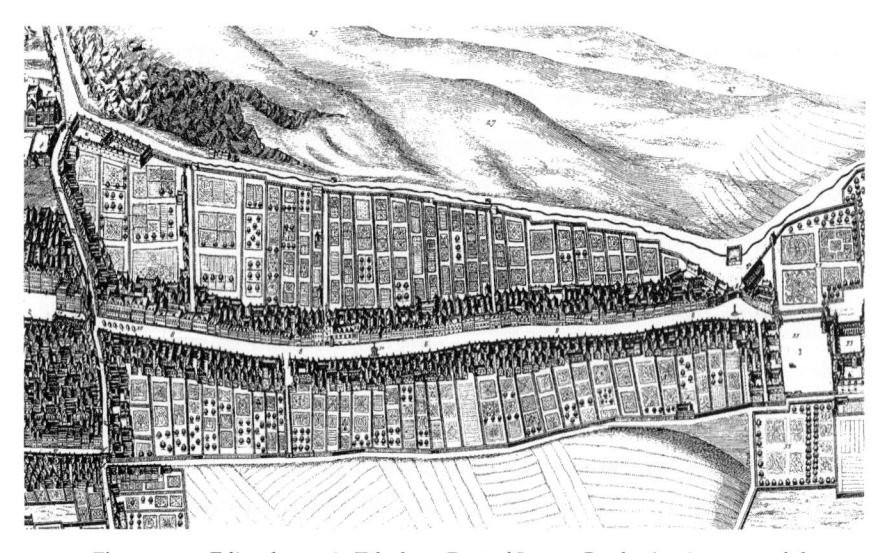

Figure 22.4 Edinodunensis Tabulam. Part of James Gordon's 1647 map of the Canongate in Edinburgh illustrates the size and sophistication of the gardens of the well-to-do. Source: Reproduced by permission of the Trustees of the National Library of Scotland.

thorn-tree [on the highest terrace] of great age, and the remains of elm bowers.'[19] It also boasted a summerhouse where the Treaty of Union was believed to have been signed.[20] James Grant notes an advertisement in the *Edinburgh Courant* of 1761 describing the Edinburgh house of Mr Grant (Baron of the Exchequer Court) for sale at the time, 'A large and convenient house, entering by a close mostly paved with flagstones, on the north side of the street near the Nether Bow ... a garden extending down the greater part of the Leith Wynd, planted with flowering shrubs ... ', and with a separate entrance.[21] The gardens of the burgh would have been more than pleasure gardens; archaeological evidence has found that cabbages, turnips, swedes and kail were grown.[22]

Many monasteries and other religious establishments were located in towns and owned large and intensively-cultivated gardens. The Palace of the Bishop of Glasgow possessed extensive gardens and orchards near the cathedral. Surrounding the palace were 32 houses, each accompanied by its own orchard and garden.[23] The closeness of burgh living facilitated the diffusion of horticultural knowledge, plants and seeds out from such centres.

NINETEENTH-CENTURY DEVELOPMENTS

The value of the gardening profession was recognised in Scotland as early as the seventeenth century. Gardeners were important figures, regarded by the public as skilled craftsmen. Many of the large gardens boasted a hierarchy of garden workers from the head gardener down to the apprentices, or garden lads. The Glasgow Incorporation, one of many societies and fraternities which developed in Scotland, has been traced back to the beginning of the seventeenth century and the Lothian incorporation to the 1670s.[24] The rise of a class of professional gardeners and nurserymen continued throughout the eighteenth century and was paralleled by the growth of provincial florists' societies which continued to increase throughout the nineteenth century. The Edinburgh Florist Society gave rise in 1809 to the Caledonian Horticulture Society, whose goal was 'to foster a taste for horticultural pursuits in every class of society'.[25] It encouraged a well-kept garden, urged experimentation to improve understanding of horticulture and published the results and other articles in its journal. Plant shows and exhibitions inspired the competitive to grow bigger and better vegetables and unusual flowers. There were no female members, however, although reports of the horticultural show of 1812 reveal women winning prizes in the currant wine category.

The uptake of ideas, knowledge and fashions increased with the publication of numerous magazines and books devoted to the study of scientific horticulture. To cater to the gardening aspirations of the expanding middle classes, Loudon wrote the *Suburban Horticulturist* in 1842. Subtitled, *An attempt to teach the science and practice of the culture and management of the kitchen, fruit and forcing garden to those who have had no previous knowledge or practice in these departments of gardening*, it was addressed to the possessors of small gardens.

Loudon was a Scotsman, a trained gardener, who wrote extensively and exhaustively on all aspects of gardening, aided by his wife Jane.

The feeling of expansiveness, confidence and wealth in Scotland, and throughout Britain, which occurred in the nineteenth century, has been linked to the return to formality in the gardens and landscapes of the nation.[26] Queen Victoria's love of Scotland meant that many families went to their estates in summer, bringing their garden designers with them. The newly wealthy built baronial halls and furnished gardens with parterres filled with bedding plants.

Victorian enthusiasm for the formal terraces and parterres of the seventeenth century suited Scotland perfectly as many of the Scottish estates had not been altered to the English Landscape style and retained their basic seventeenth-century formal geometric layout. This enthusiasm also led to the re-creation of some of the earlier formal gardens which had been lost. Drummond Castle gardens in Perthshire, first laid out in 1630, were redesigned by Lewis Kennedy in the 1820s as a Victorian re-creation of the seventeenth-century style, not a return to the original layout.

This was also a period of extensive plant exploration with individuals travelling to exotic areas of the world and sending back seeds and plants of previously unknown species. The Scots were particularly notable plant collectors. By 1830 David Douglas was sending seeds of Sitka spruce, lodgepole pine and Douglas fir from the great forests of the Pacific Coast of North America to Kew and Edinburgh for display in botanical gardens and the private gardens of the well-off. The seeds of conifers, herbaceous plants and curious specimens like the monkey-puzzle tree, plus innumerable rhododendrons and tropical palms, found new homes in specially built gardens: pineta, arboreta, woodland gardens, alpine and rock gardens and extensive glasshouses.

Gardens at Brodick Castle, Arran, and Inverewe (1860), Ross and Cromarty, proved particularly favourable for the planting of many acid-loving imports. Although difficult initially to establish, taking some fifteen years, the extensive shelterbelt of pine provided protection for Osgood McKenzie's wild garden at Inverewe. The mild climate of the west coast allowed the growing of many tender exotics (see Fig. 29.3). Crarae Glen Garden, Argyll, by Sir George Campbell in 1920, was in part designed by his mother to accommodate the new alpines, dwarf conifers and rhododendrons discovered by her nephew Reginald Farrier, the plant collector, in 1914.

The gardens of Kellie Castle and Earlshall, both Fife, were restored in the late nineteenth century by Sir Robert Lorimer. At Kellie Castle he divided the garden into compartments with hedges and walks. Within the sections fruit and vegetables grow together with the flowers. This mixed planting is characteristically Scottish and was promulgated as early as 1683 by John Reid. At Earlshall Lorimer transplanted topiary (beloved of the Victorians) in the form of twisted chessmen from a derelict Edinburgh garden in 1894. Lorimer's planting of traditional cottage plants is believed to have been influential on later twentieth-century garden designers such as Gertrude Jekyll.

THE TWENTIETH CENTURY

World Wars I and II effectively stopped ornamental garden planting with everyone being urged to grow food (Fig. 22.5). After World War II many large estates went into decline as taxation and labour costs made them increasingly difficult to run. In a bid to protect some of these great gardens and designed landscapes, the National Trust for Scotland (1931) and the Garden History Society (1965) were formed. The Scotland's Garden scheme run by the National Trust has introduced many Scots to their gardening heritage.

The end of World War I also brought the proliferation of small private homes with an attached garden. Their owners, often new to gardening, sought guidance in the design and cultivation of their plots. This came from a variety of sources. For example, the Livingston Development Corporation,

LOOK OUT FOR IT!

OUR GARDENS saved 150,000 tons of shipping in 1940. That was good. But it isn't good enough for 1941. Scotland wants 70,000 Allotments and many more food gardens next year. Every square yard put into cultivation helps to defeat the submarine and save **OUR SHIPS**

PUT YOUR GARDEN ON WAR SERVICE TO-DAY

Figure 22.5 Keeping gardens was a patriotic duty during the Wars. Source: Scottish Gardens and Allotment Committee (SGAC) *Newsletter*, 4 February 1941, 3. Department of Agriculture for Scotland.

Figure 22.6 Two-storeyed terraced houses at Meadowbank in Ladywell, Livingston, West Lothian, 1968. The twentieth century brought a garden to many people new to gardening who looked for guidance in design and maintenance. Source: Almond Valley Heritage Trust.

recognising that many of its tenants had no experience of gardening, published a *Residents' Handbook* in 1973 offering advice on how to design a new garden. This included a list of basic tools and suitable types of plants, including shrubs and trees (Fig. 22.6).

Gardening advice was also being disseminated through radio and television. One of the earliest radio productions on gardening was the BBC's 'In Your Garden', presented by Cecil H Middleton in the 1930s. Like it, 'Gardener's World', introduced in 1968 by Percy Thrower and later presented by Alan Titchmarsh on television, was broadcast Britain wide. 'The Beechgrove Garden', filmed in Scotland and aimed specifically at the Scot with a suburban semi-detached house, began in the 1970s. In addition to the increasing amount of gardening information conveyed through radio and television, the twentieth and twenty-first centuries have been characterised by a huge increase in do-it-yourself gardening books and magazines. The Internet offers still more information, on web sites from the Beechgrove Garden to the 'Virtual Allotment'.

Low-maintenance, instant gardens are characteristic of the later twentieth- and early twenty-first-century garden in Scotland as elsewhere. Most gardens are designed, built and maintained by owners whose time is limited. Other twentieth-century innovations include the increased use of machinery, for example lawn mowers, and chemicals such as herbicides and pesticides to make gardening easier.

Suburban gardens developed into a small front garden facing the public street and a larger garden behind the house. From the beginning these two gardens have had a different function. The front garden represents the public face and has to be kept neat and tidy (see Figs 29.2, 45.8). Usually consisting of a small lawn, it often includes flower beds arranged along a central walk or as a border. The back garden, as with many communal tenement back greens, is for more utilitarian needs and is not intended for display. It might include space for the washing line, the garden shed and the vegetable plot, which can often look untidy. It is also the place where children can safely play. The back garden has come to be seen increasingly as another room in the house. Conservatories attached to the dwelling are built to meet the needs of a climate that is not as amenable year round as that of California, where the idea of the outdoor room originated.

THE ALLOTMENT

Like the seventeenth-century Scottish cottage garden, the land for an allotment is rarely chosen for its site, situation or soil, usually being leftover space. Furthermore, similar to the earlier cottage gardeners, the allotment holder is often unsure how secure his or her tenure is to the plot of land. With costs increasing in several Scottish cities, land is often seen as being too valuable for its continued use for what many see as just gardening.

The Allotment Movement started with the gradual loss of common lands throughout the eighteenth and early nineteenth centuries. As enclosure took hold, many British farm labourers were displaced. The Allotment Act of 1887 in England was intended to provide land on which these 'labouring poor' could grow food. Although the 'curious fusion of philanthropic efforts and labourer land-rights activism' which lobbied vigorously for the Act in England was not present in Scotland,[27] the Allotment Act was extended to Scotland by Parliament in 1892. Under this act local authorities were required to provide allotments if there was a proven demand: representation from six or more resident registered electors. Legislation in 1919, 1922 and 1950 built on the initial Act, setting out further tenancy rights, security against arbitrary ejection and compensation for termination of tenancy.

Unlike central authorities in England, the Scottish Executive has no role in the provision or management of allotments. This is the responsibility of local authorities. The management of allotment sites varies. Those in Aberdeen are centrally managed while at the opposite end of the spectrum Glasgow has devolved all management to local site committees. Edinburgh and Dundee share management responsibilities between the local authority, the site committees and, in Edinburgh, The Federation of Edinburgh and District Allotments and Garden Associations (FEDAGA), a city-wide committee. The national organisation, previously titled the National Union of Allotment Holders, has recently re-formed as the Scottish Allotment and Garden Society (SAGS) and is attempting to co-ordinate the voice for allotments in Scotland. Its members tend to be those who have allotments under threat of closure.

Allotments, slow to gain acceptance in Scotland, flourished during the War years. Appeals to patriotism led to a great uptake and demand for new allotments with, for example, public parks being transformed. Both World War I and World War II spread the culture of allotment gardening to a wider population. No longer the realm of the poverty stricken, the unemployed and retired males, school children, youth groups, all occupations and all classes were urged to 'dig for victory' as every area of spare land was cultivated for food (Fig. 22.7).

With the end of both Wars, land, commandeered for food growing, was returned to its former use. The pressure on land after World War II in urban areas was such that housing was built on many allotment sites. At the same time the demand for allotments declined, partly as a consequence of people returning to their normal occupations and partly because many houses were being built with their own gardens attached. By the 1960s the pressure on urban land in Britain had become such that a Departmental Committee of Inquiry into Allotments was commissioned. The resulting report, submitted in 1969, was to be of great significance in the history of allotments, although it concentrated mainly on England and Wales. Professor Harry Thorpe and his committee put forward two propositions:

> The first was that the existing legislation was vague, obsolete and incomprehensible, and in urgent need of revision by one new Act. The second was the very word 'allotment' had a 'stigma of charity' about

Figure 22.7 Working in the school garden at Lilliesleaf, Roxburghshire during World War I. The Wars brought gardening to new sectors of society. Source: SLA W 610308.

it and needed to be replaced by the concept of a 'leisure garden', with sites improved and upgraded as a recreational facility for the whole family ... [28]

The recommendations were not taken up.

The demand for allotments in Britain experienced an upsurge in the 1970s and 1980s. This was largely driven by concerns about the environmental health of the planet, about the quality of the food we eat, and worry about the use of herbicides and pesticides in commercial agriculture. Although its roots were in the provision of land for growing food, cultivating an allotment plot has also come to be seen as an important source of recreation, exercise, leisure and a place for retreat. Interviews with allotment holders confirm the former stereotype of the male head of the family retiring to his allotment shed to smoke his pipe and to get away from the nagging wife and the noise of family life. This has changed only because the range of allotment holders seeking a refuge has expanded to include all ages and both sexes. Today you are likely to find a range of Scots tending their plots. In 1999, '34 per cent of plot holders in Edinburgh and 15 per cent in Glasgow were female. At Craigentinny-Telferton in Portobello, Edinburgh, 60 per cent of plot holders are women in their thirties.'[29] This is quite different from the findings of Thorpe Report in 1960s when only 3.2 per cent of allotment holders, in British allotments, were women. Members of ethnic and immigrant groups are also taking up allotments, particularly in urban areas, bringing with them different approaches to gardening which make for a changed landscape. For example, one plot in a Kelvinside, Glasgow, allotment, held by an Indian man, takes the form of a raised bed which is getting higher and higher every year as the gardener lays down a new layer of compost and soil and plants the new crop over the remains of the old.

In areas where demand for allotments is low a person may take on two plots; where demand is high some authorities divide the plots. Space in any allotment is at a premium. The traditional allotment plot is arranged with precision and neatness, with onions, potatoes and runner-bean frames lined up in rows. This is the recommended layout illustrated in the newsletters produced by the Department of Agriculture for Scotland as part of its wartime promotion of allotments, as well as most DIY books on gardening. Many plots of this type can still be seen but there are variations, generally resulting from a greater variety of gardeners. Some allotment plots are more random and chaotic in appearance, with vegetables, fruit bushes and flowers all mixed together. Permaculture, terracing and the planting of new crops has also changed the traditional allotment landscape.

Potatoes are ubiquitous on Scottish allotments. During the Wars, allotment holders were urged to plant a variety of year-round produce but many people grew only potatoes. The crops currently being grown on Scottish allotments are as varied as the gardeners who grow them but cabbages, Brussels sprouts, cauliflowers, shallots, onions, leeks, beetroot, carrots, swedes, parsnips, peas, broad beans, rhubarb and potatoes are found on most allotments throughout the country.

Cultivating an allotment is very different from cultivating a private garden. By taking on a plot one joins a community and for some allotment holders it is this aspect which is all-important. Allotments can act as a panacea for many aspects of life in a modern city. They are communities which offer a retreat to an older, slower and more basic way of life. 'Some plot holders even reflect that it is as though the allotment culture preserved patterns of social relations that have been buried in other fields of life.'[30] The exchange of seeds, seedlings, produce and advice is a significant part of these relations, provoking conversation and the sharing of flasks of tea.

GLOBALISATION OF GARDENING INFLUENCES

There is no such thing as a Scottish garden. Scotland's garden culture is necessarily linked to that of the rest of Britain (particularly England), Europe and the rest of the world. From the seventeenth century, probably earlier, wealthy landowning Scots travelled and were often educated abroad. On their return they brought new ideas, plants and seeds to their estates in Scotland. In the eighteenth and nineteenth centuries the implementation of these ideas radically changed the Scottish landscape. Improvements in agricultural and horticultural techniques filtered down the social ladder to the small landowner and then to the workers.

The growth of mass media and the relative ease of access to gardening and horticultural information worldwide in the twentieth century have accelerated this process. Global influences have never been greater on cultural phenomena. Inexpensive foreign holidays mean that people can visit other types of garden and see new plants. Web sites now allow people to communicate easily with other interested gardeners worldwide. Expert advice is readily available.

Does this mean that all gardens will look alike? This is clearly not the case. The rural, village, suburban, city gardens and allotments of Scotland are diverse; each a product, as has always been the case, of the particular gardener and the specific location.

NOTES

1. Cox, 1935, xv.
2. Parterre, literally meaning 'part earth' and indicating a pattern cut from turf, leaving bare earth to be covered in gravels or shells. Early parterres were ornate flowerbeds, usually adjoining the house, but by the seventeenth century they had become a more elaborate art form. Several types of parterres developed. *Parterres de broderies* ('embroidered parterres') have flowing plant-like designs made of box against a background of coloured earth, sometimes with bands of turf. *Parterres a l'anglais* ('in the English manner') have designs of cut turf. *Parterres de pieces coupes* ('cutwork parterres') have flowerbeds as the individual pieces of the design.
3. Loudon, 1835, 80.
4. Cox, 1935, 16.
5. Brown, 1999, 58.

6. Cox, 1935, 40–2
7. Adams, 1978, 58.
8. Fenton, 1976, 179.
9. Cox, 1935, 24, 180.
10. Fenton, 1987, 72.
11. Land Use Consultants, 1983, 21.
12. Reid, 1988, 2.
13. Cox, 1935, 84.
14. Mackay, 2001, 195
15. Fenton, 1976, 188.
16. Neill, 1813, 188.
17. Neill, 1813, 21.
18. Mackay, 2001, 90.
19. Loudon, 1835, 342.
20. Grant, 1883, 33.
21. Grant, 1883, 241.
22. Ewan, 1990, 118.
23. Cox, 1935, 27.
24. Robertson, 2000, 173.
25. Neill, 1813, 161.
26. Land Use Consultants, 1983, 23.
27. DeSilvey, 2001, 1.
28. Crouch, Ward, 1988, 7.
29. Cuthbert, McGhee, 1999, 5.7.
30. Crouch, Ward, 1988, 98.

BIBLIOGRAPHY

Adams, I H. *The Making of Urban Scotland*, London, 1978.
Brown, J. *The Pursuit of Paradise: A Social History of Gardens and Gardening*, London, 1999.
Cox, E H M. *A History of Gardening in Scotland*, London, 1935.
Crouch, D, Ward, C. *The Allotment: Its Landscape and Culture*, London, 1988.
Cuthbert, M, McGhee, P. A Survey of Scotland's Allotments and an Agenda for the Future (Food Trust of Scotland), unpublished report, 1999.
DeSilvey, C O'B. 'When Plotters Meet: Edinburgh's allotment movement, 1921–2001', MSc thesis, University of Edinburgh, 2001.
Ewan, E. *Town Life in Fourteenth-Century Scotland*, Edinburgh, 1990.
Fenton, A. *Scottish Country Life*, Edinburgh, 1976.
Fenton, A. *Country Life in Scotland: Our Rural Past*, Edinburgh, 1987.
Grant, J. *Cassell's Old and New Edinburgh, vol 1*, London, 1883.
Land Use Consultants. *A Study of Gardens and Designed Landscapes in Scotland* (Prepared for the Scottish Development Department and the Countryside Commission for Scotland), Edinburgh, 1983.
Land Use Consultants. *Inventory of Gardens and Designed Landscapes in Scotland* (Report for the Countryside Commission of Scotland and Historic Buildings Directorate), Edinburgh, 1987.
Loudon, J C. *An Encyclopaedia of Gardening*, London, 1835.
Mackay, S. *Early Scottish Gardens: A Writer's Odyssey*, Edinburgh, 2001.
Neill, P. *Scottish Gardens and Orchards* (Board of Agriculture), npp, 1813.
Reid, J. *The Scots Gard'ner* (1683), Edinburgh, 1988.
Robertson, F. *Early Scottish Gardeners and their Plants*, East Linton, 2000.

23 Food

UNA A ROBERTSON

'A dinner in the Western Isles,' observed Dr Samuel Johnson during the famous Journey of 1773, 'differs very little from a dinner in England, except that in the place of tarts there are always set different preparations of milk.'[1] Dr Johnson was just one of many individuals who recorded their observations on the food and meals of Scotland. The wealth of evidence from a wide range of written sources, both published and unpublished, is surprising when these matters might well be considered transient and too mundane to attract much attention. Evidence can also be found in pictorial representations and in the objects associated with both the dinner table and the kitchen which have been collected together in museums.

The Scottish diet aroused not only many, but conflicting reports, with some commentators expressing surprise at how well the Scots lived in spite of the commonly held perception of Scotland as a poor country.[2] The evidence from household dinner books and accounts suggests that above a certain social or economic level there was no shortage of food, or the drink to accompany it, but perhaps not great variety. Most foodstuffs were seasonal until relatively recently and there was a tendency for available items to be served on an almost daily basis, although modes of preparation might vary. Servants in these wealthier households would be fed according to their rank in the domestic hierarchy and at the lower levels this might mean monotonous feeding. However, it would at least be regular and plentiful.[3]

Those less fortunately placed existed on a diet of a few basic foodstuffs, sometimes inadequate in terms of both quality and quantity, and with famine a distinct possibility should the harvest fail. Sibbald's publication of 1699, titled *Provision for the Poor in Time of Dearth and Scarcity ... When Corns are Scarce, or Unfit for Use*, makes salutary reading.[4] The fear of famine diminished when the potato entered the general diet in the later decades of the eighteenth century. More suited to the climatic and geographical conditions of Scotland, it was a less fragile and more prolific crop than grain. In 1794 it was recorded that the food of the lower classes of Cramond, then a village outside Edinburgh, consisted principally of oatmeal for breakfast and supper, with milk in season, 'They seldom taste flesh, fish, butter or cheese; and during the winter months live chiefly on potatoes, prepared in different ways.' In addition they ate, '... cockles, muscles [sic], limpets and other shell fish'.[5] Oatmeal and potatoes were fundamental to the diet of many other Scots at this time,[6] including those within charitable institutions.[7]

The potato blight of the 1840s and 1850s thus deeply affected many lives. The diet of the poorer sections of society also worsened as a consequence of the rapid expansion of towns in the nineteenth century. The fact that so many people were now distanced from the processes of food production gave manufacturers and dealers the opportunity for malpractice – for example, the adulteration or dilution of provisions, the substitution of ingredients, or the giving of short measure – on a greater scale than ever before. The former system, whereby the market place had been regulated and supervised by the burgh, was no longer viable with the rise in population. By the middle of the century anxiety was growing over concerns of health and hygiene, laying the basis of our current legislation on such matters.

There were also serious difficulties to be experienced in supplying large conglomerations of people with sufficient food until developments in transportation and refrigeration allowed meat, eggs and many other provisions from around the world to be imported. When they did occur, such innovations, along with the development of the canning industry, allowed the wealthier sections of society to benefit from access to an increased range of foods. By being better fed, these people were better able to withstand infectious diseases such as smallpox and typhus, which ravaged many of those living in the overcrowded and sub-standard dwellings of the poorer quarters of the larger towns.

The reverberations of the outside world have always been keenly felt in the kitchen. Whether as a consequence of war, dynastic or political alliance, changing means of transport, concern over the adulteration of foodstuffs and standards of health, or imports of new foods and drinks – to list but some of the possibilities – the effects on food and its preparation have been profound. However, although specific innovations can often be dated, at least approximately, the rate of adoption of these in the home is not so easy to measure. In Scotland, both in social and geographical terms, any impact has been far from uniform throughout the country. A procedure known in medieval times, for example, may have continued until the twentieth century in some dwellings while in other households it may have been made redundant at the earliest opportunity.

One further aspect, of special relevance to Scotland, was the influence of the Reformation on the kitchen, especially at the period of celebration and feasting known as the 'Twelve Days of Christmas'. Many of the foods served at Christmas were imbued with religious significance and attracted the wrath of the reformers who deemed plain eating a virtue and extravagance and unnecessary expenditure as morally suspect. When the celebration of Christmas was banned in 1583, many individuals felt deprived of their annual period of jollification and good food and thus many aspects of the festivities were postponed until Hogmanay (New Year's Eve). As this date had little religious significance the proceedings took on a secular format.[8]

BEFORE THE NINETEENTH CENTURY

Cooking and baking
The hearth and the fuel burnt there were interdependent and influenced both the cooking vessels and the style of cookery. Peat produces a gentle heat, a similar amount of wood gives out rather more, while coal is hotter again. Dr Johnson's assertion that, in the Western Isles, preparations of milk replaced tarts gives an indication of the cooking medium since pastry needs a hot oven, difficult to achieve when baking with peat.[9]

Peat and wood will burn without an air supply underneath and in many rural dwellings the fire was set directly on the floor, with the smoke drifting upwards through a hole in the roof (see Fig. 6.2). According to the seventeenth-century traveller Celia Fiennes, 'It makes one smell as if smoaked Like Bacon.'[10] By contrast, coal requires a raised grate and a chimney or a 'hanging lum' to create the requisite draught.

Most cooking processes can be carried out over an open fire – boiling, stewing, frying, roasting and baking. Cooking pots were hung over the fire or were set beside it on a 'trivet' (see Fig. 6.4). Once the central fireplace moved to a wall and acquired a chimney cooking pots were hung from a 'swee' or 'cran' (a cran also being an iron instrument placed on or over a hearth to support a pot) or were placed on hobs – metal or stone surfaces built on both sides of the fire. Roasting was done on a spit in front of the fire which was required to be 'brisk', that is, hot. Spits were turned by hand or, in big houses, by various mechanical contrivances.[11] Meat for roasting is apt to be the most expensive and so would have been little used by the poorer sections of society. However, it is worth remembering that, until the seventeenth century, the Scots must have eaten much in the way of meat, judging from the considerable number of hides and skins exported, although consumption was probably unevenly distributed among the population. One should also question just how much meat for roasting was available on each carcase. Joints for roasting require to be marbled with fat; to put fat on an animal needs adequate grazing and feedstuffs; Scottish conditions have not always favoured such a luxury.

Relatively few dwellings had their own bread ovens. In the absence of one of these baking was done on a girdle. The girdle was a flat iron plate with a semi-circular handle and was hung over the fire.[12] The bakestone was a flat stone heated by the fire on which, when hot, items were placed, covered with an upturned pot and set among the embers.[13] Where bread ovens existed – usually only in the grandest homes – they were hefty constructions of stone or brick which were fired separately and were therefore expensive to run. On baking day logs were burnt inside the oven until it was sufficiently heated. The ash was raked out, the flat surface swabbed with a damp mop, the items requiring the most heat, such as loaves of bread, were put inside with the aid of a 'peel', a long-handled flat shovel, and the door inserted. Timing was a combination of guesswork and experience. The first batch of baking was replaced by another requiring less heat and so on throughout the day, ending with items such as macaroons or meringues which need merely gentle warmth. Finally, rags for the tinderbox or kindling for next week's baking were placed inside to dry in the residual heat.

With the increased use of coal there was concomitant use of iron for cooking pots since earthenware was unable to withstand the heat generated, unless placed in an outer container filled with water. The recipes for 'jugged hare' and 'jugged kippers' are a reminder of this technique.[14] Pots were tin coated on the inside to prevent them rusting when not in use and, later, were coated with a black varnish on the outside to reduce the amount of cleaning required. Wealthy establishments had copper pans and their popularity increased considerably after the discovery in 1763 of the Parys Copper Mine, Anglesey, sent prices tumbling.

Even the cheapest pots, along with the other ironmongery associated with cooking, comprised a hefty investment for poorer households. Here, the number of items would be kept to a minimum and each would be made to last as long as possible, being repaired repeatedly by local or itinerant metalworkers. Some people in fact managed without cooking pots altogether, as witnessed in the eighteenth century:

> I shall only at present mention one other Piece of their Ingenuity, which is, that they can boil a Quarter of Flesh, whether Mutton, Veal, Goat or Deer, in the Paunch of a Beast, which is prepared by cutting open, and turning outside in, by which it is made clean; then they affix it with Scuars on a Hoop to which they tie a String or a Thong from the Skin of the Beast, by which they hang it over the Fire.[15]

Preparation and utensils

In the above context, and no doubt some others, cookbooks were surely superfluous! Until the mid eighteenth century those found in Scotland were primarily works by popular English authors such as Markham, May, Digby, Howard and Lamb.[16] Mrs McLintock's volume of 1736 is generally considered to be the first cookery book published in Scotland, although fifty years earlier John Reid's book on gardening had included an Appendix 'showing how to use the Fruits of the Garden'.[17] The century following Mrs McLintock saw the production of Scottish publications covering every aspect of the culinary arts. The most celebrated was *The Cook and Housewife's Manual* of 1826 which preceded and overshadowed the businesslike work of Mrs Dalgairns.[18] Thereafter both English and Scottish works were readily available (Fig. 23.1).

Although many recipes, together with the methods and terminology, are common to both countries, a number reflect a Scottish element; preparations such as haggis and porridge incorporating oatmeal, marmalade as a component of breakfast, or the shortbread and black bun served at Hogmanay, for example.[19]

A plenitude of food to be prepared and cooked demands appropriate equipment. Cookbooks required their readers to slice and chop; to sieve; to

Figure 23.1 A selection of cookery books published in Scotland. From left to right the authors are Mrs Robertson, Mrs Dalgairns, Mrs Maciver, Mrs Dods and Mrs Kirk, with the Misses Kerr on the right. Source: U A Robertson.

mix, beat or knead; to roll out; to mould; to tie up or to joint; to whisk, stir or skim and many other processes still familiar today. However, considerably more preparatory work was required since poultry, game and fish came into the kitchen entire, if not actually alive, and needed gutting and skinning. Meat might arrive as 'the quarter' or 'the side', to be jointed or boned as appropriate. Moreover, far more use was made of offal than is customary for the home cook today. The necessary implements were often enumerated in domestic inventories and account books alongside the more elegant articles for serving the prepared dishes at table. Against this higher social status scenario must be set innumerable households lacking in all but a few basic utensils, in part a witness to their dependency on a handful of staple commodities (see Fig. 28.3).

Sources of supply
Markets existed where there was a high concentration of population – in urban and proto-urban settlements – for the sale of locally produced goods and were held on a set day or days in the week in designated streets. They were regulated by the local authority which laid down the conditions and hours of trading and ensured the regulations were observed. Numerous itinerant traders sold a wide-ranging variety of goods through the streets, calling their wares as they went[20] (Fig. 23.2). Hawkers and peddlers also travelled around rural areas.

Householders augmented their supplies at their local fair which, in contrast to the markets, occurred once or twice a year, beginning on dates fixed by tradition and lasting eight or 15 days. In the area around Edinburgh, for example, numerous fairs had their origins before AD 1500 and others were established later. Edinburgh itself had two fairs during the year, and Dalkeith and Musselburgh (Midlothian), Haddington, Dunbar and Prestonpans (East Lothian), and South Queensferry and Linlithgow (West Lothian) all had their own.[21] People flocked to the fairs from substantial distances as commodities from far and near would be on sale. Servants and apprentices were often granted a day's holiday to attend since the serious side was accompanied by other attractions such as quack doctors, entertainers and sellers of ready-cooked foods. Fairs dwindled in the nineteenth century until only the entertainment aspect survived, to become the fun fair of the seaside resort.

There were also those who had little need to visit markets and fairs since they produced enough of their own goods to render themselves largely self-sufficient. Elizabeth Grant of Rothiemurchus described that remote household's economy and added 'The regular routine of business, where so much was done at home, was really a perpetual amusement. I used to wonder when travellers asked my mother if she did not find her life dull.'[22] For poorer people, self-sufficiency often meant survival rather than anything more generous.

'CURDS AN' WHEY!'

Figure 23.2 Numerous traders walked the streets selling their wares. 'Curds and whey' were much esteemed by the residents of Edinburgh as a summer delicacy. Source: *Book of the Old Edinburgh Club*, vol 2, 1919, opp 203.

There was always a need to safeguard supplies, on a day-to-day basis or for longer periods. A larder or 'wet larder', in the form of a cupboard or even a small room, stored raw foods, although larger establishments would often have individual larders designated for meat, game or fish (see Fig. 4.3). A pantry, or 'dry larder', again either a cupboard or room, held cooked food and dry goods. Both types of store needed to be dry, cool and well ventilated, which generally meant they were sited away from the heat and steam of the kitchen. Similar requirements were essential for long-term storage. Country estates sometimes had an icehouse, usually for the preservation of ice for kitchen use in warmer weather (the confectioners of Edinburgh built or rented icehouses on the Calton Hill).[23] Many households, but especially those in remoter localities, preserved surplus foodstuffs against future need by a range of methods including the age-old ways of wind drying and smoking and more modern procedures using sugar.

At table
Some cookbooks advised their wealthy readers not only how dishes should be cooked but at what point in the meal they should be served and where they should be placed on the table. Diagrams were provided to show the suggested layout.[24]

During the pre-nineteenth century period the day was broken up by three principal meals: breakfast, dinner in the middle of the day and supper in the evening. It might also be punctuated at any time of day for the upper classes by glasses of something alcoholic accompanied by cakes, raisins or more solid sustenance.[25]

Breakfast was what it purported to be – food taken to break the overnight fast – and it varied widely across the years as to both timing and content. Dr Johnson sang the praises of those breakfasts he had met with:

> The breakfast, a meal in which the Scots, whether of the Lowlands or the Highlands must be confessed to excel us. The tea and coffee are accompanied not only with butter, but with honey, conserves and marmalades. If an epicure could remove by a wish in search of sensual gratifications, wherever he had supped, he would breakfast in Scotland.[26]

Others made similar observations, especially with regard to the serving of conserves and marmalades at this meal, seemingly a Scottish innovation.[27]

However, for many Scots such luxuries were unavailable. It was observed of the late eighteenth-century farm servants in the Carse of Gowrie, Perthshire/Angus, 'Oatmeal with milk, which they cook in different ways, is their constant food, three times a day, throughout the year, Sundays and holidays included; the quantity of meal allowed being 36 ounces (1 kg) to each man a day.'[28] Throughout Scotland porridge or brose was taken for breakfast. Brose is made by pouring boiling liquid over salted oatmeal and stirring. For porridge, the oatmeal is trickled into boiling water and simmered for 20–30 minutes.

Supper was the last meal of the day. It was completely variable in its composition and timing depending on whether a couple, for example, were sitting alone or whether they were entertaining friends. Many people, though, ate on getting home from work.

Between these two meals came dinner which was taken, by those who could afford to do so, with a certain formality and with servants in attendance. Originally eaten about midday, it became an early afternoon meal and then, by the end of the eighteenth century, a late-afternoon meal and it continued its onward progression thereafter – with significant effects. Meanwhile, those in employment kept to the earlier nomenclature and timing as it divided the working day more evenly.

As dinner slid towards evening, the gap between it and breakfast stretched uncomfortably and so the glass of wine or ale taken with a slice of cake mid-morning was augmented until it became nuncheon, later luncheon. This meant that no one was ready for dinner until later again and so the once-popular supper was usually superfluous.[29]

From the 1660s until well into the nineteenth century the well-to-do dined *à la française* (in the French style). A number of dishes were put on the table simultaneously to form a course. Servants were on hand to fetch and carry as required but those dining looked after each other and the host and hostess carved the meat and poultry. Carving was an accomplishment taught to youngsters, although, as Meg Dods commented, some people never became proficient.[30] In wealthier establishments or on formal occasions the first course was removed, to be replaced by a further wide-ranging selection of dishes which formed a second course.

Such elegance and formality was difficult to accomplish in the cramped conditions of many town dwellings and so much in the way of both business and entertaining was carried out in the taverns, alehouses and, later, the coffee houses that abounded in the larger urban areas. Not only was it customary for the working day to begin and end with a visit to such an establishment, but around 11.30am a break was made for a 'meridian', generally comprising ale or brandy. In addition, merchants, lawyers, craftsmen and others met with their clients in these establishments and many a bargain or other piece of business began or ended over a further drink. Favourite dishes were simple – minced collops (slices of meat), rizzared (dried, perhaps sun-dried) haddocks or tripe, a fluke or roasted skate and onions – and all were washed down with ale or claret.[31] Hot and cold food could also be bought from the numerous itinerant traders who hawked their wares through the streets: hot pies, cooked shellfish, buttermilk ('sour dook'), curds and whey and strawberries in due season were always popular.

For the rural worker, dinner, if taken at home, was likely to be based on broth, with or without a piece of meat in it, oatcakes or barley bannocks with butter and cheese, and milk, skimmed milk or buttermilk. Those taking their meal in the field would eat whatever might be available and easily carried.

After dinner
Increasingly, dinner for the fashionable was followed by the dessert course, which formed the climax to the meal. Anything which was novel, out of season or a delicacy took its place here. Items were served on the best dishes, eaten from a decorative dessert service and accompanied by expensive wines.[32] After a short time the ladies withdrew to a parlour or withdrawing room and the gentlemen might join them for tea, coffee and further sweetmeats or continue drinking in the dining room.

The dessert course originated in the medieval hall when those at the top table removed themselves after the meal until the hall was cleared for the next activity. They whiled away the time with a glass of wine and something to nibble. The room where this happened was called the parlour (from the French *parler*, 'to speak'), while what they ate was a dessert (from French *desservir*, 'to clear away'). For many years it had been customary to take dessert in a different room from the dinner, but as the formal dining room became a fashionable adjunct so dessert was taken at the dinner table. After the main course was cleared away the cloth was also removed in order to reveal the table's polished surface and the dessert course was then laid out directly on this.

In England this final course was called a 'banquet' although the term also signified a feast. It was only after the Restoration of Charles II in 1660,[33] when all things French became fashionable, that the word 'dessert' became current. However, in Scotland dessert was being used in the sixteenth century, although, confusingly, when part of a formal occasion it was sometimes termed a banquet. Dessert could also be taken informally at any time of the day, the equivalent of coffee and biscuits today.

Domestic accounts reveal that sweetmeats were increasingly purchased by the well-to-do throughout the seventeenth century. Alternatively, a confectioner could be employed for a special occasion and might at the same time teach the ladies how to make confections. The Duke of Hamilton thought such an expense justifiable in 1689, although Sir John Cochrane must have found the cost distressingly high when 'about April 1694' his 'house of Ochiltry [Ayrshire] was brunt, by the negligence of on[e] that was making confections to his Lady in it'.[34] Some cookbooks included recipes for confectionery items, although the work was considered distinct from cookery. Scottish housewives took up the new confectionery skills eagerly and went on to become proficient in making jams, jellies and marmalades as well as in baking.

After dinner the drawing-room party engaged in conversation, whether light-hearted or serious, music, singing, recitations of poetry and, later in the evening, a selection of pastimes including perhaps cards or other games. The ladies might also employ themselves with some fancy needlework or show off their artistic efforts while the men discussed recently acquired *objets d'art* or events in the newspapers.

Behind the scenes servants cleared the dining room and any adjacent service area. Food remaining from the first and second courses went back to the kitchen, to be held over for future use. Dessert items were returned to the provider, whether the housekeeper or the housewife herself, as was the precious dessert service. The butler dealt with the drinks and he or his footmen washed the glass and silverware and secured everything before the evening ended. It was the duty of the scullery or kitchen maid to wash up all the plates and dishes and to clean and scour the cooking pots and utensils, which was an onerous job when pans of water had to be heated on a range. Previously, when wooden trenchers and then pewter plates had been commonplace (to be replaced by earthenware, and then 'china', which became increasingly popular from the eighteenth century) references suggest they were merely rinsed in cold water or wiped clean with bread, bran or straw.

1800–1950

Although the domestic organisation of innumerable households remained largely unchanged until World War I, the nineteenth century transformed many aspects of kitchen life. The twentieth century witnessed change at an even greater rate, notably the virtual disappearance of the domestic servant by the 1920s and 1930s and the introduction of electricity to make a reality of 'labour-saving' devices, an important development for those middle-class housewives who now had to do their own cooking. Kitchens were brightened up and streamlined, and acquired easy-clean surfaces and floor coverings; the freestanding kitchen cabinet revolutionised storage and work space (see Fig. 12.8); distant larders were scrapped in favour of the refrigerator; hot water on tap eased clearing up; and meals themselves were simplified. Working-class housewives followed the trend as soon as they were able, although space in their homes was generally too limited for the acquisition of many modern appliances. Meanwhile the upper classes were slower to accept the necessity for change since they continued to employ servants, although they eventually came to realise that their kitchen premises would need to be upgraded if they were to keep their staff.

Cooking and baking
Cooking was increasingly women's work (see Fig. 20.6). Even in 1825 it was estimated that there were only some 400 male cooks working in wealthy households and about 40 or 50 in London hotels.[35] This reflected the considerably higher wages and prestige accorded to male servants.

The nineteenth century was the era of the kitchen range, so called because it combined a 'range' of cooking methods. There was a central firebox, open at the front for roasting and with either an open or closed top for boiling and frying, an oven at one side and a hot-water tank at the other, with the surfaces of these acting as hobs and stewing stoves (see Fig. 6.5). Cooking by gas was first tried in the 1820s but it was not a serious contender for another 50 years and its use was always limited by the fuel supply (Fig. 23.3). Electric cookers were developed in the 1890s but the supply situation again

Figure 23.3 This gas cooker of the 1890s, incorporating a tank for hot water, is recognisably a link between the coal-burning range and modern-day appliances. The tiered steamer on the right saves on both fuel and space. Source: *Cassell's Book of the Household*, 4 vols, sp edn, London, c 1893, vol 1, 9.

limited their acceptance until after World War I. The Aga cooker was a Swedish invention of the 1920s; a solid-fuel cooker, clean, efficient and thermostatically controlled, which could also provide hot water to circulate around the house. The Aga filled the large gap in the market between the old-fashioned kitchen range and the gas cooker.

Saucepans changed their shape, being given a flat as opposed to a curved base to match the ranges' flat surfaces. New finishes and materials were introduced. Vitreous enamelling protected pans from the chemical reactions produced by contact with certain foodstuffs. Aluminium was the newcomer in the 1890s and stainless steel in the 1930s. Oven-to-tableware proved popular as a labour-saving device and 'Pyrex', an early form, was available in the 1920s.

Preparation and utensils
Although many utensils were basically unchanged in shape and function (see Fig. 29.4), new materials were used in their manufacture and they became smaller and lighter. True labour-saving devices developed slowly,[36] but innumerable gadgets entered the kitchen in Victorian times.[37] Later, many of these would become far more efficient by being operated by electricity, including items such as food mixers, toasters, kettles, urns and, in due course, coffee percolators and dishwashers[38] (Fig. 23.4).

Figure 23.4 The *batterie de cuisine* of the larger kitchen. On the dresser can be seen a wooden mixing tray, an earthenware bowl and a pair of scales with its weights. Above are four copper moulds and, in the window, a tea urn. Source: Reproduced by permission of Callendar House in Falkirk, Falkirk Museums.

Sources of supply
Increasingly the housewife went 'shopping' rather than buying from markets and fairs as these lost their central place in provisioning the household, although covered markets remained popular well into the twentieth century. A wider range of goods was sold, many of the items previously being produced in the home. The concept of multiple shops with numerous branches selling uniform products was another development. Assisted by rapid rail transport and bulk purchasing, their goods were both reliable and cheap. Similarly, the Co-operative Movement's emphasis on honest trading and its dividends proved popular, particularly among working-class housewives.

An accompanying decrease in itinerant traders was also noticeable. It was recorded that Edinburgh's distinctive 'calls' were vanishing by the 1870s,[39] although fishwives from Newhaven and Leith continued to ply their trade until the 1960s. In rural areas assorted peddlers and hawkers walked or rode between villages selling their wares. The introduction of the bicycle and, later on, motorised transport increased their mobility.

With the development of railway networks from the 1830s onwards towns were no longer supplied only from their hinterland. Meat and fish

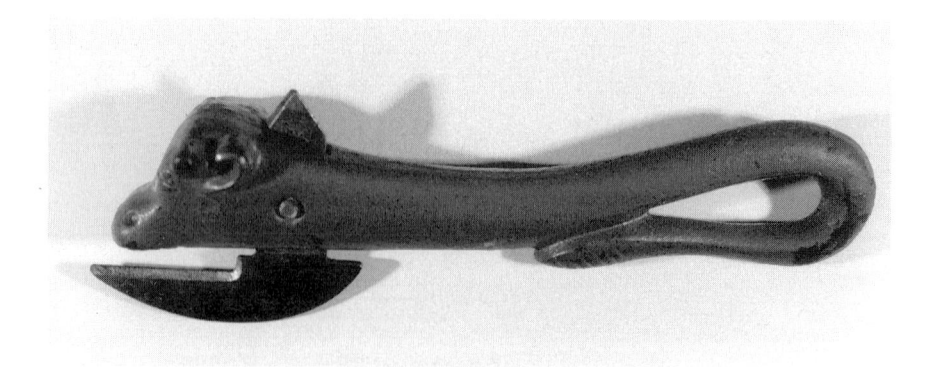

Figure 23.5 Tin openers came in about the same time as corned beef from Argentina and often took the form of a bull's head with the bull's tail constituting the handle. This one measures 7 inches long and dates from the 1890s. Source: U A Robertson.

from Aberdeen sold in London the following morning and Dumfriesshire supplied Edinburgh's milk. Steam ships, from the 1880s aided by refrigeration, imported produce from around the world and airfreight was the innovation of the twentieth century.

The importance of self-sufficiency in domestic life decreased as the evolving agricultural and industrial scene, coupled with the economic and social changes of the nineteenth century, meant that fewer people held land on which to support themselves. Notwithstanding, the Victorian housewife was encouraged to use her skills, or those of her cook, to preserve a wide range of items against future use, even although canning and refrigeration were emerging as alternative methods (Fig. 23.5). The two World Wars boosted 'self-sufficiency' as an essential element of maximising food production[40] with the corollary of encouraging food preservation. With sugar in short supply many traditional methods were resurrected and bottling in 'Kilner' jars proved popular.

For everyday storage the larder or pantry remained an essential adjunct. Country estates often had separate larders for meat, fish and game and their icehouses were in use until the 1930s. Town dwellers could purchase ice on a daily basis for an ice chest in which foods were kept cool, a method which was only superseded when refrigerators were introduced[41] (Fig. 23.6).

At table
The shifting dinner hour continued to influence the entire day (Fig. 23.7). As dinner slid towards evening, the gap between it and luncheon grew increasingly uncomfortable and during the first half of the nineteenth century 'afternoon tea' was introduced, supposedly by Anna, Duchess of Bedford.[42] Afternoon tea, as distinct from the 'tea' which followed dinner, started by being a slight refreshment of bread and butter with a few small cakes to

Figure 23.6 The cooling unit of this 1927 refrigerator was so bulky that it sat prominently on the top of the unit. Source: York Museum Trust (York Castle Museum).

accompany the drinking of tea. By the end of the century it had developed into a more substantial meal which showed off the Scottish flair for baking. Taken in drawing room or boudoir it was essentially a feminine meal in which men rarely participated, being engaged in business, sporting or their own social pursuits.

Those in employment were unable to follow this fashion and so on returning home from work they might eat instead 'high tea', an amalgamation of afternoon tea and evening meal.[43] This was often followed later on by a snack called 'supper'.

During the nineteenth century dinner in the homes of the well-to-do was served *à la russe* (in the Russian style), whereby one type of dish, whether soup or fish or meat, comprised a course and each was accompanied by a different wine (Fig. 23.8). A greater supply of cutlery and glassware was required for this style of service. Likewise more servants were necessary as dishes plus their companion sauces and vegetables had to be offered to guests in turn, so that each could serve him or herself.[44] Dessert increasingly formed an integral part of the meal and was served with the cloth still on the table, since table settings were becoming ever more complex. The twentieth-century reaction to this was to do without a tablecloth altogether, in favour

Figure 23.7 The suggested layout for a breakfast table of the 1890s. By this time it was becoming a more formal meal offering a choice of both hot and cold dishes. Source: *Cassell's Book of the Household*, 4 vols, sp edn, London, c1893, vol 1, 40.

of individual lace or embroidered place mats. In turn, the disappearance of parlour maids who washed, starched and ironed these sets after each use, coupled with the introduction of easy-care materials, further simplified arrangements.

Children rarely shared meals with their parents, generally being limited to 'nursery meals' until considered old enough to join in adult life. There is little evidence to indicate how children were eating. The general adage that 'children should be seen and not heard', coupled with 'spare the rod and spoil the child', suggests that even children belonging to the wealthier elements of society would enjoy few luxuries. What such people ate in the nursery would depend entirely on the integrity of the cook or kitchen maid who did their cooking and on the nurse in charge. Elizabeth Grant of Rothiemurchus recalled distressing scenes in her childhood in the early 1800s when a child who refused a dish at table was given nothing else until that dish had been eaten. On one notable occasion her father stood over his children with a whip in his hand. But there were happier times too in this family, when the children were allowed downstairs after dessert was set out on the parents' dinner table and they were given a taste of fruit, a little wine and a biscuit before their game of 'romps'.[45]

MENU.
3188.—SERVICE A LA RUSSE (November).

Ox-tail Soup.   Soup à la Jardinière.

Turbot and Lobster Sauce.   Crimped Cod and Oyster Sauce.

Stewed Eels.   Soles à la Normandie.

Pike and Cream Sauce.   Fried Filleted Soles.

Filets de Bœuf à la Jardinière.   Croquettes of Game aux Champignons.

Chicken Cutlets.   Mutton Cutlets and Tomata Sauce.

Lobster Rissoles.   Oyster Patties.

Partridges aux fines herbes.   Larded Sweetbreads.

Roast Beef.   Poulets aux Cressons.
Haunch of Mutton.   Roast Turkey.
Boiled Turkey and Celery Sauce.   Ham.

Grouse.   Pheasants.   Hare.

Salad.   Artichokes.   Stewed Celery.

Italian Cream.   Charlotte aux Pommes.   Compôte of Pears.

Croûtes madrées aux Fruits.   Pastry.   Punch Jelly.

Iced Pudding.

Dessert and Ices.

Figure 23.8 In a menu *à la russe* one type of food comprised a course in the style still familiar on formal occasions today. Source: Beeton, Mrs. *Mrs Beeton's Book of Household Management*, London, npd (but '140th thousand'), 954.

By mid century, the links between nutrition and health were being recognised more fully and greater attention was being paid to children's diet. Fare that was plain and simple was deemed to be the best for their needs. The majority of children, however, would have been given whatever was available. Attention was also being paid to the diet of invalids and recipe books and domestic encyclopaedias gave many directions for suitable dishes of a light but nutritious nature, as well as for comforting drinks.

After dinner
It was Prince Albert who conferred respectability on smoking by installing a smoking room at Osborne, on the Isle of Wight, to which the men could retreat after dinner. In other respects drawing-room entertainment continued along the earlier lines although by the twentieth century it might include mechanical devices such as a roller organ or an autopiano, a phonograph, polyphon musical box or wind-up gramophone. The wireless became a popular feature in family life in the 1920s. Moving pictures were created through a kinetoscope or a praxinoscope while a magic lantern and a stereoscope showed static images. Forerunners of the home movie projector were also available while more family-oriented games were marketed. Jigsaws,

which originated in the 1760s as an aid for teaching geography at Harrow, became increasingly diverse as well as popular, various card games suitable for children were created in the nineteenth century,[46] bridge and Mah Jongg were extremely fashionable during the 1920s, while board games such as Ludo (1898), Monopoly (from the early 1930s) and Scrabble (devised shortly after World War II) were also widely played.

Where servants were employed, clearing up went on as before. Little changed unless in the provision of hot water from a range or domestic boiler. Increasingly, though, the housewife had to do her own clearing away and washing up, unless she could commandeer the services of her daughters who were expected to help around the house.

1950–2000

The War years halted culinary developments. Fuel and food were curtailed and continued to be so for some time after. However once the country had returned to normality the supply networks for gas, electricity and water were expanded. The refrigerator, perceived as a luxury before the War, was considered as having both space-saving and food-saving potential. The appliances were first installed *en masse* in 'prefabs'[47] (prefabricated buildings put up initially as temporary accommodation) and were then widely taken up.

Cooking and kitchen equipment
Once post-War building restrictions were lifted, kitchens were once again subject to the processes of updating and streamlining, becoming not only technically and ergonomically efficient but, space permitting, the centre of the home as they had been for earlier generations. Cooking and an interest in food in general became fashionable pastimes, spearheaded by Elizabeth David's books in the 1950s.

The gradual acceptance of Christmas as a public holiday gathered momentum during this period, with a return to celebratory meals based on 'traditional' seasonal dishes, such as roast turkey, plum pudding and mince pies. Hogmanay had for some centuries been the focus of festivities, with its customary fare of shortbread, black bun and Hogmanay bun, assisted by various alcoholic beverages but most notably by whisky. By the end of the century, the Christmas period had merged with the New Year's festivities to give a season of celebration approaching the medieval Twelve Days of Christmas.

In terms of cooking equipment, the period witnessed numerous improvements to cookers but the principal innovation was the domestic microwave of the 1990s[48] which in fact had been used in the catering trade for several decades. Special cookware was marketed for microwave use. Pots and pans for use with conventional cookers acquired non-stick finishes in the 1960s, as did the cookers themselves, for easier cleaning. Oven-to-tableware became not only oven and flameproof but also dishwasher and freezer-proof. Electricity was increasingly applied to kitchen equipment including the tin opener, carving knife and juice extractor.

Sources of supply
Refrigerators freed the housewife from the chore of daily shopping, a trend which accelerated when the domestic freezer arrived in the 1960s. Not only did traditional methods of preserving decline but a huge market developed for frozen foods and readymade meals.

If railways revolutionised the supply situation from the 1840s onwards, airfreight did the same in the twentieth century with food arriving in Scotland from around the world. The concept of produce being 'seasonal' now scarcely exists. Smaller, individual local shops struggle to survive in most places. Supermarkets are all pervading and sell a hitherto unimaginable range of imported goods, in part reflecting an increase in foreign travel and the presence in Scotland of various immigrant communities.[49] A recent practice is computerised shopping over the Internet.

With such a proliferation of retail outlets the concept of self-sufficiency seems old-fashioned. However, there will always be those who wish to grow their own fruit and vegetables, or keep bees or hens, either to have a measure of control over their food or from a sense of creativity.

At table
Meals and mealtimes, or the lack of them, reflect the splintering of family cohesiveness, a marked trend of this period. Surveys suggest that households are less and less likely to eat together or even eat the same foods, a situation aided by the microwave and the freezer, while the number of wives and mothers employed beyond the home militates against time being spent in preparing food.[50] In addition, there has been a marked rise in the number of people living alone or whose work schedules cause them to eat on their own, and the needs of these people are catered for by numerous manufacturers and retailers who supply readymade or easy-cook fresh or frozen individual meals.

There has also been a proliferation of fast-food outlets which appeal especially to the younger generation, who often prefer to have several snacks during the day rather than to eat one, perhaps more balanced, meal. Eating out in general has increased in popularity and the last 40 years have seen a huge increase in the number of establishments where people can eat or where they can buy food to take away with them. School dinners, and sometimes breakfasts too in poorer areas, relieve the housewife of some of the responsibility of feeding her offspring. Work canteens do the same for other family members, including, of course, the housewife herself.

After dinner
When television first arrived in the 1950s whole families would watch the same programmes but nowadays, when many families rarely eat together, they are less likely to spend time together afterwards, each member preferring to follow his or her own interests.[51]

The washing up process has been somewhat eased with the marketing of liquid synthetic detergents, such as 'Teepol', from the 1940s onwards, which replaced odd ends of soap swished through the water in a small wire-mesh cage. Until the 1960s the dishwasher was considered a luxury but is now, space permitting, considered a kitchen essential.

THE YEAR 2002

Supermarket shelves are laden with produce from around the world. There is no time during the year when food is in short supply. Guidance, if not inspiration, pours daily from radio, television, newspapers and magazines. Yet there is considerable concern over our sustenance. There is worry about its nutritional content and negative impact on our health, about the potentially harmful consequences of too many over-refined and processed foods along with the effects of chemical additives, to which should be added the emerging debate over the safety of eating genetically-modified organisms.

At a time when food is available in a variety and quantity unimaginable even to our recent forebears, it is therefore ironic that it should also be a source of such anxiety to an increasing proportion of the population.

NOTES

1. Johnson, 1798, 93.
2. Hume Brown, 1891, 44, 112, 121–3; Chamberlayne, 1708, 523; Johnson, 1798, 93; for a general appraisal see Hope, 1987, 91–113; Steven, 1985, 3–10.
3. Baillie, 1911, 277–8.
4. Sibbald, 1709.
5. Wood, 1794, 114.
6. OSA, XII (1983), for example, 62, 429–30, 434; Campbell, 1966, 47–54; Sanderson, 1919, 79–84.
7. Anderson, 1867, 91. Those employed had a small allowance of meat in addition to the daily broth and bread. The infirm were given rice soup. A small proportion of individuals drank tea at breakfast and a much larger number drank it in the evening.
8. Until 1600 Scotland's New Year began on 25 March ('Lady Day'), but that year the date was changed to 1 January, thus bringing New Year into the Christmas period. Meantime, England kept to the old date, not changing it until 1752.
9. Johnson, 1798, 176. He confirms the supposition regarding the fuel for cooking, 'The only fuel of the islands is peat. Their wood is all consumed, and coal they have not yet found ... The heat is not very strong nor lasting'; see also Chapter 6 Heating and Sleeping
10. Fiennes, 1888, 219.
11. Spits could be rotated by hand, by animals, or by various devices including clockwork and smoke vanes in the chimney. The later bottlejack was independent of the fire.
12. McNeill, 1929, 49, 'The modern girdle was invented and first made in the little burgh of Culross, in Fife. In 1599 James VI granted the Culrossians the exclusive privilege of its manufacture, and this was confirmed by Charles II in 1666.'
13. Bakestones are more typical of the north of England but they were also used in

Scotland. There is a good example of one in the Highland Folk Museum, Kingussie, Inverness-shire, and a picture of two 'baking stones' can be found in Lochhead, Marion. *The Scot's Household in the Eighteenth Century*, Edinburgh, 1948, opp 192.

14. Hutchins, 1967, 115, 167–8. Items to be jugged were put into a jug or stewpan and a hot liquid (either water or stock) was poured over them. The container would then be covered and set in an outer container filled with water – a *bain-marie* – which would be placed over a low heat and left until the items were ready. Kippers needed only to be heated through rather than cooked and so merely required to be immersed in hot water for a few minutes. The hare, cut into pieces, needed three to four hours in its *bain-marie*.
15. Ray, 1749, 377; Burt, 1815, II, 253.
16. Foulis, 1894, 21, 'for markames works'; NAS, Breadalbane Muniments, 1692, 'Mays cookery and Digbys cookery'; Lamb, Patrick, Esq. *Royal Cookery* (Hopetoun Papers Trust), London, 1710; Baillie, 1911, 37.
17. McLintock, 1986; Reid, 1683, Part II, 115–24.
18. Dods, 1828. The character of 'Meg Dods' was based on Miss Marion Ritchie, landlord of The Cross Keys Inn at Innerleithen, Peeblesshire.
19. Dods, 1828, 354, 427–8, 441, 443. Although a Scottish dimension is sometimes apparent, it is instructive to note that Meg Dods considered it necessary to include a 'Bill of Fare for St Andrew's Day, Burns' Club, or other Scottish National Dinners'.
20. *Book of the Old Edinburgh Club*, 1909, 177–222; Mayhew, 1851–62, I, 4. He estimated that London had at least, '30,000 costermongers and other street folk'.
21. Banks, M M. *British Calendar Customs, vol 1 (Scotland)*, London, 1937, 181–7.
22. Grant, 1898, 180.
23. *Book of the Old Edinburgh Club*, 1953, 146–8; icehouses were also sometimes used as 'early forms of refrigerator for food and drink'. See Beaton, 2003.
24. Examples of suggested layouts can be found in McIver, 1789, 253–6; Caird, 1809, Plates 6, 7; Dods, 1828, 53–69.
25. For example, Chambers, 1912, 147–8; Grant, 1898, 22, 203.
26. Johnson, 1798, 92–3.
27. Robertson, 1998, 281–8.
28. Donaldson, 1794, 24.
29. 'Ball suppers' were served to those dancing all evening.
30. Dods, 1828, 37.
31. Graham, 1937, 103–6.
32. Faujas de St Fond, 1799, I, 254.
33. Charles II entered London on 29 May 1660 and the date became known as 'Oak Apple Day' or 'Royal Oak Day'. An Act of Parliament commanded that it be observed as a day of thanksgiving.
34. Lauder, 1840, 121.
35. Adams, Adams, 1825, 369.
36. Adams, Adams, 1825, 393.
37. Dods, 1854, 130–1, 'We have been so often taken in,' she wrote feelingly, 'with wonderful, newly-invented frying pans and infallible gridirons, that we do not venture to recommend any form. We have collected half a garretful of these and other culinary inventions, and on trial found nearly the whole useless, or little improvement on the old-fashioned utensils.'
38. As one of the most hated chores in existence many people tried to devise a mechanical device to wash the dishes. Beeton, Mrs *Mrs Beeton's Family Cookery*, London, 1923, fig between 48–9, shows 'Mechanical Crockery Washers' including no 7 which was hand operated and no 9 which was electric.
39. *Book of the Old Edinburgh Club*, 1909, 177–222.
40. The slogan was 'Dig for Victory'.

41. At the start of the nineteenth century ice was probably comparatively expensive. However in an undated but early (around 1861) copy of Beeton, npd (but 140th thousand), written for the middle-class housewife, the author speaks of taking a bushel of ice to fill the ice tub in which ice creams, etc were to be made. Ice cream was a luxury food but it appears that ice was not overly expensive.
42. Cooper, 1929, 202. This was Anna, d1857, wife of the 7th Duke of Bedford.
43. Today, many urban working-class Scots in particular call the midday meal 'dinner' and the evening meal, when taken in the home, 'tea'. Others may use 'lunch' and 'dinner' respectively instead. For all, lunch (midday) and dinner (evening) describe meals eaten 'out' in restaurants, for example. Afternoon tea and high tea is still served in some hotels, cafes and tearooms. Pubs may serve 'lunches' and 'suppers', the latter also remaining the name for the small pre-bed snack in many households. See Fenton, forthcoming.
44. Beeton, npd (but 140th thousand), 954–5.
45. Grant, 1898.
46. These included 'Old Maid' (1844), 'Happy Families' (1881) and 'Snap', sometimes called 'Snip-Snap-Snorum' (1882); see also Chapter 11 Games and Sport; Chapter 25 Childcare and Children's Leisure.
47. Women's Group on Public Welfare, 1950–2, 18.
48. In the 1980s microwaves were being introduced into homes, but only in the 1990s did prices drop sufficiently for the ovens to become 'must-have' equipment.
49. This paper cannot describe the many and varied features of food preparation and consumption among immigrant groups in Scotland. Much in fact remains to be investigated in this area.
50. Despite changing work patterns and new social trends in the last few decades, women generally continue to bear greater responsibility than men for food preparation and other domestic matters. See Chapter 17 Housework; Chapter 28 Hierarchy and Authority.
51. See Chapter 10 Television, Video and Computing.

BIBLIOGRAPHY

Adams, S, Adams, S. *The Complete Servant* (1825), facsimile edn, Lewes, 1989.
Anderson, W. *Provision for the Poor*, Edinburgh, 1867.
Baillie, Lady G (ed R Scott Moncrieff). *The Household Book of Lady Grisell Baillie, 1692–1733* (Scottish History Society, Second Series, vol 1), Edinburgh, 1911.
Banks, M M. *British Calendar Customs, vol 1 (Scotland)*, London, 193.
Beaton, E. Ancillary estate buildings. In Stell, G, Shaw, J, Storrier, S, eds, *Scotland's Buildings, Scottish Life and Society: A Compendium of Scottish Ethnology, vol 3*, East Linton, 2003, 190–210.
Beeton, Mrs. *Mrs Beeton's Book of Household Management*, London, npd (but 140th thousand).
Book of the Old Edinburgh Club, Edinburgh, vol 2, 1909.
Book of the Old Edinburgh Club, Edinburgh, vol 28, 1953.
Burt, E. *Letters from a Gentleman in the North of Scotland to his Friend in London* (1754), 2 vols, 1815.
Caird, J. *Complete Confectioner and Family Cook*, Edinburgh, 1809.
Campbell, R H. Diet in Scotland: An example of regional variations. In Barker, T C, ed, *Our Changing Fare: Two Hundred Years of British Food Habits*, London, 1966, 47–60.
Chamberlayne, E, Chamberlayne, J. *Magna Britanniae Notitia*, London, 1708.
Chambers, R. *Traditions of Edinburgh*, Edinburgh, 1912.
Cooper, C. *The English Table in History and Literature*, London, 1929.
Dalgairns, Mrs. *The Practice of Cookery Adapted to the Business of Everyday Life*, Edinburgh, 1828.

Dods, M (pseudo, C I Johnston). *The Cook and Housewife's Manual* (1828), 10th edn, Edinburgh, 1854.
Donaldson, J. *General View of the Agriculture of the Carse of Gowrie*, London, 1794.
Faujas de St Fond, B. *Travels in England, Scotland and the Hebrides*, London, 2 vols, 1799.
Feild, R. *Irons in the Fire: A History of Cooking Equipment*, Marlborough, 1984.
Fenton, A. *Scottish Country Life*, Edinburgh, 1976.
Fenton, A. *The Food of the Scots, Scottish Life and Society: A Compendium of Scottish Ethnology, vol 5*, East Linton, forthcoming.
Fiennes, C. *Through England on a Side Saddle in the Time of William and Mary*, London, 1888.
Foulis, Sir J (ed Rev A W C Hallam). *The Account Book of Sir John Foulis of Ravelston, 1671–1707* (Scottish History Society, vol XVI), Edinburgh, 1894.
Graham, G H. *The Social Life of Scotland in the Eighteenth Century*, London, 1937.
Grant, E (ed Lady Strachey). *Memoirs of a Highland Lady*, London, 1898.
Grant, I F. *Highland Folk Ways*, London and Boston, 1975.
Hope, A. *A Caledonian Feast*, Edinburgh, 1987.
Hopetoun Papers Trust, South Queensferry.
Hume Brown, P, ed. *Early Travellers in Scotland*, Edinburgh, 1891.
Hutchins, S. *English Recipes and Others*, London, 1967.
Johnson, Dr S. *Journey to the Western Islands of Scotland in 1775*, London, 1798.
Lauder, Sir J (eds A Urquhart, D Laing). *Historical Observes of Memorable Occurrents in Church and State, from October 1680 to April 1686* (Bannatyne Club), Edinburgh, 1840.
Lindsay, M. *The Discovery of Scotland*, London, 1979.
Mayhew, H. *London Labour and London Poor*, London, 4 vols, 1851–62.
McIver, Mrs. *Cookery and Pastry*, London and Edinburgh, 1789.
McLintock, Mrs. *Mrs McLintock's Receipts for Cookery and Pastry Work of 1736*, reproduced from the original, Aberdeen, 1986.
McNeill, F M. *The Scots Kitchen*, London and Glasgow, 1929.
NAS, Breadalbane Muniments, GD112/22/14, Inventory, 1692.
Ray J. *A Complete History of the Rebellion*, York, 1749.
Reid, J. *The Scot's Gard'ner* (1683), facsimile edn, Edinburgh, 1988.
Robertson, U A. Orange marmalade: Scotland's gift to the world. In Lysaght, P, ed, *Food and the Traveller*, Cyprus, 1998, 281–8.
Sanderson, W. *Scottish Life and Character*, London, 1919.
Sibbald, Sir R. *Provision for the Poor in time of Dearth and Scarcity*, Edinburgh, 1699 and 1709.
Steven, M. *The Good Scots Diet: What Happened To It?* Aberdeen, 1985.
Women's Group on Public Welfare. The effect of the design of the temporary prefabricated bungalow on household routines, *Sociological Review*, XLII (1950–2), 18.
Wood, J P. *The Antient and Modern State of the Parish of Cramond*, Edinburgh, 1794.

24 Childbirth

LINDSAY REID

INTRODUCTION

Until the twentieth century, nearly all mothers in Scotland gave birth at home, poor, homeless women being the exception. For these people were established 'lying-in' hospitals or departments attached to infirmaries which later developed into maternity units in the cities of Edinburgh, Glasgow and Aberdeen.[1] They were very primitive by present-day standards and the beds available were few compared to the numbers of women giving birth across Scotland.[2] As they were financed as voluntary hospitals, lack of funding was an on-going problem and was partly responsible for the small number of beds. Furthermore, there was a high risk of infectious puerperal fever in these institutions. Not surprisingly, mothers were very reluctant to go into hospital for their confinement or to be referred there in the case of *intra-partum* complications.[3] This fear was echoed by many Victorian professionals who therefore 'believed that there should be no hospitals at all for women in childbirth'.[4]

There were other, smaller establishments created especially for single mothers,[5] for instance the Lauriston Home in Edinburgh. There, the girls were looked after until they were sent to the Maternity Hospital for their confinement, returning to Lauriston postnatally with their babies for 14 days. In the Maternity Hospital in Aberdeen an eight-bedded room was set aside for single women. However the total number of beds in Aberdeen Maternity Hospital in 1915 was only 15,[6] which gives some idea of the large percentage of women who gave birth at home.

Lying-in hospitals in the university cities also served as teaching centres. Many of the mothers who were admitted were undernourished and physically unhealthy, and in an informal arrangement they received free care and food in return for allowing obstetricians, general medical students and pupil midwives to learn from them.[7]

By the early twentieth century most women would still not usually consider going into hospital to give birth. There were several reasons for this. Firstly, there was a strong culture of home, with women tending to conform to what their mothers and grandmothers had done before them which was to give birth in their own dwellings. There was also the perceived stigma of going into hospital. Hospitals were for the very poor, for those women who had no one to look after them and for unmarried mothers. In addition there

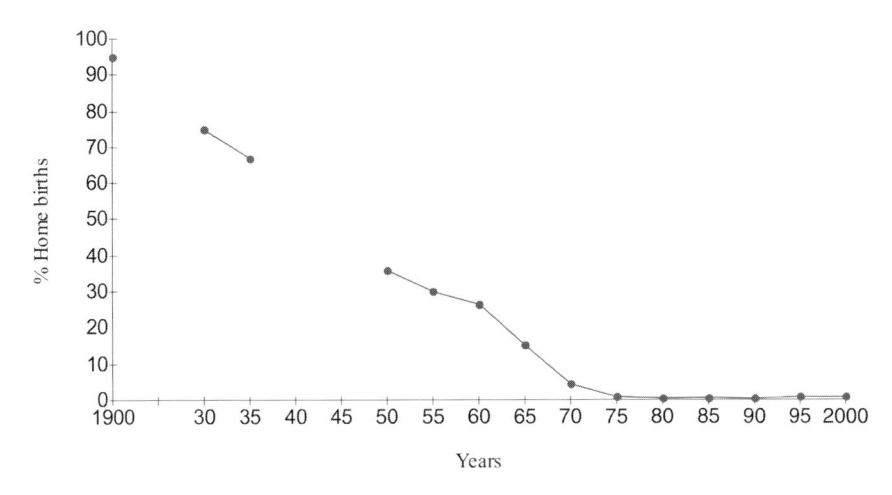

Figure 24.1 Graph showing the decline in home births in Scotland in the twentieth century. Source: Lindsay Reid/Scottish Multiprofessional Maternity Development Group.

was often a low quality of care. While infection was considered a hazard of birth wherever it took place, hospital figures for *puerperal sepsis* were considerably higher than at home largely because of poor hygiene and the close proximity of patients in the wards. Further, as the number of maternity beds in Scotland bore no relation to the number of births, going into hospital would simply not have been an option in most cases. Finally, as Marjorie Spring Rice highlights in *Working-Class Wives*, some mothers felt they could not leave the responsibilities of home to the care of others.[8]

Mothers giving birth at home, particularly those who were undernourished and badly housed, were not immune to problems. Mackenzie, reporting in 1917 on the physical welfare of Scottish mothers and children, comments on the conditions in which many mothers gave birth. He acknowledged that the high level of puerperal fever was partly to do with the transmission of infection from mother to mother, particularly by careless doctors who dealt with all types of patient. He also recognised that poor social conditions, with frequent occurrences of whole families living in one room, intensified the incidence of infection.[9] Even so, Munro Kerr, an eminent Scottish obstetrician working in Edinburgh in the early decades of the twentieth century, recommended institutional treatment for only eight to 12 per cent of deliveries.[10] Despite this, the change away from the home as the main place of birth began to gather pace in the early decades of the twentieth century and reached a peak in 1981 with 99.5 per cent of Scottish babies being born in hospital[11] (Fig. 24.1).

WHO DELIVERED THE BABIES?

Before the mid eighteenth century, women in childbirth were cared for and delivered by midwives, the majority of whom were untrained and as such known in many areas as 'howdies'. Male medical practitioners would be called to a birth in an emergency only. From the mid eighteenth century the presence of male practitioners became more common even for normal births,[12] while a greater number of midwives undertook a form of training under their supervision. There was, however, no uniformity of midwifery training throughout Britain until the twentieth century and nothing to regularise it. In addition, apart from the special case of the Diocese of Glasgow where the Faculty of Physicians and Surgeons had the power to license and un-license midwives from 1740 (the date of the Faculty's Act anent Midwives) until around 1820,[13] any woman could practise midwifery at will. The untrained midwife was, in fact, the more popular option, especially amongst less well-off women. In 1902 the Midwives Act for England and Wales was passed and was followed by the 1915 Midwives (Scotland) Act which did much to regularise midwifery. Legally, midwives were expected to enrol with the newly formed Central Midwives Board for Scotland within five years of its start date in February 1916. In practice, it took much longer, with howdies occasionally being seen as late as the 1950s.[14]

General practitioners have a long history of involvement in home births. Scotland's medical practitioners received a wider training than their English counterparts and this affected their attitude to the field of maternity care. From the seventeenth century, students within the Scottish university medical schools were customarily trained in medicine, surgery and midwifery. By 1880 about 66 per cent of medical personnel practising in Scotland were graduates with this broad education.[15] However Ferguson writes that the proportion of confinements in Scotland attended by medical practitioners dropped in the first decade of the twentieth century.[16] This was probably because of a gradual improvement in midwifery practice. In 1907 midwives in Scotland were still under no supervision. However, according to A K Chalmers, Medical Officer of Health for Glasgow, medical practitioners in Scotland observed the improvement of midwifery practice in England and Wales after the implementation of the 1902 Midwives Act for those countries. There was a concerted effort to raise standards in Scotland by increasing courses of instruction even though the Scottish Midwives Act was not passed until 1915. With higher standards of midwifery it is possible that medical practitioners felt able to place more trust in the skills of their midwifery colleagues.[17]

The Maternity Services Schemes in Scotland,[18] which developed from the 1915 Notification of Births Act, strengthened the part the medical profession could play in midwifery in Scotland by giving power to local authorities and their Medical Officers of Health to organise maternity care. Under the Schemes medical practitioners could take a more active part in the care of childbearing women than their counterparts in England[19] and were nominally in charge of all home confinements, although they were not necessarily present at the birth. In fact it was not uncommon for a doctor to be unaware of a pending birth until called. This was especially evident in rural areas where, in the early twentieth century, trained midwives were seldom available and most of the babies were delivered at home by untrained midwives,[20] although the Highlands and Islands Medical Service was to play an important part in rectifying this. Howdies could, if necessary, call for medical aid, 'There are no trained midwives in the district [Unst, Shetland], but there are four very capable women who attend most of the confinements, the doctor being called in when labour seems for some reason unknown to the midwife to be unduly prolonged.'[21] The howdie went everywhere, at all times and in all weathers as David Rorie writes:

> A' gate she'd traivelled day an nicht,
> A' kin o orra weather
> Had seen her trampin on the road,
> Or trailin through the heather.[22]

However, in the decades following the 1915 Midwives (Scotland) Act supervision of the practice of midwifery became tighter under the auspices of the Central Midwives Board. The number of certified midwives practising in Scotland grew and the number of howdies diminished.

PREPARATIONS FOR HAVING A BABY AT HOME

In the first half of the twentieth century many women received little or no antenatal care from either certified midwife or doctor and pregnancy was a very private, secret state.[23] In some cases a howdie might have been booked and taken up residence beforehand to help with the housework, other children and even the chores conducted outside, but the family doctor might not have been aware that a baby was imminent. 'Then, ye see, ye didna tell the doctor aforehand, the doctor jist cam fan ye needed them.'[24] In this particular instance, the 18-year-old who was to become an established howdie in Aberdeenshire had never seen a baby being born before. The doctor was late because of snow and she had to perform a very difficult delivery with the cord wound twice round the baby's neck.[25]

Antenatal care in Scotland, eventually offered by municipal clinics in the cities employing their own midwives and doctors and by general practitioners, developed slowly. It was aided by the comprehensive ideal of the 1918 Maternity and Child Welfare Act which enabled municipal authorities to use their funds to provide salaried midwives and health visitors, free or cheap food for mothers and children, and antenatal clinics and day nurseries.[26]

In general, however, in the 1920s and 1930s the amount and type of antenatal care offered across Scotland varied widely.[27] Local authority ante natal clinics functioned with limited success. A large number of rural areas lacked them altogether, and even where they were available, mothers often did not bother to attend. Nor did they visit their doctor. Midwives delivered women who may have booked them but who had received no antenatal care, or women who had not booked at all. Many mothers preferred to book a howdie, not just because they performed many other tasks in the home as well as looking after the mother and her new baby, but because they were cheaper than qualified midwives. There is also evidence that some general practitioners preferred working with howdies.[28]

Ann Lamb, in midwifery training in Edinburgh in 1927, makes the link between lack of antenatal care and poverty.[29] This is reinforced by Molly Muir training in the 1930s in Gorgie, one of the poorest areas of Edinburgh at that time:

> Working in Gorgie at that time there were very few antenatal clinics. Sometimes when the mother went into labour you found it was a face presentation. That was the worst I ever came across. Other things happened that you didn't know anything about until the mother was in labour. There was very little antenatal care ... I think they had all made arrangements even though there were so few antenatal clinics. I think that when somebody was pregnant and they knew they were going to get the midwife in for the delivery, it was arranged quite a long time in advance. Then, when the time came she would phone the hospital [for a midwife] from a public telephone.[30]

Yet, working on Unst, during and after World War II, Mima Sutherland told a different story, highlighting the fact that not all places offered the same quality of care:

> In my time the mothers were all supervised. You could see them and they could get their urine tested, you see, and then you would know if they needed to see the doctor. They were seen regularly. Then you would wait for them to send for you [when they went into labour].[31]

After the implementation of the NHS Acts in 1948 there was a gradual increase in the uptake of antenatal care performed by general practitioners in their surgeries, and, in time, by obstetricians in hospital antenatal clinics. This accompanied a growing medicalisation of childbirth and a decrease in the number of home births in Scotland.

From a practical point of view, preparing for a home birth depended on who the mother was, where she lived and how many babies she had already had. In the days before the common use of contraception large families were the norm and an extra mouth to feed was not always welcome, 'The poor mothers they really didn't – after they had had nine or ten children – they didn't really want another baby.'[32] The birth rate fell 'first and fastest' in middle-class families and the difference in family sizes in 1911 ranged from those of physicians, averaging 3.91 children, to those of crofters, averaging 7.04 children.[33]

Many poverty-stricken mothers made no preparations at all. One midwife from South Africa spoke about working in Glasgow during the 1930s:[34]

It was very difficult ... and of course we always took newspaper with us because it was cleaner than the sheets or whatever they had on the beds. There were these beds in the wall where they all slept and you had to turn father and all the children out whilst mother produced her thirteenth. There were huge families. I was at Rotten Row six months and I never went into a house where the man was working. The 'Queen Mary' was up on the stocks. Unemployment was terrible and there was such a kind of despair about it. Glasgow Corporation produced a layette for the new baby if the woman had been to the antenatal clinic and got a form saying she was pregnant. They had to go and get this layette ... and very often they hadn't even bothered to go and do it. There was such a kind of feeling of I don't know what – apathy.

And then things became a little better, for some:

Whilst I was there they started on the 'Queen Mary' again ... They had stopped building altogether and then they re-started whilst I was there. It would have been the summer of 1934 and the bands all played the workers on to the job and some of the wifies' husbands were working and they were so superior and so pleased. 'My man's working.' It was all so sad and pathetic and these awful closes and tenements. I had never seen people living in those sort of conditions. I couldn't imagine it. Things may have been bad in our slums in South Africa; they did have slums in Johannesburg and places but nothing as bad as that.

Not all poor families lacked preparations for the new baby. Anne Chapman working with a Traveller family said:

I delivered the baby in the tent on the floor. I remember the lady said, 'We've got a clean sheet nurse.' And I used it for the delivery and it was a clean blue sheet. And they always had clothes for the baby, not new but they were always washed, maybe not ironed but always clean, [and] a shawl or a blanket or whatever, to wrap round the baby afterwards.[35] (Fig. 24.2)

Another important part of preparation for a home birth was the delivery pack or box, which local authorities were required to supply. It contained sterile equipment for the birth and the midwife usually took it to the house about a month before the expected date of delivery. No one was supposed to touch anything until it was required:

The most common thing that happened everywhere, after 36 weeks [of pregnancy], you went in and collected your brown box [and took it to the house]. This was the delivery box and in it was waterproof paper, cotton wool, sanitary towels and so on – and a bottle of Dettol. Well, the Dettol in Glasgow was always white! 'We got it like that Nurse. That was how it came.' Of course they had used the Dettol and filled

Figure 24.2 Living room prepared for a homebirth in the 1980s. Source: Liz Andrew.

it up with water and of course when you put water into Dettol it turns white. So we all finished up with this white Dettol.[36]

Sometimes the mother collected her own delivery pack. Mary McCaskill recalled:

If it was a booked case, if the mother was booked to have a home confinement, well, she got the pack [from Central Office] at about 36 weeks ... She would go with a chitty from me and collect that pack. [I took it] only if it was an unbooked or emergency delivery. You kept say four or maybe six packs in your own home so that you could produce them during the night when the Central Office would be closed.[37]

Midwives also carried old newspapers to most births, or mothers and their neighbours supplied them. They were commonly used to protect the mattress from damage, and in very poor homes they also took the place of sheets, 'One woman had nothing but a pile of Radio Times and when I went downstairs the wifie gave me some clothes for the baby.'[38] Newspapers were also used for wrapping up the placenta before the midwife put it on the fire, and even served to protect midwives' coats:

I remember seeing bugs on the wall, various things on the wall. I don't remember fleas because if there's a flea about I get it. I didn't know quite where to hang my coat in the houses. I would wrap it up in some of the newspaper – we went bulging with newspapers. It was used for everything. But I don't ever remember being bitten.[39]

NEIGHBOURS AND FEMALE RELATIVES

There were almost always other women around when a woman was in labour and giving birth; grannies and other female relatives of the mother, neighbours and friends. They were there to give encouragement and advice to the mother and the midwife or the howdie (who was sometimes a neighbour as well), and to give practical help and join in the celebrations after the baby was safely born.

As time went on, the number of women within the birthing room grew smaller, but neighbours were still offering support, albeit in the background:

> Most of them had a neighbour who was very good to them ... A neighbour would probably be in just helping her to get everything ready ... if it wasn't granny it would be somebody from upstairs or downstairs or something. They wouldn't necessarily be in the room. No, no they wouldn't. But you knew, they would say, 'I'm just next door if you want me.' You know the sort of thing. But obviously she had been in helping to get everything and the efforts some of these people made.[40]

Molly Muir also had reason to appreciate neighbours:

> We called the neighbour and no matter what the neighbour was called we called her Maggie. She made tea half a dozen times – she was there the whole time, sort of sitting in the background. She was there making the tea. We seemed to drink an awful lot of tea and [eat] scones and things. [She also got] hot water to bath the baby – you know there was no hot water in the tap in these houses ... so, Maggie was in the background and usually they were the nicest people and very important ... and when the baby was born they were so excited.[41]

Neighbours were there for the bad times also. Of one sad occasion in Shetland in 1930, when both mother and baby died, Chrissie Sandison said, 'When I went in the house there was a neighbour there. There are always that kind of folk that go to people in distress.'[42] Neighbourly support in times of adversity is also testified to by Ella Banks who delivered many babies in very poor homes.[43]

FATHERS AT THE BIRTH

Over the years there has been a marked change in the attitudes of professionals, and of the fathers themselves, to the suggestion that men might attend the births of their babies. Fathers used to take no part in the preparations for a birth, nor were they present at the birth itself, 'In these

CHILDBIRTH • 447

days [the 1940s] it wasn't considered the thing to have a husband present, husbands were shooed out the door.'[44] Gradually the presence of a father at the birth became not just acceptable, but expected. Fathers in Scotland, from the 1970s onwards, have regularly attended parentcraft and preparation for birth classes.

POSTNATAL CARE IN THE HOME

The immediate postnatal period, once known as the 'lying-in' period, begins with the birth of the baby and is considered to last for about ten days. The full *puerperium* lasts for six to eight weeks while the woman's body returns to its pre-pregnant state. Since the 1915 Midwives (Scotland) Act, it has been a statutory requirement for midwives to attend upon and examine mothers and their infants during the postnatal period.[45] Postnatal physical examinations of the mother and baby by a midwife remain the same today as they were in the first half of the twentieth century and as laid down by the Central Midwives Board rules.

For many years mothers were expected to stay in bed for the full ten days of the lying-in period. Help was usually available from neighbours and female relatives, and husbands if unemployed. However, many mothers got up against advice and got on with work, 'The mothers weren't supposed to get up out of their beds till about the tenth day and you very often would go [to visit] and the mother had skipped out for messages.'[46] One midwife described how she always knew if the mother had been out of bed by the state of her feet, and another recalled:

> ... one time calling at this house. I did the mother and bathed the baby and then I said, 'You know you've got on awfully well. I think you could get up this afternoon for an hour.' I carried on further up the street to another lady and she said, 'Hasn't Mrs S done well. She was up having tea with me yesterday afternoon.' You never knew what was happening when your back was turned.[47]

As time went on midwives and doctors came to realise that mothers fared better without prolonged bed rest, whether they had been delivered at home or in hospital, and now it is customary for them to be up and about soon after the birth.

Care of the cord or *umbilicus* has seen many changes over the years. It is said that on the islands of St Kilda, off the western coast of North Uist, in the late 1800s the midwife dressed the cord stump with a paste of fulmar oil resulting in many infantile deaths from *tetanus neonatorum*, a condition common enough elsewhere. Yet in Banffshire in the early twentieth century one howdie had the right idea. Her daughter, Ann Lamb explains:

> She told us once how she dressed the babies' cords. She washed and boiled old white linen rags, dried them and lit a candle. Then she cut a hole in the centre of each square of linen and held it over the candle

until it was burning, then blew it out and rolled them all up in a clean towel. That was sterilisation.[48]

There were also phases of using 'cord ties' which the midwives made themselves and, depending on the custom of the area, changed at regular intervals until the cord fell off. A big innovation was a cord tie which came ready wound up in a sterile bottle and the midwife only had to pull out and cut off what she needed. Sometimes the baby's umbilical area was swabbed with spirit, sometimes dusted with sterile powder, sometimes dressed and bound with a broad bandage which was stitched, stuck with tape or safety pinned. Now, the usual procedure is to wash the *umbilicus* when the baby is being bathed, dry it carefully and leave it.

For at least 2000 years there has been an association of the event of childbirth with psychiatric disorder. This was recognised and noted by Hippocrates who suggested that it could be caused by milk diverted from the breast to the brain. The mental aspect of postnatal care was not routinely given much attention, unless the mother was acutely ill, until well into the latter part of the twentieth century. Now it is part of the midwifery curriculum and taken very seriously. Up to one in seven women in Scotland may suffer from some form of postnatal mental illness. There is no evidence to show that women having a planned homebirth in Scotland fare any differently from those having a planned hospital birth. However, where women have wanted a homebirth, and for some reason this has not been possible, or when women have had a particularly unhappy experience in hospital, then place of birth may be a factor in the occurrence of postnatal depression.

For any woman, becoming a mother, particularly for the first time, requires psychological adaptation. Some people find this easier than others; much depends on the individual's personality and background. Men too have to adapt, and the stress of becoming a new father is now recognised. The present government is acknowledging the importance of the family, whatever form it takes, with increased maternity pay and allowances, and – for the first time ever – paternity leave.

BIRTH CELEBRATIONS

Having a baby is usually the cause of much happiness and rejoicing. In west-coast homes in the past this celebration could take the form of a banquet or 'bangaid', although customs varied very much from place to place.[49] In the north east the feast was:

> called the merry meht, part of which was the indispensible cheese, or cryin kebback. In some districts a cryin bannock made of oatmeal, milk, and sugar, and baked in a frying pan, was served up. Each one present carried off a piece of the cheese to be distributed among friends, and everyone who came to see the mother and baby also carried away a piece for the same purpose.[50]

Figure 24.3 The christening party of Donald Ban MacNeill, Oronsay, Argyll, c 1930. Source: SLA W301421.

The custom of giving 'pieces' was also common on the way home from baptism across Scotland. Until comparatively recently most babies in Scotland were baptised (Fig. 24.3). The mother prepared something to eat, a 'christening piece', and carried it to the church. After the baby was baptised and they were walking home the mother would give the lucky piece to the first person she met.[51]

Midwives have long been associated with baptism. In the Middle Ages many carried the baby they had delivered to church for the ceremony. In Scotland if the minister were not available, the midwife sometimes baptised the baby herself, 'in cases of necessity'[52] to protect him or her from evil spirits 'by letting three drops of clear cold water fall on the tablet of his forehead'[53] in the Name of the Father, the Son and the Holy Spirit. This lay baptism has its origins in pre-Christian 'Celtic' times and was considered to be temporary until the 'real' ecclesiastical baptism could take place. Without baptism some believed that the baby would have had to be carefully guarded in case the fairies took it away. Once baptised, however, the baby was safe from such hazards.

Sometimes the midwife was given the honour of dressing the baby for its first outing to church:

> [I was] met with granny ... and the great-granny of the new baby all smiling. They said, 'Oh how nice nurse', and she says, 'I hope you don't mind but', she said, 'we just left the baby for you to dress because this is its first outing to go to the church ... we thought that perhaps you would like to [do it].' And I said, 'That would be nice to do that.'[54]

THE DECLINE OF CHILDBIRTH IN THE HOME

From a slow start, the decline in homebirths gathered pace in the second half of the twentieth century. Medicalisation included the increasing use of pharmacological pain relief such as chloroform, introduced by Sir James Y Simpson in Edinburgh in 1847, and 'twilight sleep' (morphine and scopolomine), introduced in the early decades of the twentieth century. Midwives were not allowed to administer these drugs, which meant that a doctor had to be present. Constant observation of the mother was also necessary. In the 1920s private maternity homes came into vogue for the wealthy and it was much easier to admit the mother to one of these where both equipment and attention were on hand.

Hospital confinements increased during World War II, as mothers, particularly those from urban areas, encouraged to have their babies in Emergency Maternity Hospitals, found that they enjoyed the break from home responsibilities.[55] In the early months of the War the proportion of institutional births in Glasgow rose by 20 per cent and it is probable that the organisation of the wartime maternity services was a contributory factor in the sharp post-War upturn in the institutional delivery rate.[56]

The implementation of the NHS Acts in 1948 also contributed to the change in place of birth and a decline in the status of the midwife. The organisation of maternity services changed with the development of a tripartite system of maternity care: hospital services dealing with obstetricians and hospital births, general practitioner services dealing with antenatal care and providing GP cover for homebirths if necessary, and local-authority services providing district midwives for home deliveries and postnatal care. Thus, care of mothers during the childbearing episode was fragmented between the three systems. Women could now go to their GP free of charge when they became pregnant and midwives were no longer the first point of contact. Under the NHS, GPs received extra payment for undertaking maternity care and the proportion of them providing more antenatal care than hitherto rose quickly.[57] From full medical contact in the antenatal period, it was a short step to encouraging mothers to go into hospital for the birth. Many mothers were happy to comply as they were told childbirth was safer in hospital with the added bonus that they were able to look forward to a ten-day rest from housework. The provision of enough beds did not happen overnight, however, and there was a time when mothers took for granted a hospital delivery but did not in fact receive it.[58]

There were further developments in the second half of the century. Selection criteria for appropriate types of care evolved in the 1950s and 1960s which steered many women into hospital.[59] The field of obstetrics, a relatively late entry into the widening groups of different medical specialties, developed rapidly in the twentieth century and thus obstetricians, a powerful group, were influential in the production of the 1970 Report of the Standing Maternity and Midwifery Advisory Committee under the chairmanship of J Peel (the 'Peel Report').[60] For the first time a government report advocated 100 per cent hospital delivery on the grounds of safety, even though there was no supporting evidence for this.[61] Although the Peel Report did not apply to Scotland, the Scottish Home and Health Department's *Maternity Services: Integration of Maternity Work* (the 'Tennent Report') agreed with its main conclusions and recommendations.[62]

THE CHALLENGE

In the 1980s the cry for change in maternity services grew louder in the UK. Pressure groups, such as the National Childbirth Trust, combined with the media, mothers, midwives and some doctors to challenge the increasing medicalisation of birth, the apparent decreasing level of choice for women and the assumption that hospital is the safest place for childbirth. In 1991, the government made a commitment that the NHS in Scotland should deliver services, including maternity care, responsive to the needs of the user.[63] In April 1992, representatives of the Clinical Resource and Audit Group and the Scottish Health Management Efficiency Group (CRAG/SCOTMEG) formed a Framework for Action Working Group on Maternity Services in Scotland. Working closely with CRAG/SCOTMEG was a government-led group which published the *Provision of Maternity Services in Scotland: A Policy Review* in 1993.[64] This document promoted 'woman-centred care' and the importance of informed choice for women as to how and where their maternity care should be given, including place of birth. The policy review group did not anticipate a great upsurge in the numbers of women asking for homebirths. Nevertheless, the fact that the option is there for women to investigate is important (Figs 24.4, 24.5).

In 2001 the Scottish Executive published *A Framework for Maternity Services in Scotland*.[65] It reiterates the philosophy of the 1993 *Policy Review*, 'Women have the right to choose how and where they give birth.' The official homebirth rate in Scotland in 1999 was around nine births per 1000[66] (however this figure masks local variations with some areas showing higher levels than others).[67] Mothers, knowing that they have a choice, can plan a birth with their partner, family and with professionals and find what for them is the best and safest option. The excitement and joy of today's planned homebirth is a real team effort.

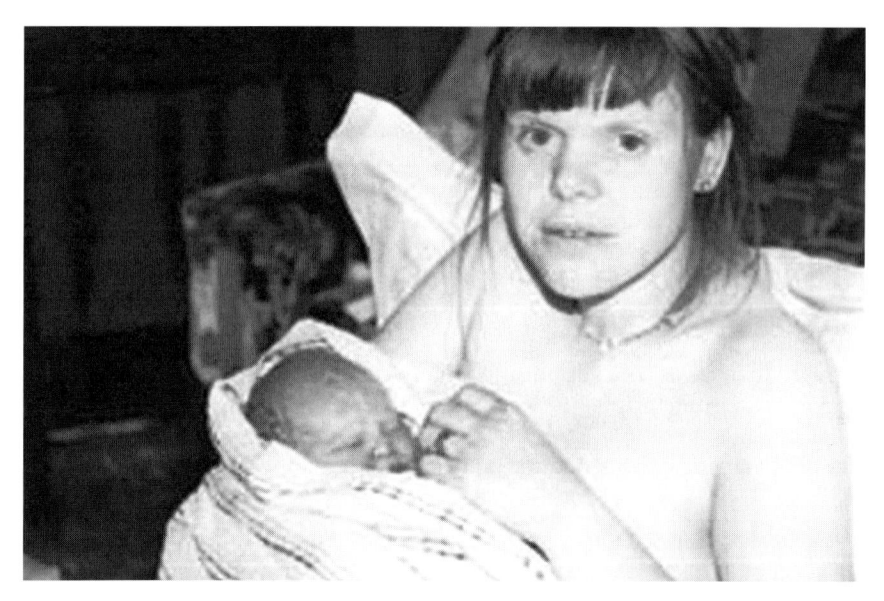

Figure 24.4 Homebirth. Source: Liz Andrew.

Figure 24.5 Homebirth. Source: Liz Andrew.

NOTES

1. Munro Kerr, 1954, 3.
2. Catford, 1984, 36.
3. Munro Kerr, 1954, 4.
4. Checkland, 1980, 185.
5. Ferguson, 1958, 518.
6. Mackenzie, 1917, 64.
7. Tew, 1995, 46.
8. Spring Rice, 1939, 101.
9. Mackenzie, 1917, 55.
10. Munro Kerr, 1933, 255.
11. General Register Office for Scotland, Edinburgh.
12. Donnison, 1988, 34.
13. Cameron, 2000.
14. Reid (LR), oral testimonies between 1997 and 2001, numbered 1–124, 108.
15. Hogarth, 1987, 166.
16. Ferguson, 1958, 510.
17. Chalmers, 1930, 204.
18. Mackenzie, 1917, 535.
19. Cumberlege, 1948, 28.
20. Ferguson, 1958, 510.
21. Mackenzie, 1917, 485.
22. Rorie, 1935, 64.
23. Oral testimony, LR, 61. 'The women in the early twenties they never had any examinations or care beforehand. I remember Jimmy's mother telling me about when she was young when she was having her children it was kept a secret until she was

sure and then the mother who was expecting would speak to some old neighbour, somebody she knew would help and they kept it a secret ... Lots of those expectant mothers of all those years ago would have had no pre-natal examinations. An expectant mother *at da faain-fit*, which meant getting near to the time of delivery, would have "spoken for a midwife."'
24. Oral testimony, LR, 101.
25. Oral testimony, LR, 101, 'So I sort a pushed the heidie back a wee bittie and I got one finger in below the cord and I got it ower – the second bit o the cord wis easy, it wis the first ye see. An eventually, the baby, efter I got at daen, he wis born, nae doctor, nae hubby, naebiddy. I had ti wrap it – bit at the same time the baby wis motionless, an, an like, blue, ye ken. An I thocht oh well, the baby's deid ... I wrapped it up an it sort a gaed a kin a half cough ... ye ken, sort of spluttered a bittie and I thocht, well it's still here ... An then aa at aince it gid a yell and I thocht, O God, I wis never so pleased to hear a bairn.'
26. Oakley, 1984, 54.
27. Douglas, McKinley, 1935, 25.
28. Oral testimony, LR, 61. 'The doctor that I remember when I was a child was a good doctor and the women liked him. But he would never fetch the midwife [Certified Midwife] until she was required. This was my aunt's second baby. She said he sat down the stairs reading a book and he never went upstairs until he knew that the midwife had gone to the lavatory. This was a good bit from the house. It was a wee house across a burn, there was no bucket just a seat across a burn and I don't know about toilet roll – likely it was a bit of newspaper and no buckets to empty. Well anyway when she had to go, he went up to inspect [how the mother was progressing] and then when she came back he would go downstairs again. When she had her next baby – that would be in 1929 – she employed another woman, one who had had a big family but who wasn't a certified midwife. Then the doctor was quite happy. He didn't like working with the trained midwife.'
29. Oral testimony, LR, 47. 'There was no antenatal care. Some people didn't even know they were having a baby till they delivered ... Sometimes you would be going down the street to see a woman and a head would pop out at the window, 'Oh nurse come and help me.' You had never seen her before, you didn't know her name, and up the stairs you went and delivered the baby. She would know who you were. There was no antenatal care and poverty was very bad, very, very bad, but cheery people.'
30. Oral testimony, LR, 46.
31. Oral testimony, LR, 56.
32. Oral testimony, LR, 50.
33. Smout, 1986, 174.
34. Oral testimony, LR, 50.
35. Oral testimony, LR, 11, pupil midwife, Glasgow, 1940s.
36. Oral testimony, LR, 91.
37. Oral testimony, LR, 29.
38. Oral testimony, LR, 56, pupil midwife, Aberdeen, 1930s.
39. Oral testimony, LR, 50.
40. Oral testimony, LR, 35, midwife, Leith, Edinburgh, 1940s.
41. Oral testimony, LR, 46.
42. Oral testimony, LR, 61.
43. Oral testimony, LR, 2, midwife, Motherwell, Lanarkshire, 1940s. 'We delivered babies in houses where there was real poverty. They didn't have anything ready for the baby either in the way of a cot or a blanket or a garment of any kind. Literally nothing. I've wrapped many a baby in a towel and left it with its mother. There was always a friend to help. Although they were so poor they were so friendly and

helpful to one another. The neighbour would go round the other neighbours who had had babies and get clothes for the new baby. That was their lot and they accepted it. The houses were very poor but they were so hospitable – you got tea out of jam jars and bowls, and a biscuit or whatever they had to give you. When you delivered the baby, there was usually a nice neighbour there to hold the baby while you did what you had to do to the mother.'

44. Oral testimony, LR, 27.
45. Maternity Services (Scotland) Act, 1937, section 1 (1).
46. Oral testimony, LR, 93.
47. Oral testimony, LR, 46.
48. Oral testimony, LR, 47.
49. Oral testimony, LR, 56, Raasay: 'Then as soon as possible if the mother was feeling well enough they had a function ... All the women folk were invited and the father made a party for them you see ... Everyone went upstairs and spoke with the mother who was in bed. And they brought presents for the baby and the midwife was there too. I would leave laden with presents – homemade jam and the like. They always had new headgear. I remember one time a young girl came with a bargain book, you know, a catalogue to choose what she was going to wear. Before this, in the old days the women all got new mutches for this party'; Oral testimony, LR, 99, Outer Hebrides: 'The friends went before she got up, to see the mother and the baby and there is an old Gaelic word for it: "Bangaid", which is like banquet. It is pronounced *banget* with a grave accent on the first "a". They went at night to these bangaids. It was war time with the blackout and you had to carry a torch upside down. No roads down to some places. They got a glass of wine ... I think it might have been port ... I wasn't invited to the bangaids which weren't big gatherings. It was just a small community and it was mostly the older married ladies who went. They came with a wee present and got tea and wine. It was an old, old custom, and usually at night.'; for further information on birth customs see Chapter 30 Custom.
50. Gregor, 1992, 6.
51. For further information on 'christening pieces' and other food eaten around the time of birth, see Fenton, forthcoming.
52. Edgar, 1886, 213.
53. Carmichael, 1900, 166.
54. Oral testimony, LR, 20.
55. MacGregor, 1958.
56. Oakley, 1984, 118.
57. Robinson, 1990, 74.
58. Oral testimony, LR, 27. 'I was [a Green Lady] for five years. Now, at that time ... it was just post-War and hospital beds weren't plentiful ... and when a mother or a relative phoned in to say that Mrs A or Mrs B was in labour – could she go into hospital? If the answer was no, if there were no beds, we were sent out. And often you got a hostile reception at first because a family was unprepared for this. They hadn't been prepared for a home confinement and they maybe didn't have a lot in the way of bed linen and towels and even the minimum of baby clothes.'
59. Scottish Home and Health Department, 1993.
60. Campbell, Macfarlane, 1990, 217.
61. Campbell, Macfarlane, 1990, 218.
62. Scottish Home and Health Department, The Tennent Report, 1973, 38–40. The main thrust of the recommendations of the Tennent Report was to unify maternity services in each area of Scotland within a Maternity Services Division (recognising the problems of fragmentation of maternity care caused by the NHS tripartite system), and to make institutional confinement 'available' to all pregnant women.

63. NHS in Scotland, 1991.
64. Scottish Home and Health Department, 1993.
65. Scottish Executive Health Department, 2001.
66. General Register Office for Scotland, Edinburgh.
67. For instance, in 2000, Highland Acute Hospitals and Highland Primary Care Trust had a 2.1 per cent homebirth rate while West Lothian Healthcare NHS Trust had a 0.2 per cent homebirth rate. National Childbirth Trust, 2001, 6.

BIBLIOGRAPHY

Bennett, M. *Scottish Customs from the Cradle to the Grave*, Edinburgh, 1992.
Cameron, A. Licensed to Practise within their Bounds: The Faculty of Physicians and Surgeons and the regulation of midwives in eighteenth-century Glasgow. Lecture presented at History of Nursing Research Colloquium, Oxford Brookes University, 18 March, 2000.
Campbell R, Macfarlane, A. Recent debate on the place of birth. In Garcia, J, Kilpatrick, R, Richards, M, eds, *The Politics of Maternity Care*, Oxford, 1990, 217–37.
Carmichael, A, ed. *Carmina Gadelica*, Edinburgh, 1900.
Catford, E. *The Royal Infirmary of Edinburgh, 1929–79*, Edinburgh, 1984.
Chalmers, A. *The Health of Glasgow, 1818–25*, Glasgow, 1930.
Checkland, O. *Philanthropy in Victorian Scotland*, Edinburgh, 1980.
Cumberlege, G. *Maternity in Great Britain: A Survey of Social and Economic Aspects of Pregnancy and Childbirth undertaken by a Joint Committee of the Royal College of Obstetricians and Gynaecologists and the Population Investigation Committee*, London, 1948.
Donnison, J. *Midwives and Medical Men*, New Barnet, 1988.
Douglas, C A, McKinley, P L. *Maternal Morbidity and Mortality in Scotland* (Report for the Department of Health for Scotland Edinburgh), Edinburgh, 1935.
Edgar, A. *Old Church Life in Scotland: Lectures on Kirk-Session and Presbytery Records*, Paisley, 1886.
Fenton, A. Celebration food on personal and work occasions. In Fenton, A. *The Food of the Scots, Scottish Life and Society: A Compendium of Scottish Ethnology, vol 5*, Edinburgh, forthcoming.
Ferguson, T. *Scottish Social Welfare, 1864–1914*, Edinburgh, 1958.
Fildes,V. *Breasts, Bottles and Babies: A History of Infant Feeding*, Edinburgh, 1986.
Gregor, W. North East, 1874. In Bennett, M. *Scottish Customs from the Cradle to the Grave*, Edinburgh, 1992, 54–6.
Hogarth, J. General practice, part II. In McLachlan, G, ed, *Improving the Common Weal: Aspects of Scottish Health Services, 1900–84*, Edinburgh, 1987, 163–212.
MacGregor, A. *Public Health in Glasgow, 1905–1946*, Edinburgh and London, 1958.
Mackenzie, W L. *Report on the Physical Welfare of Mothers and Children, vol 3: Scotland*, Dunfermline, 1917.
Maternity Services (Scotland) Act, 1937, 1 Edw 8 and 1 Geo 6, ch 30.
Munro Kerr, J. *Maternal Mortality and Morbidity: A Study of their Problems*, Edinburgh, 1933.
Munro Kerr, J. The maternity services. In Munro Kerr, J, Johnstone, R, Phillips, M, eds, *Historical Review of British Obstetrics and Gynaecology, 1800–1950*, Edinburgh and London, 1954, 3–8.
National Childbirth Trust. *Home Birth in the United Kingdom*, London, 2001.
NHS in Scotland. *Framework for Action*, London, 1991.
Oakley, A. *The Captured Womb*, Oxford, 1984.
Reid, L, ed. *Scottish Midwives: Twentieth-Century Voices* (Flashbacks, no 12), East Linton, 2000.

Robinson, S. Maintaining the independence of midwives. In Garcia J, Kilpatrick, R, Richards, M, eds, *The Politics of Maternity Care*, Oxford, 1990.
Rorie, D. The howdie. In Rorie, D. *The Lum Hat Wantin the Croon*, Edinburgh, 1935, 90–1.
Scottish Executive Health Department. *A Framework for Maternity Services in Scotland*, Edinburgh, 2001.
Scottish Home and Health Department. *Maternity Services: Integration of Maternity Work* (The Tennent Report), Edinburgh, 1973.
Scottish Home and Health Department. *Provision of Maternity Services in Scotland: A Policy Review*, Edinburgh, 1993.
Smout, T. *A Century of the Scottish People, 1830–1950*, London, 1986.
Spring Rice, M. *Working-Class Wives*, Harmondsworth, 1939.
Tait, H. Maternity and child welfare. In McLachlan, G, ed, *Improving the Common Weal, Aspects of Scottish Health Services, 1900–84*, Edinburgh, 1987, 411–40.
Tew, M. *Safer Childbirth: A Critical History of Maternity Care*, 2nd edn, London, 1995.

25 Childcare and Children's Leisure

LINDSAY REID

INTRODUCTION

Childhood is a comparatively recent concept. For many centuries the individual's early years were not perceived by society as constituting a distinct phase of his or her life. The Reformers, for example, regarded infants as simply being at the beginning of a solemn life-long journey, one from the 'wicked ignorance in which they had been born to the perfect knowledge of God in which they ought to die'[1] (Fig. 13.6). Yet with the Reformation's strict moral upbringing came increasing concern with learning and a rise in the number of years of education. From this gradually emerged the idea of childhood as a stage of life with a standing of its own. The authoritarian upbringing of the early eighteenth century gave way to a time of enlightenment towards children's care, influenced by the work of William Cadogan, who, in his writing of 1748,[2] concentrated on the nurture of infants in their own right. This slowly led towards late nineteenth and early twentieth century psychological theories of upbringing and a more child-centred approach.

The beginning of the twentieth century also witnessed increased awareness of the value of healthy children for society. As late as the end of the nineteenth century, the infant mortality rate, as well as that of mothers, was very high. This, along with a falling birth rate, suggesting the possibility of a population decline, brought maternal and child health into the political arena.[3] In addition, the growing acknowledgement of the poor physical state of children in Britain, highlighted by the rejection of many army recruits for the Boer War, stimulated interest in the importance of maternal and infant welfare at the beginning of the twentieth century. In 1905 a delegation from Glasgow and Edinburgh attended the *Congrès Internationale des Gouttes de Lait* [International Congress on Milk Depots] in Paris which resulted in the First National Conference on Infant Mortality in Britain, in London in 1906.[4] This conference was the forerunner of a permanent national body working for the Infant Welfare movement.

Ideas about the upbringing of children varied considerably over the course of the twentieth century. In the 1920s and 1930s Truby King[5] gained notoriety with his strict routines, countered in the 1950s by Dr Spock and very much more permissive advice. Today bookshops are full of publications which together present a range of views on the bringing up of babies and children. This chapter explores changing aspects of caring for children, as opposed to adolescents, in Scottish dwellings; and also children's play in and around the home.

WORKING CHILDREN

Being a child in the past did not provide an excuse from working long hours (see Fig. 17.4), especially on farms and other holdings where children of both sexes were expected to help from an early age[6] (Fig. 25.1). In the majority of Scottish households girls were given domestic responsibilities as soon as they were able to cope, and by the beginnings of adolescence – around the age of 12 or 13 – these could constitute quite a burden. Today most children, in normal circumstances at least, are expected to help minimally around the home, their chores often being limited to tidying their bedrooms and perhaps caring for family pets.

Figure 25.1 Helping with what may be pet lambs, Shetland, 1950s. Many Scots children have had to work long hours on farms and other landholdings to help the household economy. Source: Lindsay Reid.

Children were also involved in paid work, for example in factories, such as those at Alva, Clackmannanshire, where 12-year-old girls were expected to work 14 hours a day. Various Factory Acts were passed in the 1800s 'fixing the hours of employment for women and children in most factories at 56½ per week' and eventually prohibiting children under ten from undertaking any factory work.[7] In the north, 'crofters' daughters who were not needed at home often had little choice but to go into domestic service ... They could continue to contribute to the household economy by sending or bringing goods home if they could afford to do so.'[8] Childhood offered no choice but compliance with parental rules. Naomi Mitchison noted in her Wartime diary, 'She [her friend] ... said that they were treated as objects of ownership by the parents.'[9] However children also felt a sense of responsibility towards contributing to the household economy, despite the restrictions which working placed on their rest, play and education. Patrick McVeigh, writing about the 1920s and 1930s in *'Look after the Bairns': A Childhood in East Lothian* illustrates this clearly in his chapter on family economics.[10]

Government action to protect people in Scotland from working for an income at too early an age developed from the 1833 Factory Act, which forbade the employment of children under nine years of age and limited the hours of work for those under 18.[11] This Act was largely ineffective at the time and the progress of reform was very slow. At last, in 1890, a committee appointed by the House of Lords revealed the appalling working conditions endured by men and women, children and adolescents. A further Factory Act was passed in 1891 to deal with these and amended in 1906. Now if a newsagent, for example, wants to employ young people to deliver newspapers, they must be over 12 years of age – adolescents rather than children. Even so, their conditions and hours of work are, in theory at least, carefully controlled.

TOO MANY BABIES

Contraception was not commonly used in the first half of the twentieth century. There were various methods available at the time – sponges, condoms, pessaries, douches, the diaphragm or cervical cap and the withdrawal method.[12] However the whole issue was cloaked in secrecy and many looked on the idea of family planning as sinful and 'against nature'. Consequently most families were large. Some babies were wanted and cared for, while others were not. For many mothers becoming pregnant yet again was the cause of considerable worry. Space could be at a premium and Anne Bayne recalls a birth in one particularly overcrowded home:

> I think this woman was para ten [eleventh pregnancy]. She had this one room. She was thrilled because they had actually the sink – the jaw-box in the window. There was a fireplace with a mantelpiece and it was a little open fire, a range where she did all her cooking. That

night the father and the oldest boy went [away]. There was a box bed ... There was a sheet over this bed and there were five children of varying ages in behind that sheet and so there were five little peep-holes in the sheet. And then she had her bed and at the end there was a heap of coal and there was a cot and there were two in the cot. Then there was another cot across at the other side. It was like a Moses basket. There was another bath there and the last baby was in it and now we were delivering the new one. There wasn't room for the doctor and the midwife and myself [as pupil midwife] to stand on that floor. We took it in turns to sit on the bed. There was a wee paraffin lamp which was all the light we had and we took turns to hold that depending on who was doing what.[13]

The bringing up of numerous children in a confined space was far from uncommon and made matters of food preparation, cleanliness, privacy and physical comfort problematic. How could the family sleep in peace? How was the baby to be comforted when teething, not only for his or her own sake, but also for that of the rest of the family? Mothers, perhaps with other female relatives in the household taking a turn, would have been up many a night, walking the floor and possibly resorting to the 'cures' deplored by the professionals: teething powders, soothing syrups, patent medicines and, as a last resort, rubbing sore gums with a whisky-coated finger.

CLOTHING

Babies' bodies were formerly tightly tied or 'swaddled'. Frightened that flexed limbs would permanently disable a growing child, the midwife would bind the arms and legs in swaddling bands, sometimes tying the infant on to a board which could be hung on the wall, to ensure straight growth. The custom of swaddling gradually declined from the eighteenth century.[14] Babies are still wrapped at birth to keep them warm, but not so tightly that they cannot exercise their limbs (see Fig. 24.4).

Early twentieth-century babies were dressed according to the class and income of their parents. Most, however, wore long gowns day and night and were not 'shortened' until after they had been baptised. Even after shortening, boys still wore dresses for many years (Fig. 25.2). This practice continued occasionally in some parts of Scotland until at least the 1960s. One mother recalled, 'I put him out in his pram with rompers on and she [a friend] came along and said in a scandalised voice, "Ye're nae shortenin him afore he's christened?"'[15] Chrissie Sandison of Aith, Shetland, recalls how some babies there were dressed in the 1920s:

> Then they had a winter thing, a winceyette barry coat, sleeveless and all tapes, and you could open it to change the baby's nappy. It was long enough to cover the feet. Then they had a long cotton frock. I think maybe in winter it wisna cotton ... what I mind seeing when

Figure 25.2 Edith and Andrew Inkster, Shetland, c 1907. Edith has been carefully groomed and dressed up for a very special photograph. Her brother has been 'shortened' (his feet are visible) but he is still very much in a frock. Source: Lindsay Reid.

> I was going round looking at babies, they had the frock turned up around the baby's head ... to stop it getting wet. You see there were no plastic pants.[16]

As children grew older the distinction between boys' and girls' clothing became apparent. Little boys were put into sailor suits or knickerbockers (Fig. 25.3). From the 1910s many boys wore wide shorts, thick knitted stockings and 'tackety' boots with a knitted jersey which might have a collar and occasionally was worn with a tie (see Fig. 15.5). To own a jacket which matched the trousers was considered very up-market. Girls from reasonably well-off families had a female version of the sailor suit but the trend soon changed to frocks and pinafores with frilled edges. In many families, not always the poorest, there was a lot of cutting down of adult clothing and making do. Then, as Anna Blair says in *Tea at Miss Cranston's*, 'the day of the gym tunic dawned'.[17] It was to take about 60 years before the sun set on these very economical and versatile garments, which could be taken in or up and let out or down.

For many years in the first half of the twentieth century both boys and girls had to suffer a scratchy all-in-one undergarment known as combinations or 'combies' which came in two styles. One type was crotchless; the other had a flap at the back and was very difficult to undo without help.[18] Both varieties were designed to keep the kidneys in particular warm. Another common and warm undergarment was the liberty bodice. This was kinder to the skin and was used only by girls, being worn on top of a woollen vest and with suspenders attached, front and back. Both combies and liberty bodices were necessary evils in the days before central heating.

Not all present-day Scottish homes have central heating, or even heating which is adequate, of course, and this, combined with fashions inappropriate for the Scottish climate, results in some children being cold for long periods at a time and suffering from a variety of infections as a consequence of lowered resistance.

'Best' clothes were usually kept for Sunday and going to the 'kirk'. Most girls possessed a Sunday coat and hat even if it was the same hat summer and winter with a different trimming. The kilt, everyday wear for some boys in nineteenth-century highland Scotland, evolved into the formal Sunday garb of many lads across the whole country in the twentieth century. Children sometimes wore their best clothes for important occasions such as hospital visiting. To be in one's best could be very useful as Anne Bayne recalls:

> She [my mother] took me to visit my aunt and I remember this visit very well, in Aberdeen Royal [in Woolmanhill] and it was the time [mid 1930s] ... when it was all voluntary donations and the nurse was made to go round [the visitors] with the plate and ... to get a little person [to help] ... I remember I had ... this little fur jacket and leggings and a little bonnet and they came and handed me the plate to take round.[19]

Not every parent could afford the luxury of 'best' clothes for their children as Anna Blair notes:

> And while [he] was photographed gazing into a goldfish bowl, long hair curling about his neck and shoulders and wearing a cotton dress

Figure 25.3 Andrew Inkster with his sailor suit on, Shetland, c1911. Source: Lindsay Reid.

with lace inserts, white socks and boots, his small contemporary on inner-city streets was toddling barefoot alongside his shawlie mother.[20]

Even in relatively wealthy households garments were recycled if at all possible by being handed on to the next child. Crofters' children could enjoy the benefits of wearing home-knitted clothes created from the wool of their own sheep:

> When we clipped the sheep in the summer time my mother used to send the wool to a mill and she would receive back wool, spun, in place of the sheep's wool. And she knitted our jerseys and our skirts and our knee socks, our hats and scarves and gloves and things for the winter time ... it was rather itchy. I don't believe I would wear it today.[21]

Modern children are usually highly fashion conscious, sometimes from a very early age, and expect to have a say in their choice of clothing, their tastes largely shaped by advertising and peer pressure. Although children of immigrants may sometimes wear clothing traditional to their parents' homelands, the majority of young people choose to wear sports and casual clothes whenever possible, although specific fashions may last for only a season. For this reason, and because few of the cheaper manufactured garments last long enough to be re-used, most children now possess garments previously unworn by others, though some handing down of items still goes on within families, usually for wear in the home or at least away from school.

In terms of footwear, in the late nineteenth and early twentieth century button boots were worn by both boys and girls and not always with happy memories, 'I've still got my button-hook I had when I was a girl, and I can feel yon wee tweaky feeling yet, when you nipped your leg getting hold of the button.'[22] Since then, boots have moved to shoes, sandals, and back to boots again, although in the last two decades fashions have fluctuated from season to season and the number of pairs of footwear in an individual's possession has greatly increased. Two unisex staples of the late twentieth century were Doctor Marten's style boots and training shoes, the latter still the most popular option and often very expensive.

There have been big changes over the years in nappies and keeping babies dry. At one time nappies or 'hippens' were homemade, clumsy and rough. By the 1930s, the advice for mothers who could afford them was, 'Harrington's absorbent napkins are best, being easy to launder and very soft to the skin.'[23] When rubber pants appeared, mothers were told, 'Never put mackintosh drawers on a baby as they are most injurious to its tender skin as well as to its general health.'[24] With the development of finer and more sophisticated materials, widespread use of plastic pants, along with nappies of differing shapes, sizes and composition, was inevitable. Then came the disposable nappy. As early as around 1940 Mary Truby King advised mothers to use disposable cellulose wadding inside the napkin,

'The squares dissolve instantly in water after use.'[25] By the late 1950s the popularity of disposable nappies was growing. They became increasingly sophisticated and today's most popular brands are complete with a throw-away plastic outer layer. However, in the latter decades of the twentieth century debate arose over the damage to the environment caused by discarding disposable nappies, many homes now lacking open fires on which to burn such waste. In families where both parents work, there is little time for washing nappies and people seek alternatives. In the last few years this has led, especially in the cities, to the development of nappy-washing and delivery services.

PRAMS, CRADLES AND COTS

As well as being warmly dressed and kept dry, babies also need somewhere to sleep and a means of being safely transported. Infants in the past could be carried within a shawl wrapped around both mother and baby and this was customary until the late nineteenth century and even until the mid twentieth century amongst the poorest families. Also in the late nineteenth century, and sometimes earlier, people were beginning to put their babies in pushcarts and perambulators.[26] Babies have been transported in a variety of such items. Most early prams had wicker and then wooden bodies, and, with new materials, hoods developed which could go up and down depending on the weather (Fig. 25.4). The coach-built high pram, a status symbol which reached its peak in the middle decades of the twentieth century, declined in popularity as car ownership increased. Such prams were, and are, of little use on tenement stairs and in many country contexts and it may be no

Figure 25.4 200-year-old pram, photographed in Stornoway, Lewis. It is thought to be of German origin. The wooden sides have engraved patterns. Note the brake on the handle. Source: Lindsay Reid.

surprise that today fashion has come full circle with the 'baby carrier' or sling which fulfils the same functions as a shawl. These are complemented by new types of pram and push-chairs, which are made of fabric, fold and are versatile, and which fit into twenty-first century lives where babies accompany their parents in cars and public transport, into supermarkets and shops, and to and from work and the child minder.

Within the house, rocking wooden cradles with wooden hoods, which kept both evil spirits and draughts away,[27] were replaced in the nineteenth century, at least by those who could afford it, by frilly creations. However, Mary Truby King said, 'Frills, bows and drapings ... in old-fashioned cots used to harbour dust and germs and prevent the free circulation of air', and she recommended a simple wicker cradle and stand.[28] In the 1960s came carrycots, echoing earlier devices such as the Moses basket. These can be lifted in and out of house and car with ease. Some families, though, have had to make do without any specialised equipment:

> The folk had no money. I mind a neighbour house [where] ... they had a few children. They had this creel and they had a flour bag from the shop, probably held 70 pounds of flour and it was bleached white and it was stuffed with straw. It was soft Shetland oats so it wasn't stiff like Scots oats and that was the bed and the pillow was something soft like wool inside a small pillow case. In one house where they were not very well off, a grey or fawn woman's underskirt cut up ... and spread out for a blanket and before the baby was laid in it, it was held up before the fire and sometimes it had a strong smell, and the baby was fine and cosy and then they rocked the baby to sleep. They then would have a quilt made out of old bits and pieces.[29]

And, as Linda Stamp recalls, many babies' first resting place was a drawer, 'I've had to put a baby just in a wooden drawer because there was nowhere else for him ... Glasgow was poverty stricken then.'[30] This practice was repeated in homes all over Scotland. Sometimes there was not even a drawer. Molly Muir[31] delivered a mother of her thirteenth child and remembers, 'This mother didn't have a cot for the baby; he just stayed in bed with the mother and ... another half dozen children climbed in as well.' Once a baby grew out of the first cradle, the more affluent procured a cot with bars. However, it was a common occurrence until well into the twentieth century for children of all ages to sleep in bed together. Today most children past the stage of infancy have a bed to themselves, if not always a separate bedroom. Where rooms are shared, bunk beds are a common solution to lack of space.

FEEDING

Breastfeeding is generally accepted as the best way to feed a new baby. In Scotland today, a mother who, for whatever reason, does not breastfeed, will usually give her child formula milk. Until the late nineteenth century the option of wet nursing existed instead, a practice used throughout the world from at least Biblical times.[32] In Britain, the fact that Queen Victoria employed a wet nurse probably helped perpetuate the practice for so long.[33] A mother of sufficient means might hand over her infant to the care of a poorer woman. She, by breastfeeding the other woman's baby, augmented her income, although doctors and moralists repeatedly stated that the wet nurses' own infants suffered as a result.[34]

Although wet nursing had gone out of fashion by the beginning of the twentieth century, the feeding of infants in Scotland was still considered 'insufficient and improper'.[35] Breastfeeding remained unpopular and was probably not available to many infants because of the unhealthy physical condition and poor nutrition of their mothers. A large number of women in both urban and rural areas went back to work soon after giving birth and their babies were bottle fed on condensed milk.[36] What fresh milk was available was of inferior quality and often unclean, while many homes did not possess adequate storage facilities for fresh products. The resultant poor nutrition and diarrhoea killed hundreds of babies every year.[37] Not surprisingly, some parents chose instead to give their babies a little of their own food, 'even bits of tinned salmon, if it was thought that "the wean's stomach is fit for stronger food"'.[38]

For a mother to breast feed satisfactorily she must herself have a reasonable supply of good food. In 1906 the Scots followed the example of the French and established restaurants for nursing mothers, in Dundee and then in Glasgow, in an attempt to improve maternal and baby health and decrease mortality rates.[39] Mothers, along with toddlers, could attend for dinner from three months of pregnancy onwards and for nine months post-partum, provided they were breastfeeding. Nowadays the number of Scottish mothers breastfeeding their babies from birth is rising although the number breastfeeding after six weeks is still lower than the UK average.

A good-quality supply of safe cow's milk was a necessity and again the French led the way, setting up special milk depots in Paris which were copied in Leith, Edinburgh, in 1903 and in Dundee and Glasgow in 1904. Subsequent developments included an official acknowledgement of the connection between poor nutrition and disease. This highlighted the need for a pure milk supply for the general public and for further investigation into the nutrition of Scottish children with the long-term aim of better health.[40] Specific mention should be made of Lord John Boyd Orr's important work in this field.

With World War II came rationing and a fairer distribution of food. The government further acknowledged the importance of good food for children by giving free cod-liver oil, orange juice and school milk (1/3 pint per day), as well as subsidised school dinners which were free to those unable to pay. Immediately post-War, school dinners were 5d, 4d, and 3d on a sliding scale from eldest to youngest in a family. Despite this, the 'jeely' piece, or jelly or jam sandwich, remained a lunch favourite. This was made with the traditional Scottish 'plain loaf' with the top crust known as 'Curly

Kate',[41] and the bottom as 'Plain Geordie'. Mothers saved up rationing 'points' for enough sugar to make jelly with brambles picked in the autumn. It is probable that rural children fared better than those in towns during the War, with home-grown vegetables, eggs and other foods which were otherwise strictly rationed. Fruit was virtually unavailable although there was the occasional other treat which mothers would divide up by saying:

> Neevie, neevie nick, nack,
> Which han will ye tak?
> Tak een, tak twa,
> Tak the best abeen [above] them aa.

After the War and the eventual cessation of rationing, the supply of food escalated.[42] We now have more and better food available than ever, but advertising has created tastes for fast food and other foods of a low nutritional content. Less food is home cooked and the likes of porridge, homemade soups and stews and fruit and vegetables are increasingly unpopular. Tooth decay and deterioration in long-term overall health are the consequences. To try and reverse this trend, some schools in Scotland are currently supplying free fruit to their pupils, and offering carefully thought out school dinners with optimum nutritional content.

KEEPING CLEAN

It is hard to imagine today with modern bathrooms, perhaps en-suite and maybe even incorporating jacuzzis and bidets, what keeping a family of children clean was like not so long ago. As late as 1951, 43.1 per cent of all Scottish households did not have access to a fixed bath,[43] and even so, bathing was a weekly affair, often on a Friday or Saturday night with clean clothes laid out for Sunday. For many children the weekly bath was in a zinc tub in front of the fire. The same water did for all, with the cleanest child going in first and the dirtiest last. Anna Blair's *Up to the Neck in the Dolly Barrel* describes the struggles mothers had to keep their offspring clean and nit-free in very difficult circumstances.[44] She also highlights the problem of lack of private space. At puberty, some urban boys at least could have weekly visits to the public baths. However the novelty of being in a full-sized bath vied with the continuing lack of privacy:

> It was a completely new experience to lie stretched out in the bath, half floating in the water. He forgot for a long time to soap himself, but lay there looking up at the single electric bulb on the ceiling ... he could clearly hear the man in the next bath, grunting and scrubbing and snorting ... He saw that he could stand on the end of the bath and look over, but decided not to do so.[45]

Women and girls continued using the tub in front of the fire, with the men and boys going out on a Friday night to allow them some privacy.[46]

New babies have had a variety of bathing receptacles including baking bowls or 'just the sink'. One man has a particular memory of such make-shift measures:

> My grand-aunt was a nurse and midwife ... She was born in Glasgow in 1869. I was born in 1922 and my aunt's career would be drawing to a close when I was still a boy. I remember her saying some did not even have a basin for water, but an old corned-beef tin cut in the middle. I know of the tins; I haven't seen one in many a year.[47]

Even if a house had a bathroom, the bath was not always put to the use for which it was intended. Stories of coal in the bath abound but one midwife working in Leith in the 1930s had a surprise of a different kind:

> I remember going upstairs thinking – oh marvellous, running water. There'll be a bath ... we'll get the sheets [washed]. I remember the awful disappointment when I went into the bathroom. It was a fisherman's family and the lobsters were in the bath ... You never know what you'll find.[48]

It was still common, well into the twentieth century, to have an outside lavatory and although girls often had an unfair amount of work when it came to washing days and cleaning, there was one job which was left for a boy in 1940s rural Ayrshire:

> My father was called up and we had to move from a tied house to a very small house nearby. Before he left my father constructed an outside lavatory at the foot of the garden. This consisted of a seat over a galvanised bucket, lined with grass which I had to empty every few days in a nearby fast running burn, wash the bucket out, re-line it and replace it. This was my task until at the end of the War we moved again – this time to a house with an inside lavatory and bath.[49]

CHILDREN'S PLAY IN AND AROUND THE HOME[50]

On a recent BBC 2 TV programme, 'HomeGround', a Scotswoman said of changes in childhood play:

> We *had* to get out to play. Now they have TV, videos and computers. There's not the rush to get outside. Summer was the favourite time with long evenings. We still went out to play in the rain ... the water was like a wee stream and we had paper boats ... we always came back dirty after being up the bing playing. [We played] with skipping ropes, and 'ring a roses' and 'the farmer wants a wife'[51] (Fig. 25.5).

In the past, play was seen as more of a privilege, to be fitted in between more serious activities (see Figs 34.6, 42.2). Today children are expected to play and learn through play from an early age.

Children's toys have changed in form and purpose. Many in the past were simple and made by household members and most were gender

Figure 25.5 Coming home from minnowing, Perthshire, late 1940s. Playing outside was the norm for most Scots children before the late twentieth century. Source: Lindsay Reid.

specific. A little girl cuddled her rag doll, for example, while a toy bow and arrow provided her brother with enjoyment. These toys, common until the second half of the twentieth century (and still to be found), along with usually shop-bought items such as dolls' houses, model trains and toy soldiers, expressed parental toleration of the child's freedom in play while reflecting the adult that the child was to become. Modern toys, colourful, often kinetic and noisy and less gendered, rather than prepare children for adult social roles are more concerned with inviting them into a fantasy world within the imagination. To some extent the toys reflect changes in theories of upbringing.[52] Instead of children being 'seen and not heard' they are now allowed, even encouraged, to express themselves.

Certain games and activities played in the environs of the home in the twentieth century were also either for boys or for girls. For instance, girls skipped, although occasionally they would get a reluctant boy to 'ca the ropie' (turn the rope) when the numbers were not enough. They also jumped over chalk-drawn 'beds' or 'beddies', and played houses, hospitals and ball, the last alone or together. But boys came into their own in relation to games like cowboys and indians. Girls definitely did not get involved in this, in most cases anyway. Neither did they make paper aeroplanes, play football, nor, strictly speaking, play 'bools', with a 'kypie' dug into the ground with the heel of a tackety boot. Another male activity (but often invaded by the girls as it was so much fun) was riding on a homemade 'hurlie'. Depending on where one originated, a hurlie was also known as a 'bogie', 'piler' or 'cairtie', and was usually a four-wheeled contraption made of wooden planks, boxes and old pram wheels. Other modes of transport enjoyed by all were, and are, tricycles, bicycles, scooters, skateboards, roller skates and rollerblades, and now luges have made their appearance for travel even without snow. Sledging has always been a part of the Scottish winter scene for both boys and girls, with sledges gradually becoming more sophisticated. For those who did not manage to obtain a sledge there was always a tray borrowed from the kitchen or even, in later times, the black bin bag. Ever-popular are slides made by smoothing a length of ice or snow for sliding along on foot.

To this day both boys and girls play games like 'hide and seek' and its variation, 'sardines', in gardens and other open spaces in residential areas. In sardines, if a 'seeker' finds the 'hider', instead of proclaiming victory as is usual in hide-and-seek, he or she changes roles and joins the hider. This goes on until all are hiders except the last player. By this time all the hiders are squashed in the hiding place like sardines. There are other challenging running games like 'tic and tac', or 'tig', where the one who is 'it', or, more commonly in Scotland 'het', has to chase the group until he or she catches someone who then takes over the role of chaser. Whether they are playing this kind of game at home, in the street or in the playground, before they begin, to find out who is going to be out, children say a 'pye', like:

> Eetle ottle black bottle,
> Eetle ottle out,
> If you want a piece and jam,
> Please step out.

Another favourite is when all the children in the game hold out their fists and one counts on all the fists:

> One potato, two potato, three potato, four,
> Five potato, six potato, seven potato, *more*.

The fist which receives *more* is put down and the game goes on until the last fist up is 'it'.

Before the days of television there was plenty of household entertainment involving members of all ages (see Fig. 11.1). Early twentieth-century families played card games like Snap and Happy Families, or board games such as Ludo, Snakes and Ladders, or, from the 1930s, Monopoly, all of which continue to enjoy some popularity. The latter were joined by the likes of Scrabble and then Trivial Pursuit. Family listening to the wireless as a group was popular in the 1930s, 1940s and 1950s, especially to programmes such as the 1950s The MacFlannels, along with gatherings round the piano to sing favourite songs. Music lessons and their popularity seem to go in cycles. Until comparatively recently, schools provided singing as part of the curriculum but tuition in the playing of instruments was an extra which had to be paid for by parents. Lessons were fashionable only among families who could afford them, although there were some households in which children could receive free instruction from family members or neighbours. Now, many more children have the opportunity to learn to play an instrument at school, often starting with the recorder, and some may even go on to join youth orchestras. Much depends on the individual's dedication and also the tolerance of other household members who have to endure the practice sessions at home.

Children of parents who could afford it also attended dancing classes throughout the twentieth century with highland dancing, for example, being taught in village halls and practised at home (Fig. 25.6). Earlier, fiddlers, sometimes notable figures such as James Scott Skinner in the late nineteenth century, would come and teach the steps and fiddle for dancing.[53] In later years, increasing quantities of music in the home were provided by 78s then LPs, singles and cassettes and now compact discs.

Children still play the old games but they also spend substantial amounts of time watching television and their leisure is increasingly under the influence of media advertising. In the 1980s little girls combed the synthetic hair of My Little Pony and played with mass-marketed dolls such as Barbie (boys had and have an equivalent in Action Man). These have been joined by numerous sophisticated toys with batteries and moving parts, plus calculators and personal organisers which may have a recreational element. The fact that a large number of children also have a computer, computer games and TV in their own room (some may also have their own mobile

Figure 25.6 Children Highland dancing at Banchory Games, Kincardineshire, c 1952. Source: Lindsay Reid.

phone largely used for text messaging, although these are more common among adolescents) has perhaps lessened the togetherness of family life and increased solitary activity. Yet reading, very much a solitary activity, has been enjoyed by children since they first had access to books. The success of modern children's writers, such as J K Rowling, shows that there are still many children who are sometimes glad to be alone with a book.

Organised group activities and their popularity have waxed and waned, depending on the thinking of the time. These include the Boy Scouts, Girl Guides (now known as The Guides), Girls' Guildry, Boys' Brigade and Life Boys to name but a few. The first experimental scout camp, led by Robert Baden-Powell, was held at Brownsea Island, in Poole Harbour, Dorset, in 1907.[54] From this small beginning, and with the encouragement and example of William Smith of the Boys' Brigade,[55] the scouting movement, including Scouts, Rangers, Wolf Cubs and now Beavers (a later addition for very small boys), has evolved world wide. In 1910, Baden-Powell and his sister Agnes founded the female equivalent of the Scouts, the Girl Guides, with Brownies and now Rainbows for younger girls. Currently, there are 69,000 Scottish Guides. For those who cannot attend regularly, they can participate through the Internet as a NetGuide.[56] The scouting movement was based on a philosophy of mutual understanding between people and thoughtfulness to others demonstrated by the giving of practical help. From the latter evolved the well-known habit of doing a 'good turn' every day. Now, although Scouts and Guides still meet separately, girls may, if they wish, join the Scouts.[57]

In 1883, almost 25 years before Baden-Powell held his first Scout camp, William Smith started the Boys' Brigade in Glasgow,[58] a new idea for the Christian development of adolescent boys. It was part of a busy programme of mission activity taking place at North Woodside Mission and the surrounding area at the time. Its object was 'The advancement of Christ's Kingdom among Boys and the promotion of habits of Reverence, Discipline, Self-respect and all that tends towards a true Christian manliness.'[59] Thus began a movement which was to welcome and influence boys of all classes across the world.

Other organisations have existed, derived from the Boys' Brigade, and all designed exclusively for boys. It was Dr William Francis Somerville, a medical practitioner and Free Church member in Glasgow, who saw the possibilities of extending the Brigade idea to girls 'from the leisured class' and also from the tenements[60] and in 1900 he founded the Girls' Guildry, an organisation which spread throughout Scotland with similar ideals to that of the Boys Brigade.

FROM HOME TO SCHOOL

Both children and parents, since at least 1872 and the introduction of compulsory education, have viewed going to school as a significant milestone. Before the days of playgroups and nurseries children might always have been at home with their mothers or other close carers. Suddenly they were out in the world with schoolbags on their backs. Many of today's adults will remember the first day, with crying five-year-olds within and weeping mothers at the school gate. One child was horrified when she learned she had to go back the next day; she thought the first day was it. Another small boy from the north east, at the end of the first day, and obviously thinking about his place at home, turned to the teacher and said, 'Jist ti let ye ken. A'll nae be back the morn. A've learnt aa A need. Wi me here, ma faither's a man short.'[61]

Today the first day at school is not nearly so traumatic for all involved. Since the 1960s there has been a proliferation of kindergartens, playgroups and nursery schools both privately and state-run for under five-year-olds. The number of these institutions has risen as more women work outside the home and the value of pre-school education is increasingly recognised. Children, socialised into groups from an early age, walk into school with barely a backward glance and parents, by this time accustomed to seeing their offspring go, meet the milestone with equanimity.

NANNIES AND OTHERS

Before the social upheavals associated with the World Wars many Scots employed help in their homes. This varied from a few hours a week of housework and laundering, to, in very large establishments, the full-time labour of a fleet of servants including nannies (or head nurses)[62] and nursery maids. The living-in nanny of the past was an important person. Highly trained in childcare, often by working up through the ranks, and sometimes from the late 1800s and early 1900s college trained,[63] she usually had full charge of the children, carried a status which matched her responsibilities and looked on her post as permanent, at least until the last baby was at school and sometimes longer. 'In the really old days nannies didn't leave. They would still live upstairs in the nursery and their food would still come up on a tray and if nanny wasn't that old she would go on to the next generation'.[64] To leave before the children were at an appropriate stage was unusual enough to be remarked upon by M C Moncrieff who noted:

> Nurse left, I do not know why, when I was four [about 1902] ... In her place came Lucy Abbot. Lucy was to serve as nurse and sewing maid in our house for the next seven years ... and after she left, to remain a faithful correspondent and affectionate friend all the rest of her life.[65]

Yet, while nannies in many cases became lifelong friends of the family, Gathorne-Hardy in his comprehensive history of the British nanny describes and analyses situations where nannies were cruel, authoritarian and sacked from their posts, as well as brave, loyal and loving.[66] He also highlights the significant element of snobbery and competition between nannies working for families of different standing, which must surely have rubbed off on their charges.[67]

Now, in the twenty-first century, many children have both parents working outside the home, and nannies are making a comeback. Not always live-in, sometimes wearing jeans and tee-shirt instead of a starched uniform, today's nanny bears just as much responsibility, if not more, than yesterday's. The extra responsibility is a product of both parents being out of the home for long periods, the reduced input of grandparents, and the lack of backup by other servants. Twenty-first century parents are perhaps also more demanding, being especially fearful for the safety of their children with high levels of road traffic and increased publicity about child abductors and paedophiles.

Nannies are not the only child carers. Mother's helps and au pairs also take charge of other people's children with greater or lesser degrees of responsibility. Mother's helps may be full- or part-time but are not trained nannies. The au pair lives in, is often from another country and usually has a primary objective of learning to speak good English. Thus her hours with the children have to be negotiated so that she can attend classes and socialise.

Another option for having children looked after when both parents are working is childminding. Today's childminder usually works in her own home, is officially registered with the local authority, by whom she is inspected once a year, and must abide by certain rules and fulfil various criteria. Children may be looked after from babyhood on a full-time basis with hours negotiated with the parents, or, later, on an 'after-school' basis. For information parents can contact their local social services or a Childminding Association.[68]

School hours seldom conform exactly to parents' working hours. During the 1980s and early 1990s, as the number of families with both parents working increased, so did the number of children who went home to an empty house. There was little or no provision made for these 'latch-key kids' after school and sometimes at lunchtime. One informant described a situation where a neighbour's children let themselves into the house at lunchtime, boiled a kettle and made an instant lunch in a plastic pot before going back to school. They were then alone again in the house after school.[69] Now, many local authorities provide well-staffed after-school clubs, and in these children may play games, do homework or just pass the time until their parents come to collect them.

BOARDING OUT

From the beginning of the nineteenth century until the 1960s, the state boarded out children. This was a specifically Scottish solution to the problem of large numbers of usually extremely poor, sometimes homeless, children without parents capable of looking after them.[70] Children who were not living with their parents were either orphans, deserted, or separated from their parents by court order. Boarding out was a practice which suited the ideas of Scotland's child-welfare professionals who advocated placing a child with a family rather than in an institution. Few argued with this principle and in 1917 Mackenzie said, 'The boarding-out system is one of the best departments of Poor Law work. The primary intention of the system is to secure to the child an opportunity for growing up in a normal social environment. In most cases the only alternative is some institution.'[71]

Yet Abrams[72] highlights many problems associated with the boarding-out system. Many children were uprooted from cities, Glasgow especially, often without warning, and put with families in rural areas such as the Highlands and Islands and the north east. There was little or no vetting of prospective foster parents or homes and there was no idea of 'matching' a child with a family. Some parents had no idea of how many children they were receiving and felt able to reject a child they did not like the look of. There was an element of financial dealing, indicating that the interests of the child were not always put first; it was cheaper for the placing authority to board out a child than put him or her in an institution. Foster parents were paid for taking in children and there is evidence that some children did not receive the full benefit of this. Sometimes they were not fed as well as other household members and not treated emotionally as part of the family. In addition, if a child earned any money, it often had to be handed over to the foster parents. There was also the problem of mixing cultures. The system was not viewed favourably by many local communities, especially when the numbers of boarded-out children and their non-rural ways looked like provoking change in the local schools and society. It was not all one sided however. One girl, sent to Barra, said, 'I couldn't speak Gaelic and I felt different from the others … I made sure I learned and worked out the meaning of every word so that I could speak Gaelic and be like the others at school.'[73]

Not all boarded-out children were unhappy, ill-treated, poorly educated or emotionally starved, of course. Yet recent research appears to show that there was much in the system of the time with which to find fault, despite the apparently well-meaning attitudes and objectives. Fostering in a family home is still considered to be a better option than placing a child in an institution. However, fostering is now much more likely to be temporary, with carefully chosen and dedicated foster parents, and to allow access to the child's natural parents.

CONCLUSION

This chapter has attempted to show some ways in which childcare and children's leisure in the Scottish home have changed over the years. Children were born whether wanted by their parents or not and, despite high infant mortality rates, families were often large. The fear of losing a baby or child might be one reason why there sometimes appeared to be few open displays of love between parents and children. Cultural inhibitions, possibly reflecting Scottish Presbyterian disciplinarianism, also prevented some parents from overtly valuing their children and demonstrating much affection. In recent decades, however, there has been a shift to a more open culture, accompanied by low child mortality rates. Family planning has led to smaller families, and parents are encouraged to 'enjoy' their children and to bond with their babies. A hundred years ago a woman spent much of her adult married life pregnant and looking after babies. No sooner was one child weaned than she might be pregnant again. Now, the average age of a mother having a first baby in Scotland is around 28 and she may go on to have only two or three children. With fewer children in the family it is easier for them to be treated as individuals by the adult members. Children's opinions are listened to, they often call adults by their first names, and they probably have more confidence than children even a few decades ago did.

The concept of the family has also changed. It is rare in present-day Scotland to see various related couples with their children living together as an extended family under the same roof, although this is sometimes the case for members of certain ethnic minority groups. There was a time when this happened not infrequently among the general Scottish population. More recently, related families commonly lived in the same street or town. With modern mobility that too has become rarer and children may no longer see their grandparents and other more-distant relations very often. Many families are 'nuclear', that is, a unit consisting of husband and wife (or other type of partnership) and dependent offspring or, occasionally, adopted or fostered children.[74] In addition there has been recognition of single-parent families as a distinct and viable family type[75] with 1998 figures giving an estimate of 162,000 one-parent families in Scotland.[76] The big change here has not been in the numbers of single-parent families, but in the reasons why they have come to exist. In the past the main reason for single parenthood was bereavement, while illegitimacy was a prominent issue and widely condemned. Today, however, the majority of lone parents are separated or divorced, while 38 per cent of lone mothers and 13 per cent of lone fathers describe themselves as single.[77] In 1999, 89 per cent of all single-parent households were headed by a woman and 11 per cent by a man.[78] In addition, in 1999, 70 per cent of single-parent households in Scotland had a net annual income of less than £10,000. This compares with 25 per cent of small two-parent families and 29 per cent of large two-parent families.[79] Combined with greater financial insecurity (79 per cent of single parents have no savings), it might be assumed that there is more likelihood of children in single-parent families going without necessities. However, research shows that single parents appear to spend almost as much as two parents do on their children, but they spend considerably less on themselves, sometimes going without food.[80]

As the wider concept of the family has changed, so has the concept of childcare within the home. For a long time the responsibility for bringing up children rested primarily with the mother, often assisted by female members of the extended family. Fathers away at paid employment or war saw the day-to-day business of raising children as women's work whether the mother was working outside the home or not. Now, in Scotland as elsewhere, many fathers are much more ready to take an active part in the upbringing of their children. Changing roles in childcare in the home are all part of social change in Scotland today.

NOTES

1. Smout, 1985, 91; for further discussion of changing attitudes towards children and childhood see Chapter 28 Hierarchy and Authority.
2. Wilson, 1979.
3. Loudon, 1992, 206.
4. Tait, 1987, 415.
5. King, 1938.
6. Jamieson, Toynbee, 2000.
7. Ferguson, 1958, 65.
8. Jamieson, Toynbee, 2000, 18.
9. Sheridan, 1985, 48.
10. McVeigh, 1999, 20.
11. Muir MacKenzie, 1941, 250; Whatley, 1988, 246.
12. Wood, Suitters, 1970, 142, 158; see also, Carter, Duriez, 1986, 201–44.
13. Oral testimony, LR, 91.
14. de Mause, 1974, 38.
15. Oral testimony, LR, 123.
16. Oral testimony, LR, 61.
17. Blair, 1998, 16.
18. Many will empathise with the cry emanating from lavatories up and down the land, 'A'm needin somebiddy ti help haud oot ma combies'.
19. Oral testimony, LR, 91.
20. Blair, 1998, 16.
21. Jamieson, Toynbee, 2000, 21.
22. Blair, 1998, 12.
23. King, 1938, 34.
24. King, 1938, 36.
25. King, c 1940, 35.
26. McGregor, 1994, 13, 17, 31, 43.
27. de Mause, 1974, 30.
28. King, c 1940, facing page 40.
29. Oral testimony, LR, 61.
30. Oral testimony, LR, 115, pupil midwife in Glasgow in the early 1940s.
31. Oral testimony, LR, 46, pupil midwife in Edinburgh in the 1930s.
32. *Holy Bible*, King James Version, Exodus, 2, 9, 46.
33. Williams, 1992, 109.
34. de Mause, 1974, 34; Marshall, 1984; Gordon, 1992, in ch 11, Morals of the People, describes how wet nurses in the 1700s who were unmarried were handled by the Kirk Sessions of the day; further information on wet nursing in Scotland and France is supplied by Diack, 1999, ch 5; see also Fildes, 1986; Golden, 1996.
35. MacGregor, 1958, 106.
36. Mackenzie, 1917, 463.
37. Tait, 1987, 414; Ferguson, 1958, 533.
38. MacGregor, 1958, 110.
39. Mackenzie, 1917, 155; MacGregor, 1958, 108.
40. Tait, 1987, 415.
41. Curly Kate and Plain Geordie: The top of the plain loaf was rounded and usually

over-fired in the oven. Children were persuaded to eat it by being told that if they did so, their hair would grow curly.
42. See also Chapter 23 Food.
43. Anderson, 1992, 41.
44. Blair, 1987, 149–53; see also Chapter 16 Personal Hygiene.
45. Montgomerie, 1985, 16.
46. Anderson, 1992, 41.
47. Personal Communication, LR, 63.
48. Oral testimony, LR, 35.
49. Oral testimony, LR, 124.
50. See also Chapter 8 Music; Chapter 10 Television, Video and Computing; Chapter 11 Games and Sport; Chapter 21 Reading and Study.
51. BBC2, 'HomeGround', 14 June 2001, 7.30pm.
52. Cross, 1997, 9.
53. Purser, 1992, 238.
54. Baden-Powell, 1952, 13.
55. Springhall, Fraser, Hoare, 1983, 15.
56. Randall, 2001, 12–15, website: www.guides.org.uk
57. Personal communication.
58. Springhall, Fraser, Hoare, 1983, 28.
59. Springhall, Fraser, Hoare, 1983, 39.
60. Springhall, Fraser, Hoare, 1983, 71.
61. Personal communication.
62. Beeton, 1892, 1540; for further information on servants with childcare duties see Chapter 20 Work in the Homes of Others.
63. Gathorne-Hardy, 1985, 184.
64. Smith, Grunfeld, 1993, 12.
65. Moncrieff, 1984, 16.
66. Gathorne-Hardy, 1985, ch 6, 170–208.
67. Gathorne-Hardy, 1985, 199, 'The Pryce-Jones Nanny had wheeled herself behind the Memorial and sat down on an empty bench. After a while an older Nanny appeared, pushing a pram on which was painted a small gold coronet. She sat down too, and they eyed one another. At length the older Nanny turned to the younger one, coughed, and said, "Excuse me Nanny, is your mummy a titled mummy?"
68. Scottish Childminding Association, Room 7, Wellgreen, Stirling, FK8 2DZ. <http://www.childminding.org>.
69. Personal communication.
70. Abrams, 1998, 35; Ferguson, 1958, 521–30.
71. Mackenzie, 1917, 108; see also Chapter 37 Children's Homes.
72. Abrams, 1998, 35–77.
73. Abrams, 1998, 65.
74. Gordon, 1978, 159.
75. O'Donnell, 1992, 52.
76. One Parent Families Scotland, 2000.
77. One Parent Families Scotland, 2000.
78. One Parent Families Scotland, 2000.
79. One Parent Families Scotland, 2000.
80. One Parent Families Scotland, 2000; Middleton, Ashworth, Braithwaite, 1997.

BIBLIOGRAPHY

Abrams, L. *The Orphan Country*, Edinburgh, 1998.
Anderson, M. Population and family life. In Dickson A, Treble, A, eds, *People and Society in Scotland, vol III, 1914–90*, Edinburgh, 1992, 12–47.
Baden-Powell, R. *Scouting for Boys*, 27th edn, London, 1952.
Beeton, I. *The Book of Household Management*, London, 1892.
Blair, A. *Croft and Creel*, London, 1987.
Blair, A. *Miss Cranston's Omnibus, Including Tea at Miss Cranston's and More Tea at Miss Cranston's*, Edinburgh, 1998.
Carter, J, Duriez, T. *With Child: Birth Through the Ages*, Edinburgh, 1986.
Cross, G. *Kids' Stuff*, Harvard, 1997.
Diack, H L. 'Women, Health and Charity: Women in the poor-relief systems in eighteenth-century Scotland and France', unpublished PhD thesis, University of Aberdeen, 1999.
Ferguson, T. *Scottish Social Welfare, 1864–1914*, Edinburgh, 1958.
Fildes, V. *Breasts, Bottles and Babies: A History of Infant Feeding*, Edinburgh, 1986.
Fontanel, B, d'Harcourt, C. *Babies: History, Art and Folklore*, New York, 1997.
Gathorne-Hardy, J. *The Rise and Fall of the British Nanny*, 2nd edn, London, 1985.
Golden, J. *A Social History of Wet Nursing in America*, Cambridge, 1996.
Gordon, A. *Candie for the Foundling*, Edinburgh, 1992.
Gordon, M. Industrialisation and the family. In Worsley, P, ed, *Modern Sociology*, 2nd edn, Harmondsworth, 1978.
Hurst, J, Hubberstey, S. *Working with the Professionals to Get the Best for your Child*, London, 2001.
Jamieson, L. Toynbee, C. *Country Bairns: Growing Up, 1900–30*, Edinburgh, 2000.
King, M. *Mothercraft*, London, c1940.
King, T. *Mothercraft*, London, 1938.
Loudon, I. *Death in Childbirth*, Oxford, 1992.
MacGregor, A. *Public Health in Glasgow, 1905–46*, Edinburgh, 1958.
Mackenzie, W L. *Report on the Physical Welfare of Mothers and Children, vol 3, Scotland*, Dunfermline, 1917.
Marshall, R. Wet nursing in Scotland, 1500–1800, *ROSC*, 1 (1984), 43–51.
de Mause, L, ed. *The History of Childhood*, London, 1974.
McGregor, I. *Bairns: Scottish Children in Photographs*, Edinburgh, 1994.
McLachlan, G, ed. *Improving the Common Weal: Aspects of Scottish Health Services, 1900–84*, Edinburgh, 1987.
McVeigh P. *'Look After the Bairns': A Childhood in East Lothian* (Flashbacks, no 8), East Linton, 1999, 20.
Middleton, S, Ashworth, K, Braithwaite I. *Small Fortunes: Spending on Children, Childhood, Poverty and Parental Sacrifice*, York, 1997.
Moncrieff, M. *Yes, Ma'am! Glimpses of Domestic Service, 1901–51*, Edinburgh, 1984.
Montgomerie, W. Coals in the bath. In Kamm, A, Lean, A, eds, *A Scottish Childhood*, Glasgow, 1985, 12–16.
Muir MacKenzie, A. *Scotland in Modern Times*, London and Edinburgh, 1941.
O'Donnell, M. *A New Introduction to Sociology*, Walton-on-Thames, 1992.
One Parent Families Scotland. *Fact Sheet*, 13 Gayfield Square, Edinburgh, EH1 3NX, August 2000 <http://www.opfs.org.uk>.
Purser, J. *Scotland's Music*, Edinburgh and London, 1992.
Randall, L. Carry on Guiding, *The Scotsman Magazine*, Saturday, 6 October, 2001.
Reid, L (LR). Oral testimonies between 1997 and 2001, numbered 1–124.
Sheridan, D. *Among you Taking Notes … : The Wartime Diary of Naomi Mitchison, 1939–45*, London, 1985.

Smith, J, Grunfeld, N. *Nanny Knows Best*, London, 1993.
Smout, T. *A History of the Scottish People*, 1560–1830, London, 1985.
Springhall, J, Fraser, B, Hoare, M. *Sure and Stedfast: A History of the Boys' Brigade, 1883–1983*, London and Glasgow, 1983.
Sussman, G. *Selling Mothers' Milk: The Wet-Nursing Business in France, 1715–1914*, Illinois, 1982.
Tait, H. Maternity and child welfare. In McLachlan, G, ed, *Improving the Common Weal: Aspects of Scottish Health Services, 1900–84*, Edinburgh, 1987, 413–40.
Whatley, C. The experience of work. In Devine, T M, Mitchison, R, eds, *People and Society in Scotland, vol 1, 1760–1830*, Edinburgh, 1988, 227–51.
Williams, E. The rise and fall of the wet nurse, *Professional Care of Mother and Child* (April 1992), 108–9.
Wilson, A. The Enlightenment and Infant Care. Paper presented at the 9th SSHM Annual Conference, *Medicine and the Enlightenment*, University of Hull, July 1979, published in *Society for the Social History of Medicine Bulletin*, 25 (December, 1979), 44–7.
Wood, C, Suitters B. *The Fight for Acceptance*, Aylesbury, 1970.

26 Infirmity

LINDSAY REID

INTRODUCTION: CARING FOR THE INFIRM

The word 'infirm' has various connotations. The infirm have been defined as people who lack strength or vitality, for example because of illness or advanced age.[1] Also to be included are those who are mentally and/or physically impaired as a consequence of congenital conditions or accident.[2] Those who are infirm may require care or support, on a short- or long-term basis, in an institution or looked after at home by a variety of different people. This chapter focuses on aspects of caring in the Scottish home for individuals, who, for whatever reason, have needed help in their day-to-day living.

Attitudes towards caring for the infirm have in some respects gone full circle. From an era when nearly all who needed it were looked after at home, we moved to a period of intense hospitalisation and then recently to a policy of admitting people to hospital only when absolutely necessary, sometimes for as short a time as a day or even less. The object now is to get individuals 'out into the community' as fast as possible and this usually involves the need for care and support in the home.

OVERVIEW OF HISTORICAL ASPECTS

Until the nineteenth century it was common for people who were ill, old or disabled to be looked after at home (Fig. 26.1). Hospitals were considered a last resort, and were very much associated with the poor.[3] They were also dirty, offered low-quality care and were the scene of much cross-infection. Furthermore, if home conditions were reasonable and hospital beds were few, the women in a family were expected, possibly aided by the servants of the house, to look after their sick at home,[4] and would perhaps encounter censure from the local community if this did not happen. Hospitalisation would certainly be a more costly business than home care, while from the infirm person's point of view going into hospital could seem like the end of the road.

The practice of caring for the infirm in the home was supported by a culture of independence and a very strong sense of family cohesion and duty, perhaps enhanced by people not moving so far afield as they do or can today. Linked to this was the custom of unmarried daughters remaining at home to look after ageing parents, which, according to Cline in *Lifting the Taboo*, still occurs.[5]

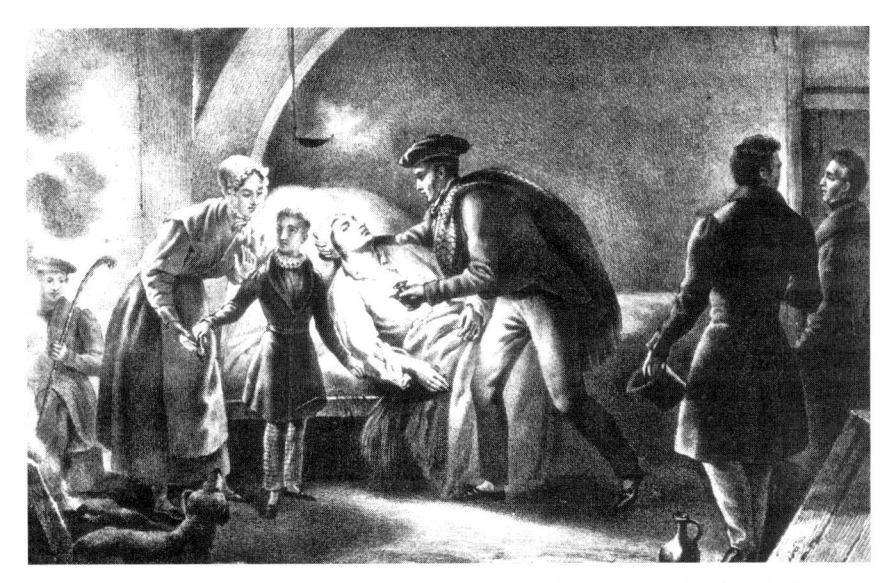

Figure 26.1 '*La Chaumière Ecossaise*' (from d'Hardiville, *Souvenirs des Highlands 1832*, Paris, 1835). A cruisie hangs above the patient in the alcove bed. Source: SLA 128A(1) C.65/B S.9689.

During the nineteenth and early twentieth centuries the number of hospitals in Scotland increased. Many were specialist hospitals, catering for particular types of condition, such as the Glasgow Eye Hospital. Care within all hospitals improved and the work of Macewen and Lister in introducing asepsis and reducing infection greatly improved outcomes.[6] All in all, hospitalisation became a more acceptable idea. New hospitals were built in Scotland during World War II in readiness for battle casualties. When the expected numbers did not materialise, the extra beds they provided were used for civilians. These institutions gave a stimulus to post-War specialist surgery and medicine, and they formed the basis of the hospital service provided by the NHS, formed in 1948.[7]

Despite these advances, in the early decades of the twentieth century there existed many parts of the country where people continued to be cared for largely at home, mainly by their relatives, friends and neighbours (Fig. 26.2). This was particularly evident in rural areas, such as much of the Highlands and Islands, where hospitals, doctors and nurses were few and far between. Even in better-served areas, and with an evolving custom of 'medical clubs' acting as an insurance off-setting the costs of medical help, calling in a doctor for a home visit could be an expensive affair. Some doctors gave their services free in extreme cases, but generally medical treatment was beyond the means of many families, who had to cope as best they could themselves.[8] The National Health Insurance Act of 1911[9] provided medical insurance for some. However, crofters, for example, were not 'employed' and they did not meet certain other criteria of the Act. Another factor in the low

Figure 26.2 Andrew Minto, Coulterhaugh, Biggar, Lanarkshire, being cared for at home, c 1910.
Source: SLA 98B 54/57/A.

level of professional intervention in some areas was a fatalistic disposition and a refusal to call the doctor as 'the hour had come'. This also led to lack of certification of nearly 60 per cent of deaths in some parishes.[10] Altogether, as Brotherston notes, 'In certain remote areas of the country, and in the poverty-stricken western Highlands and Islands in particular, healthcare was totally inadequate.'[11]

In 1912 a government committee 'was set up under Sir John Dewar to advise on how the special healthcare needs of the crofting counties might be met'.[12] The main recommendation of the Dewar committee resulted in the Highlands and Islands Medical Scheme, which aimed to encourage doctors to work in the region. The government agreed to pay doctors an enhancement on condition that they would attend, at a reduced fee, crofting families, dependents of insured persons and those of similar income. Although the Scheme could not begin until after World War I, once it was under way it proved to be successful in improving medical services across the Highlands and Islands. Even so, Dr James Durham, who started medical practice in the parish of Tingwall, Shetland in the mid 1930s, noted:

> It was obvious that very often the doctor was only called out as a last resort and when he made his diagnosis and supplied medication, he then assumed, in the absence of another call, satisfactory results. Quite simply, many people could not afford to pay even these fairly moderate charges.[13]

The problem of paying for medical services was removed in 1948 with the promise of the NHS that these would be 'free at the point of need'.

Although the rest of the population was relatively well served in terms of medical coverage before the outbreak of World War I, this was to diminish drastically when the War began, with the problem escalating as the hostilities continued and more doctors were needed to serve in the forces. Extra funding was found to supplement the incomes and cover the mileage and practice expenses of hard-pressed doctors in sparsely populated rural areas and at the end of the War demobilisation of doctors began immediately.[14] The post-War years were a time of new beginnings not least of which was the establishment of the Scottish Board of Health in 1919, 'which became the umbrella unit for all the existing components of Scottish health administration'.[15] To have an administration in Scotland, separate from the rest of the UK, helped to take into account the needs of the geography and demography of Scotland, where 80 per cent of the population live in 17 per cent of the land mass.

Today the customs and expectations of the people of Scotland are often very different from those of a hundred years ago, not least because of the arrival of new groups of immigrants. When considering the care of the infirm in the home such recent variety should be taken into account. While those of differing ethnic and racial backgrounds acquire statistical differences in some medical conditions, Professor Raj Bhopal of the University of Edinburgh recommends that care planning should be based on specific data, not generalisations. However, there is currently very little information about Scotland in regard to the care needs of recent immigrants and their descendants and researchers are largely reliant on data from England and Wales. From this, it appears that particular attention needs to be paid to language, cultural differences, educational background, level of health knowledge and socio-economic status. For a better service to evolve, it is crucial to acknowledge differences in outlook, improve communication and plan services with the needs of recent immigrant groups in mind.[16]

POSSIBLE REASONS FOR INFIRMITY

One of the principal causes of long-term poor health in Scotland was overcrowding in homes, which was widespread in both cities and rural areas. In 1911, 50 per cent of Scots lived in one or two-roomed dwellings[17] while the proportion of people in England and Wales living in similar accommodation was just 7 per cent. Overcrowding was usually accompanied by poverty, and together they contributed to malnutrition, lack of hygiene, infectious disease and cross-infection. Matters were exacerbated by the domestic and industrial use of polluted water which gave rise to epidemics of diseases like cholera. The long awaited opening of the Loch Katrine water supply for Glasgow in 1859 was to be completely vindicated in the cholera epidemic of 1865–6 when relatively few people in the city died. Formerly, most fatalities from cholera in Scotland had occurred in Glasgow.[18]

People living in the Highlands were generally healthier than their urban counterparts, but they were subject to high rates of tuberculosis.[19] The disease was contracted by those with poor resistance, a consequence of inadequate diet, poor personal and domestic hygiene and high exposure to infection. In the nineteenth century if tuberculosis was contracted there was little chance of recovery.

Industry created its own category of disease. Before health and safety measures were introduced, work-related diseases which caused long-term disability were prevalent. These included chest conditions like asbestosis, related to working on building sites and in shipyards,[20] and miners' pneumoconiosis. Accidents at work could also result in long-term disability (although health-and-safety regulations are much more strictly enforced now, these still occur) and there were in addition diseases specific to living in large urban areas, such as a range of chest diseases provoked by inhaling smoke and smog.

In the past, many people with congenital or life-long disabilities, both physical and mental, were able to stay at home even though they needed long-term support and care. However, those who required it, particularly those who were mentally unbalanced or described as insane, were admitted to what was deemed to be the appropriate institutional care. Their numbers appeared to be very great. Ferguson, for instance, reports that in 1914 only 17.2 per cent of 'the lunatic poor' lived in private dwellings, although this might include those who had been 'boarded out' by the authorities. The remainder were in asylums and institutions for imbeciles and licensed wards of poorhouses.[21]

CHANGES IN CLASSES OF INFIRMITIES AS TIME ADVANCES

Illness
Throughout the twentieth century, and into the twenty-first, there have been changes in the types of illness requiring care. Better public health measures have resulted in cleaner water, purer air in cities, stricter food regulations, improved housing, more education on food preparation and hygiene, and a greater understanding of the principles of nutrition. Thus epidemic diseases like cholera are no longer with us, although the outbreak of typhoid fever in Aberdeen in 1964[22] was a warning against complacency. Prevention of illness therefore covers social and economic as well as personal and medical aspects. With better conditions and nutrition the ability of people to withstand some illnesses has also improved. Medical advances, such as the discovery of antibiotics and the use of blood transfusions, vaccination, aseptic techniques and X-rays, along with improved education for all professionals involved in the care of the ill, have eradicated or reduced certain conditions like diphtheria and small pox, and made others, for example those caused by bacterial infections such as tuberculosis, easier to treat and therefore short- rather than long-term illnesses.

However, while the incidence of some diseases has fallen, and some have disappeared completely, there has been a rise in the incidence of others.

These include smoking-related diseases such as cancer of the lungs, bronchus, oesophagus and breast, coronary artery disease and other heart conditions, raised blood pressure, and cerebral vascular accidents. Although the diet of the population is much better than it used to be, a worrying trend is an increase in the number of obese Scots. According to a recent report in *Scotland on Sunday*, the proportion of seriously overweight men rose from 16 per cent to nearly 19 per cent between 1995 and 1998 and among women over the same period, the increase was from just over 17 per cent to nearly 21 per cent.[23] Obesity contributes to the rise in heart disease and to an increase in the number of Scots developing diabetes and its accompanying long-term problems.[24]

Illnesses, little known or unheard of a few decades ago, now have a high profile. One of epidemic proportions, especially in Africa, but also affecting Scotland, is the human immunodeficiency virus (HIV) which can develop into full-blown acquired immune deficiency syndrome (AIDS) over time. There are other examples: Creutzfeldt-Jacob disease, new bacterial infections which fight off antibiotics, neurological diseases such as multiple sclerosis, and chronic fatigue syndrome or myalgic encephalomyelitis.

It is difficult to say whether mental illness has increased over the decades or not. A century ago, mental illness was a forbidden topic. Today, people are much more open about it and more ready to seek help. Currently, one in five adults is affected by mental-health problems in any one year and at least 30 per cent of GP consultations have a psychological component.[25] Sadly, the number of young men committing suicide in Scotland has risen recently. Six hundred people kill themselves in Scotland each year. Of these the incidence is more than three times higher among men than women.[26]

Infirmities related to old age
Old age does not always imply infirmity. Nevertheless Scotland's ageing population has begun to put more pressure on care demands. The proportion of old people in Scotland has increased as life expectancy has grown longer. In 1861 a man could expect to live for 40.3 years and a woman for 43.9 years. In 1911 the expectation rose to 50.1 years for a man and 53.2 years for a woman.[27] The reduction in infant mortality will have played its part in improving the life-expectancy figures. However there are other factors which should be taken into account. These include lifestyle and nutrition of the individual and improved medical and nursing care along with developments in drugs such as antibiotics, and modern surgery. Now, life expectancy for men in Scotland is 76.6 years and for women 82 years,[28] and today, one in five people in Scotland is over 60.[29] It is hoped that the majority can lead independent lives, and with improving health levels this should increasingly be possible. However, a recent article in the *Scotsman* highlighted the need for more research into diseases of the elderly if this optimism is to become a reality. A recent General Household Survey 'showed that 50 per cent of over-75s said their illnesses prevented a full and active life, and that at least two-thirds of pensioners over 75 had a long-standing illness'. However, with more

high-quality research these figures and the quality of life for elderly people should improve.[30] Whatever happens, there will remain a need for care to be provided for at least some old people. Many individuals hope that this can be given in their own home if possible.

In the sixteenth century 'our Jacobean ancestors believed that society had a Christian duty to maintain those whom the 1574 Scottish Poor Law referred to as the "puyr, aigit and impotent [ie, infirm]"'.[31] This maintenance, designed to be performed in the ageing person's own home, was gradually overshadowed by the development of poorhouses. These, however, were so unpleasant that ill, old people, who normally might have welcomed relief, avoided the indiscriminate care on offer as long as they could. It was only when the Poor Law authorities put pressure on those administering poorhouses that the standard of medical care started to improve. Serious difficulties remained, though, in caring for the sick old person, and these only began to be resolved with the growth and official acknowledgement of geriatric medicine, although even after the start of the NHS in 1948 there were still large numbers of elderly sick not obtaining the medical care they required. In 1953, Dr Nairn Cowan, the Medical Officer of Health for Rutherglen, Lanarkshire, demonstrated the need for healthcare teams to attend upon the elderly in their own homes with input from geriatric consultants at local healthcare centres. He recommended that the teams should include general practitioners, district nurses, health visitors, social workers and para-medical staff, especially chiropodists[32] (Fig. 26.3). It is interesting to note that chiropody was included in NHS home services for

Figure 26.3 Nurse Mary Morrison and Annie MacLeod at Mrs MacLeod's door. Carloway, Lewis, 1952. Reproduced by kind permission of Carloway Historical Society and Mary Morrison. Source: SLA 98A(3) 59/2/13/Carloway Historical Society 8/1990.

the elderly in Scotland but not, at the time, in England.[33] There was further progress and acknowledgement of geriatrics (today known as gerontology or, more generally, as 'care of the elderly') as a specialism in 1965, when the first clinical chair of geriatric medicine in the world was endowed at the University of Glasgow, along with accompanying research facilities.[34]

Care and treatments in medicine in general are changing. Alongside what are seen as more conventional medical developments, the NHS in some areas of Scotland now offers forms of complementary medicine such as homeopathy, acupuncture, aromatherapy and reflexology. These slower, more gentle approaches lend themselves particularly well to home care and hopefully 'have helped pave the way towards the kinder, more open climate of the close of the millennium'.[35]

WHO LOOKS AFTER THE INFIRM IN THE HOME?

As noted above it was usual for the infirm at home to be cared for by women relatives, aided by servants if they had them. However there is some evidence that men did the caring too. In December 1701 Lady Polwarth died of consumption at Polwarth House after a few months of illness. Before her death, 'Lord Polwarth had been passionately devoted to his wife and had nursed her during her illness.'[36] Very often the main burden of caring would have been placed on an individual member of a family. Even children had to, and still have to, look after a parent needing care at home. In the nineteenth century, Angus Maclellan aged 13 had to look after his family by himself in South Uist during a fever epidemic:

> The doctor was sent for and he said that Hector had the fever. The doctor went away. Only a few days after he had gone two of my sisters fell ill with the fever. Three of us were ill then. Then my father fell ill. I was happy enough as long as my mother was up, but unfortunately she herself fell ill and the five of them were ill together and there was no-one to look after them but myself. No-one was coming near them or near the house. I had to fetch peat from a good distance and I had cattle to look after too and I had no-one to help me.[37]

Another Scots boy looked after his mother for years during World War II:

> My mother had always been an invalid but when war broke out in 1939 and I was only nine years of age my father was called up. This left myself and my sister and my mother at home, say, three miles [5 km] from the nearest village. I had to learn to give her injections daily and to attend to the general chores and make sure that we got ourselves off to school. Often my mother could not be left alone so I had to be absent from school frequently. The school as far as I remember took no notice of this despite the fact that I sometimes had more absences than attendances. The doctor came willingly when called; I had to go to the village and get him. This continued till the end of the War.[38]

HELP FROM OUTSIDE THE HOME

Informal recruitment of help for those who care for the infirm at home has always existed. Relatives of the main carer, in the same or in other households, would often lend a hand. Particularly in areas of cities typified by tenement housing, families lived close by each other and neighbours and friends could rally round and give help as and when it was needed. Although it must have been difficult in many rural areas because of the distances involved, especially in the days before cars and telephones, neighbours were there to help too. Chrissie Sandison recalls one particular occasion in Shetland in the 1920s:

> There was another very small baby and the mother died. Four or six weeks after the birth the baby took whooping cough from his brothers and sisters. And he was kept by the fire in a cradle and the women took turns to sit with him – they had their knitting you see – and every time that child began to cough they got him up. If ever anybody was in any kind of trouble, the neighbours rallied around. That baby lived.[39]

In addition to family, friends and neighbours, immediate help might be given by the local minister with his medicine chest,[40] wise women with 'legendary lore', quack doctors with their potions and the 'howdies' who practised midwifery[41] and laid out the dead. Bonesetters also had a part to play, frequently a blacksmith working from his smiddy, practising a craft often handed down from father to son. They were seen by medical men as humbugs and yet they were often preferred to doctors by the public as 'a body kens doctors ken naethin aboot banes'.[42]

Professional help with caring for the infirm at home gradually developed. General practitioners aided where they could. However there were areas where distance as well as finance precluded the constant attention of a GP and sometimes the local laird would finance the engagement of a resident nurse instead. Through this, there evolved a system of nursing associations, partly privately funded, partly helped by Local Authorities and strengthened by the provisions of the 1897 Public Health Act and the 1892 Local Taxation Account (Scotland) Act.[43]

Nursing associations were further reinforced by the formation of the Queen Victoria's Jubilee Institute. In Queen Victoria's Jubilee Year, 1889, the women of Great Britain and Ireland collected money for a fund as a jubilee offering to her. The Queen wished to devote £70,000 of the Fund for the promotion of home nursing and was advised that the best way to do this was to found an institute to train and maintain nurses who would thus be qualified to nurse people at home. The three main centres of the Institute were in London, Edinburgh and Dublin. Its functions were to give extra training in district work to registered nurses (who, after the Midwives Acts, could be certified midwives as well), to supply nurses to Local District Nursing Associations and to arrange for the work of each Queen's Nurse to be inspected.[44] The District Nursing Associations continued until the late 1940s. One dual role nurse/midwife described how they worked in Angus:

> And we had, you know the Nursing Association. Each district had that ... It was made up of the ladies of the – good standing as [they say] and we used to have to go to their various houses and have my book signed once a month and give account of what I'd been doing ... to these ladies. [They] called themselves the Local Nursing Association. They stopped when the NHS took over.[45]

Other sources of outside help have evolved over the years. One well-known example is the home help. The first home helps in Scotland started informally after the disappearance of the howdie who, when attending a mother in childbirth, was prepared to do housekeeping duties as well as care for the mother and baby. In 1943 women employed as home helps were given official recognition and the system developed from there.[46] The home-help system, now administered by Social Services, is widely used to offer assistance with, for instance, housework and shopping, to people in their own homes. Other support to help the old, disabled and very ill remain at home includes health visitors, social workers, meals on wheels, 'tucker-ins' (nurses who settle people down for the night), MacMillan and Marie Curie cancer nurses and other community services such as community psychiatric nurses and physiotherapists. In addition, the Report (the Sutherland Report), published in 1999, of the Royal Commission on Long-Term Care of the Elderly[47] recommended further help in the form of free personal care as well as nursing care for those who require it. Personal care includes help with feeding, washing, dressing and going to the lavatory. In 2000 the Scottish Executive agreed to implement the recommendations.

There are many financial implications involved in caring for a person with long-term illness or other infirmity in the home. While the government helps with the appropriate support and allowances, looking after a family member may well mean the loss of income of the main carer, which can represent a considerable weakening of the household economy. To this might be added further costs of transport, extra equipment and adaptations to the home. However, the Scottish Executive demonstrated its commitment to help by saying:

> Care of older people accounts for 40 per cent of social-work and 40 per cent of health-service budgets in Scotland. We need to ensure that these resources are managed effectively to deliver effective integrated services. We need to demonstrate by our actions that people will get the care and treatment they need, when they need it and where they want it, irrespective of their age. There is no place for ageism ... in any part of the NHS in Scotland.[48]

And, at the other end of the age spectrum, 'The Executive is committed to improving the provision of services for children with special needs. Wherever possible and appropriate, we will deliver this care at home.'[49]

Although some individuals prefer, for various reasons, to go into an institution, for an infirm person of any age or diagnosis, to be looked after at home can often seem a much preferable option. Yet, the whole household has to adjust to this, and not just in relation to primary caring. Relationships can be affected. Children may not get the attention they need because a mother is occupied with a family member who needs a greater amount of care. Siblings of a child with special needs may have to grow up more quickly than usual. Their social lives may be curtailed, and they may feel guilt at being angry about this. They may even wonder why they are not 'infirm' too.[50] Carers and other household members also need to be cared for, and here, the system of respite care, for example, is an important asset. A change of carer and scenery are important to enhance the quality of life, the dignity and independence of all involved. To help with this there are provisions such as day centres and lunch clubs which enable those cared for at home to get out and do something different, see new faces, and talk to other people (Fig. 26.4). There are other networks, groups and projects, for example the Craigmillar Health Project, Edinburgh,[51] which provide a wide variety of supportive and healing care for people of all ages.

For many infirm people, living at home equates with self-respect and independence. Staying in one's dwelling during infirmity should be viewed as a challenge, not a burden, by all involved – the infirm person, the

Figure 26.4 'Milan' Day Centre for Older Asians, Pilrig Street, Edinburgh, 1992. Source: SLA 98B E/93/36.

household and the integrated services as promised by the NHS in Scotland, the last respecting needs and views on how and where care should be provided.

Living alone should not preclude remaining at home if that is what an individual wishes. The Scottish Executive is committed to making this happen through providing additional money to deliver extra services throughout the country. These include practical help, for instance 'a local service in every part of the country for shopping, laundry and minor household repairs, helping 10,000–15,000 people to preserve their independence at home'.[52] In addition, a person living alone who needs help, for example, with getting up in the morning and going to bed, can receive this, along with a home help, meals on wheels and other services if necessary.

THE MATERIAL CULTURE OF THE INFIRM IN THE HOME

Those who are infirm and at home often require special equipment to help them, and when appropriate, those who look after them cope, with daily living. The nature of, and reliance on, such equipment depends of course on the class of infirmity. Usually a team of professionals will work in partnership with the patient and/or carers to assess what is the best response to personal and environmental needs and to supply, if possible, what is necessary. Volunteer and support groups and societies also loan and maintain or help to buy expensive items.

The temperature of living space must be at a comfortable level for those who are physically disabled or very elderly and unable to move with ease, and thus central heating may have to be installed in a home. Someone who is bedridden may also require warm but lightweight bedclothes to minimise pressure, special pillows or an adapted design of bed, possibly with cot sides. Articles for increasing comfort include sheepskins to lie upon and heel pads to reduce the effects of constant rubbing. A bedpan and/or a commode are necessary items for many while some bed-ridden individuals may need to use draw sheets and incontinence pads. A number of common household articles are not appropriate in this context: electric blankets are not an option if incontinence is a problem and hot-water bottles may burn frail skin.

Most infirm people at home are up and about during the day, although they may still make use of equipment to help with washing, dressing, going to the lavatory, eating and preparing meals. In the bathroom, for example, metal supports may be fixed on either side of the water closet or attached to the wall side of the bath. There also exist bath seats which can be electrically driven or worked by the weight of the person, depending on their level of disability. A bath hoist is possible in some homes while in others showers can be fitted with handles and seats.

Articles for the kitchen include utensils with specially shaped or thickened handles and jar openers for those people without full movement in their fingers. Plates which keep food warm may be appreciated by slow eaters. A cooker guard is also useful for some infirm people, along with fireguards for use in any room with an open, electric or gas fire.

If mobility is a problem there are various measures which can be employed. These include the use of walking sticks, with rubber tips to prevent slipping, and Zimmer frames, which with four legs are very stable. Some Zimmer users find it helpful to attach a bag or basket (such as a bicycle basket) to the front of the frame so that small items like spectacles, books or knitting can be carried around. Wheelchairs are available in different styles and sizes, including models with removable sides which can fit through narrow doors. The difficulties associated with going upstairs can be avoided by having the infirm person's bedroom downstairs or all the rooms of the dwelling on one floor. Homes built for the elderly or physically disabled usually adopt the latter format. Where the use of purpose-built accommodation is not practicable stair lifts can be installed. This is not a cheap option but it can avoid the need for moving home and enable infirm members to participate more fully in the life of a household. Going out of the dwelling, the local authority may provide a ramp with a side rail from the front door. Some support societies offer assisted public transport or help provide vehicles for individual use which can have an orange disability badge for easy parking.

Those with sensory disabilities may also require special measures and equipment in the home. Blind people usually prefer landmarks throughout the dwelling to remain unaltered. They may make use of talking books, and for some time now these, along with books in large format for the partially sighted, have been supplied by libraries. For deaf people, a loop induction system can be installed for television and radio. British Telecom, in association with the Royal National Institute for the Deaf, launched a new typetalk telephone system in Autumn 2000 called TextDirect which connects telephone users to those to whom they wish speak straightaway, rather than their going through what was formerly a slow call set-up process.[53] Also available are lights which flash when the door bell or phone rings and alarms that can be put under a pillow and wake the sleeper by vibrating.

Those who are infirm can feel particularly vulnerable regarding their personal security. Thus the Social Work Department may supply personal alarms which automatically contact the emergency services when activated.[54] For further peace of mind, chains and other devices can be fitted to outer doors.

It is important for infirm people and their carers to know how to obtain extra help and equipment. General practitioners, district nurses, health visitors and social workers can all advise. Even the local phone book will yield contacts which can develop into tangible assistance and support.

DYING AT HOME

Peter Pan said, 'To die will be an awfully big adventure.'[55] A hundred years ago most people would have expected to face this adventure in their own home, usually with one or more members of their family with them. Now, more and more people are dying in institutions: hospitals, hospices (see Figs 36.1, 36.2), nursing homes and residential care homes. There is evidence that the majority of people would rather die in their own home.[56] However medicalisation of death developed in the second half of the twentieth century along with a medical tendency to see death as a failure. Medicalisation includes intervention, specialisation, organisation, time-tabling and control. All of these preclude dying at home and encourage dying in hospital, the idea being that death in hospital is tidier and easier to manage. The public have become used to institutional death and indeed have come to expect it. As Nick Bass says in his article 'Where to die', the fewer deaths there are in the community, the less acceptable it becomes, 'adding a public push to the professional pull to hospital'.[57]

Dying at home used to mean the relative absence of professionals. Female members of a deceased person's family could wash and dress the body themselves with the privilege sometimes being granted to the oldest female relative.[58] In some areas, before the 1915 Midwives (Scotland) Act, the local howdie did this. The Statutes of the Act however attempted to put a stop to the practice, stating that certified midwives were not allowed to lay out any body except that of a mother they had attended. Occasionally, however, the order was sidestepped, particularly by midwives who were also employed as Queen's Nurses in the dual role of nurse and midwife. Anne Bayne recalled working in Tullibody:

> You werena supposed to do midwifery and lay out bodies. Well, I did. They were people I had nursed. You couldn't say 'no' to a request like that. You just accepted that you wanted to be a district nurse and this was what it was like. You were often with them when they were dying and you had attended them and become friends of the family and supported the family and it was a very special position you were in, and it gave tremendous support. [I remember one] old man ... he died. I stayed with her [the daughter-in-law] that night. It was quite late on and he died and I dressed him and I sat with them.[59]

In some areas of Scotland, when a death occurred the family sent for a wright[60] who came with his 'streakin boord' upon which the body was laid, then washed and dressed and covered with a white sheet to await the next important ceremony, the coffining.

A body would not have been put in a coffin immediately. Probably this originated in the coffin having to be made to measure by the local joiner. One or two days might elapse before the kisting, coffining or chesting of the body. This was an important rite performed by a privileged few close friends and family members, as highlighted by David Marshall from Kinross, who wrote in his diary of the coffining of his father two days after his death at home, 'My father was coffined tonight about 40'ck [sic]. Aunt Mary, Alexr Bennet Snr, Mrs Blackwood and Aunt Helen, Agnes, Mother and I.'[61]

Viewing the body is another important ceremony which used to take place at home and now often occurs elsewhere. Formerly the coffin and body would have been placed in the best room and for the following few days those who wished to pay their respects would visit. This is still done, but often in the undertaker's parlour. Similarly, most funerals took place from the deceased person's home. Now the body is usually taken from the undertaker's premises to church or crematorium and it appears that the undertaker and his funeral parlour have largely superseded the neighbour or the district nurse and the best room. Dying in Scotland, as elsewhere, has become the province of the professional.

Not everyone is happy with this state of affairs. In the last decades of the twentieth century acceptance of death seemed to have disappeared in the need to 'strive to keep alive', even when it was known that someone was dying. Now there is a slowly growing trend to turn this round and, especially for the old and terminally ill, dignify dying as a natural inevitability. To die at home may yet return to being the norm.

CONCLUSION

Not all infirm people require to be looked after but for those who do customs of caring for them at home in Scotland have changed greatly. We have seen that there was a time when home care was the regular practice even for the very ill and disabled. With large modern hospitals and medical advances this tradition broke down and a new culture of hospitalisation developed. Now 'care in the community' is a standard phrase and becoming normal practice. Many patients go into hospital for treatment but this is for as short a time as possible.

Yet being 'discharged into the community' does not always mean returning to one's own home. Options are sometimes possible which allow people to move into home-like situations where they can retain some independence (see Fig. 2.1). These include sheltered housing with a warden at hand or a 'granny flat' supplied by the family. For those without these possibilities there are residential and nursing homes. In 2002 there was formal recognition that care here and elsewhere must be of the best, and that personal care as well as nursing care should be paid for by the state.

It is also often considered desirable that those who are disabled in some way should live in their own home. In order that they may live as independent and full a life as possible they may also need support, sometimes given by the appropriate society in conjunction with the state. A good example of this is the Scottish Society for Autism which administers care, support according to individual need, day centres and housing for those who are in the autistic spectrum. There is now in Scotland a practice of client, purchaser and provider, where the Social Services (as the purchaser), the provider (for instance the Scottish Society for Autism) and the client (the person who requires care) work together as a team to provide the best outcome.[62] A new tradition of caring for the infirm in the home is emerging, with all involved working together.

NOTES

1. Encarta World English Dictionary, London, 1999.
2. Clarke, Kofsky, Laurol, 1989.
3. Cameron, 1998, 1.
4. Gaffney, 1982, 139.
5. Cline, 1995, 87.
6. Hamilton, 1987, 223.
7. Hamilton, 1987, 256; Chapter 36 Hospices and Nursing Homes provides information about infirmity in the modern hospital and residential-care setting.
8. Ferguson, 1958, ch 8, 436–508.
9. Pringle, 1912.
10. Ferguson, 1958, 439.
11. Brotherston, 1987, 36.
12. Brotherston, 1987, 54.
13. Durham, 1988, 17.
14. Hogarth, 1987, 177.
15. Jenkinson, 1998.
16. Bhopal, 2001.
17. Smout, 1986, 35.
18. Smout, 1986, 43.
19. Crowther, 1990, 284.
20. Johnston, McIvor, 2000.
21. Ferguson, 1958, 499.
22. Diack, 2001.
23. Fracassini, 2001, 7.
24. Clinical Standards Board for Scotland, 2001, 21. In 2001 it is known that approximately 120,000 people in Scotland have been diagnosed with diabetes and there may be as many as 90,000 more who are as yet undiagnosed.
25. Scottish Executive, 2000, 71.
26. Gilchrist, 2001, 6.
27. Ferguson, 1958, 32.
28. Registrar General Office (Scotland), 2000.
29. Scottish Executive, 2000, 63.
30. Morgan, 2001, 8.
31. Anderson, 1987, 369; for further information on poorhouses and the Poor Law see Chapter 37 Children's Homes; Chapter 47 Homelessness.
32. Anderson, 1987, 376.
33. Webster, 1991, 185.
34. Anderson, 1987, 378.
35. Pallister, 1998, 55.
36. Kelsall, Kelsall, 1986, 65.
37. Smout, Wood, 1990, 220.
38. Oral testimony, LR, 124.
39. Oral testimony, LR, 61.
40. Rodger, 1893, 253. The Rev Dr Alexander John Forsyth, minister of the parish of Belhelvie, Aberdeenshire from 1791 to 1843, was the parish vaccinator; see Chapter 27 Medicine for further information about professional and non-professional medical practitioners.
41. Ferguson, 1958, 459.
42. Buchan, 1994, 28.
43. Ferguson, 1958, 454.

44. Ferguson, 1958, 457.
45. Oral testimony, LR, 70.
46. Tait, 1987, 424.
47. Scottish Executive, 1999.
48. Scottish Executive, 2000, 63.
49. Scottish Executive, 2000, 61.
50. Featherstone, 1980, 82.
51. Craigmillar Health Project, 1999. The Craigmillar Health Project has been up and running since 1989. 'It has grown to become such a crucial part of the Edinburgh health network providing a framework for active participation in health by so many Craigmillar residents' (Introduction). The challenge for the project is to reduce the gross inequalities in health caused by poverty and disadvantage. It operates within the community, from a five apartment semi-detached house which has often been called 'The Healing House' by local people. Some of its many services include a drop-in crèche, workshops, courses, 'away-with-stress' groups, counselling, aromatherapy massage, acupuncture, reflexology, mid-years group, men's group and a dementia group.
52. Scottish Executive, 2000, 64.
53. The Scottish Council on Deafness publishes a newsletter which keeps deaf people in Scotland up to date with recent developments in the field: ScoD Newsletter, Clerwood House, 96 Clermiston Raod, Edinburgh EH12 7UT
54. McKay, Patrick, 1995, 8. The person using the alarm should keep it with him or her at all times.
55. Barrie, J. Peter Pan. In *The Concise Oxford Dictionary of Quotations*, London, 1981, 16; for further information on death practices see Chapter 30 Custom.
56. Surman, 1988, 289; Cline, 1995, 33.
57. Bass, 2001, 9.
58. Bennett, 1992, 192.
59. Oral testimony, LR, 91.
60. Bennett, 1992, 190.
61. Marshall, 1986, 29.
62. McKay, Patrick, 1995, gives an account of the system and the legal rights of the individual.

BIBLIOGRAPHY

Anderson, F. Geriatrics. In McLachlan, G, ed, *Improving the Common Weal: Aspects of Scottish Health Services, 1900–84*, Edinburgh, 1987, 367–82.
Bass, N. Where to die, *Health Summary* (May 2001), 9–10.
Bennett, M. *Scottish Customs from the Cradle to the Grave*, Edinburgh, 1992.
Bhopal, R. Minority Ethnic Health Inequalities, paper presented at a Health Inequalities Seminar, Royal College of Physicians, Edinburgh, 6 March 2001.
Brotherston, J. The development of public medical care, 1900–48. In McLachlan, G, ed, *Improving the Common Weal: Aspects of Scottish Health Services, 1900–84*, Edinburgh, 1987, 35–102.
Buchan, D, ed. *Folk Tradition and Folk Medicine in Scotland: The Writings of David Rorie*, Edinburgh, 1994.
Cameron, A. The Origin and Experience of Hospital Patients in Britain, c.1800–1948, M Phil in History, Wellcome Unit for the History of Medicine, University of Glasgow, 1998.
Clarke, P, Kofsky, H, Laurol, J. *To a Different Drumbeat: Parenting Children with Special Needs*, Stroud, 1989.
Cline, S. *Lifting the Taboo*, London, 1995.

Clinical Standards Board for Scotland. *Diabetes Draft Standards*, Edinburgh, 2001.
Craigmillar Health Project. *Working Together Towards Health and Wellbeing*, Edinburgh, 1999.
Crowther, A. Poverty, health and welfare. In Fraser, W H, Morris, R J, eds, *People and Society in Scotland, vol II, 1830–1914*, Edinburgh, 1990, 265–89.
Diack, L. Myths of a beleaguered city: Aberdeen and the typhoid outbreak of 1964 explored through oral history, *Oral History*, 29/1 (Spring, 2001), 62–72.
Durham, J. *Night and Day: Recollections of an Island Doctor*, Lerwick, 1988.
Featherstone, H. *A Difference in the Family*, New York and London, 1980.
Ferguson, T. *Scottish Social Welfare, 1864–1914*, Edinburgh, 1958.
Fracassini, C. Obese patients put strain on hospitals, *Scotland on Sunday*, 24 June 2001.
Gaffney, R. Women as doctors and nurses. In Checkland, O, Lamb, A, eds, *Healthcare as Social History: The Glasgow Case*, Aberdeen, 1982, 134–48.
Gilchrist, J. A most desperate measure, *The Scotsman*, S2, 6 November, 2001, 6; Help for those who need it: The Samaritans, 0131 557 4444; Theatre Nemo, 01355 331483; Papyrus 01706 214449; Penumbra 0131 475 2380.
Hamilton, D. *The Healers: A History of Medicine in Scotland*, Edinburgh, 1987.
Hogarth, J. General Practice. In McLachlan, G, ed, *Improving the Common Weal: Aspects of Scottish Health Services, 1900–84*, Edinburgh, 1987, 165–212.
Jenkinson, J. The Scottish Board of Health and the Department of Health for Scotland: Their creation and impact on healthcare, 1919–48, paper presented at the Wellcome Unit for the History of Medicine, the University of Glasgow, 3 November 1998.
Johnston, R, McIvor, A. *Lethal Work: A History of the Asbestos Tragedy in Scotland*, East Linton, 2000.
Kelsall, H, Kelsall, K. *Scottish Lifestyle, 300 Years Ago*, Edinburgh, 1986.
Leighton, A, ed. *Mental Handicap in the Community*, New York and London, 1988.
Marshall, D. *A Victorian Countryman's Diary*, Kinross, 1986 (Original diary from 1847).
McCrae, M. The Great Highland Famine of 1846: The lack of medical aid, *ROSC*, 14 (2001), 58–73.
McKay, C, Patrick, H. *The Care Maze* (Enable and the Scottish Association for Mental Health) (1995), review and update, Glasgow, 1998.
Morgan, H. Don't get too hopeful for a healthy old age, *The Scotsman*, Edinburgh, 9 July 2001.
Pallister, M. Soothing balm of alternative medicine. In, *NHS in Scotland, 1948–98*, Edinburgh, 1998, 55.
Pringle, A. *The National Insurance Act, 1911*, Edinburgh and London, 1912.
Registrar General Office (Scotland). *The Health of the Population of Forth Valley* (Forth Valley Health Board), Edinburgh, November 2000.
Reid, L (LR). Oral testimonies between 1997 and 2001, numbered 1–124.
Rodger, E. *Aberdeen Doctors*, Edinburgh and London, 1893.
Scottish Executive (Chair Sir Stewart Sutherland). *The Report of the Royal Commission on Long-Term Care of the Elderly*, Edinburgh, 1999.
Scottish Executive. *Our National Health: A Plan for Action, a Plan for Change*, Edinburgh, 2000.
Smith, R. *Life, Death and the Elderly*, London, 1991.
Smout, C. *A Century of the Scottish People, 1830–1950*, London, 1986.
Smout, C, Wood, S. *Scottish Voices, 1745–1960*, London, 1990.
Surman, L. Care of the dying. In Wright, S, ed, *Nursing the Older Patient*, London, 1988, 263–75.
Tait, H. Maternity and child welfare. In McLachlan, G, ed, *Improving the Common Weal. Aspects of Scottish Health Services, 1900–84*, Edinburgh, 1987, 411–40.
Webster, C. The elderly and the early NHS. In Pelling M, Smith, R, eds, *Life, Death and the Elderly*, London, 1991, 165–93.
Wright, S, ed. *Nursing the Older Patient*, London, 1988.

27 Medicine

JOYCE MILLER

HISTORY AND BACKGROUND

Since the dawn of the Scientific Age and the concept of rational thought – somewhere around the eighteenth century – medicine and the treatment of illness have been gradually removed from the domestic sphere and the concern primarily of family and local communities to increasing control by physicians and surgeons trained in the allopathic tradition of Western medicine. The powerful hegemony of the medical profession has encouraged the rest of society to believe that the medical approach employed now has always been used. This ignores the fact that medicine has had a long history of development. Ideas which were once acceptable – such as the doctrine of signatures, sympathetic medicine or the theory of the four humours[1] – have either been replaced by new theories and practice or have evolved into something quite different. What is also often forgotten is that although in the twenty-first century there appears to be one dominant system of medical theory and practice in Western societies, in the past this was not so and in the future it may not be so again. Indeed we are already seeing evidence for imminent plurality in the increasing acceptance and use of complementary or alternative medicine such as homeopathy, herbal medicine, acupuncture and chiropracty.[2]

Before the advent of the NHS in Britain in 1948, there were relatively few university-trained medical professionals and their services were, for the majority of the population, expensive.[3] The medicine offered by these people was based on classical medical texts such as those of Hippocrates and Galen, the fathers of ancient medicine. Some of the treatments recommended were unpleasant and often unacceptably painful to a modern-day audience. The basic principle was to restore humoral balance by purging the patient using expectorants, enemas, emetics and bleeding, followed by restoration with tonics and other supplements.[4] Patients could survive these practices, and some may even have improved as a result of their treatment, but it was not until the nineteenth and twentieth centuries that real advances and successes in medical and surgical treatments occurred, with the development of anaesthetics, antibiotics and improved surgical methods.

WHY USE DOMESTIC MEDICINE?

In the period before the Scientific Age, in addition to university-trained professional physicians, and professional 'quacks', there existed a wide range of non-professional healers, including folk healers.[5] Folk or domestic medicine was both more accessible and cheaper for the majority of the population. This is not to imply that it was a second-rate option. For centuries it was often the best available and indeed the only option. People diagnosed and treated themselves and others according to what was understood within the culture of the time, using either natural remedies – incorporating herbs and other plants as well as animal and mineral substances – or magico-religious systems of belief involving rituals and charms, or both in combination. Indeed there was little distinction made between natural and magico-religious medicine. Many of the rituals associated with the collection or use of natural substances could be termed magico-religious. It was, for example, believed that the power of herbs or other substances was improved by the timing of their collection – perhaps at sunrise or on a particular day – or when they were administered.

The scientific argument has often dismissed folk or domestic medicine on the grounds that it is backward, whimsical and based on superstition rather than on empirical knowledge, but in many aspects it evolved through man's capacity to observe and deduce. The fact that many so-called 'modern' chemical medicines are based on plant originals supports this view, as does the length of time for which some practices in folk medicine are recorded as having endured.[6] Research undertaken on medieval documents, including material related to magic and witchcraft, has revealed that some natural folk remedies being used in the twentieth century were based on medieval practices passed down through generations.[7] Other measures may have been adapted over time. Certainly a large number of the practices used in Scotland were to be found in the rest of Britain or Europe or in America. The problem is that many practices were based on a system of belief and understanding about diseases and disease causation that was, for the most part, replaced by scientific orthodoxy. As the twentieth century advanced, the rationale and explanation for these 'cures' were forgotten and it appeared that they were based simply on superstition. The study of folk medicine has attempted to reclaim its position as a valid and traditional method of treatment.

HOME HEALERS

Before the medicalisation of disease and treatment, the focus of care was the home and the local community, so medical advice could often be obtained from family, friends and neighbours. Domestic medicine was important not only for poorer members of society who could not begin to afford the fees of physicians and surgeons but also for the wealthy. Literate élite members of society would record the ingredients and efficacy of treatments in their diaries and often circulated them among friends and relations. An example from 1674 demonstrates the kind of home medical recipe that was all the rage in certain social circles:

> That medicine, which my mother used to give for a stitch in the side, is now the only thing prescribed by the king's physicians, and it is stoned horse dung, steeped in white wine some hours then drank, and I can assure you it is of such essence, for wholesomeness as ... drink it every year, all May long.[8]

More unpleasant than recommending vitamin C or cod-liver oil for colds, perhaps, but word-of-mouth or personal endorsement appears to have been just as common in the seventeenth century as it is now. Clergymen's diaries, and sometimes farmers' diaries, also provide information about domestic medicine. Ministers and others often recorded treatments and medical anecdotes, both folk and scientific, for posterity and were consulted for advice owing to their status as educated men.[9] There were also families who were particularly associated with healing and medicine. In the Highlands, the Beatons formed a distinguished medical dynasty whose members combined the learned medicine of Greek tradition with local folk tradition. Some of their number were employed by the Stuart monarchs.[10] However, members of ordinary households or local healers provided most folk with domestic medical advice. What made the latter special, if indeed they were?[11] They were individuals who possessed a reputation for having special knowledge or skills. They included the informally trained 'howdie'[12] or 'skeely' (a wise woman or man), plus those who had received a training in some non-medical field, such as blacksmiths and gardeners. In other instances, it was simply an accident of birth, a physical deformity, some thing or circumstance which marked the individual as different. Having red hair was sometimes sufficient. People who had breech births or were born with the caul[13] attached were usually regarded as special. The power of the seventh son of the seventh son is internationally recognised, and in Scotland such people were particularly associated with curing scrofula, the 'King's Evil' or lymphatic tuberculosis.

Bonesetters are a well-known type of local healer and appear to have obtained their role through inheritance rather than through any training in anatomy. Unlike the chiropractors of today who manipulate joints very carefully, bonesetters were often ineffectual in their 'putting in' of bones. Indeed, their treatment of fractures could be dangerous, and could lead to deformity or worse. Bonesetters were being consulted occasionally in Fife as recently as the 1980s, and there may still be some that claim to have the expertise, but it is hoped that their treatments have become more responsible.

The most common performers of folk or domestic medicine were ordinary household members, especially mothers, or more commonly, grandmothers, who made use of personal experience – sometimes with a measure of accident thrown in – and inherited knowledge. One example of the latter is the practice of the 'ballan', applying a cattle or sheep horn over an infected or inflamed site to reduce swelling, which was recorded in the

Hebrides and the Northern Isles,[14] knowledge of the remedy appearing to have been passed on from generation to generation. In addition, in many Scottish dwellings the housewife possessed a household book, which contained cooking and household management advice, as well as basic medical recipes. Other important sources of advice and information were lay manuals, some of which were available for literate members of society from as early as the sixteenth century.[15] Later self-help manuals included David Buchan's *Domestic Medicine*, which ran to many editions. Buchan first published his work in the eighteenth century, claiming that, 'Diffusing medical knowledge among the people would not only tend to improve the art to banish quackery, but likewise to render Medicine more universally useful, by extending its benefits to society.'[16] Indeed he pointed out in his introduction that it was medicine and the physicians themselves which were to be regarded as suspicious, even as quackery and quacks, in the late eighteenth century.

REGIONAL AND LOCAL VARIATION IN DOMESTIC MEDICINE AFTER THE EIGHTEENTH CENTURY

The practice of domestic medicine did not cease with the arrival of the Scientific Age. There was a lack of formally qualified practitioners in many more remote areas, and where present, few households could afford to make much use of their services.[17] Even after the creation of the NHS, folk medicine continued, and continues to be practised to some degree.

It has been held that folk or home medicine was unique to Gaelic-speaking areas of Scotland.[18] This is far from true. Much information about folk medicine and domestic cures was recorded in Gaelic-speaking areas by medical doctors in the nineteenth and twentieth centuries as part of the burgeoning interest in folklore.[19] Although such documentary evidence suggests that folk medicine may have continued to be widely practised in these areas for longer, it may also be the case that fewer people collected information from other areas of the country. Very similar practices were found in non-Gaelic speaking areas, as the material recorded by David Rorie from Aberdeenshire and Fife attests.[20]

Another commonly held misapprehension is that this form of medicine was found only in rural communities. Certainly there was longer retention of some practices in the countryside, probably because of the greater absence of doctors and other health workers. There have been assumptions, made especially in recent times, that rural or agricultural communities are more in touch with their surroundings and the cycles of life than urban populations. In fact both types of community had their own preferred treatments, many of which were appropriate to the local environment. In urban areas a child with a cough or cold would be told to follow men mending a road as the fumes from the molten tar were believed to encourage expectoration. In rural communities, taking some earth from a newly dug grave or taking the patient and laying him or her on the ground beside a grave was recommended for a chest condition. On Islay in the twentieth century, a local treatment for whooping cough involved taking the child over three boundaries or 'crìochan'.[21] Another industrial, rather than urban variant of this remedy was to lay the patient next to a mineshaft so that he or she could breathe in the fumes from the mine. Inhaling the fumes from a gas works was an alternative recommendation.[22]

As a health worker in former mining communities in Fife, the author, in the late 1980s, was told of what people used to treat their families. For children who were suffering from nausea, boiled and slightly cooled lemonade was the best remedy. If someone had been sick but was recuperating, then some 'brae o' the broth' – the liquid of a soup with no vegetables or pulses – was recommended. And indeed this treatment was one that the medical profession during the twentieth century would have recommended, when easily digested slightly salted stock, 'beef tea', was a popular form of nourishment for convalescents. Until the second half of the twentieth century, and before the development of more sophisticated antibiotics, in order to draw the infection from a boil or wound a poultice made of sugar or bread could be used. These practices were common knowledge, and people felt no need to go to a doctor to seek alternatives.

THE MATERIA MEDICA OF HOME MEDICINE: TREATMENTS AND SUBSTANCES THROUGHOUT THE CENTURIES

Home treatments have been diagnosed and administered by the patient or by another person, or a combination of the two. The use of rituals such as visiting holy wells or using charms, for example using rowan (mountain-ash) twigs to ward off witches, were popular forms of self-treatment and were often regarded as protective rather than curative.[23] Thomas Pennant noted in 1709 that farmers placed twigs of rowan and honeysuckle in their byres to protect their livestock from harm, often from witches or fairies, and sometimes humans.[24] Crosses made of rowan bound with red wool were still being used in the north east in the nineteenth century, and even at present rowans can be found in many gardens, perhaps 'just in case'.[25] Material charms or rituals which accompanied treatments could be deployed for any number of conditions or symptoms. Amber beads, worn around the neck, as well as being decorative, were believed to help eye diseases. There are other talismans, including silver heart brooches or special stones, which were believed to bring good luck or protection when worn by an individual (Fig. 27.1). The reputation of these talismans, such as the Lee Penny or the Clach Dearg (the Red Stone of Ardvorlich, Dumbartonshire), was often widespread and people might travel great distances to be treated with them[26] (Fig. 27.2).

Other charms were believed to cure specific ailments rather than provide general protection. These included a cure for toothache, which consisted of some words from the Bible written on a piece of paper. The paper was then folded, without being read, and worn around the neck until

Figure 27.1 Amulets. Top left: the Maclean Leug [Jewel], ridges and silver mounted from the Ross of Mull. Top right: the Glenorchy Charm Stone (in the possession of the Earl of Breadalbane). Centre: rock crystal, mounted in brass, Oban, Argyll. Bottom left and right: charm stones of rock crystal, silver mounted, which belonged to the family of Mackenzie of Ardoch. Source: SSS IIX 4 4033.

the pain receded. If the writing was read, then the healing power of the words would be lost. In the journal of Robert Landess, a minister in the seventeenth century, a cure for toothache consisted of frogs boiled in water and vinegar.[27] In Buchan's *Domestic Medicine* the recommended treatment for toothache was first to use purgatives, then scarify, or bleed the gums by applying leeches to the affected area, at the same time bathing the feet in warm water. He also suggested warm figs or poultices of herbs applied to the gum to relieve the pain. Another option was to apply a blistering plaster between the shoulders.[28] Since it was believed that toothache was caused by a worm in the tooth, a further cure involved carrying a worm in the mouth over the boundary between two parishes.[29] In the light of these cures, and although extraction of a decayed tooth was still available as a last option, the use of a piece of paper and some writing does not seem so bizarre, and certainly far more pleasant (Fig. 27.3).

Many other remedies have focused on natural treatments, alone or in combination with magico-ritual elements. Some are simple, others fairly complicated, some contain quite innocent ingredients, others much more unpleasant substances. From listing some of them it might seem that anything and everything could do: cold tea and bread poultices for boils;

Figure 27.2 The Lee Penny from Lanarkshire. The 'Talisman' of Sir Walter Scott's novel, in the possession of Major S F Macdonald Lockhart of Lee. Source: SSS IIX 8 4035.

onions round the neck tied in an old sock or a spoonful of sugary butter for sore throats; bathing sore feet or chilblains in urine. Wart lotion can be purchased over the counter but other ways to cure warts included spitting on them for seven or nine consecutive mornings or rubbing them with a potato, which would be left at a crossroads in the hope that whoever picked it up would develop the warts instead. Alternatively the sufferer might be able to persuade someone to buy them from him or her, although it is hard to understand why anyone would wish to do so. For a 'stye' or an eye infection a doctor would probably prescribe an antibiotic eye drop – for example chloromycetin – or bathing the eye with a proprietary product such as Optrex, but plain salted, boiled water is a practical, sensible and cheaper home remedy. Another, more unusual method was to rub the eye with a gold wedding ring, although that might seem both less hygienic and less acceptable to us.

There are numerous proprietary cures for indigestion which can be purchased without prescription, but it can also be treated with the much cheaper option of a spoonful of bicarbonate of soda. The traditional aphorism 'starve a fever, feed a cold' is not an empty saying, but is quite sensible advice. During fever a patient cannot tolerate heavy food, but it is vital to maintain an adequate fluid intake, especially if there has been a lot of sweating. During the early modern period other measures were used with fevers, some to induce sweating. On Skye if a patient had not managed to sweat, then his or her shirt was boiled in water and then put back on. The patient was also encouraged to take dulse, a type of seaweed, boiled in water, which would not only encourage sweating but also ensure that the patient maintained urinary output. Another drink, given after a fever, was whey in which violets had been boiled. This was both cooling and refreshing.[30]

Whooping cough is a distressing and dangerous illness. In past times a child suffering from whooping cough could be passed under the stomach of a donkey, taken to a mine shaft to breathe the air from the mine or placed over a newly dug grave.[31] The reason behind these measures was to ensure a change of air, albeit that the 'new' air might not seem particularly healthy to modern observers. In the case of the donkey cure, another explanation is that it was hoped the disease would transfer itself to the donkey. In Scotland, as elsewhere, it was commonly believed that illness or disease was the result of bad luck or the evil eye. Illnesses were transferred onto a person either by accident or deliberately and in order for a cure to be effected it was necessary for the illness to be transferred to someone, or something, else.

Another remedy for whooping cough was to visit special healing wells (Fig. 27.4). In Scots whooping cough was known as 'chincough' or 'kinkhost', and there were several wells around the country which were associated with the disease. The Kinkhost Well near Comrie in Perthshire and the Chincough Well near Glasserton in Dumfriesshire are two examples. The Spring of the Minister (Fuaran a' Mhinisteir) at Petty near Inverness was a sea well and the Dripping Well (Tobar nan Dileag) at Tillichuil in Perthshire was also believed to cure whooping cough. In Buchan's *Domestic Medicine*

Figure 27.3 Another way of curing toothache. A nail hammered into the stone is thought to take away toothache. The Toothache Stone, Islay. Source: SSS JIX 4 7304 Photo I Fraser.

Figure 27.4 Certain wells were believed to cure other diseases. This is chalybeate well at West Linton, Peeblesshire, which was thought to cure tuberculosis. Note the cup. Source: SSS JIX 7b 6944 Photo A Ross.

recommended treatments for the chincough concentrated on fluid intake and maintaining regular bowel openings. However, the volume also contained the advice that millipedes or wood lice, infused in white wine, should be strained before being given as a drink.[32] Buchan further suggested that garlic rubbed on the soles of the feet, or spread on a rag and applied as a plaster, would help the condition. Another method used in the twentieth century was to expose children to infectious diseases in the hope that they would develop immunity. Children were therefore brought together when there was an outbreak of measles or chickenpox, which were regarded as children's illnesses and a normal part of childhood. There is no doubt that in the case of whooping cough and measles these home remedies would be no substitute for the modern option of immunisation by inoculation.

For aches and pains of the rheumatic type, carrying a potato in the pocket was regarded as a useful cure as it was thought to draw pain out of the blood. Thrashing painful joints with nettles was also believed to reduce the pain, perhaps as a form of distraction from one type of pain to another. Further treatments included wearing red flannels around the affected joints or rubbing into the skin the oil from boiled slugs.[33] Sulphur flowers or brimstone sewed into a cloth and worn near an affected joint was believed to reduce rheumatic pains or prevent cramp, and was especially useful for lumbago when used with a blue flannel.[34] For sciatica Martin Martin described applying spirewort or *Flamula-jovis* to the thighbone in order to create a blister. The blister would be drained three times and the cure would be complete.[35] A modern remedy, which echoes some of the older ones, is the use of copper bracelets to prevent or reduce joint pain.

Although the domestic treatment of mental health is perhaps more difficult to investigate, a number of practices have been recorded, demonstrating that recognition of emotional or mental illness occurred well before Freud. As with physical conditions, families and households bear the brunt of caring for emotionally or mentally disturbed members. This was no different in the past and before the development of lunatic asylums, from the eighteenth century, and especially psychiatric hospitals in the twentieth century, families attempted to seek cures for afflicted relatives. These cures took a range of forms, and to the modern reader may seem rather shocking. For example, on his tour of the Western Isles, Martin Martin noted a cure for 'faintness of the spirits'. The blacksmith in Kilmartin parish on Skye would lie the sufferer on the anvil, facing upwards. He would then pick up a large hammer in both hands and draw it back, as if he intended to give the patient a blow on the head. As he brought the hammer down he would direct it away at the last moment. It was claimed that this struck so much terror into patients that they were cured of their depression.[36]

There are also several locations, some associated with Christian saints, which seem to have been used in the curing of lunatics. A healing well on an island in Loch Maree, Ross and Cromarty, which was dedicated to Saint Maelrubha, was thought to be particularly effective in helping the insane. St Fillan's pool and chapel at Strathfillan, Perthshire, were also believed to help those suffering mental or emotional distress. The sufferers were immersed in the pool, then carried to a nearby chapel, where they

Figure 27.5 St Fillan was associated with other cures, such as those involving these healing stones kept at Killin, Perthshire. Source: SSS I X 5c 6891.

were secured with ropes or tethers. The bell from the chapel was then placed over their head, and they were left alone overnight. In the morning if the sufferer was found with the ropes untied then there was some hope of recovery. However, if the ropes were still in place then the future looked less bright[37] (Fig. 27.5).

Even relatively mild emotional and mental disturbances, such as being troubled by 'one's nerves', appear to have been recognised by society in the past and attempts made to ease suffering and generally help people cope. The rationale behind the attempted cures is perhaps more challenging for us to understand than some other domestic medicinal treatments, and we do not know how successful they were, but for those involved they at least offered some hope in the midst of distress.

THE RENEWED INTEREST IN TRADITIONAL TREATMENTS AND SUBSTANCES

Domestic remedies may strike us as being somewhat quaint and maybe even humorous. Some of the medical profession argued, and still do argue, although opinion is very much divided, that they are irresponsible. But little significant harm could be done by drinking slightly warm, flat lemonade or beef stock, for example, although this might not always have been the case with all home remedies. Household or domestic remedies have also been rejected as old wives' tales or placebos, with no scientific evidence to prove their efficacy. And clearly some did not work. However, modern medicine has started to recognise that a number of older treatments do prove effective. Leeches for example, associated with excesses of bleeding in the medieval and early modern period, are being applied to areas of venous congestion or clotting in order to re-establish blood circulation. Dressings of sugar or honey are once again being applied to wounds and burns in order to create a low-water environment which inhibits bacterial growth. One of the fathers of Renaissance surgery, Ambroise Paré, noted in the sixteenth century that injured soldiers applied maggots to their wounds in order to keep them clean. Sterilised maggots were applied to wounds in the 1930s and 1940s, but their use declined as antibiotics were developed and introduced. Now, with the overuse of the newer, extremely powerful antibiotics, some organisms are largely resistant to treatment with drugs and maggots are once again being used to treat necrotic tissue, eating the dead flesh from wounds while their enzyme production has been shown to have antimicrobial or protective properties.[38]

Interest in herbal medicine is currently increasing and it is acceptable today to use, for example, camomile for poor digestion, bramble for bronchitis, eyebright for eye washes, fennel for stomach complaints and urinary problems, garlic for urinary troubles and protection of the heart and circulation, or onion for toothache, all as a first option rather than going straight to a doctor.

The acceptability or efficacy of treatments may not be the only reason that some people continue to prefer to use domestic medicine. There is the matter of convenience, and the sense of comfort which traditional knowledge provides. Or it may be the case that the individual lacks confidence in the efficacy of modern medical services, or simply that he or she finds going to the doctor or nurse embarrassing. Going to see medical professionals or calling them out, even after the introduction of the NHS, can induce a range of anxieties. In the past, hesitation in approaching practitioners was clearly related to the cost, as a doctor's visit was an expensive matter, but it has also been to do with the relationship between doctor and patient. Doctors and other health professionals including nurses – like priests or ministers – have an almost unquestioned access to the innermost aspects of an individual's life. Over the years an unequal relationship has developed, in which the doctor leads and advises and the patient is expected to follow unquestioningly. Although this may well be changing now to an ambience of mutual participation, as patients become confident in demanding to know more about their illnesses and treatments, it has nevertheless influenced society's attitude to professional medicine. The consultation with the doctor or other medical professional can be an unsatisfactory experience as the language, or jargon used often deliberately excludes the patient and his or her family. Some questions are not asked and answers given may not make things any clearer. The professional medical practitioner still makes home visits, giving the patient the advantage of being on 'home territory', but at the same time perhaps creating a source of stress among those who feel it necessary to tidy up and generally make the dwelling 'presentable' before receiving visitors. In all events, household members and relatives have done the majority of caring for those not sick enough to require hospitalisation. It is therefore not surprising that some people regard professional medicine as external to daily life, beyond their control, and view it almost as a last resort.

Clearly domestic medicine has constituted an acceptable form of medical care and advice to much of the general public. Despite scepticism from the medical profession it has survived. It is still an important, and probably the original, alternative 'complementary' form of medicine. Very few people go to their doctor with minor to moderate problems without trying something at home first, perhaps a suggestion from a home medical book, a web site (including the NHS' own), a phone line such as 'NHS Direct', or some advice from the local pharmacist, a friend or a relative. Domestic medicine may have changed considerably over the centuries, but not entirely. The treatments for some conditions are now very different, while others continue to preserve something of the essence of the past. What is pleasing is that domestic medicine's very survival demonstrates that it is a dynamic and vigorous option.

Domestic medicine also reflects the cultural diversity of Scottish society. Chinese medical practitioners are becoming well established and increasingly consulted, not only by members of Chinese communities but by those from other parts of society as well. Health advice, including healthy-eating campaigns which can be carried out at home, supported by the professional medical establishment, also reflect these cultural changes. A Mediterranean-style diet, including plenty of fresh vegetables, olive oil and garlic, is seen as healthier than a more traditional Scottish diet involving large quantities of starch and animal fat. Scotland is a multicultural nation. It has always drawn on ideas and practices from other cultures and, with changing population demographics in the twenty-first century, this cross fertilisation is bound to continue. Domestic medicine has, therefore, many reasons to remain in a healthy state.

NOTES

1. The four humours were blood, yellow bile, black bile and phlegm and were regarded as fundamental to the working of the body. It was believed that any disorder in their balance would result in illness and treatments were designed to restore harmony between them.
2. There is a vast range of complementary medical treatments on offer; some of these are available on the NHS but others need to be paid for directly by the patient. There is also a proliferation in the number of self-help manuals which can be purchased. A glance at the 'Mind, Spirit and Body' or 'Medicine and Self-Help' sections in any bookshop can be quite overwhelming at times.
3. An important exception to this is to be found in the Highlands and Islands Medical Service, created in 1913, 'the first comprehensive and free medical service in Britain and the forerunner of the NHS 35 years later', McCrae, 2001–2, 71.
4. Expectorants encouraged coughing, enemas cleared the bowels and emetics caused vomiting. See Dingwall, 1995, 123–8, for a discussion of early modern medical practice.
5. Hamilton, 1987, 21, states that 'it would be a mistake to look for an organised medical profession in medieval times ... [as] everyone was a doctor.'
6. Hand, 1971–3, discusses international healing customs and beliefs.
7. There are a number of works which cover the history of medicine in Scotland. These include Hamilton, 1987; Dingwall, 1995; and see also Beith, 1995.
8. NAS, GD 406/2669.
9. See for example, NLS, Acc 6605, John Knox's medical recipes, 1688–9; MS 548, Journal of John Landess; MS 1668, Memoirs of John Brand, minister.
10. Bannerman, 1986.
11. Hand, 1971.
12. See Chapter 24 Childbirth; Chapter 25 Childcare and Children's Leisure; Chapter 26 Infirmity.
13. If the caul or amniotic sac was still attached to a baby at birth it was believed to give special healing powers or was kept as a talisman to protect against drowning.
14. Mitchell, 1897–8 describes the use of a ballan in Northmavine in Shetland. It was placed over the affected joint after having been heated. As the air cooled, a vacuum was created and a large blister would appear which exuded blood. It was still being used in the islands of Skye and Lewis in the 1940s and 1950s. See also Cheape, 1996–7.
15. *Ane Gude Boke of Medicines*, from 1595, which includes a transcription of Petyt, Thomas. *The Treasure of Poor Men*, Edinburgh University Library, Dc.8.130; see also Comrie, 1832, 188–9.
16. Buchan, 1800, xxiii.
17. McCrae, 2001–2.

18. An excellent and up-to-date account of the history and materia medica of Gaelic medicine is found in Beith, 1995.
19. Masson, 1887–8; McPherson, 1929.
20. Buchan, 1994.
21. SSS, SA 1969/28.
22. In the southern states of the USA during the early twentieth century a leather string tied around the neck of the child was used as a treatment for whooping cough. See Digby, 1997, 294.
23. Miller, 2000, 27–52, discusses and lists many of the healing wells which were visited in Scotland.
24. Pennant, 1774, 171.
25. Dalyell, 1834, 9. The NMS has examples of crosses of rowan which were made in Pitsligo in the nineteenth century.
26. The Lee Penny is still in the possession of the Lockhart family. It was used to cure disease, particularly animal illness, well into the seventeenth century. It is claimed that it was used during an outbreak of plague in Newcastle in 1645. The Clach Dearg is now held in the collection of the NMS, Edinburgh.
27. NLS, MS 548, Journal of Reverend Robert Landess.
28. Buchan, 1800, 358–9.
29. Rorie, D. In Buchan, 1994, 112.
30. Martin, 1994, 223.
31. Rorie, D. In Buchan, 1994, 105.
32. Buchan, 1800, 287.
33. Rorie, 1994, 106.
34. Rorie, 1994, 39. Brimstone was widely recognised in relieving cramps and rheumatics.
35. Martin, 1994, 225.
36. Martin, 1994, 148.
37. Miller, 2000, 34.
38. Church, 1996; Courtneay, Church, Ryan, 2000.

BIBLIOGRAPHY

Bannerman, J. *The Beatons: A Medical Kindred in the Classical Gaelic Tradition*, Edinburgh, 1986.
Beith, M. *Healing Threads: Traditional Medicines of the Highlands and Islands*, Edinburgh, 1995.
Buchan, D, ed. *Folk Tradition and Folk Medicine: The Writings of David Rorie*, Edinburgh, 1994.
Buchan, W. *Domestic Medicine*, 17th edn, London, 1800.
Cheape, H. Cupping, *ROSC*, 10 (1996–7), 135–8.
Church, J C T. The traditional use of maggots in wound healing, and the development of larva therapy (biosurgery) in modern medicine, *Journal of Alternative and Complementary Medicine*, 2/4 (1996), 525–7.
Comrie, J. *History of Scottish Medicine*, London, 1832.
Courtneay, M, Church, J C T, and Ryan, T J. Larva therapy in wound management, *Journal of the Royal Society of Medicine*, 93/2 (2000), 72–4.
Dalyell, Sir J G. *The Darker Superstitions of Scotland*, Edinburgh, 1834.
Digby, A. The patient's view. In Loudon, I, ed, *Western Medicine: An Illustrated History*, Oxford, 1997, 291–305.
Dingwall, H. *Physicians, Surgeons and Apothecaries: Medical Practice in Seventeenth Century Edinburgh*, East Linton, 1995.
Gregor, W. *Notes on Folklore of North-East Scotland*, London, 1882.
Hamilton, D. *The Healers: A History of Medicine in Scotland*, Edinburgh, 1987.

Hand, W. The folk healer: Calling and endowment, *Journal of the History of Medicine*, 26 (1971), 263–78.
Hand, W. Folk curing: The magical component, *Béaloideas*, 39–41 (1971–3), 140–53.
Loudon, I, ed. *Western Medicine: An Illustrated History*, Oxford, 1997.
McCrae, M. The Great Highland Famine and the lack of medical aid, *ROSC*, 14 (2001–2), 58–73.
McPherson, J M. *Primitive Beliefs in the North-East of Scotland*, London, 1929.
Martin, M. *A Description of the Western Islands of Scotland c.1695* (1698), facsimile edn, Edinburgh, 1994.
Masson, D. Popular domestic medicine in the Highlands fifty years ago, *TGSI*, 14 (1887–8), 298–313.
Miller, J. *Myth and Magic: Scotland's Ancient Beliefs and Sacred Places*, Musselburgh, 2000.
Mitchell, A. Description of some Neo-Archaic objects, *PSAS*, XXXII (1897–8), 181–94.
Pennant, T. *A Tour of Scotland in 1709*, 3rd edn, Warrington, 1774.
Rorie, D. Disease and its cure. In Buchan, 1994.
Rorie, D. Some Fifeshire folk-medicine. In Buchan, 1994.

28 Hierarchy and Authority

LYNN JAMIESON

INTRODUCTION

It is tempting to characterise the history of family life as the gradual softening of hierarchies of gender and generation. The claim would be that, as the centuries progressed, the rules of traditional loyalty and authority which stipulated the deference of younger generations to older, of women to men and the subordination of individual needs to 'family honour' have eroded away. It is as if the space of traditional values has been progressively reduced by the ascendancy of democratic individualism, the expression of individual and equal rights. But social reality is not so simple, history is not so linear and social change is rarely mono-causal.

The variation in the meaning and experience of 'family' at any one point alerts us to the potential complexities involved in understanding its change over time. 'Family' can be experienced very differently by two people in the same family household, and divergent understandings of how to be 'a family' can live side by side in the same society. In any case, Scotland has never been a completely homogeneous society in terms of everyday cultural practices. In some historical periods social divisions have included very marked differences between social classes. Social, cultural and economic variance has always been associated with topographical and regional differences, for example between Highlands and Lowlands, Northern Isles and Western Isles, between the north-east Lowlands and the Borders, and between the cities. It may be that such variation rules out the possibility of ever talking about a 'typical Scottish family' in any period. There is certainly much cause for being wary of that expression in the present. We nevertheless have good reason to continue to explore what we know about family life and domestic environments in the past.

This chapter focuses on hierarchy and authority in family life in Scotland, and in particular hierarchy and authority in family households. In public debate and anxiety about changes in family life, the issues of equality between partners and of parental authority over children are recurrent themes. For some, the perceived erosion of traditional authority in these domains causes not only the demise of 'the family' but also the weakening of the moral fibre of our society. However, for others equality in these aspects of life is proof of our progress to a more humane and just world. Claims about the past are implicit and sometimes explicit in both positions. The two

Figure 28.1 Group family portraits sometimes depict family hierarchies as well as particular occasions. Consider who is in the centre of the picture, who is standing and who is sitting. In this picture the oldest son is in the centre with his father sitting to his left and his mother standing in the row behind. Source: Lynn Jamieson.

Figure 28.2 Golden Wedding, Turriff, Aberdeenshire, 1910. Source: SLA 95(1) 56/28/20.

main sections of this chapter discuss relationships between partners, usually husbands and wives, and relationships between parents and children. While the dynamics of gender equality in family households have some bearing on our attitudes to children and their position in the family, it is not unreasonable to treat these two central aspects of family life as having distinct histories. There is much that could be said within a larger chapter about other relationships within households – those between co-resident grandparents and younger generations, the significance of visiting aunts and uncles to their nieces and nephews, the impact of non-kin such as lodgers or servants living in the household, and relationships of fosterage[1] and adoption, for example. However, not all can be covered adequately here. The aim is to give a flavour of what we know about change in hierarchy and authority in Scottish family households from the seventeenth century to the present, although, as the first two sections briefly explain, our knowledge of the earlier part of the period is rather thin. It is more descriptive than analytical, arranging food for thought rather than digesting it for the reader[2] (Figs 28.1, 28.2, 28.3).

Figure 28.3 Cottage interior with figures, by William Shiels, 1841. Source: SLA 128A(3) C.199775.1699.

HOUSEHOLDS AND FAMILY HOUSEHOLDS

Notions about the disadvantages and benefits children might gain from being brought up in institutions have waxed and waned historically.[3] When average life expectancy at birth was around 30 and welfare systems as we now understand them had barely developed, the numbers of orphaned, ill-cared for and informally fostered children were no doubt very significant. Nevertheless, across the period of this study the majority of children in Scotland have grown up in family households. It is important to state what is meant by 'family household'. Households are about people living together, not simply sharing a dwelling but also to some degree pooling resources, such as eating together or being dependent financially upon a common purse. In this sense there are many types of household which are not families. Students sharing the rent, putting money in a kitty and taking turns at shopping and cooking are a household, albeit in a temporary arrangement.[4] Migrant labourers sharing a bothy might become a household if they pool food, cook and eat together. Moving between a family household and temporary accommodation shared with other migrant workers was a common seasonal pattern for many adults working in farming and fishing in previous centuries.[5]

Family households are people living together who would be regarded as a 'family' or who think of themselves as 'family', whether or not they would be legally recognised as such. In the first sense this might include a range of kin living together, for example, brother-and-sister households, children being brought up by older siblings, grandparents, aunts or other relatives and elderly people cared for by a resident child or another relative, as well as more conventional groupings composed of married couples, and parents and their dependent children. The second sense includes people not related by blood or marriage, who nevertheless define themselves and live as a family, for example those involved in informal adoption or foster arrangements. All of these family household arrangements have existed and continue to exist, although more conventional family households have been by far the largest group.

The dominance of the family household has fluctuated historically. The Scottish Household Survey, a research tool used in the twenty-first century, lists the following eight types of households: single adult and single pensioner (one adult, respectively above or below retirement age), small adult and small older (two adults, respectively both below or at least one above retirement age), large adult (three or more adults), single parent (an adult and at least one child), small family and large family (two adults and, respectively no more than or more than two children). Only households with two adults and children are labelled 'family' in this categorisation but it is likely that many more contemporary households think of themselves as families. For example, research indicates that most adults and children in single-parent households define themselves as a family, and some couples living together think of themselves as a family despite the absence of children. However, it remains common to see dependent children as the defining characteristic of a family, although 'large adult households', made up of a couple and their adult children, are unlikely to stop seeing themselves as a family just because the children have grown up. Nevertheless, the proportion of Scottish households defining themselves as families has probably declined since the mid twentieth century. Longevity and lower birth rates have combined to create many more small households and households without children, and people living alone[6] are a growing proportion of households across the adult life course. The average age at which mothers now have a first child is high and the period of continuing to have children is relatively short. In 2003, it was only in the age group 35–44 that the majority of Scottish adults were living with children. Only 27 per cent of Scottish households contained children, and only 4 per cent had three or more children.[7]

The composition of family households has always been varied but the nature of the variation has changed historically. There have been historical periods in which lodgers, servants, apprentices and others who were not kin were more common in households. In the nineteenth and into the early twentieth century, it was the norm for the growing middle classes to have at least one resident domestic servant and in earlier times apprentices might live with their masters.[8] At the same time some forms of co-resident kin beyond the nuclear family of parents and children may now be more uncommon than in previous periods, with stepfamilies or 'blended family' of partnered adults and their children by other relationships perhaps the exception as this is a growing form of family household.

Three-generational households, however, appear to be less common than in some previous periods (see Fig. 31.1), except among certain ethnic minority groups, which form a small proportion of Scotland's present population. Three-generational families have never been the dominant form of family household. Michael Anderson's work on nineteenth-century Preston, Lancashire, suggested that older people living with their adult children became a somewhat more common phenomenon with the development of the old-age pension, which meant poorer people could afford to care for their elderly relatives. In the oral histories which the author has been involved in gathering, the clearest account of a three-generational family was that given by Mrs Gillies, born in 1899 on the Isle of Lewis to a relatively well-off crofter household and the oldest of eight children. She describes her grandfather as the head of the household and comments, 'They didn't throw their old people out, ever ... there was this great respect for old age. Not chucked into homes.'[9] Although her grandfather was the formal head of the household and Mrs Gillies' mother had moved into the house of her parents-in-law on marriage, in practice the grandparents may have stepped back and allowed their oldest son and daughter-in-law to run their affairs as they wished. Mrs Gillies' relationship with her grandmother suggests a closer and less deferential relationship than she had with her mother. 'I just adored her. When I wanted to ask any awkward questions, like when your periods came ... I remember going to my granny.' To have two grandparents still alive at puberty would have been relatively unusual in this period, particularly for children who were not the first born. A substantial proportion of the population surviving to older old age, beyond 85 years old, is a relatively recent phenomenon. As this age group has grown in number across the twentieth century, so the proportion who are co-resident with their children has declined. Few children in Scotland live with both a parent and a grandparent. The loss of a parent from the family household, usually the father, is far more common than living with a grandparent.

The most common cause for children's experience of family household change in the twenty-first century is divorce or separation. But children's exposure to radical change in the composition of their household is unlikely to be greater now than in previous centuries when death provided very frequent disruptions to family life. A radical recent change has been in a shift away from marriage as the starting point for creating a new household. It is now normal for a couple to live together before they marry and this sometimes results in children being born prior to marriage and marriage being permanently deferred. In 2003, 46 per cent of births in Scotland were to mothers who were not married, but in the majority of cases the child's mother and father lived at the same address. In only 6 per cent was this not the case.[10] Children brought up by unmarried parents are no longer stigmatised as illegitimate and the notion that the sins of parents are somehow stamped on their children no longer has potency.

FAMILIES, HOUSEHOLDS AND SOURCES OF DATA

Unfortunately only very partial insight can be gained into the everyday quality of household dynamics and tenor of familial relationships in centuries remote from our own. Historians have long lamented the fact that most 'ordinary' people left no personal records in their seemingly poor, short lives. Many had no means to write biographies or even short accounts of themselves, and if they did manage to write letters (or biographies) these have usually not survived. The technologies for tape-recording voices and capturing images on film and the systematic social-science survey are relatively recent phenomena. In general, those who have left written accounts of everyday life are atypical. The wealthiest and most privileged were the most likely to write and also preserve personal records about themselves, and those who additionally had a particular reforming mission were the most likely to provide written commentary on others.

The most solidly evidenced based insight of a more general kind is provided by demographic work, a painstaking building up of facts about births, marriages and deaths, and patterns of residence. Religious registers long pre-date the public records and the censuses of populations adopted by the state in the nineteenth century. While historians can skilfully interpret trends in ages of marriage, patterns of childbirth, household composition and migration to build a picture of family life, the snap shots which result can also be misleading, missing much of the fluidity and flux of everyday life. Information on many topics remains elusive. For example, a great deal about the nature of power and authority in family households among the 'ordinary folk' or among wider kin in early periods is simply not known.

FAMILY LIFE UNDER THE CLAN SYSTEM

The Gaelic term 'clann' has the primary meaning of 'children', but it is the secondary meaning of 'kindred' which has given its name to the clan system. A clan is a patrilineal descent group with a ruling family or lineage at its core but the term is used more loosely to mean individuals and groups affiliated to the family, although not by blood.[11] It is a system of alliances and feuds which dominated the Highlands until the eighteenth century, in which the chiefs were men and privileges passed through the male line. Little has been written about the personal dynamics of hierarchy and authority across gender and generation during the clan system. Such issues have not been at the centre of the interests of historians undertaking much of the work in this area and information about relationships between husbands and wives or domestic partners, parents and children is scant. We do not have much sense of whether and how everyday distinctions were made between family households, 'close family' and wider kindred and clan, and it is not clear how notions of family mapped onto the household arrangements in which people lived.

That said, there is some reason to presume that a degree of separation between close family and the wider clan was part of the experience of life of the time. A sense of collective rights and loyalties as a clan or kindred seems to have co-existed with more narrowly familial privileges, which were patrilineal, passing from father to son.[12] MacGregor notes two factors important for the ascendancy of a new clan: expansion into new territory; and growth in the population of the kindred. The control of women was important to both of these, with likely effects on the dynamics between husbands and wives, fathers and daughters. Territorial expansion could be achieved by raids on neighbours but marriage alliances were also an important means of consolidating control of land. With respect to the latter, MacGregor refers to 'Gaelic Scotland's relaxed attitude to marital and sexual matters, which permitted rapid multiplication rates'.[13] This is said without any comment on the consequences for men and women. Other commentators indicate circumstances which make it unlikely that women enjoyed as much freedom as men to pursue sexual pleasure in spite of marital arrangements. Stiùbhart notes that 'Even more so than in Ireland, the culture of the early modern Scottish Gàidhealtachd was imbued with a strong patriarchal ideology.' It was not only that power and land were in male hands,[14] there existed also a masculine culture which celebrated male violence.[15]

At the same time commentators on clan culture have emphasised the familiarity which co-existed across the social hierarchy. It seems unlikely, however, that this recognition of common cause between unequals was a form of egalitarianism which moderated social divisions between men and women, parents and children. In describing how those outside the clan system could be puzzled by familiarity between unequals, Grant notes that 'The snobbish Boswell wondered to find that the country people of Coll shook the hand of the son of the Laird.'[16] Boswell does not make it clear, however, whether men and women, old and young had an equal right to shake the laird's hand. The co-existence within the clan system up to the eighteenth century of a clear patrilineal hierarchy and more collective claims of kinship had its parallel in an earlier period of lowland culture. In the Lowlands of the Middle Ages, although control of land was dominated by a feudal system, the kin of the lord, even if 'little more than peasants ... had a right to family protection'.[17] However, this recognition of kindred across social boundaries was clearly not an ethos of equality which threatened to undermine hierarchy. Rather loyalty was generated for the hierarchical system by a sense of the right to share the benefits gathered by the clan chief or the feudal lord.

HUSBANDS AND WIVES: HIERARCHY AND GENDER EQUALITY

The power play between men and women living together as couples can only be understood in the context of how wider society denies, creates or supports equality between men and women. In early centuries, many powerful institutions gave support to a hierarchy in which men were above and had authority over women. Much of this institutionalised sexism has now been dismantled. Even when women were legally the lesser sex, everyday realities were often rather more complex and sometimes turned this order on its head. The traditional authority of men over women was undermined by the combined forces of economic change – transforming opportunities for making a living and the economic independence available to men and women – and ideological/political change, including the demand of people for universal suffrage and of rights for women[18] (these forces played out rather differently for men and women in different social and economic positions). However, even in the twenty-first century, domestic equality between men and women is more of an ideal than a reality for many. This section aims to illustrate these claims (Fig. 28.4).

Until the twentieth century, across many societies, Christian teaching and legal systems typically sanctioned men's rights to exercise power and authority over women and children, particularly as heads of households, husbands and fathers. In the period 1500–1800, married women in Scots law had more legal protection as individuals than in many other parts of Europe but, nevertheless, had fewer legal rights than their husbands, widows (who had not remarried) or single women over the age of 21.[19] For example, if a widow married, her husband gained power over property which was previously hers. He could dispose of as many of her goods as he saw fit and

Figure 28.4 Husband and wife on an afternoon jaunt to Cove, near Cockburnspath, Berwickshire, c 1900.
Source: SLA 95(1) C.19472.

although he could not do the same with her land without her permission, she no longer had the right to dispose of it without his say. Leneman describes how, in the late seventeenth century, Margaret Seaton, daughter of a skipper in Leith, applied to the courts for an 'adherence' order requiring her deserted husband to return from Holland, although her motive was not to get him back but to register his absence. When she married she was a merchant in her own right but as a married woman business transactions required the permission of her husband. Leneman concludes, 'The decree of adherence from the Commissary Court (in 1696) could not have had any force in Holland, but presumably in some way entitled her to carry on trading without her husband.'[20]

From 1560 in Scotland, men and women formally had equal access to divorce on grounds of adultery (a situation of gender equality not achieved in English law until 1923) and from 1573 for desertion.[21] Prior to the Reformation marriage was regarded as a sacrament which could never be broken. Leneman states that for the architects of the Scottish Reformation, 'The right to divorce had to be the same for both sexes for, in Calvin's words,

"A man may hold the primacy in other things but in bed he and his wife are equal."'[22] This formal equality did not counter-balance the legal and financial inequalities between men and women which left the latter with much less room for escape from marriage. Indeed few women in oppressive marriages had any possible legal or financial means of escape. From the eighteenth century onwards, suffering from violence or cruelty could be grounds for a legal separation but was not grounds for divorce until 1938. A wife suffering abuse could seek legal separation from the necessity of sharing bed and board and be granted a yearly aliment for the rest of her life, but not a divorce. However, only exceptional upper-class women with economic support from their natal families would have the resources to make a complaint. Legal options were in any case desperate measures, not only because of the costs involved, but also because scandal, shame and stigma remained attached to separation and divorce until divorce rates reached contemporary levels in the last decades of the twentieth century.

Husbands also had a customary right, sanctioned by law and Church, to control their wives and others in their households by chastisement, and this included physical punishment.[23] How severe this could be without attracting condemnation no doubt varied according to time and place. There are records of women bringing successful legal complaints of severe and repeated beatings by their husbands as early as the seventeenth century and in the eighteenth century the courts granted legal separations in a series of cases of such violence.[24]

In Scotland the law never defined rape in a way which denied that description to sex violently forced on a woman by her husband. However, this made little difference in terms of women's recourse to the law, as it was unknown for wives to bring rape charges against their husbands until the twentieth century, and even then they typically failed to acquire conviction. For the majority of women who suffered abuse and violence and could not afford recourse to the law, abandonment of the relationship was only an option if they could find another household – that of family, neighbours or friends – willing and able to take them in despite their limited ability to contribute economically and the associated risk of trouble from their husbands. While local communities sometimes attempted to discourage violent men from abusing their women through such shaming and intimidating practices as 'rough music' (a crowd gathering outside the house to create a disapproving din), there are also many historical reports of communities turning a blind eye to domestic violence.[25]

How much does the fact that, for most of this period, the Church and the law presumed men's authority over women tell us about the everyday balance of power in family households in practice? The fact that some, and probably many, men treated women callously and otherwise abused them need not mean that this was the typical tenor between husbands and wives. There is autobiographical evidence[26] that many relationships were caring and affectionate and reason to suppose that these were more common than violent and abusive relationships. Making this point, Clark[27] quotes the

Paisley weaver Robert Tannahill's poem 'When Jonnie and Me were Married' which includes the lines 'sae eydent [ardent] aye were we. The lowe [flame] o' love made labour light'. Even without such strong amour, the need to share the burdens of life modified the formal authority which men had over women. In the case of households engaged in subsistence farming and in many of the trades, businesses and other occupations associated with a predominantly agricultural society, making a living required co-operation between household members.[28] Husband and wife were effectively economic partners mutually dependent on each other and the man could not afford to lord it too much over the woman. In the case of some occupations, employers made contracts with men which spelled out requirements concerning the productive work of their wives or another female dependent, sometimes a sister or daughter.[29] The following account is of the duties expected of the wife of a hind, or farm worker, from the Lothians in 1656:

> The Wives of Hinds, whether whole or half Hinds, are to Shear dayly in Harvest, while until their Masters Corn be cut down. They are also to be assisting with their Husbands in winning their Masters Hay and Peats, setting of his Lime-kilns, Gathering, Filling, Carting, and spreading their Masters Muck and all other sorts of Fuilzie [manure], fit for Gooding and Improving the Land. They are in like manner, to work all manner of work at Barnes and Byres, to bear and carry the stacks from the Barn-yards to the Barnes for Threshing, carry meat to the Goods [livestock], from the Barnes to the Byres, Muck, Cleange, and Dight [clean out] the Byres and Stables, and to help to winnow and dight the Cornes.[30]

While in many occupations men needed women to ensure their survival or prosperity, women's economic opportunities to live independently – without the potentially controlling hand of a husband, father or other male relative – were undoubtedly limited. For example, there are instances from the sixteenth and seventeenth centuries of burgh councils enforcing regulations preventing women who had never been married from living alone and setting up businesses on their own or otherwise working in the towns except in domestic service.[31] Houston argues that restrictions on single women were more to do with powerful burghers maintaining local labour supplies favourable to themselves than men imposing authority on women. However, he concedes that there were no such restrictions on single men living alone and explains this with reference to fears about the 'immorality and scandal' associated with women doing the same. DesBrisay argues that 'Scottish men were just as convinced as their English and European brethren that women were the root of temptation, and the burgesses among them were especially apprehensive about the small army of female servants occupying their cities.' Widows had more freedom to carry on businesses and trade in their own right and 'only widows and respectable spinsters and their female servants were allowed to live without the benefit of a male head of household'.[32]

In the sixteenth and seventeenth centuries restrictions on women's independent economic activity did not yet presume that their place was exclusively in the domestic sphere of their own homes. The idea of a 'housewife', a married woman whose sole occupation is looking after her home and family, gained currency among the Scottish urban middle and upper classes in the eighteenth century.[33] By the nineteenth century the concept had become the middle-class norm and was increasingly idealised in the early twentieth century by all classes as the division of labour most enabling of good family life, although the reality was always more complex.[34]

The interpretation of these shifts remains contested but they certainly involved new forms of restriction on women's opportunities for economic independence. The fight for a 'family wage', which would enable a male worker to keep a wife and family, was simultaneously an acceptance of wage differentials between men and women. The idea that women did not require the same level of wages as men persisted well into the twentieth century and in the 1900s women in Scotland earned on average 45 per cent of average male wages. Marriage-bars, which had excluded married women from many professions in the nineteenth century, persisted into the twentieth century.[35]

Divisions of labour arguably became more gender segregated as husbands earned money by long hours of work conducted outside the home and without any assistance from their wives, while wives continued to manage households with little or no assistance from their husbands. Although this became the conventional division of labour between married man-the-earner and woman-the-housewife, it was never universal. Many married women took part-time or full-time paid work in response to the death, absence, ill health or the unemployment of their husbands or to supplement the household income where the husband's wages were very low (see Fig. 19.2).

The conventional division was increasingly common in working-class families in the cities and towns of Scotland in the early twentieth century, but the testimony of oral histories[36] suggests that the trend may have had less impact on the balance of power in some households than others. The division did not always or automatically increase working-class men's power and authority over women. In poorer households women had more control over the finances than men, despite being cut off from earning money. The common pattern was for the man to hand over his intact wage packet to his wife. Children also gave any earnings they made to their mother and, if they were lucky, like their father, received back a small amount of pocket money. This system was necessary for the greater good of the household. It was taken for granted that every penny was needed to 'make ends meet'. Elderly informants speaking of their working-class childhood in the early 1900s generally noted their mothers' skill, determination and endless effort in keeping them fed and clean on the available income.[37] Although the background culture of the time increasingly encouraged respect and deference for 'the man as the worker' and 'the earner', the domestic worker at home was clearly essential to the survival of all members of the household. Wives and mothers spent many hours in visible toil on behalf of their households. The lack of domestic technologies we now take for granted (such as hot water on tap, electrical appliances and means of heating and cooking which do not rely on tending fires) made keeping the house and its occupants clean and the family fed a physically demanding business.[38] Very limited incomes also required labour-intensive strategies for making ends meet, like mending and foraging. Rikki Fulton's description of hierarchy and authority in his family household in the mid twentieth century satirically captures the strong character of one such mother:

> The system under which we lived was matriarchal and *my* Mother ruled her four slaves [Rikki's father and the three sons] with a rod of iron ... I've always wondered if it was significant that when talking to me, my brothers referred to Her as '*your* Mother'.[39]

It was, perhaps, not until the post-War period that the majority of Scots began to benefit from the wealth of industrial society. In the context of relatively high male wages and improved standards of housing and domestic apparel, a view of the marriage arrangement as one of a male earner and his economically dependent, privileged and less than equally contributing wife was more possible. Pahl's[40] study of the distribution of money in twentieth-century marriages was conducted in England, but a broadly similar pattern is likely to have been typical of Scotland also. The 'allowance system' or 'housekeeping system', often exercised in ways that emphasised women's subordination, came to predominate in the 1950s and 1960s. Men decided the distribution of 'their wage' between their own personal spending and the 'housekeeping' and many husbands deliberately concealed their total earnings from their wives. The concept of 'housekeeping' left women with no legitimate personal spending power: if they wanted to buy something for themselves they had to ask for additional money:

> My housekeeping isn't for me; it is to run the house and the children, not for me. I've always had my own money up to the time I was married. I don't like being dependent.[41]

> I'd like to work then I wouldn't have to spend your money, I could spend my own ... You do your own goddam shopping from now on ... All your precious money would be gone in no time.[42]

By the 1970s and 1980s, the conventional division of labour of the mid twentieth century was being radically modified by an increased proportion of married women, including married women with children, in both full-time and part-time paid employment. Couples were less likely to use the allowance system and forms of pooling money, most commonly by using joint bank accounts, were starting to predominate. Partly as a consequence of the Women's Movement, but also of wider socio-political change, norms of equality and 'fairness' were more commonly embraced than traditional ideas proclaiming a 'women's place' as the home and domestic and caring work as 'women's work'. Most of the legal apparatus supporting men's control of women and quashing the independence of single women had long been dismantled.

However, a number of factors continued to encourage divisions of labour which involved women carrying more of the burden of domestic work than men. These included inequalities between men's and women's earnings and earning prospects, a lack of institutional support for combining paid employment with bringing up children and the persistence of the idea that women are more suited to and responsible for bringing up children. The 'hands on' father who is directly and heavily involved in all aspects of childcare was not yet a feature of popular culture; 'good' fathers were primarily providers although there was also the expectation that they would be interested 'helpers' in the bringing up of their children. In a study of married couples living in Aberdeen at the beginning of the 1980s, Askham[43] explored the extent to which men and women felt their individual freedom constrained by marriage. Men discussed losses of freedom including the freedom to travel, to have sexual relationships with other women, and to go out drinking, while women talked of the limits placed on their free time or time for themselves by the responsibilities and demands of household tasks. These compromises and constraints might now sound anachronistic and unattractive to some young people in the early twenty-first century but many may come to make broadly similar compromises in their efforts to sustain long-term relationships within which to bring up children.

THE SOFTENING OF PARENT–CHILD HIERARCHY

Ariés' work, *L'enfant et la vie familial sous l'Ancien Régime*, first published in 1960 and subsequently published in English as *Centuries of Childhood*, has become a respected classic text. Although writing about France, his history of childhood is often assumed to have a wider European relevance. Ariés argues that, excepting a brief period of indulgence towards children in their infancy, medieval parents and adults in general were unsentimental about the young, treating them more or less as little adults from the age of about seven. Beyond that age there was no childhood in the sense of a protected existence in which people were specially nurtured and shielded from aspects of the adult world.[44]

Being treated as 'little adults' meant individuals engaging in the business of earning their keep, knowing their place in the social order and generally entering into a world in which many were subjected to harsh discipline and toil all their lives. Other historians have confirmed that across Europe children were often sent away from home to be servants or apprentices in the households of others at a very young age.[45] Using depictions of children in paintings, records of dress, toys and games, and religious and educational writings about the young, Ariés concludes that children were not given a childhood as such until the seventeenth or eighteenth century and then it was only those among the more privileged sectors of society – first the sons and then the daughters of the highest ranking families – who received this, the idea trickling down much later to the majority.

Ariés places much emphasis on the effect of changes in religious teaching, writing about the softening of religious pedagogical views in a country with a very different religious history from that of Scotland. Scotland emerged as a largely Protestant country from the Reformation of the sixteenth century. Some faces of Protestantism have been accused of being particularly bleak and stern, encouraging harsh and dour family life. For example, writing of eighteenth-century Scotland, Graham declares:

> ... life wore a grave and sombre aspect. In Presbyterian families especially was this the case; for the taint of a grim creed and the rigid spirit of the Church was still over the land. It was an age of austerity and probation. Severity was the characteristic of school discipline, which often amounted to brutality, and rigour was the note of all family training, in which the Solomonic maxim against sparing the rod and spoiling the child was orthodoxly followed ... 'My children from the youngest to the eldest loves me and fears me as sinners dread death. My look is law.' These words of the vigorously minded Lady Strange express the hard, austere spirit prevailing in many a household and the dismal discipline of every nursery, the memory of which was burned into many minds that lived to see more genial times. In the household the head of the family was regarded with awe as at table he presided with his hat on, and as he sat in his exclusive seat at the chimney corner.[46]

However we do not know for certain if the renewed religious vigour of the Reformation resulted in a general heightening of hierarchy and harsher discipline in families, nor do we have much detailed evidence to tell us if there was a growing contrast between family life in Catholic and Protestant homes. Graham suggests that 'In the homes of lairds of the Episcopalian persuasion a more genial atmosphere was found, less religious austerity, less Sabbatarian rigour.' But note that the 'vigorously minded' Lady Strange was a Jacobite supporter of Catholic Bonnie Prince Charlie. Protestantism and Catholicism have both been used in particular times and places as the legitimate instrument with which both to eradicate 'indulgent' or 'undisciplined' ways and to encourage greater moderation (see Fig. 13.3). Steel's[47] account of the course of Presbyterian Protestantism in the most outlying of Scottish inhabited islands, St Kilda, off the coast of North Uist, paints a particularly bleak picture of the potential of religion to extinguish joy from family and community.[48]

Smout has suggested that, for the majority of upper-class Scottish society, parents became more indulgent and affectionate towards their children during the period which he refers to as 'the revolution in manners' accompanying the greater affluence, status consciousness and Anglicisation of the gentry following the Union of 1707 and particularly post-1745:

It was perhaps something of a paradox that even as polite behaviour became more necessary there was a relaxation in the internal formality of the family and in the strictness with which a child of the upper classes was treated by its parents. On such things it is always hard to generalise, but people later on referred with wonder to the instances of domestic sternness in the early part of the [eighteenth] century, when for example children were not allowed to sit down in the presence of their parents and girls wrote to their mothers as 'Dear Madam', signing themselves 'Your affectionate daughter and humble servant'. Even by 1760 there was more familiarity between parents and children, much less stiffness, and less fear of the father in particular.[49]

Smout goes on to quote from the family papers of Mrs Elizabeth Mure.[50] Reminiscing in the 1790s she notes that, unlike in former times, fathers now treat their sons like friends and mothers seek intimacy with their children and scold nurses and ministers alike for frightening them with stories of witches or Hell respectively (Fig. 28.5).

The late eighteenth century may have been a more indulgent time among Scottish upper-class families but, as always, rather less is known about the lot of poorer children. By the early twentieth century family life in Scotland was certainly not yet typically child centred, although there were always exceptional parents who wanted to be more like a friend than a higher authority to their children and attempted to mask their greater power. Oral histories and biographies of the early 1900s indicate that many children, from rich and poor backgrounds, still described their parents as emotionally

Figure 28.5 Dressing up, Dalhousie, Midlothian, 1888. Source: SLA 124B 52/2/11.

distant figures and sometimes harsh and frightening ones. Christina Miller, who was brought up in the early twentieth century on her father's inherited highland estate, describes the deference she and her siblings were required to show to both their parents, and her fear of her father:

> by both training and inclination a cavalry soldier ... an implacable disciplinarian dealing with us in a strictly military manner. No matter what we were doing, if either he or my mother came into the room we had to spring up and stand to attention beside our chairs, and at meals we sat bolt upright, like puppets suspended by strings from the ceiling, afraid to open our mouths except to put in food.[51]

This is followed by accounts of her father's use of the cane and her mother's ineffectual remonstrations against this. Alasdair Alpin MacGregor's various biographical accounts of his father, the Colonel, are also littered with details of military-style discipline. The detailed and demanding codes of conduct his father attempted to enforce were often justified by asserting the general overarching need to preserve the honour of the family and clan name, which he referred to as 'the Old Name', claiming ancestry through Rob Roy MacGregor from King Kenneth MacAlpin. The Colonel was no doubt an extraordinary character. Indeed his son likened him to the legendary highland chiefs[52] whom the Colonel may have hoped to personify. Oral histories suggest that while most fathers may not have been as colourful as the Colonel, and many children did not fear their parents, emotionally distant parents were not very unusual. For some middle-class and upper-class children, their mothers and fathers were more akin to benevolent visitors and background benefactors than 'hands-on' parents:

> Father didn't seem to take a great deal to do with us but when he did see us he just put his hand in his pocket and gave us another penny or two pence or something. He was very good, that way.
>
> ... We spent a good deal of time with the maid ... I didn't see such a great deal of my parents, because if we were playing in the house we were in the huge kitchen with the maid, mostly there, so that I didn't really see such a great deal of my parents from that point of view ... If we had visitors, in those days, of course, it was a case of afternoon teas perhaps. Mother had her circle of friends and each one had an afternoon 'at home' and we were dressed up to go in and meet the visitors and have our tea with them. But they did a lot of entertaining in the evening ... of course, we were not allowed at table with them in the evening if they were entertaining. We usually had our meals in the kitchen with the maid under these circumstances ... I don't remember getting a lot of attention from my parents. I really can't remember my mother ever, you know, taking me in her arms or anything like that but we were happy somehow or other. We were left pretty much to amuse ourselves and get on with it. (Catherine, born 1902, commercial traveller's daughter)

> My mother, you know, never had charge of us [children]. She liked to see us but then when we got rather obstreperous we were banished again ... Then at the age of eleven I went to boarding school. We all went to boarding school. My mother said it was the happiest day of her life when we went to boarding school. (Caroline, born 1910, mill owner-manager's daughter)

In the first half of the twentieth century, the maxim 'Children should be seen and not heard' was still practised in many households, as the quotations below illustrate,[53] and many working-class children still had very deferential relationships with their parents:

> I don't know if you know the golden rule, 'Children should be seen and not heard'. You were not to talk back to elders. You never joined in the conversation at the table. You were never allowed at the tea table if there were visitors in. Children were kept apart. (Angus, born 1902, tailor's son)

> We weren't allowed to sit at the table. We had our tea over a big blanket chest that sat in the corner. That's where we got our tea. (Jimmy, born 1903, mason/builder's son)

After his mother's death, Jimmy's sister took over the role of housewife, which meant not simply doing the housework but also acting as a servant to her father. He was an extreme version of the patriarchal father, being too 'above' his children to consider having a conversation with them and preferring to issue commands without uttering words. 'He never spoke to us in the house. Never heard his voice. And if he wanted a cup of tea, he rattled his spoon in his saucer for my sister to pour out his second cup of tea.' It was only after Jimmy had become a tradesman and happened to encounter his father on a building site they experienced something approaching a conversation, when Jimmy was treated as 'one of the men'. Working-class children were also still expected to be at the service of their parents. In other words, parents considered it right and proper to make demands of their children for errands, housework and odd jobs that would help with 'making ends meet'.

Parents' requirement of respect and obedience from their children did not necessarily mean children were beaten or regularly smacked. Meg, an illegitimate child informally fostered by an elderly couple who spent most of their income on alcohol, suffered more neglect than many children but not exceptional physical punishment. 'I never used to have a pair of shoes on my feet. If anybody ever gave me anything they pawned it. They were old. They never should have got me.' However she also noted, 'They never ill-used me', adding, 'I used to get a skelpit backside if I needed it' and 'I loved them because they were all I had.'

Many of the people the author interviewed about growing up in Scotland during the 1900s explained that a look from their parents was enough to bring them into line and that their parents never needed to hit them. However, certain offences, such as being disobedient to teachers or truancy without parental consent seemed to result in automatic physical punishment in some households. Several descriptions of working-class childhoods involved the explanation that if you were belted at school it was important to keep it quiet at home or you would be belted again.

The expectations and demands made of children were rather different for daughters and sons and sometimes for older and for younger children. Working-class mothers struggling with large families, small incomes and heavy burdens of housework often made more demands on daughters for assistance than they did of sons. The conventions of the time were that certain types of domestic work, including heavy cleaning and clothes washing, were predominantly seen as 'women's work'. Therefore, when a mother was unable to manage unaided, her oldest daughter was often her first choice of assistant, as Betty explained:

> I sympathise with a person that's the oldest girl because she does get it put on her. She [younger sister] would get out [to play]. When she come out of the school my mother would say, 'Get the wee one [the baby of the family] ready'. Well she would take her [the wee one] oot and she would leave her at the side of the road and play at jumping ropes. I had to go in and scrub the floor and wash [clothes]. I was standing at the green, hanging up clothes [standing] on a tin box or something because I couldnae reach the [washing] line. (Betty born 1905, miner's daughter)

Over half of the working-class men and almost half of the working-class women interviewed by the writer had taken part-time jobs when school children. The most common job was delivering milk. In all cases their earnings were handed over to their mothers in order to help towards 'keeping the house'. The greater involvement of boys in paid work was a reflection of girls' greater involvement in housework. For example, when Maggie was asked if she had a part-time job at school she said, 'No, no, no. I never had a chance.' Cause my mother was oot working, I had to be in the house with the bairns. [I was] washing and everything at twelve and thirteen years old.' (Maggie, born 1899, docker's daughter)

Many working-class school children also contributed to subsistence by foraging for food and fuel. This sometimes meant hunting and gathering wild food, but more often in cities it involved being in the right place at the right time to get cheap or free merchandised goods such as stale bread and broken biscuits. Partly because of their lesser involvement in housework, working-class boys generally had more freedom to roam than girls and were more likely to be involved in these activities. Part-time employment and foraging were often undertaken by children in a more voluntary way than housework, which was usually demanded of daughters by mothers. It can be argued that rather than a conscious effort to make their daughters into dutiful housewives, this typically reflected mothers' preoccupations with meeting their housekeeping responsibilities. Nevertheless, in the early 1900s some mothers did explicitly teach their daughters to be subservient to men and boys. For example, Mary noted that not only were her brothers exempt from housework but:

> We used to have to wash out their white gloves and clean their patent leather shoes to let them go to the dancing. They were the apple of my mother's eye. Nothing could be wrong with the boys. (Mary, born 1897, stonemason's daughter).

The overwhelming majority of working-class children left school at the minimum school-leaving age or earlier and went straight into full-time employment. From handing over their wage packet to their mothers for 'the house' and getting back a small amount of pocket money, many made the transition to 'keeping themselves' or 'paying their keep', that is buying their own clothes and other expenses over and above bed and board for which they paid 'digs money'. While paying your keep was associated with more freedom to come and go, the privileges of being an earner were always less for young women than young men. Young women often continued to help with domestic work, despite being earners, and parents remained more watchful of their daughter's comings and goings than their sons, for example by checking more carefully that they were home by the agreed time and intervening if they lingered with a young man outside the house when being 'walked home' from a dance.

Gender differences were even starker in some nineteenth- and early twentieth-century middle-class households. In such families, girls and young women were significantly more constrained and chaperoned than young men. For example, when Robert was asked if he had a set bedtime when he was at school he replied:

> Not me. The girls in the family, yes. I mean I used to go to the Literary Association meeting at school on Friday night. If I got home before twelve o'clock, jolly good. But the girls had to be in by nine o'clock or else, which I thought was fair enough. (Born 1898, son of an owner/manager of a tree and plant nursery)

There was no expectation of either sons or daughters contributing to the standard of living of the family household and in upper-class households only boys were being brought up to anticipate earning a living. Upper-class daughters had many opportunities to socialise and to travel but within carefully supervised worlds in which they were almost never unchaperoned.

In the early twentieth century, the possibilities of 'family time' were constrained for working-class people by the necessity of long hours of paid employment and domestic work and the relative lack of space and comfort in many working-class dwellings. Class differences remained very acute in terms of both family size and standard of housing. Many large working-class families were still accommodated in one, two or three rooms. The 1901 Census indicates that 58 per cent of Scottish families were living in dwellings of no more than two rooms. A few of the 64 adults interviewed by the author who were born in the early 1900s into working-class urban families could remember home-based leisure for the whole family in their childhood. But these were typically activities which parents enjoyed and which children could be part of rather than play organised for the children's benefit. They reported evenings structured around music (the piano, organ, accordion, fiddle or gramophone) and sometimes games such dominoes or ludo.[54] Most working-class children eagerly escaped from their home to a world of play in the streets or, for rural children, the fields and woods, with other children and without any adult involvement.[55]

School-age middle-class children were not always allowed to play in the streets. Girls in particular were supervised and chaperoned to a greater degree, often by servants rather than parents. They too had a childhood world of play but more commonly it involved being placed in 'suitable' environments, such as the tennis club and parties at friends' houses (Fig. 28.6). It is interesting that many upper-class parents were not markedly more engaged in having fun with their children, despite having more leisure time and far greater resources than others. Many of the 23 adults interviewed about their upper and middle-class childhoods in the early 1900s reported musical evenings, trips to the pictures and theatre and family holidays but only exceptional parents were involved in their children's everyday play or in routinely encouraging conversation with them.

The idea that children and parents should be more like friends was not more pervasive until 1950s[56] (Fig. 28.7). By the 1970s, when Backett-Milburn conducted her study of middle-class Scottish parents,[57] family life

Figure 28.6 Games, Corstorphine, Edinburgh. Source: SLA 124B C.1752.

was clearly child centred and 'understanding the child' – that is being in touch with his or her fears and other feelings, interests and pleasures – was the main task of parenting, not simply keeping children clean, well-fed and providing a 'good' (clean, healthy, moral and perhaps educational) environment. Many factors contributed to the change. Middle-class households were the first to have smaller families but by the late twentieth century all classes were having fewer children. In middle-class families, the virtual end of domestic service coincided with a greater engagement in making a profession of full-time mothering while in working-class homes help from children was less essential in the struggle to maintain a decent standard of living.

The demise of the notion that children should be at the service of adults perhaps also reflects some more general changes in the social structure of post-War Britain. The social upheaval of the two World Wars did much to undermine many forms of traditional authority exemplified by the class system. The welfare state represented a more developed and inclusive

Figure 28.7 Charles (Chic) Rooney and his father Charles Rooney senior (and dog), from Dundee, on excursion to the Carse of Gowrie, Perthshire /Angus, 1936/7. Source: Catherine Smith.

form of citizenship consistent with a more democratic and egalitarian society. As the view of the world in which the working classes should 'know their place' increasingly became an unacceptable anachronism, so denigration of other forms of authority, hierarchy and privilege which were legitimated solely by tradition became more vigorous.

The child-centred household in which parents organise themselves around their children was not a general norm until the late twentieth century. As family sizes have shrunk, the dynamics between siblings have changed. Older children are less likely to be required to look after much younger children. Not only has the sense of responsibility for younger siblings on behalf of older children diminished but also the experience of hierarchy between older and younger siblings. Gender hierarchy within family households has also changed radically. Girls are no longer likely to be required to perform domestic service for their brothers because of their respective genders. Now parents are more likely to see themselves as at the service of their children than to demand service from them and many parents strive to be more akin to friends with their children. Although mothers and fathers are now likely to believe that they should both be very involved with their children, in practice mothers often tailor their employment and leisure more around children and therefore know their children better than the fathers do, as shown in the couples studies by Backett-Milburn.[58] A recent study of Scottish children's perceptions of their family life[59] finds that they are sometimes worried by their parents' preoccupations with work and finances, and sometimes believe that they do not have enough time to listen to their problems, but on the whole are happy with their families, seeing parents, and mothers in particular, as a key source of support. Research also suggests that grandparents and other kin beyond the family households who maintain contact with children can also play an important role in their lives. The practical care and emotional support these individuals can provide has been shown to be particularly important to children when stresses make mothers and fathers less accessible, or when parents are damaged or debilitated by illness, drug or alcohol abuse.[60] In such cases, grandparents are often playing the role of substitute parent more in the mode of an adult friend than a patriarch or matriarch authority figure.

NOTES

1. For information on fosterage in the form of 'boarding out' see Chapter 25 Childcare and Children's Leisure; Chapter 37 Children's Homes.
2. For the author's discussion of the debates about the causes and timing of social change in family life see Jamieson, 1987. For her comment on more contemporary debates see Jamieson, 1998.
3. Abrams, 1998.
4. See Chapter 39 Shared Flats and Houses.
5. See Chapter 33 'Time Out' Work.
6. See Chapter 43 Living Alone.

7. Scotland's People: Results from the 2003 Scottish Household Survey Annual Report <http://www.scotland.gov.uk/library5/housing/shsar03-05.asp>
8. See Chapter 17 Housework; Chapter 19 Working from Home; Chapter 20 Work in the Homes of Others.
9. Mrs Gillies quoted in Jamieson, Toynbee, 2000, 46.
10. Scotland's Population 2003, The Registrar General's Annual Review of Demographic Trends, <http://www.gro-scotland.gov.uk/grosweb/grosweb.nsf/pages/03annual-report>
11. MacGregor, 2001, 93.
12. Clan chiefs usually controlled land which was parcelled out to, managed, inhabited and lived off by kin in a complex chain of social gradations, involving varying access to land with different rights and obligations which often passed from father to son. MacGregor, 2001.
13. MacGregor, 2001, 94.
14. See also Coutts, 1999; Reddington-Wilde, 1999.
15. Stiùbhart points to the evidence left in the dominant form of Gaelic poetry which was created in praise of 'the hunter-warrior chief as defender of the clan' and reinforced 'notions of an assertive masculine independence based upon physical prowess and violence'. Stiùbhart, 1999, 233.
16. Grant, 1995, 31.
17. Smout, 1972, 35.
18. See Gordon, 1991; Clark, 1995 for accounts of the struggle for gender equality which itemise the interaction of economic and political change.
19. Coutts, 1999, 176; Houston, 1989.
20. Leneman, 1998, 234.
21. Leneman, 1998.
22. Leneman, 1998, 6.
23. Dobash, Dobash, 1980; Clark, 1995.
24. Leneman, 1998.
25. Historical work on rough music and 'riding the strang' is summarised by Clark, 1995, 84, as is turning a deaf ear, 85.
26. Burnett, 1984; see also the discussion of this point by Clark, 1995, 64–6.
27. Clark, 1995, 65.
28. See Chapter 19 Working from Home.
29. For the evolution of the demand that the hind supply female labour as part of his contract, whether or not it was that of his wife, see Robertson, 1990.
30. Firth, 1899.
31. Houston, 1989; DesBrisay, 2002.
32. DesBrisay, 2002, 139; see also Dingwall, 1999.
33. 'For these sections of society, female leisure was increasingly an indication of a man's social status. For the lower orders it was an unthinkable luxury.' Houston, 1989, 122.
34. Gordon, 1990.
35. Devine, 1999, 535.
36. Paul Thompson's study *The Edwardians: The Remaking of British Society*, St Albans, 1977, is the classic oral-history study of family life in Britain. But for oral histories giving direct insight into divisions of labour and authority between husbands and wives in England see Gittens, 1982; Roberts, 1984; in Scotland, Jamieson, 1983; Jamieson, Toynbee, 2000.
37. Jamieson, 1986.
38. See Chapter 17 Housework.
39. Fulton, 1985, 111.
40. Pahl, 1989; see also Morris, 1990; Vogler, 1994.
41. Mrs Cox. Quoted by Pahl, 1989, 116.

42. Argument reported by Luxton, M. *More than a Labour of Love*, Toronto, 1980, 164. Quoted in Morris, 1990, 121.
43. Askham, 1984.
44. See Chapter 25 Childcare and Children's Leisure.
45. Gillis, 1974.
46. Ariés, 1973, 2–25.
47. Steel, 1975.
48. He states 'After the Disruption in 1843 within the Church of Scotland, the stern faith of the Free Church, in the manner of its application and of its acceptance, made slaves of the people of St Kilda.' The claimed effects included reform of the islanders' social structure and culture. New and more extreme hierarchy was invested in the social structure as the Free Church gave some men authority over others and encouraged more extreme inequalities of gender and generation. The cultural transformation meant that 'The St Kildans lost their sense of gaiety and their love of song and dance. Their way of life, previously governed solely by wind and tide, was thereafter subject to the demands of regular church-going.' The pleasures and freedoms of childhood were presumably significantly reduced by these changes. Steel, 1975, 93, 94.
49. Smout, 1972, 269–70.
50. Smout, 1972, 270; *Selections from the Family Papers Preserved at Caldwell*, Maitland Club, Glasgow and Edinburgh, 1954, I, 270.
51. Miller, 1981, 15–16.
52. MacGregor, 1945, 7.
53. Unless otherwise indicated, quotations are from 'A Case Study in the Development of the Modern Family: Urban Scotland in the early twentieth century', unpublished PhD, University of Edinburgh, 1983, which is also the work drawn on in the cited single-authored articles by Jamieson. Also quoted in Jamieson, 1983, 26; 1986, 61. A number of other quotations cited in this chapter can also be found in one or other of these articles.
54. See Chapter 8 Music; Chapter 11 Games and Sport; Chapter 23 Food.
55. See Chapter 25 Childcare and Children's Leisure.
56. Jamieson, Toynbee, 1990.
57. Backett-Milburn, 1982.
58. Backett-Milburn, 1982.
59. Borland, et al, 1998.
60. Bancroft, et al, 2002; Dench, Ogg, 2002.

BIBLIOGRAPHY

Abrams, L. *The Orphan Country: Children of Scotland's Broken Homes from 1845 to the Present Day*, Edinburgh, 1998.

Anderson, M. *Family Structure in Nineteenth-Century Lancashire*, Cambridge, 1971.

Ariés, P. *Centuries of Childhood*, Harmondsworth, 1973.

Askham, J. *Identity and Stability in Marriage*, Cambridge, 1984.

Backett-Milburn, K C. *Mothers and Fathers: A Study of the Development and Negotiation of Parental Behaviour*, London, 1982.

Bancroft, A, Carty, A, Cunningham-Burley, S, Backett-Milburn, K *Support for Families of Drug Users* (Scottish Executive), Edinburgh, 2002.

Borland, M, Laybourn, A, Hill, M, Brown, J. *Middle Childhood: The Perspectives of Children and Parents*, London, 1998.

Burnett, J. *Destiny Obscure: Autobiographies of Childhood, Education and Family from the 1820s to the 1920s*, Harmondsworth, 1984.

Clark, A. *The Struggle for the Breeches: Gender and the Making of the British Working Class*, London, 1995.
Coutts, W. Wife and widows: The evidence of testaments and marriage contracts, c.1600. In Ewan, E, Meikle, M M, eds, *Women in Scotland, c.1100–c.1750*, East Linton, 1999, 176–86.
Dench, G, Ogg, J. *Grandparenting in Britain: A Baseline Study* (Institute of Community Studies), London, 2002.
DesBrisay, G. Twisted by definition: Women under Godly discipline in seventeenth-century Scottish towns. In Brown, Y G, Ferguson, R, eds, *Twisted Sisters: Women, Crime and Deviance in Scotland since 1400*, East Linton, 2002, 137–55.
Devine, T M. *The Scottish Nation, 1700–2000*, Harmondsworth, 1999.
Dingwall, H. The power behind the merchant? Women and the economy in late seventeenth-century Edinburgh. In Ewan, E, Meikle, M M, eds, *Women in Scotland, c.1100–c.1750*, East Linton, 1999, 152–62.
Dobash, R, Dobash, R. *Violence against Wives: A Case against the Patriarchy*, London, 1980.
Dodgshon, R A. 'Pretence of blud' and 'place of thair dwelling': The nature of Highland clans, 1500–1745. In Houston, R A, Whyte, I D, eds, *Scottish Society, 1500–1800*, Cambridge, 1989, 169–98.
Firth, C H, ed. *Scotland and the Protectorate* (Scottish History Society), Edinburgh, 1899, 405–8. Quoted in Fenton, A. *Scottish Country Life* (1976), East Linton, 1999, 166–7
Fulton, R. Happy birthday. In Kamm, A, Lean, A, eds, *A Scottish Childhood: 70 Famous Scots Remember*, Glasgow, 1985, 110–112.
Gillis, J. *Youth and History: Tradition and Change in European Age Relations, 1770 to the present*, London, 1974.
Gittens, D. *Fair Sex: Family Size and Structure, 1900–39*, London, 1982.
Gordon, E. Women's sphere. In Fraser, W H, Morris, R J, eds, *People and Society in Scotland*, vol II, 1830–1914, Edinburgh, 1990, 206–35.
Gordon, E. *Women and the Labour Movement in Scotland, 1850–1914*, Oxford, 1991.
Graham, H G. *The Social Life of Scotland in the Eighteenth Century*, London, 1928.
Grant, I F. *Highland Folk Ways*, London [1961], 1995.
Houston, R A. Women in the economy and society. In Houston, R A, Whyte, I D, eds, *Scottish Society, 1500–1800*, Cambridge, 1989, 118–47.
Jamieson, L. Growing up in Scotland in the early 1900s. In Glasgow Women's Studies Group, *Uncharted Lives: Extracts from Scottish Women's Experiences, 1850–1982*, Glasgow, 1983, 17–34.
Jamieson, L. Limited resources and limiting conventions: Working-class mothers and daughters in urban Scotland, c.1890–1925. In Lewis, J, ed, *Labour and Love: Women's Experiences of Home and Family, 1850–1940*, London, 1986, 49–69.
Jamieson, L. Theories of family development and the experience of being brought up, *Sociology*, 21 (1987), 591–607.
Jamieson, L. *Intimacy: Personal Relationships in Modern Societies*, Cambridge, 1998.
Jamieson, L, Toynbee, C. Shifting patterns of parental control. In Corr, H, Jamieson, L, eds, *Politics of Everyday Life: Continuity and Change in Work and Family*, London, 1990, 86–113.
Jamieson, L, Toynbee, C. *Country Bairns: Growing Up, 1900–39* (1992), Edinburgh, 2000.
Leneman, L. *Sin in the City*, Edinburgh, 1998.
Lynch, M, ed. *Oxford Companion to Scottish History*, Oxford, 2001.
Mackillop, A. Clans of the Highlands and Islands, 2: 1610 onwards. In Lynch, M, ed, *The Oxford Companion to Scottish History*, Oxford, 2001, 95–6.
MacGregor, A A. *The Turbulent Years*, London, 1945.
MacGregor, M. Clans of the Highlands and Islands, I: To 1609. In Lynch, M, ed, *The Oxford Companion to Scottish History*, Oxford, 2001, 93–5.
Miller, C. *A Childhood in Scotland*, Edinburgh, 1981.
Morris, L. *The Workings of the Household*, Cambridge, 1990.

Pahl, J. *Money and Marriage*, London, 1989.
Reddington-Wilde, R. A woman's place: Birth order, gender and social status in Highland houses. In Ewan, E, Meikle, M M, eds, *Women in Scotland, c.1100–c.1750*, East Linton, 1999, 201–9.
Roberts, E. *A Woman's Place: An Oral History of Working-Class Women, 1890–1940*, Oxford, 1984.
Robertson, B. In Bondage: The female farm worker in south-east Scotland. In Gordon, E, Brietenbach, E, eds, *The World is Ill-Divided: Women's Work in Scotland in the Nineteenth and Early Twentieth Centuries*, Edinburgh, 1990.
Smout, T. C. *A History of the Scottish People, 1560–1830*, London [1969], 1972.
Steel, T. *The Life and Death of St Kilda*, Glasgow, 1975.
Stiùbhart, D U. Women and gender in the early modern western Gàidhealtachd. In Ewan, E, Meikle, M M, eds, *Women in Scotland, c.1100–c.1750*, East Linton, 1999, 233–49.
Vogler, C. Money in the household. In Anderson, M, Bechhofer F, Gershuny, J, eds, *The Social and Political Economy of the Household*, Oxford, 1994, 225–66.

29 Status

CRISSIE W H WHITE

SILENT WITNESS

Concepts of status form entities which, along with other values, are explained, negotiated, imposed, rejected or accepted – in short distilled – as different generations interact with each other. Scotland's *Statistical Accounts* present striking evidence of the extent to which 'ranks and orders' have been built into the social structure and language of the nation. In Edinburgh 'people of quality and fashion' were set apart from 'tradesmen and people in humble and ordinary life'. In Kirkcaldy, Fife, there were 'men of rank and fortune.' In Perth, there were, 'opulent inhabitants ... [who live] genteelly ... [and are] of better rank'. Only 'proper company' was allowed to attend the assemblies in Montrose, Angus. In Glasgow, 'rank and consequence' marked social divisions and we find here early use of the term 'class', when educational provision is being made for the 'lower classes' and the 'higher classes'.[1]

An understanding of status and the other frameworks of social life is essential for the individual's survival, as is the ability to 'look and laugh' at these and to work with and round them. Indicators of status abound. Not least are material possessions, including dress (particularly on special occasions) and dwellings, which make important statements about who and what the possessors are in their community. Domestic display, allied with the residents' behaviour, sends a message to the immediate community and neighbouring ones, which interpret this information and make their own assessment of households and individuals. There are many and varied contributory elements within the symbiosis of dwelling and status, at every level of society. This chapter seeks to explore, in context, some of those which have been particularly salient in recent centuries in Scotland.

ACHIEVING STATUS IN SCOTLAND

Status land and property
For 91.8 per cent of Scots in 1750,[2] no matter if they were 'landed gentry', tenants, sub-tenants or cottars, survival depended in some way on working the land for food. Ownership of property, and to a lesser extent tenanting of land and buildings, was and remains an important signal of status in society. In cattle-rearing country, like the Highlands, prestige and wealth were

measured largely by the number of cattle kept, indicating the amount and/or quality of land owned or tenanted.

The position of those lower down the social hierarchy could be precarious. Where communities were dominated by a landlord, a casual word or action out of place could lose a household its tenancy and the living tied to it at a stroke or at the next 'quarter day'. The fear of such an occurrence influenced patterns of behaviour and speech in both public and private. Winning and maintaining the good opinion of a landlord was fundamental while dire consequences could result, for example, from adding variety to a meagre diet:

> In 1829 [James] Hogg was caught poaching on the Duke of Buccleuch's land. His nine-year lease on Mount Benger was up for renewal at the time, and ... the Duke chose to evict him and demand the rent that was owing. James and Margaret were ruined, losing most of their possessions in a forced auction, selling their sheep at a loss, and moving back into the four-roomed cottage at Altrive. Here they crowded together with their four children (a fifth would arrive shortly) and their new lodger Mr Brooks, 'a gentleman of weak mind'.[3]

The status of feu superior carried great power. The man or woman of independent mind might 'look and laugh at a that' behind closed doors. Independent means were required to do it openly.

The power of feu superiors became particularly apparent during the period of agricultural improvement when radical tenurial reforms had profound impact on much of the countryside's population. As improvement proceeded in lowland areas in the late eighteenth and early nineteenth centuries many former tenants lost all stake in the land. At the same time, for those fortunate enough to be able to secure one of the reduced number of tenancies:

> ... distinctions of function and status were firm and at any one time, clear ... [yet] families were not locked permanently in unchanging positions. Some holders of land would be bettering themselves, perhaps by adding to an existing holding perhaps by taking over a new and bigger place.[4]

In nineteenth-century highland areas, however, there was less of such opportunity and landowners who changed established patterns in the interests of sheep farming and hunting and fishing often had a severely detrimental effect on the lives of their tenants and their cottars. In the crofting townships created at this time, the ownership of a cow provided food in terms of milk and milk products and represented health and wealth, money on legs, the difference in status between dependence and independence. Yet many households were denied this:

> We feel very much the want of milk for our families. Many of us would be very glad if we could get a cow's grass ... at a reasonable

rent, which we would pay. The rent of cow's grass ... is £3 a year, which we consider an intentional discouragement against anyone aspiring to the dignity of keeping a cow. Our landlords say, 'How much more valuable is, not even a sheep, but a game bird than a man.'[5]

Reputation
Reputation can be a signal of status in its own right. In Scotland reputation could rarely be achieved by purchase in a society significantly poor in coin. However opportunities to 'rise in the [material] world' have often depended on being of 'good character', while the status of 'bad character' is difficult to shed. Recognition of 'worth' or 'uprightness' appears to have meant a great deal, oiling the wheels of life with or without property or other wealth. Moral qualities, which Robert Burns identified to be of 'higher rank' or standing than material wealth, are honest poverty, an independent mind, 'pith o sense and pride o worth'.[6]

The character reference remains an element in an employer's armoury when deciding whom to hire. Characteristics sought are often the same as those recorded on the memorial inscriptions and in the obituaries of past generations; honest, reliable, responsible, punctual, industrious, dignified, an upstanding member of the community.

Ways of rising in the world
The ability to generate wealth through trade grew apace in the eighteenth century, gathering momentum in the nineteenth century and beyond and witnessing the rise of the middle classes. Some of the income was spent in acquiring land in the belief that higher status would accrue from it. But there were also new means of enhancing status:

> The flourishing clubs and societies – literary, scientific, professional – provided an arena for conviviality, and enhanced the sense of identity ... With an emphasis on intellectual improvement, rational enquiry and the positive use of time, this culture was distinct. When linked to a positive espousal of active Christian observance and the evolving notion of respectability, the intellectual identity of the middle class was further refined and differentiated from other groups.[7]

The possibility of improvement in material conditions and status occurred for those in lower classes at this time also, but effecting the change from subsistence living to life with a surplus required great effort (Fig. 29.1). Going to sea to fish or sail with the Navy or merchant fleet was one way to augment a family income and improve the household's position. Educating children for a trade or profession was another. Finlay J Macdonald of Harris relates how his grandfather saved money from his days at sea to build a 'solid core of respectability'. He moved to better housing, built up a stock of sheep, became an expert on sheep rearing and saw his elder daughter

Figure 29.1 Blackhouse, corrugated-iron dwelling and stone-and-lime-mortared house together. These illustrate the three stages of Highland dwelling improvement. In the nineteenth century the 'tin' house was seen as an improvement upon the blackhouse but was considered a temporary dwelling to be used only until the stone-and-lime-mortared house was built. It later became a store and extra accommodation for summer visitors, as did the blackhouse until such time as it collapsed through neglect. Castlebay, Barra. Source: SLA C.12684.

through school and college, while his elder son became a master cabinet-maker. Thus he made himself a pillar of the community.[8]

Education opened opportunities to new ways of life. Much is owed to the providers of schools and to the skill and dedication of teachers in the nineteenth century that urban and single-teacher country schools alike achieved success in preparing pupils from very humble homes for trades, professions and university entrance. For many, scholarships were essential precursors to study at secondary school let alone university or college:

> This was accomplished many years before the Welfare State was ever dreamt of … when the words pensions and grants had as yet no meaning for those folk who worked and saved so that their children might get a better start in the world than they themselves had enjoyed.[9]

STATUS AND HOMES: EXTERIOR ASPECTS

The external aspects of homes are the outward face of the people within and usually much can be deduced about the residents' status from the type of dwelling lived in, its location and the alterations made to its exterior and environs. Arriving at such an assessment is often complex and never easy.

Dwelling type
HOUSING FOR THE WORKING CLASSES

Poverty and the single-storeyed dwelling were once largely synonymous. The association of cottages with the 'lower orders' ensured that charity houses and nineteenth-century homes on estates, servants quarters' and other dwellings for workers showed a continuous stylistic link with the form of vernacular buildings. Nineteenth-century working-class tenement dwellings of one and two rooms (see Fig. 3.1), in plan, arrangement and occupation of rooms, as well as status, parallel these precisely and in essence are stacked cottages, made possible by complex gable chimneys, communal stairways, internal plumbing and gas light, with later innovations being electric light and piped hot water. Such flats had toilets outside the dwelling, at ground level or on the half landing of the stairs. The kitchen-living room had a range for cooking, a dresser and a built-in bed – with storage or a hurlie bed underneath – and a cold-water tap at a sink, which had to serve as wash-hand basin also. The 'room', in cases where the dwelling possessed two rooms, was kept for best, and allowed the residents to make visitors welcome without embarrasment.

Not all nineteenth- and twentieth-century tenement homes were intended for working-class people, however. Flats also served the needs of members of the rising middle class, and in this context the building form was developed *par excellence*. Internal and external features associated with villas were employed to establish a sense of importance. Bay windows, a spacious panelled hall, a drawing room, dining room, multiple bedrooms, a maid's room, etc, combined a villa's features with a convenient city location. Stained-glass windows, a 'wally' (tile) close with a door on it, a tiny front garden, cast-iron railings and fancy front steps were prerequisites in the highest-quality tenements. An own or main-door flat with control of the front garden was, and remains, a most sought-after home of this type.

Though graduations in status between examples of tenements, and other dwelling types, at this time were generally apparent in material terms, other expressions of worth were to be found at all levels which cut across more formal ideas of hierarchy. A warm welcome, good manners and a clean well-kept house spoke powerfully of their residents' quality in cottage, tenement and mansion alike.

Nevertheless physical differentiation of housing on class lines continues to be a powerful concept. The Glasgow Garden Suburb Tenants Ltd, Westerton, 1913–15, a visionary co-partnership concept, was geared to improving living conditions for tenants, and the first of its type in Scotland. It aimed to provide 'healthful, cheerful houses with good gardens and pleasant surroundings'[10] for rent. The style of the terraced cottages, with private front doors and an area of garden, was inspired by English vernacular examples, yet follows the Scottish tradition of compact dwellings for workers. Real improvements to living standards were thus achieved for working people while the continuity of vernacular style ensured that such households were identifiably housed according to an allotted place in society.

Post World War II reconstruction in Scotland produced new housing quickly and relatively cheaply, both in the private and public sector. Within the latter, a number of New Towns and very large council schemes were created (see Fig. 22.6), some with success. Others without sufficient planning of community and transport facilities courted disaster and at times contributed to a marginalisation of the residents. Built at a point of decline in the use of open fires and before the advent of central heating, some flatted homes failed to keep dwellers warm and dry. There was a general increase in frequency of bathing in the post-War period and the washing of clothes in the dwelling rather than in a washhouse, steamie or laundrette. This, combined with poor insulation and lack of ventilation and resources adequate for efficient heating to make walls of inadequately constructed flats mildew or run with condensation. Modernist concrete flats in the Gorbals, Glasgow, designed by Sir Basil Spence, were considered by some to be among the more significant and irredeemable failures and have since been demolished. Many makeovers of similar areas of housing have been made at vast expense in recent years in an attempt to improve living conditions and help communities thrive, as is the case in Drumchapel, Glasgow.

Small groups of semi-detached houses and four-in-a-block flats in villa form were constructed within the public sectors in many parts of the country before and after World War II. Set within existing communities, these dwellings have generally continued to be popular with tenants. The blurring between private and public has increased at this level as mixed communities have developed since the 1970s as council homes built by local authorities are sold. These become sought-after, affordable purchases, particularly for first-time buyers, purchasing after the original tenant-owners cease to require them.

Local-authority controls on the exterior of tenanted property can be seen to confine free expression in contrast to those homes adapted by the freedom of ownership, with, for example, new doors and windows. While the gardens of rented homes offer some freedom of expression, shared areas leading to a tenement home are difficult to use with pride when systems of care, such as rotas for cleaning the close, break down and vandalism occurs. Setting the tone for visitors through dressing entrance doors and windows and the approaches to a dwelling is an important way of communicating status at all levels of society. Until the advent of Christmas lighting provided a form of demonstration available to the highest multi-storey flat dwellers, expression for tenants of rented flats might be confined to a choice of bell push, identifying label and door mat. An example of an own-door flat with a dedicated front garden in a local-authority tenement demonstrates the potential for display which tenants might make for a visitor given the opportunity (Fig. 29.2).

Figure 29.2 An 'own-door' flat in a council block of the 1960s. The common three-, four- or five-storeyed design was used in many towns in Scotland. In contrast to the other flats in the block, and no matter how small the patch of garden available, self-expression is possible and the resident can give notice of their view of the world to the visitor. Maryhill Road, Glasgow. Source: Crissie White.

HOUSING FOR THE MIDDLE CLASSES

As cottage and castle had evolved through economic dependence on land, the wealth acquired through a profession or trade triggered the growth of new types of urban dwelling for the middle classes. Stylistic borrowings from the dwellings of those of 'rank and fortune' are to be seen in the variety of design in middle-class dwellings, in the development of the villa and its extension, suburbia.

The magnificent eighteenth-century terraces of Edinburgh's New Town created a radical shift in the concept of gracious living in Scotland. By their elegance and proportion they helped to inspire the building of other, Georgian dwellings, usually in villa form. Publications such as *Rudiments of Architecture*[11] inspired builders throughout the country, in the same way that Chippendale and Sheraton inspired local furniture makers and cabinet-makers. Villas of various types appeared in large numbers in both town and country when the rise in wealth and associated status of the middle classes was at its most accelerated, in the eighteenth and nineteenth centuries. In towns they were often associated with suburbs. Victorian villas tended to be on a grander scale than their Georgian antecedents and were judged socially by the number and arrangement of rooms and the type and quality of their architectural detail, as well as the servants' quarters and the number and type of outbuildings (Figs 29.3, 29.4). They were built in many styles. A form of small castle inspired by Scottish tower houses was a notable example developed in the wake of Queen Victoria's passion for Deeside. Many other antiquarian approaches to design were displayed in Gothic Revival, Greek, Egyptian and Roman styles as well as English and Scottish vernacular.

The villa as a signal of independent living has had a strong hold on perceptions of status for the middle classes. Reluctance to leave it behind, as costs demanded, spawned the double villa in the late nineteenth and early twentieth centuries. The Victorian detached villa form was sustained while dividing ownership between two households, and these extended to four-in-a-block villas built for the middle classes in Kilmacolm, Renfrewshire. This continues in vernacular style in the four-in-a-block council housing built between the two World Wars, previously discussed.

Much of the detailing of buildings offered in suburbia up to the present day is derivative of other countries – Olde English, with imitation timbered gables, Colonial America, the open porch with pillars, or Spanish Hacienda, with rough plaster coatings and patios, for example. *Art deco* style dwellings were an attempt in the 1930s to break from this and look afresh at the medium of harled double brick walls combined with metal window frames with which to introduce large areas of glass.

Figure 29.3 Late eighteenth-century Georgian marine villa with Victorian update, in c1925. The status of a Victorian villa is sought by the replacement of six-pane sash windows with plate glass, a t-piece extension at the back with a bedroom and bathroom above and a dairy/scullery and maid's room below. The hall floor is rendered in decorative ceramic tiles and a cast-iron porch with lead canopy completes the picture. Seafield, Lamlash, Isle of Arran. Source: Crissie White.

Figure 29.4 The sale of furniture from Seafield House, Lamlash, Isle of Arran, on 26 May 1885 lists the furnishings of the middle class at that time. Advertisement, *Ardrossan and Saltcoats Herald*, May 1885. Source: Isle of Arran Heritage Museum.

> AT SEAFIELD HOUSE, LAMLASH, ARRAN,
> On TUESDAY, 26th MAY
>
> PUBLIC SALE OF
> HOUSEHOLD FURNITURE,
> CONSISTING OF
>
> MAHOGANY TELESCOPE TABLE, Mahogany Chiffonniere, Sideboard, Mahogany Sofa in Damask, Mahogany Rocking Chair, Mahogany Chairs, Rosewood Cottage Pianoforte, by R. Allison & Son; Small Commode with Mirror, Mahogany Easy Chair in Leather, Couch in Cretonne, Small Tables, 4 Chairs in Blue Damask, Painted Drawers and Press above, Mahogany and Painted Dressing Glasses, Mahogany and Painted Chests Drawers, Mahogany, Birch, and Painted Basin Stands; Painted Wardrobes, Towel Rails, Brussels, Tapestry, and Scotch Carpets; Iron Bedsteads, Bedding, Grates and Tiled Hearths, Fenders, Irons, Kitchen Range, Dresser, Table, Chairs, Hamper, Kitchen Utensils, 2 Fire-Clay Vases, Garden Seat, Mat, Sofa, Water Barrel, Lawn Mower, Barrow, 2 Small Boats and Oars, &c.
>
> BY AUCTION
>
> P. BURN, SON & DOUGLAS (of 97 Bath Street, Glasgow) beg to intimate that they are instructed to Sell the above, which belonged to a Lady deceased.
> On View on Morning of Sale.
> Sale to begin at 11.30 o'clock.

An important form of twentieth-century middle-class housing was the speculatively built bungalow. In arrangement it was effectively a middle-class tenement flat transposed to the ground, roofed and detailed to ensure no association with the cottage and poverty could be inferred. This dwelling type of four or five apartments built in the 1930s for the aspiring middle class

Figure 29.5 The speculative bungalow. A restrained and well-proportioned example of a 1930s bungalow, rough-harled, with arched entrance, the brick quoins in stone proportions, a brick base course and central steps. Its original windows remain in place. Bearsden, Glasgow. Source: Crissie White.

has been described as a miniature French château.[12] Here, adherence to the concept of a high-status dwelling on a tiny scale, with a central front door approached by a flight of steps and public rooms on each side regardless of aspect, sacrificed convenience in pursuit of an image of grandeur to impress buyers. The home as castle is confirmed by details such as bands or quoins around the front door (Fig. 29.5).

The complexity of homeownership today
Aspiration to the status of homeowner has risen steadily in recent decades. Owners of all sorts of dwellings are generally viewed as of a higher status than, and distinct from, those who rent. At a smaller scale, however, matters are more complicated, now more so than in the past.

The change from travel by foot and horse to rail, bus and car has increased choice about where to live, while the variety of dwellings available today is greater than ever, with new building complementing the old. Rising property prices, especially in many central urban locations, have encouraged people of reasonable means to consider homes in formerly run-down areas of towns and cities. Dwellings in more out of the way places or some redundant non-domestic buildings which have been converted to residential use may now be chosen in preference to others.[13] The status which the owners of such properties seek is difficult to interpret. Recognising the perceptions of potential clients, an artist weaver looking for a home decided that a cottage in the Highlands was essential. Whoever heard of anyone weaving in a bungalow?

There are new types of complexity. A large home may not always be desirable, even where there is sufficient wealth comfortably to purchase and maintain it. Such a dwelling requires a great deal of housework and other maintenance to keep it in good condition, and few people today have the time or servants to carry this out. Modest purchases may therefore represent a desire to be free to spend time, as well as money, elsewhere, for example on a boat or a second home. Small is not only convenient, it can also be beautiful. In the twenty-first century, unlike the eighteenth, ownership of a cottage or servants' quarters may be an attractive prospect and accrue a status higher than anything dreamed of by previous occupants. Driving these choices is a desire for a dwelling with 'character' or 'charm', as well as a 'good' location – that is, one of pleasant and safe physical surroundings, socially acceptable neighbours and accessibility. Old dwellings in general tend to occupy better-aspected and more generous sites. Assessing status and wealth by property size and style has become a tricky business.

Retention and alteration in dwellings: Windows
In each succeeding generation, alterations to and decoration of the exterior of homes give out messages of status achieved or status aspired to. Much can be deduced from the way windows are dressed and doors arrayed, the addition of porches, decorative ceramic tiled entrances, steps and pathways, for example, as well as the alteration of doors and windows. Such

adaptation to signal new status, or new status aspirations, was as frequent in old houses as it is today.

The Georgian terraced house and villa, with its vertically orientated twelve-pane sash-and-case windows, inspired window types in rural and urban areas and at most levels of society, although, with the panes hand-made by the laborious, expensive and size-limiting process of spinning molten glass onto a flat surface and glass at times subject to a heavy tax, humbler homes were less generously fenestrated than those of the well-to-do. The development of plate glass in the later nineteenth century created a new marker of status – windows with large panes. Astragals were removed and windows were fitted with this new material, a move equivalent to the fitting of picture windows in the 1960s. Bay windows, dormers, porches and glazed double doors could be added to Georgian houses to mimic the new Victorian dwellings. Dormers also began to feature in many cottages formerly equipped with skylights. Metal window frames introduced before World War II allowed for whole walls of glass to be incorporated into some dwellings. Today double-glazed windows dominate Scottish homes except where their fitting is prohibited by planning rules, as in conservation areas and listed buildings. Status clearly, like fashion, is not all. Stylistic details change, yet many well-proven forms remain as icons of successful design, whether for practical reasons, out of habit or a more conscious desire to maintain contact with the past.

Name and number
Owning a named, rather than a numbered dwelling adds an element of status. It suggests property with land. However, names given where numbers are sufficient for the postal service may also signal aspiration, identity, homesickness, romantic association, family connection or local history. Although the standard of property division has risen dramatically in the last 30 years, doing much to allay former associations of division with multiple occupancy in areas in decline, ambivalence about owning a dwelling numbered A, B or C remains. Identifying flats by their position within a building is a way around the problem of low status; 'main door', 'garden' or 'studio' are all usually quite acceptable to the status-conscious dweller.

A good address
The location of a dwelling is an important factor in the assessment of its status. But as well as living in the correct area or street, possessing a suitable street name may be of significance. A street address once signalled higher status, being paved and thus an advance on a dirt road. Living in a 'Square', or 'Place', or on an 'Avenue' or 'Crescent' can often still imply grand living, although in many modern suburban developments crescents are no more than roads with a bend and avenues are usually without trees. Nevertheless the connotations of gracious living may entice purchasers, as can pretty, rustic-sounding names like 'Elm Grove' and 'Heather Dene', which demonstrate a lack of understanding of the meaning of words and context. Elms,

for example, are as unlikely to grow in groves as heather would in denes. Contradictions and complexities abound. In the twenty-first century living in a 'lane', for example, may carry prestige or the opposite depending on location, in city, town or country.

Local authorities generally have not employed to the same extent the type of street names used by private developers. Street names with historical or political references and streets named after leading local figures and councillors are often found in areas of local authority housing. Keir Hardy Street in Methil, Fife, is a case in point.

STATUS AND HOMES: INTERIORS

A Victorian villa

Status and aspiration to status, relating to both households and the individuals within them, are demonstrated in most aspects of the interior of dwellings: decoration, furnishing and plenishing,[14] and the use of space. In some homes the expression is covert, in others status is boldly, sometimes extravagantly stated.

Melfort House, a substantial villa of 1877, was built near the new Bearsden Railway Station, Glasgow, of 1875. Each element and detail of this Victorian villa, inside and out – entrances, hallways, rooms and the decoration of these – was proportionate to the status of the people expected to use them. The device was common to all such properties and ensured that the owner and family, household servants and guests would literally 'know their place'. Principal apartments for the use of the owner, from the drawing room with its bay window to the master bedroom, were spacious and situated with the finest aspect. They had the highest and most decorative ceilings and cornices and were furnished accordingly. The front porch and conservatory off the dining room were equally substantial (see Fig. 29.6).

The lower ceilings and dormer windows in the attic bedrooms were appropriate to the immature status of the children of the family. Kitchen, scullery and maid's room formed an adjunct occupying a low-roofed extension to the rear. The smells of cooking could be entirely isolated from the main house and a toilet at the back door ensured that the needs of servants and tradesmen were met without their having to enter the main house. Behind the house, and across a narrow courtyard of glazed ceramic brick, stood 'offices', a stable and coach house and washhouse. The two-roomed flat and hayloft above were the most modest parts of the ensemble, details only distinguishing the exterior of the human accommodation from the fodder storage area. The stable and coach house appear to have been equipped to a high standard with gas lighting, a stove and piped central heating, greenhouse-fashion, for the horses (Fig. 29.7).

Melfort House was formally divided in 1974 into three dwellings, the house into two homes and the stable block converted to form a third. The occupants benefited from being part of a small community, accepting shared responsibilities and valuing the mixed messages the property gave as a

Figure 29.6 Victorian suburban villa of 1887. The five windows and single door opening and the raised site echo the Georgian villa. Every element on the exterior has a unique feature drawn from Scottish and classical sources to indicate the importance of interior spaces. Note the château-inspired roof with decorated dormers and a turret-like end, the substantial covered porch with granite pillars, the large windows and the semi-circular bays, the crow-stepped gable, and cast-iron rail on the balcony of the master bedroom. Also, the plate-glass windows, cast-iron finials, knobs and rails, decorative barge boards and chimney pots. Melfort House, Bearsden, Glasgow, photographed, 1960s. Source: John Morrison.

burglar deterrent. In the 1980s the main-door flat was sold. Entered at the principal door, it occupied all but one of the prime apartments. The new owners had difficulty in adjusting to the idea of shared occupation. The flat radiated messages of detached status and a mantle of ownership over the entire house. The owner of the upstairs flat observed to the owner of the stable block, 'R thinks he owns the house, and we are the upstairs servants, and you're the gardener and the chauffeur.' Unhappy, they moved on. The protocols of owner occupation of part of a divided dwelling and ground with shared access are less explicit than in a factored tenement property.

Signals of hospitality: The dresser and sideboard[15]
The offering of hospitality accords status regardless of dwelling type or contents; witness the renowned highland hospitality to strangers, displayed in homes from great houses to cottages. Certain objects and features have long been used to facilitate hospitality and the entertaining of guests, including seating, the hearth and the dresser.[16] The sideboard and dresser, and their descendent the fitted kitchen (essentially a fixed, elongated dresser), are silent witnesses to a dwelling's potential to offer hospitality.

Figure 29.7 Melfort House, Bearsden, Glasgow, 'offices' on the rear boundary wall. A pavement of glazed ceramic brick lies in the courtyard between this and the main house. The two-roomed flat in the upper storey, built to accommodate a married couple acting as coachman and servant or housekeeper, has windows in the crow-stepped gables for the bedroom and kitchen-living room. The hayloft is to the right. The coach house and stable were converted to garages and in 1976 these and the washhouse were converted into a two-bedroomed dwelling with sitting room, kitchen and bathroom. Source: Crissie White.

The dresser was ubiquitous (see Fig. 12.8). Timber versions could possess a raised shelf and small drawers or a separate or integral plate rack, while increased affluence introduced the buffet niche in the eighteenth century. The associated sideboard, which flourished in the dining rooms of the nineteenth and twentieth centuries, could be large and imposing and made of splendid materials. But no matter how ostentatiously the object featured as a mark of affluence, the basic form and function was sustained, the four compartment dresser in the prehistoric home seen at Skara Brae in Orkney presages precisely the proportion of the ubiquitous oak sideboard of the Victorian era. Both the dresser and sideboard exist to contain materials required for serving. In a larger house, the 'best' will be stored in the sideboard and the second best, perhaps the discarded 'best' of a previous generation, will be demoted for use in the kitchen before obsolescence and relegation to spaces such as the garden shed.

Like the mantelpiece, the surface of the dresser or sideboard has been used to display objects, some ornamental, some useful, most special in some way. As an example, a marriage gift of a pair of glazed ceramic meal barrels marked Mrs Munro, Tarrel 1893, made in the Seaton pottery near Aberdeen, took continuous pride of place on a dresser and then a sideboard in Ross-shire from 1893 to 2002 (see Fig. 4.5). The skill of the homemaker could be represented in hand-embroidered cloths, lace or crochet work, which protected the top of the dresser or sideboard from damage from the bowls, vases, special plates and fancy jugs which were among the items displayed. Jugs and ornamental vessels 'too good' for daily use held odds and ends.

Bedrooms and visitors
Personal beds in the past were generally a rarity. Usually a bed was shared between several household members. When individual beds and bedrooms became more common in the eighteenth and nineteenth and especially the twentieth centuries, rooms in more substantial houses were usually identified by location ('back bedroom' for example), user or status (the 'best bedroom'). Best bedrooms were usually for parents and guests of importance; that is adults of an equal or superior social position to that of the parents. Where space has been limited, as in many modern homes, guests change the pattern of occupation and the household has to adapt for the duration of their stay.[17] They may expect to have to fit themselves around the household's habits. Teenage visitors today, for example, may sleep on the floor of any room, an activity made practicable by fitted carpets, central heating and sleeping bags.

The kitchen-living room
Each space within the home has its own relationship with the concept of status. The kitchen-living room, present since prehistory in the homes of the majority, has defied attempts to replace it. The first effort towards its eradication was by the middle classes in the nineteenth century, when the 'separate spheres' of workplace for men and home for women were fully established and servants were the norm, even if they did not always reside in their workplace. Eating in the parlour or dining room became a signal of membership of the middle classes while cooking, eating and living in one room were associated with poverty. In large villas the kitchen, sometimes with integral box bed, was deemed the suitable social space for servants.

Within the speculative bungalow, compliance with this idea was achieved by creating a kitchenette too small to eat in. The bungalow kitchenette cartooned in 1936 by W Heath Robinson[18] was in effect a scullery washhouse with a cooker, winning over purchasers by being well equipped.

Its diminutive size ensured there was no backsliding from eating meals in the dining room. This arrangement encapsulated a serious attempt to alter the pattern of living of families aspiring to be middle class and effectively consigned the primary status of women from wife and mother to servant of husband and family. The rest of the household sat in comfort in the living or dining room waiting for meals to arrive while the housewife toiled alone.

The kitchen-living space reappears in middle-class homes at the end of the twentieth century, having once again achieved the status of respectability. Its revival followed the return of middle-class women to the workplace in large numbers. The addition of extensions on 1930s bungalows and villas since World War II to create family kitchen-living rooms is testimony to the triumph of household needs over the readymade dwelling.

STATUS AND THE HOUSEHOLD

Within the household: Age, size and status
Creating a clear framework for family members ensures a secure place for everyone in the home in which confidence can be acquired and attitudes formed and revised. Age and size are tools used to define status within the household, to exert authority and claim rights. For children size is often a synonym for age. I am big/old enough now; to stay up late, to ride a bike, to go to school myself. You are too wee/young; to climb trees, go out on your own. Thus is the process of growing up negotiated. Parents and children in turn assert their authority in the move towards independence.

Key objects signal rites of passage on the path from childhood to adulthood. Each is associated with a particular age or stage of development. Feet were usually shod first with bootees, then shoes or tackety boots, with bare feet common for summer until World War II. From the War, until trainers came along in the 1980s, it was rubber boots, then shoes, sandals or sandshoes in summer, often with the toes cut out for growth. Loss of status did not accrue if every child in a community was the same. 'September was the month of the shoes,' Findlay J Macdonald writes movingly of the year there was no money and there were no new boots, so that he went reluctantly to school on the first day of term in sandshoes.[19] Once there he took comfort from being one of many. Trainers, the over-designed cousin of the sandshoe, are a key element of current dress. In the 1990s failure to own a labelled pair of trainers brought to one 8-year-old the ignominy of being called 'No Name' at school.[20]

A teddy bear, a doll and pram, a trike (tricycle), a fairy cycle (the smallest two-wheeled bicycle on the market), and then a full-sized bicycle also marked steps on the road to adulthood within the twentieth century, as did being trusted to own and take care of a fountain pen or a watch. The latter might appear at the age of twenty-one. For girls, the first wearing of a bra and lipstick, for boys the first wearing of long trousers and beginning to shave, were among the seminal moments.

Size as a feature of status is used affectionately, humorously or threateningly in and beyond the home depending on circumstances. Big Peter and Young Peter distinguish father and son. The use of diminutives like 'wee', 'small', or 'peerie' in the family take the sting out of criticism. 'Big man' or 'the big yin', are expressions local to Glasgow and the west of Scotland. They identify importance, power, someone 'in charge'. The designation may be benevolent or malevolent, admired or feared; big-hearted, big-personality, big-strong, big-bully.

Reinforcing status: Cutting down to size and embarrassment
Conflicts arise when roles are confused. Andrew Melville forcefully reminded King James VI of his role in the church at Falkland Palace in 1596. He called him, 'God's sillie vassal', and said, 'There are twa Kings and twa Kingdoms in Scotland; there is Christ Jesus and his kingdom the Kirk, whose subject King James the sixth is, and of whose kingdom, not a king, nor a lord, nor a head, but a member.'[21]

In the home, when individuals seek a superior status not agreed upon by others, various expressions originating in real and imagined perceptions of kings, princes and landed and upper-class society in the past are used. Ideas such as owning a horse, wearing expensive clothes or a top hat, and putting on airs are a few among many which live on from previous centuries. Individuals setting themselves above others in skill, knowledge, looks or importance are cut down to size. Poking fun is one way to curb delusions of grandeur, using expressions such as, 'she's high falutin', 'father is on his high horse' and so on. 'His Lordship' or 'Her Ladyship' and 'wee madam' are used to describe and mock people getting above themselves. Standing to mock attention and saluting or bowing and touching an imaginary cap, saying 'yes sir' and 'no sir', are used to alert anyone perceived as bossy. The infliction of embarrassment in any form is an important tool in the maintenance of ideas of status. Being laughed at or talked about is a fate children and adults alike go to great lengths to avoid.

Departing from and working around ideas of status: Daftness and the comedian in the family
Insanity, like death, remains feared and rarely talked about, yet there have been many words and phrases to do with insanity and commonly associated phenomena in everyday usage. Descriptions like 'simple', 'daft', and 'weak minded' up to the nineteenth century were genuine attempts to describe the condition of an individual's mental faculties. In loving and affectionate families, words like 'daft' and 'mad' are now used to express a temporary and tolerated aberration from normal behaviour. Households make their own rules and know by word and tone of voice what is serious and what is not. Flying in the face of family convention prompts actions like pointing a finger to the head and screwing it round ('a screw loose'), or comments like 'don't be mad', 'silly', 'stupid', 'daft' or 'idiotic'. 'Mad' and 'daft' also express incomprehension of others' behaviour or enthusiasms, be it mad or daft

about boys, girls, football or motorbikes. According to John Currie,[22] 'mad about football' means more than enthusiasm. It is identified in a 7-year-old son knowing the names of goalkeepers of obscure football teams.

One way of liberating oneself from censure when failing to conform to the rules of household life, including those relating to status, is to play the clown. Adopting the role of comedian in the family may grant a certain immunity and also act as a defence mechanism.

Mixed messages
Both within and beyond the home, signals of status of one generation survive and overlap with new ones and the behaviour and standards of one generation may inadvertently offend another. Dress can be particularly sensitive. In 1990, at a wedding the author attended, a fellow guest remarked, 'My daughter has decided to get married in a red dress. I don't know how I'm going to tell my mother. She will surely have a heart attack!' A white dress, as a symbol of purity associated with marriage, was for much of the twentieth century the expected garb of a bride, while red was not at all respectable. Yet John Wallace, farmer in the Parish of Rosskeen in Ross-shire writes of his father's time, 1779–1800, that, 'No young ladies covered their head until married. Their hair was their pride ... At the marriage ceremony, the bride was always covered with a scarlet plaid, and if she had not one of her own, got the loan of one.'[23] The fashion, and thus status, statements of one generation are redesigned in the next, and in return revised to offend another generation.

The household within the local community: The definite article
The use of the definite article is a signal of singularity in an object, role or persona projected. It accords recognition and implies a relationship of value or respect to individuals or their roles within a locality: the laird, the doctor, the minister or the banker, for example. There is also the 'big house'. Status by association may be a source of pride or a burden and problematic, as 'the doctor's niece', 'a son of the manse' or 'the schoolmaster's wife' can often testify. Being associated with someone of importance in a community may help or hinder attempts at advancement.

CONCLUSION

Questions of status in the dwelling have always been varied, sometimes in conflict and often complicated and difficult to interpret. As new technology and economic and social circumstances have arrived in recent centuries, so has this complexity increased. In the last few decades television has brought the trappings of wealth and power, the style of dwellings and other possessions and the behaviour of the rich or famous from around the world into almost every home. Those living in Scotland today find more complex signals of status than ever to witness to their aspirations and to identify in others.

NOTES

1. Nenadic, 1988, 120.
2. Devine, 1988, 28 (after de Vries, Jan. *European Urbanisation, 1500–1800*, London, 1984).
3. Groves, 1988, 143–4; Hogg, J. A very ridiculous sermon, *Fraser's Magazine*, 1895.
4. Grey, 1988, 54–5.
5. Yeoman, 2000, 371–2, extract from the Royal Commission led by Lord Napier into conditions of crofters and cottars in the Highlands, 1884, statement of grievances by crofters and cottars in Lochaline, the District of Morven, Argyll.
6. Burns, R. 'A Man's a Man for a That'.
7. Nenadic, 1988, 121.
8. Macdonald, 1986, 109.
9. Sillars, M. Old Arran-Bogary, *Ardrossan and Saltcoats Herald*, c 1960.
10. Westerton Garden Suburb Tenants Ltd, 1913, 2.
11. 'Inadequately constructed' is not an entirely accurate description. These buildings were securely built and of their time. The use of concrete walls for flats proved inappropriate for families used to coal fires and paraffin heaters for reasons of economics. Lack of insulation affected the comfort of all double brick dwellings in the public and private sector from World War I onwards where money or servants were lacking.
12. Described by Cluny Rowell, in a lecture on architecture to students at Glasgow School of Art in 1958.
13. See Chapter 46 Idiosyncratic Homes.
14. See Chapter 12 Artistic Expression and Interiors.
15. See Chapter 4 Storage for further discussion of dressers.
16. See Chapter 6 Heating and Sleeping; Chapter 7 Rest.
17. Woe betide anyone waking in the morning who forgets the visitor and inadvertently enters the wrong room; in a large manse one morning a minister's wife walked into and across the spare bedroom to look out the far window. Dressed in corset, Celanese knickers and stockings, she had forgotten that a guest, a bachelor minister friend of her husband, was in the bed. He was awake. She paused, retreated. No words were exchanged, and the incident was never mentioned again (except against herself years later to her grandchildren). Mrs Archibald Hunter, Abbey Manse, Kilwinning, Ayrshire.
18. King, 1984, 167.
19. Macdonald, 1983, 103; see also Chapter 25 Childcare and Children's Leisure for further information on children's clothing.
20. James Beattie, East Kilbride, 1995.
21. Burleigh, 204. The original appears in Pitcairn, R, ed, *Autobiography and Diary of James Melville* (Wodrow Society), Edinburgh, 1842, 370.
22. John Currie, Drumchapel, Glasgow, 2002.
23. Munro, 1992, 21, ms of John Wallace died aged 93 in 1875.

BIBLIOGRAPHY

Blair, A. *Miss Cranston's Omnibus* (1985), Edinburgh, 1991.
Burleigh, J H S. *A Church History of Scotland*, Oxford, 1960.
Devine, T M. Urbanisation. In Devine, T M, Mitchison R, eds, *People and Society in Scotland, vol 1, 1760–1830*, Edinburgh, 1988, 27–52.
Dickson, J, Elliot, C. *Rudiments of Architecture: The Young workman's Instructor* (2nd edn Edinburgh, 1778), reprinted Whittinghame, 1992.
Duncan, J. *My Friends the Miss Boyds*, London, 1968.

Faley, J. *Up Oor Close*, Glasgow, 1990.
Fenton, A. *Scottish Country Life*, East Linton, 1999.
Grey, M. The social impact of agrarian change in the rural Highlands. In Devine, T M, Mitchison R, eds, *People and Society in Scotland, vol 1, 1760–1830*, Edinburgh, 1988, 53–69.
Groves, D. *James Hogg: The Growth of a Writer*, Edinburgh, 1988.
King, A D. *The Bungalow*, London, 1984.
Phillips, A. *My Uncle George*, London, 1986.
Macdonald, F J. *Crowdie and Cream*, London, 1982.
Macdonald, F J. *Crotal and White*, London, 1983.
Macdonald, F J. *The Corncrake and the Lysander*, London, 1986.
McVicar, A. *Salt in My Porridge*, Glasgow, 1975.
McVicar, A. *Capers in My Kirk*, London, 1988.
Munro, W. *Averon: Tales and Legends of Alness District*, Alness, 1992.
Nenadic, S. The rise of the urban middle class. In Devine T M, Mitchison R, eds, *People and Society in Scotland, vol 1, 1760–1830*, Edinburgh, 1988, 109–26.
Smout, T C. *A Century of the Scottish People, 1830–1950*, London, 1986.
Westerton Garden Suburb Tenants Ltd. *Prospectus*, Glasgow, 1913.
Whitelaw, M H. *A Garden Suburb for Glasgow*, Glasgow, 1992.
Worsdall, F. *The Glasgow Tenement: A Way of Life: A Social, Historical and Architectural Study* (1979), Edinburgh, 1991.
Yeoman, L. *Reportage Scotland*, Edinburgh, 2000.

30 Custom

GARY J WEST

INTRODUCTION

In the everyday goings-on of the Scottish home, each space was allotted its function. In the grand houses of the well-to-do, this demarcation would tend to be at the level of specific rooms for specific purposes – dining, sleeping, receiving guests – while in the more modest dwellings inhabited by the majority of the population, the one or two rooms which usually comprised the home were by necessity multi-functional.[1] These spaces retained a significance which went far beyond mere utility, however, for various internal parts of the building were imbued with meanings which were treated with reverence by the householders. The hearth, for instance, was a symbol of great importance as the beating heart of the home, while doors and windows, as points of entry and exit, represented boundaries which had to be viewed with caution during certain key periods and events.

It is these periods and events that shall be the focus of this chapter. At particular points in the annual cycle, and at times of great significance in the lives of families and individuals, the home has become the setting for rites and customs. These represent the manifestations of a complex set of beliefs and cosmological blueprints that might be dismissed by some as 'superstition' but which nonetheless retain a potent legacy right down to the present.

LIFECYCLE

In Scotland, as indeed appears to be the case universally, various stages of the human lifecycle are considered to be of particular significance and are therefore marked in special ways. Some of these are based on biology – birth and death, for example – but others, such as coming-of-age ceremonies and betrothals have more social or cultural roots. Termed *rites de passage* by Arnold van Gennep in 1909, these lifecycle markers attract ritual celebration and behaviour which feed into the traditional custom and belief patterns of local communities throughout the nation. The places within which this activity takes place are highly significant, for movement through physical space is often used as a symbol of the metaphorical 'passage' of the individual from one life stage to the next. Some events require the participation of the wider community and as such are held in public space, while others are acted out within the confines of the private home. This chapter concentrates on the latter.

Basing his analysis on a wide range of cultures, van Gennep concludes that each ritualised rite of passage within the human lifecycle has three distinct phases: 'separation', 'transition' (also known as 'liminality') and 'incorporation' (or 'aggregation'). Rites of separation mark the period in which the individual is removed from the normal social order. This is followed by a period of transition in which he or she is considered to be in a state of liminality, isolated from the former social role, but not yet integrated into the new one. The final passage involves rites of incorporation through which the initiate is accepted into the new status.[2] Holden succinctly captures the essence of the liminality phenomenon:

> Each transition in the lifecycle involves a period during which the initiate has no distinct status and is in a state of danger as well as being dangerous to others. The danger is alleviated by ritual abstentions and taboos, often involving segregation from the community and dietary restrictions.[3]

Van Gennep's model has withstood the rigorous examinations of subsequent generations of scholars reasonably well, and as such, it is adopted here as the most useful template against which to discuss lifecycle customs relating to the Scottish home.

Birth
The creation of new life is naturally a time for celebration and rejoicing, and is of such fundamental concern to all societies that it is of little surprise that pregnancy, childbirth and the first weeks of life tend to be demarcated particularly vividly in many cultures. So it is, and was, in Scotland, where both religious and secular belief systems have thrown up a great many rites and customs associated with this primary rite of passage. While the clinical procedures of pre- and post-natal care are nowadays the domain of formally trained professionals – as well as the birth itself, of course – the spiritual associations of pregnancy and childbirth continue to display strong elements of traditional belief, most often articulated simply as 'superstition'.[4] These may be predictive (of the baby's sex, stature, health, complexion, etc) and have been protective, aimed at counteracting the evils of otherwordly creatures, diabolic powers or even the jealous gaze of an unfriendly neighbour. Indeed, they may even be ritualised offers of practical advice, based on the collective experiences of generations of mothers and their carers. The following is a selection of those rites and practices which have centred specifically upon the home.

The confinement of the expectant mother was an important marker of the beginning of the birth process and can be said to represent van Gennep's 'separation' stage. A highly significant part of this separation was the removal of men from the scene, and preferably from the house altogether. The comment made by Alice Hick of Leith, Edinburgh,[5] that 'the men would be out of the house – we rarely saw the husbands' is something that many will recognise, although this tradition does now seem to have been consigned to the past. 'She's for the neuk' was a common saying in Orkney, a reference to the 'neuk beds' which were often built into the stone wall of Orkney homes,[6] and once bedded, it was considered important to ensure that no harm befell the mother-to-be while in her confinement. This was a time in which she was vulnerable to the goings on of the otherworld, and so various protection rites were carried out, with the Bible and some form of metal object being among the more common items involved:

> … it used to be quite common among women, at the time of their confinement to stick a darning needle in their bed, and place a Bible beneath the bolster, and this was thought to keep the fairies away.[7]

The fairies were considered a potent threat during the whole birthing process, the risk lying in their legendary liking for human milk. The protection of the new mother, and her milk in particular, was paramount therefore, and this was reflected in the large number of rites and practices which relate to lactation:

> Neither could a woman giving suck seat herself on the edge of the bed of the lying-in woman, from the belief that such an action stopped the flow of milk of the lying-in woman.[8]

If this taboo was ignored, and the new mother later had difficulty in feeding the infant, an elaborate procedure was carried out involving the clandestine 'borrowing' of the offending woman's child which was passed under and over the apron to encourage the flow of milk. Such talk of fairy hosts and secret rites may appear strange to us today, but it is surely easy to understand the preoccupation with the potential loss of milk in a nursing mother. Without the back up of modern formula alternatives, a mother's milk was virtually the sole life-giving force for the newborn, and without it the child's survival would be very unlikely.

During the labour and birth, and indeed for the days following, both mother and child were seen to be in a state of liminality, again a most dangerous period, with the mother not yet 'churched' nor the child baptised.[9] Certain precautions could, however, be taken in the meantime, with both mother and child being 'sained'. Regional traditions varied on the exact form of this protection rite, although in the north east in the late nineteenth century the procedure went as follows:

> A fir candle was lighted and carried three times round the bed, if it was in a position to allow of this being done, and if this could not be done, it was whirled three times round their heads; a Bible and bread and cheese, or a Bible and a biscuit, were placed under the pillow, and the words were repeated, 'may the Almichty debar a' ill fae this umman, an be aboot ir, an bless ir and ir bairn'. When the biscuit or the bread and cheese had served their purpose, they were distributed

among the unmarried friends and acquaintances, to be placed under their pillows to evoke dreams.[10]

The final part of Gregor's description hints at another important signature of the celebration of rites of passage; the close involvement of other members of the community, and not just the family or household concerned. The home, in this respect, while obviously affording a degree of privacy, was also a centre of community gathering. There were, however, strict conventions regarding the concept of 'visiting', none more so than in relation to another major life marker – marriage.

Marriage
As he approaches the marital home for the first time since the wedding ceremony, the modern-day groom is probably aware of the tradition which suggests he should carry his bride 'over the threshold'. Should he rise to the challenge, he is embracing a general concept which appears to be of great antiquity, and one that lies at the heart of many of the rites and customs associated with marriage in Scotland, for he is negotiating a conceptualised boundary. During the whole marriage process – conceived of as a journey rather than a single event – those involved have to pass through various stages of ritualised behaviour, each representing a change in their status. The metaphorical boundaries marking the beginning and end of each phase have usually been represented by physical boundaries, some relating to public space, and others to the home. The threshold of the marital home is just one example and can be said to represent the boundary between these public and private spheres. Martin has closely examined marriage ritual in relation to Scotland and he explains the point thus:

> The tripartite structure of separation, transition and incorporation which characterises the marriage ritual indicates a symbolic conception of the change in the central figures' social position, a step-by-step movement towards the public acknowledgement of new status, identities and responsibilities. This symbolic movement is linked with actual spatial passage ... [11]

This 'spatial passage' has often been accompanied by ritual dialogue and as Martin clearly shows, the Scottish tradition in this respect is just one localised manifestation of a very widespread international phenomenon.[12] The homes of the bride and the groom have both become significant spaces during the marriage journey, and specific physical features within these homes have also taken on special associations. One or two examples can perhaps illustrate the point.

At one time, the 'betrothal' usually took place in the bride's home, and in highland society normally involved two stages, the 'rèiteach beag' or 'còrdadh' (small betrothal) at which the permission of the girl's father was first sought, and the 'rèiteach mór' (big betrothal), a more formalised ceremonial occasion attended by close relatives and friends of both parties. In the former, the suitor would often be accompanied by a matchmaker, usually a strong, healthy local male who had the wit and verbal skill to conduct his duties properly.[13] These duties included taking part in an allegorical dialogue with the girl's father in which permission was sought for the marriage to take place. To make the request in overt terms would be dangerous, as in the event of a negative response there would be no way for the father to couch the refusal in a gentle and polite manner. The use of allegory got round this problem though, as '... one of the functions of this device is to provide a method of refusal which would preserve the dignity of the suitor, regardless of the outcome.'[14] The usual practice was to refer to the girl in metaphorical terms:

> When the company has had a dram out of the bottle, tea follows, after which the friend contrives to introduce the subject of their visit, in the best form possible. If he is a man of wit, or eloquence, he has the advantage in the use of these gifts in discharging his delicate task. Here is an example of the business part, of what is reported to have been said on such occasion:– After a few suitable words by way of introduction, the friend went on thus:
>
>> 'We have been building a house, and have got it all ready for the roofing, but we find we are short of the leg of a couple, to match another we have already got. We know you have got such a thing to spare, and as we are wishful to have the best that can be had, and being assured of the good quality of those you have got beside you, we have come to request the favour from you. If you can see your way to oblige us, you will contribute greatly to our house and to our happiness.'[15]

The space within the home in which this ritual drama was acted out was significant, with the hearth and the table together providing a key focus for the action. Both are associated with the basic components of hospitality – warmth and food – but there is strong evidence to suggest that these focal points of the home took on more subtle meanings during the acting out of customs such as the 'rèiteach'. According to one South Uist account,[16] the representatives of the prospective bride and groom were spatially divided into two camps, their respective positions leaving no room for doubt that it was the girl's family which was in the position of power within this relationship:

> The hosts' 'solemn decorum' is the first indication that this is no ordinary visit, but a clearly structured ritual event. The spatial separation of the two 'camps' is clearly delineated; the girl's family are gathered round the fire, physically aligning themselves with their representative who occupies the 'superior' end of the table. The table is a 'male' space which becomes the 'duelling ground', and the visitors are conducted by their hosts to their 'inferior' place at the end of the table furthest away from the hearth.[17]

Another place of great metaphorical significance within the home, particularly in relation to marriage, is of course the bed. Consummation was an essential part of the marriage journey but the exact stage of the process at which it took place could vary. In Shetland, the betrothal was centred around the 'speiring night' and 'the couple were expected to sleep together that night as a seal of the contract, but not to have intercourse again until after the wedding'.[18] With the 'bedding' of the bride taking place in front of many witnesses, we can see the inversion of normal social order when the private sphere of the bedroom is brought fully into the public domain.[19] This might be done in the company of a large gathering of friends and family, and in Shetland was even accompanied by music from a fiddler. John Irvine provided the music on many such occasions:

> The bride was put to bed, and the whole lasses, the whole women went into the bride's hoose and they put her to bed. And there was no man allowed in there at all – unless the fiddler – and I was always it so ... I had to step up and play the fiddle ... There were no room to dance you see, the hoose was a full as she could ha'd you see, there were fifty or sixty lasses'd be within. Then after they got her in, aa the men cam in wi the bridegroom.[20]

There are many other examples of the importance of particular spaces within the home in relation to marriage, but this rite of passage does in fact share a good deal of common ground with the final journey in the lifecycle, and it is to this that we now turn.

Death

> There was no privacy, no play space, no work space, no place to get out of the tensions of family life, to think, relax or sulk. There was not even space to die.[21]

Christopher Smout here paints a most striking picture of the problems associated with overcrowding in many of Scotland's urban areas, particularly in the 'single-end' tenement flats comprising only one room. For those living in two-roomed houses, the 'good' room (the 'ben' end in rural dwellings) at least provided the chance to define the use of the limited space more specifically. Indeed this demarcation was adhered to almost religiously, the kitchen accommodating most of the everyday activities such as cooking, cleaning and much of the sleeping, leaving the second room relatively free for special activities. It was to the good room that an expectant mother might retreat for her confinement, and there that friends and family might gather to bring in the New Year. And until the twentieth-century specialisation of commercial undertaking took hold,[22] it was there too that the bereaved family would lay out its dead; it was this 'liminal' space which was denied the single-end family. Referring to the high level of infant mortality in nineteenth-century Glasgow, one Medical Officer of Health commented 'Their little bodies are laid on a table or on a dresser so as to be somewhat out of the way of their brothers and sisters, who play and sleep and eat in their ghastly company.'[23]

Yet lack of space would certainly not stand in the way of the plethora of rites and actions which had to be undertaken within the household immediately following the death. Well into living memory, in rural and urban communities alike, families knew exactly what was expected of them, this knowledge being based entirely on unwritten tradition. As with marriage, space, liminality and boundary seem to lie at the heart of much of the activity. Doors and windows were opened as soon as the death occurred, 'to give the departing spirit full and free egress, lest the evil spirits might intercept in its heavenward flight'.[24] Clocks were stopped – a clear sign that the household was now entering a liminal time zone between the occurrence of the death and the final incorporation into the afterlife following the funeral. Mirrors were covered or removed, perhaps to avoid confusing the departing spirit or, as some adherents to the tradition suggest, to prevent the family from seeing the face of the deceased appear in the glass.[25] Blinds or curtains were drawn as a signal that the normal order of life had been upset, although this might also serve to communicate the death to the members of the wider community who would then draw their own curtains in turn.[26]

This involvement of the wider community is essential, for it serves to reinforce the sense of belonging which membership demands. Indeed it is manifested in several ways in relation to a death. After the body was laid out or 'straiked' (stretched) on a board or table and dressed in the 'deid claes', usually by the local howdie wife,[27] the room in which the body lay became public space as relatives, friends and neighbours came in to pay their respects (see Fig. 7.2). For the two or three nights before the funeral was held neighbours might also take turns to sit with the body to 'watch' it, essentially to protect it from evil spirits. In many instances this took the form of a 'lykewake', a communal gathering held in the room in which the corpse lay and at which much food and drink was taken:

> ... the great attraction of the old-style lykewakes was the hospitality shown at them, all part of 'decent burial', and no-one minded how long the interval between death and burial might be. There was sure to be a plentiful supply of tea, beer and bread, with perhaps cheese, while new pipes and tobacco as well as snuff were provided for the men.[28]

Representatives of the community would also attend the 'kisting', the placing of the body into the box or coffin ready for removal from the house for the funeral service and the burial. And just as arrival into the world was marked in particular ways within the home, and the first entry into the dwelling following marriage, so too was the final exit from the house another boundary, one which had to be crossed at the end of life's journey. The body was removed feet-first so that it could not see the way back into the house, while there was also a long-standing tradition in parts of rural

Scotland that the corpse should not be removed through the doorway at all, leading to some very complex extractions through smoke holes, windows or even through temporary breaks in the wall made for this specific purpose.[29] In more recent memory, the use of windows for extracting the coffin was most often associated with suicides.[30]

It is clear, then, that there are common links between all of these main lifecycle rites of passage – birth, marriage and death – in that, in each, elements within the home take on special significance. Particular spaces are clearly 'set apart', underlining the liminality which is a central feature of these journeys from one status form to the next. These themes are by no means peculiar to lifecycle events, however, for they also appear in activities and beliefs associated with seasonal custom.

CALENDAR CUSTOMS

The Scottish home has for long been the site for the concentration of rituals and customs aimed at marking the passing from one phase of the year into the next. This may relate to the seasons, to religious observance, to important events in the economic production cycle such as the harvest, or indeed to a host of other contexts and purposes. What scholars normally refer to as the four quarters of the old Celtic year have, however, provided the foundation stone for much of the custom associated with calendar celebrations within the Scottish home.[31] 'Samhainn', 1 November, and 'Bealtainn' (Beltane), 1 May, are said to have marked the beginning of winter and summer respectively, while two other divisions occur in between, 'Lunasdal' (or Lammas) on 1 August and 'Imbolc' or St Bride's Day on 1 February (the latter having become closely bound up with Candlemas on 2 February). In Gaelic tradition the night is considered to belong to the following day rather than to that just gone, and so the 'eve' of some of these markers attracts celebration too; Samhainn festivities have for long been replaced with those associated with Halloween on 31 October for example.[32]

Other dates brought yet more cause for domestic celebration. Yule or Christmas was marked enthusiastically in certain areas, especially those regions where Episcopalian or Catholic adherence predominated, yet it was virtually ignored by large sectors of the rest of the population who preferred to wait for New Year before breaking from the normal routines of daily life. Here, of course, is another example of the emphasis being placed on the 'eve', Hogmanay attracting more popular interest in Scotland than New Year's Day itself.[33]

The manner in which events such as these were marked (and still are in many cases) has varied through time and space, but certain common traits can be identified. As with lifecycle rituals, movement from one calendar period to another involves a transition of status, introducing a state of liminality which can be exploited for good and for bad. It is a time for foretelling the future, for disguising identity, for undertaking role reversals, for commemorating that which has gone and embracing that which has yet to come.[34] People come together to perform rites aimed at celebration, purification, protection and charity. Some of these practices are communal and public, others personal and confined to the family and the private home, while the concept of 'visiting' unites the two. The following sections provide a brief 'taster' of a selection of these traditions as they have pertained to life within the Scottish home.

Protection
The use of fire and water and the employment of verbal sayings, chants or songs are among the most common methods of offering protection to the householders at these special times of year. Fire was of course to be found in the hearth, the 'beating heart' of the home, and so it was the hearth which acted as the focus of much of the ritualised procedure relating to the calendar. For those who could afford the necessary fuel it was normal practice to keep the fire alive in the hearth at all times as it was considered bad luck to let the flame go out.[35] Thus, the fire was 'smoored' or smothered at night, the embers being covered with a large peat or log, ready to be coaxed back into full life the next morning.[36] Keeping the fire alive was of particular importance through the transition from one period to another, and so on Hogmanay in particular great care was taken to 'rest' the fire correctly:

> On Hogmanay the fire was 'ristit' by the 'guedeman'. The sign of the cross was made on the peat that was put among the burning ashes. The peat was then covered up by the ashes, and after the whole was smoothed, the sign of the cross was made over the whole. No woman was allowed to take a hand in this work, though it was her province to do this work at all other times.[37]

This passage succinctly captures several elements which are commonly found in relation to such liminal periods: the use of fire, the grafting of Christian symbolism onto older pagan beliefs, and the gender role reversals, which are all common signatures of the Scottish tradition (Fig. 30.1).

While many of the practices associated with the marking of calendar customs took place within the domestic sphere, individual households did not exist in isolation and were linked to the wider community in various ways. One symbolic act which served such a purpose was closely associated with Beltane. To celebrate the beginning of summer, a large community fire was built ('tein-eigin' in Gaelic), often on a prominent hillock, and cattle were driven past this as a ritual purification to protect them for the coming year. It was also a chance for the community members to re-assert their common sense of belonging, for all the household hearths were extinguished and relit anew from the communal source, rendering them 'people of one fire'. The symbolism was powerful and could be used as a social sanction as well as a sign of common identity:

> In Rannoch the tein-eigin was produced within living memory. My

Figure 30.1 People of One Fire. The essence of the need fire tradition survives with the annual burning of the Clavie at Burghead. Source: SSS, HIII 6 8661 Photo: Ian Mackenzie, 1986.

father, who died at the age of 82, remembered taking part as a boy in the production of tein-eigin in a distant glen on the borders of Inverness and Perth shires. It was then the custom for each family in the district to receive a brand from the sacred fire to kindle the domestic hearth. But those who were in arrears of rent, had failed to pay their just debts, had been guilty of theft or meanness, or were known to have committed certain offences against good morals were deprived of the privilege.[38]

The links between households might also be strengthened at other times through various forms of visiting, often formalised through certain rituals, and a task commonly bestowed upon the young men of a community. 'Guising' belongs in this category (the word simply means disguising), and while certainly associated with Halloween, as at present, it was New Year which attracted the most enthusiastic sartorial frolics. In the Highlands, bands of youths, some dressed in animal skins, would roam from house to house welcoming in the new year and offering the householders protection from evil spirits for the months ahead. This protection was achieved through a combination of fire-based rituals and chants and was offered in return for food and drink[39] (Fig. 30.2).

CUSTOM • 571

Figure 30.2 Guisers, Iochdar, South Uist, c1983.
Source: SSS, V 3b 8265
Photo: Peter Cooke.

Ritual hospitality
It would be misleading to imply that guising was aimed solely at offering protection to the householders, however, for often no protective rites were present, the emphasis being placed instead on the request for charity. Groups of young men would wander from house to house requesting meal, cheese and other foodstuffs which would then be offered to those in need within the local community. 'Thigging' as this was called, was often accompanied by the singing of a song requesting alms:

> Rise up gueedwife, and dinna be sweer [lazy],
> B'soothan, b'soothan.
> An deal yir chirity t' the peer,
> An awa b' mony a toon.[40]

The dramatic tendency which was often integrated within these guising traditions reached its high point in the mummers' plays, common in many parts of England in the eighteenth and nineteenth centuries and surviving in Scotland as 'The Galoshins'.[41] This folk play, based on a death and resurrection motif, was widely known in lowland Scotland and the

Borders through the nineteenth century, and survived in pockets into the second half of the twentieth century. One early account from Abbotsford captures the essence of the play succinctly:

> Yesterday being Hogmany there was a constant succession of Guisards – ie boys dressed up in fantastic caps, with their shirts over their jackets, and with wooden swords in their hands. These players acted a sort of scene before us, of which the hero was one Galoshin, who gets killed in a 'battle for love', but is presently brought to life again by a doctor of the party. As may be imagined, the taste of our host is to keep up these old ceremonies. Thus, in the morning, yesterday, I observed crowds of boys and girls coming to the back door, where each one got a penny and an oaten cake. No less than 70 pennies were thus distributed – and very happy the little bodies looked with their well-filled bags.[42]

While the houses of the landed gentry were no doubt viewed as a lucrative source of revenue by the guisers, the performance of the play might take place in any household. Given that their presence was seen to encourage fertility and prosperity, coupled with the widespread belief that it was bad luck to let the lads leave empty handed, they could usually be sure of a reward of some kind wherever they went.[43]

Decoration

The celebration of seasonal custom involves not only the dressing up of people but also of buildings, for the decoration of both public and private space remains a central marker of these traditions. Today the arrival of Christmas, or a birthday perhaps, may well find us adorning our walls, doorways, windows and floor space with trees, lights, banners and balloons, while localised festivals, parades and galas often produce a less homogeneous set of decorative traditions. The community emphasis at the latter events tends to result in the bulk of this decoration being situated within the public or municipal sphere, but in order to demonstrate a sense of belonging to their community individual householders often get in on the act too. The homes of those individuals who are key players within the celebrations often receive special treatment, such as that of the elected 'Queen' in the Grangemouth Children's Fair. Founded in 1906, the fair was simply intended as 'a day of mirth and cessation from scholastic labours',[44] although it soon began to adopt certain traditions associated with May Day including the election of a local school girl as 'Queen'. By the 1920s the Queen's house had become the focus for an ostentatious annual makeover:

> ... the entrance to the home of the Queen-elect was a bower of flowers. A deep arch of rhododendrons and greenery above the entrance trailed off artistically in streamers of marguerites, and there was not a ledge of the building upon which flowers were not heaped, while all manner of flags and embellishments transformed the place.[45]

As Burnett and MacCallum point out, these localised festivals are highly flexible in form and function 'because they have no rules as such, merely conventional forms such as floats, costume, decoration and processions'.[46] They are therefore very adaptable and can tap into changing fashions of popular taste and culture, resulting in a dynamic decorative tradition which owes as much to the present as to the past. Today the decorative arches which adorn the Grangemouth Queen's home incorporate themes of 'Disney Cartoon characters, popular music, film, sports stars or television favourites'.[47] The Grangemouth festival serves as just one example of many which continue to link individual homes with the rest of their community in towns and villages across Scotland.

Divination
The liminality resulting from the progression from one calendar division to the next was believed to create a window on the future, and so divination rites were commonly performed at such times, especially by the young. The specific form and function of these varied enormously from place to place, but predictions of marriage and death seem to have been something of a universal, with Halloween acting as the focus for most of this prognostic activity. Fire, water and salt all served as agents of divination. Nuts thrown into the fire might predict a future marriage, one nut representing the thrower and the other an eligible potential partner: the manner and direction in which the nuts 'popped' were indicative of their compatibility. The placing of saucers on the floor was another common method of predicting the future, one being filled with earth, another with salt, a third with clean water and the last with dirty water. The subject of the test was blindfolded, and then placed a hand on one of the saucers, thus discovering his or her fate, which might be good (salt or clean water) or bad (earth or dirty water).[48]

THE NEW SCOTS

Those members of immigrant or black or minority groups who have made their lives in this country have of course brought with them a plethora of customs and traditions which have added whole new dimensions to the culture of Scotland.[49] To do justice to the richness and diversity of the ritual celebrations of these groups in relation to the domestic sphere would require a whole new study of its own, but it is clear that in many respects the underlying structures of these customs have a good deal in common with those discussed above. Much of this celebration centres around the life-cycle and the calendar, for instance, and in both religious and secular activities the concepts of boundary and protection are very often evident. Fire and water are commonly brought into service as agents or catalysts, and strict divisions between the private and public use of space are often retained or re-emphasised in ritualistic ways, although these boundaries are crossed at certain times and in particular forms in order to celebrate the links between the individual, the family and the wider community. A few examples of some of these underlying structures in practice can serve to highlight this common ground.

The liminality concept is present in rites associated with birth and death within many of Scotland's immigrant groups. Within the Islamic tradition, the newborn child has the *Adhan* or call to prayer recited in the ears soon after birth and is named on the seventh day.[50] At the other pole of the lifecycle, a deceased Muslim must be buried as soon as possible after death, perhaps revealing a fear of this particular form of liminal state. In an echo of the highland 'churching', in traditional Sikh society the newborn child is taken with his mother to the temple as soon as possible in order to be named and to afford the family and community a formal opportunity to give thanks.[51] In some families such traditions may well become diluted following migration, but as Baldev Singh who came to Leith in 1958 points out, this is by no means inevitable:

> Here in Britain the community feels that they don't want to let the traditions die because they've been brought from India to a different country. Here they would get blamed for being Westernised and forgetting all their traditions, so in that sense we've stuck to them, more so than our relatives in India.[52]

For Scotland's Jewish communities, the weekly liminality of the Sabbath brought a time for reflection and family communion, the boundary between ordinary time and the set-apart period being Friday evening. For Howard Denton growing up (as Hyman Zoltie) in the 'Happy Land' of St Leonard's in Edinburgh, the Sabbath eve brought a cherished calmness despite his rejection of the Orthodox principles which lay at its roots:

> ... I still have my good memories of our Friday evenings at home, after we'd been to the synagogue. We rarely saw father all week but he was always back from his travels early on Fridays. We all ate together and there would be readings from the Jewish Bible. The tranquillity was bliss after a week of squabbling and shouting which sometimes got so bad that I wanted to pull a blanket over my head. On Friday evenings, even a slightly flippant comment by anyone would immediately be shushed by Father. I grew to love that atmosphere of calm.[53]

This ritualistic 'time out' – and indeed the predominance of custom in general within the Jewish faith – was also appreciated by those of a higher social standing within the Edinburgh Jewish community. David Daiches, son of the Chief Rabbi, tells a similar story:

> ... memories of light and warmth and festivity predominate in my recollections of childhood. Jewish family life makes much of the domestic festival, and the lighted candles, the large clean white tablecloth, the abundant food, made every Sabbath and festival eve a celebratory occasion.[54]

CUSTOM • 575

The message which emerges from both of these passages is surely that customs, as well as being a manifestation of the belief systems which underpin them, also serve important functions in their own right, for they afford the opportunity for individuals to interact with their families, and families to interact with their communities with an intensity and to a depth that is much more difficult to achieve within the social or domestic norm.

CONCLUSION

The Scottish tradition is extremely rich in customs and beliefs associated with seasonal and lifecycle activities, and it has been possible to present only a very small sample of the huge corpus of material within this brief overview. Indeed, there are many aspects of the tradition which have been omitted altogether here – birthdays, coming-of-age celebrations, anniversaries, graduations and a wide range of surviving or revived seasonal customs – while the author is also very conscious that this account has been biased towards the past rather than the present. This should not be taken as a sign that these traditions have been lost or forgotten, but rather that contemporary ethnological studies of custom in practice, particularly within the domestic sphere, are still few and far between, and it is an area of Scottish ethnology which remains ripe for closer examination.

Despite these omissions, however, it is certainly possible to identify a number of patterns relating to seasonal custom and rites of passage within the Scottish home, and to suggest some conclusions accordingly. For instance, van Gennep's assertion that all such custom is based upon a tripartite system of separation, transition and incorporation does appear to be backed up by the Scottish evidence, for in both personal and seasonal passages the 'journey' taken by those involved is marked out in a highly structured way in most cases. And while certain elements of these journeys require the involvement of the wider community and as such are situated within the public sphere, much of the action is centred within the family or household unit and is therefore acted out within the private space of the home. Even this relatively limited space, however, is demarcated very carefully in order to provide a physical environment which reflects the various ritualistic stages through which people and time must travel. Division of space requires the establishing of boundaries, and it is the crossing of these boundaries which lies at the heart of much of the activity highlighted in this chapter.

It is clear, then, that the home, as both concept and place, is a highly complex phenomenon indeed, and one that plays its part in the cosmological outlook of the people of Scotland, past and present. We have some way to go yet, however, before our understanding of the full range of meanings we bestow upon our own homes is in any way complete.

NOTES

1. Even in 1861, 64 per cent of the Scottish population were living in one or two-roomed houses. Smout, 1986, 33.
2. Van Gennep takes pains to point out, however, that the relative importance of each phase varies according to context. 'Rites of separation are prominent in funeral ceremonies, rites of incorporation at marriages. Transition rites may play an important part, for instance, in pregnancy, betrothal, and initiation; or they may be reduced to a minimum in adoption, in the delivery of a second child, in remarriage, or in the passage from the second to the third age group.' Van Gennep, 1960, 11.
3. Holden, 2000, 267.
4. See Chapter 24 Childbirth.
5. Clark, Carnegie, 2003, 82.
6. As reported, for instance, by Stephen Bichan, Deerness, Orkney, SSS, SA 1967.115.
7. Maclagan MSS, p7396. Recorded from Miss Maclachlan, Ardrishaig, Argyll. Published in *Tocher*, 38 (1983), 50.
8. Gregor, 1874, 84.
9. The first attendance at church by the mother can be seen to represent her 'incorporation' into the Christianised version of this tripartite model of a rite of passage. Similarly, the child's journey into life is considered safely complete only once he or she has been formally baptised.
10. Gregor, 1874, 87.
11. Martin, 1998, 284.
12. The list of countries from which Martin cites evidence includes Wales, Ireland, Brittany, Germany, Austria, Norway, France, Russia and Japan. Martin, 1998.
13. Martin suggests that the common requirement for the matchmaker to be physically strong is indicative of the idea that he was not just a representative of the suitor, but also a physical manifestation of him and that 'he was, after all, embodying the young man's youth, strength, capability and determination'. Martin, 1998, 194.
14. Martin, 1998, 201.
15. MacLeod, 1978, 375–6. The specific metaphor employed could vary, although often it related to the occupation of the girl's father, for example a lamb for a crofter, a boat for a fisherman, etc. Where the father was keen to add to the verbal drama he may well have taken the metaphor further; he might explain, for instance, that the boat 'has never yet been sailed in'!
16. Rea, 1964, 141–3.
17. Martin, 1998, 225.
18. Smout, 1981, 205.
19. It was the placing of the bride in the bed which was carried out under the gaze of those present, not the actual act of consummation itself.
20. SSS, SA1961.27. Quoted in Cooke, 1986, 85–6.
21. Smout, 1986, 34.
22. See Chapter 26 Infirmity.
23. Dr J B Russell. Quoted in Smout, 1986, 34.
24. Gregor, 1874, 137. The opening of doors and windows was not confined to customs associated with death, for time, as well as spirits, had to be allowed free movement when appropriate. The author recalls from his own childhood the practice of opening the front and back doors of his family home to allow the old year out and the new year in.
25. Many of these practices remain current in urban as well as rural Scotland. Hugh Hagan of Port Glasgow witnessed all of these beliefs in action following a family death in 1990. SA 1991.15. Quoted in Bennett, 1992, 243–6; for a detailed

anthropological account of marriage ritual in contemporary Scotland see Charsley, 1991.
26. See, for example, Faley, 1990, 137. This serves to explain why some people still become alarmed if they see a neighbour's curtains drawn during the day.
27. The howdie took on the responsibility of helping to see members of her community both into and out of this world, providing practical help at both births and deaths.
28. Gordon, 1984, 24-5.
29. Gordon, 1984, 9.
30. Martin, 1998, 287.
31. A re-assessment by Hutton, 1996, however, casts doubt on the appropriateness of the 'Celtic' label, in that the practices and beliefs associated with such festivals as Beltane, for example, were neither universal within the Celtic language areas of Europe, nor peculiar to them; Lyle, 1990, 75-85, also examines the four-part division of the Celtic year in some depth, and suggests that it may have developed from an older three-point structure.
32. Hutton, 1996, 370, argues that the characterisation of Samhainn as having been a 'festival of the dead' is unfounded. Rather, 'its importance was only reinforced by the imposition upon it of a Christian festival which became primarily one of the dead' [All Souls' Day]. The modern 'Halloween' probably owes as much to the Christian tradition as it does to the pagan Samhainn.
33. The term 'Hogmanay' may derive from the medieval French *aguillanneuf*, a word used to refer either to New Year itself or to a New Year's gift. Hutton, 1996, 33; an alternative but unlikely suggestion is that it relates to 'hugman' bread given to beggars at the end of Yule. Hayward, 1992, 16. The earliest use of the word within a Scottish context dates to the beginning of the seventeenth century; see also *SND* s v Hogmanay.
34. See Lyle, 1998.
35. For a more detailed discussion of this practice see West, 2001.
36. In some areas, the smothering of the hearth flame was accompanied by the recitation of a blessing. Alexander Carmichael included several of these in his *Carmina Gadelica* collection of hymns and incantations of the Highlands. Carmichael, 1992; see Chapter 13 Spirituality; Chapter 6 Heating and Sleeping.
37. Banks, 1939, 29.
38. MacDonald, 1893-4, 273.
39. The archives of the SSS contain many recorded accounts of such activity. See for example, SA 1981.79, Finlay Maclean, Lewis; SA 1971.43, Donald Alasdair Johnson, South Uist; SA 1980.106, Donald MacDonald, Lewis. The chant commonly sung was the 'Duain Challuinn', the New Year rhyme, which as Alan Bruford points out, could take any one of a number of forms '... in many places several were known, ranging from ancient Ossianic lays to newly improvised farragos with satirical verses about local people and a good deal of pure nonsense'. Bruford, 1993, 102.
40. This was just one verse of a song collected in the north east of Scotland in the 1880s. Gregor, 1881, 161.
41. The origin of the name 'Galoshins' is obscure, although as Brian Hayward shows in his analysis of this tradition, a number of suggestions have been made over the years. None seems convincing, however, and Hayward himself believes that the term may derive from middle English 'galosh' (Old French, *galoche*) meaning a wooden shoe or clog. Hayward, 1992, 72-88.
42. Captain Basil Hall's 'Journal' (MS), Abbotsford, 1 January 1825, printed in Lockhart, 1837, 385. The 'host' referred to was Sir Walter Scott.
43. At Polwarth, Berwickshire, the players took on 'an almost idealised progress from

the Manse and the "laird's big house" to the farmhouse and the humble cottage'. Hayward, 1992, 43.
44. *Falkirk Herald*, 24 June 1905, 26 June 1907. Quoted in Burnett, MacCallum, 2001–2, 51.
45. *Falkirk Herald*, 25 June 1924. Quoted in Burnett, MacCallum, 2001–2, 54.
46. Burnett, MacCallum, 2001–2, 56.
47. Burnett, MacCallum, 2001–2, 54.
48. This was the process recalled by Nan MacKinnon, Vatersay, in 1958 (SSS, SA 1958.152.); the details could vary a good deal, however. One source describes the use of six plates rather than four: pure water (predicting an unexceptional husband); salt (marriage to a sailor); meal (a farmer); earth (a death); dirty water (a disreputable husband); an empty plate (no husband at all). Goodrich-Freer, 1902, 53.
49. The following works provide a good starting point for further reading on this theme: Clark, Dick, Fraser, 1996; Cowan, 1974; Daiches, 1987; Denton, Wilson, 1991; Edensor, Kelly, 1989; Maan, 1992; Storrier, 1993.
50. Maan, 1992, 188.
51. Maan, 1992, 197.
52. Edensor, Kelly, 1989, 91; for a detailed analysis of the theme of acculturation in relation to Edinburgh's Jewish community see Storrier, 1993.
53. Denton, Wilson, 1991, 34.
54. Daiches, 1987, 26–7.

BIBLIOGRAPHY

Banks, M. *British Calendar Customs: Scotland*, vol II, London and Glasgow, 1939.
Bennett, M. *Scottish Customs from the Cradle to the Grave*, Edinburgh, 1992.
Bruford, A. Festivities and customs, seasonal. In Daiches, D, ed, *The New Companion to Scottish Culture*, Edinburgh, 1993, 100–4.
Burnett, J, MacCallum, M. Grangemouth Children's Day, Bo'ness Fair and Lanimer Day at Lanark, *ROSC*, 14 (2001–2), 43–57.
Carmichael, A. *Carmina Gadelica: Hymns and Incantations Collected in the Highlands and Islands of Scotland in the Last Century* (1900), Edinburgh, 1992.
Charsley, S R. *Rites of Marrying: The Wedding Industry in Scotland*, Manchester, 1991.
Clark, H, Dick, L, Fraser, B. *Peoples of Edinburgh: Our Multicultural City. Personal Recollections, Experiences and Photographs*, Edinburgh, 1996.
Clarke, H, Carnegie, E. *'She was Aye Workin': Memories of Tenement Women in Edinburgh and Glasgow*, London, 2003.
Cooke, P. *The Fiddle Tradition of the Shetland Isles*, Cambridge, 1986.
Cowan, E. *Spring Remembered: A Scottish Jewish Childhood*, Edinburgh, 1974.
Daiches, D. *Two Worlds*, Edinburgh, 1987.
Denton, H, Wilson, J. *The Happy Land*, Edinburgh, 1991.
Edensor, T, Kelly, M, eds. *Moving Worlds: Personal Recollections of Twenty-One Immigrants to Edinburgh*, Edinburgh, 1989.
Faley, J. *Up Oor Close: Memories of Domestic Life in Glasgow Tenements, 1910–45*, London, 1990.
van Gennep, A. *The Rites of Passage*, London, 1960.
Goodrich-Freer, A. More folklore from the Hebrides, *Folklore*, XIII (1902), 30–62.
Gordon, A. *Death is for the Living*, Edinburgh, 1984.
Gregor, W. *An Echo of the Olden Time*, London, 1874.
Gregor, W. *Notes on the Folklore of the North East of Scotland*, London, 1881.
Hayward, B. *Galoshins: The Scottish Folk Play*, Edinburgh, 1992.

Holden, L. *Encyclopedia of Taboos*, Oxford, 2000.
Hutton, R. *The Stations of the Sun*, Oxford, 1996.
Lockhart, J G. *Memoirs of the Life of Sir Walter Scott, Bart*, Edinburgh, 1837.
Lyle, E. *Archaic Cosmos: Polarity, Space and Time*, Edinburgh, 1990.
Lyle, E. *Scottish Customs in a Wider Context* (SSS Occasional Papers, no 1), Edinburgh, 1998.
Maan, B. *The New Scots: The Story of Asians in Scotland*, Edinburgh, 1992.
MacDonald, Rev J. Stray customs and legends, *TGSI*, XIX (1893–4), 272–86.
MacLeod, M. Rèiteach, *Tocher*, 30 (1978), 375–99.
Martin, N. 'Ritual Dialogue in Marriage Custom with Special Reference to Scotland', unpublished PhD thesis, University of Edinburgh, 1998.
Rea, F G. *A School in South Uist: Reminiscences of a Hebridean Schoolmaster, 1890–1913*, London, 1964.
Ross, A. *Folklore of the Scottish Highlands*, Stroud, 2000.
Smout, T C. Scottish marriage regular and irregular, 1500–1940. In Outhwaite, R B, ed, *Marriage and Society: Studies in the Social History of Marriage*, London, 1981, 204–36.
Smout, T C. *A Century of the Scottish People, 1830–1950*, London and Glasgow, 1986.
Storrier, S. Jewish cuisine in Edinburgh, *SS*, 31 (1993), 14–39.
West, G. Oral Testimony and the Hearth: Custom, belief and superstition in the Scottish tradition. In Wood, M, ed, *The Hearth in Scotland* (SVBWG Regional and Thematic Studies, no 7), Edinburgh, 2001, 86–99.

Archival Sound Recordings
SSS:
SA 1958.152, Nan Mackinnon, Vatersay, Outer Hebrides.
SA 1961.27, John Irvine, Whalsay, Shetland.
SA 1967.115, Stephen Bichan, Deerness, Orkney.
SA 1971.43, Donald Alasdair Johnson, South Uist.
SA 1980.106, Donald Macdonald, Lewis.
SA 1981.79, Finlay Maclean, Lewis.
SA 1991.15, Hugh Hagan, Port Glasgow.

31 Animals

CATHERINE SMITH

> Pussy at the fireside
> Suppin up brose
> Doon came a cinder and burnt pussy's nose
> Och, said pussy
> That's no fair
> Weel, said the cinder
> Ye shouldnae been there
>
> (traditional rhyme)

INTRODUCTION: ANIMALS IN THE HOME

From at least the Neolithic Age in Scotland, people have lived alongside and benefited from keeping domesticated animals. Whilst cattle, sheep, goats and pigs were originally kept primarily to supply meat, milk, wool and hides, other species, primarily the dog, and later the cat, had roles as working animals in the service of man. Generally, the flesh of dogs and cats was not eaten, except possibly in cases of food shortage, although their skins were often used.[1]

Some authors have argued that pets are animals which are neither eaten nor do any work.[2] However, from the prehistoric period, to perhaps early modern times in Scotland, as in other parts of Europe, it is unlikely that any animal had the single function of non-working companion. Food and shelter were more likely to have been provided for those animals with multiple roles, including acting as working beasts and as a source of meat and raw materials for human use. It is perhaps fairer to say that an animal distinguished as a 'pet' is one which is allowed into the home.[3] A pet is also an animal which has been given a personal name,[4] although of course it is not possible to know whether individual animals had their own unique personal names in times lacking written records. The *Carmina Gadelica*, a late nineteenth-century collection of incantations and songs derived from oral tradition, possibly reflects earlier attitudes. Here cows are given names relating to their characteristics, such as 'Mineag' [gentle].[5]

Pets need not be mammals. At present a wide variety of non-mammalian species are kept in the home, and these, it is claimed by their owners, are indeed pets. They may be birds, reptiles, fish, insects or arachnids. In addition, the number of mammal species which are kept in Scottish homes is ever increasing. It is becoming more common to find animals previously viewed only as a source of meat, for example the Vietnamese pot-bellied pig, snuggled up on the hearthrug. During a devastating outbreak of foot-and-mouth disease in 2001, media attention focused on a woman who turned her living room into a refuge for a flock of Dutch Zwartbles sheep in a vain attempt to save them from slaughter. It availed little that she argued that these were not domesticated animals, but pets, each with its own name.[6] Ironically, a calf accidentally saved from a funeral pyre lost its status of domestic animal intended for slaughter by acquiring the name Phoenix and thus becoming a pet.[7]

Today, we tend to consider as pets those animals on which particular affection is bestowed. Often if town dwellers think of domestic livestock such as cows and sheep at all, it is with indifference, or perhaps guilt at the plight of battery hens and veal calves kept in crates. The distinctions between pets and domestic animals may, however, sometimes be blurred, especially in rural communities. In the Highlands of Scotland, until the late nineteenth and early twentieth centuries, it was common for the home to be divided into two parts: a room for human occupation, with an adjoining area set aside as a byre for cattle and sheep[8] (sometimes the layout comprised a good room and a combined kitchen-cum-byre). The byre was usually set on lower ground, so that animal waste could run out without contaminating the human living area. Excavations of late medieval dwellings on Eilean Mór on Islay have found evidence of byre dwellings which may confirm Pennant's eighteenth-century observations of milk cows and horses being housed in winter.[9]

On Orkney, houses were fitted with plenishings which allowed hens to roost in the kitchen end, calves to be tethered and geese to nest inside, alongside the humans who might be sleeping in the neuk bed.[10] Sometimes young pigs or litters of puppies were also to be found there, the 'pet' animals alongside the purely 'domestic'. Small openings were sometimes let into the outer walls to let cats and hens in and out, as at Appiehouse in Stenness.[11] Good stockmanship involves factors such as careful handling of animals, a basic knowledge of their requirements for food and shelter, recognition of departures from normal behaviour of individual animals and above all patience and empathy with the stock,[12] all of which could be more easily satisfied by keeping livestock in the house.

The types of animals kept in nearest proximity to the beds of humans in the house were probably young or newborn beasts which might have needed special attention, sick animals and those about to give birth. Livelihoods depended on these animals and this manner of housing them would have allowed quick access in case of a nocturnal veterinary emergency, as when a cow was having a difficult time calving. Milking cows and stirks were more likely to be kept in the byres attached to the main living areas.

That such animals could be treated almost as members of the family is again shown by the *Carmina Gadelica*.[13] There is real affection here in the songs the women sang to their cows as they milked them, and an indication of value with which they were accorded, 'Jewel of the white cows art thou'.[14] Significantly, an orphaned lamb which has lost its mother at birth is often still kept in front of the kitchen fire, bottlefed and referred to as a 'pet lamb', the term often applied by extension as an endearment to children or a beloved. Successfully reared lambs are put out to the fields with the rest of the flock and it can be disconcerting to hear an Angus farmer's wife rounding up lambs to go to market, calling out 'come on, Petty'.

THE PREHISTORIC PERIOD

Since the so-called Neolithic 'revolution', when people began to settle in one location, cultivate crops instead of gathering wild plants, and keep domestic animals rather than hunting wild game for meat, our relationship with animals has changed. Domestic animals are dependent on human beings for shelter and for food in a way that wild animals are not.[15] Neolithic peoples first began to farm in what is now Scotland around 6000 years ago,[16] bringing their domestic animals with them from mainland Europe. There is archaeological evidence of these animals, in the form of butchered bones, from habitation sites such as Buckquoy, Orkney.[17] We cannot know if the bones of dogs from such early sites were those of pet animals, but at least we can say that it is from these beginnings that the notion of pet keeping must have developed.

Animal bones have been found in other archaeological contexts. Burial sites, such as the chambered tombs of the Neolithic, Bronze Age barrows and the cist burials of the Iron Age, have occasionally yielded glimpses of people's relationship with animals. For instance, the chambered tomb of Quanterness, Orkney, contained both dog and puppy bones, as well as the bones of domestic livestock.[18] It is thought that ritual meals may have been eaten within the tombs. In the Bronze Age, both in continental Europe[19] and in Scotland, it was customary to deposit a joint of meat alongside many burials. Legs of pork were often favoured and examples in Scotland are known from East Lothian, Kinross-shire, Aberdeenshire and Perthshire.[20] A curious burial from a cist at Lop Ness, Orkney, contained the skeletons of two newborn or foetal lambs, positioned over the feet of the buried body.[21] Perhaps the lambs were a special food offering for the journey to the underworld, or they were there for other reasons, impossible now to know.

Animals also feature in hunting and riding scenes on the enigmatic carved stones of the Pictish (late Iron Age) period. In other examples of Insular art, such as the Book of Kells, cats and mice are commonly, and often humorously, depicted.[22]

THE MEDIEVAL PERIOD

Both written records and archaeological evidence survive for the medieval period. Reconstructing the medieval attitude to animals from these sources is, however, problematic. It is easy to view the past in the light of modern prejudice and this may include making the assumption that medieval animals were customarily treated with mindless cruelty. For instance, there is abundant evidence from medieval archaeological assemblages from Scottish burgh towns that the bones of horses, dogs and cats were frequently butchered,[23] a practice of course now unknown in Scotland in relation to these species. The economy of medieval Scotland was, however, based chiefly on animal products. The Exchequer Rolls of Scotland are the main documentary records of goods exported from Scotland, and it is clear that the revenues raised from the sale of hides, wool, woolfells (sheepskins) and the skins of other animals were of prime economic importance.[24] Dogs, cats and horses thus played their part in the export economy of the nation. But not all the marks on the bones can be simply ascribed to removing the animals' skins. It is obvious that some of the horse bones were butchered in exactly the same way as cattle were. Thus it may be inferred that horse meat was eaten by medieval Scots, in much the same way as it is today in continental Europe. Some of the horse meat may also have been intended as dog food.[25]

Attitudes towards animals at this period were shaped by contemporary understanding of the Christian Church's teaching. There was widespread belief that animals had been created by God for man's use, in the words of Genesis IX, 3, 'so that every living thing may be meat for you', or, as interpreted by the poet Henryson, 'all creature he maid for the behufe, of man and his supportatioun'.[26] Thomas Aquinas (1225–74), an influential thinker of the medieval Church, considered that, unlike humankind, animals possessed no immortal souls,[27] which conveniently allowed people to dispose of them in any way they thought fit. However, there is evidence for animals clearly occupying the role of pet during the Middle Ages. A very small dog skeleton was recovered from the excavation at 75–77 High Street, Perth. This dog was only about nine inches high (23.4 cm), was suffering from osteoporosis and was probably fairly elderly when it died. Most likely it had been a cosseted lapdog, of the type known in Scotland as a 'messan', often favoured by well-to-do ladies.[28] A story told of the execution of Mary, Queen of Scots, relates how, after the death of the Queen, her little dog was found hiding amongst her petticoats and 'could not be gotten forth but by force'.[29] Other dogs were not so fortunate, spending their lives outside the home, perhaps as watch dogs, herding dogs or even pariahs, scavenging on the burgh middens for food.

THE POST-MEDIEVAL PERIOD TO THE NINETEENTH CENTURY

Amongst the concerns of James VI of Scotland when succeeding to the Crown of England was the suppression of the practice of witchcraft. A coven of witches from North Berwick was accused of trying to wreck a ship in which James was sailing, their power having come from a spell involving the maltreatment of cats.[30] The supposed existence of witches led to many trials of women accused, amongst other things, of having communion with familiar spirits, supposed to have taken on the shape of earthly animals and most commonly that of a cat. Cats were also required for a ritual known as the 'taigheirm', which survived in the Hebrides until the mid eighteenth century. This custom involved roasting a succession of cats on a spit, the purpose being to conjure spirits and thus obtain benefits such as prosperity or courage.[31]

Nor was the recreational use of cats confined to roasting. A popular pastime, probably a survival from pre-Christian midsummer ritual, which continued in Perth and Kelso until the late eighteenth century, required a cat to be hung up in a bag or barrel of soot. The object of the 'game' was to break the barrel open and thus release the animal, soot and all, to the hilarity of the crowd.[32] Other pastimes requiring animals ranged from cock fighting to bear baiting and dog fighting. Archaeological evidence of cock fighting has been found in a fifteenth-century deposit at 75–77 High Street, Perth.[33] Here, the bony spur on the lower leg of a cockerel was sawn off in order to fix on a more deadly metal attachment.

Although some fifteenth- and sixteenth-century Scots poets such as Henryson (fl 1450) had shown a sympathy towards birds and animals, this does not seem to have been typical of the period, and it is not until the later eighteenth and the nineteenth centuries that we begin to see the modern attitude to pets developing. The sentimentality of the late Victorians is probably typified by the tale of Scotland's most famous dog, the nineteenth-century Greyfriars Bobby, whose life was mythologised by the American author, Eleanor Atkinson, and also by several filmmakers. Forbes Macgregor, a twentieth-century headmaster, writer and journalist,[34] investigated the story of the dog who would not leave his master's grave and found that the real Bobby was originally the pet of a policeman named John Gray of Edinburgh. Bobby did indeed keep watch in Greyfriars Kirkyard after Gray died in 1858 and was famous for having learned to come for his dinner at the sound of the one-o'clock gun at Edinburgh Castle. He was adopted by various people in turn, including the Traill family, who were threatened with a fine because they did not have a dog licence. The Lord Provost of Edinburgh was so impressed with Bobby that he paid for the licence himself. Eventually the story, somewhat embroidered, spread throughout Scotland and abroad. Today a statue erected in Bobby's memory stands in Candlemaker Row in Edinburgh (dogs may not be commemorated in a graveyard intended for humans), and a memorial in Greyfriars Kirkyard to Bobby's late master was paid for by the dog's American admirers. These tributes are as much a celebration of human sentimentality as canine faithfulness.

The animal paintings of Sir Edwin Landseer demonstrate another aspect of Victorian attitudes to pet animals. Although not a Scot himself, Landseer drew much inspiration from his visits to the Highlands and painted many scenes of pathos set there such as the 'Old Shepherd's Chief Mourner' (1837), in which a collie rests his head on his master's coffin in an attitude of complete sadness.[35]

Unwelcome visitors
As well as the animals which people deliberately keep in their homes, uninvited guests can creep in. Stored cereal grain invites small rodents like field mice, house mice and even voles to enter barns and houses. On many Orkney archaeological sites is found a particular sub-species of vole (*Microtus arvalis orcadensis*) which is thought to have travelled by sea, hidden in stored seeds, with Neolithic farmers from continental Europe. Such small mammals have the potential to cause famine if allowed to overrun, and for this reason cats were essential companions in the home and around the barns. In Henryson's poem 'The Two Mice', a sixteenth-century telling of Aesop's fable of the 'Town Mouse and the Country Mouse', the mice encounter 'vitell grit plentie; baith cheis and butter ... and flesche and fisch eneuch ... sekkis full of grotis, meill and malt' – a feast of cheese, butter, meat, fish, oats and barley;[36] all goes well enough until 'Gib Hunter, our jolie cat' puts in an appearance.

Cats were also essential in keeping down the rat population. No archaeological evidence for the black rat has yet been found in medieval Scotland, although the devastating outbreak of plague, known as the Black Death, in the fourteenth century implicates *Rattus rattus* as a carrier. Close proximity between rats and humans is indicated, and it seems likely that rats were present in medieval dwellings. In 1703 Martin Martin saw 'a great many rats in the village Rowdil [Rodel, Harris] which have become very troublesome to the natives'. Apparently, the village cats were quite overcome by the encounter, and had to be revived with warm milk, successfully, it has to be said.[37] The invasive rodents were black rats, since the brown rat (*Rattus norvegicus*) did not reach Scotland until after 1728, the date of its first arrival in England.[38] This was a troublesome, voracious species which could carry a variety of diseases and was welcome neither in home nor barn.

Other vermin besides rats and mice can, and did, invade the dwelling. Fleas specific to humans (*Pulex irritans*), as well as cat and dog fleas (*Ctenocephalides* spp.), and body and head lice (*Pediculus humanus*) have been unwelcome visitors in the home, as have bed bugs (*Cimex lectularius*). Lice were well known to everyone in Scotland, Robert Burns famously describing one impudent individual which he saw 'on a lady's bonnet at Church'. All of these insect home dwellers feed on human blood and at the very least leave red puncture marks on the skin, usually provoking itching, scratching and sometimes secondary skin infections; more seriously they also act as carriers of micro-organisms which can cause fatal infectious diseases. Lice in particular carry the organisms which cause typhus, a major epidemic disease, relapsing fever and trench fever, the last causing further loss of life during the Great War.

Old people who remembered the Edwardian fashion for mantelpiece frills later condemned them as 'herbours o' dirt', as they could shelter troublesome parasites. Textile hangings around beds, as well as bedlinen and mattresses, could also harbour bed bugs which were in addition reputed to live under wallpaper in Victorian tenement buildings. When many of these buildings became derelict in Dundee and Glasgow in the 1960s and 1970s, firemen and demolition workers complained bitterly of being viciously attacked by bed bugs lying dormant in the cracks between the floorboards. Bed bugs were virtually eradicated in the later twentieth century, but in the last few years they seem to have made something of a comeback, probably because of their increasing resistance to modern pesticides.

PETS OF THE WORKING CLASS IN THE TWENTIETH CENTURY

Throughout the twentieth century and into the twenty-first, pets have played an important part in Scottish home life. The types of animals kept as companions have necessarily been influenced by the dwellings in which their owners have lived. In the first half of the century, the homes of both rural and urban working-class people were often small and cramped.[39] In Dundee, the commonest type of working-class dwelling was the two-roomed tenement flat, often known as a 'room and kitchen', most examples of which were built in the latter half of the nineteenth century.[40] Conditions in these tenements could be difficult. Apart from the crowded nature of the living accommodation, sanitary arrangements were often unsatisfactory and must have contributed to outbreaks of cholera and typhus.[41] Open middens in the back green were the norm until the 1930s[42] and into these went, along with household rubbish, what was euphemistically termed 'night soil'. In the matter of refuse disposal, little had really changed since the medieval period, and it was still possible for cats and stray dogs to forage through waste for food.

Charles (Chic) Rooney, born in 1927 in Court Street, Dundee, remembers spending his childhood in the company of various animals (see Fig. 28.7). His first dog, a collie named Laddie was 'a bra' doagie'. Sent away to live with grandparents when his mother Annie was hospitalised in the 1930s, Chic was heartbroken to return home to find Laddie had been 'gien awa'. Never long without a pet, he found a substitute in a white rat, named Confucius, which he transported to school every day inside his jumper. At night, Confucius slept in the kitchenette of the house in Wedderburn Street where the Rooneys now lived.

In contrast to this peaceful coexistence with rodents, rural rats were still seen as harmful agricultural pests. Edith Ferguson, who lived in a two-roomed clay-built and thatched cottage at Tayfield, St Madoes, in the Carse of Gowrie in the early 1950s, kept a pet fox terrier called Toby.[43] It was Toby's job to defend Edith's baby, Donald, from the rats which scrabbled in the thatch at night. Toby therefore slept inside, rather than outside. At threshing time, the local farmer often borrowed Toby's mother to clear the rats from the barns. These particular dogs were probably pets first and working dogs second.

In town, dogs were less likely to have a role in rodent control, being kept primarily as pets, while it was the cats' job to keep down mice. Dora Buchan, born in 1920 in the Hawkhill area of Dundee, well remembers her family's pet spaniel, Beauty. She was a gentle, brown and white dog who was allowed the privilege of sleeping on the bed. Beauty's party trick was to 'lie down and die for the Labour Party'. Her diet, as was the case with most urban dogs in the 1930s, was much the same as the humans in the house, and if she was lucky it might often consist of mince and tatties. This was obviously more than adequate for Beauty, who became sadly overweight. In old age, when it was obvious that she was on her last legs, Dora's brother Peter paid another youth a packet of Woodbines to 'tak her to Spreull's the vets, tae be pit doon', because he had not the heart to do it himself.

Dogs need space, which in the typical tenement flat of the past was at a premium. Betty Gerrard, born in 1929, lived in and around Dundee's Princes Street, an area dominated by jute mills, where she remembers cats were kept far more often than dogs. When Betty's mother Cathy came home from her shift as a weaver, their cat, Darkie, would jump on her shoulder and purr in her ear until Cathy's 'piece and jam' was shared with her. During wartime, food rationing also played a part in influencing whether dogs or cats were kept. Cats ate less and when hungry would forage for mice and birds. While food for humans was scarce, cats were fed on the more unpalatable bits of fish, such as boiled cod heads. Whiting, one of the cheaper white fish, was also considered suitable for cats. Dora Buchan's cat, Tam, was supplied with fish heads from Mrs Knight's fish shop in the Hawkhill.[44]

As for sleeping arrangements, cats can doze almost anywhere, and frequently do. The most common place for the cat to be found was in front of the coal fire. Betty Gerrard's Auntie Nellie was fanatical about black-leading the grate of her fire in Princes Street and it was no coincidence that her black cat was named Zebo after the grate polish of the same name. In Dundee at this period some of the two-storeyed cottage dwellings still had solid-fuel kitchen ranges for cooking.[45] At night, when the fire had gone out or been banked down, the range would still give off heat, and it was common for cats to sleep on top of it. Chic Rooney remembers a friend's house in Court Street where the kitchen oven was filled with drying kindling when not in use, and often a cat too.

Only occasionally were cats allowed to sleep on beds. Dora Buchan's cat Tam was one of the few allowed this privilege. Since most people had a fear of infestations of fleas or bed bugs[46] the habit was not often tolerated.

Besides cats and dogs, caged birds such as budgerigars and canaries were popular as they took up little space and, apart from the initial expense of buying a cage to house them, were relatively inexpensive to keep (Fig. 31.1). A song popular in Dundee from the nineteenth century to the present day, 'Auld Maid in a Garret' illustrates one of the reasons why people kept caged birds – they kept feelings of loneliness at bay, 'If I cannae get a man then I'll shairly get a parrot'.[47]

In real life, a parrot called Loretta (a red and blue Macaw) was one of the Buchan family's more unusual pets. She (or perhaps he) had been acquired in the 1930s from a fellow member of Mrs Buchan's church congregation, a ship's chief engineer. Unfortunately, Loretta was of an uncertain

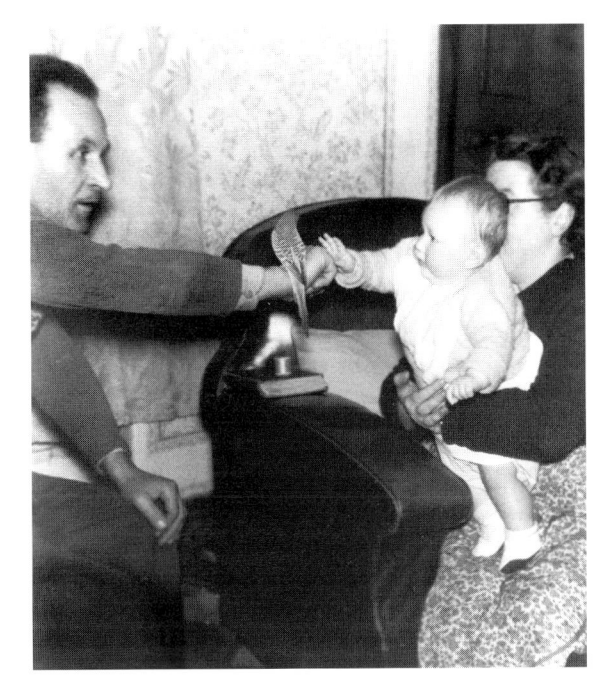

Figure 31.1 Micky the budgie, with Charles Rooney senior, Annie Rooney, and granddaughter Catherine, 1955, Wedderburn Street, Dundee. Source: Catherine Smith: Photo Charles Rooney.

temperament and had a wholly unsuitable vocabulary, given her God-fearing previous owner. 'Go tae Hell' was apparently one of her stock phrases. Loretta's food consisted of peanuts, bought from McVie's Mealstore in the Hawkhill, of which Dora often ate the 'maist pert' on the way home. The parrot eventually outlived her welcome since the local bairns entertained themselves by continually banging on the room window so that she would screech and swear. This was too much for Mrs Buchan, who took her to a man who had an aviary and swapped her for one of his much quieter budgies.

In the post-War period, working-class housing in Dundee improved dramatically. Many families moved from the centre of the town out to newly built schemes such as Charleston and Fintry. The new homes were far more spacious than the tenements of Hawkhill and Princes Street had been and provided more room and opportunity to keep pets. Greater prosperity allowed people to choose pets they could not previously have afforded to feed, and dogs became more common. Tinned dog and cat foods became more readily available and were now relatively cheap to buy.

In 1955 Chic Rooney moved his young family to a terraced house in the Charleston housing scheme. Here he was to keep a succession of gun dogs. Employed at the nearby National Cash Register factory, he became a member of the NCR Wildfowling club, which met at weekends. His gun dogs were trained to retrieve game birds, rabbits and hares (Figs 31.2, 31.3, 31.4). The dogs' reward for their hard work was the rabbit meat, cooked for them by Betty Gerrard, now Mrs Rooney, although when that was not

Figure 31.2 Kim, with the NCR Wildfowling Club, c 1965. Source: Catherine Smith: Photo Charles Rooney.

Figure 31.3 Kim, retrieving pheasant in snow, c 1965. Source: Catherine Smith: Photo Charles Rooney.

Figure 31.4 Dundee NCR Wildfowling Club, c1965. Source: Catherine Smith: Photo Charles Rooney.

available they were fed on tinned food and Vims or Winalot dog biscuits from the Co-operative Store in the scheme. Chic's dogs were undeniably animals with a dual function. Apart from their weekend work, they were also family pets. Being labradors, both Kim and his successor Thunder were canny by nature. Kim in particular allowed small children to sit on him while he slept (Fig. 31.5). Thunder lived to the extremely old age of eighteen, a much-loved companion.

Chic and Betty's children carried on the pet-keeping tradition. In the 1960s, the fashion for keeping hamsters had spread from the United States, and the young Rooneys followed suit. Hamsters have a similar advantage to budgies in that they take up little space. The disadvantage for parents is that cages need regular cleaning if they are not to smell, and for children, that hamsters are more active at night than during the day. Sharing a bedroom with several small creatures running around in a squeaky metal wheel is not conducive to a good night's sleep and eventually the Rooney hamsters were rehoused in the garden shed. There were other pets including rabbits, guinea pigs, canaries and, most of all, budgies.[48] The Rooneys did not continue to keep cats, however. There was a new pride in gardening in the housing schemes, and cats were less well tolerated than they had been in the tenements because of their habit of digging in newly turned soil and upsetting rows of plants and seeds. In addition, the new houses were free of mice and rats and so cats were not as necessary as they had been in the crowded urban centre of Dundee.

At the beginning of the twenty-first century cats have overtaken dogs

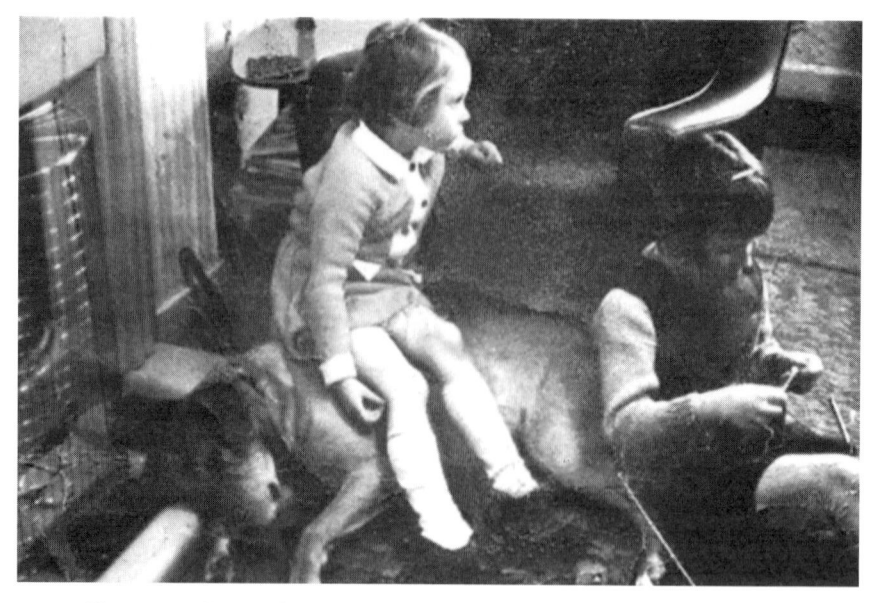

Figure 31.5 Kim with the Rooney sisters, Dundee. Maureen is sitting on the dog while Allison is knitting, resting a needle on the dog's hind legs, c1963. Source: Maureen Rooney.

in popularity simply because they do not take up much space and do not require to be taken outside for exercise. They therefore fit more easily into both their owners' cramped modern living accommodation and their busy lives. Within the European Community as a whole, there are now approximately 47 million pet cats compared with 41 million pet dogs.[49]

Pet keeping in modern Scotland is not confined to those with fixed homes or indeed a roof over their head. For some homeless men and women living on the margins of society, close relationships are formed with pet dogs, who by their nature are uncritical, rather than with other people, who very often are.

PET PARAPHERNALIA

In past times, pets required little other than perhaps a comfortable place by the fireside. They may have had their own bowls to eat and drink from, but were just as likely to have scavenged their food wherever they could find it. Although no evidence survives to show what they looked like, dog collars must have existed in the medieval period since we know that hunting dogs often ran in paired 'leashes'[50] and collars or perhaps harnesses would have been needed for attachment. Collars may have been leather straps, or maybe braided textile. Nowadays, a whole industry is built on selling pet accessories such as dog baskets, coats, combs, nail clippers, shampoos and toys, not to mention medicinal items such as flea collars and worm tablets.

Prepared, often tinned, dog and cat food is the norm and books such as *The Good Cat Food Guide*[51] are available as a reference source for pet lovers.

Such paraphernalia would have seemed superfluous in earlier times, but perhaps they demonstrate the changing role which pets play in the lives of modern Scots. In an age where families are shrinking in size and some couples choose to remain childless, dogs and cats may become substitutes for children or other dependants, encouraging a feeling of being needed. The health benefits which pet ownership can bring are well documented. Medical studies have shown that pet owners are less likely to suffer heart disease as well as lesser health problems such as colds.[52] The reasons for better health may include the increased exercise which 'walking the dog' entails, but emotional factors are also part of the story. Other studies have shown that pets can enhance the quality of life for frail older people; simply introducing a friendly dog cheered the occupants of one nursing home to a remarkable, and scientifically measurable, extent.[53] The description 'companion animal' rather than 'pet' may be more apt in these circumstances.[54] Dogs have long been used to protect homes, both the possessions they contain and the people who live in them. In modern urban areas, especially in the context of those who live alone, perceptions of the need for such protection are perhaps stronger than ever.[55]

There are, however, some drawbacks to keeping pets. Leaving aside the health risks to humans from animal parasites such as tapeworms and fleas, there is the inevitable sadness involved in the disparate life spans of animals and their owners. Because pets can become members of the family with such ease, the death of a beloved dog or cat can be as painful as losing a human relative. Yet in spite of this, there is little doubt that keeping pets continues to enhance the lives of many people in modern Scotland (Fig. 31.6).

Figure 31.6 Robin Spark and his cat greeting each other, Bruntsfield, Edinburgh, late 1990s. Source: Robin Spark.

ACKNOWLEDGEMENTS

Grateful thanks are extended to the Rooney family (Betty, Chic, Allison and Maureen), the Buchans (Dora and Cathie, now Mrs Smith and Mrs Wells respectively) of Dundee and to Edith Ferguson of St Madoes who allowed me to question them relentlessly about their homes and pets from the 1930s to the present day.

NOTES

1. Harcourt, 1974, 171–2.
2. Fiddes, 1991, 133–7; Thomas, 1983, 115.
3. Thomas, 1983, 112.
4. Thomas, 1983, 113.
5. Carmichael, 1992, 344.
6. *The Times*, 2 May 2001.
7. *The Times*, 25 April 2001.
8. Grant, 1998, 75–6.
9. Caldwell, McWee, Ruckley, 2000, 62–4.
10. Fenton, 1978, 147–9.
11. Fenton, 1978, 125.
12. English, et al, 1992, 15.
13. Carmichael, 1992, 335–52.
14. Carmichael, 1992, 348.
15. Serpell, 1996, 1–5.
16. Ashmore, 1996, 22.
17. Noddle, 1983, 92–100.
18. Clutton-Brock, 1979.
19. Green, 1992, 107–8.
20. McCormick, 1991, 113–4; Cowie, Ritchie, 1991, 98; Harman, 1977, 90; Stewart, Barclay, 1997, 43.
21. Smith, 2001, archive animal bone report.
22. Alcock, 1998, 522.
23. Smith, 1998, 876–8.
24. Guy, 1986.
25. Smith, 1998, 876.
26. Henryson's poems may be found in the editions of Bawcutt, Riddy, 1987; 1992.
27. Serpell, 1996, 153–4.
28. Smith, 1998.
29. Lindsay, 1989, 75.
30. Lindsay, 1987, 49.
31. Howey, 1989, 114–7.
32. Penny, 1986, 135; Hazlitt, 1905, 96.
33. Smith, 1998, 881.
34. Macgregor, 1990.
35. Illustrated in Ormond, 1981, 110.
36. Bawcutt, Riddy, 1987, 142.
37. Martin, 1999, 34.
38. Lever, 1979, 89.
39. Butt, 1983, 233.
40. Lenman, Lythe, Gauldie, 1969, 90.

41. Whatley, Swinfen, Smith, 1993, 105.
42. Dow, 2000, 35; also information from Chic Rooney; see in addition Chapter 32 Paying for Dwellings.
43. The clay cottages at Tayfield (once known as Newlands) were built prior to the 1850s and were thus a century old when Edith lived there. They were abandoned in the 1950s, but some of the walls are still standing.
44. Tam was noted for his routine of following Dora's mother every morning on her way to work, as far as the Miller's Wynd washie. Every day, when she returned, he would still be there. Whether he stayed there all day is a mystery, as yet unsolved.
45. Some tenement buildings in Dundee still had kitchen ranges in the 1960s. Alice Fox, a very old lady who lived in Ferguson Street, off Princes Street, taught Scottish dancing in her 'room', while the kitchen contained a very old, black-leaded range.
46. Tarrrant, 1998, 71.
47. This song and many other Dundee gems are contained in Gatherer, 1986, 87.
48. The favourite budgie was one which landed on the kitchen windowsill one day, was captured, and because its original owner could not be found, stayed for many years.
49. FEDIAF website information, 2001
50. Smith, 1998, 869; Rogers, 1880, 107.
51. Gasson, Gasson, 1992.
52. Serpell, 1996, 101–2.
53. Serpell, 1996, 95–6.
54. Serpell, 1994, 129.
55. See Chapter 2 Security.

BIBLIOGRAPHY

Alcock, L. From realism to caricature: Reflections on Insular depictions of animals and people, *PSAS*, 128 (1998), 515–36.

Ashmore, P. *Neolithic and Bronze Age Scotland*, London, 1996.

Bawcutt, P, Riddy, F. *Longer Scottish Poems, vol I, 1375–1650*, Edinburgh, 1987.

Bawcutt, P, Riddy, F. *Selected Poems of Henryson and Dunbar*, Edinburgh, 1992.

Butt, J. Working-class housing in the Scottish cities, 1900–50. In Gordon, G, Dicks, J, eds, *Scottish Urban History*, Aberdeen, 1983, 233–67.

Caldwell, D, McWee, R, Ruckley, N A. Post-medieval settlement on Islay: Some recent research. In Atkinson, J A, Banks, I, MacGregor, G, eds, *Townships to Farmsteads: Rural Settlement Studies in Scotland, England and Wales* (BAR British Series, 293), Oxford, 2000, 58–77.

Carmichael, A (introduction by J MacInnes). *Carmina Gadelica. Hymns and Incantations Collected in the Highlands and Islands of Scotland in the Last Century* (1900), Edinburgh, 1992.

Clutton-Brock, J. Report of the mammalian remains other than rodents from Quanterness. In Renfrew, C, ed, Investigations in Orkney (Society of Antiquaries of London), London, 1979, 112–35.

Cowie, T, Ritchie, G. Bronze Age burials at Gairneybank, Kinross-shire, *PSAS*, 121 (1991), 95–109.

Dow, B. Dundee tenements, *Tay Valley Family Historian*, 55 (2000), 31–5.

FEDIAF European Pet Food Industry Federation website, 2001 <http://www.ipsofacto.be/fediaf/>.

English, P, Burgess, D, Segundo, R, Dunne, J. *Stockmanship: Improving the Care of the Pig and Other Livestock*, Ipswich, 1992.

Fenton, A. *The Northern Isles: Orkney and Shetland*, Edinburgh, 1978.
Fiddes, N. *Meat: A Natural Symbol*, London, 1991.
Gasson, R, Gasson, A. *The Good Cat Food Guide*, London, 1992.
Gatherer, N. *Songs and Ballads of Dundee*, Edinburgh, 1986.
Grant, I F. *Highland Folk Ways* (1961), Edinburgh, 1998.
Green, M. *Animals in Celtic Life and Myth*, London, 1992.
Guy, I. The Scottish export trade, 1460–1599. In Smout, T C, ed, *Scotland and Europe, 1200–1850*, Edinburgh, 1986, 62–81.
Harcourt, R. The dog in prehistoric and early historic Britain, *Journal of Archaeological Science*, 1 (1974), 151–75.
Harman, M. The bones from Uppermill. In Kenworthy, J B. A reconsideration of the 'Ardiffery' finds, Cruden, Aberdeenshire, *PSAS*, 108 (1977), 89–90.
Hazlitt, W C. *Dictionary of Faiths and Folklore* (1905), facsimile edn, London, 1995.
Howey, M O. *The Cat in Magic, Mythology and Religion*, London, 1989.
Lenman, B, Lythe, C, Gauldie, E. *Dundee and its Textile Industry* (Abertay Historical Society Publication, no 14), Dundee, 1969.
Lever, C. *The Naturalised Animals of the British Isles*, London, 1979.
Lindsay, M. *The Scottish Cat*, Aberdeen, 1987.
Lindsay, M. *The Scottish Dog*, Aberdeen, 1989.
Martin, M. *A Description of the Western Isles of Scotland c.1695*, Edinburgh, 1999.
McCormick, F. Animal bones. In Dalland, M. A short cist at Grainfoot, Longniddry, East Lothian, *PSAS*, 121 (1991), 113–4.
Noddle, B. Animal bone from Knap of Howar. In Ritchie, A. *Excavation of a Neolithic farmstead at Knap of Howar, Papa Westray, Orkney, PSAS*, 113 (1983), 92–100.
Macgregor, F. *Greyfriars Bobby: The Real Story at Last*, Edinburgh, 1990.
Ormond, R. *Sir Edwin Landseer*, London, 1981.
Penny, G. *Traditions of Perth* (1836), Edinburgh, 1986.
Rogers, C R. *Rental Book of the Cistercian Abbey of Coupar Angus*, vol II, London, 1880.
Serpell, J. *In the Company of Animals: A Study of Human-Animal Relationship*, Cambridge, 1996.
Serpell, J, Paul, E. Pets and the development of positive attitudes to animals. In Manning, A, Serpell, J, eds, *Animals and Human Society*, London, 1994, 127–44.
Smith, C. Dogs, cats and horses in the Scottish medieval town, *PSAS*, 128 (1998), 859–85.
Smith, C. Report on the Animal Bone from Stenchme, Lop Ness, Orkney, archive report for Historic Scotland/GUARD/SUAT, 2001.
Stewart, M E C, Barclay, G. Excavations in burial and ceremonial sites of the Bronze Age in Tayside, *Tayside and Fife Archaeological Journal*, 3 (1997), 22–54.
Tarrant, N. *Going to Bed* (NMS), Edinburgh, 1998.
Thomas, K. *Man and the Natural World: Changing Attitudes in England, 1500–1800*, London, 1983.
Whatley, C, Swinfen, D B, Smith, A. *The Life and Times of Dundee*, Edinburgh, 1993.

32 Paying for Dwellings

PETER A KEMP

The dwellings in which we live are so expensive that most of us cannot afford to purchase one outright from our income. Consequently most people must rent their home from a landlord or, alternatively, buy it with a mortgage from a bank or building society. Even then, the rent or mortgage repayments usually represent the largest single item in the household budget, and there are always at least some costs associated with furnishing, decorating and maintaining the home, the payment of service bills or charges for certain municipal services. Although in the past it was not uncommon in rural areas for people to build their own homes, this practice became less common in many places during the agricultural-improvement period, when farmers began to provide accommodation for their workers, and elsewhere after 1919 when local authorities increasingly became involved in meeting housing needs. It is because dwellings are so expensive and provide one of the most essential of human needs – shelter from the elements – that, since World War I, governments have intervened extensively in the housing market to help people afford their accommodation.

THE LATE NINETEENTH CENTURY

The experience of paying for dwellings has altered fundamentally for the Scottish people since the late nineteenth century. At that time, approximately nine out of ten households in Scotland rented their home from a private landlord. Almost all other Scots were occupying dwellings which they owned, had built or were buying with a mortgage. In 1913 only about 1 per cent of the population in a total of 11 local authorities were living in houses let by the council. Renting from a private landlord was thus the normal way for many Scottish people, whether poor or better off, to pay for a dwelling.

In order to buy their property, private landlords and owner occupiers typically took out a mortgage using the dwelling as security for the loan. In Glasgow at the outbreak of war in 1914, as much as 90 per cent of property owned by private landlords was said to have been purchased in this way. In general, owners put forward one third of the cost of the dwelling and borrowed the remaining two thirds. Private mortgagees, and to a lesser extent building societies, were the principal sources of mortgages, but banks were rarely a source for such funds.

Private mortgages were generally organised by solicitors. In the absence of an organised capital market they brought together money from a variety sources, such as private individuals and trust funds, and arranged for it to be loaned on mortgages to owner occupiers and landlords. Unlike building-society lending, these private mortgages were interest-only loans which could be recalled at relatively short (usually six months') notice. At the time, lending money on a mortgage was seen as a relatively attractive investment as it was underpinned by the security of the property. Meanwhile, there were few alternative investments available to small investors as most companies were privately rather than publicly owned and not listed on the stock market. Building societies were not seen as the safe haven for deposits that they became after World War I, and because pension provision was relatively underdeveloped, the interest on mortgages could provide a 'nest egg' for old age.

In the Highlands it was relatively common for people to build their own homes. Under agricultural leases, it was the landlord's responsibility to provide farmsteads, but crofters had to provide their own dwellings. Built to traditional styles and methods which varied from locality to locality, these cottages were often relatively primitive and tended not to be long lasting. Building standards began to improve from around the 1880s or even later in some areas. In Lewis, the money for building such cottages was said to have come from the earnings of people who went elsewhere for seasonal work and from money sent home by those who had left to work in the armed forces or the colonies. Notwithstanding these improvements, in 1917 the Royal Commission on the Housing of the Industrial Population of Scotland described housing conditions on the island of Lewis as 'deplorable'. After World War I, local authorities and the Agricultural Department of the Scottish Office provided loans and grants to enable homes to be improved or new ones built to higher, more modern standards.

Although paying rent was an everyday occurrence in late nineteenth-century Scotland, in rural areas especially it was not uncommon for certain types of tenant to occupy their home rent free by virtue of their employment. Farm workers in particular often lived in what was called 'tied accommodation', owned by their employer and provided rent free or at a low rent.

In general, employers provided rent-free or low-rent accommodation not for charitable reasons, but because their workers were paid very low wages, out of which it was difficult to afford accommodation in the normal housing market. The need for workers to live close to the workplace, especially in more remote areas, was also an important reason for the provision of tied housing. Accommodation in return for seasonal help on the farm was common in some areas. For example, in south-east Scotland the cottage was in some cases paid for by the provision of a woman who would undertake farm work when required.

In urban Scotland, the amount of rent which people paid to the landlord – or the landlord's managing agent, known as a 'house factor' – varied according to the size, quality and location of the dwelling. In late Victorian

Scotland building costs and hence rents were generally higher, and earnings lower, than in England and Wales. The result was that north of the border rents accounted for a larger proportion of household incomes. In general, poorer households paid a larger share of their income on rent than better-off households. For the poorest Scots, paying for their dwelling was a struggle at the best of times. Many working-class households lived in overcrowded, unsanitary conditions in very poor or even slum properties, for that was all they could afford in the unsubsidised private housing market.

During the late nineteenth and early twentieth centuries, seasonal and cyclical fluctuations in employment meant that rent arrears were a not-infrequent experience among the poorest tenants. In fact, arrears could be so common during trade depressions that landlords were sometimes obliged to allow their poorer tenants to run them up in the expectation – or at least the hope – that the individuals would be able to repay the debt when trade improved. It was usually housewives who had to deal with the rent collector or landlord as part of their role of manager of the household budget.

At this time, under the common law of hypothec, Scottish landlords had the right to sequestrate (ie seize) their tenants' work tools, furniture and furnishings as security for the rent. In practice this occurred if the tenant refused or was unable to pay the agreed rent for the year. Landlords were not slow to avail themselves of this right, especially in the city of Glasgow which accounted for between a third and one half of the annual value of sequestration for rent in the period 1899–1914. Ultimately, failure to pay rent led to eviction from the dwelling.

Evictions were also very high in Glasgow. Contemporary evidence indicates that there were on average over 5,000 evictions per year in Edwardian Scotland excluding Glasgow, but within that city there were three times that amount in 1902 alone. At this time, in order to repossess the dwelling landlords or their factors had to apply to the courts for a warrant for ejection of the tenant. While elsewhere in Scotland such warrants were normally granted, in Edwardian Glasgow more than a third of them were contested, mainly in order to gain a few extra days in which to remove.

For the poorest of the poor, 'flitting' from the dwelling – usually under cover of darkness – was one, not infrequently used form of rent evasion. To tackle this problem, some landlords and factors insisted upon a character reference from prospective tenants' previous landlord or factor (which was often called a 'factor's line'). Those unable to secure a factor's line could find themselves living in 'farmed-out' accommodation sub-let by 'house farmers' who had rented it from the landlord. This was run-down property in older parts of the city, let with rudimentary furniture, and occupied by people with very limited means. A census of farmed-out accommodation in Glasgow in 1902 found that one in six occupants gave the lack of a factor's line as their reason for living there (Fig. 32.1).

Letting part of the dwelling to a lodger or lodgers was also a very common, but more respectable, method of economising in a trade recession or if times were hard for some other reason. Conversely, moving in with

Figure 32.1 House flitting, Edinburgh, 1934. Source: SLA 128K S.10288/© The People's Story Museum, The City of Edinburgh Council, 1994 (ref B579).

another household – 'doubling up' as it was often called in official reports – was a widely used strategy for economising on rent among the poorest tenants. The sharing, and indeed overcrowding, of accommodation was widespread. At the Census of 1871, for example, 23 per cent of all Glasgow's families contained lodgers. One in five of these families lived in one room with their lodger, while half lived in two rooms.

Another way in which a household could reduce its rent burden was to move to a cheaper dwelling.[1] Tenants' ability to move was, however, restricted by the system of long lets which prevailed in Scotland at this time on working-class property.[2] While landlords and factors accepted that the very cheapest dwellings had to be let on weekly or monthly tenancies, for more expensive working-class homes they insisted that tenants sign up for yearly leases.[3] Moreover, tenants were commonly required to sign a 'missive' (a legally binding written agreement) to renew their tenancy four months prior to the end of the term. Thus, in effect, tenants had to commit themselves to the dwelling for a year and a quarter. Almost all removals were required to take place on 28 May, with the result that this time of the year witnessed a frenzy of activity in the urban housing market as decoration, outstanding repairs and removals all took place.[4]

The system of long lets infuriated tenants and was especially stressful for women since the home was usually seen as their particular responsibility in the division of labour. Moreover, if the household had to move mid year and was unable to find a sub-tenant to replace it, the system of long lets could also create financial hardship as a result of having to pay rent on two dwellings at once. It was not until 1911 that the law in Scotland was changed – much against the will of most landlords and factors – to allow tenants to take out monthly tenancies instead of yearly ones. By way of compensation, the same legislation gave landlords power to evict their tenants at a mere 48 hours' notice.

For better-off tenants, paying for their dwelling was one among many financial transactions, but for the poor it could be highly problematic for at least some or even all of the time. Loss of household goods, or even the home itself, could result from an inability to pay rent, as a result of trade depressions, reduced earnings, or the death of the main breadwinner.

These were difficulties of the market economy in an era when state provision of social security was minimal. What particularly caused resentment among working-class tenants in Scotland, and especially on Clydeside, was the law of hypothec and the missive system. Neither was an inherent feature of a market economy. Tenants viewed them not simply as unjust, but as a product of undue power on the part of landlords and factors within the urban housing market.

THE DEMISE OF MARKET RENTS

The lingering resentment of tenants against their landlords and factors reached new heights following the outbreak of war in 1914. The outcome was a highly significant and long-lasting change in the nature of the rent nexus between landlords and factors on the one hand and tenants on the other, mediated by the intervention of the state.

At the beginning of the War the rapid influx of workers into areas of munitions production led to a sharp reduction in the number of vacant dwellings, to which landlords responded by increasing rents. In turn, the rent increases led to a great outcry from tenants, fearful of being evicted at a time of acute housing shortage. The response by tenants, especially in the Clydeside shipbuilding districts of Partick and Govan, was to refuse to pay the *increases* in rent.[5] These rent strikes were organised on a close-by-close basis, often by women, who thus played a pivotal role in initiating and sustaining what many saw as a class struggle. Landlords and factors were portrayed as 'Huns' seeking unpatriotically to profit from the War while tenants were away fighting at the front (Fig. 32.2).

When the rent strike in Glasgow threatened to spill over into an industrial strike by tenants engaged in vital munitions production, the wartime coalition government of Lloyd George stepped in and introduced rent-controlling legislation.[6] This was a quite unprecedented intervention in the housing market, albeit one that was mirrored in other countries involved in World War I.

The control of rents was initially intended to last until six months after the cessation of hostilities. But by the end of the War the localised housing scarcity in key centres of munitions production had become a nationwide shortage and a pressing political issue. In 1919 and again in 1920, the government reluctantly extended the period for which rent control was to remain in place and expanded the number of dwellings subject to the legislation such that all but very expensive homes came within its ambit.[7] The coverage of the legislation gradually reduced during the inter-War years, but the bulk of working-class dwellings remained subject to rent control throughout this period.

The introduction of rent control in 1915 was a major landmark in the history of the dwelling and the role of the state in the housing market. It meant that the amount of rent tenants paid for their dwellings in the controlled rental market was no longer just the outcome of market forces. Instead, the rent was determined by the amount which happened to be charged in August 1914 plus what governments subsequently decided was an appropriate increase over that sum. It effectively substituted politics for the market in the rent-setting process.

The continuation of rent control after the War was one factor behind the introduction of state subsidies for council housing in 1919 and the subsequent expansion of this new form of housing thereafter. The introduction of exchequer subsidies for council housing was partly prompted by genuine fears of Bolshevik revolution over the housing shortage. Indeed, the Royal

Figure 32.2 Rent Strike photograph. Source: Image courtesy of *The Herald & Evening Times* picture archive.

Commission on the Housing of the Industrial Population of Scotland, which reported in 1917, noted that 'bad housing conditions may fairly be regarded as a legitimate cause of social unrest'. However, it argued that the housing problem was in fact a long-standing one, but had been exacerbated by the shortages and disruption caused by the War. It found that investment in rental housing by private enterprise had dried up even before the War, largely because (for a variety of reasons) tenants could not afford to pay an economic rent. The Royal Commission concluded that this situation was unlikely to change at least for the immediate future and recommended that local authorities be given responsibility to provide housing at subsidised rentals.

Initially built to provide homes 'fit for heroes to live in' the new council dwellings were built to high standards. However, despite the subsidy from the exchequer and from local taxes (the 'rates'), the rents of the new homes were beyond the pockets of the poorest Scottish households. Better-off workers and lower middle-class households – such as those of clerks, teachers and policemen – were the predominant occupiers of these dwellings rather than those living in the squalid slums of urban and rural Scotland.

Falling building costs during the course of the 1920s eventually made it possible to build council dwellings which had more affordable rents. This allowed a growing number of working-class tenants to move into a modern dwelling, rented from a landlord whose primary interest was to meet housing need rather than to earn profit. While the rent could be expected to increase every year it was no longer subject to the vagaries of market forces. The provision of rented dwellings was coming to be seen as more of a social service and less of a market commodity. Even so, council tenants who failed to pay the rent were evicted just as in the privately rented housing market.

In the 1930s, when the focus of council house building had switched to slum clearance and relief of overcrowding, council rents became more affordable but at the expense of lower housing standards. Nevertheless, the tenants who were rehoused moved into new dwellings which were still built to a significantly higher standard, and with better amenities, than the slums they had left behind, although the rents were often higher. During the industrial depression, especially the early 1930s, it was not unknown for the families of some unemployed workers to move from their municipal housing scheme back to cheaper slum property.

By the outbreak of World War II in 1939, council housing had become firmly established. It was also now the largest supplier of rented dwellings: two thirds of all new dwellings constructed between April 1924 and December 1938 were built for local authorities. However, Edinburgh was an exception to this general pattern as private-sector house building exceeded local-authority output between 1918 and 1939. Few privately rented homes were constructed during the inter-War years, especially the 1920s. Meanwhile, unlike England and Wales, owner occupation was slow to develop in Scotland (although, again, Edinburgh was an exception). Indeed, during World War II a government committee of inquiry was set up to consider and advise on the measures required to encourage the provision of houses for owner occupation in Scotland.

THE POST-WAR YEARS

In September 1939, in order to avoid a repetition of the events of 1915 and to prevent profiteering by landlords, dwelling-house rents were frozen at the level then being charged. Meanwhile, house building and renovations came to a halt as resources were diverted to the war effort. By the end of the War there was a considerable excess of households over dwellings, so much so that tackling the housing shortage was at the top of the political agenda.

After the War, the new Labour Government gave councils the dominant role in house building as they sought to tackle the housing shortage in a planned way. Council rents were kept at historically low levels relative to earnings, making them affordable to most working people. At the same time, private-sector rents remained controlled at their 1939 level, despite the substantial increases in retail prices, earnings and building costs which had taken place since that date. For tenants, whether renting from a private landlord or the local council, this meant that the amount of their household budget devoted to rent was historically low relative to earnings.

Meanwhile, the post-War commitment to full employment and the new system of social security (inspired by Beveridge's famous report of 1942) meant that tenants had never been so financially secure and better able to afford their rental payments.[8] Under Beveridge's Plan for Social Security, national insurance was to be the basis of protection against the risks of unemployment, incapacity to work and old age, supplemented by child benefit for families with children. Flat-rate pensions and benefits for incapacity and unemployment were provided at subsistence levels in return for flat-rate national insurance contributions. A separate system of means-tested national assistance was intended to provide a safety net for the minority of people who had been unable to make sufficient national insurance contributions. Whether dependent on national insurance or national assistance, people were much better placed to afford their rent or mortgage repayments under the post-War system of social security than they had been before World War II.

Despite the ecstatic reception with which the Beveridge Report was received, one problem it failed to tackle was what Beveridge called 'the problem of rent'. This referred to the fact that rent varies markedly from one area to another and between households of the same size in the same parts of the country; and that expenditure on rent cannot be reduced during temporary interruptions of earnings (for example, due to unemployment or ill-health) unlike expenditure on fuel, energy and clothing. The system of flat-rate national insurance benefits set at subsistence level and incorporating a notional element for housing costs was ill-equipped to cope with these variations in rents. In fact, social insurance recipients whose rents were above 10p per week in 1948 prices would have been better off claiming the means-tested national assistance since this covered all reasonable housing costs. It was not until 1972 that a system of rent allowances was introduced to provide help with housing costs for people who were not receiving means-tested benefits (see below).

In the post-War years, the large council house building programme meant that a growing number of Scots were renting from their local authority rather than a private landlord. In fact, the size of the private rented sector slowly declined year on year, as existing landlords sold vacant dwellings into the expanding owner-occupier market and few new ones were produced. But while owner occupation increased, its rate of growth was much slower than south of the border. In Scotland, local authorities dominated house building for the three decades following the War.

By the 1970s council housing had become the largest tenure, followed by owner occupation and then private renting. In fact, in many districts, council housing accounted for more than half, and in some districts more than two thirds, of all dwellings. Scotland had become largely a nation of council tenants. Not only were tenants renting their home from the council and not a private landlord, they were also renting from a landlord which owned most of the dwellings in the district. A plurality of competing private landlords had been largely replaced by a municipal oligopoly.

Meanwhile, successive local authority re-organisations generally involved the amalgamation of local authorities into a smaller number of larger councils. One consequence of these mergers was that the number of dwellings owned and managed by each council increased substantially. Concerns began to emerge that council housing was becoming increasingly bureaucratic, remote and unresponsive to tenant wishes, and this was one of the reasons why, from the mid 1970s, governments sought to expand the role of housing associations at the expense of local authorities. Although the rents which tenants of housing associations had to pay were higher than those paid by council tenants, they were usually less than those paid by private tenants.

At the same time, new methods of rent collection, such as Post Office 'giros' and latterly standing orders and direct debits, resulted in the declining use of rent collectors calling at the door to collect the rent every week or month. The changing scale of municipal landlordism and the new methods of rent collection meant that the payment of rent had become less personal and more distant. Increasingly, tenants were not handing cash to the landlord, either to the rent collector at the door or to the cashier at the council offices; instead money was being transferred from one account to another. Paying the rent, to borrow the language of information technology, was becoming more 'virtual' than 'real'. It was a more efficient, but much less personal, way of doing things.

RESERVED FOR THE POOR

From World War II until the 1970s the amount of rent that tenants were charged was low relative to earnings. Council rents were kept low by subsidies from central government and from local ratepayers, while rent control meant that private tenants generally paid rents which were well below what would have been the market price for the dwelling. Unlike local councils – and after 1974 the growing number of housing associations – private landlords received no subsidies from the state, which gave them little incentive to continue letting property; it also reduced their ability to undertake repairs and maintenance. Consequently many private tenants were paying a sub-market rent for sub-standard or un-modernised dwellings. Not surprisingly, council housing, built to modern standards and with subsidised rents, looked a more attractive proposition to most Scottish tenants.

The 1970s marked the beginning of the end of the low-rent housing market in Scotland. At this time there was talk about 'over-subsidised council tenants' and 'indiscriminate' subsidies from local ratepayers and national taxpayers to tenants who did not need assistance with the rent. Successive governments sought to raise rents, initially in order to reduce subsidies to councils and housing associations; and, from the late 1980s, to substitute private for public-sector finance. In 1989, rents on new lettings were deregulated in the private sector, partly as a way of helping to revive this part of the housing market.[9]

In order to help the poorest tenants cope with these rent increases, means-tested rent rebates (for council tenants) and rent allowances (for private and housing-association tenants) were introduced in 1972. These schemes became known as housing benefit from 1982/83 and were merged with the *rent* component of means-tested income-support benefits in 1988. Since 1982/83, housing benefit has been administered by local authorities on behalf of the social-security ministry (currently called the Department for Work and Pensions). For owner occupiers, help with *mortgage interest* payments has remained part of income-support benefits.

The introduction of means-tested housing benefit has effected a major transformation in the payment of rent for low-income tenants in Scotland. In the first place, whereas bricks and mortar subsidies and rent controls keep rents low for *all* households living in the subsidised or controlled dwellings, housing benefit is available only to the *poorest* tenants. To use the jargon of the policymakers, help with housing costs has been increasingly 'targeted' on those most in need. Tenants who are not among those most in need are expected to spend a larger share of their income on rent and this has encouraged considerable numbers of people to consider buying their home instead of renting it.

Secondly, this new form of personalised subsidy involves means testing to establish whether or not a household is entitled to receive such help. As such, it singles out as poor those tenants who apply for it. Long and complicated forms (up to 20 pages long) have to be completed, enquiring about all manner of personal details, such as income, savings and the relationship between the various members of the household. Even quite minor changes of household circumstances have to be reported to the local authority in case it affects housing-benefit entitlement. The inconvenience of claiming and the stigma attached to receiving means-tested housing benefits can in fact be so great as to deter some tenants from applying to their local authority. Elderly people are especially likely to be deterred from applying for housing benefit because of the perceived stigma and notion of charity attached to means-tested benefits.

Because the housing benefit scheme is so complicated, it is difficult for officials to administer quickly and accurately. Consequently tenants can find themselves having to wait weeks, and in some cases months, to get the help to which they are entitled. This can make it difficult to pay rent in the meantime and potentially jeopardise their relationship with the landlord. It also makes some private landlords reluctant to let their accommodation to tenants receiving housing benefit. Tenants whose housing-benefit claim is stuck in the system, or incorrectly assessed, can find themselves having to contact the local-authority benefit office to speed things up or get the claim corrected, in order to be able to pay their rent to the landlord. Housing officers working for housing associations or local authorities often have to spend some considerable time helping to sort out their tenants' housing-benefit problems, not the least in order to prevent rent arrears from developing.

Finally, the introduction and growing importance of housing benefit also has implications for *how* rent is paid. Council tenants who have made a successful claim are entitled to a rent rebate, in which the housing benefit office transfers the money into their rent account, while the housing department bills them for the remainder of the rent. Since more than half of all council tenants have all of their rent covered by housing benefit, they no longer have to hand over any rent money. 'Rent direct', as it is known, saves the tenant from having to pay the rent and the landlord from having to collect it. But it may also, to some extent, have undermined the tenant's bargaining position vis-à-vis the landlord, as he or she is no longer handing over money in exchange for the right to occupy the dwelling.

In theory, low-income private and housing-association tenants are supposed to receive a rent allowance from the housing benefit office to enable them to pay the rent to their landlord. But increasingly – and often at the landlord's insistence – the benefit is paid directly to the landlord, bypassing the tenant altogether. Again, if the benefit covers all of the rent, the tenant effectively no longer pays any rent to the landlord. If the benefit covers only part of the rent, the tenant is supposed to hand over the remainder to the landlord or agent. But many landlords complain that tenants who have only small residual amounts to hand over to them frequently fail to do so.

For private tenants living in deregulated dwellings, however, housing benefit frequently does not cover all of the rent. This often occurs either because the rent the tenant has agreed to pay the landlord is considered to be too high or because the accommodation is deemed to be overlarge for their needs. The outcome of these 'rent restrictions', as they are called, is that private tenants on housing benefit do not know how much rent will be considered by the local authority to be reasonable. As a result tenants cannot be sure that they will be able to afford the dwelling until after their housing-benefit claim has been processed, which can be some weeks after they have signed the tenancy agreement. Private tenants on housing benefit can thus find themselves having to make up the shortfall between what they are contractually bound to pay the landlord and the amount that the local authority housing benefit office regards as reasonable for the tenant to pay in rent. This problem particularly affects young single people under the age of 25, for whom the benefit system is less generous than for other types of tenant. Such people can find themselves struggling to pay a substantial shortfall and ending up in rent arrears or having to move to a cheaper property in the hope that the rent on that dwelling will be considered reasonable by the local authority.

A further difficulty, especially for private and housing-association tenants, is that the rent and benefit payment cycles are often out of step. Housing benefit is generally paid fortnightly in arrears, but landlords commonly require tenants to pay the rent weekly, four-weekly or monthly in advance. The result is that so-called 'technical rent arrears' can arise in between the period between the date that the rent is due and the date at which housing benefit is received. For private landlords especially, this lack of synchronicity is an additional irritation in letting property to housing-benefit recipients.

PAYING FOR HOME OWNERSHIP

Since the 1980s, owner occupation has become much more widespread in Scotland. The expansion of this tenure was accelerated by legislation introduced in 1980, which gave council tenants the right to buy their dwelling at a very considerable discount from its market value. In addition, from the mid 1980s easier access to mortgage finance has made it possible for households on relatively modest incomes to buy their homes. By the late twentieth century, owner occupation had become the tenure of choice, or of aspiration, for the great majority of Scottish households.

Inevitably, as owner occupation has expanded and moved down the income scale, the number of house buyers who run into difficulties with mortgage repayments has also grown (although of course it fluctuates to some extent in line with the economic cycle). The small but significant problem of mortgage arrears is exacerbated by certain social and economic trends which emerged in the final decades of the last century. The rise in unemployment from the late 1970s and the subsequent growth of less-secure forms of employment are important factors which have helped to precipitate arrears. The growth in divorce and relationship breakdown since the 1970s onwards has led to an increase in the number of low-income households, particularly lone parents with children, who are likely to get into arrears.

The amount which homebuyers have to pay their lender depends not only on how much they have borrowed, but also *when* they borrowed it. Because of inflation – the annual rate of which in the last third of the twentieth century was often in double figures – the real value of mortgage repayments falls over time. This means that the proportion of household expenditure devoted to the mortgage falls commensurately as well, provided the household does not move and buy somewhere else with a larger loan.[10] Consequently, it is often low-income, first-time buyers in the early years of their mortgage who have the most difficulty in affording the repayments.

However, since interest rates can fluctuate up or down, so too can the monthly mortgage repayment, even if the household does not move. If interest rates increase at the same time as unemployment is rising, the result can be an increase in mortgage arrears and repossessions. Mortgage repayments can therefore fluctuate with the economic cycle, with significant consequences for the budgets of low- and moderate-income households.

Whereas low-income tenants can apply for housing benefit, owner occupiers receive no such help unless they are claiming means-tested income-support benefits. Yet many people in the lower half of the income distribution are now owner occupiers; it is no longer solely the preserve of middle- and upper-income households in Scotland.

Despite the growth in the number of low-income home buyers, the social safety net for owner occupiers was cut back in the 1980s and 1990s in an attempt by governments to cut back the cost of social security and to encourage people to take out private mortgage-protection insurance. Unemployed home buyers now have to wait nine months before they get any help with their mortgage interest, during which period arrears are likely to develop if the household has insufficient savings or does not have private insurance cover. Inevitably, it is low-income buyers – the group of people most at risk of unemployment – who are the least likely to take up private mortgage-protection insurance.

Once the mortgage is paid off (normally after 25 years), the outgoings of owner occupiers are reduced and they have only to pay for items such as repairs and maintenance (and perhaps service charges if they live in a flat). Since this often occurs when households are near to retirement age – a period when incomes generally reduce considerably – older owners are usually better placed than tenants, who still have to pay rent unless receiving housing benefit for some or all of it. Thus, whereas the amount of rent that tenants pay remains a more or less fixed proportion of household income over time, for owner occupiers there is a lifecycle pattern to paying for their dwelling.

Compared with other assets, home purchase receives favourable tax treatment from the Inland Revenue. Although mortgage interest payments are no longer tax deductible, the notional income that owner occupiers receive from living in their home (by paying rent to themselves instead of a landlord) has also been exempt from income tax since 1963. Likewise, owner occupiers do not have to pay tax on capital gains on their main residence.

However, owners do pay stamp duty (on all but the cheapest dwellings) when they buy their home, as well as other costs such as fees for solicitors and estate agents. These house-selling and purchase costs can be quite considerable, but if house prices have risen significantly since the last purchase, they can be paid for out of the capital gain.

Inheritance tax is paid on the most expensive homes when they are passed on after the death of the owner. With the growth of owner occupation, an increasing number of (mostly middle-aged) people are inheriting homes or monies from the sales of inherited homes. Two out of five privately rented dwellings in Scotland in 1993 had been acquired through inheritance. For these and other landlords, the dwellings on which tenants pay rent provide a source of income (or, if they are sold, the prospect of a capital gain). In 1999/2000, approximately 1 per cent of Scottish households were receiving income from property or sub-letting.

CONCLUSION

Over the past century, the Scots have moved from being a nation of private renters to a nation of owner occupiers and social renters. Prior to World War I, most households secured a dwelling by paying rent to a private landlord within an uncertain and, for the poorest third of households, often harsh economic climate. Following the introduction of rent controls and exchequer subsidies for council housing, the market was to some extent replaced by politics in determining how much rent Scottish tenants had to pay for their dwellings.

The rapid growth of council housing, and from the mid 1970s housing associations, meant that tenants in Scotland were increasingly renting within an administered housing system. Their dwelling was allocated to them by officials, rather than chosen by themselves (subject of course to their purchasing power, preferences and the vagaries of the market place). This meant that the officials allocating the dwellings determined how much of their budget tenants had to spend on accommodation rather than the tenants themselves. Nevertheless, because council, and latterly housing-association, rents were subsidised, tenants had to pay considerably less than if they had been renting a comparable dwelling in the private market. Although council and housing-association rents have risen in real terms over the past two decades, they are still below private-sector rent levels.

Meanwhile, the proportion of Scots buying their home with a mortgage from a bank or building society, rather than renting from a landlord, has gradually increased. Home ownership is now the largest tenure, accounting for over two thirds of all households in Scotland. Paying rent to a landlord is widely perceived as 'dead money', whereas repaying a mortgage is seen as a secure investment for the future. Although owner occupation has many financial advantages over renting, it also has more pitfalls for the small but significant number who are unable to keep up with their mortgage repayments.

As a result of the expansion of owner occupation, renting has increasingly become the preserve of low- and moderate-income households, although not to the same extent as south of the border. The majority of tenants in Scotland are paying their rent with the help of means-tested housing benefit. A major source of financial anxiety has thus been removed.[11] For these people, paying for their dwelling is a very different, and considerably easier, experience than it ever was for the poor in the late nineteenth century housing market.

NOTES

1. In fact, moving was a common experience for middle- and working-class tenants, though especially for the latter. It enabled them to adjust their housing circumstances and corresponding rental commitments – whether up or down as the case may be – to their changing financial circumstances.
2. In England and Wales, working-class tenants commonly rented their accommodation on weekly tenancies. By the turn of the century landlords and their managing agents began to complain that many lower-middle-class tenants now preferred to rent their dwelling on a monthly tenancy instead of the more common yearly or three-yearly lease.
3. In common lodging houses rents were paid nightly. In 1902, 1 per cent of Glasgow's population was living in lodging houses.
4. In Scotland (unlike England and Wales) tenants' freedom to move had been limited since the seventeenth century. Under the Terms Removal Act of 1886, removal was restricted to Whit Sunday and Martinmas, but in practice urban removals were largely confined to the former (which was defined as 28 May). Yearly tenants were obliged to sign a 'missive' or written agreement when they took on, or renewed, a lease.
5. There were also rent strikes in centres of munitions production elsewhere in Scotland and in England.
6. The 1915 Act restricted rents to the amount charged at the outbreak of war in August 1914. The rate of interest on mortgages with terms of less then ten years was also restricted by the Act, which benefited landlords – most of whom had purchased their dwellings with a mortgage – at the expense of lenders. The exclusion of loans granted for terms of ten or more years effectively exempted building societies from mortgage-interest restrictions, which instead mainly affected private lenders.
7. The 1919 legislation allowed landlords to increase the rent by 10 per cent above the 1914 level; the 1920 Act allowed them to increase it by 40 per cent above the 1914 level. Agitation against these rent increases took place on Clydeside and especially in the shipbuilding district of Clydebank.
8. Although the inquiry was established to draw up proposals for streamlining the fragmented and overlapping system of social insurance, Beveridge's report set out a vision for the Welfare State as a whole. He argued that five 'giants' had to be tackled on the road to reconstruction: Want, Disease, Squalor, Ignorance and Idleness. This would require, not just a new system of social security to tackle Want, but also a commitment to full employment and a free national health service. The report, which was greeted with widespread acclaim, rapidly sold out and had to be reprinted. Within two years, over 200,000 copies had been sold worldwide.
9. By 1996, private tenants living in furnished rental accommodation in Scotland were spending 35 per cent of their net (after tax and national insurance) income on rent,

while those in unfurnished rental accommodation were spending 20 per cent. This compared with 23 per cent among public-sector tenants, 26 per cent among housing-association tenants, and 15 per cent among people buying their house with a mortgage.
10. And provided also that they do not remortgage their existing dwelling for a larger sum than the original loan.
11. However, this removal of anxiety assumes that their housing benefit is paid correctly and on time by the local authority, and they remember to report any changes of circumstances as soon as they occur. In fact, a substantial minority of local authorities fail to process benefit claims within the required 14-day period. Meanwhile, failure to report relevant changes in household circumstances is not uncommon, which can result in overpayments of benefit, which the claimant has to repay.

BIBLIOGRAPHY

Beveridge, W. *Social Insurance and Allied Services*, London, 1942.
Burrows, R, Wilcox, S. *Half the Poor: Home Owners with Low Incomes*, London, 2000.
Butt, J. Working-class housing in Glasgow, 1815–1914. In Chapman, S D, ed, *The History of Working-Class Housing*, Newton Abbot, 1971.
Cairncross, A K. *Home and Foreign Investment*, Cambridge, 1953.
Damer, S. *Property Relations and Class Relations in Victorian Glasgow*, Glasgow, 1976.
Damer, S. *From Moorepark to 'Wine Alley': The Rise and Fall of a Glasgow Housing Scheme*, Edinburgh, 1989.
Damer, S. 'The Clyde Rent War!': The Clydebank rent strike of the 1920s. In Lavalette, M, Mooney, G, eds, *Class Struggle and Social Welfare*, London, 2000, 71–95.
Daunton, M J. *A Property-Owning Democracy? Housing in Britain*, London, 1987.
Deacon, A, Bradshaw, J. *Reserved for the Poor: The Means Test in British Social Policy*, Oxford, 1983.
Englander, D. *Landlord and Tenant in Urban Britain, 1838–1918*, Oxford, 1983.
Gauldie, E. *Cruel Habitations: A History of Working-Class Housing, 1780–1918*, London, 1974.
Gibb, A. *Glasgow: The Making of a City*, London, 1983.
Grant, I F. *Highland Folk Ways*, London, 1961.
Harloe, M. *The People's Home*, Oxford, 1995.
Holmans, A E. *Housing Policy in Britain*, Beckenham, 1987.
Johnstone, C. Housing and class struggles in post-War Glasgow. In Lavalette, M, Mooney, G, eds, *Class Struggle and Social Welfare*, London, 2000, 139–54.
Kemp, P A. Some aspects of housing consumption in late nineteenth-century England and Wales, *Housing Studies*, 2 (1987), 3–16.
Kemp, P A. *Housing Benefit: Time for Reform*, York, 1998.
Kemp, P A, Rhodes, D. *The Lower End of the Private Rented Sector: A Glasgow Case Study*, Edinburgh, 1994.
Kemp, P A, Rhodes, D. The motivations and attitudes to letting of private landlords in Scotland, *Journal of Property Research*, 14 (1998), 117–32.
Macgregor, A. *Public Health in Glasgow*, Edinburgh, 1967.
Melling, J, ed. *Housing, Social Policy and the State*, London, 1980.
Melling, J. *Rent Strikes: People's Struggle for Housing in West Scotland, 1890–1916*, Edinburgh, 1983.
Merrett, S. *State Housing in Britain*, London, 1979.
Merrett, S, with Gray, F. *Owner Occupation in Britain*, London, 1982.
Morgan, N J, Daunton, M J. Landlords in Glasgow: A study of 1900, *Business History*, 25 (1983), 264–83.

O'Carroll, A. The influence of local authorities on the growth of owner occupation: Edinburgh and Glasgow, 1914–39, *Planning Perspectives*, 11 (1996), 55–72.
Orbach, L. *Homes for Heroes*, London, 1977.
Pryce, G, Keoghan, M. *Safety Nets for Mortgage Borrowers*, Edinburgh, 1999.
Rodger, R G. The Victorian building industry and the housing of the Scottish working class. In Doughty, M, ed, *Building the Industrial City*, Leicester, 1986, 151–99.
Rodger, R G, ed. *Scottish Housing in the Twentieth Century*, Leicester, 1989.
Swenarton, M. *Homes Fit for Heroes*, London, 1981.
Timmins, N. *The Five Giants: A Biography of the Welfare State*, 2nd edn, London, 2001.
Treble, J H. The market for unskilled male labour in Glasgow, 1891–1914. In Macdougall, I, ed, *Essays in Scottish Labour History*, Edinburgh, 1978, 115–42.

PART TWO

•

Less Common Types of Scottish Household and Home

33 'Time Out' Work

ROGER LEITCH

This chapter is intended to draw attention to the nature and conditions of those who, for economic reasons, have been away from their homes for shorter or longer periods. It brings together examples from different parts of the country, but it is clear that the subject is a large one, substantially unexplored, and the present notes are merely a 'taster' which, it is hoped, will act as a source of encouragement for further research.

Many people in Scotland have been required to leave their homes and reside elsewhere so as to undertake some form of work.[1] This has been part of the round of subsistence agriculture or has constituted self employment, piecework or employment for a fixed wage or salary, and some individuals have taken 'time out' from their normal domestic life to be involved in several of these categories of work. The period of residence can be a matter of a few days or weeks, for a season (however defined), or for longer. It may be a single experience or one which is repeated throughout one stage or at various points within a person's working life. Being away from home can be fraught with feelings of isolation and other forms of hardship and may be endured only for the material gains. Or it may provide the opportunity for positive experiences, including variety, companionship and an increased measure of liberty. Much depends on the individual's personality and experience, his or her age and home circumstances and the nature of the work to be done. The accommodation provided or occupied can also have a considerable effect on morale. It may be self-built or an adaptation of a natural feature, a moveable manmade object or a building with a previous function. It may be purpose-built by an employer and rented to workers or form part of their employment contract. It may be rented or hired from, or lent by, third parties, who may or may not have created the building for purposes of accommodation. It may be a room in a guest house, a bothy, a cabin in a ship, a dormitory in an oil rig or many other things.

LEAVING HOME FOR A SEASON'S WORK

Scotland has a long history of seasonal migration. At the smallest scale, it was sometimes necessary to leave one's main dwelling for limited periods to utilise fully landholdings within the context of subsistence agriculture, cutting peat and carrying it home, or, more commonly, grazing cattle at the summer pastures or 'shielings'. Commercial agriculture has required extra hands for the harvesting of crops such as grain, hay, potatoes or soft fruit, plus other seasonal tasks. The sea has taken people from their homes at certain times of year in the search for food, produce to sell or paid employment. In the last century or so tourism has required an influx of workers to service accommodation or staff leisure activities during the summer months. Although some of the Scots and people of other nationalities involved in seasonal work in Scotland may have moved from job to job without having a base as such, the majority have returned to a home. Men and younger women have played a prominent role in migrant or semi-migrant work, which can require people to be away from their family and home community for several months at a time, although it sometimes has been the case that individuals from the same family or locality have banded together to seek and undertake work.

At the Uist shieling
The practice of transhumance was at one time common throughout much of upland Scotland. In South Uist, the cattle shielings, or summer grazings, were described by Seton Gordon (1886–1977)[2] who recorded that in the 1880s those far up Glen Dorchaidh or upon the lower slopes of Stulaval were still being used for the transhumant movement of dairy cattle from the beginning of July until the end of August. The sites of the shieling huts or 'bothain àiridh' can still be traced close to running water, now marked only by yellow-flowering irises indicative of soil enriched by the droppings of tethered cattle. Older shieling huts in Scotland were generally circular in form (Fig. 33.1). Those built from the nineteenth century until the practice died out in the early twentieth century (transhumance surviving longest in Britain on the island of Lewis) were rectangular and more like the permanent township houses.

An unpublished account of Uist shieling life is to be found in the MS notebooks of the Swedish ethnologist, Åke Campbell. Calum MacLean, who acted as his guide, interpreter and scribe during their 1948 fieldtrip to the Outer Hebrides and the island of Canna, took down a lengthy passage of oral testimony from Annie MacDonald, living on Canna, but originally of South Uist,[3] who recalled going to the shielings about 1903.

In this informant's account, Tuesday is said to have been the favourite day for flitting. Initially all the family went to put everything in order, but only the older girls and younger women would stay on.[4] Two or three families would go to one locality, taking as many as 24 cows with them and the same number of calves. As a rule their huts were close to one another and also near to running water, essential for the cleansing of dairy utensils as well as for personal use. Equipment for milking and milk processing was stored inside the shieling hut along with basic supplies and food utensils, the latter including a kettle and a cooking pot.

Each girl or woman brought a milking cog with her. Sieves for straining milk into basins to let the cream separate out were also taken and the cream was converted into butter using a churn, or 'cuinneag', butter

Figure 33.1 Remains of a circular shieling hut, Uig, Lewis, c1987. Source: SLA 99(3) 53/55/4A.

being churned every day, cream permitting, except for Sunday. Cheese making was another important activity carried out at the sheiling[5] and required the use of a press.[6] The finished butter and cheeses were stored on built-in shelves in the shieling huts until the girls and women were ready to go home.

The working day at the shieling began at 6 o'clock, with the cows being rounded up with the help of dogs. When each cow was milked, half of the milk was given to the calf in its own basin. Milk, along with bread, also provided the girls' breakfast. About midday a meal of potatoes and fish, either salted herring or home-smoked and dried saithe, was eaten. The evening fare was milk and porridge. There was home-ground barley meal as well as oatmeal in the shieling hut and porridge, brose and bannocks could be made from both, the latter baked before a brisk peat fire by being placed against stones. Fresh supplies of foodstuffs were brought from home on Sundays, after the girls had gone to church.

In spare moments the girls and young women knitted stockings for themselves and their menfolk. Recreation was also provided by the receiving of visitors, some of them young men making the most of an opportunity to visit without parental interference. Gaelic song makes repeated reference to such visits, as it does to shieling life in general.[7] In Annie MacDonald's time, a dance to celebrate the end of the summer pasturing was held in the town ship. The 'Bàl Stocainn' or Stocking Ball took its name from the pair of stockings given to the piper in lieu of a monetary fee, with, on occasion, all the girls offering him such a gift.

Going to the shieling, in South Uist as elsewhere, was usually regarded as a very enjoyable activity. The weather could be at its best, and most of the people involved were young and relished days spent in the open air, in a break from the usual routine and with good company and a measure of freedom.

Berry pickers

The farms around Blairgowrie, Perthshire, and also in parts of Angus, Fife and Lanarkshire, provided many urban dwellers with something of a working holiday during the six weeks or so of the soft-fruit season (Fig. 33.2) The cash earned from the piece work of picking mainly raspberries and strawberries – often destined for Dundee jam factories – was needed to augment the income of impoverished folk from Dundee and Glasgow, but a few weeks at the summer berry fields also provided the chance for fresh air and good company. These benefits were especially appreciated by the younger pickers, in some respects similar to the school-pupil potato pickers or 'tattie howkers' who enjoyed a holiday of field work in many parts of the country until relatively recently.[8] The accommodation for the urban pickers – the so-called 'respectable' pickers – took the form of a range of

Figure 33.2 Enjoying a break and a 'piece' at the berry picking. Aberlemno, Angus, 1957. Source: SLA 65 L987.

corrugated-iron huts provided by the farmers; in places these makeshift ensembles were known as the 'Tin City'.[9]

The city dwellers were joined at the berry fields by large groups of Travelling People who converged for the season's work and a chance to see old friends and indulge in merrymaking along the lines of an impromptu people's carnival. Hamish Henderson describes his experience of the scene in the 1950s:

> At camp fires in the berryfields of Blairgowrie ... members of the Travelling fraternity, young and old, sang rare Child ballads, lyric love songs, execution broadside ballads, kid's rhymes, contemporary pop songs, you name it, they sang it.[10]

A holiday atmosphere prevailed among the pickers in general as did a sense of community, although the Travelling folk and the townspeople tended to keep their distance from each other. The Travellers for example shunned the iron huts and preferred their own caravans or tents, or makeshift accommodation. Sandy Stewart, a member of the Travelling community, paints a colourful picture in his autobiography of an array of tents and caravans at Gowthens Muir (Gothens), three miles (5 km) to the south of Blairgowrie, straddling the A93 road to Perth.[11] The description no longer stands as the Muir was cleared of Travellers in the early 1960s, at the same time as many Travellers and others were first able to claim social security. With state benefits berry picking and its low wages fell out of favour and in many respects a chapter of folk history closed.

In recent decades some farms have offered to let the public 'pick their own' soft fruit. People drive out for a few hours' picking, enjoy the countryside and take home fruit at lower prices than those charged by retailers. Picking as employment continues, however, although the pickers are likely to be students now, often from abroad, or other foreign workers. Like their predecessors, students especially may go to the berry fields hoping not only to earn a little cash but also enjoy a change of scene, new company and a taste of fresh air.

Fishwives and herring gutters

Before vans were used by fish merchants, the women of certain east-coast fishing communities walked often prodigious distances carrying produce in a creel, perhaps taking advantage of trains if these were available. Along with others, Christian Watt (b1833) from Fraserburgh, Aberdeenshire, travelled on foot each August with a summer cure of dried fish deep into the isolated settlements of the Grampian Mountains where she slept in barns[12] and exchanged fish for butter and other products of the land. If creels had been heavy on the outward journey, they were often heavier on the return. The women, however, would have been used to strenuous physical effort. Many helped in the several processes associated with baiting their menfolk's fishing lines, and some even carried the men on their backs to the boats so that they should start their working day dry. When the men returned, women would help process the catch then market it locally; they could also lend a hand with the repairing of nets and other equipment. All of this in addition to the normal responsibilities of housewife and mother. Perhaps then, it is not surprising that the summer trek inland, for all the weight and rough sleeping involved, was seen as something of a break.

The predominantly teenage gutting girls who came from most Scottish fishing communities, but especially those of the Outer Hebrides, were usually relatives of the men who crewed the herring boats.[13] Known as 'gutting quines' in the north east, they migrated round the British and Irish coast, to gut and pack herring; the 'king of fish', offered employment to thousands in the boom years at the end of the nineteenth century and the beginning of the twentieth century (Fig. 33.3). For processing and curing the fish many extra hands were needed. Initially the fishermen themselves had undertaken the work, the girls who accompanied them cooking and generally tending to the men's needs, with in turn an older women, known in some cases as 'the Mistress', responsible for the girls' wellbeing as well as their moral welfare.[14]

Annie MacDonald went to the gutting after she ceased going to the shieling in South Uist. Like others she worked a 12-hour shift, sometimes longer if there was a glut of fish. The work required the teams, of two gutters

Figure 33.3 Former fisher girl, Isa Ritchie, aged 91, outside her home in Whitehills, Banffshire, in May 1991. Source: R Leitch.

and one packer, to work together with great efficiency. During the first stage of gutting, the herring was sprinkled with salt to enable the gutter to obtain a good grip. It was held usually in the left hand with the throat of the fish exposed. The knife was then inserted through the gills and with a sharp upward turn was drawn outwards over the herring's head and in one deft action the gills and guts were removed. To protect their hands from the salt, the women wore 'clouts' or rag strips bound round their fingers, often torn from old cotton flour sacks. Girls who packed the gutted herring into barrels were usually taller as this was a job which required less bending, although it too was arduous. The gutting girls were visually distinctive in other ways – muscular as a result of their work and usually dressed in oilskin wraparound protective 'quoits', or apron and bib, and a headscarf.

In Shetland the girls were housed for the season in utilitarian wooden huts at the fishing station. These, however, were kept spotlessly clean and the women would add a homely touch by papering the walls, sometimes using newspaper. Their week was a busy one, not only with long hours of paid work, but also because all housekeeping, including the preparation of meals, had to be carried out by themselves. Despite this the Sabbath was unfailingly observed. Limited as their leisure time was, the women might be able to meet men from the boat crews and perhaps be entertained by music from a button-key accordion or a melodeon. The Gaelic-speaking girls were themselves well known for their musical ability, primarily in the form of singing.

Gutting girls also travelled to English ports such as Great Yarmouth, Norfolk and Lowestoft, Suffolk, when the surrounding seas were in full harvest. What amounted to a Scottish invasion took place about the beginning of October for the autumn fishing, 'There used to be something like 2,000 Scotch girls come [to Lowestoft] for gutting the herring and Scotch girls used to come down from Fraserburgh, Peterhead, Lerwick to Lowestoft.'[15] The English ports offered attractions such as the cinema and shops and again allowed young women and men to get to know each other, often through the activities of missions and other religious institutions. The rare moments of free time were also spent in knitting, despite raw hands.

In Great Yarmouth the women were housed in digs arranged by the curers, usually in an area known as 'The Rows', where they were crowded into garrets and other rooms, frequently with all three of a gutting crew accommodated in one bed. Children in Lowestoft were sent to stay with grandparents or might even be consigned to garden sheds to make room for Scots girls as boarders, the extra income being much appreciated by the householders. Because the girls' oilskin clothing was impregnated with grease and littered with fish scales, stories abound of landladies pinning up brown paper or oilcloth to protect their walls, particularly in the hallway where the girls first entered.[16]

The enormous exodus of fishermen and gutting girls effectively came to an end in the 1930s with depleted herring stocks and the onset of World War II. Memories of the Scots fisher girls in England are now beginning to fade, although some retired fishermen remember them clearly and with nostalgia, recalling how, in the evening, the Scots girls would walk down the main street of Lowestoft speaking in different dialects of English, and in Scots and Gaelic, all equally impenetrable to the English listeners' ears.

LONGER SEASONS

A number of jobs with a seasonal basis have provided the main livelihood, rather than a seasonal supplement, for those engaged in them. Usually, but not always, the season has occupied the larger part of the year but despite this most workers have retained a home base, usually defined by family, elsewhere. Commonly men's jobs, the work may be approaching the habitual, but the associated domestic life is not normally seen in the same light. Those involved in this type of work have included timber floaters working on the River Spey, deep-sea fishermen, and the salmon netsmen described below.

Salmon netsmen

The salmon from the River Earn and especially the 'Silvery Tay' have enjoyed an unrivalled reputation for succulence, and the netting of salmon by sweep net and river coble goes back to a very early period, on the River Tay in particular. Cobles are flat-bottomed vessels capable of being launched from a river bank or sandy bay. Twentieth-century examples were built of fibreglass and motorised but followed the form of the older larch-built and oared boats. The procedure for netting the fish is described as follows:

> One end of the net is kept near the shore by means of a tow rope while the remainder of the net is swept out over the river from the coble ... A huge floating bag is thus formed. The net is then drawn ashore at the 'haling', a specifically prepared landing place, free from obstructions and usually covered with gravel.[17]

The section of a river where cobles could be launched or landed, usually with a gravel inlet or shore and a netsmen's bothy or 'lodge', is known as a netting or fishing station. At one time there were over 100 such stations on the River Tay alone, extending as far up river as Stanley, Perthshire.

Salmon netsmen on the Tay and Earn relied on fishing to provide the larger part of their annual income, during a season which commenced in early February and finished on 20 August. They were often housed in very poor conditions in bothies, known as 'lodges' on the Tay, which were at one time cramped affairs, their earthen floors often riddled with rat holes. One of the lodges on Mugdrum Island off Newburgh, Fife, was so badly infested with rodents that all food had to be kept in glass jars and it was not unknown for a rat to spring up on an upturned fish box used as a table when the men were at their 'piece' or snack.[18] In the main, those bothies on the banks of the Earn were smaller structures, previously just dens buried into the riverbank and covered with reeds. Conditions improved on both rivers with the erection of stone or brick-built buildings with slated or pantiled roofs and bunk or double beds on which chaff-filled 'tykes' or mattresses were placed. There were often fireplaces at opposite ends of a single room, which were fuelled by coal and had to be stoked overnight lest the water in the gunmetal kettles – a feature of the bothies in many instances – froze (Figs 33.4, 33.5). On the Tay, two generations of lodges stand side by side bordering the estuary east of Newburgh in Fife and these reveal re-use of building stone. Some of the structures were quite remarkable, such as the example at Lower Taes, with its classical design dimensions (probably taken from a pattern book), a fireplace in each of the two rooms and intervening storage space for a coble.

A hierarchy existed amongst the workers, from the foreman of the seven-man crew on the Tay, down to the 'ropey' or lad who coiled the ropes and made sure the kettle was boiling for the often drenched men coming in off a shift. The workers slept in pairs, the foreman sleeping with the ropey. With long and irregular hours of work, dictated by tides, cleaning the lodge was not a priority and leisure time rare. One fisherman, Angus Morrison, recalls that the only form of entertainment within his bothy, the Pyerod, was that derived from a piano accordion. Meals were fitted around shifts, the frying pan serving up a near-continuous diet of ham and eggs, supplemented by other foodstuffs obtained from mobile shops and the more homely fare prepared by the womenfolk of some of the fishermen – items such as potted head, girdle scones and jam.

Figure 33.4 Late nineteenth-century Pitmillie salmon fishers' bothy, Hillhead, by Boarhills, on the east coast of Fife, with its pantile roof intact. Ten years after this picture was taken (1989) the roof had completely collapsed. Source: R Leitch.

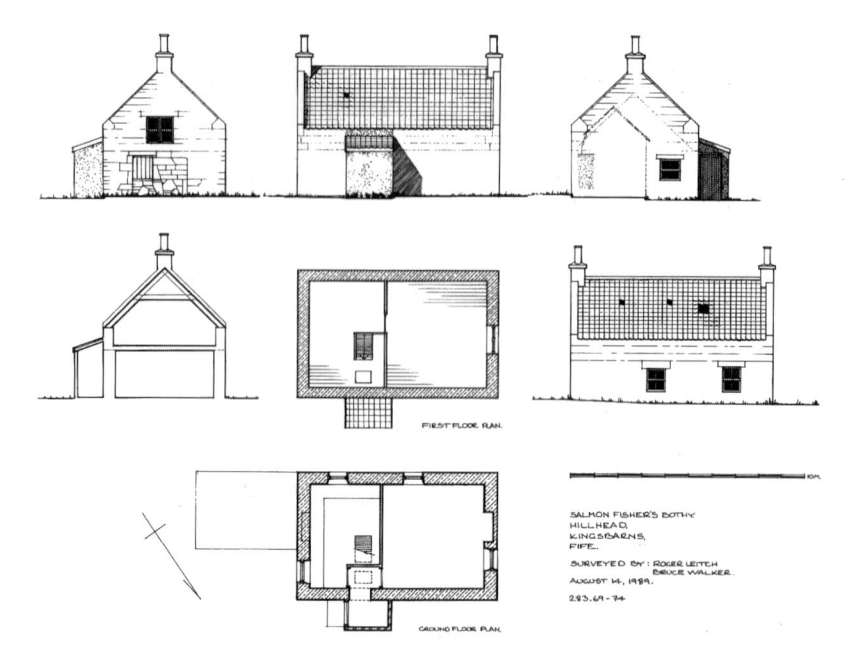

Figure 33.5 Salmon fishers' bothy, Hillhead, by Boarhills, Fife. Surveyed by R Leitch and B Walker, 14 August 1989. Source: R Leitch, B Walker.

For the Hebridean men who migrated to work the Tay salmon fisheries such luxuries from home were not possible. From the 1920s it was the custom, especially on the island of Scalpay, for young single men to migrate to the Tay fishery. Held in high esteem for their boat-handling skills, the Hebridean squads were hired by the main gaffer, at one time a Mr Scotland from Newburgh, who travelled out to the Outer Isles and was, in the words of the late Tom Jarvis Jnr, 'treated like a god', returning with webs of tweed from his visits.[19] Men whose homes were within close proximity to their lodges might return at the weekend as there was no fishing on Sunday, but for the Hebrideans it was a matter of remaining at the fishing station for the duration of the season, save the occasional weekend trip to Perth to buy small luxuries.

For much of the twentieth century the main netting interests regarding salmon were controlled by the Tay Salmon Fisheries Company, instituted in May 1899.[20] Around 1920, MacFisheries Ltd acquired complete control of the Tay industry and by arrangement with the Tay Salmon Fisheries Company, the entire output was dispatched from their shops and distributing centres. Rapid delivery of wild salmon to the customer base was essential to ensure the product reached its destination in first-class condition. A steamer collected boxes at various fishing stations and delivered them to Perth, Newburgh, or Dundee. Salmon taken up to 8pm could be dispatched by rail and arrive in London the following morning. The collection station at Perth consisted of offices, a fish house capable of holding 2000 fish, ice houses of 800 tons' capacity, a box-making shop, stables and a private railway siding.[21]

Seasonal workers such as the salmon netsmen were for long unable to receive state benefits out of season, and where there was no landholding at home, and sometimes where there was, it was necessary to do any type of supplementary work available, such as harvest work, rabbit trapping or jobs in the orchards and reed beds of the Carse of Gowrie, Perthshire/Angus.[22] Towards the end of the twentieth century, more and more salmon netting was undertaken by students on vacation working alongside a greatly reduced number of permanent workers.

The decline of the industry occurred in the last decades of the twentieth century, a commercial tradition ended by pollution, poaching and competition from rod fishermen. The last view netting as having a detrimental effect on catches by rod and line and, usually in the form of angling syndicates, have now bought up the netting interests on the Tay.

YEAR-ROUND TEMPORARY RESIDENCE

In a range of occupations workers, usually male, are away from home on a habitual basis, returning whenever possible. These have included Scotland's wide variety of itinerant traders and salesmen, travelling preachers and missionaries, peripatetic musicians and teachers, and, today, some professionals, oil-rig and construction workers, lorry drivers and certain types of transport worker.

Itinerant traders
Migrant workers have offered the communities they pass through not only new faces, but often also a source of news and entertainment. This was especially the case with the wide variety of itinerant traders to be found at one time throughout Scotland.

Until the nineteenth century pedlars or chapmen served most of the country and carried everything and anything possible. The tradition was maintained into the twentieth century by those individuals who sold a miscellany of items: almanacs, shirt-collar studs, chanter reeds, shaving brushes and trumps, to name just a few lines. Generally, though, chapmen were eclipsed by specialist dealers – travelling hardware merchants, itinerant drapers and stationers – who appeared on the rural scene as early as 1820. The heyday of the great rural fairs was largely over by the 1840s, however, and their demise saw the disappearance of the itinerant trader from the mainstream of the country, relegating him to outlying areas.[23]

As late as the 1950s, the more remote corners of the country were still the stamping ground of packmen, some of whom had been born far from Scotland's shores; there were a number of Asian and Jewish traders, for example.[24] Amongst the packmen were colporteurs whose stock-in-trade was Bibles. There were many 'Heather Jocks' who sold besoms and pot reenges, although hawking wares such as tin goods, horn spoons and clothes pegs, as well as willow baskets, was the traditional preserve of the Travelling People. There were also individuals such as Irishman Johnny Morgan, who came to south-west Scotland for almost 50 years until his death in 1901. He sold agricultural and domestic wares such as hones, keel for reddening the hearthstone and bags of scythe sand (white granite sand used in sharpening tools).

Such men, and occasionally men and women in partnership, travelled the country on regular beats, much in the manner of the Travelling folk, where they became acquainted with the local people, securing a cup of tea, a 'piece', and a place to lay their head. A higher class of the old chapmen, who travelled the countryside on horseback, had been welcome to reside for the night in farmhouses.[25] Humbler individuals might have found accommodation in a cot house or have been shown to an outbuilding, or perhaps slept in a cave. The author Gavin Maxwell, for example, in 1960 mentions that a category of pedlar was well received by local people in parts of western Inverness-shire, the men living rough in caves round the coast near Glenelg.[26] Traders coming into the towns and cities from outlying areas might secure lodgings in formerly run-down parts like Edinburgh's Canongate. Indeed those individuals with no home base would make regular use of such accommodation, or live in lodging houses, when not working.

A number of travelling traders became reasonably successful businessmen, and used the profits and experience of their peripatetic days to set up shops, for instance. One example is Alasdair Steven, born 1915 in Clunie, near Blairgowrie, Perthshire, who started his antiquarian and second-hand book business travelling around the countryside on a bicycle with panniers and a rucksack on his back, carrying a stock of about 125 hardback books. He visited areas which had minimal or no public transport and where people were fairly isolated from the outside world. He brought something different into their lives with his store of books, news and chat. As he came to know regular customers they would provide him with food and put him up for the night, although he also slept in barns and in youth hostels.

Such work must often have been lonely, frustrating and exhausting, and it must have made family life very difficult to accommodate, yet it also offered much that was positive: opportunities for social contact, variety, fresh air, independence, challenge and, occasionally, a fair profit.

Oil-rig workers
With the exception of the British armed forces, the largest present-day migratory or quasi-migratory workforce in the UK is composed of the men and women who work in the offshore oil industry in over 150 different installations amidst the hostile environment of the Atlantic and North Sea. Domestic conditions on the installations vary so much that it is difficult to generalise even at the broadest level. Some have two-man cabins with adequate soundproofing, air conditioning, television, video and hi-fi systems and the obligatory kettle since there is a strictly enforced no-alcohol

policy offshore. But one can still encounter dirty, noisy accommodation with poor-quality furnishings, a lack of ventilation, inadequate heating and communal showers at a distance from the sleeping area. Four men might have to live in such a confined space for a fortnight at a time, seeking rest there after 12-hour shifts. The industries have been able to let such conditions develop because the regulations governing offshore accommodation are very lax.[27] In terms of recreation, the workers mainly watch television, listen to music or practise hobbies, model making or ornithology perhaps. Others simply cannot adjust to the routine and different life pattern involving flights to and from the rigs by helicopter, sometimes over rough seas and through storms.[28]

Working offshore has positive aspects too, though. Some of the workers enjoy the environment as it gives a degree of privacy, the opportunity for solitude and reflection and a strong sense of camaraderie. At one time the men could also enjoy luxurious meals. More recently these have been replaced by a more health-conscious diet. Another important change is that today around 1 per cent of the offshore workforce (estimated at 36,000 in 1990, but only 23,000 in 1997 as a consequence of increased automation and the decommissioning of certain platforms) are now women. Women work in technical fields as well as at more traditional female tasks such as catering. The first women began to work offshore in 1984. A Alvarez described a visit offshore with a female company representative, likening it to 'Following Helen around the ramparts of Troy ... there were no bawdy comments, no wolf whistles, only a wave of incredulous silence, as though the men were overwhelmed by this dazzling reminder of life elsewhere.'[29]

Lorry drivers

Drivers of articulated lorries are often considered 'the kings of the road', not least because of their skill in manipulating heavy vehicles over very large distances. Before the worsening of problems in the Middle East, some Scottish truckers were involved, for example, in haulage to Iran, each trip involving a considerable time away from their homes, families and friends.

There has always been a strong sense of brotherhood among such drivers, as witnessed by the fuel protests of September 2000. The work is dangerous and difficult, in no small measure due to the heavy traffic to be encountered on many principal roads; and often lonely, although drivers generally contrive to maintain a good measure of communication between each other through using CB radio.

Physical comforts can be scarce. If they have bunks in their cabs, drivers often sleep in their vehicles, parked in lorry parks. Most, however, would prefer to have a bed to sleep in. Today there is reliance on in-cab cooking, whereas even ten years ago there was greater dependence on roadside cafés and restaurants. Again most people would probably prefer to get out of the cab and have a sit-down meal. For all such workers leisure time and the activities which can be undertaken within it are very limited. For many the attitude is just to get through the job as well as possible.[30]

INDEFINITE TEMPORARY RESIDENCE

Spending a number of years, usually as a young adult, away from home is an experience in which many Scots have participated. For these people residence may be less 'Time Out' as constituting a temporary home, although 'returning' is often a pertinent, but not always realised idea. Included within this category are certain types of agricultural worker and domestic servant and the modern equivalents of the latter such as au pairs, as well as students.

Bothymen

Unmarried male farm servants in many parts of eastern Scotland were hired at feeing markets twice a year, unlike the married men who were hired only once a year[31] (Fig. 33.6). Farm servants were essentially employed as ploughmen, each man responsible for a pair of horse, apart from the 'orraman' who had to undertake any task that came his way. The men and adolescents were usually accommodated in the north east in 'chaumers', rooms in the steading, but with meals served in the farmhouse kitchen, while further south they lived in 'bothies', rooms in the steading or sometimes separate buildings where they ate and slept (see Fig. 7.1). Carter writes that 'as early as the 1850s improved bothies were being built with the cooking area separated from the sleeping area, and with single beds in cubicles. These improvements started in the great bothy districts south of the North East.'[32]

Figure 33.6 Feeing Saturday, Arbroath High Street, Angus, in 1918. Source: SLA W 28/06/18.

After a long day's work the lads in the bothy could look forward to less than comforting rations. Their diet – prepared by themselves – was largely composed of oatmeal, in the form of brose or porridge, with an occasional Sunday treat of sausages, or bread and jam, or a pilfered egg or two and sometimes the hen as well. Domestic arrangements were, unsurprisingly, basic. A weekly rota for undertaking chores was the order of the day. Cleaning did not have a great part in this, partly because there was little to clean, the bothies being sparsely furnished. Like the salmon netsmen, bothymen slept in pairs in built-in wooden beds with straw-filled mattresses. Kists stored meal, and clothing and other personal possessions, and also served as seats. There was a recognised custom of seating with the foreman and second lad having their places nearest to the fire, and this hierarchy was strictly adhered to in other contexts. Honesty was another important concept within the bothy. Stealing, apart from the farm eggs, was virtually unknown amongst the men and when it did occur the perpetrators could experience severe reprisals.

Drink did not tend to play a large part in the lives of the bothymen, except at the feeing markets, where it would indeed be taken. Recreational activities in and around the bothy were limited by the lack of leisure time. Rarely did the men read books, the weekly papers being more popular during winter nights,[33] as were card games. Dominoes and draughts were also part of the scene, and music. Born in 1913, Andrew Bruce migrated from his native Aberdeenshire to secure better money on farms in Angus and Perthshire. As a single lad he was mostly fee'd on farms around Forfar, Angus. As for bothy entertainment, Andrew recalled, 'Heavens, I've seen the bothy, a dozen lads in it at night, havin a bit concert or a singsong, tellin jokes and playin moothies [mouth organs], melodeons, fiddles an aathing – great.'[34]

Outdoors, during the longer summer evenings, football was played, as well as a test of strength called the 'sweir tree' or 'pullin the swingletree'. Two lads entered into a contest to try and lift each other off the ground, the soles of their boots being placed together and both gripping the shaft of the plough's swingletree.[35] Similar activities included tossing the caber, or throwing a 56 lb (25 kg) weight over a line. Quoits, a throwing game, was a further leisure pursuit.[36]

Friday-night dances were the highlight of the week for the men, who would often cycle miles to attend a dance. Indeed it was not unknown for busloads of young women from towns to arrive at a village dance.[37] These events must have been fairly successful; many bothy men eventually married and, if they continued in agricultural work, they usually raised their families in the comparative comfort of a tied cottage.

Bothies were still in existence in Angus until World War II, although on a much-reduced scale. Sprott states that 'in large parts of the east coast the bothy system was in full swing from early last [the nineteenth] century to Hitler's War, when it was broken up largely by powered mechanisation and the weekly wage.'[38] (Fig. 33.7).

Figure 33.7 Joe Tindal, born 1918, Kirkton of Monikie, Angus. Former ploughman. Pictured on 16 March 1993 at Inchture, Perthshire. Source: R Leitch.

IN CONTRAST: TEMPORARY RESIDENCE AS LONG-TERM HOME

For some workers, temporary accommodation is their habitual domestic scene and constitutes their only real home. This way of life may be chosen by an individual or may be imposed by circumstances such as the lack of family or of sufficient funds with which to maintain a base. Certain jobs may demand that the personnel involved are almost constantly on site. This includes some domestic servants and many of those in the armed forces. Exactly where and how these people position their home is often part of a complex personal equation.

Those in the armed forces
Members of the armed forces who are married are usually provided with family accommodation which allows them to make their section's base as much their home as it is for single men or women. When the section moves base, so do the families, in many respects taking their social world with them. When off duty, the various ranks find recreation in the different Navy Army Airforces Institutes, or 'NAAFIs', from the 'squaddies' to the Sergeants' Mess and the Officers' Mess. The NAAFI is where people go to relax, socialise and bond with each other, enjoy different types of recreation and escape from the pressures placed on modern servicemen and women. There are also numerous clubs and activities such as adventure training

which members of the forces and their family members living on the base can enjoy. Indeed there is always some activity going on.

However, service men and women sometimes have to experience extended periods in action away from their families, perhaps under emotionally highly-charged circumstances. Sgt Major Jimmy Wright of the Black Watch comments that his regiment has served all over the world and like most others offers welfare and recreational support while people are away in action.

Itinerant craftsmen and labourers

In addition to members of the Travelling People,[39] Scotland has possessed many types of itinerant craftsman and labourers, some of whom have had a base, others not. There were itinerant tailors, who returned to their homes between bouts of work, for example, but there were also those who survived by performing small tasks and services around the farm in return for a night in the barn. The late Wull Halliday of Meikle Kirkland, by Crocketford, Dumfries and Galloway, recalled:

> They all had their trades. Pat Moaney swept chimneys, another auld fella mended dishes wae staples, Bob Hanley did a bit o' droving, Kellie played the tin whistle for a living, and Auld Cassidy hawked naphthalene balls.[40]

Like travelling traders, such individuals may well may have been welcomed for the society they provided as much as the services they offered. Hector Urquhart mentions that:

> In my native place, Poll-ewe, Ross-shire, when I was a boy, it was the custom for the young to assemble together on the long winter nights to hear the old people recite the tales or 'sgeulachd', which they learned from their fathers before them. In these days tailors and shoe-makers went from house to house.[41]

Alan Bruford describes how, along with 'quarterers':

> 'a class of respectable beggars' who were given lodgings in return for the news they carried, and sometimes their skill in music, medicine or repairs ... chapmen or pedlars who stayed the night; and travelling tailors, who used to live in their customers' houses while they made up garments ... [are] often mentioned in stories.[42]

They would certainly have told stories themselves.[43]

The demise of the itinerant craftsman and labourer occurred at different times in different locations. Improvements in transportation allowed industrially produced goods and machinery to find their way into the most remote areas, while the creation of the Welfare State meant that many people were given an alternative to ekeing out existences in this manner. In some cases the authorities felt it their duty to remove what they saw as the less-desirable itinerants from local communities.

ACKNOWLEDGEMENTS

The writer is indebted to Martha Stewart for gathering new material from oral sources, to Jake Molloy of OILC in Aberdeen for his contribution on offshore oil workers and truckers and to Sgt Major Jimmy Wright of the Black Watch. Catherine Walker of Aberdeen Maritime Museum supplied invaluable material. The author also acknowledges the contribution and help of Susan Storrier and Professor Alexander Fenton.

To the memory of the 167 men who perished as a consequence of the Piper Alpha disaster on 6 July 1988 and to the 61 survivors.

NOTES

1. It is not possible in this short article to deal with Scots working abroad nor with occasional temporary residence in relation to work, such as that undertaken by many present-day professionals. Information relating to further groups of people working away from home is included in the Bibliography.
2. Gordon, 1923.
3. Extract from the MS Field Notebooks of Åke Campbell, Uppsala, Sweden, 1948. Transcribed and translated from the original Gaelic by Calum I MacLean.
4. The women might be up to around 50 years of age. Annie MacDonald was 'about 15' when she first went to the shieling.
5. Annie MacDonald has left a description of the process. Cheese was made once a week. A good pot of milk was warmed slightly over the fire. Rennet was added, made from the 'deasguinn' (inner membrane of the stomach) of a sheep. To make rennet the deasguinn was placed in water and left to ferment. The liquid was then poured off and bottled. One spoonful of rennet was sufficient to curdle a pot of warm milk which would then be left until it had cooled. By now a soft cheese, crowdie, had formed. This was taken out of the pot and placed in a clean cloth. The wrapped cheese was then immersed in pickle to salt it and the cloth removed before being placed in a basin and mixed by hand to distribute the salt evenly throughout the cheese. The cheese was then pushed into the press – usually positioned on stones – and a weight placed on top. The whey would run off through the holes in the base of the press to be given to the calves or used in baking and the production of a kind of porridge ('brochan miùg'). When the cheese was fully pressed, the lid and stones were removed and it was left outside to dry.
6. In other areas cheese could be consolidated by simply hanging it up in a cloth.
7. See, for example, Shaw, 1986.
8. See Holmes, 2000.
9. For detailed information about living conditions in the raspberry district of Perthshire and the strawberry areas of Lanarkshire see Leitch, 1999.
10. Henderson, 1980, 80.
11. Stewart, 1988, 51.
12. Watt, 1988, 29; see also Chapter 42 Fishing and Homes.
13. For information about the fisher girls see Bochel,1996; Kidd, 1992.
14. Oral information from Mary Murray, Cellardyke, Fife, recorded by Roger Leitch in 1983 (SSS, SA1983/123).
15. The Roaring Boys (produced by Nick Patrick), *The Archive Hour*, BBC Radio 4 broadcast on 9 June 2001.
16. The Roaring Boys (produced by Nick Patrick), *The Archive Hour*, BBC Radio 4 broadcast on 9 June 2001.

17. MacFisheries Ltd, 1920, 7–8. The author is grateful to Iain MacRae, Head Librarian, A K Bell Library, Perth, for bringing this information to his attention and to Jeremy Duncan, Local Studies Librarian of the same institution.
18. See Leitch, 1998–9, 7–32; the unsanitary nature of the lodges was described by Alexander Carmichael. NAS AF56/1422, Fishing Lodges on the Tay and Earn, 11 May 1889; see also *Dundee Advertiser*, 8 August 1889, 7.
19. The men who could not find places on a west-coast fishing boats were in general those who had to migrate to the Tay, at least this was the custom from the early 1920s on Scalpay. Oral information obtained from Angus Morrison, Blairgowrie, Perthshire, formerly of the Isle of Scalpay, Harris. RWL/101 (author's personal tape reference).
20. The Tay Salmon Fisheries Company also had stake-netting interests at Lunan Bay, Angus, a distance of 70 miles (123km) from Stanley, giving an impression of their interests in netting wild salmon, both on river and at sea.
21. MacFisheries Ltd, 1920, 9–10.
22. Working at the reed beds was an onerous job when the sickle was used. Today the process is mechanised. See Leitch, 1998–9.
23. Leitch, 1990.
24. See Edensor, Kelly, 1989.
25. Penny, 1836.
26. Maxwell, 1969, 17.
27. Oral information from Jake Molloy, General Secretary, Oil C, interviewed by Martha Stewart on 6 November 2001 in Aberdeen.
28. Correspondence from Catherine Walker, Assistant Keeper (Maritime History), Aberdeen Maritime Museum, dated 31 October 2001.
29. Correspondence from Catherine Walker, 31 October 2001.
30. Oral information from Jake Molloy, recorded by Martha Stewart on 6 November 2001 in Aberdeen.
31. See Carter, 1979, 121.
32. Carter, 1979, 122–3; in Allan, 1999, the so-called 'improved chaulmer' – which he describes as a bothy – was where the men slept only, taking their 'meat' in the farmhouse.
33. See Chapter 21 Reading and Study.
34. RWL/129 Andrew Bruce (1913–), retired ploughman, Inverurie, Aberdeenshire, recorded by Roger Leitch at Perth on 8 March 1993.
35. Adams, 1992, 42–3.
36. Adams, 1992, 42–3.
37. Interview with Bob Smith (1919–), former orraman, recorded by Roger Leitch at Dundee on 20 March 2001; verbal communication from Mrs Margaret Scroggie, by Bridgefoot, Angus.
38. Sprott, 1996, 102; for the story of other types of migratory agricultural worker, specifically tattie pickers, and their dwellings see Holmes, 2003.
39. See Chapter 44 Communities on the Move.
40. RWL/8 William Halliday (1910–89), retired farmer, Crocketford, Dumfries and Galloway, recorded by Roger Leitch at Meikle Kirkland on 24 January 1982.
41. Campbell, 1994, I, 5.
42. Bruford, MacDonald, 1994, 1–31; for information on quarterers see Fenton, 1987, 6.
43. Fenton, 1987, 6.

BIBLIOGRAPHY

Adams, D G. *Bothy Nichts and Days: Farm Bothy Life in Angus and the Mearns*, Edinburgh, 1992.
Allan, J. Chips from a Deeside lumberman's log, *The Deeside Field* (1925), 35–8.

Allan, J. *Farmer's Boy* (1935), reprinted Edinburgh, 1999.
Baldwin, J R. The Berries, *The Scots Magazine* (October 1972), 10–14.
Bell, J. Donegal women as migrant workers in Scotland, *ROSC*, 7 (1991), 73–80.
Bochel, M. The fisher lassies. In Kay, 1996, 147–57.
Bruford, A, MacDonald, D A, eds. *Scottish Traditional Tales*, Edinburgh, 1994.
Campbell, J F. *Popular Tales of the West Highlands* (1860–1), reprinted Edinburgh, 3 vols, 1994.
Carter, I. *Farm Life in North-East Scotland, 1840–1914*, Edinburgh, 1979.
Cheape, H. Shielings in the Highlands and Islands of Scotland, *Folk Life*, 35 (1996–7), 7–24.
Devine, T M. Temporary migration and the Scottish Highlands in the nineteenth century, *Economic History Review*, 2nd series XXXII (1979), 344–59.
Devine, T M. Highland migration to Lowland Scotland, 1760–1860, *Scottish Historical Review*, LXII (1983), 137–49.
Devine, T M. *The Scottish Nation, 1700–2000*, Harmondsworth, 2000.
Eden, R. *Going to the Moors*, London, 1979, 190.
Edensor, T, Kelly, M. *Moving Worlds: Personal Recollections of Twenty-One Immigrants to Edinburgh*, Edinburgh, 1989.
Fenton, A. The currach in Scotland with notes on the floating of timber, *SS*, 16 (1972), 61–85.
Fenton, A. *Wirds an' Wark 'e Season Roon on an Aberdeenshire Farm*, Edinburgh, 1987.
Fenton, A. *The Northern Isles: Orkney and Shetland*, East Linton, 1997.
Fenton, A. *Scottish Country Life*, East Linton, 1999.
Fenton, A, Laurenson, J J. Peat in Fetlar, *Folk Life*, 2 (1964), 3–26.
Gordon, S. *Hebridean Memories*, London, 1923.
Henderson, H. The ballad, the folk and the oral tradition. In Cowan, E J, ed, *The People's Past: Scottish Folk, Scottish History*, Edinburgh, 1980, 69–107.
Holmes, H. *'As Good as a Holiday': Potato Harvesting in the Lothians from 1870 to the Present*, East Linton, 2000.
Holmes, H. Housing for seasonal agricultural workers. In Stell, G, Shaw, J, Storrier, S, eds, *Scotland's Buildings, Scottish Life and Society: A Compendium of Scottish Ethnology, vol 3*, East Linton, 2003, 236–48.
Kay, B, ed. *The Complete Odyssey: Voices from Scotland's past*, Edinburgh, 1996.
Kidd, D I. *To See Oursels: Rural Scotland in Old Photographs*, Edinburgh and Glasgow, 1992.
Leitch, R. 'Here chapman billies tak their stand': A pilot study of Scottish chapmen, packmen and pedlars, *PSAS*, 120 (1990), 173–88.
Leitch, R. Working lives of the Tay and Earn salmon fishermen, *ROSC*, 7 (1991), 67–72.
Leitch, R. Fishermen's bothies and other shelters, *Vernacular Building*, 14 (1997), 4–19.
Leitch, R. The Tay and Earn Salmon Fisheries, *Folk Life*, 37 (1998–9), 7–32.
Leitch, R. 'Seasonal Workers: Their dwellings and living conditions, 1770–1970', unpublished PhD thesis, University of Dundee, 1999, 241–9.
MacFisheries Ltd. *Tay Fisheries*, London, 1920.
MacLean, C I. Hebridean traditions, *Gwerin*, 1 (1956–7), 21–33.
Martin, A. *Fishing and Whaling*, Edinburgh, 1995.
Maxwell, G. *Ring of Bright Water*, London and Harlow, 1969.
Michie, M F (comp and ed A Fenton, J Beech, with a chapter by C MacKenzie). *Glenesk: The History and Culture of an Angus Community*, East Linton, 2000.
Miller, J. *Salt in the Blood: Fishing Communities Past and Present*, Edinburgh, 1999.
Orr, W. *Deer Forests, Landlords and Crofters*, Edinburgh, 1982, 112–13.
Penny, G. *Traditions of Perth*, Perth, 1836.
Rosander, G. *Gård Farihandel Inorden [Farm Travelling Traders in the North]*, Stockholm, 1980.
Shaw, M F. *Folksong and Folklore of South Uist*, 2nd edn, Edinburgh, 1986.
Sprott, G. A weel plou'd rig. In Kay, 1996, 99–111.
Stewart, S (comp, ed, with introduction and notes by R Leitch). *The Book of Sandy Stewart*, Edinburgh, 1988.
Watt, C (ed D Fraser). *The Christian Watt Papers*, 2nd edn, Aberdeen, 1988.

34 'Time Out' Leisure

ROGER LEITCH

This chapter seeks to outline some aspects of temporary residence in relation to leisure in Scotland. Not surprisingly, given the size of the subject area, no single work on the history of leisure exists for Scotland.[1] However, the beginnings of the country's tourist industry have been shown to lie in a 'wave of curious visitors'[2] on what were essentially holidays to the Highlands, inspired by the Ossianic trail, Sir Walter Scott's *The Lady of the Lake* (1810), and the English clergyman William Gilpin's guide to the Picturesque. These visitors, who were mainly, but not exclusively English and usually of the wealthier classes, came in increasing numbers from the 1790s and left some 80 published travelogues of their journeys between 1792 and 1812.[3]

It was the cult of the Picturesque, largely focused on the Trossachs, Stirlingshire, but also on other dramatic sites, such as Fingal's Cave on the Isle of Staffa, off Mull, which drew most visitors to the Highlands. Many, however, were not fully prepared for the wild scenery of the area, nor what was often seen as the basic nature of the accommodation. Numerous commentators, not least the Wordsworths and Samuel Coleridge, and sometimes Samuel Johnston, judged the country by the standards found in their native England. Dorothy Wordsworth's journal of 1803, for example, includes comments which were none too complimentary, such as references to 'the dirt usual in Scotch public houses'.[4] Of a different background was James Hogg, 'The Ettrick Shepherd'. He was, however, similarly unimpressed. In 1801 he stayed at the government-built inn[5] at Dalnacardoch, Perthshire, the finances for which were partly raised through money from the Forfeited Estates.[6] Hogg amused himself in his room by reading the graffiti on the plasterwork, much of it to the discredit of the inn (Fig. 34.1). Further north at the inn of Invershiel, Kintail, Wester Ross, he slept in a vermin infested bed and was rudely wakened from his sleep by drunken vagabonds who rifled through his coat pockets and stole letters of introduction to gentlemen in Sutherland, a county Hogg failed to explore as a result.

However, T Garnett in his observations of the Highlands (1811) states that the inn at Dalnacardoch was in fact 'a very good one'. Other written sources point to a range of standards, including the comfortable. For example, the village of Inver, near Dunkeld, Perthshire, an important crossing point for drovers and others on the main north – south route prior to the construction of Telford's bridge over the Tay in 1809, boasted two inns. According to the inventory for the older and less well-known Ferry Inn

Figure 34.1 The frontage of Dalnacardoch Inn, Perthshire, in 1994, since restored. Originally a change house on the main road north, 10 miles (17km) from Blair Castle and 13 miles (22 km) from Dalwhinnie, the inn closed in 1865. Source: R Leitch.

(no longer extant), the rooms were generally of a high standard, being wallpapered and having feather beds, linen sheets and table cloths with napkins.[7]

STALKING AND SHOOTING

As well as trips simply to gaze at Scotland's 'ominous' mountains, pursuit holidays were established in the late eighteenth century[8] for a rich minority seeking sport in the hills; an aspect of Scottish tourism which has been examined by Willie Orr in respect of the Western Highlands.[9] Local people were taken off large parts of the inland Highlands in the eighteenth and nineteenth centuries to make way for sheep.[10] However, the flock masters themselves and their dogs experienced a second 'clearance' to make way for deer forests, vast tracts of rough treeless land at altitude devoted to deer stalking and the prestige of the sportsman hoping to obtain an antlered head to adorn a trophy wall.[11] Stalking at this time was a strenuous sport, far removed from the modern version and its four-wheel-drive hill vehicles, telescopic sights and ease of access to the hill.

Red deer were not the only game. The red grouse finds an ideal habitat in open moor land and the deer forests were also used for shooting. R Eden points out that, 'The growth of organised shooting in Scotland is reflected in the fact that game books first made their appearance towards the end of the eighteenth century.'[12] By the 1840s, deer forests had received royal patronage with visits to highland estates by Queen Victoria and Prince Albert. Deer stalking became highly fashionable amongst the wealthy, inspired by Scrope's book on *The Art of Deer Stalking* (1838),[13] the works of Scott, and such icons as Landseer's oil paintings depicting the noble stag. The Scottish aristocracy were happy to accommodate the *nouveau riche*, often brewing magnates such as Lord Burton of Bass, Sir E C Guinness, S Whitbread and others of substantial fortune (Fig. 34.2).

Areas associated with deer forests originally housed the sportsmen in unpretentious cottages but this changed to a more grandiose form of accommodation completely at variance with that of the majority of local people.[14] As the factor for Lord Seafield's property in Strathspey, Inverness-shire, explained to the Napier Commission in 1883:

> I remember when people taking shootings went into the next cottage and lived six weeks there, and went away again. Now everybody who has a shooting wants a country house and we have to build them.

Figure 34.2 Brewing magnate Lord Burton (of Bass) in 1903, then owner of the 25,000 acre Glenquoich deer forest whose shooting guests included the King and the Tsar of Russia. Source: Courtesy of J S Simnett, photographers, Burton on Trent.

'TIME OUT' LEISURE • 639

They bring their families and visitors, and spend a great deal of money in the place.[15]

Although the last point is debatable, it would appear that there was an upsurge in the building of large lodges in the late nineteenth century. Such buildings accorded prestige and extravagant sums were spent constructing dwellings furnished with every amenity available at the time; £10,000 for example, in the case of Invermark Lodge, Angus (twice visited by Queen Victoria).[16] The Seafield estate spent £12,517 on Balmacaan House, Inverness-shire, complete with gas lighting, nine bedrooms, three dressing rooms, adjoining stabling for ten horses, and two coach houses.[17]

The tied homes of the estate workers, the deer ghillies, ponymen and deerstalkers vital to the success of any stalk, often correspond to a distinctive individual estate architecture with flourishes echoing the details of the lodges and their outbuildings. In contrast, the deer watchers occupied extremely rudimentary and remote bothies close to the marches of adjoining estates, where the movement of herds could be controlled using dogs.

From the nineteenth century onwards, stalking and shooting provided at least part of the livelihood for a considerable number of people, with the 'Glorious Twelfth' heralding the start of a short but profitable season for certain communities.[18] The number of gamekeepers in Scotland, for example, rose from 608 in 1836 to 1,050 in 1868.[19] Road making for the purpose of allowing sportsmen access on Sir John W Ramsden's Ardverikie deer forest in Inverness-shire extended to over 20 miles (32 km) of carriage roads and 18 miles (29 km) of pony tracks, many of the labourers migrating from the west coast and being housed in barrack accommodation.[20]

However, opposition was vehemently directed against the sports and associated lifestyle by those such as J M MacDiarmaid, Labour candidate for Kinross and West Perthshire, in his 1926 booklet *The Deer Forests and How they are Bleeding Scotland White*.[21] This was the year of the General Strike, with millions living in poverty while those able to indulge in stalking and shooting were effectively part of another world, their sporting interests generally benefiting few in terms of employment and making a questionable addition to the economy of the glens. For example, in the seasons of 1921 to 1928, the Nairn family of linoleum fame stayed at Glenshirra Lodge, Inverness-shire. The retinue of these affluent industrialists comprised children's nannies, a lady's maid, two pantry maids, two kitchen maids, two housemaids and two chauffeurs.[22]

FISHING

Many lodges have been built and maintained in Scotland to facilitate not just the hunting of birds and mammals, but also fishing. For those without the extreme wealth necessary to purchase an estate there has existed the possibility of leasing fishing rights on a short-term basis. The country's major rivers – the Tay, Spey and Tweed – for the most part flow through large estates which have derived a substantial income from letting beats to well-off rod fishermen.[23] The salmon season on the River Tay, for example, starts on 15 January each year in the headwaters near Loch Tay, Perthshire, complete with the ceremonial blessing of the boats, an anglers' march with pipe band and the Kenmore Trophy for the first fish to be landed. After years of declining spring runs of salmon, Scottish anglers enjoyed their best sport since the 1980s in 2006.[24]

Anglers of this type usually require accommodation and wish it to be of a good standard. Half a century ago R C Saunders' *A Guide to the Fishing Inns of Scotland* was published,[25] which provided a list of popular anglers' haunts. Angling holidays in Scotland during the 1930s represented substantial business for hotels and inns close to the main rivers and lochs and often the more prestigious hotels held permits for bordering stretches of water. The Garve Hotel in Ross-shire, for example, had three miles (5 km) of salmon beat in the Blackwater River. Around 1937, the hotel advertised its amenities as including electric light, hot and cold water in all 12 bedrooms, and fresh butter, eggs and cream from the hotel dairy. Fishing was free to residents, the tariff for a week's stay from 1 July to 30 September being five guineas per person.[26] Today, many small rural communities still find the business generated by visiting anglers offers a much-needed supplement to the local economy.

Salmon fishing is primarily a sport for the middle and upper classes, and it is not surprising that poaching has existed for those of more limited means, although often it was a case of 'one for the pot' rather than large-scale hauls. There is a wider range of options in relation to trout fishing, in loch, river or stream, as seen in the writings of John Buchan who immortalised his recollections of hill-burn fishing days on the Upper Tweed.[27] Only some of these fishermen, however, require overnight accommodation, most returning home at the end of the day.

HILL WALKING AND MOUNTAINEERING

The latitude of the Scottish hills and their climate in winter gives them an alpine character, making them particularly attractive to hill walkers and mountaineers (see Fig. 28.4). Access to Scottish mountain areas was opened up in the later nineteenth century as a result of a more comprehensive railway network, and gave rise to climbing clubs. The oldest of these, the Cairngorm Club, was formed in 1887, followed by the prestigious Scottish Mountaineering Club (SMC) two years later.[28]

The SMC drew its members from the more leisured classes largely of Scotland's two principal cities and, as with the two male clubs, the Ladies' Scottish Climbing Club (founded in 1908) was in essence a middle-class preserve. Two of the members of the SMC, both men of the cloth, are noted as the first recorded Munroists; the Rev A E Robertson and the Rev A R G Burns, the latter recognised as the first 'Compleat Munroist' of the 558 tops over 3,000 feet (914 m). Landowners formed part of the early membership of

Figure 34.3 An isolated, windowless bothy in Upper Glen Almond, Perthshire, left to decline to discourage poachers. Source: R Leitch.

the SMC and the club had various huts in climbing areas, although earlier times saw meets linking train journeys to hotels synonymous with climbing, such as the Kingshouse Hotel under the shadow of Buachaille Etive More at the gateway to Glencoe, Argyll; The Loch Awe Hotel, Argyll; The Sligachan Inn, within reach of the Black Cuillin on the Isle of Skye; and the Cluanie Inn, Kintail, Ross and Cromarty, which was a famous stopping place serving numerous journeys and traverses of the South Shiel ridge and the Five Sisters of Kintail.

In the far north-west, the influence of the motor car was gradually felt in the 1920s and previously unexplored areas, such as parts of Wester Ross and Sutherland, were opened up to climbers for the first time. Skye, however, remained the chief attraction. The 1920s also witnessed the publication of a series of mountain guides which stimulated interest in the hills and mountaineering, an indication that the tide was beginning to turn in Scottish mountaineering circles after a rather stagnant period. The formation of the Junior Mountaineering Club of Scotland in 1925 proved a landmark in this revival. Many routes on the Cairngorm cliffs were pioneered by its members who were a *tour de force* from the next decade onwards, when the standard of both rock and snow-and-ice climbing was consistently raised.

The predominance of climbers from the middle classes lasted longer in Scotland than in England and Wales, but the climbing and walking scene was set to change come the 'Hungry Thirties':

Increasing numbers of youths and young men, poor, tough, and in many cases from the working-class areas of the big cities, first took to the sport of mountaineering. Glasgow was the centre of the movement ... Those who took part in it lived hard, toughly, and cheaply, reaching the mountains by exercising the art of hitch-hiking ... and sleeping in caves, barns, howffs and deserted bothies. Moreover they climbed extensively and well.[29] (Fig. 34.3).

Theirs was an escape from the like of the shipyards of Clydeside or unemployment, as documented in the oral history series 'Odyssey', edited and presented by Billy Kay for BBC Radio Scotland,[30] and in Alastair Borthwick's nostalgic 'Always a Little Further'.[31] To listen to the voices of men like Jock Nimlin, Sydney Scroggie[32] and Tom Weir is to listen to life experienced at the edge[33] (Fig. 34.4). Jock Nimlin, a working-class west-coast trailblazer, was one of those who often slept rough in caves, minus blankets or sleeping bags and whose experiences centred on roughing it and camaraderie.[34] Tom Weir in turn states in his autobiography:

> The '30s and '40s were decades of cheap and literally plentiful public transport because car ownership was rare. To meet the sudden surge

Figure 34.4 The well-known climber, writer and poet, Sydney Scroggie, pictured at home in 1997 prior to becoming an Honorary Doctor of Laws from the University of Dundee. Source: R Leitch.

'TIME OUT' LEISURE • 643

towards the great outdoors in the 1930s the railway companies ran special rambler tickets, entitling you to put your bicycle on the train. I did that. I got off at Inverness, cycled north to Little Loch Broom, camped, and climbed the Fannichs, An Teallach, then moved round the coast of Loch Maree and on home from there by Achnasheen with a stop off at Aviemore to climb Cairngorm. The total coast for a fortnight was £5.[35]

Not everyone viewed mountaineering in such a positive light, however. Styles encapsulates these grumbles, 'Up to 1939 climbing was unpopular with Press and public. It was regarded by the great majority of people as a foolish and dangerous sport pursued by folk who were mildly mad!'[36]

Gone are such days and with them something of the magic of discovery and the pioneering spirit arising from limited funds and the desire to spend time in a worthwhile manner. The working-class mountaineers of the 1930s and 1940s used boat, cycle, train and foot to reach the hills, undertaking journeys of a far more arduous and possibly more adventurous nature than can be experienced today with widespread ownership of cars and the existence of good quality roads. Yet in many ways mountaineering remains a philosophy and perhaps the ultimate escape. Despite costing numerous lives, among them those of some of the giants of the climbing scene such as Dr Tom Patey, whose simple loss of concentration during an abseil of The Maiden, Sutherland, led to his death in 1970, mountaineering continues to offer to some an irresistible contrast to everyday life.

YOUTH HOSTELS

For town and city dwellers, from the inter-War years onwards, escape to the outdoors has been greatly facilitated by a network of accommodation opened and run by the Scottish Youth Hostels Association (SYHA). The hostels were initially concentrated in the Borders, the first being founded in 1931 at Broadmeadows in the Yarrow valley, Selkirkshire. By the end of that year another eight were also in operation.

The early hostels were promoted by a group of enthusiasts who recognised the need for a network of simple, inexpensive accommodation to cater for town dwellers – usually but not always young – who wished to explore the countryside on foot or by bicycle.[37] The structures used were usually custom-built wooden huts or re-used buildings such as large, stone-built farmhouses or lairds' houses bequeathed to the Association. For a time, in the late 1930s, Hoddom Castle in Dumfriesshire, for example, served as a hostel. Retaining all its splendour, the spacious accommodation in the castle included a common room with padded sofas and chairs as well as a piano.[38] In contrast, most other hostel common rooms – the only area apart from the kitchen where the sexes could mix – were spartan affairs, as were the dormitories. Despite the segregation of sleeping space, however, the hostels allowed a largely unprecedented level of social interaction between the sexes. Almost as remarkable was the ability of 1930s hostelling to encourage equality between the classes in a decade of extreme class consciousness.

Hostels in more remote parts of the Scottish countryside served the needs of hardier walkers. They included Craig, Ross and Cromarty, which was reached by a lengthy walk in from the nearest road, and that in Glen Affric, Inverness-shire, which also lacked vehicular access and, like many other less accessible hostels, was left unlocked and unmanned during the wintertime (Fig. 34.5).

Hostels were a particular boon to cyclists. Sydney Scroggie from Dundee recalls arriving at the hostel at Newbigging, Glen Clova, Angus, late one night in the 1930s:

> When Colin Brand, my brother Jack and I got off our bikes at eleven o'clock at night and chapped on Mrs Harper's door – she was the warden – she signed us in. We fumbled over a field through some silver birches and into the hostel itself. It was lit by a paraffin lamp, so we got that going and there were some sticks for the fire so we got it going and 'drummed-up' and had a meal cooked on a sort of range, which had hot plates, an oven and a lum which went up through a hole in the roof.
>
> It was bare. There were two forms to sit on and pigeon holes to put your grub in. There was nobody else in the hostel so we dragged straw palliasses through from one of the dormitories and laid this beside the stove, which was red-hot by this time and we had a very cosy night. All that was provided were huge iron kettles, big iron frying pans and

Figure 34.5 Interior of Glen Affric hostel, Inverness-shire, under natural light with Iain Smart sitting to the left of the fire, where a sack of coal had been generously left for hillwalkers and hostellers in winter. Source: I Smart/R Leitch.

large iron griddles – that was about the lot. You had to bring your own crockery and cutlery.[39]

The hostel at Newbigging was one of the earliest hostels, a chalet-type wooden building which had been opened by the Earl of Airlie. It was replaced around 1959 by a former shooting lodge in Glen Doll, which was better positioned in relation to Jock's Road and the Capel track over the Mounth. In 2001 this hostel was sold by SYHA.

To some extent the original hostelling ethos has been lost. The great rise in car ownership, along with an increased number of foreign visitors, has transformed the distribution of hostels, which are now frequently situated in cities and substantial towns of historical interest and near tourist attractions of international renown. The older style of hostel, such as that described by Sydney Scroggie, is now rarely found. Today most SYHA hostels offer facilities largely on a par with cheaper hotels, with central heating, hot showers, TV lounges and, in some instances, such as the Braemar hostel, Aberdeenshire, Internet access. Dormitories remain the norm, but in many places they take the form of four-bedded rooms which can be booked for family or group use (not always single sex), the beds supplied with fresh linen and duvets, though still of bunk form. Well-equipped kitchens are available, some with microwaves, and the great bane of mornings in the old hostels – the chore – has been done away with for some time, the wardens' responsibilities increasing correspondingly.

SYHA bed-night prices have risen in line with the increased facilities, but the hostels remain an inexpensive form of accommodation, especially for those under 18 years of age, for whom reduced fees are charged. Slightly cheaper are a considerable number of 'independent' hostels which have grown to popularity in the last decade or so. These establishments appeal not only to holiday makers seeking accommodation with fewer rules attached (the SYHA takes a firm stance on matters such as the consumption of alcohol, smoking, closing and opening times and lights out, although this is relaxing somewhat, especially in the big-city operations), but are also important initial bases for those arriving from abroad hoping to undertake an extended working holiday in Scotland.

Of the Scots who use hostels, walkers remain an important element. Indeed hill walking has experienced a further rise in popularity in the last few decades, amongst other things leading to increased land-use conflict. Instances in the past of conflict between hill walkers and gamekeepers have been described in *Highland Days*[40] in which Tom Weir recounts his experiences of particularly hostile treatment from keepers in Knoydart, Inverness-shire, and Torridon, Ross and Cromarty. Generally speaking, the impasse has been overcome by an unwritten code of conduct whereby walkers avoid stalking areas during the season; usually August to November, but sometimes later. In addition particular estates, such as Letterewe, Wester Ross, have worked with the Mountaineering Council of Scotland to create more harmonious relations, which will hopefully serve as a model for elsewhere.

CYCLING

Young and old and both sexes alike have been able to enjoy this form of recreation and exploration of the Scottish countryside. Nevertheless, despite the bicycle's popularity, not enough attention has been paid to cycling within the literature and histories of travel.[41] We know, however, that many cycling clubs were established in the 1870s,[42] and by the turn of the twentieth century the bicycle was relatively common on Scotland's roads. Specific regional pocket road books were published by the Cyclists' Touring Club (CTC). The author possesses a weather-stained example for the Borders and Galloway dating from 1898 (first edition 1885), complete with small-scale route maps and short descriptions of off-road antiquities such as tower houses, as well as mileages between destinations.

The development of youth hostels and the use of bicycles went hand in hand and cycling experienced tremendous popularity and growth in the inter-War years, along with rambling and hill walking. In 1920, 385,000 bicycles were sold, and this figure had risen to 1.6 million in 1935. In 1938 the CTC and the National Cyclists' Union (NCU) had 60,000 members between them.[43]

Long before mountain bikes as such came into being, hardy cyclist-mountaineers were accessing the hill areas on two wheels.[44] Cyclists could travel further afield by making use of trains and steamers, and were accommodated not only in hostels, but also hotels and similar establishments, as witnessed by the old cast-iron symbols of the CTC and NCU.[45]

Dundee in particular has a long and proud tradition of cycle-touring and cycle-racing clubs. Early SYHA yearbooks list such bodies as the Charles Star Cycling Club, the Dundee Western Cycling Club, and the Heatherbell (Ladies') Cycling Club. Now resident in New Zealand and well into her eighties, Mrs Ida Jacob (nee Bloy), describes in correspondence with the writer a cycle journey she undertook in 1942 from Bridge of Earn in Perthshire to the Lake District. En route she stayed in hostels, open in spite of the War. Her route took her via Scotland's highest village, Wanlockhead, Dumfriesshire, where the hostel was a spartan timber hut with minimal amenities. Ida remembered her bicycle was a Sun Wasp (the makers Sunbeam) purchased from Jimmy Christie, cycle merchant, King Street, Perth, on hire purchase at 2s 6d per week. The full amount of £6 was paid off in just one year. Ida's daily mileage might extend to 96 miles (154 km) on roads which, though lacking heavy traffic, were made difficult to navigate by the defacement and removal of milestones and wayside signposts as a wartime precaution. Crossing the Lanarkshire moors, she saw POW camps surrounded by barbed-wire fences housing German and Italian prisoners of war, 'groups of dark-skinned men wearing long brown overcoats'.[46]

Tom Weir's early explorations of his native land display the tenacity of some cyclists.[47] Leaving the grime of Glasgow city life he reached the outdoors by any means possible, including over-laden bicycles, a form of transport which could lead to the 'knock' – a trembling of the lower limbs common among hard-pushing cyclists.[48] For those who did not have very much in the way of financial resources there was a camaraderie built on determination and hardship. Cyclists, in the same way as mountaineers, would always have time for a 'bylie-up' on a stick fire by the wayside, using a tin billy formed from a redundant Lyall's Golden Syrup or Fowler's East India Treacle tin with a wire handle.

Sydney Scroggie, like Tom Weir, ignored the 1930s dance halls and cinemas, in this case of Dundee, and took to 'floating around' the Clova hills in Angus and the Cairngorms, reaching them with like-minded companions on cycles which were laden to the maximum. Once unloaded in the upper stretches of Glen Clova the bicycles could be left unlocked and still be *in situ* upon their owners' return from nights spent in bothies and caves[49] and sometimes the Newbigging youth hostel.[50] Sydney recalled that after a day on some perilous rock climb, he and his climbing friends would return to the hostel to be afforded the best seats by the fire and often served with a mug of hot tea by the girls of the Heatherbell Cycling Club.

Changing hostel provision has altered the experience of cycling in Scotland. Likewise, bicycle design has been revolutionised since the penny farthings of the 1881 Dundee Bicycle Club,[51] and there is now an emphasis on lightweight and brightly-coloured clothing and accessories and specially designed high-energy foodstuffs and drinks. Modern roads are often without the extreme gradients of former days, but at the same time most are subject to heavy traffic. The conveyance of bicycles on modern trains can be extremely difficult as the rail network has been severely cut back and machines can no longer always be transported free of charge. Occasionally, however, it is still possible to enjoy something of the former style of cycling. Cycling holidays in the Outer Hebrides using independently-run hostels are a remarkable step back in time, although the pleasure can often be curtailed by the frequent strong winds.

BOARDING HOUSES, BED AND BREAKFASTS, ETC

As the twentieth century progressed, more and more people other than members of the leisured middle and upper classes were able to afford a holiday, although many workers, such as farm workers and women in the textile industries, continued to have little free time to spare and money to spend until the century was well through.

The east coast with its drier and sunnier climate had a ring of seaside resorts, extending from Angus, through Fife, to East Lothian. The professionals of Edinburgh took holidays in places like North Berwick, East Lothian, or they might go to Whitby, North Yorkshire, throughout the first half of the century, while those of Glasgow often went to centres such as Elie and St Andrews, Fife. Glaswegians and others in west central Scotland would, however, go first and foremost 'doon the watter' to the resorts and golf courses of the Ayrshire coast and the Firth of Clyde. Those of less ample means also went to such centres, but mainly to enjoy the outdoor bathing,

Figure 34.6 One of the sea-side pleasures for the very young. A one-man band, 1930s. Source: Alex Wilson, East Wemyss, Fife.

either in constructed swimming pools or the sea, although an equally popular attraction was the Pierrots, or variety artists (Fig. 34.6).

Railway companies ran 'specials' to the resorts during the summer. These returned in the evening and avoided the need for an overnight stay, thus making the pleasures of the sea available to a far greater number of people. For those with the time and money to stay overnight, several options were available. The richest might possess holiday homes. Others might rent an entire, staffed dwelling. A modest version of the latter is described by one 87-year-old lady who recalls going on a family holiday from Perth to St Andrews and taking over a family dwelling. The house belonged to a widow who cooked the food the mother of the family bought each day. The family had the run of the parlour and bedrooms, while the lady of the house was relegated to the kitchen, where she presumably slept in the bed recess,[52] in an arrangement similar to that which occurred in highland stalking areas before the building of shooting lodges.

Hotels, guesthouses and boarding houses were usually very important in seaside resorts. The last accommodated perhaps ten to 20 visitors with full or half board of breakfast, possibly lunch and high tea. Otherwise a landlady might let out just a few rooms in her home to provide seasonal additional income,[53] a full or half-board precursor to the bed and breakfast.

Bed and breakfast establishments (B&Bs), now commonplace stop-overs for both home and foreign visitors, came into their own with the end of petrol rationing in the 1950s and have burgeoned in the last 20 years. They are found mainly, but not exclusively, in rural locations and together offer a comprehensive network of usually good quality accommodation at moderate prices, at least for couples or small family groups (single rooms are not common and are often relatively expensive). Crofters and farmers, and most often their wives, for example, may run a B&B business only in the short summer season to supplement their income, while others operate a year-round enterprise from which the main livelihood is derived. All offer a cooked breakfast and usually an evening cup of tea and coffee or at least tea and coffee-making facilities. Some may also provide an evening meal for a separate charge.

The Scottish seaside resorts have suffered from increasing pollution of sea and sand, and from the irresistible competition of foreign package holidays, which not only offer a guaranteed sunny break, but are also often cheaper than vacationing at home. However, Scotland as a whole remains, and is increasingly, a popular holiday destination, both for those living within and beyond the country, and it continues to offer a wide range of accommodation. Full- or half-board can still be found in hotels, guesthouses and some B&Bs. Despite the problems they may cause in areas of limited housing stock, holiday homes – whether conventional dwellings, purpose-built chalets and cabins, mobile homes, static caravans or eclectic conversions such as railway carriages or even the former Debtor's Prison in Inveraray, Argyll – are bought and sold in large numbers every year, or are rented by individuals, couples, families or groups of friends, maybe for just a week or two. Those also seeking to enjoy a private dwelling for holidays may occasionally choose to build their own structures. Groups of self-build holiday dwellings have existed for some time at Glentress, Peeblesshire, for example.[54] Timeshare holiday homes, where the individual buys the perpetual right to use serviced accommodation for a number of weeks per year, is a relatively recent phenomenon and one usually more popular with older holidaymakers. For those prepared to do without private sleeping accommodation, hostels have been augmented by bunkhouses and groups may use outdoor or educational centres. Camping has its devotees, both those who favour campsites and those who prefer to 'rough it in the wilds'. Other alternatives for the hardy include the use of bivvy bags (sacks, formerly made of plastic but now of sophisticated materials such as Goretex, in which a sleeping bag can be laid out), and, like the early mountaineers and hill walkers, caves.

Whatever the nature of it, one's holiday accommodation provides, in optimum conditions, the backdrop to an escape from the routine of everyday life and many of its constraints. This relative freedom, coupled with the stimulation of new surroundings and activities, means that holidays often have a disproportionate prominence in our memories and importance in our lives.[55]

NOTES

1. Simpson, 1997 is a very useful starting point; a pictorial publication of relevance to this subject is Durie, 1988.
2. Lynch, 1992, 362.
3. Calculated from Mitchell, 1902.
4. Wordsworth, 1981.
5. Scottish inns are believed to date back to the seventeenth century when they were mainly established in the larger burghs and served the function of drinkers' dens and were patronised by clubs. They were not far removed from basic public houses, offering few alternatives to drink, but also served as change houses in line with the development of stagecoach travel, which occurred when Scotland was opened up by better roads. Inns were also to be found at ferry points, coaching halts, sailors' and smugglers' howffs and the overnight stances used by drovers. In many instances board and lodging were provided at the more reputable establishments, but travellers from England often found the inns to be inadequate in their provision of victuals, other than that of whisky. In fact the comfort and provisions varied in Scottish inns and it is dangerous to generalise, especially when travellers' descriptions are based on comparison with another culture. See Walker, 2003.
6. Garnett, 1811, vol II, 41.
7. NAS CC7/6/6. Testament Dative of James Johnston who Died in 1771 (Boat of Inver or 'The Ferry Inn', Perthshire); Garnett, 1811, vol II, 41.
8. Colonel Thomas Thornton is the earliest example of the wealthy Englishman pursuing sport on the Scottish moors and lochs, each autumn in the 1780s. He kept a pack of foxhounds, raced on both horse and foot, fished, hawked and shot. See Eden, 1979, 19–20; see also Suffolk, Peak, Aflalo, 1897; Ellangowan, 1889; Fittis, 1889.
9. Orr, 1982.
10. Donnachie and Hewitt, 2003, 53, point to the fact that the Clearances – a latter day term – have been the subject of 'greater controversy and more myths than perhaps any other theme in Scottish history (discounting perhaps the Reformation and the Union)'.
11. The concept of deer forests was entrenched in the Highlands by the latter part of the nineteenth century and the proportion of hill land devoted to this form of recreation was substantial. See Orr, 1982, 28.
12. There are several game books in the NLS (eg MSS15152–59), partially printed volumes relating in various ways to Ellices' estates in Glengarry, Inverness-shire, including the partly submerged lands of the Glenquoich estate, Visitors' book, see Dep. 356. Glenquoich Lodge was lost when the waters of Glenquoich were raised after a dam was constructed in the mid twentieth century; Eden, 1979, 39.
13. Scrope, 1838.
14. A photograph by George Washington Wilson reveals the sumptuous surroundings of the Billiards Room, Glenquoich Lodge, Inverness-shire, rented for many years to Lord Burton, who took 100 stags a year from the famous forest of 46,000 acres. Durie, 1988, 34.
15. Napier Commission, Minutes of Evidence, 43, 243–4
16. Edwards, 1876, 117–8.
17. Orr, 1982, 97.
18. In 1907 the *Sportsmen's and Tourists' Guide* referred to the annual expenditure on shooting as 'a golden stream of no small magnitude and one which benefits all classes alike'. Alternatively, the *Oban Times* (12 October 1895) was more strident in its assertion that any wealth derived from shooting immediately went south where the lairds passed their summer.

19. Eden, 1979, 39.
20. Napier Commission, Minutes of Evidence, 43, 243–7.
21. MacDiarmaid, 1926.
22. Nairn, 1996, ch 4.
23. Correspondence from James Mair, Newport-on-Tay, Fife, 18 July 2001.
24. *The Courier and Advertiser*, 22 June 2006, 14.
25. Saunders, 1951.
26. Castle, 1937, advert opp 126.
27. See Buchan, 1896; Buchan, 1901.
28. Simpson, 1997, 66.
29. Clark, Pyatt, 1957, 195.
30. Two books accompany the radio series. Kay, 1982.
31. Borthwick, 1939.
32. See Scroggie, 1989.
33. The writings of the late W H Murray, 1947, who penned his experiences in POW camps only to have the drafts seized by the Gestapo, also give a vivid impression of the way of life.
34. See MacLean, 1980.
35. Weir, 1994, 104.
36. Styles, 1964, 18.
37. Smith, 1981, 8.
38. Correspondence from Professor Emeritus J P Duguid relating to 1939.
39. Sydney Scroggie (Tape 11), recorded by Roger Leitch on 14 August 2001.
40. Weir, 1948.
41. Eric Simpson, in his short history of vacationing in Scotland, highlights that the 'Cycling in Scotland' section of Black's *Shilling Guide to Scotland* (1906) occupies 11 pages of 167. Quoted in Simpson, 1997, 65.
42. Durie, 1988, 14.
43. Howkins, Lowrenson, 1979, 46, 55; see also Stevenson, 1990, 27, 110, 392.
44. See Douglas, 1897. In the 1930s such cyclists were known as 'pass stormers'.
45. Simpson, 1997, 65–6.
46. Correspondence from Ida Jacob, Ohaupo, New Zealand, 14 September 2000 and 4 November 2000.
47. Weir, 1948.
48. Weir, 1948, 23.
49. Anyone wishing to savour Sydney Scroggie's encyclopaedic knowledge of the Clova hills and the Cairngorms should see his collection of poems, *Give me the Hills*, Dundee, 1978.
50. The warden of the Newbigging hostel, 'Ma Harper', is revered in the poem of the same name in Scroggie, 1978, 46–7.
51. See *The Courier*, 2 June 2001, 10.
52. Information from Helen Jackson (1915–), Perth.
53. Information from a discussion with J J Robertson, Senior Lecturer in Roman Law at the University of Dundee and also correspondence from Professor Emeritus J P Duguid, formerly of the University of Dundee.
54. See Chapter 45 Domestic Life in Temporary and Self-Built Dwellings.
55. The writer would like to acknowledge the assistance of James Mair of Newport-on-Tay in pointing him to certain sources, along with the kindness and similar help provided by Goronwy Owen, Dundee, and R D Soutar, Dundee.

BIBLIOGRAPHY

Borthwick, Alastair. *Always a Little Further*, London, 1939.
Buchan, J. *Scholar Gypsies*, London and New York, 1896.
Buchan, J, ed. *The Compleat Angler*, London, 1901.
Castle, P. *Angling Holidays in Scotland*, Edinburgh and London, 1937.
Clark, R W, Pyatt, E C. *Mountaineering in Britain*, London, 1957.
Donnachie, I, Hewitt, G. *Collins Dictionary: Scottish History*, Glasgow, 2003.
Douglas, W. Mountaineering with Cycles, *Scottish Mountaineering Club Journal*, 4 (1897), 279–83.
Edwards, D H. *Historical Guide to Edzell and Glenesk*, Brechin, 1876.
Ellangowan, E. *Outdoor Sports in Scotland*, London, 1889.
Fittis, R S. *The Sports and Pastimes of Scotland*, Edinburgh, 1889.
Garnett, T. *Observations on a Tour through The Highlands and part of the Western Isles of Scotland*, 2 vols, London, 1811.
Howkins, A, Lowrenson, J. *Trends in Leisure*, London, 1979.
Kay, B, ed. *Odyssey: Voices from Scotland's Recent Past* (1980), Edinburgh, 2 vols, 1982.
Lynch, M. *Scotland: A New History*, 2nd edn, London, 1992.
MacDiarmaid, J M. *The Deer Forests and How They are Bleeding Scotland White*, 2nd edn, Glasgow, 1926.
MacLean, I. Mountain men. In Kay, 1980, I, 78–87.
Mitchell, A. A list of tours etc, relating to Scotland, *PSAS*, XXXV (1902), 522–42.
Murray, W H. *Mountaineering in Scotland*, London, 1947.
Nairn, A. *Silver Spoon*, Edinburgh, 1996.
Orr, W. *Deer Forests, Landlords and Crofters: The Western Highlands in Victorian and Edwardian Times*, Edinburgh, 1982.
Saunders, R C. *A Guide to the Fishing Inns of Scotland*, London, 1951.
Scroggie, S. *The Cairngorms Scene and Unseen* (The Scottish Mountaineering Trust), npp, 1989.
Scrope, W. *The Art of Deer Stalking*, London, 1838.
Smith, R. *Outdoor Scotland* (Scottish Youth Hostels Association), Stirling, 1981.
Stevenson, J. *British Society, 1914–45*, London, 1990.
Styles, S. *Modern Mountaineering*, London, 1964.
Suffolk, Earl of, Peak, H, Aflalo, F G, eds. *The Encyclopaedia of Sport*, npp, 1897, 2 vols.
Walker, D. Inns, hotels and related building types. In Stell, G, Shaw, J, Storrier, S, eds, *Scotland's Buildings, Scottish Life and Society: A Compendium of Scottish Ethnology, vol 3*, East Linton, 2003, 127–89.
Weir, T. *Highland Days*, London, 1948.
Weir, T. *Weir's World: An Autobiography of Sorts*, Edinburgh, 1994.
Wordsworth, D. *Recollections of a Tour Made in Scotland, AD 1803*, facsimile reprint, Edinburgh, 1981.

There is an abundance of existing bibliographies and source lists associated with this subject:
Brown, D, Mitchell, I. *Mountain Days and Bothy Nights*, Barr, Ayrshire, 1987.
Brown, H M. *Hamish's Mountains Walk and Climbing the Corbetts*, London, 1997, 346–7, 371–2.
Devine, T M. *The Scottish Nation*, Harmondsworth, 2000, detailed index.
Durie, A J (comp). *George Washington Wilson: Sport and Leisure in Victorian Scotland*, Aberdeen, 1988.
Eden, R. *Going to the Moors*, London, 1979, 202–6.
Simpson, E. *Going on Holiday* (Scotland's Past in Action Series, NMS), Edinburgh, 1997, 82–3.

35 Prisons

ROD MacCOWAN

LIVING IN PRISON

As at March 2002, home for around 6300[1] people in Scotland is prison, in which anything from a few days to 20 or more years can be spent. If you have never been in prison it is impossible fully to understand what it means to have one's liberty taken away and to lose commonly taken-for-granted control over many aspects of life. At this moment you can stop reading and it is likely that you can go out of the room, make a cup of tea, phone a friend or go for a walk. In prison each of these activities may well need someone else, first of all to allow it, and then to supervise it. While there may be some parallels, for example with domestic service in the past or with the infirm or elderly in residential settings, life in prison is dominated by considerations of security and supervision which arise from the coercive relationship between the prisoner and the prison.

Home in prison may be a Victorian cell shared with a stranger, without electrical power points for a television or radio and without access to sanitation except when the prisoner is let out of the cell. Or it may be a modern cell with in-cell toilet and washing facilities and a TV. Home may be an inner-city prison or an open prison in the countryside.

A prison is:

> ... a hotel function, it is a medical function, it has got its own industries, it's got its own college, its own chaplaincy, its own welfare workers, reception staff ... it is more a total institution ... than anything you could describe. There are almost no total institutions in this country. There is a ship at sea, there is an enclosed religious order and there is a prison ... difference is ... only half of [those in] the prison ... are volunteers.[2]

It is a remarkably self-contained world which can offer accommodation, work, education, leisure, social work and medical services. In fact it may provide all the services required by its residents, from the opportunity to have visits from family, to counselling, to courses to help the individual with drug or alcohol problems or with offending behaviour. In addition it has its own rules and its own internal policing of these and punishments for breaking them. All of this can be without the prisoner once leaving the prison during his or her sentence (Fig. 35.1).

Figure 35.1 HMP Edinburgh Gatehouse. Source: Rod MacCowan.

HISTORY OF THE SCOTTISH PRISON SERVICE

There is no one history of prisons and imprisonment and thus choosing any starting point is merely one way to begin a particular interpretation. However, 1877 seems as reasonable a place as any to begin ... for in that year, with the passing of the Prison Act, the foundations of a modern, centralised Prison Service were laid.[3]

Before 1835 there is scarcely any justification for referring to a *prison system* in Scotland. Almost every prison was run under an independent process financed by the relevant local authority.[4]

The first piece of major legislation affecting prisons in Scotland was passed in 1597, requiring each burgh to build and run a prison, 'Prison Houses suld be bigged within all Burrowes.'[5] The same legislation also stated that prisoners were to be maintained at their own expense. Concern at the state of prisons grew, particularly during the late eighteenth century, and led to the Prisons Act of 1835, which among other things set up an inspection process to give central government oversight of prisons. This provoked recommendations that the entire control of prisons come under central government control. Further Prisons Acts followed in 1839 and 1877:

> By 1877, the number of prisons in Scotland had been reduced to 56, their administration- by the Prisons (S) Act of that date – transferred from the local authorities to the state, and their maintenance provided for by Parliament.[6]

In Edinburgh, prisoners were held in the Old Tolbooth (the 'Heart of Midlothian') until a new prison started in 1808 on the Calton Hill replaced it, Lord Cockburn being moved to declare that, 'It was a piece of undoubted bad taste to give so glorious an eminence to a prison.'[7] A site at Saughton was purchased for £10,000 in 1913 for a prison to replace the Calton Jail. World War I hampered construction, however, and it was not until an Order of 1919 that the new prison was appointed as a legal place of detention. Although building work continued until 1930, all prisoners had been transferred from the Calton Jail by 1925. Substantial additions have been made to the prison since the 1930s.

THE SCOTTISH PRISON ESTATE TODAY

The Scottish Prison Service (SPS) operates prisons on 15 sites from Dumfries in the south-west to Peterhead in Aberdeenshire, providing accommodation for some 6300 prisoners. In addition, a private provider operates one prison at Kilmarnock, Ayrshire, under contract to the SPS.

These prisons have a variety of functions and are divided into six main types. More than one function may take place at any one site.

Local prisons
Barlinnie (Glasgow), Aberdeen, Perth, Edinburgh, Inverness, Greenock (Renfrewshire) and Kilmarnock. These prisons hold remand prisoners (that is, those awaiting a court appearance or awaiting sentence) and short-term convicted prisoners (prisoners serving up to four years). Local prisons normally serve the courts and the population of a defined geographical area of Scotland. They may, however, also have additional functions, such as holding special categories of long-term prisoner (sentenced from four years to life imprisonment). Greenock and Edinburgh, for example, both have units holding long-term prisoners who are nearing the end of their sentences. As part of a sentence management plan, they will qualify to work or attend college in the local community, on a day release basis, up to five days per week.

Long-term prisons
These are Shotts (Lanarkshire), Glenochil (Clackmannanshire) and Peterhead, plus certain accommodation halls at Edinburgh, Perth, Kilmarnock and Greenock. As the name suggests, these hold prisoners serving from four years to life. Some units within these prisons specialise in dealing with certain groups of prisoners. For example, those starting their sentences will normally go to Shotts (where the National Induction Centre is located), Glenochil or Perth, while long-term prisoners preparing for release will move to Edinburgh or Greenock.

Short-term prison
HMP Low Moss (Glasgow) holds convicted prisoners serving up to 12

Total			6172
Untried (includes those convicted but awaiting sentence)			
Male:	Adult		871
Female:	Adult		62
Male:	Under 21 years		211
Female:	Under 21 years		11
			1155
Convicted			
Male:	Adult		4288
Female:	Adult		166
Male:	Under 21 years		530
Female:	Under 21 years		16
			5000
Others			17

Figure 35.2 Prison population, 8 March 2002. Source: Rod MacCowan.

months and selected prisoners serving up to four years who are in the last nine months of sentence.

Open prisons
Castle Huntly (Perthshire), Noranside (Angus). These hold a variety of prisoners, short- and long-term, who are not considered to pose a risk and who can be held in conditions of lesser security.

Young-offender institution
Polmont (Stirlingshire). This institution deals with all convicted males under the age of 21.

Women
Cornton Vale Prison (Stirlingshire) holds most of the female prisoners in Scotland (under 21, remand, short-term and long-term). Some local prisons have small units for females.
The SPS also ran a prison at Zeist in the Netherlands until March 2002, which was set up as part of the arrangements for the Lockerbie Aircraft Bomb Trial (Fig. 35.2).

CLASSIFICATION OF PRISONERS

For the purposes of managing the prisoner population there are four main areas of classification, which will normally decide where a prisoner will serve his or her sentence and the degree of security required.

Age
As far as possible young offenders (under 21 years old) are kept separate from adults, either in separate institutions or separate areas within, usually, local prisons.

Gender
While small units exist at Aberdeen, Dumfries and Inverness, mainly for women remand prisoners or those serving short sentences, the majority of women go to Cornton Vale Prison near Stirling.

Sentence length
Normally those serving less than four years undertake their sentences at local prisons, usually the prison nearest to their home or to the court they were convicted in. This group will also be considered for short-term or open prisons. Long-term prisoners (serving over four years) will normally start off in a long-term prison, but may progress to other prisons, including open prisons, with facilities geared to preparing them for release.

Security category
There are currently five security categories:

Category A	Those most likely to pose a risk to the public and who need the highest levels of supervision.
Category B	Those posing a significant risk, including all remand prisoners.
Category C	Those who pose some risk but can be held under less strict security conditions.
Limited Category D	Those, particularly long-term prisoners, who pose little risk and can take part in activities which allow their suitability to be assigned to Category D to be tested.
Category D	Those who pose minimal risk (these include prisoners in open prisons or those on Training for Freedom and similar schemes).

SCOTTISH PRISON SERVICE: PURPOSE

The SPS Mission Statement clearly defines its purpose:

The Mission of the SPS is: to keep in custody those committed by the courts; to maintain good order in each prison; to care for individuals with humanity; and to provide all possible opportunities to help prisoners to lead law-abiding and useful lives after release.[8]

Much of the response of the SPS to managing prisoners stems from the policy document *Opportunity and Responsibility*, published in 1990. This focuses on such matters as developing the role of the prison officer, involving prisoners in the decision-making process (including sentence planning) and the concept of 'the responsible prisoner'. The last views the prisoner not so much as someone who requires a variety of things to be done to and for him or her to assist his or her rehabilitation, but instead regards the prisoner as 'a person who is responsible for his actions and who should be encouraged to accept his responsibilities by providing him with a range of opportunities for personal development'.[9] In dealing with those in prison the SPS has developed a number of opportunities, ranging from the possibility of improved contact with friends and families to sophisticated programmes aimed at addressing the offending behaviour of those who come into prison.

Currently the SPS is engaged in refining its vision – as described in *Protecting the Public and Reducing Offending* – to meet the continued and changing demands and expectations placed on it. This will build on what is currently done, and will seek to involve the SPS in the wider Correctional Agenda while concentrating on aspects of service delivery, providing value for money and the further development of staff professionalism. The Correctional Agenda involves the SPS addressing to a greater degree prisoners' offending behaviour, creating partnerships with other agencies, and developing throughcare for prisoners and managing their transition back into society.

HM PRISON EDINBURGH

> Edinburgh Prison serves the Sheriffdoms of Lothian and Borders and Fife. The prison holds male remand prisoners, both adult and under 21. It holds short-term convicted adult prisoners; under 21 and long-term convicted prisoners awaiting transfer; provides a facility for very long-term prisoners and a Training For Freedom Hostel.[10]

Known locally as Saughton Prison, Edinburgh Prison is situated in the west of the city. The prison currently comprises eight accommodation units of different design, size and function. Construction of a new houseblock commenced in January 2002 and when complete this will replace three of the original houseblocks (Fig. 35.3). The prison holds some 720 prisoners, about 25 per cent above its number of single-cell places.

As with most prisons, Edinburgh is virtually a self-contained community, trying to provide within its confines all the prisoners' requirements. This reflects a wish to ensure that peoples' needs are adequately met and also recognises that, for reasons of security, it is better to provide a service to prisoners within a secure perimeter than to take them outside the prison. The prison is home to a prisoner for the duration of his sentence and all the types of services and activities which happen in the wider community are reflected inside.

Figure 35.3 Exterior of Glenesk House Accommodation Hall, HMP Edinburgh. Source: Rod MacCowan.

Accommodation
Accommodation tends to reflect one or more of the four areas of classification referred to above. The Prison and Young Offenders Institutions (Scotland) Rules 1994 state:

> Rule 11
> Every Prisoner may be classified by the Governor according to:
> age
> sex
> offence
> period of sentence or committal
> previous record.[11]
>
> Rule 14
> The Governor shall, so far as reasonably practicable, keep civil prisoners, untried prisoners and young prisoners apart from other categories of prisoner.[12]

In Edinburgh Prison untried prisoners will go to Glenesk Hall. Under–21 prisoners are held as far as possible in a separate section of the Hall from adults. Glenesk is a two-year-old, three-storeyed building, where each cell has a toilet cubicle, electric power points, kettle and television (Fig. 35.4).

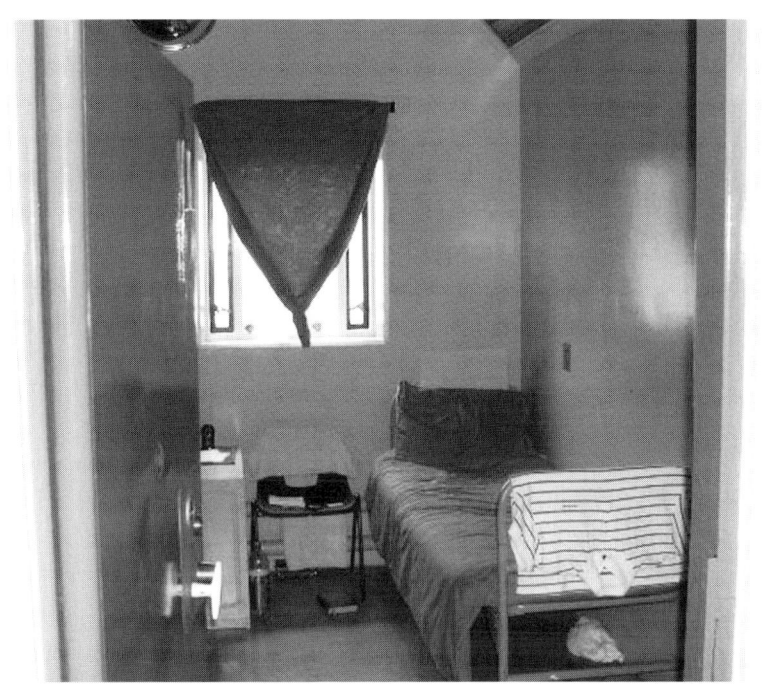

Figure 35.4 Cell, HMP Edinburgh. Source: Rod MacCowan.

Televisions are part of the equipment in each cell. As a component of a system of privileges, they can be removed as part of a scale of punishment or loss of privileges. Within the privilege system, a prisoner can have a variety of things in use, including radios and cassette or CD players. There is no restriction on the playing of these, although they are subject to security checks. Where the items have recording facilities, these are disconnected. While this may seem odd to the outsider, the precaution prevents taped messages of a threatening or distressing nature being sent to, for example, victims.

The Hall itself has a range of recreational and fitness equipment on each of its three levels, including a mini gym, pool table and table tennis. There is a food servery on each level, there are payphones in each section and a medical consulting room. A typical day in the Hall is shown in Fig. 35.5.

Once convicted, a prisoner is moved to one of a number of other halls. Each hall has its own routine and a progression system is in place which allows prisoners who are co-operative, and in particular those who do not wish to be involved with drugs, the opportunity to have additional privileges (Fig. 35.6).

Some of the halls have old Victorian-style cells without sanitary facilities or in-cell power. While regular opportunities to use the toilets exist during the hours of unlock, when locked up for the night (from about 9pm until about 7am) there is no access to toilets. Plastic pots are available in each

Residential Hall Daily Routine
Monday to Friday
Remand

Time	Activity
7.15	Numbers check
7.45	Unlock, showers, etc
8.00	Breakfast
9.00	Prisoners to activities as appropriate
11.30	Exercise
12.45	Lunch
13.30	Visits
14.00	Recreation/activities
16.00	Lock up, numbers check
17.30	Tea meal
18.00	Visits, recreation, kit changes, etc
21.00	Lock up, numbers check
21.30	Night shift take up duty

Residential Hall Weekend Routine
Remand

Time	Activity
7.30	Numbers check
8.00	Unlock, showers, etc
8.30	Breakfast
9.00	Lock up, church service, domestic duties
10.00	Exercise
12.15	Lunch
12.45	Lock up and patrol period to allow staff break
13.45	Unlock, recreation, visits, PE, etc
16.45	Tea meal
17.30	Lock up, numbers check

Residential Hall Daily Routine
Monday to Friday
Convicted (D Hall)

Time	Activity
7.15	Numbers check
7.30	Unlock for showers, if sick, etc
8.00	Breakfast
8.30	Prisoners move to work, education, PE, etc
11.30	Prisoners return from work, education, etc
11.45	Exercise
12.45	Lunch
13.30	Prisoners move to work, education, programmes, etc
16.45	Prisoners return from work education, etc

17.20	Tea meal and lock up/numbers check
18.00	Evening recreation, visits, etc
21.00	Lock up, numbers check
21.30	Night shift take up duty

Residential Hall Weekend Routine
Convicted

Time	Activity
7.30	Numbers check
7.45	Unlock, issue breakfast then showers, etc
9.00	Lock up, church service, domestic duties
10.15	Exercise
12.15	Lunch
12.45	Lock up and patrol period to allow staff break
13.45	Recreation, visits, PE, etc
16.45	Tea meal
17.30	Lock up, numbers check

Figure 35.5 Residential Hall daily routines for remand and convicted prisoners, HMP Edinburgh. Source: Rod MacCowan.

Figure 35.6 'A' Hall west elevation, HMP Edinburgh. Source: Rod MacCowan.

cell. This leads to the very unpleasant procedure of 'slopping out' which occurs each morning when the waste of over 100 men is taken from the cells and disposed of in the hall toilet area.

To pass the time, prisoners have battery-powered radio cassettes. Recognising that these prisoners are disadvantaged by not having power in their cells, rechargeable batteries and chargers are available. During lock-up periods prisoners can read, although the lights are turned on and off from outside. Papers are supplied or can be sent in, and books can be obtained from the prison library or can be handed in. Prisoners do jigsaws, write, draw, paint and some undertake in-cell craft activities such as modelmaking.

Prisoners are supplied with prison clothing which now usually comprises polo shirts, sweatshirts and sweatpants. These are colour co-ordinated, each hall having its own colour both for supervision purposes and for stock control. Additionally, jeans, tee-shirts, socks, pants, pyjamas, etc, are supplied, as is protective clothing for work. Prisoners can wear their own clothes during recreation periods (usually the evening) and at the weekend. Beds in cells are equipped with sheets and duvets. All clothing and linen is laundered in the prison laundry.

Food

Food is very important in prison. Catering staff and prisoners work in the prison kitchen providing three meals a day, 365 days per year for around 700 prisoners; in total some three quarters of a million meals per year.

The meals are mostly served from a hotplate in the hall and taken back to the cells to be eaten. The crockery and cutlery is collected and washed centrally. With a menu choice system in place similar to that run by hospitals, most dietary needs can be met. Choices include vegetarian and vegan options. Regard is given to religious dietary requirements and the options available will usually provide for these, failing this additional dishes are supplied. A typical day's menu can be seen in Fig. 35.7.

Arrangements are made for religious observance, such as during Ramadan when Muslim prisoners fast during the daylight hours and suitable food is provided in the evening. At Christmas and New Year catering staff supply a range of special meals to celebrate the festive season. Alcohol is not available.

In addition, prisoners can purchase food and other items from the prison canteen. Goods on sale include tobacco, sweets, fruit and toiletries. This is paid for both from personal funds and from wages earned while working in the prison. Prisoners are not allowed cash; instead a computerised credit system is used. There are limits placed on the personal money which can be handed in and that available. In addition to any wages earned, a prisoner can have up to £5 of his own money credited for use each week.

Staff

Between 400 and 500 staff work at Edinburgh, most of whom come into contact with prisoners in some capacity. The Governor has a Deputy Governor

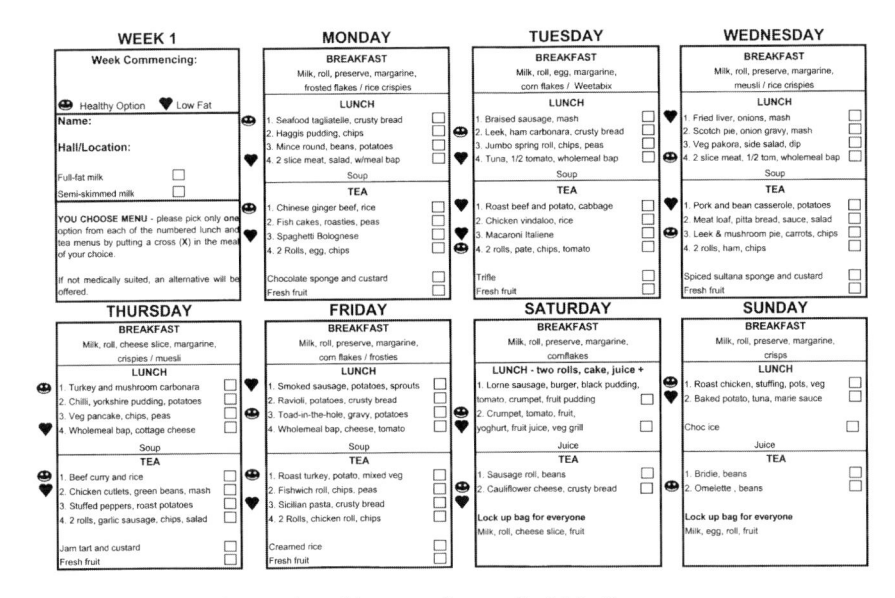

Figure 35.7 A typical week's menu. Source: Rod MacCowan.

and a senior management team to manage the prison. This team's responsibilities are divided into various areas:

RESIDENTIAL
Responsible for prisoner management, casework, running and managing the accommodation areas.

OPERATIONS
Security, escorts, intelligence, visits.

PRISONER ACTIVITIES
Work, throughcare, prisoner programmes, drug strategy, catering, healthcare and managing Education, Social Work and Chaplaincy.

FINANCE AND ADMINISTRATION
Budget, resource management, prisoner wages and money, legal compliance, stores.

HUMAN RESOURCES
Staff issues.

ESTATES
Maintenance, energy and contract supervision.

During his time in prison, a prisoner will see both prison officers and specialist staff, the SPS encouraging as much meaningful engagement as possible between staff and prisoners. Prisoners interact with a range of people on a daily basis. Within the hall a prisoner has a personal officer who is his first point of contact. This officer works as part of a team assigned to run a particular hall. Serious matters may be referred to a Hall Manager, leading one of the teams, or to a Unit Manager, who is in charge of two or three halls and who deals with very serious management issues surrounding the prisoner and his casework.

If a prisoner is ill he will be seen by a nurse and referred to one of the medical officers (doctors) employed in the prison. A multi-disciplinary Drug Referral Team member or a Mental Health Team member may also see him. If at risk of self-harm, a multi-disciplinary case conference will be held in the hall to draw up an action plan for his management.

On an average day a prisoner goes to work in one of the jobs detailed below, the wages from which can be spent on a range of items ordered from the prison canteen. He may also go to an education class or to physical education (PE) during the day. If he is on a programme (for example, Anger Management, Cognitive Skills, or an alcohol or drugs programme) he may spend all or part of the day there. With unresolved personal issues he may attend throughcare and see one of a range of community agencies present, or he may speak with a social worker or chaplain. He is entitled to a period of exercise, usually at lunchtime, and usually has free time (recreation) in the evening where again he may go to classes or take part in organised competitions, etc. He is also entitled to visits supervised by staff.

If a prisoner commits an offence against prison discipline he may be removed to the purpose-built Segregation Unit and within 24 hours be seen by a senior manager in a Prison Orderly Room. Here a number of actions may occur: dismissal of the charges, or imposition of sanctions, such as the loss of privileges, or a period of confinement to cell.

Work
Prison Rule 68 states that all convicted prisoners are required to work. Within Edinburgh there is a variety of employment. The prison has woodwork, textile and metalwork manufacturing workshops and there is a large paintspray workshop. Prisoners can also learn a range of skills including painting and decorating and industrial cleaning, perhaps leading to national qualifications. They also engage in a range of domestic duties: cleaning, cooking, gardening and the maintenance and cleanliness of the prison grounds.

A board meets twice a week to allocate prisoners to work parties. This is done on the basis of the vacancies available, the prisoner's experience and his preferences. Other issues, surrounding security classification and intelligence, associates or affiliations, will also influence allocation to work parties. Prisoners are paid an average of about £7 per week for their work. Some long-term prisoners who are on a programme leading to release work under supervision in the community from one to five days per week.

Education
Education is provided by a local college in the prison's own Education Unit. The Education Department tries to meet individual needs, from basic literacy

and numeracy classes to Open University courses. Prisoners can apply to take classes during the working day. Evening classes are also available.

Exercise and leisure
Rule 75 stipulates that all prisoners are entitled to take exercise for not less than one hour per day and to have the opportunity to spend time in the open air daily. While this is usually time spent in an exercise yard, the prison also has a small gymnasium and trained PE instructors who undertake a range of games and fitness activities with those prisoners who wish to take part. The prison has two all-weather five-a-side pitches.

There are limited recreation facilities in or near each accommodation hall. Typically this includes a TV, pool or snooker tables, and electronic games. Additionally the prison has an excellent library, which is a branch of Edinburgh City Council's own library service. It provides a range of services similar to those of any of the city libraries.

Health
The prison has a Health Centre, which includes in-patient rooms and small wards, consultation rooms, treatment rooms, a dispensary and dentist's surgery. Healthcare staff see every prisoner on admission to the prison. Many prisoners, particularly those admitted with drug problems, present a wide variety of needs which are dealt with by a team of doctors and nurses employed in the prison along with a number of specialists. Prisoners with serious injuries or conditions are referred to outside hospitals. Most minor complaints and convalescence are provided for in the Health Centre. A number of clinics and health-care initiatives are also run from here. Healthcare staff in addition form part of a number of multi-disciplinary teams within the prison, including those which work with drug users, prisoners with mental health issues and individuals at risk of self-harm.

Programmes
A team of specially trained officers provide a variety of risk and needs assessment and a range of programmes and courses designed to address prisoners' offending behaviour. These programmes are also formulated to develop skills which will be of help upon release. In the Throughcare Centre prisoners receive an induction into prison, assessments are carried out, and it is possible to meet with representatives of outside agencies to help resolve problems, such as those connected with housing and benefit. Prisoners can also undertake courses aimed at acquiring job skills with agencies such as APEX, which works in partnership with the prison. Prisoners who have specific needs will be encouraged to undertake specific programmes. These include Anger Management, Cognitive Skills (thinking, problem-solving and dealing with others), Drug Awareness, and Drug Relapse. These programmes and courses are delivered by prison staff – who are also part of wider multi-disciplinary teams – supported by specialists, for example, psychologists. These activities are central to the work being carried out with

Figure 35.8 Interior of the chapel, HMP Edinburgh. Source: Rod MacCowan.

the prisoners by the SPS. In his last formal inspection, HM Chief Inspector of Prisons identified the work of the Throughcare Centre as one of eight examples of best practice at Edinburgh, noting that, 'The Throughcare Centre enabled the adoption of an integrated rational approach to mainstream prisoner management by identifying risks and needs.'[13] Unlike many prison services, the SPS trains the prison officers who deal with the prisoners on a daily basis to carry out this work.

Religion
Religious needs are met by a Chaplaincy Team, most of whom work in the prison on a part-time basis and provide a range of religious activities for prisoners including Prison Fellowship and the Hope Group. The team will arrange provision for any religious faith (Fig. 35.8).

Contact
It is important to remember that when someone is in prison his life still goes on. Most prisoners have families, many have children. School problems still happen, family members get ill, get married, leave and grow old. They need support and advice. Being in prison does not remove these responsibilities from the individual. Contact is crucial in allowing people to keep in touch and to manage the life in the community they left and will return to.

There are formal agencies to help with this. The prison has its own Social Work Department, which deals with statutory requirements and exceptional problems. However, as far as possible, prisoners are encouraged to try to deal with their own issues.

Prisoners can receive and send mail and also have access to payphones in all the accommodation units. Mail cannot be opened if intelligence as to the content suggests that there may be contraband contained. In such a case the mail will be examined in the presence of the prisoner. Payphones can be listened to and conversations recorded. Notices on all payphones advise prisoners of this.

Visits

Of greatest importance to prisoners are visits. These may be from legal advisers before and during any trial or appeal. Just as significant, however, are visits from family and friends. Edinburgh has a new Visitor Centre built outside the prison which provides a bright and welcoming environment for visitors to book in and wait for their visit. The Centre is run by the Women's Royal Voluntary Service and has a café, play area and meeting room and offers a wide range of information to visitors. In the period 1 April to 31 December 2001 there were over 13,000 visits to untried prisoners and over 11,000 visits to convicted prisoners (see Fig. 35.9).

After the Visitor Centre visitors then enter the prison, visits being taken in an extensively refurbished Visit Room, frequently around coffee tables. Those who take advantage of the arrangements, usually by trying to smuggle contraband, will go on closed visits which take place behind glass screens. The emphasis is on making the visits as comfortable and enjoyable as possible, while having regard to security.

Untried prisoners can ask for daily visits on weekdays. The prison

Figure 35.9 The Visitor Centre interior. HMP Edinburgh. Source: Rod MacCowan.

rules allow convicted prisoners a 30-minute visit once a week or two hours in any four-week period. This is the minimum allowance and many prisoners are able to exceed it. Provision is also made for visits at the weekend. Long-term prisoners can qualify for a scheme which allows a number of visits to take place at home, leading eventually to unescorted visits home as part of a release programme.

Prison visits can be difficult for everyone. Families on low incomes may find the costs involved high (although there is an assistance scheme). Visitors sometimes have to travel long distances or use infrequent and inconvenient public transport, often with young children. Pressure is frequently put on families to provide cash, clothing, etc for the prisoner, which they may be ill-able to afford. Difficulties in relationships do not disappear when someone is sentenced, indeed separation can intensify them. A crowded Visit Room may not be the best place to resolve family matters.

Visits are a source of tension for the prison. The more relaxed and informal they are, the greater the opportunity for their misuse, especially in terms of smuggling drugs. However, the preventative measures used, which include searching, drug 'sniffer' dogs and staff surveillance, can intimidate visitors. Despite this, most people find visits a good chance to maintain family contact and help the prisoner continue to feel part of his social network outside.

Discipline and complaints
Prisoners are subject to the Prison Rules. Those who break the rules are dealt with daily in the Prison Orderly Room where a senior manager will hear the charge, take evidence and, if the individual is found guilty, give an award. Remission (time off for good behaviour) no longer exists. There was an award of up to 14 additional days added to the sentence for serious offences. However SPS withdrew that punishment in 2001. Punishments are mainly withdrawal of privileges, monetary penalties from wages and confinement to cell.

Edinburgh, with over 700 men resident 24 hours a day, in the period from 1 April to 31 December 2001 had an average of three offences per day by prisoners deemed serious enough to lead to prisoners being charged under the prison discipline procedures. Complementing the Governor's power to impose penalties is a series of complaints procedures which seeks to ensure that prisoners are treated fairly. Prisoners can use an internal procedure, allowing the Hall Officer or Hall Manager to address their complaints. Complaints can go on to an Internal Complaints Committee and, if unresolved, go to the Governor. Any prisoner can also formally complain directly to the Governor. All of these avenues of complaint must be recorded and replied to within set periods. Prisoners can further complain to an independent Complaints Commissioner or request to see a member of the Prison Visiting Committee. Additionally, they can complain to their legal agent, the courts, The European Court of Human Rights, their MP or MSP, the Chief Constable or to anyone else they wish.

CONCLUSION

A sentence of imprisonment is a serious matter. It is the most serious sanction our society places on its citizens. Prison is home for the individual while he or she is on remand or serving a sentence. However prisoners come from the community and will return there. The SPS tries to provide the individual prisoner with appropriate links with the community and offer opportunities to acquire skills which will help him or her lead a useful and law-abiding life upon return to the community.

NOTES

1. Prison numbers fluctuate throughout the year, this figure is taken 8 March 2002.
2. Willmott, 2001.
3. Bryans, Wilson, 2000, 1.
4. Coyle, 1991, 27.
5. Coyle, 1991, 26.
6. Cameron, 1983, 128; see Chapter 37 Children's Homes for information on children's imprisonment in the past.
7. Cameron, 1983, 55.
8. SPS, Mission Statement.
9. SPS, 1990, 63.
10. HMP Edinburgh, 2001.
11. Prison and Young Offenders Institution (Scotland) Rules, 1994.
12. A civil prisoner is one who is imprisoned for such things as contempt of court or breach of interdict. At the time of writing there was one civil prisoner in Scottish prisons.
13. HM Inspectorate of Prisons for Scotland, 2000.

BIBLIOGRAPHY

Bryans, S, Wilson, D. *The Prison Governor*, 2nd edn, Wooton-under-Edge, 2000.
Cameron, J. *Prisons and Punishment in Scotland*, Edinburgh, 1983.
Coyle, A. *Rethinking Scotland's Prisons*, Edinburgh, 1991.
HMP Edinburgh. *Strategic Plan, 2001–2* (Internal SPS Management Document), Edinburgh, 2001.
HM Inspectorate of Prisons for Scotland. *Report on HM Prison Edinburgh*, Edinburgh, 2000
MacGregor, C A. Buildings of administration. In Stell, G, Shaw, J, Storrier, S, eds, *Scotland's Buildings, Scottish Life and Society: A Compendium of Scottish Ethnology, vol 3*, East Linton, 2003, 249–75 (tolbooths and early prisons).
SPS. *Opportunity and Responsibility* (SPS Document), npp (Edinburgh), 1990.
Willmott, Y. Governing women's prisons, *Prison Service Journal*, 136 (July 2001), 42–9.

36 Hospices and Nursing Homes

DEREK B MURRAY

THE HOSPICE MOVEMENT

The modern Hospice Movement began in the United Kingdom in the 1960s and has spread rapidly throughout many parts of the world, taking various forms, each reflecting the culture and healthcare structures of the countries where it has been adopted. Building on the experience of the older Homes for the Dying – which took their name and concept from medieval inns on pilgrimage routes – and impelled by strong Christian convictions, Dame Cicely Saunders founded St Christopher's Hospice in south London in 1967. The primary aim was to care for adult patients of any age with advanced cancer who had been treated with curative medicine and were now in the final stages of the disease. Other conditions such as end-stage heart and lung disease and motor neurone disease were also cared for, but much smaller numbers of individuals were, and are involved. In the next 30 years many similar units were opened throughout the United Kingdom, reflecting the ideals of St Christopher's.

The first such hospice in Scotland was St Columba's in the north of Edinburgh which received its first patient in 1977. Soon Strathcarron Hospice was opened near Denny, Stirlingshire, to serve Central Region, closely followed by units in Glasgow, Greenock, Inverness and other centres. In Aberdeen and Dundee previously existing units within the NHS, both called Roxburghe House, became part of the movement, and over the years the Marie Curie centres in north Glasgow and south Edinburgh became identified as hospices, the catchment areas in each city being adjusted. More recently hospice wards have been incorporated within larger hospitals, at Dumfries, Kirkcaldy, Fife and Aberdeen, and a few smaller units have been opened in outlying areas such as Cowal, Argyll and Stornoway, Lewis.

In other countries the Hospice Movement has adapted to different models of healthcare and funding, and in some, such as the USA, has concentrated on voluntary help for people at home. In China palliative care is mainly for older people, and in southern Europe it is largely hospital based.

Many hospices have a day-care facility, to which patients come once or twice a week for observation, diversion and company (Fig. 36.1). The whole of the United Kingdom is now served by specialist palliative care nurses, familiarly known as Macmillan nurses, who often operate out of hospices and provide supportive care to people in their own homes.[1] Most hospices have an education unit where healthcare professionals are trained and where staff are kept in touch with the latest developments in palliative care.[2]

Typically in Scotland hospices are partly funded for their running costs by Health Boards and rely heavily on charitable giving and community support for a proportion of their day-to-day expenses and for all capital costs. In Edinburgh, Milestone Hospice for the care of people affected by the human immunodeficiency virus (HIV) or acquired immune deficiency syndrome (AIDS) has been run by a trust and has suffered from lack of adequate funding and also from the changing parameters of the whole AIDS scene. There is a children's hospice at Kinross, Rachel House, begun by the efforts of a few parents and other interested people, now so successful that a second children's hospice, Robin's House, has recently been opened at Balloch, Loch Lomond.

Nursing homes, which have long been part of the range of 'less common types of homes', have enjoyed a renaissance in recent years and many custom-built units have taken their place beside the conversions of large dwelling houses. Co-operation with hospices has resulted in training for the staff of such homes in palliative care. As most residents suffer from the slow deterioration due to old age, with a fair proportion having forms

Figure 36.1 Day Hospice – a homely place.
St Columba's Hospice, Edinburgh. Source: Derek B Murray.

Figure 36.2 St Columba's Hospice exterior view. Source: Derek B Murray.

of dementia, the whole ethos of the nursing home significantly differs from that of the hospice. There is, however, a two-way traffic between nursing homes and hospices, in patients and in the exchange of knowledge, to their mutual benefit.

Hospices of all kinds, and to a lesser extent nursing homes, set out to provide a homely atmosphere (Fig. 36.2). This varies depending upon the sort of care given and the individuals and families being cared for. In an AIDS hospice, families of the often young patients are encouraged to live on the premises and many clients have multiple admissions, spending the intervening time in their own homes. In children's hospices there is also provision for parents and other siblings to be accommodated and the care is often for short breaks. Play areas, well-arranged grounds, and a very flexible routine make the atmosphere as homelike as possible. For hospices where most of the patients have cancer, and many are older people, the provision of homeliness is different, as space can be restricted and relatives stay overnight only in a minority of cases. Proposals to mix types of hospice, such as having AIDS patients cared for in 'ordinary' hospices, or mixing children and adults, have run into difficulties of perception and of practical care. Therefore, at least in Scotland, the various types have tended to remain separate.

A further distinction of the Hospice Movement is its spiritual or religious orientation. Dame Cicely originally saw her hospice as a religious community and only as reality overtook theory did the present St Christopher's evolve into a home for those of any creed or culture. The chapel is still central to the building. Something similar happened in St Columba's, Edinburgh, where the chapel has an important place and the initial financial and personal support came from the churches of the city. Other hospices have had similar origins, while many are secular foundations. The Marie Curie centres for long resisted explicit religious commitment, but now both centres in Scotland have full-time chaplains, as have several of the other hospices. So pervasive has been the idea of religious origins that there are stories of patients in other hospitals resisting transfer to a hospice on the grounds that they are not religious enough to qualify, or that they would feel excluded by the excessive religiosity they anticipate. It has also been noticed that few people of other faith communities use hospice facilities, and there is anecdotal evidence that they are deterred by the overtly Christian nature of some units.

THE HOSPICE AS HOME

Basic to the idea of the hospice is its very name. Hospice derives from the Latin *hospitium*, which has the meaning both of host and guest. The hospice is designed to be a place of hospitality, practised in mutuality. Just as in our own homes there can be a welcoming atmosphere and careful preparation, and it can all be wasted on inconsiderate and ungrateful guests, so in a hospice there must be both the evident welcome and a willingness to accept the need for care. It is glib to talk of any institution as 'just like home'. In the author's experience, people in nursing homes often say, 'It's very nice and the staff are kind, but it's not home.' In a hospice, where many die, the talk of home is at times a little hollow. Yet some patients and their families speak appreciatively of the homelike atmosphere, so it is worthwhile attempting to illustrate this perception.

Theory has been provided by Katherine Froggatt, who has written in the journal *Mortality* of the hospice as standing on the 'limen' between life and death. Hospice care speaks of a journey, and the crossing of a threshold. The dying person and his or her family are welcomed and cared for, the individual is encouraged on his or her life journey, and the family receives bereavement follow-up and aftercare. Liminality implies absence of status and the patient and his or her family are encouraged to see the hospice as 'family'.[3] But the images of family in hospice are often fragmentary and selective. The hospice can scarcely promote the more negative aspects of family life, the anger and division all too common in our experience. Perhaps within that mysterious area between living and dying it is necessary to overstress the positive aspects of family life. Nursing homes are less able to mirror the life of the family home and indeed some recently-built establishments resemble a hotel environment rather than a home, with uniformed waiters as well as identifiable nurses.

What attributes of family life can be reproduced in a hospice? In our own homes we dress as we please and there is no 'uniform' appropriate to the normal household. There has been some argument in hospices about the necessity of uniforms. Doctors seem to require a white coat in which to keep their stethoscope, that universal badge of medical office, and nurses prefer a uniform to protect their ordinary clothes from the unavoidable stains and marks of illness. The old-fashioned hats worn by nurses from the time of Florence Nightingale have been abandoned without much regret. All staff and volunteers must wear badges, and the hospice differs from home in its rigid insistence on security. There are self-locking doors, alarms, digital entry devices and all the trappings of institutions, although many of the patients will have come from sheltered housing, with similar precautions, or from private homes with an alarm system. Patients are encouraged to wear their own clothes and to make themselves as comfortable as possible within the constraints of shared accommodation. There can be difficulties when patients disagree about which television programme, if any, they should watch; about the volume of such a programme; and, although in a diminishing degree, about the use of tobacco. Smoking areas have contracted noticeably over the years, although there is no question of a total ban. Sometimes visitors can be seen to interfere with the lives of other patients, but similar situations can occur at home!

The hospice can be too luxurious for some of its residents. A patient may be cleaned and shaved against his better judgement, and afternoon tea served in china cups and saucers may reflect the middle-class nature of the volunteers rather than the habits of the patients. In shared rooms privacy is impaired and even in single rooms the door may well have to be kept open for access by staff. Signs saying 'Do not enter' may be ignored by officious flower arrangers, for example, and for those of retiring disposition there may be altogether too much emphasis on sharing in conversation and entertainment. It is difficult for those who are used to their own company to adapt to life in any kind of small hospital, and in many nursing homes the communal sitting room is crowded and seats and positions are jealously guarded. On the other hand visitors can be controlled in number and length of stay, there are private interview rooms and locks on toilet doors. But in shared accommodation conversations inevitably can be overheard, especially as many residents have difficulty in hearing normal speech, and residents can become intrusively interested in their neighbours.

ROUTINES, CELEBRATIONS AND PASTIMES

Even in the smallest institution mealtimes must be at set hours and a daily routine develops. Patients may be allowed to sleep as late as they please but they risk missing breakfast. Nevertheless the kitchen, usually being close to the nursing wing, can be flexible, diets and preferences can be respected and a choice of food made available. The doctors' rounds, the arrival of the drug trolley, the coming of food from the kitchen, are all daily and expected events, following a reassuring pattern. Variety is provided by volunteers in their ordinary clothes bringing afternoon tea, and changing waterjugs and rubbish bags, and by the daily visit of the cleaners, uniformed indeed but often more approachable than more 'professional' staff.

In the day hospice there is a definite but unobtrusive routine, from arrival and coffee to time for chat and appointments with the doctor, to entertainment of various kinds, interspersed with food, rest and programmed relaxation, all in a room set out, with fireplace and bookshelves, to resemble a home. The need of some for solitude is recognised.

Visiting entertainers, both in the day hospices and less often in the nursing wings, bring the images of the television programme to life, and residents have the occasional opportunity to meet celebrities who kindly give time to hospital performances. Authors, actors and musicians are warmly welcomed, and patients can become objects of envy to their families when they boast of shaking hands with performers who are household names. At Christmas there are special meals and programmes, and on Christmas Day itself there are opportunities to hear carols and share dinner with family and friends. But such celebrations often have an edge of sadness or even desperation as many deaths occur around the festive season. It is difficult to feel festive and at home when you are aware of the very sick person in the next bed or up the corridor, and there are evidently distressed relatives mingling with those trying to look seasonal.

Events such as a Royal Jubilee are celebrated appropriately. They become useful occasions for reminiscence and nostalgia. Very occasionally a visit from a royal or other socially eminent personage occurs and there is a general lightening of mood, mixed with some apprehension. Patients' birthdays are celebrated, and the kitchen provides a cake. Anniversaries can be marked by a meal, eaten in a quieter area of the building, and some patients reveal that they have not celebrated any such event for many years. More specifically religious festivals such as Good Friday and Easter seem to hold significance only for those involved in church life. Although Christmas can produce Santa Claus, Easter is purely Christian.

One of the great enemies of people in small hospitals, as in large, is boredom. When pain is under control, and all decisions seem to be made by someone else, when there is a measure of physical comfort, how can one pass the hours? Some try to read, and spend many hours over the same page because concentration is difficult. Others measure their mental acuity by their ability to solve the daily crossword. And some listen to music or talking books. But for many people there are long stretches of time to think backwards and forwards, or just to do nothing. Not even the birds and the squirrels in the garden provide amusement or interest. Visitors are a burden, with their bright chatter, or gloomy silence. But attempts are made to keep interest alive. Volunteers will read with patients or play draughts. Large-print books are provided. Parties are organised. Television brings sports events into the room and indeed some doctors observe that a good football match can enable patients to need less pain relief. There are always a few patients who pursue their interests, such as translating and writing, or organise their own business, until almost the end. But for many the problem of what to do with the waiting time is paramount, and seemingly insoluble.

The hospice projects itself as an extended family. It has been suggested that the Medical Director is father and the Matron mother,[4] but for most patients that is a rather fanciful concept. What matters to them is their own immediate family, who have fairly free access at most times. Family members may be involved in care, or they may just sit and talk and listen, or watch as the patient sleeps. They may have come from afar, and not have visited for several years. They may bring guilt as well as joy. They deliver news from the outside world, as do members of the bowling club, or the minister or the flower distributor from the church. Visitors reconnect the patient with the world outside.

It is easy to expect that family and friends will visit frequently and expertly, but there are many factors which militate against this. Not all visitors are made welcome, and not all are able to appreciate each other. Bedside fights are not unknown and old grudges can be rekindled even in hospital settings. Some feel they have to fortify themselves with alcohol to visit, which can lead to unpleasant or embarrassing scenes. Some have to make long, expensive and complicated journeys, perhaps entailing going home in the dark, and so visit seldom. Visitors are a mixed blessing, yet their presence can often be of paramount importance for the patient. Sometimes longstanding grievances and silences can be ended and people reconciled. In hospices whose catchment areas include remote rural communities, special arrangements can be made for transport and for the accommodation of relatives in the vicinity of the hospice. Much depends on the work of palliative care sisters in the community, whose support can allow patients to spend a longer time at home.

Alongside visitors are numbers of volunteers who make up a significant part of the workforce in hospices and also play a part in the life of nursing homes and wards of long-stay hospitals. They serve as flower arrangers, coffee bringers, receptionists, drivers who make it possible for frail spouses to visit, and in a multitude of other roles. They become friends, if not virtual family, of the patient, and must be guarded against over-involvement. If well managed they work smoothly with salaried staff while enabling the immediate neighbourhood to feel some ownership of the hospice. They chat and even gossip, and add to the atmosphere of comfort which approximates to home.

Children of all ages have a particular part to play. Babies are brought in to see their great-grandmother, for example, almost as soon as they are born, and the other patients can share in the joy of new birth. Staff and volunteers also appreciate the liveliness and hope of babies. Small children on the whole do not show much embarrassment in the presence of the old and the sick and will happily go from bed to bed and chair to chair and enliven a few moments. Some older children and young adults can be very constrained, upset and uncomfortable in the presence of illness, and may refuse to come and visit, while others are very much at their ease. Students, working as nurses for the summer vacation, bring their own freshness and friendship.

Pets are important to many, and some people's chief anxiety when they leave home is the care of their cat or dog. Health and safety regulations have made it more difficult for resident pets in institutions, especially indoors, but dogs which come to visit have a special place in many people's affections. 'Therapets', often retired guide dogs, can tour the wards and are graciously pleased to be patted and admired.

The use of Christian names seems to be the rule in hospices. This is in part a result of the decline of formality in society, and in part because of the 'homely' approach of staff and volunteers. Usually patients welcome being called by their familiar names, which are not always the names which appear on their official documents. Any complaints seem to come rather from the relatives who feel that their honoured father, for example, is not being treated with due dignity. Sensitivity dictates that people such as the retired schoolmistress who wishes still to be 'Miss' is obeyed. But at home we are called by our given or indeed by our 'pet' names, and we are certainly not 'the tumour in bed three'.

THE CHALLENGE OF MINORITY ETHNIC COMMUNITIES

The author is aware that this is potentially a very sensitive subject, writing as he is from a white Christian perspective. There is anecdotal evidence that amongst Islamic and Hindu communities especially, hospices can at times be perceived as middle class, white and aggressively Christian. Certainly they can look like that to a casual observer. In some places quiet rooms or prayer spaces are replacing chapels, and symbols of religions other than Christianity are displayed. But so long as the prevalent religion is Christian it is not surprising that Christian iconography and assumptions prevail. It has also been assumed that minority ethnic communities are more skilled in and committed to caring for members of their families at home, but this is debatable.

There is no doubt that members of all communities are welcomed in hospices, and offered hospitality, but misunderstandings persist. Staff must be told about customs surrounding the dying and practices concerning dead bodies. They might have to be persuaded that the patient may in some cases wish to have twelve or more relatives constantly at the bedside. Dietary provisions are not always honoured, and many people are too polite to refuse food which is not in accordance with religious rules. Continuous education is required to transform goodwill into real acceptance. There are also patients who do not have an interest in any religion, whose wishes must be honoured, especially around Christian festivals.

SPIRITUALITY

Spirituality is a slippery term, much used in nursing literature and sometimes seen to be the good counterpart of the out-dated and dangerous concept of religion. Of course the reality is much more complicated. In both hospices and residential homes there will be residents who seek the consolations and comforts of their own faith community, and access is given to ministers and other practitioners. The person who brings the church flowers after the Sunday service is providing a link with the worshipping and caring community. The warden from the synagogue does a similar task. Religious services can be provided, and several hospices have designated quiet rooms or chapels where regular daily prayers and Sunday worship is conducted. A stock of religious texts is usually available, and hymnbooks can be useful when planning one's funeral service. Indeed this is one of the coded ways in which patients talk about approaching death.

But spirituality goes far beyond religious belief and observance. Spiritual needs include the need to be valued, to find meaning in strange experiences, to be aware of love and care, to be listened to, to be treated as more than a sick or worn-out body. Spiritual pain is very real, and it can be addressed by many members of staff. The doctor can listen to the imprecise talk of symptoms, the nurse can respect the patient's wish for privacy, the cleaner can chatter and share experiences, and the chaplain can sit in silence.

Theoretically, hospice and care-home staff are aware of spiritual needs. In practice it can often be easier to attend to physical emergencies and to 'send for the chaplain' whenever there is a deeper question raised. Nurses are often reticent about sharing their own beliefs or experiences, and it is not so long since such exchanges were forbidden between carer and patient. It is easier simply to note the patient's religious affiliation and not to probe further. Hospices should be better at meeting spiritual needs than busy hospitals. But there is still much to learn in this field.

FINANCIAL MATTERS

This is too complex a subject to be examined in any depth in this chapter. The funding of hospices is subject to many pressures. Public generosity has been remarkable, but the multiplication of institutions, with the opening of AIDS and children's hospices, has spread that support more thinly. Patients are not expected to pay for care, and many families are generous in their recognition of what the hospice has done. Collections at the church or crematorium have become quite usual in the last few years. But it has occasionally been evident that the family is happier that a relative is kept in the hospice rather than being transferred to a nursing home, where assets and savings will be taken into account. The whole area of caring for the dying patient who is becoming a little stronger with good nursing and diet and is no longer appropriately placed in a hospice is a vexed one, with moral and financial aspects which are not easily solved.

There have been suggestions that hospices should all be incorporated into the NHS so that the state can take responsibility for the care of all dying people and charitable resources are not channelled into what is seen as the privileged minority of people who gain access to hospice care. Care homes are also under severe financial restraints as the state contribution does not always meet the shortfall between patients' resources and running costs. The Church of Scotland has been forced to curtail its provision of care for older people for these reasons. However, at the time of writing (2002) free personal and nursing care for older people is in the process of being introduced. This radically changes the previous financial arrangements, the ability of individuals to pay and thus the decisions they make about care.

SOME CASE STUDIES

The first concerned one of Edinburgh's homeless people. He had held a good position in a bank, but had lost it many years before. He may have been married at one time, but this had ceased to have any meaning. He valued his solitude, but spent Wednesday evenings in our church hall, playing dominoes. Visiting him in hospital as a minister, the author noticed his neatly-rolled school tie at the end of the bed – his last link with 'respectability'. A few years later he was admitted to the hospice with end-stage cancer. He remained quiet and self-contained, and he still had his tie. In the hospice he managed to hide himself, in odd corners, in the visitors' shower-room, sometimes in the chapel. He maintained his reserve until he was too weak to leave his bed. Then he admitted that he had a sister nearby. She was found and some sort of reunion took place. He died soon after.

This man had rejected any conventional idea of home. Cheerful chat and a shared watching of the football match on the television were not for him. Nor were the blandishments of nurses or social workers. Yet he found a home of a sort, and when he was reunited with his sister some memories of past happiness must have been stirred.

Another case was a busy lady in her seventies who had been severely slowed down by her illness and entered the hospice seemingly bedridden. After a few days of coaxing, feeding and general care, she was up and about, taking an interest in all around her. She became a ward monitor, and began to regard herself not as a patient or even a volunteer, but as a member of staff, involving herself in the care of other patients; 'busybodying about', as a nurse described her activities. She was obviously much better, and too well to be occupying a hospice bed. Her closest relative was a niece, who was willing to visit but not in a position to do much more. It suited her to have her aunt close at hand in 'such a nice place'. So when a nursing home was suggested the niece was horrified.

Nevertheless, after the inevitable delay, the woman went to a highly recommended nursing home, where she was miserable. A few weeks later she was back – to die. Her niece and her church friends were angry because she had been 'put out' of the hospice against her will. But the hospice is designed for fairly short-term care, and those who improve are a challenge.

In contrast another woman of 70 decided to sell her flat and go into a home for the elderly run exclusively for members of the large church to which she had belonged all her life. Thirty years ago that was not so unusual as it is now, when older people are kept at home by many means of community support. The woman was happy in her new home. She could still go to church and to all her meetings, and her fellow residents and staff were for the most part in agreement with her on all religious and social matters. She flourished, and became quite proprietorial. 'Matron and I have decided which room will belong to a new resident,' she confided, and this was not a problem. She lived peacefully into the new millennium, when she died aged 97.

But for how many will this seamless transition from home to very sheltered home be possible? The last example is important because the lady concerned was so atypical. For many old people a care home will be the last resort, in every sense, and they will have no chance to choose their company or the ethos of the home. Where they go will depend on what is available. Such a person may well find a new home, or may be permanently discontented and upset. Dwellings have dysfunctional aspects, and neither hospices nor care homes suit all who enter them.

NOTES

1. See Chapter 26 Infirmity.
2. 'Palliative' is the word used since the 1980s to denote that care which is not curative but which relieves the symptoms of far-advanced disease. Its exact definition is still a matter of discussion.
3. Froggatt, 1997.
4. Graves, 1982.

BIBLIOGRAPHY

Cassidy, S. *Sharing the Darkness*, London, 1988.
Cobb, M, Robshaw, V, eds. *The Spiritual Challenge of Healthcare*, Edinburgh, 1998.
Froggatt, K. Rites of passage and the hospice culture, *Mortality*, 2: 39 (July 1997), 123–36.
Graves, D. Models of hospice management. In St Christopher's Hospice, *Hospice Management* (Factsheet, no 9), London, 1982, passim.
Hockey, J. *Experiences of Death*, Edinburgh, 1990.
Lawton, J. *The Dying Process: Patients' Experience of Palliative Care*, London, 2000.
Murray, D B. *Faith in Hospices*, London, 2002.
Walter, T. *The Revival of Death*, London, 1994.

37 Children's Homes

JOE FRANCIS

From the mid nineteenth century through to World War II the childcare services played a significant and active role in Scottish family life. Families battered and sometimes torn apart by death, illness, poverty and cruelty were no strangers to the parish inspector, 'the cruelty' and the local children's home.[1]

Childcare services in Scotland have affected the lives and experiences of many individuals for almost two centuries. Despite an enduring preference throughout this period for 'boarding out' Scottish children, it is perhaps true to say that it is children's homes, more than any other facet of childcare services, which have affected the collective conscience as well as the lives of individual families. For a large proportion of the population the idea of institutional life has always been regarded with repugnance[2] and for those experiencing separation from their families, whether on a temporary or permanent basis, life in a children's home had, and continues to have, a profound impact.

Acquiring some insight into these perceptions and experiences requires an understanding of how residential childcare has developed in the past and how those developments have influenced the present. It is only through examining the external social pressures which have shaped and formed them over time that we can begin to appreciate the contribution of children's homes as places of care, control, treatment, custody and substitute parenting. In particular, a historical perspective gives us an understanding of the structure and functioning of childcare institutions in Scotland and of the images surrounding them today. It also helps us to consider how, throughout time, people have come to view care in residential establishments as an experience which bestows stigma and handicaps on young people rather than providing them with an adequate compensatory experience or, indeed a head start in their adult life.[3]

FROM LOCAL COMMUNITY TO POORHOUSE: THE POOR LAW RESPONSE TO CHILDREN IN NEED.

Childcare practices and expectations in contemporary Scottish society are significantly different from those of earlier times. Children's homes, like the concept of childhood itself, are relatively modern phenomena. During the medieval period care and concern for children was normally a matter for local communities. Orphaned, abandoned and destitute children were usually looked after by their neighbours or kinsmen. It was not until the sixteenth and seventeenth centuries, when rural society in Britain began to change rapidly and the process of urbanisation started to accelerate, that any organised measures of public provision of residential accommodation became necessary.[4]

One of the consequences of the increased urbanisation of society was that there were greater numbers of destitute people and, in turn, many more neglected and abandoned children began to appear on the streets. The main response to this situation was the enactment of Poor Law legislation (1579 in Scotland; 1601 in England and Wales) and with it the establishment of the first forms of residential homes, poorhouses, where children were accommodated along with destitute adults, the sick and old people.

The main function of poorhouses was to train the people who lived there to become independent and not to be a burden on the state. For many young children this initially meant receiving some basic training but eventually it led to their farming out as cheap labour in the mills, factories and mines which were burgeoning in the early stages of the Industrial Revolution. In country districts, children from the poorhouse might be sent to work on the land. This philosophy of 'training' and 'independence from the state' underpinned the legislation for almost three and a half centuries and was still predominant when the Poor Law was reformed in the mid nineteenth century.[5]

The reformed legislation, the 1845 Poor Law (Scotland) Act, was even more draconian than that which had hitherto existed and, in an effort to minimise expenditure on the many impoverished people, workhouses were introduced. Under the terms of the new legislation, destitute people were now required to enter these workhouses if they wished to receive help or, as it was known, 'relief'. In Scotland, destitute people were entitled to relief only if they were unfit for work, so in this respect the poor of Scotland were even further disadvantaged than the destitute in England.[6]

Initially, the process of urbanisation was slower in Scotland than in England and consequently very little institutional care existed outside Glasgow or Edinburgh until the middle of the nineteenth century. Until then most care for children was usually organised within local communities. From that time, however, the speed of industrialisation and increased urbanisation in Scotland created intense pressures on the social structures of both rural and urban communities. The resultant poverty and hardship was particularly harsh on lower-working-class families and their circumstances were further exacerbated by the influx, mainly into the towns and cities in the central belt, of significant numbers of immigrant Irish, desperate to escape the effects of the potato famine of 1846. Thus poor working-class families found themselves competing for unskilled labouring work and for the scarce cheap accommodation which was available at the time. The impoverished families of urban Scotland, suffering from poor sanitation, overcrowding and unstable employment, found it increasingly difficult to provide for their children, resulting in many seeking assistance from the authorities.

Illegitimate children were often the most disadvantaged and, in the urban areas in particular, mothers of such children were frequently unable to provide care owing to a lack of kinship support and also because they were not eligible for unconditional poor relief. In these circumstances women were forced to place their children in care to allow themselves to find work or to continue in work. Abrams gives an account of one such case,[7] describing how a hotel chambermaid implored Aberlour Orphanage in Banffshire to take her 10-day-old baby as she was unable to find anyone to care for the child while she worked. Such occurrences were not uncommon during this period. Indeed, by 1884 parishes had the authority forcibly to remove illegitimate children from their mothers and place them in care. By doing so they were seen as both rescuing the child from poverty and also

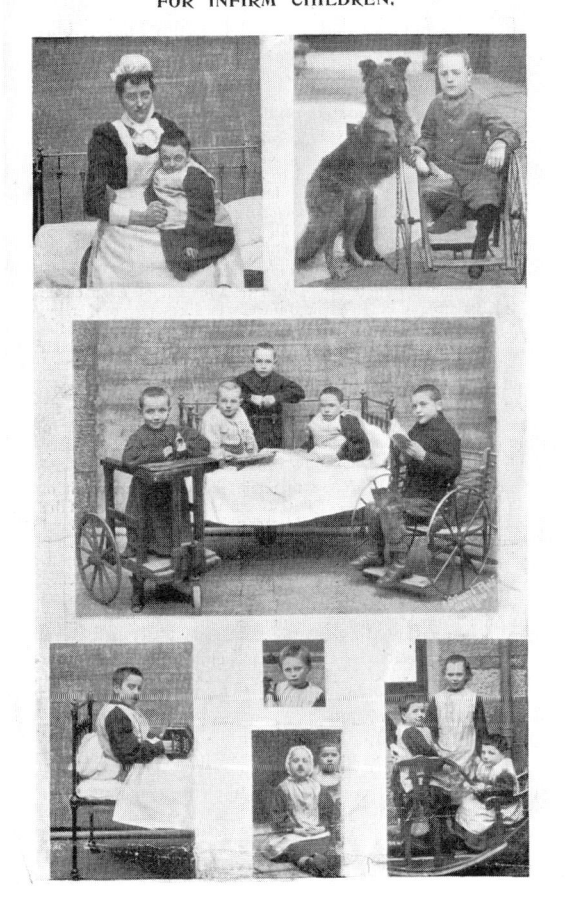

Figure 37.1 East Park Home for Infirm Children. Source: Reproduced by permission of Heatherbank Museum of Social Work, Glasgow Caledonian University.

releasing the mother from the responsibilities of caring for the child, thus permitting her to return to employment.[8]

Widows and widowers also frequently found themselves placed in such circumstances. Even in situations where relatives and extended family were on hand, they were often unable to accommodate the children as a consequence of their own family commitments and the hardships and privations that they themselves were experiencing. Sometimes the children of widowed parents were placed for short periods to allow the family time to make alternative arrangements but all too often, partly because of a policy of actively discouraging contact between the parent and the children, the links were permanently severed.

Life for many families was exceptionally harsh during this period and children were often the most disadvantaged and badly affected by the poor conditions. One of the foremost Scottish reformers of the time, Thomas Guthrie, describing one of his pupils in 1847, wrote that he had 'neither shoes nor stockings; his naked feet are red swollen, cracked, ulcerated with the cold'.[9] Indeed, such was the deplorable and unsanitary condition of many of the children, with their clothes in rags and lice-infested hair, that the Edinburgh Ragged School took to shaving their heads and scrubbing them from head to foot (Fig. 37.1).

THE BOARDING-OUT PHILOSOPHY

The situation in England in the mid nineteenth century was somewhat different from that in Scotland. There, destitute children were more likely to receive care in the workhouse. In Scotland, however, parishes (which were responsible for funding and administering poor relief) favoured the policy of 'boarding out' children; that is, removing them from the destitution of urban families to new homes with families in rural communities.

To some degree, this policy reflected the traditions of clan and highland lifestyle where fostering children had been quite commonplace.[10] However, it also reflected a desire to find a cheaper alternative to residential provision, a characteristic which still influences current policy. Perhaps more importantly though, it demonstrated a more critical attitude towards the less-desirable effects of institutional care. Such attitudes were reflected in the report of the Merchant Company of Hospitals or Schools in 1886 which indicated that residential care was not desirable because of:

> '... the fixed hours for every act, the unvarying round from day to day, the necessity for acting in masses, the regimental routine in which every duty is embedded', these features, the report argued, 'inevitably tend to the destruction of individuality: and where individuality disappears, morality is either a habit of fear or one of external imposition'.[11] (Fig. 37.2)

Given this level of antipathy towards residential care, the boarding-out policy was reinforced in the Poor Law (Scotland) Act of 1876 which required

Figure 37.2 'Feeding the Children'; Children in a workhouse in London. Source: Reproduced by permission of Heatherbank Museum of Social Work, Glasgow Caledonian University.

that no children should be retained in any poorhouse for longer than three months. Consequently, by 1880, more than 5,000 children out of a total of 8,000 dependent on poor relief in Scotland were boarded out and by the start of World War I, 86 per cent of Scotland's 8,873 poor-relief children were placed with foster families.[12]

However, despite the overwhelming preference for boarding out, Scotland did retain a level of provision in workhouses. The conditions in workhouses were extremely harsh and comfortless, reflecting the political intention of discouraging dependence on the state and reducing public expenditure on the relief of destitution. Known as the principle of 'less eligibility', this meant that conditions were made as stringent and repellent as possible so that only those who were truly in desperate need would apply for assistance. For many children, life on the streets was preferable to the rigorous and dispiriting regimes of the workhouses, which soon became hated symbols of repression. Those who had no alternative but to seek relief in the workhouses became figures of contempt and were regarded as carriers of the 'workhouse disease', which was probably typhus fever.[13]

To many reformers of the day the workhouse was completely inappropriate for a child, prompting one commentator to describe them as places where:

> ... children were degraded by hideous uniform, shuffling boots and prison-cropped hair. With overcrowding, vermin and skin disease

spread like wildfire; purulent opthalmia was so common that it was proposed to open schools for eye cases only. No child had ever handled money; some had never eaten with a knife and fork. Workhouse children grew up permanently unfitted for normal life.[14]

It was not long before it became apparent that workhouses were not enabling children to develop economic independence and so a new system of District Schools and Industrial Schools, on the model founded by Sheriff Watson of Aberdeen in 1841, was established to provide education and training for children and young people. Initially children were sent daily to these schools from the workhouses but they soon became established as separate residential institutions for children. This, then, was the beginning of separate public provision of residential care for children and young people.

CHILDREN IN PRISON

At about the same time as these developments, a number of changes were emerging in the treatment of juvenile offenders. At the beginning of the nineteenth century, children who broke the law were subject to the same punishments as adults and those who were imprisoned were placed in the same institutions as prisoners of all ages. Such a policy reflected the reality that the concept of childhood, as we understand it today, was alien to society at that time. Consequently the needs of children were frequently overlooked when they came before the authorities because of petty offences. For example, describing the backgrounds of nine child prisoners in Preston Gaol, Lancashire, in 1844, the prison chaplain wrote:

> Two of the six were orphans, a third had lost one parent by death and the other by desertion, the fourth was altogether neglected by his father and mother, the fifth and sixth had parents who encouraged them to steal. Only two of the nine have well conducted parents.[15]

The anonymous journals of an official at Maybole Ragged School in Ayrshire also record the circumstances of such children. His notes for one 12-year-old boy state that he:

> ... was in the habit of begging and stealing. For the latter of which he was sent to the county prison. Altho I may here remark that poor James was compeled either to beg, steal or starve; this much I learned from his mother's door neighbours.[16]

Confining children in prisons gradually came to be regarded as counterproductive and harmful to the children and a movement led by Mary Carpenter eventually resulted in the establishment of reformatory schools in 1860. The reformers argued, on moral and social grounds, that children who were incarcerated with hardened criminals were likely to become hardened themselves and that children who were idle were more likely to commit crimes. Children who committed crimes were henceforth sent to the reformatory schools where it was intended that they would be rehabilitated into society through a regime of training and education.

The parallel systems of residential training provision for destitute children and reformatory schools for young offenders co-existed until they were amalgamated in 1933 and became known as Approved Schools. Children who were in need of care and protection were placed in these schools alongside children who had committed offences and children who were regarded as being beyond the control of the local authority (for example, because of failure to attend school or being difficult to control in the workhouse). These schools were managed both by local authorities and by voluntary organisations and they were characterised by rigorous regimes of activity, education, trade training and physical fitness.

THE 'RESCUE MOVEMENT'

Alongside public-sector developments in the expansion of the workhouse system and separate institutions for young offenders, the second half of the nineteenth century witnessed the rapid expansion of the work of philanthropic and Church organisations. This period established and promoted a considerable growth in the number of charitable children's homes and later became known as the 'rescue period'.

Many people were unhappy with the stringency and severity of the Poor Law system and, in particular, its harsh impact on the lives of children. Public debates about the conditions of destitute children, the need for separate provision for young delinquents and for educational provision for all children, resulted in the concept of the 'child' emerging in the latter part of the century. Consequently, a number of religious groups and concerned individuals began to develop voluntary alternatives to the public provision.

Among the best known of the larger philanthropic institutions which were founded in Scotland were Quarrier's Homes at Bridge of Weir, Renfrewshire (the Orphan Homes of Scotland established in 1873 by William Quarrier), and Aberlour Orphanage on Speyside (founded in 1875 by the Reverend Jupp). Quarrier's housed almost 1000 children by the end of the century and the Aberlour Orphanage cared for more than 2000 children before 1914.[17] Both of these institutions created large communities for children which were separated from the local communities and comprised living quarters, church, school and training facilities. Children placed in these establishments were thus rescued from, and protected against, the corrupting influences of their parents and the urban environment (Figs 37.3, 37.4, 37.5, 37.6).

In addition to these large-scale institutions a number of smaller children's homes operated throughout Scotland, many catering for the needs of specific groups of children. The Smyllum Orphanage in Lanark, for example, offered provision for Roman Catholic children while the Magdalene Institutes sought to rescue wayward girls.[18]

Figure 37.3 Aberlour Home. Source: Reproduced by permission of Heatherbank Museum of Social Work, Glasgow Caledonian University.

Figure 37.4 Aberlour Home. Source: Reproduced by permission of Heatherbank Museum of Social Work, Glasgow Caledonian University.

Figure 37.5 Quarrier's Children's Home. Source: Reproduced by permission of Heatherbank Museum of Social Work, Glasgow Caledonian University.

Figure 37.6 Homelea and Paisley Home. Source: Reproduced by permission of Heatherbank Museum of Social Work, Glasgow Caledonian University.

The motives of these philanthropic organisations were largely evangelical and they sought to 'rescue' children from the damaging influences of immoral parents and of the Poor Law system itself. The majority of the residents were orphans, destitute children, abused and neglected children or children who were the victims of family crises. Almost all of the workers were Christians, often poorly paid and unpaid women, specifically recruited by the societies, and they believed that children should be raised in circumstances where they would experience sound moral influences. The administrators believed also in the value of education and training. These twin features characterised most of the Church and voluntary services, distinguishing them from public provision by the additional emphasis on religious instruction. Thus their aim was to educate the children in the 'Four Rs'; reading, writing, arithmetic and religion.

Citing the annual reports of Whinwell Children's Home in Stirling, Abrams notes that institution's evangelical mission to rescue children from their 'paths of vice and wickedness ... '.[19] Further insight into the religious zeal which motivated many at that time can be gleaned from the descriptions of children attending the Maybole Ragged School. Of one boy it is recorded:

> ... 13 years of age, father dead; was brought to school the 29th day of January 1849. This boy was well known to be a notorious thief and a town pest. He was ignorant as a heathen, neither knowing a letter nor a God. Could not tell who made him; did not think it a sin to steal or tell a lie; did not know where the wicked go after death; could not say the lord's prayer; never heard of Jesus of Heaven or of Hell except when pronounced with an oath ... I despaired of breaking in this boy ... When he left school he was a good reader in the bible and I am happy to say that James is no longer a thief nor a beggar ... I find he has attended both church and sabbath school and is very obedient.[20]

In furtherance of the aim of rescuing children from the corrupting influences of their families, parents who admitted their children to many children's homes were required to give the institutions the power to send the children to any situation within Scotland or abroad or otherwise to dispose of them in whatever way thought best. For many children this resulted in them being sent to Canada, Australia and other parts of the Empire. Indeed in the latter part of the nineteenth century and the early part of the twentieth century, thousands of Scottish children were sent overseas in the belief that this was the best way of cutting the ties with their families and giving them a new start in life.

Not all children's homes supported this practice, however, and the Aberlour Orphanage, for example, sent only around 50 children overseas throughout its history. The administrators there felt that they were able to provide a sound upbringing within the institution, a policy which resulted in its continued expansion to accommodate the increasing number of children for whom admission was sought.[21]

Most homes were run on strict lines and were characterised by regimes of discipline, religious instruction, harsh physical punishment and domestic chores and duties. Children were required to attend frequent religious services in the majority of the homes and usually came to regard this activity like other chores. Food, clothing and other aspects of personal daily life were extremely institutionalised. The children had very little free time and normally spent substantial proportions of their day engaged in routine activities that were designed to prepare the girls for domestic service and the boys for manual work or farm labouring. Abrams' interview with a former resident of Bethany House in Aberdeen furnishes a glimpse of this existence:

> Well [we] wouldnae really had much time to talk because eh at breakfast time after we'd the dishes to wash before we went to school, there was scrubbing to do, er the dormitories to be cleaned out and er you never got time to talk about things and then you were in bed at 6 o'clock at night.[22]

These features of life in Scottish children's homes remained relatively constant until 1930 when the Poor Law service became the responsibility of public-assistance departments and children's homes began to come under more rigorous scrutiny from government inspectors.

Many of the reports of the time were extremely critical of the inappropriateness of the eating and sleeping arrangements, of the over-emphasis on cleanliness, orderliness and discipline, and of the general austerity and dullness of the environments in children's homes. More importantly, critics were greatly concerned about the lack of parental affection or of conditions which were designed to promote the emotional development of the children. Evidence from the institutionalisation of children during the War years suggested that the development of their individuality and sense of self was impaired by the experience of long-term residential care. Such views were highly influential in shaping childcare policy in the post-War period and, in particular, in the enactment of the 1948 Children's Act and the subsequent establishment of Children's Departments.

POST-WAR PROVISION

Bowlby's classic work on maternal deprivation[23] had a strong bearing on the views and decisions of many post-War policy-makers. He argued that children were profoundly disturbed by separation from their mothers and also that children placed in residential establishments suffered serious personal and social losses as they were emotionally stunted as a consequence of prolonged residence in these circumstances. His report to the World Health Organisation in 1951 concluded that a deprived child's ability to make relationships, other than superficial ones, was impaired; that a child would show inaccessibility in allowing others to relate to him; that his social relationships would be characterised by deceit and stealing; and that his intellectual development would be impaired.

Goffman's (1961)[24] research into life in institutions also adversely influenced attitudes towards residential childcare. His concerns echoed those of the Scottish critics almost a century earlier who had highlighted many of the worst features of institutional care. He found, for instance, that various aspects of work, leisure and rest were all carried out in the same place and under the same authority. Residents were depersonalised through batch treatment and all activities were highly organised, leaving no free time for the individual. Goffman argued that people in institutions experienced loss of individual identity symbolised by a depersonalising admission procedure, a lack of personal clothing and a lack of privacy. They suffered humiliation, typified by non-reciprocal forms of intimate address and insistence on deference, and their areas of personal choice were severely limited through lack of power. Many of Goffman's criticisms centred on features of life in residential establishments which were the vestiges of Victorian institutions.

The concerns raised by Bowlby and Goffman resulted in many people coming to regard institutional care as an anathema, particularly in relation to the provision for children and young people. Successive policies during the second half of the twentieth century have reflected these views and have been based on the principle of admission to residential care as a last resort. The impact on provision in Scotland during the final decades of the century was marked. For example, from 1976 to 1990 the number of children in residential homes fell by two thirds from 6336 to 2161, and the number of children's homes was reduced from 294 in 1980 to 154 in 1990.[25]

At present there appear to be two polarised perspectives about the place of children's homes in modern childcare services. Following his study of 20 children's homes in England and Wales and his analysis of the role of residential childcare in Britain as a whole, Berridge[26] suggests that there are, on one hand, those who might be described as optimists while, on the other, there are those who fall into the category of pessimists.

The optimists argue that the introduction of the 1948 Children's Act heralded appreciable changes in residential care which were both 'rapid and radical'. In their view, the situation since that time has improved for a number of reasons. First, there has been a general reduction in the size of children's homes and a corresponding diversity in the functions that they serve. For example, 65 per cent of children's homes in Scotland now have fewer than ten beds and only 12 per cent (mostly schools) have more than 25 beds.[27] Second, there has been increasing attention paid to the value of developing sharper, more focused and more realistic aims and objectives. Third, greater emphasis has been placed on maintaining children's family and social links while they are in care. Finally, policy-makers and managers now stress the importance of integrating children's homes within their local communities.

Alongside these developments, they argue, there is an increasing recognition of the complexity and demanding nature of the task undertaken by staff in children's homes and this is gradually leading to improvements in training and conditions of service for this group of workers. In Scotland, the recent establishment of the Scottish Institute for Residential Childcare, the implementation of the residential-care pathway on the Social Work degree (the current professional award) and the development of a post-qualifying Master's degree in residential childcare, has highlighted this. The protagonists would thus conclude that residential care for children and young people has improved considerably in the post-War era and is most certainly a great deal more responsive to their needs than it was before then.

The opposing view, however, is that placement in residential care does nothing to improve the quality of life of our children and, indeed, it may be damaging for them. This more pessimistic perspective has dominated much of the thinking and planning around the development of childcare services during the last 50 years. Two major reports (Curtis, 1946 and Clyde, 1946)[28] had a significant impact on the development of post-War childcare legislation. Both were very critical of public residential childcare, advocating that children should be placed in children's homes only as a last resort and that strenuous efforts should be made in every instance to find them an adoptive or foster home. They argued that it was undesirable to place children in residential establishments for long periods of time and their views, which were broadly reflected in the 1948 Children's Act, resulted in children's homes being seen as places offering temporary care rather than substitute parenting.[29]

As already indicated above, the work of Bowlby and Goffman fuelled the belief that residential care was undesirable where children were concerned. At an ideological level, policy-makers were influenced by Goffman's critique of institutions and this led to closing, or minimising the use of children's homes. Alongside this, a preference for family-based care was pursued in good faith and on the strength of an evolving understanding of Bowlby's concepts of attachment.[30] The argument that children's homes should be used purely as a last resort continues to have an influence on policy today and, coupled with the fact that foster care is also a much cheaper alternative to children's homes, has resulted in a shift of resources towards the latter, leaving residential provision as the Cinderella of children's services.

Moreover, while some would suggest that there have been appreciable changes in training provision for residential-care workers, this has been countered by those who argue that the reality is that qualification rates among staff working in children's homes has remained unacceptably low. Although government targets were set in the early 1990s[31] to improve the qualification levels of staff in residential childcare (aiming for 90 per cent of senior staff and 30 per cent of care staff to be professionally qualified in social work and a further 60 per cent to be qualified to Higher National Certificate level) these targets have not been met and, if anything, qualification rates appear to be declining.

The concern that institutional care is potentially damaging for children can be seen in the changed age profile of children in residential care in

Scotland during the second half of the twentieth century, with adolescents now making up the vast majority of residents in children's homes. This trend has been amplified by the introduction of recent legislation, the Children (Scotland) Act 1995, which also reflects many of the established concerns about the potential harmful effects of residential care. The guidance accompanying the 1995 Act states that most children can best be looked after in a family placement and that children under the age of 12, in particular, require the comfort and care that a family can offer. The guidance emphasises that children under 12 should be placed in residential care only in exceptional circumstances and then only after an assessment of their needs indicates that such a placement is appropriate.[32]

Consequently, recent government statistics indicate that approximately 80–90 per cent of children in residential care are aged between 11 and 16.[33] The majority of these children are boys (approx 65 per cent) and relatively few of them remain in care for lengthy periods, with about 75 per cent leaving within one month of placement and only 17 per cent staying for more than six months.

The number of children in residential care is now less than in the first half of the twentieth century but the turnover rate of admissions and discharges is higher. The function of residential care in the second half of the twentieth century and early twenty-first century has therefore shifted towards that of temporary respite care rather than a permanent home or a processing centre for children to be shipped overseas. Children's homes now act as safety nets for families experiencing temporary crises and there are clear legal and professional expectations that family links be actively sustained.

Despite the fact that there has been little recent research into children in residential care in Scotland, we know from government statistics that most children in children's homes now enter care for the first time as teenagers.[34] We also know from research which has been conducted elsewhere in Britain[35] that the circumstances leading to admission are extremely complex and varied. In some cases the young people have experienced neglect, abuse (physical, sexual or emotional), or rejection by their parents. Others have been involved in activities such as offending, school truancy or substance misuse. Some are prone to self-harm or violence and often have experienced family breakdown. Many young people experience a combination of these difficulties and it is widely recognised that there is an interplay of personal and socio-economic factors in most cases. Certainly, the majority of children in children's homes are from socially deprived backgrounds, nearly all have emotional disorders and there is a high incidence of conduct disorders. The children who now reside in children's homes are characterised by their chaotic behaviour and poor impulse control, their proneness to harming themselves or others or to destroying property, their school problems and persistent offending, their inappropriate sexual behaviour and their difficult relationships with parents.[36] While it is difficult to generalise, nevertheless it can be said that children in residential care have usually experienced very traumatic and disadvantaged circumstances and consequently have very complex and demanding problems that present many challenges.

The majority of children's homes in Scotland continue to be provided by local authorities and are normally managed by the social work departments of those authorities. A few are still provided by voluntary organisations such as Barnardo's. The homes are usually located in local residential neighbourhoods and are staffed by a unit manager, senior residential-care staff and care workers. Some homes also employ cooks and domestic staff but this varies depending on the size and ethos of the home.

Admission to children's homes can be voluntary, that is by agreement between the parents and the local authority, or compulsory following a decision by a children's hearing or the courts. Where admission is voluntary, parents can normally remove their children at any time but those admitted under a compulsory order require the decision of the court or children's hearing before discharge.

In most circumstances, children placed in residential care have a local-authority social worker who has responsibility for the overall case management. This includes assessment of the child's needs and circumstances, care planning with a range of professionals and agencies, and regular review of all the plans and decisions taken about the child's care arrangements. The paramount priority of the social worker is the welfare of the child and this must be reflected in all plans and decisions. The social worker also has a duty, wherever possible, to support the upbringing of the child in his or her own family. This means that contact between child and parents is normally actively promoted.

Links with other family members and with friends and community are also usually sustained and it is expected that the young person will continue to go to his or her own school or attend another local school. Similarly, staff try to involve young people in a variety of social and leisure activities in their local communities, using the range of available facilities such as youth clubs, cinemas and leisure centres.

Many young people return to their families for occasional overnight stays or weekend visits as part of a planned process of reunification. Equally, parents and family members are encouraged to have contact with the young person in the home. Birthdays and other important occasions such as Christmas are celebrated within the placement but where possible these are spent in the family home.

Where return to their own family is unlikely, young people are helped to prepare for permanent placement within a foster or adoptive family. Alternatively, older children are helped to prepare for independent living and are given support with accommodation and other practical matters once they leave care.

At the start of the twenty-first century, the role of children's homes has increasingly become one of supporting families in times of difficulty or crisis rather than acting as providers of substitute care. Children are no longer placed 'in care' but are 'looked after' by the local authority. This change in terminology in recent legislation denotes a shift in policy and philosophy towards a partnership with parents who are seen to retain parental responsibilities for their children.

CONCLUSION

One of the prevailing attitudes towards children's homes is that they retain vestiges of the Poor Law principle of 'less eligibility' and many argue that the visible, public nature of residential care is such that it sustains the stigmatising effect conferred on the residents by a society which views admission to care as an indication of failure by the immediate or extended family. Furthermore, policy continues to support the perception that residential care is positively detrimental, especially to children who need the warmth and emotional security which can only be provided in a family setting.

Children's homes in Scotland therefore remain a controversial alternative to foster care and continue to be the poor relation of the childcare services in the eyes of both public and professionals. Despite this, it is recognised that residential care has served the needs of many children for more than two centuries. The Skinner Report (1992)[37] demonstrated that for many Scottish children who have experienced family difficulties, residential care continues to serve a useful purpose. Such a view was strongly endorsed in a more recent report[38] when it was stated that placement in a children's home might be preferred to placement in the wrong foster family or, worse still, in a series of foster homes. While it is unlikely, therefore, that we will witness the extinction of children's homes in Scotland in the near future, their role and function in childcare services will undoubtedly continue to be the focus of much heated debate.

NOTES

1. Abrams, 1998, Preface vii.
2. Wagner, 1988, 8.
3. Triseliotis, 1988, 4.
4. Stroud, 1973, 99.
5. Stroud, 1973, 100.
6. Abrams, 1998, 10.
7. Abrams, 1998, 13.
8. Abrams, 1998, 13.
9. Cited in Seymour, 1995, 12.
10. Fosterage in highland society had often been for the purposes of alliance building. See Curle 1895–6; for further information on boarding out see Chapter 25 Childcare and Children's Leisure.
11. Cited in Triseliotis, 1988, 5.
12. Abrams, 1998, 37.
13. Stroud, 1973, 101.
14. Hill, 1889.
15. Clay, 1861.

16. Cited in <http://www.maybole.org/history/articles/mayboleraggedschool.htm>
17. Abrams, 1998, 26.
18. Abrams, 1998, 27.
19. Abrams, 1998, 82.
20. Cited in <http://www.maybole.org/history/articles/mayboleraggedschool.htm>
21. Abrams, 1998, 92.
22. Abrams, 1998, 97.
23. Bowlby, 1951.
24. Goffman, 1961.
25. Kendrick, 1995.
26. Berridge, 1985.
27. Scottish Executive Statistical Bulletins, 1998–9 and 1999–2000.
28. *Report*, Cmd.6922, 1946; *Report*, Cmd.6911, 1946
29. Triseliotis, 1988.
30. Smith, 2002.
31. Skinner, 1992.
32. Scottish Office, 1997, 70.
33. Scottish Executive Statistical Bulletins 1998–9 and 1999–2000.
34. Department of Health, 1998.
35. For example, Berridge, Brodie, 1998; Brodie, 2001; Packman, Hall, 1998.
36. Whitaker, Archer, Hicks, 1998, 16.
37. Skinner, 1992.
38. Kent, 1997, 23.

BIBLIOGRAPHY

Abrams, L. *The Orphan Country: Children of Scotland's Broken Homes from 1845 to the Present Day*, Edinburgh, 1998.
Berridge, D. *Children's Homes*, Oxford, 1985.
Berridge, D, Brodie, I. *Children's Homes Revisited*, London, 1998.
Bowlby, J. *Maternal Care and Mental Health: A Report prepared on behalf of the World Health Organization as a Contribution to the United Nations Programme for the Welfare of Homeless Children* (World Health Organization), Geneva, 1951.
Brodie, I. *Children's Homes and School Exclusion: Redefining the Problem*, London, 2001.
Clay, W L. *The Prison Chaplain*, Cambridge, 1861.
Curle, A O. Notice of four contracts or bonds of fosterage; with notes on the former prevalence of the custom of fosterage in the Scottish Highlands, *PSAS*, 30 (1895–6), 10–22.
Department of Health. *Caring for Children away from Home: Messages from Research*, Chichester, 1998.
Goffman, E. *Asylums: Essays on the Social Situation of Mental Patients and Other Inmates*, New York, 1961.
Hill, F. *Children of the State*, 2nd edn, Edinburgh, 1889.
Kendrick, A. *Residential Care in the Integration of Childcare Services*, Edinburgh, 1995.
Kent, R. *Children's Safeguards Review*, Edinburgh, 1997.
Packman, J, Hall, C. *From Care to Accommodation: Support, Protection, and Control in Services*, London, 1998.
Report of the Care of Children Committee (Curtis), PP, 1945–46, Cmd.6922, X, 559–754.
Report of the Committee on Homeless Children (Clyde), PP, 1945–46, Cmd.6911, X, 755–96.
Scottish Executive Statistical Bulletins. *Children in Residential Accommodation*, Edinburgh, 1998–9, 1999–2000.
Scottish Office. *Scotland's Children: The Children Scotland Act 1995 Regulations and Guidance, vol 2*, Edinburgh, 1997.

Seymour, C. *Ragged Schools, Ragged Children*, London, 1995.
Skinner, A. *Another Kind of Home: A Review of Residential Childcare*, Edinburgh, 1992.
Smith, M. Stands Scotland where it did? Perspectives and possibilities for child and youth care, *Cyc-online: The International Child and Youth-Care Network*, 47 (2002), <http://www.cyc-net.org/cyc-online/cycol–1202-smith-scotland.html>
Social Work Services Inspectorate, Scottish Office. *Another Kind of Home: A Review of Residential Childcare* (Skinner Report), Edinburgh, 1992.
Stroud, J, ed. *Services for Children and their Families: Aspects of Childcare for Social Workers*, Oxford, 1973.
Triseliotis, J. Residential care from a historical perspective. In Wilkinson, J, O'Hara, G, eds, *Our Children: Residential and Community Care* (National Children's Bureau, Scottish Group), London, 1988.
Wagner, G. *Residential Care: A Positive Choice*, London, 1988.
Whitaker, D, Archer, L, Hicks, L. *Working in Children's Homes: Challenges and Complexities*, Chichester, 1998.
www.maybole.org/history/articles/mayboleraggedschool.htm

38 Residential Schools and Student Residences

IAN MORRIS

Although the tradition of residential education in Scotland is not particularly strong, unlike, for example, that of England, it does exist in a range of options and at most social levels. Residential institutions have been established to meet perceived needs and have changed or even closed when those needs have ceased or a different value system has arisen (see Figs 41.5, 41.6). Some of the earliest schools were founded in the seventeenth century as charitable institutions, usually catering for poor children or orphans. It was in the nineteenth and twentieth centuries, however, that most schools were created.[1]

Despite the variety of schools and post-school institutions with residential provision, there are many common features of life, just as there are others specific to the particular establishment. Most schools, for example, witness a tension between specificity and social inclusion (a belief that togetherness is best for the individual and for society). This applies especially to those institutions catering for pupils with particular forms of disability, for whom, in more recent times, the local authority by statute has to make provision.

Many other difficulties and numerous benefits are associated with the residential-school environment, and this chapter hopes to outline some of these. Boarders of all sorts in Scotland today form a small group, but apparently they are a well-supported and contented one.

INDEPENDENT SCHOOLS

The independent sector in Scotland
Parents who believe in the value of residential education and who have adequate financial means are able to choose from a number of independent schools with residential provision. Today Scotland has 73 independent schools registered with the Scottish Council of Independent Schools (SCIS), including some with specific functions. About 18 per cent are in Edinburgh, yet 19 of the 32 Scottish local councils have independent schools within their area. Some 35 independent schools have residential provision, with, at 2001, a total of 3,710 boarding pupils. The number of boarders has been falling for the last four years, although much of this can be accounted for by the general

decline in family size among the Scottish population. In contrast, the total number of day pupils is 27,625, or about 4 per cent of pupils in Scottish schools, a figure which has been relatively constant for the last 15 years, and one which thus represents a real increase.[2]

Nomenclature
What is in a name? The term 'independent' is only a partially accurate description. In fact such schools get help from public funds in kind, this being in the training of teachers. Notwithstanding, 'independent school' remains a useful classification. Earlier nomenclature for independent schools can be confusing. In England, independent schools were called 'private' schools while those provided by the local authority were 'public' schools, but in Scotland both independent and state schools were termed 'public'. The anomaly was addressed in the early 1950s with the adoption of the 'independent' label.

Reasons for choosing an independent school
Families who choose independent education often defend the decision on the basis that they wish their children to receive the best possible education. This is seen as being achieved by the superior quality of the teaching, freedom from state intervention as social engineering, and the provision of a wider than otherwise range of extra-curricular activities. They may also feel that their children will make connections supporting a future network likely to give advancement and preference. In addition, parents can associate at school functions and forge their own advantageous acquaintanceships. Many families believe that pupils behave better at independent schools and regard the institutions as fostering a greater attitude of study in which the keen worker is not discouraged by a peer culture of studied idleness. Certainly there is a widening gulf between independent and state schools in terms of achievement.

Other factors can be at play. Boarders may be sent to a school because of its alleged character-building qualities. Parents may consider themselves unable or may be unwilling to care for a child at home on a full-time basis, or they may be working abroad and wish their children to be educated at home. In some instances there is not a suitable school within daily travelling distance of the family home and boarding presents the only option.

Images of boarding schools: Boys' weeklies
Despite the limited tradition of independent residential schools in Scotland the majority of Scottish school pupils in the first half of the twentieth century were familiar with boarding-school life from weekly magazines. The main magazine, English-based like most of the others, was *The Magnet*, edited and almost wholly written between 1908 and 1940 by Frank Richards. There, at a fictitious school named Greyfriars, Harry Wharton, Bob Cherry and other good English chaps reigned supreme. They were in 'The Remove',[3] a concept almost totally foreign to Scottish pupils, but accepted without demur. Other

stereotypical characters included Wun Lung and Hurree Jamset Ram Singh, foreigners from wealthy families, the latter being the son of a Rajah.

The most successful stereotype was Billy Bunter, who became an eponym. Fat because of greed, he possessed weak eyesight and pebble-glass lenses in his spectacles so that he was called 'the Owl of the Remove'. His efforts to raise money to buy food for himself usually failed, while his school cap was too small and figured in his being bullied. Bunter always protested against the bullies in terms of fair play but with little effect, yet the bullying was relatively gentle.[4]

Images of boarding school: Biography and autobiography
Because of England's longer and more extensive tradition of boarding-school education it is easier to find personalised accounts of this way of life there, usually in the form of biography and autobiography. Simon Schama,[5] for example, has compared the lives of Churchill at Harrow and Orwell at Eton for the BBC.[6] C S Lewis is one of a number of illustrious individuals who have written descriptions of their own experiences of boarding-school life. His account, like many, highlights the difficulties and indeed the oppression of the residential-school environment for some pupils, especially those unable to achieve social success there.[7]

Life at Christ's Hospital, past and contemporaneous, is discussed in detail in the writings of former pupil Norman Longmate.[8] Christ's Hospital was founded by Edward VI, son of Henry VIII, in 1553 in London, later moving to the west. Uniforms were extremely important at the school; these had been initially designed to identify the pupils as poor. Boys wore a blue coat or gown of ankle length with a belt and buttons. They had knickerbockers of yellow and stockings of similar colour. The name Bluecoat Boys in time was shortened to Blues, with Old Blues for former pupils, these having a magazine of the same title.

There was a separate school for girls. When the two schools came to be placed on the same campus at Hertford in 1784, a high wooden fence was erected to keep the groups apart and only at church on Sunday were girls and boys able to view each other. Girls too wore an identifying uniform, of green or brown skirt with a blue bodice and white cap. Both rather ridiculous-looking uniforms were changed in 1875, leading to a raft of other reforms such as changes in curriculum and the easing of some petty restrictions. In 1980, by merger of institutions, girls became boarders for the first time at Christ's Hospital.

Longmate writes that, from the eighteenth century, senior pupils who studied Greek as well as Latin were known as Grecians and went by right to the Universities of Oxford or Cambridge at the age of 19. They were above prefects in the school hierarchy but, unlike them, took no part in discipline. Discipline was left not only to prefects, but also to monitors, usually 15-year-old boys with a penchant for bullying (the roles of prefect and monitor also dating from the eighteenth century). The monitors learned their brutality in turn from the masters, who inflicted punishment by caning both boys and girls, the custom of beating girls continuing until the mid twentieth century. Faults considered worthy of punishment included inelegant Latin prose, bedwetting and fraternising with children from another House, even when the school was on vacation. No effort was made to screen out or reprimand unsuitable masters, even when their activities were obscene, and parents appear not to have criticised the actions of the school despite their children sometimes speaking of being sent 'into exile' or of 'ten years of penal servitude'.

The commonest complaint in all such schools was of a lack of privacy. Cubicles in toilets had doors which were open at top and bottom and shared beds were in use until the mid twentieth century. Equally as distressing for some was the lack of a place to keep treasured articles hidden from view. The painting by Millais displayed at the Royal Academy in 1867, 'The Bluecoat Boy's Mother', captures the anguish experienced by more sensitive parents over the conditions in which their children had to live.

The school gloried in its Old Blues: Lamb, Coleridge, Leigh Hunt, Barnes Wallis and Blunden,[9] for example, show the range of former pupils. Charles Dickens is another, who seems to have modelled some of his characters on Christ's Hospital staff, Wackford Squeers, the schoolmaster of Dotheboys Hall in *Nicholas Nickleby* (1839), being the most obvious.

A modern Scottish example: Dollar Academy
The concept of modern Scottish boarding education is probably best grasped by looking in some detail at one school. The most popular schools are

Figure 38.1 This beautiful building is McNabb House (named after Dollar's founder), Dollar Academy, Clackmannanshire. Source: John Robertson, Rector.

Figure 38.2 Chatting in a bedroom, promotional material, Dollar Academy, Clackmannanshire. Source: John Robertson, Rector.

non-denominational, away from a main city, yet central, with reliable transport links, all age, co-educational and – like any business – financially sound. The school may be 'mixed', that is with both day pupils and boarders, with these groups associating freely or choosing to mix only within their own category. The families of prospective pupils will also wish to consider the school's academic, sporting and cultural facilities and records.

One school which meets many families' criteria is Dollar Academy, Clackmannanshire (Figs 38.1, 38.2, 38.3, 38.4), a member of the bench mark 'Headmasters' Conference'.[10] The school has over 1240 pupils, of whom 88 are boarders, with such a discrepancy between the numbers of day and boarding pupils being common in present-day Scottish boarding schools.

Dollar Academy is probably the oldest co-educational school in Britain, founded in 1818 from the estate of Captain John McNabb. Its logo, a ship, signifies the work of the founder, and metaphorically the journey through life of the pupils. The Academy's academic performance, the extra-curricular achievements of its pupils and its ethos make it a powerful independent voice in Scotland, as well as an important part of the local

Figure 38.3 Female pupils and others at a barbeque, Dollar Academy, Clackmannanshire. Source: John Robertson, Rector.

Figure 38.4 Male pupils at a formal social event, Dollar Academy, Clackmannanshire. Source: John Robertson, Rector.

scene. In a small town of 3000 people, the school has a high profile and is aware of its need to maintain a good reputation.

The school is led by Rector John Robertson, who is responsible to a Board of Governors, and it may be examined by statute by HM Inspectors of Education. A key concept in the running of the school is communication, which is effected by holding meetings, including social occasions, having one-to-one discussions and providing comprehensive and easy-to-read material.[11]

Duties for the Rector, senior staff, house parents and others are very wide and are not confined to school hours. After classes come the supervision of sports and intellectual pursuits such as chess and debating clubs, as well as the offer of support to those pupils encountering problems. Individuals must possess a strong sense of commitment to work in such an environment. The care and welfare of pupils is largely achieved through preventative measures. Within the four boarding houses the staff are *in loco parentis*, and must oversee the provision of well-maintained rooms of adequate size with secure lockers and security systems for entry, ensure there is identification of strangers and that pupils have access to responsible adults when necessary. Quiet rooms for study must be available along with common rooms where boarders may meet friends (these are provided separately for the junior and senior age groups), kitchen facilities and the opportunity for students to personalise their own rooms with pictures and so on. Modern technology and media should be at hand; Internet access, telephones, television, audio and video recorders, play stations for the younger children and for the older pupils newspapers and books.

Outside of the classroom and the boarding houses the children need exercise with safe play areas for younger pupils and an opportunity for all to participate in school sports teams. Also required are school buses to transport pupils to external functions.

The lists of needs and provisions are endless. Parents and pupils are encouraged to state their expectations but must adhere to a 'statute of limitations', what lawyers would call 'the reasonable man' test, and the system is supported by a 'Suggestions' book. Boarding schools tend to expect a higher level of religious observance than might otherwise be the case today in Scotland and the chaplain or a local priest may promulgate views at variance with those of other members of staff – this is especially significant when the staff have guidance duties. The views held are unlikely to be doctrinal but alcohol, drugs and sex are often trip wires and conflicting opinions can bewilder pupils. All boarding schools are at risk of being accused by parents and pupils alike of inappropriate action and commonly the 'Three Rs', the concepts of Rules, Rituals and Relationships, are employed as a preventative measure. However, an overwhelming set of rules and a policy of zero tolerance not only are no guarantee against mistake, mishap and perhaps litigation, but may have a seriously negative effect in that they curb exploration and risk taking and thus learning. It is necessary therefore that staff must be assured of a right to use their discretion in many circumstances.

Brochures for all schools with boarding facilities emphasise the value of education, their records in examinations and their ability to 'develop the full potential of every pupil'. Boarding education is promoted as a 'family-style' enterprise. The ancient universities of Europe had cloisters and quadrangles where small groups of students could walk about and talk. The spacious and delightful grounds of Dollar Academy are designed to lend themselves to such activities.

A SCHOOL FOR THE CHILDREN OF THOSE IN THE SERVICES[12]

A unique school in Scotland is Queen Victoria School at Dunblane, Perthshire, which is wholly boarding and accommodates children of Scottish service personnel, even when they are posted abroad. It became co-educational in 1996/7. The Ministry of Defence provides most of the finance, while parents contribute a modest sum. Accommodation here is more spartan than in other residential schools, with houses being named after service heroes such as Haig, Trenchard, Cunninghame and Wavell. Daily chapel attendance is expected. The ambience of the school tends to steer pupils towards a career in the armed services.

PHILANTHROPIC INSTITUTIONS AND STATE SCHOOLS FOR CHILDREN WITH SPECIAL NEEDS WITH RESIDENTIAL PROVISION

Non-state schools
There are a number of institutions which have been provided for pupils by various religious groups and philanthropic organisations such as Barnardo, Quarrier and Steiner. They tend to be primarily homes rather than schools and the pupils may be educated on the premises or travel daily to a local authority school. Dr Barnardo's homes have shoots throughout Scotland; Quarrier's home in Bridge of Weir, Renfrewshire, which is now closed, had its own school on the premises. Steiner[13] has schools in Edinburgh and at one time a school complex near Aberdeen, this having boarders and accommodating children with emotional and behavioural problems and some who were blind. Staff not only must be very committed to the regime but also must be trained in specialist 'colleges' run by the organisation, with head teachers sometimes possessing medical degrees. Teaching methods at times are controversial.

The Royal Blind School
The Royal Blind School, a state school, is based in Edinburgh but serves all of Scotland. Visually impaired pupils may opt out of mainstream state education for a range of reasons, including deterioration in their eyesight or the family moving to a new area where the local school cannot adequately accommodate them. The current roll of the Royal Blind School is 131 pupils,

from 3 to 19 years of age, of whom 86 are resident. Most pupils begin residency at 5 years of age, although a few with multiple disabilities may become resident when 3 years old. Among the advantages of being in residence is the possibility of extra-curricular activities in the evenings.

The main aim of the school is to develop independent living skills in its pupils, including mobility skills. The learning may be cumulative, co-operative and progressive. However, as skills increase, so too does controversy. Is a proposed activity a step too far or not far enough? In a litigious society, safety tends to be the determinant of action over the acquisition of skills, although parents, in the abstract at least, encourage staff to seek the far horizons. The three participating bodies – pupils, parents, and staff of the Royal Blind School and Firrhill School, a neighbouring school – have different agendas and ideas, but work closely together.

Boarders are collected from the school by parents on Friday afternoon and return on Monday morning. Thus the link with home is maintained and the risk of institutionalisation minimised. Whilst in the school the physical safety of pupils is ensured by an environment in which cars are restricted and security doors on the buildings are fitted with intercoms. In addition, the Suzy Lamplugh Trust[14] provides personal attack alarms for all female pupils and staff.

The Scottish Executive favours social inclusion for all pupils. Any decision to end grant aid for the Royal Blind School, Donaldson's School for the Deaf, also in Edinburgh, and similar institutions could mean pupils being spread across Scottish schools. In a class of 30 children, any pupils with sensory impairment are likely to have things done for them, if only to save time. Thus they miss an opportunity to learn by doing. However, administrators appreciate economies of scale. There are 105 residential care staff in the Royal Blind School, for example. Replicating this level of care when pupils are spread through schools across the country would undoubtedly be more expensive.

BOARDING FOR GEOGRAPHICAL REASONS

Scotland possesses many islands grouped under the title Inner and Outer Hebrides and Northern Isles, with a number of additional islands and some very isolated settlements on the mainland. Primary schools can function on many of the smaller islands and in modest-sized settlements but secondary education has to be undertaken either at a school on a larger island, such as Portree High School in Skye, or mainland schools such as Oban High School, Argyll. Roman Catholic pupils, mainly from Barra and South Uist in the Outer Hebrides, were at one time accommodated at Lochaber High School, Fort William, Inverness-shire, with some at Oban High, where a priest attended regularly.

The lack of, and restrictions on the availability of sea and road transport is such that it is impossible for many pupils to attend secondary school on a daily basis and return home each evening. Only parents in villages near to towns with a secondary school can have their children brought home daily by bus. Even returning home at the weekends can be impossible or at least very difficult for those who must travel by sea. Ferries to the islands are usually run by commercial companies[15] and sailing times are rarely geared to ensuring that pupils can leave school on Friday, take an afternoon boat home and pick up an early return sailing on Monday morning. Some ingenious methods have been used by schools to get children home using non-public transport, but all are constrained by prohibitive insurance costs.

The only feasible solution at times has been to accommodate children in town, in specially provided hostels and sometimes other types of accommodation such as relatives' homes. Oban High School, for example, had in the late twentieth century around 200 boarding pupils with a range of transport requirements (in 2003 there were some 80 boarders). Although pupils may spend all week in a hostel, it is essentially a Monday to Friday experience, closely related to the school, but maintaining links with the pupils' own home life. This link to home life becomes more apparent at the weekend when greater freedom is given to the children. In the Oban High hostel, for example, all pupils attend a daily study period from 6pm to 8pm, a pattern which some former pupils have maintained in adult life. Opportunities to visit the town are limited and staff aim to keep their charges fully occupied. There is the usual complaint about the quality of the food. Many hostels have had their accommodation upgraded in recent years, but with limited finance available to schools, such investment is not always regarded as a priority.

For pupils from remote areas, arriving in larger towns to attend secondary school can be quite an experience; not only do they have to become accustomed to being away from home for considerable periods of time, but also the size, scale, noise and bustle are very different from what they are used to and can provoke both delight and 'culture shock'. They also have to find their place among the local children. Even where the pupils are second-generation boarders, their parents cannot entirely prepare them for what to expect, because the gap between the generations is sometimes too great and there are considerably changed circumstances.

Parents and teachers do not wish to have pupils divorced from their roots. Regular visits home, even if not on a weekly basis, are essential. The local authority in Argyll, for example, eventually decided (in the 1970s) that pupils would be allowed home once a month. The Rector[16] of Oban High School later changed this to one visit every three weeks but arranged the timetable so that pupils would not have to miss the first two hours of classes on a Monday morning.

The 1973 Local Government Reorganisation (Scotland) Act created 53 District Councils and nine Regional Councils, plus three Island Councils. Of the last, one was the Western Isles Council whose creation resulted in Ross and Cromarty and Inverness-shire no longer controlling separate parts of the Hebrides. The structure of Councils was reassessed in the early 1990s, leading in 1994 to a reversion to 32 Councils but retaining the three Island

Councils. The consequent new financial arrangements affected boarding pupils by creating new schools, such as the community school at Liniclete, Benbecula, and new catchment areas, and generally resulted in fewer pupils having to be away from home for long periods in an environment markedly different from that of their home. Many of the islands, along with very remote parts of the mainland, are part of a fragile economy. Depopulation is a constant threat, provoked by lack of facilities and scarcity of jobs. Not sending children away for secondary education is an important factor in helping retain population.

POST-SCHOOL STUDENTS: ST ANDREWS COLLEGE, GLASGOW

Post-school students in Scotland have been able to choose from a range of accommodation types whilst undertaking their studies.[17] Although taking lodgings was for long preferred by Scottish students unable to live at home, opportunities to board in purpose-built residences have been available at certain institutions for centuries.

Many more-recently founded institutions have provided residences for their students. The Roman Catholic Hierarchy was reinstated in Scotland in 1878 and by 1894 had established Notre Dame College in Glasgow for training women teachers, with residential accommodation available on site to students in the form of halls of residence. The Robbins Report in 1963 criticised the notion of single-sex and monoculture colleges for students and it was in the late 1960s that Notre Dame began admitting male students, at the same time as the institution moved to the suburb of Bearsden. After merger with Craiglockhart, Edinburgh, in the 1970s, Notre Dame became St Andrews College of Education. In late 1998 this joined with Glasgow University, as did boarders from each institution. Student teachers are now expected to live with, for example, chemists, sociologists and engineers to mutual benefit. At the same time they have exchanged the suburban environment of the outskirts of Bearsden for life in the city centre. At a superficial level, the mix of backgrounds, possible vocations, religious persuasion, ages and city anonymity has worked well, both in academic and personal terms.

University education tends to be frozen in instructional mode with education through discussion with one's peers being greatly undervalued. A place to talk with fellow students in pleasing surroundings is helpful. With project-based assignments and assessment, having classmates living nearby can be very useful. Other advantages of living in residences include ease of access to on-site activities such as libraries, sports facilities and cultural events. Students also form good relationships with those staff who act as residence wardens. In addition, halls of residence or 'halls' offer an assured minimum standard of comfort, freedom from many domestic chores and responsibilities, sometimes lower costs and an enhanced degree of security.

Then there are social opportunities, ranging from general companionship to more intense friendships. On the negative side there is often more restriction than when living in alternative types of accommodation. When treated as younger than one's age in an institutional setting people may react in an infantile and negative way.

Today most students arriving at residences attached to post-school educational institutions have not come from a boarding-school background. Nevertheless, they may encounter much which is familiar to boarding-school pupils. Although boarding-school rituals are not echoed in Scottish undergraduate life, as they are in certain English universities,[18] the everyday domestic routine may be similar and groups of young people in higher education are ultimately the responsibility of staff in their institution. They are also exposed to similar risks and problems, albeit in a different guise, to those of boarding and other school pupils. These include issues of personal safety, adequate nutrition, drug and alcohol use and safe sex. The skill, from the institution's point of view, is in providing support while not crowding the student and preventing growth of independence.

DISCUSSANTS

The author wishes to thank the following individuals for sharing their knowledge of boarding-school life during the preparation of this chapter:

Professor Bart McGettrick, formerly Principal of St Andrew's College, Glasgow; presently at the Faculty of Education, University of Glasgow; Dr Farquhar McIntosh CBE, pupil, teacher and Head Teacher at Portree High School, Isle of Skye, and Rector of Oban High School, Argyll, later Rector of the Royal High School, Edinburgh, presently Chair of the Board of the Royal Blind School, Edinburgh; John Robertson, Rector of Dollar Academy, Clackmannanshire; Dr Dorothea Sweeney SND, formerly Warden of Residences and Vice Principal of St Andrew's College Glasgow; Kevin Tansley, Principal of the Royal Blind School, Edinburgh.

NOTES

1. Paterson, 1999, 94.
2. See Scottish Council of Independent Schools for yearly figures concerning pupils in Scottish independent schools.
3. The equivalent of Primary 7 in Scottish schools.
4. The character of Billy Bunter was resuscitated in the 1950s for television. The series ran for several years yet did not achieve the success of the magazines, partly because ideas on lifestyles had changed. George Orwell, the pen name of Eric Blair, novelist, essayist and Old Etonian, examined this phenomenon in an essay, Boys' weeklies, in the collection, *Inside the Whale* (1940), discussed in Crick, 1980, 623.
5. Simon Schama has been Professor of Art History and History at Columbia University for 20 years. He returned to England to make a television series, 'A History of Britain', for the BBC which was published in 2002 by BBC Worldwide, Ltd.
6. Televised under the 'History of Britain' banner, BBC2, 7 December 2002.
7. Lewis, 1955.
8. Longmate, 2001.

9. Charles Lamb, essayist and literary critic; Samuel Taylor Coleridge, poet; James Henry Leigh Hunt, poet, essayist and reformer; Neville Barnes Wallace, engineer (created 'bouncing' bombs used in World War II); Edmund Charles Blunden, World War I poet, writer, critic.
10. A badge of quality.
11. Publications include booklets titled *Information for Pupils*, *Information for Parents* and *Extra-Curricular Activities*, plus regular newsletters addressed to parents, former pupils and friends of the school. There is also an extensive website with up to date HMI and Care Commission reports on boarding.
12. Established to cater for children of soldiers killed while serving in the South African Wars, c 1903.
13. Steiner, 1861–1925, was an Austrian philosopher who founded anthroposophy, the science of spirituality. He linked it to educational and therapeutic work.
14. Susie Lamplugh was on the staff of an estate agency in England and disappeared after meeting a client. Her parents set up a trust in her memory to provide help for young people in danger.
15. The population of the islands and adjacent mainland have a love-hate relationship with MacBrayne. A poem explains it:
The earth belongs unto the Lord
And all that it contains
Except the Western Isles
For it belongs unto MacBrayne's.
16. Dr Farquhar McIntosh.
17. See Chapter 39 Shared Flats and Houses.
18. The University of St Andrews, alone of Scottish universities, has red gowns for undergraduates, something in keeping with their Kate Kennedy and Raisin Monday celebrations. Kate Kennedy, Bishop Kennedy's daughter, heads a procession where women are excluded and there is much cross-dressing. Raisin Monday is where a bejant gives his or her mentor a bag of raisins in appreciation of help given.

BIBLIOGRAPHY

Clark, A T, ed. *Was it Only Yesterday?*, Edinburgh, 2002 (recollections of former public school boys who served the British Empire in northern Nigeria).
Crick, B. *George Orwell: A Life*, Harmondsworth, 1980.
Emerson, R W. *Journals*, Boston, 1849.
HM Inspectorate of Education has reported on aspects of boarding facilities of independent schools from 1999. Available on the Internet.
Holmes, H, ed. *The Institutions of Scotland: Education, Scottish Life and Society: A Compendium of Scottish Ethnology, vol 11*, East Linton, 2000.
Lewis, C S. *Surprised by Joy: The Shape of My Early Life*, London, 1955.
Longmate, N. *The Shaping Season*, Cirencester, 2001 (life as a boarder in Christ's Hospital).
Morris, I. Independent educational institutions. In Holmes, H, ed, *The Institutions of Scotland: Education Scottish Life and Society: A Compendium of Scottish Ethnology, vol 11*, East Linton, 2000, 237–53.
See Pearson, K. The independent sector. In Bryce, T G K, Humes, W M, eds, *Scottish Education*, Edinburgh, 1999, 93–101.
Pinkerton, R M, Windram, W J. *Mylne's Court: 300 Years of Lawnmarket Heritage*, Edinburgh, 1983.
Scottish Council of Independent Schools. *Which School? A Directory of Independent Schools in Scotland*, Edinburgh, published annually.

39 Shared Flats and Houses

SUSAN STORRIER

Sharing has always marked home life, provoking strong feelings and complex strategies. We share non-tangible phenomena, such as experience, thought and emotion with the other members of the household, as well as the physical resources of the home, in the first instance space. In present-day family households of ordinary means, rooms are commonly shared by all members of the household with the only private space being bedrooms, which children may in fact share. In the case of sexual partners it will be expected that they will share not only a bedroom but a bed also.

Even those who live alone may find themselves sharing conversation with their neighbours as well as resources such as boundary walls, communal stairways and drying areas; for implicit in most domestic life is sharing not only with household members but also with the local community, if only on an infrequent basis or in times of emergency. Our dwellings may also be shared, temporarily, with those from beyond – non-locally living acquaintances, friends and family – and with those who work in homes.[1] At a further level, the media and the authority of the public world, in the form of byelaws, tenancy agreements, mortgage requirements, planning policies, etc, may impinge to such an extent on the workings of our domestic life that the notion of sharing our homes with these may not be entirely inappropriate.

SHARING SPACE IN HOMES BEFORE THE TWENTIETH CENTURY

In the humble homes of the past it was common in the crowded conditions for individuals to possess no part of the dwelling, no matter how small, for their own exclusive use, the exceptions being the heads of the household who might have their own chairs to sit upon and the very elderly or ill who could have a bed to themselves.[2] Sharing beds was the norm for most 'ordinary' people in Scotland until into the twentieth century and for some of those in better-off households too.

At the neighbourhood or local community level, sharing one's dwelling often had considerably more significance than it does today. This included general social interaction and the heightened form seen at times of individual and community rites of passage,[3] mutual support in times of crisis and in the use of services. In the context of urban dwellings especially, households of limited economic means could expect to share a far wider variety of domestic resources with those living nearby than we would accept today. They might make use, for example, of communal wells, pumps and piped water supplies, sinks, washhouses, ash pits and other sanitary facilities.

Unmarried workers provided with accommodation by their employers would also usually be expected to share intensively. The men housed in agricultural bothies, for instance, were often required to share beds in a form of dormitory and made use of a communal living/cooking area,[4] while domestic servants in large households might also have to share both bedrooms and beds.

If sharing within a household was regarded as normal and respectable, and sharing the home with the community at certain levels was quite acceptable, the sharing of one dwelling by different households – especially if there were children present – was often regarded in a poor light. Such was often the reaction to 'doubling up', meaning two households (or more) sharing a dwelling but usually with some private space for individual household use, even if just one room, although in the most extreme cases households shared a single apartment.[5] The authorities tried to prohibit the mixing of unrelated adults of different sexes within accommodation, especially sleeping accommodation. They were not always successful. One report from Mylne's Court, Edinburgh, in the nineteenth century described a 'house' (apartment) of three rooms, 'Sunday night there slept fourteen souls of different sexes and no family connections', although the reporter noted that, 'the inhabitants appear healthy and are extremely civil'.[6]

It was not only indefinitely that dwellings could be intensively shared. At the most temporary level, the custom of strangers of the same sex sharing both rooms and beds was followed at inns for a long time. Those engaged in seasonal work might, like year-round agricultural workers, be housed in bothies or equivalents.[7] Sharing bedrooms or dormitories, and often beds, as well as washing and dining (or cooking facilities) also typified lodgings, lodging houses and hostels of various sorts, which accommodated those away from home or those with no permanent base.[8] The experience of living in these types of accommodation varied enormously, as university students (a group likely to make frequent use of such housing) regularly discovered. Writing of Edinburgh in the 1870s 'Alisma' states that:

> In our day we had no splendid hostel wherein we might live with the minimum of care or bother. Every student had to cast about for lodgings, and of course these varied in all manner of shades and degrees.[9]

Until the eighteenth century, Scottish universities provided lodgings for some of their students in serviced and often shared rooms (two or three to a room) within the college, with dining taking place in the Hall, as occurred at Glasgow University.[10] Later, private lodgings or living at home were exclusively relied upon until the idea of halls of residence came into being in the nineteenth century. Although some of these 'halls' took the form of hostels with dormitories, or were supplied with double rooms (but no shared beds), others pointed more firmly to the future by providing single study bedrooms. All might be accompanied by communal dining and washing facilities and often tutorial or even lecture space and a library.[11] Such accommodation was considered especially suitable for female students and was thus particularly restrictive and closely monitored. However, residences at Scottish educational institutions were far from widespread and did not become so until after 1945.[12] Then they became not only more common, but more relaxed and accommodated more male students.[13] Similar types of residence can be found for those with special needs of various kinds.

SHARING HOMES IN THE TWENTIETH CENTURY

As the twentieth century progressed, ideas about acceptable minimum standards of privacy and space, and the implementation of these by the authorities, coupled with a reduction in the size of households and a general decline in community interdependence and interaction at neighbourhood level, were making both very intensive (within the household) and extensive (beyond the household) sharing of domestic space less common. There were fewer informal visits and visitors. Most dwellings were supplied with their own water supplies and other facilities for the household's exclusive use and increasingly they possessed enough space to ensure that parents could enjoy a room to themselves and that bedrooms were shared only by children of the same sex, who would now have their own individual beds. By the end of the century, and into the twenty-first, children's bedrooms have often become in effect bedsits supplied with their own television and other recreational – as well as study – facilities. Sometimes meals are eaten here, in solitude in front of the TV or computer.

A reduction in sharing occurred in other accommodation. That provided for employees still required bedrooms to be shared sometimes, but not beds. The twentieth century continued to witness the taking in of lodgers (although this became less common as time progressed) but the lodger would now almost certainly have a room to him- or herself. In guest houses and hotels the norm is now one room per person and although double and twin rooms are very widely available these are not for sharing with strangers. Increasingly, bathroom facilities are private also (as they are becoming in student and other residences).[14] Although hostels for recreational use maintained the idea of single-sex dormitories until recently,[15] other types of hostel, such as those for homeless people, divided dormitory space into semi-private cubicles within single-sex sleeping accommodation. Such a move towards self-containment within domestic space is even being seen in prisons where, in a few instances, en-suite bathroom facilities are being installed.[16]

The overall trend, then, through the twentieth century and into the twenty-first has been towards an increase in personal privacy and a decrease in sharing within homes. However there have been moves in the opposite direction which have flourished in recent decades. Two of the most prominent involve unrelated adults of both sexes sharing domestic space. Sheltered housing provides self-contained flats for older or infirm residents but with a shared common- or day room. More profound, and more radical, is the sharing of houses and flats by unrelated single people, often men and women together.

SHARED FLATS AND HOUSES

For the purposes of this short article, 'shared houses and flats' describes the homes of those households in which, with no more than average imposition of authority from beyond (thus excluding institutional, 'care' and commercially based households) residents pertain to different families and have, in theory at least, equal status. Those households where membership is determined by a common ideology or belief are dealt with elsewhere in this volume, as are shared homes associated with specific types of work. Of course, distinctions are not always clear, nor are categories discrete. It could be argued, for example, that adult siblings living together have more in common with 'sharers' than conventional family households.[17] And there is the question of bedsits.[18] Space does not permit their inclusion here, yet in many instances they operate more as a shared home than a collection of essentially self-contained units within a building. This chapter limits itself to houses and flats shared by student households and those formed by, usually, young workers and, to a lesser extent, by those in receipt of state benefits.

Information has been gathered from bibliographic sources – notably Nicholson and Wasoff's detailed study of student housing experiences at the University of Edinburgh in the late 1980s[19] – and a brief survey of 24 students, carried out at the University of Aberdeen in January 2005 (this took the form of a face-to-face questionnaire composed of 25 questions, asking about the facts surrounding the students' present accommodation and household and their views about these; see the Appendix for an analysis of some salient results), supplemented by the author's 16 years of experience of living in shared accommodation.

SHARED FLATS AND HOUSES: THE 1960S ONWARDS

By the 1960s, the growth of higher education[20] was putting a strain on the existing forms of accommodation available to students away from home; that is lodgings and residences. This occurred as new liberal ideas – including those concerning the role of education and greater personal freedom, especially sexual freedom – came to prominence. Until the 1960s the emphasis had been on young adults staying in the parental home until marriage, unless study or work forced them to go away from home, but now independence was to be valued instead. Lodgings became not only scarcer but, with their rules and regulations, less desirable to many.[21] The grant system was introduced to universities which allowed students to choose accommodation for themselves:

> ... the preference for living away – on campus or in shared student flats – was also part of a more complex set of changes in student culture ... a shift from university years as an apprenticeship in the social mores of the middle classes to concept of student life as an autonomous interlude free of the pressure of class and status.[22]

Campus accommodation was popular for a while, with single-sex self-catering halls of residence becoming prominent at some Scottish institutions.[23] However, expansion was not sufficient to meet the growing housing needs of students and in any case halls remained to some extent supervised environments. The greatest demand for many years now has been for shared furnished flats,[24] either university supervised, or more commonly, those provided within the private rented sector, although residences are still popular with first-year undergraduate students[25] and of course many students live at home and some alone.[26]

The growth of higher education has continued, but funding has not increased commensurately.[27] By the end of the 1990s, 40 per cent of the 'student' age group were in higher education.[28] The provision of university-controlled accommodation (either halls or student flats and houses) has not been able to keep up with this expansion, so that even more emphasis has been placed on the private rented sector, of which very few properties can be afforded by a single tenant.

Recent changes in the student accommodation scene include the number of mature students, who may have specific housing needs (for example, they might have dependants who require to be housed with them), and the increasingly popular option open to better-off students (or their parents) of buying flats in which they live and rent rooms to fellow students.[29] The expansion of the latter has been fuelled largely by the massive rise in property values in most of Scotland in the last decade or so which makes flat-buying likely to considerably increase one's capital as well as reduce living costs for the owner during his or her studies thanks to the rents of co-resident tenants.[30] Mortgage interest rates are relatively low and many lenders are now more than willing to offer mortgages to students. This is reflected in the findings of the University of Aberdeen survey where one third of respondents lived in a flat or house with a resident student landlord.

The practice of young people living together in shared flats, often in mixed-sex groups, is not confined to students. Many young workers and others have followed the trend of leaving home and living independently,[31] a vogue brought to wider notice by the media, especially the television comedy series 'The Liver Birds', which ran for nine series from 1969.

As with students, lodgings and other traditional forms of temporary accommodation became both harder to find and increasingly unpopular with these people, often being seen as restrictive and impersonal, occasionally threatening and certainly unfashionable domestic environments for those with 'get up and go'. At the same time there was less rented housing available as, from the 1960s, more and more property was being taken into owner occupation. Of that on the market, much bore an inflated price-tag, especially in popular locations, and individual tenancies became beyond the means of many people.[32] Once the way of life became generally regarded as acceptable (at least for people before they embark on marriage and children), sharing a flat or house became an important option in areas of housing pressure, especially for those with constrained incomes and/or with a sociable outlook and a wish to postpone taking on board the full responsibilities of having a home of one's own. Cementing matters further were high interest rates in the 1980s, making property purchase more difficult, and decades' worth of policies associated with public-sector housing which effectively exclude many single people from the system.[33]

The greatest demand for rented accommodation among young employed people tends to be in areas where there is a high concentration of students, the inner areas of cities and large towns, so there is much competition.[34] Different in some respects are those single young people who depend on state benefits yet who may also desire or be forced to share a home, usually on a more indefinite basis. They too may wish to obtain a room in a central flat so as to be nearer employment possibilities and the facilities and general buzz of the city centre, or they may share a home in less popular areas. Sharing is sometimes the only option for those living in certain parts of rural Scotland where there is a lack of housing.

A flat may be shared by students, workers and benefit recipients together. The groups tend to have a fair amount in common in terms of financial constraint and ideas about domestic lifestyle. Indeed an individual may experience life in all three categories whilst sharing, even sharing the same home. Sharing is not the only form of domestic life that people can experience at one time – those working or studying away from home may choose shared accommodation for 'through the week' and return to their families or other base at weekends and during holidays.

There are many motivations behind sharing homes and usually more than one factor prompts any individual's decision. Amongst these, reducing costs is clearly one of the most important. Expenses may be divided among the residents and effectively reduced for the individual as a consequence. But just as significant can be the sense of independence and the freedom from regulation and formality as well when sharing a rented flat or house, perhaps '... the first real introduction into adult social life freed from the constraints of parental control and neighbourhood and family ties'.[35] In common with much rented accommodation, minimal maintenance of the dwelling is required on behalf of the tenants, who are not responsible for furnishing or decorating the home (important status signals elsewhere).

There is the additional boon that they are usually not expected to adhere to the same standards of cleaning displayed among other types of household and that which *is* done can be shared, in theory at least, among a number of people. Shared homes also provide company, a very important factor for some people. Indeed they frequently offer a highly sociable and fun environment and may provide an increased measure of security in comparison with living alone.

Nowadays not only young people and students can find themselves sharing flats, although the way of life is stereotyped as 'young'. Individuals may also choose to take a room in a shared house or flat later in life, for example after divorce or separation. Sharing accommodation can be one of the more attractive alternatives for getting life re-established as a single person. It may be difficult to adjust to not 'having one's own space', but the costs are relatively low and there is the potential for emotional support from other household members. For many people finding themselves without an address – after illness, unemployment, upon return from living abroad or release from imprisonment, etc – sharing is the first step towards establishing or re-establishing their own home. It is also a useful housing form for those on temporary work contracts or those who otherwise have a high degree of mobility. Working in a location for a few months may be too long to spend in hotel-type accommodation, or restrictive lodgings, but not long enough to merit taking a tenancy for an entire dwelling (which often have fixed lengths), let alone purchasing a property when this is otherwise a possibility.

For landlords, student or other shared households can be advantageous. The rent charged on a property can be higher,[36] people move on frequently[37] and they may be willing to accept lower standards of furnishing and maintenance than other types of tenant.

For long, renting a home, in any form, was seen as a short-term measure for students and those in work; both groups would be expected to buy their own home as soon as possible. However, the surge in property prices in recent years, coupled with substantial improvements in the quality of rented accommodation, has made renting a longer-term option.[38] Many of those who rent into their thirties will be in a partnership, and perhaps have children and will not live in a shared home as such. But others will continue sharing with non-related adults, a considerable number sharing with the landlord. A further, occasional (but increasingly popular) present-day option, in the central and very expensive areas of the cities of Glasgow and Edinburgh especially, is for shared owner-occupation, where two or more friends club together to take on the mortgage of a shared home.

BUILDINGS USED AND WHAT IS SHARED

The buildings used by sharing households can include purpose-built houses and flats such as some student accommodation service properties. Most commonly though, they are dwellings designed for conventional single

Figure 39.1 Tenements in South Edinburgh, near the University of Edinburgh. Many of these flats are rented and sometimes bought by students. Source: Susan Storrier.

household use, occupied either in adapted or unaltered form (Fig. 39.1). Disused buildings have at times been used as 'squats', usually for relatively large sharing households.[39]

Most shared flats have three or four bedrooms, but much larger dwellings can be found. Those sharing a flat or house will usually expect to have their own private bedroom with at least a bed, a chair and some type of storage furniture or fitting (and often a desk or other study facilities too) and to share the use of a hall, kitchen and bathroom. At times a room may be a windowless boxroom rather than a bedroom as such (the resident usually pays a proportionately smaller share of the rent as a consequence). Often there is also a shared living room, although in circumstances where rents are high and the incomes of the residents very low, the lounge may be used as a bedroom for a further housemate instead (Fig. 39.2). Bedrooms are occasionally shared in the form of twin rooms[40] and, more rarely still, where two of the residents are in a sexual relationship together, double rooms. Conversely, in shared new-build flats – many of which have just two bedrooms and are occupied by those in professional employment, or students, one of whom is the landlord – the principal bedroom has an en-suite bathroom, leaving the other resident the exclusive use of the main or 'family' bathroom.

In shared homes not only space and the rent (or the mortgage) are shared. Other costs are divided among the residents – where appropriate

Figure 39.2 Living room of a shared flat used as a bedroom. The bed, which is placed against the wall, is laid out as a day-bed/sofa. Source: Susan Storrier.

this includes the Council Tax (Community Charge plus Water and Sewerage),[41] as well as utility bills (such as gas, electricity, telephone) and sometimes other recurring expenses such as Internet line rental, digital or satellite television connections and television licences. Usually most housekeeping costs are also shared (cleaning materials, soap, toilet paper, etc) and sometimes full or partial food costs.[42] A number of these expenses are commonly paid via a kitty or in turn (housekeeping costs and fresh foods such as bread and milk), whilst others are usually divided up equally or proportionately between the number of residents. Responsibilities, from cleaning to payment of bills to liaising with the landlord or agent, are similarly shared out.

Conversation and life experience are also shared widely in most of these homes. Often this is to a much more profound degree than in family households, for example, where the presence of a hierarchy, a wide age range and other differences may limit somewhat the interaction between members.

LIFE IN SHARED FLATS AND HOUSES: SOME IMPORTANT THEMES

Finding a Home

For those on low incomes, certainly in many urban areas, it is usually far easier to find a shared home than a dwelling to oneself. Yet obtaining a shared flat, especially the right one for one's needs, can be problematic. Housing markets change for worse or for better and generally seem to increase in complexity. There is discrimination against students by some landlords and an occasional tendency for owners and agents to dislike mixed groups of males and females, or a preference for female residents over male. Individuals might have to consult several sources before finding a suitable home[43] and this is particularly the case for non-students who are without access to student accommodation services and the notice boards of institutions (although many non-students will consult the latter). Reliance may be placed on a card in a shop window, a newspaper advertisement, a web site, a solicitor or an estate/accommodation agency. Most people, however, seem to prefer a personal recommendation. For those leaving their parental or other main home for the first time it is a daunting matter deciding if a dwelling, or a room in a pre-existing household, is really what one needs and wants. Students again have the advantage in that their institutions usually provide advice including web sites with checklists of important points.[44]

The most successful flats in social terms are often composed of friends who look together for a place to live. Sometimes one person takes on a property alone and then looks for others to share with him or her. In either case, early commitment to a good property is essential and those wishing to secure a home may immediately have to provide a parental guarantee, bank references and up to two-and-a-half month's rent (to cover a deposit, perhaps an agency fee and the first month's rent).[45]

When a flatmate leaves, the other residents may have to cover his or her portion of the rent until a replacement can be found. Not surprisingly great efforts are made to find a new person as soon as possible. Most households would prefer to take on-board someone already known to and vouched for by at least one of the residents. Failing that, it is hoped that a good number of people will come to look at the flat and that a satisfactory choice can be made after an informal interview has been conducted with each. However, so great is the urgency that many households have little choice about who comes to join their number. With luck the replacement flatmate will fit in, but sometimes this is not the case. As long as differences are not very extreme and no serious offences are committed, it is a case of everyone being tolerant and enduring the situation with as much good grace as possible.

Those leaving a shared home and looking for a room in another may also have very little time in which to decide about a flat and they will often be expected to move in almost immediately, paying rent and a share of the deposit up front. People wishing to change flat need a good measure of luck in finding someone to take on their old room straightaway and in finding something suitable at an appropriate point. The alternative can be paying double costs for a number of weeks.

Paying for Accommodation
One of the great advantages of shared flats is the sharing of costs. By this means it is possible for people on low or even very low incomes to live in well-built, spacious flats in expensive areas.[46] Yet although shared accommodation may seem like a cheaper alternative, the rent can still account for a large part of the income of many residents and this is especially the case for students. In 1996 Callender and Kempson estimated that students in Great Britain spent on average 23 per cent of their income on housing costs, exceeded only by their expenditure (28 per cent) on food, household items and bills together.[47] In some popular university cities, such as Edinburgh, the percentage paid on rent may be far higher.[48]

Almost all sharing of finances in shared flats is done on trust, with a strong ethic of honesty amongst flatmates supporting this and making it successful in the vast majority of cases. Although individual tenancies for each resident in a shared flat may occasionally be the case, more often there is one tenancy which is signed by all, or the tenancy is taken on by one flatmate who effectively sub-lets to the others.[49] In the latter two contexts, a flatmate unable or unwilling to contribute his or her share of rent will have the effect of increasing the financial burden on the others in the home. Defaulting on any shared cost is the foremost 'offence' against shared flat life and that most likely to provoke an insistence that the person at fault find somewhere else to live.

Status
Shared flats and houses usually offer freedom from some of the strictures of normal adult life including the obligation to 'keep up with the Joneses'. It can be a relatively carefree, liberated existence, tolerated by others perhaps because it is seen as being a temporary phase which will let young people get unconventional ideas 'out of their system'. At the same time, those in shared flats and houses may be stereotyped as undesirables, in terms of noise, lack of cleanliness, sexual behaviour or drink or drugs taking, and they may find relations with the more settled members of the immediate community affected as a result. The rapid turnover of personnel in shared homes often contributes to the lack of knowledge and understanding between 'sharers' and their neighbours.

Equality and Hierarchy, Conflict and Resolution
One of the essential features of shared-dwelling life is a lack of authoritarian or hierarchical structure. All should be equal and living with as few rules as possible. In fact this is a highly sophisticated domestic system which demands a significant number of rules, but these are largely unstated. Commonly the interpretation is that everyone should be free to do as they wish within reasonable bounds – that is, paying shares of rent, bills and

other expenses promptly and in fair proportion, generally keeping noise down after certain hours, and not making use of others' property without permission. Having partners living in the dwelling but not contributing to the household, especially in economic terms, can also be badly viewed. Any conflict between interests is dealt with by an emphasis on tolerance.[50] It is better to endure than criticise or complain and when matters are genuinely intolerable negotiation is tried before the last – and unusual – resort of asking someone to leave the home. Although the notion of tolerance usually works well, it is open to abuse. More dominant personalities quickly assume an authoritarian role and a 'boss' can also be created from, for instance, the longest-standing resident, the person who has negotiated the tenancy, or even the tenant with the biggest bedroom, leaving the other residents with little option but to fall quietly into line or look for somewhere else to live.

Whatever the level, the process of sharing can work well or with difficulty or it may prove impossible. For it to work smoothly those involved have to judge deftly where individual need must give way to the needs of others. Where there is not the ability to make this distinction, or the will, or where – most commonly – the boundary is drawn differently by those involved, conflict occurs. Our ideas about what constitutes the good, the reasonable or the unacceptable in domestic terms come in large part from our own early training and experience, usually lie deep and are not quickly changed. Shared dwellings and conflict, therefore, can accompany each other to a particularly marked degree and the latter is more likely the more crowded a home is.

In flats where tenants share the accommodation with the owner there tends to be less crowding and consequently less tension from that source. Owners can decide at what point they want to balance financial gain against lack of space and loss of privacy. Other advantages of living with the owner can include a lower rent, the lack of a deposit to pay, better furnishings and maintenance and generally a more homely atmosphere. However, the sense of equality among those living in the flat may be less and there may be more regulations – especially in terms of cleaning and care of the property – and less chance for personal expression on behalf of the tenants. These downsides are more commonly encountered when the home is in the possession of a working resident landlord rather than a student owner.

Security
One of the motivations behind sharing a dwelling may be the additional security that the way of life offers. A household of several people all coming and going at different times is less likely to be subject to burglaries, for example, and the presence of good company may help relieve various forms of mental or emotional anxiety. Mutual support and understanding are salient features of successful shared homes, and not just recent ones. 'Timber girl' Jessie Mackay, describing life in a shared hut during World War II war work says, 'The girls were a grand bunch and true friends. We shared our joys and sorrows and helped one another through difficult times.'[51]

At the same time, shared-flat life may be very insecure. Tenancies can be fragile, despite legislation,[52] and for those living with the owner it may be especially easy to find oneself without a home after very little notice. The people one shares with and who visit the home can very occasionally pose a threat to one's possessions[53] or oneself. More commonly, tensions abound in shared dwellings and not all flatmates may be able to control adequately the resultant feelings of anger and frustration. In addition, people may be sharing at particularly vulnerable periods in their life – having just left home, entering student and working life, after separation, etc – and may bring some of this to bear on their companions. In non-student shared dwellings the range of experience may be very wide and great and differing demands may be placed on individuals when they are out of the dwelling, for example at work, adding to the potential for trouble at home.

Privacy
Shared flats and houses may counter the trend of decreasing interaction within Scottish homes, yet privacy remains an important notion. Some people, notably students in the first year of their studies, or those with particularly marked financial constraints, may share a bedroom and have no private room to speak of. Areas of a shared bedroom may be demarcated by unspoken or spoken agreement for the exclusive use of one person, but of course all activities in these areas can be witnessed by the other resident. Timetabling, usually by informal and unspoken agreement, allows a measure of privacy in such circumstances. A factor helping to offset the effects of a lack of privacy is the choice of roommate. Sharing with a friend, or at least someone in similar circumstances (similar courses for students, for example) and with a compatible personality and in whom one can trust, can reduce friction – or result in the end of the friendship! Sharing a room is only really practicable for very young students who have similar lifestyles and experiences. As individuals establish themselves in their new surroundings, and in independent life in general, the demands made on living space increase. A wide variety of relationships form and increasingly important examinations need to be prepared for. It is not surprising therefore that few people share a bedroom for more than one academic year.

Most people sharing a dwelling do have exclusive use of one room however, sometimes provided with a lock (although employing this can be interpreted as a lack of trust in one's flatmates). Entering these spaces without the invitation of the resident is generally regarded as untoward, depending on how close to each other the people in the household are, and certainly is disapproved of if the resident is not present and more so still if his or her possessions are disturbed. Private rented accommodation tends to be older and reasonably well soundproofed but more modern flats can make it easy to overhear the sounds in other rooms. In any circumstances, giving the appearance of not listening to others' private goings-on is expected. The number and size of public spaces for the use of the entire household vary. Where there is one available, a lounge may be used as a 'community' room, where the household meets up and relaxes together, or, with older students and others, its use may be geared more towards each member entertaining his or her visitors separately, an extension of individual privacy into public areas.

Where the kitchen is large enough, it can often act as much more of a community room than the lounge. Certainly it is here that most interaction between household members occurs. But here too privacy is a concern. Some households may buy and prepare all food in common, which certainly is more economical, but many others operate on the basis of each member provisioning him- or herself according to taste and means, perhaps just with fresh goods such as milk or bread bought in common. Thus most kitchens require room for each resident's food to be stored, a cupboard or at least a shelf, and a space in the fridge for every person. Removing and consuming food from others' store areas is greatly disapproved of. Likewise it is expected that any toiletries in the bathroom not bought in common are not used by others.

Housework and Food Preparation
Housework is often a bone of contention. Standards vary considerably from person to person and few people are prepared to compromise too far on this most personal of domestic matters. For some dirtiness and untidiness are unbearable, for others the undertaking of chores, especially at others' insistence, is contrary to the ethos of relaxed, tolerant living which shared-flat life represents. The pressure usually increases with the number of residents. When flatmates are being chosen, it may be indicated that the successful candidate should be 'tidy' or 'relaxed', but the interpretation of these adjectives varies greatly.

In practice, it is often the case that standards are maintained towards the lower end of the range within the household, with some individuals contributing far more than others, although the less active on the housework front may take on other tasks such as overseeing the payment of bills. In a bid to ensure fairness rotas may be drawn up or areas of the home and particular tasks may be appointed to individuals, but these measures can be resented by some who dislike being thus regimented and the frequency and manner of executing chores can be flexibly and not mutually interpreted. Bedrooms are the responsibility of the occupants, but all public areas, including the entrance and stair within a block of flats, may be involved in rotas or apportioning, which can also include taking out rubbish, undertaking shopping for communal items and cooking.

Not surprisingly it is in the kitchen that conflict over housework is felt most. Dishes, work surfaces and appliances may be left uncleaned by those who insist on eating in a relaxed manner, to the annoyance of those who cannot prepare their food and eat it without cleaning them first. Cooking together on a rota basis may seem to make kitchen matters more straightforward.[54] Certainly it avoids a scramble for the cooker, but some have greater culinary skills or flamboyant tastes or bigger appetites than others, some may not like or cannot eat certain foodstuffs and some may resent having to be at home at a certain time on a certain day to prepare a meal, or even eat it.

The Physical Characteristics of the Property
For those with access to it, the shared accommodation provided by higher-education institutions tends to be less popular as it is seen as having uniform furniture and fittings, poor soundproofing, small rooms and few amenities.[55] In contrast, much private rented housing used for sharing in Scotland is essentially of good quality and fairly generous room proportions, being often of Victorian or Edwardian construction (such as the student flats of Marchmont, Edinburgh),[56] upgraded in the late twentieth century by grants and more recently in response to the increasing value of property. Most now have central heating – usually gas-fired or in the form of storage heaters – and double glazing (see the Appendix). There are also stringent standards laid down in law concerning many aspects of all types of 'houses in multiple occupation'.[57] The results of the University of Aberdeen survey certainly indicate that the days of widespread grotty student housing are gone.

Yet problems remain. Lack of maintenance on behalf of landlords still occurs. While one flat was described in the survey as 'luxurious' and others as 'very comfortable', in other instances homes can be cold and draughty, inadequately furnished or have unfashionable, drab or damaged décor. Even if flats are supplied with central heating, the residents may not be able to afford to use this sufficiently to keep the dwelling warm and damp free.[58] The greatest problems, however, are overcrowding and lack of space and facilities. Although minimum standards are laid down by law, landlords and the tenants themselves do not always respect these. Very large flats may not have the legally requisite number of bathrooms, cookers and sinks.[59] In smaller flats the living room and boxrooms may become bedrooms to reduce costs and the pressure on the rest of the dwelling can be very considerable as a result. The University of Aberdeen survey makes it clear that the lack of a living room is a concern for a considerable number of people. Nevertheless, it seems that many in shared dwellings, particularly students, are remarkably easygoing about the physical drawbacks of their homes, possibly because they see them as being temporary.[60]

Study
Study provision in student flats and houses, in the form of a desk and chair, table lamp, etc,[61] may be echoed in other types of shared dwelling. Homes shared by people who are all studying offer the opportunity for mutual support and understanding and, if the residents are undertaking the same courses, the chance to pool intellectual and other resources in preparation for submitting work or sitting exams. There may be a healthy degree of competition.

The converse is that stress levels can grow to unbearable proportions as household members feed off each others' study anxieties and excessive competition creates a lack of confidence in some. Most students can study in their institutions' libraries, for example, and many do so at exam time to reduce these problems. Some who prefer to study at home may insist on absolute quietness there which prevents others from relaxing in their customary ways. They may also wish to use heating more than others do as a consequence. Others may wish peace to study but find it impossible to achieve in a busy flat. Such problems are particularly likely in those instances where students and non-students live together.

Leisure and Socialising
Another of the major advantages of sharing dwellings is the opportunity to socialise and take part in leisure activities with one's companions. Very strong and lasting friendships can be formed between those sharing domestic space. This may be particularly important for those with little alternative support network nearby. Counselling, practical help and fun are often exchanged among flatmates, both within the home and, to a lesser extent, when out together, at for example the cinema, clubs or pubs.

Visiting and Receiving Guests
Shared flats usually witness a far greater number of guests and general interaction with those beyond the household – especially friends and study or work colleagues – than most other types of present-day Scottish home (Fig. 39.3). The sociable lifestyle is generally greatly appreciated by the

Figure 39.3 Flatmates and guests enjoying a barbeque in the drying green of a tenement, Edinburgh, 1990s. Source: Susan Storrier.

residents. Of course, for the more introverted or those who require their home to be a quiet one for whatever reason such sociability can be less of a positive trait. It is also a sociability which, in contrast to family life, pertains more to the everyday than special occasions. Although birthdays may be celebrated with one's flatmates, most residents go somewhere else for Christmas, for example.

Sexual Relationships
With lots of people visiting shared homes and single men and women sometimes living together in these, it is possible to get to know potential sexual partners in a relaxed and friendly atmosphere, quite different from dating, for example. It is not surprising, therefore, that for many people important sexual relationships and experience may come about through sharing a flat, with flatmates offering support and advice on sexual and relationship matters in the same way as on other topics.

The downside includes the number of residents at times effectively doubling as partners unofficially move in or at the least are round very often. This can put tremendous strain on the space, resources, sense of privacy and occasionally the level of security within the home. Cooking facilities may be overstretched by couples preparing more elaborate and 'family-like' meals for themselves than unattached flatmates do, while officially non-resident, but unofficially resident, partners sometimes do not contribute in any way to the costs of running the home or cleaning it and may not feel the same need to respect the fabric of the dwelling and the norms of the household. The finishing of sexual relationships between flatmates may cause a very uncomfortable atmosphere in the home and those relationships which go from strength to strength may cause other residents to feel up against an excluding 'club of two', quite against the notion of sociable equality. People living together with very different belief and moral systems may be offended by the sexual actions, or the sexual disengagement, of others. Although people usually seek to be as easygoing as possible about such matters, it is certain that sexual relationships can sometimes be a source of stress and conflict within shared homes.

CONCLUSION

It may seem that there is potential for those in shared flats and houses to encounter numerous negative experiences, yet most of the problems seem to be of relatively little consequence to the vast majority of people. Those difficulties which can undermine shared-dwelling life are few and occur relatively rarely, thanks to an unwritten, yet highly developed, code of good behaviour.

Indeed it appears that most people are extremely happy with their shared homes and households and content to live in this way for a considerable period of time. The lifestyle is commonly portrayed and reported to outsiders in a positive light – youthful, sociable, carefree, tolerant – and this is especially the case for students.[62] Workers and those on benefits may be older or facing the prospect of living in this way indefinitely, and they may have less choice over their companions and experience a lesser degree of compatibility with them. For these people, although the ethos of tolerance will help and the positive qualities of sharing will often still be appreciated, living in a shared home may sometimes be relished less. For a few it may become unbearable. Most people do eventually move on to other types of home, although few do so because they have become unhappy sharing.

Domestic life in Scotland is increasingly private and the concept of sharing, although still of importance, has been lessened for more than a century in many types of home and household, indefinite or temporary, family or otherwise. Yet the growth in shared flats and homes since the 1960s is in stark contrast to this. These homes are more about inclusion than exclusion, communality than individuality, exchange than isolation. The prime factor is perhaps to keep down living costs, but sharing can be regarded as a very positive experience in many other ways. And it seems that it is an ever more common domestic arrangement, not just for more people, but for longer and at more points in their lives.

APPENDIX: QUESTIONNAIRE ANALYSIS

Number of informants = 24
Informants living with the landlord: 33 per cent
Flats with central heating and double glazing: c84 per cent

What is shared among the flatmates (excluding rent)
utility bills (gas, electricity, phone): 66 per cent
food: 33 per cent (alternate shopping/communal eating)
household cleaning materials: 25 per cent (alternate shopping/kitty)
other costs: Sky TV c17 per cent; TV licence c8 per cent

What is especially liked about the present home and way of life there
car-parking space
close to town
close to university
independence
large bedroom
large kitchen
large sitting room
modern and comfortable
often play games and talk with flatmates
one's own home with one's own character
on the first floor
secured entry
sociable

What is especially disliked about the present home and way of life there
carpet needs replacing
colder than family home
flat is on the fourth floor
informant does not feel very comfortable and at home in the flat
kitchen
no living room

What the informant would change about the present home and way of life there
better cleaning
better heating
bigger rooms
convert loft
have a lounge: c17 per cent
nothing: 50 per cent
would prefer a ground-floor property

How shared-flat life compares with other forms of domestic life

POSITIVE COMMENTS (PARAPHRASED)
chance to practise and improve English
everyone the same age with similar tastes and no complaints about each others' lifestyle
very enjoyable way of life
more freedom
more fun
more privacy (locked bedroom)

NEGATIVE COMMENTS (PARAPHRASED)
can be less enjoyable than in a family home if one does not have the correct flatmates
food not so good as in a family home
harder to keep tidy
less intimate than in a family home
more chores to do

NOTES

1. See Chapter 20 Work in the Homes of Others.
2. See Chapter 7 Rest.
3. See Chapter 30 Custom.
4. See Chapter 33 'Time Out' Work.
5. See Rodger, 2001, 279.
6. Pinkerton, Windram, 1983, 59.
7. See Chapter 33 'Time Out' Work; also MacDougall, 1993, 23; Robertson, 1996–7, 70.
8. For information on these types of accommodation see Burnett, 1970; Gauldie, 1974; Rodger, 2001, 279; also Blair, 1991, 26, 65–6; Catford, 1975, 101; Easton, 1988, 23; MacDougall, 1993, 56–7; MacDougall, 1995, 31.

9. Alisma, 1918, 15; see also Grant, 1884, 488–9, for information on student lodgings in nineteenth-century Edinburgh; Brown, Moss, 1996, 16, for lodgings in Glasgow.
10. Coutts, 1909, 181, 333; Brown, Moss, 1996, 5, 12, 15, 16; for Edinburgh, Harrison, 1884, 56; for Aberdeen, Henderson, 1947, 64.
11. See Chapter 38 Residential Schools and Student Residences; Moore, 2000, 325.
12. Moore, 2000, 325. In 1919 only 14 per cent of women in Scottish universities lived in such accommodation, one third in lodgings, half at home; Anderson, 2000, 16.
13. See Footman, 1983, 75–6, for an outline of the twentieth-century residences built by Edinburgh University; Pinkerton, Windram, 1983, especially 69–73, for Mylne's Court in the Lawnmarket, Edinburgh; Brown, Moss, 1996, 40–1, 63 for residences in Glasgow; also the various universities' accommodation service web sites; see Reid, 2000, 36, 72, for experiences of living in nurses' homes.
14. See the University of Dundee web site.
15. See Chapter 34 'Time Out' Leisure.
16. See Chapter 35 Prisons.
17. For attitudes by communities in the past towards adult siblings living together see Blair, 1991, 34; Holmes, Finkelstein, 2001, 53.
18. Shared flats and houses are Houses in Multiple Occupation (HMOs). Today HMOs also comprise bedsits, lodgings and hostels. Nicholson, Wasoff, 1989, 12.
19. Nicholson, Wasoff, 1989.
20. The new definition of 'higher education' embraced non-university institutions. See Anderson, 2000, 167; this accelerated form of post-1945 expansion derived from spontaneous social demand and government policy, Anderson, 2000, 169, 171. Not only in the UK was this happening. Since the end of World War II substantial expansion of higher education in many industrial and developing nations had occurred. It was seen as necessary for economic and national growth and also drew strength from the idea of the individual's right to education: Morgan, McDowell, 1979, 2.
21. Morgan, McDowell, 1979, 15.
22. Anderson, 2000, 167.
23. Brown, Moss, 1996, 41.
24. For example, for the University of Edinburgh, Nicolson, Wasoff, 1989, 20, 71, state that in 1989 32.3 per cent of students lived in rented furnished accommodation, while 79 per cent of undergraduate students regarded sharing in the private rented sector as their first choice of accommodation. This was followed in popularity by living on one's own in a self-contained flat, then in a university-owned flat. There was a strong preference for communal independent living and a strong dislike for institutional or 'digs'-style accommodation, seen as a form of social control.
25. Nicholson, Wasoff, 1989, 21; halls of residence are also more likely to be used by science students than arts or social science students: Morgan, McDowell, 1979, 26, 28.
26. Brown, Moss, 1996, 64.
27. Nicholson, Wasoff, 1989, 14.
28. Anderson, 2000, 171.
29. Callender, Kempson, 1996, 52, found that 10 per cent of all students in Britain were home buyers while 53 per cent lived in private rented accommodation.
30. See Nicholson, Wasoff, 1989, 29–31, for such tenant-landlord relationships.
31. Morgan, McDowell, 1979, 16.
32. See Rivers, et al, 1992, 152–3, for example.
33. Morgan, McDowell, 1979, 57.
34. Morgan, McDowell, 1979, 16.
35. Morgan, McDowell, 1979, 14.

36. Nicholson, Wasoff, 1989, 31.
37. Nicholson, Wasoff, 1989, 15.
38. See *Press and Journal*, 11 January 2005, 'What has changed over recent years is that there is no longer any social stigma in living in rented homes. Landlords provide good, decent accommodation at a fair price where tenants are happy to lie for as long as it suits their personal circumstances, or until they can afford to buy a home.'
39. See Anning, et al, 1980, 122–3, 160, 197, 199, for information about the nature of and motivations behind squatting.
40. Nicholson, Wasoff, 1989, 31, write that only 7 per cent of University of Edinburgh students who shared flats in the late 1980s shared a room with a non-related adult.
41. Generally speaking the Council Tax bill – which is usually very substantial – for a property is shared among those in employment. Unemployed people often receive benefit to cover this and exempted are some people under 18, students, student nurses, apprentices, school leavers, skillseekers (someone under 25 receiving government training) and people with severe mental disabilities and their carers. People on low wages may receive some help but this does not include the water and sewerage charge. Difficulties occur when mixing these categories of people within one dwelling, resulting in those who are entitled to exemption sometimes having to pay a share of the tax. They may be prepared to pay the extra if it means they can live where they want or with whom they prefer. Sometimes the Council Tax is 'included' in rent charged by a (usually resident) landlord, which is often more palatable. For information on the Council Tax see the Scottish Executive web site, *Council Tax in Scotland: A Guide for Students, Student Nurses, Apprentices, Skillseekers and School Leavers*.
42. See the Appendix for shared costs among students at the University of Aberdeen.
43. Morgan, McDowell, 1979, 39, 40.
44. For example, the University of Dundee Student Accommodation Service web site, *Welcome*, advises looking on the Internet for accommodation and also at the City Council, solicitors' offices, etc. Properties should be inspected, especially in regard to personal security. One should choose flatmates carefully, sort out money matters, discuss housekeeping, insurance, fire safety, TV licence (if there is a joint tenancy only one is needed per flat, otherwise there must be one licence for each set), check the tenancy agreement and discuss what will happen if one person leaves and be clear about inventories and inspections.
45. See, for example, the University of Edinburgh Student Accommodation Service web site, *A Guide to Student Housing: Finding Student Flats in Edinburgh*.
46. See Nicholson, Wasoff, 1989, 28.
47. Callender, Kempson, 1996, 46.
48. Nicholson, Wasoff, 1989, 18, state that in late 1980s students in Edinburgh on average paid 42 per cent of their income on rent; since then the relative financial position of many students has worsened through the introduction of the Council Tax and first the reduction in student grants then their replacement with student loans. Today the rent for a room in a shared flat in central Edinburgh (excluding bills and Council Tax) can be as high as £90 per week. In contrast accommodation is available in Dundee for as little as £40 per week. See the University of Dundee Accommodation Service web site.
49. Nicholson, Wasoff, 1989, 39, 50–2, 55, report that in Edinburgh in the 1980s joint and several tenancies were most frequently used for shared flats (54 per cent). A main tenant agreement granted to one person who later sub-let accounted for 24 per cent.
50. See Chapter 40 Communes.
51. Gray, 1998, 121.

52. There is much trust, and a lack of knowledge of the law, involved in these tenant-landlord relations. Many tenants not only have insecure tenancies, which may leave them open to being evicted at will by the landlord, or have no written tenancy agreement at all, but may in addition receive no type of receipt for rent payments made. Despite this, most tenant-landlord relations appear cordial. See Nicholson, Wasoff, 1989, 39, 50–2, 55.
53. The University of Dundee Accommodation Service web site, December 2004, states that one in four students is the victim of theft and recommends that all students in shared homes obtain insurance covering their possessions.
54. Such rotas are far from new. See Gray, 1998, 108 for accounts of timber girls living together mostly in huts during World War II. Margaret Fraser describes the rota for cooking at the weekends.
55. Nicholson, Wasoff, 1989, 79.
56. Nicholson, Wasoff, 1989, 28.
57. See the Scottish Executive web site, *Houses in Multiple Occupation: A Guide for Landlords*; North Lanarkshire Council web site, *Multiple Occupation: Useful Information*.
58. Nicholson, Wasoff, 1989, 31.
59. Nicholson, Wasoff, 1989, 32–6.
60. Nicholson, Wasoff, 1989, 76.
61. See the Scottish Executive web site, *Houses in Multiple Occupation: A Guide for Landlords*.
62. See Nicholson, Wasoff, 1989, 74.

BIBLIOGRAPHY

Alisma (pseud). *Reminiscences of a Student's Life at Edinburgh in the Seventies*, Edinburgh, 1918.
Anderson, R D. Scottish universities. In Holmes, H, ed, *Institutions of Scotland: Education: Scottish Life and Society: A Compendium of Scottish Ethnology, vol 11*, East Linton, 2000, 154–74.
Anning, N, Wates, N, Wolmar, C. *Squatting: The Real Story*, London, 1980.
Blair, A. *More Tea at Miss Cranston's: Further Recollections of Glasgow Life*, London, 1991.
Brown, A L, Moss, M. *The University of Glasgow, 1451–1996*, Edinburgh, 1996.
Burnett, J. *A Social History of Housing, 1815–1970*, London, 1970.
Callender, C, Kempson, E. *Student Finances: Income, Expenditure and Take-Up of Student Loans*, London, 1996.
Carruthers, A, ed. *The Scottish Home*, Edinburgh, 1996.
Catford, E F. *Edinburgh: The Story of a City*, London, 1975.
Coutts, J. *A History of the University of Glasgow: From its Foundation in 1451 to 1909*, Glasgow, 1909.
Currie, H, Third, H, Satsangi, M, Brown, A. *Good Practice in the Use of Licensing Schemes for Houses in Multiple Occupation in Scotland*, Edinburgh, 1998.
Easton, D. *By the Three Great Roads: A History of Tollcross, Fountainbridge and the West Port*, Aberdeen, 1988.
Footman, R. *Edinburgh University: An Illustrated Memoir*, Edinburgh, 1983.
Gauldie, E. *Cruel Habitations: A History of Working-Class Housing, 1780–1918*, London, 1974.
Grant, Sir A. *The Story of the University of Edinburgh: Its First 300 Years*, vol 1, London, 1884.
Gray, A (eds U Roberston, J Robertson). *Timber!* (Flashbacks no 7), East Linton, 1998.
Harrison, J. *Oure Tounis Colledge: Sketches of the History of the Old College of Edinburgh*, Edinburgh and London, 1884.
Henderson, G D. *The Founding of Marischal College, Aberdeen*, Aberdeen, 1947.

Holmes, H, Finkelstein, D, eds. *Thomas Nelson and Sons: Memoirs of an Edinburgh Publishing House* (Flashbacks no 14), East Linton, 2001.
The Lord Provost's Commission on Social Exclusion. *One City: The Lord Provost's Commission on Social Exclusion, Final Report*, Edinburgh, 2000.
MacDougall, I, ed. *'Hard Work, Ye Ken': Midlothian Women Farmworkers* (Flashbacks no 2), East Linton, 1993.
MacDougall, I, ed. *Mungo Mackay and the Green Table: Newtongrange Miners Remember* (Flashbacks no 4), East Linton, 1995.
Moore, L. Women in education. In Holmes, H, ed, *Institutions of Scotland: Education: Scottish Life and Society: A Compendium of Scottish Ethnology, vol 11*, East Linton, 2000, 316–43.
Morgan, D, McDowell, L. *Patterns of Residence: Costs and Options in Scottish Housing* (Research into Higher Education monographs), Guildford, 1979.
Nicholson, L, Wasoff, F. *Students' Experience of Private Rented Housing in Edinburgh* (Department of Social Policy and Social Work/Student Accommodation Service, University of Edinburgh), Edinburgh, 1989.
Pinkerton, R M, Windram, W, J. *Mylne's Court: 300 Years of Lawnmarket Heritage*, Edinburgh, 1983.
Reid, L. *Scottish Midwives: Twentieth-Century Voices* (Flashbacks no 12), East Linton, 2000.
Rivers, T, Cruickshank, D, Darley, G, Pawley, M. *The Name of the Room* (BBC), London, 1992.
Robertson, F W. The working life of Scots gardeners between the Wars, *ROSC*, 10 (1996–7), 67–85.
Rodger, R. *The Transformation of Edinburgh: Land, Property and Trust in the Nineteenth Century*, Cambridge, 2001.
Wood, S. Education in nineteenth-century rural Scotland: An Aberdeenshire case study, *ROSC*, 7 (1991), 25–33.

Web sites
University of Edinburgh Student Accommodation Service, *A Guide to Student Housing: Finding Student Flats in Edinburgh*, <http://www.accom.ed.ac.uk/>
North Lanarkshire Council, *Multiple Occupation: Useful Information*, <http://www.northlan.gov.uk/living+here/housing+and+housing+advice/housing+advice/houses+in+multiple+occupation+useful+information.html>
Scottish Executive, *Council Tax in Scotland: A Guide for Students, Student Nurses, Apprentices, Skillseekers and School Leavers*, <http://www.scotland.gov.uk/library3/localgov/ctst-00.asp>
Scottish Executive, *Houses in Multiple Occupation: A Guide for Landlords*, <http://www.scotland.gov.uk/library5/housing/hmogfl-00.asp>
Scottish Executive, *Housing and Anti-Social Behaviour: The Way Ahead*, <http://www.scotland.gov.uk/library2/doc10/hasb-00.asp>
Scottish Executive, *Scottish Household Survey Bulletin no 8*, <http://www.scotland.gov.uk/library5/housing/shs8-00.asp>
Scottish Executive, Homelessness Task Force, *Single-Room Rent: The Research Evidence Since 'Benefit Shortfalls'*, Paper HTF 4/1/00b, <http://www.scotland.gov.uk/Topics/Housing/homeless/htf>
The Scottish Office, *Students in Further Education Colleges, 1997–8: Provisional Figures*, <http://www.scotland.gov.uk/news/releas99_2/pr0282.htm>
University of Dundee Accommodation Service and Student Advisory Service, <http://www.dundee.ac.uk/adviceguidance/accommodation.htm>

40 Communes

DAVID RICHES

Throughout Western history, living in communes has generally accompanied the rejection of the mainstream. A 'commune' implies a group sharing access to common resources, and this signals opposition to *individualism*, the social value which has been dominant in the West from the dawn of Christianity. Living communally also generally entails intensive and mutually supportive social interaction. People who shun the dominant society need this both to justify their commitment to an 'alternative' way of life, and to defend themselves from being treated badly by those unsympathetic to their cause. In non-Western societies communal living is commonly the prevailing way of life. But in the West, where esteem generally goes to people who independently make their lives from their own personal resources, it is unusual and arouses suspicion.

The historical bases of communal living in the West are varied.[1] In North America, for example, three movements can be discerned in the past. The Hutterite and Amish communes, their lineages dating back to the eighteenth century, stem from religious persecution in Europe. The Shakers, and similar 'communities' based on a peasant livelihood of small-scale production and sale, arose in the nineteenth century in response to a sense of individual dislocation stemming from rapid industrialisation, while the 1960s wave of commune formation can be seen as connected to people's experience of disenchantment relating to increased secularisation and globalisation. There is a great deal of variety, but communes from all historical periods commonly include the sharing of land, and also the use by members of at least some communal buildings, typically for eating or sleeping.

In the post-1960s period in the West, two main types of experiment in communal living have been seen, though some examples seem to straddle both categories. Firstly there are ideological communes, which are marked by cultic beliefs and normally manifest a fairly rigid and hierarchical internal organisation and clear social barriers between the members and the outside world. The much-reported apocalyptic communes such as the Branch Davidians, led by David Koresh (Waco, Texas, USA), and the People's Temple (which removed to Jonestown, Guyana) are extreme instances where members were involved in terrible violence. Most ideological communes – whose beliefs typically aggregate elements from a mixture of various world religions – are in intent relatively benign, the Unification Church (The Moonies) or Bhagwan (The Rajneeshees) being examples.

This chapter will concentrate on the second type of experiment in communal living, of which there are several instances in Scotland. This is the holistic commune, which seeks to harness economic and social behaviour to a principle of ecological ethics commonly expressed in spiritual terms. Ecological ethics upholds the idea of a living environment, which subsumes all other things, human and non-human, organic and inorganic. Reversing the mainstream Western notion of humans dominating the environment, this celebrates the concept of the environment caring for people, providing its resources are used modestly and with respect. The spiritual principles accompanying such ethics are mainly drawn from pre-Christian (pagan) and tribal (especially North American Indian) beliefs, and are perhaps most famously voiced through the concept of Gaia, which expresses the notion of a divine and all-embracing universe.[2] Holistic communes tend to have rather egalitarian and fluid social organisations, and be relatively open to the wider world, towards which many consider they have an educational duty.

Holistic communes, which see themselves in the vanguard of the utopian dream of environmental sustainability, may be found in various parts of Scotland, mostly in rural areas where they bear witness to a oneness with the environment through a non-exploitative, sympathetic ecological practice known as permaculture. The Diggers and Dreamers *Handbook* for 2000–1, which describes most examples of such communities in Britain, lists eight holistic communes in Scotland, although there exist others which are not included.[3] Findhorn, close to the Moray Firth, is internationally renowned as one of the first and most successful experiments in holistic communal living and is featured frequently in the *Handbook*, while Monimail in north-east Fife, the other commune examined in this chapter and one more typical of the commune movement, despite functioning rather similarly to Findhorn, albeit on a much smaller scale, is not mentioned.

The Diggers and Dreamers *Handbook* is especially useful in exposing the variety of holistic communes in Scotland. It tabulates each commune in terms of its size, its social composition (especially the ratio between adults and children) and whether or not it welcomes new members; most, in fact, do. Also listed is an indication of the nature of a group's communality, for example, whether the members share income or partake of communal meals. Substantial variation can be noted here. In addition there is a summary of the ethical principles by which each commune lives, ranging from the nature of its dietary regime – vegetarianism is popular – through to the type of spirituality it holds, if any. It is obvious from the photographs in the book that there is a very wide range of dwelling structures occupied: the Laurieston Hall group with its imposing mansion in Kirkcudbrightshire; Findhorn, which began as a single caravan but now incorporates architecturally impressive, ecologically sustainable houses; and many other communes with more rudimentary and improvised dwellings, including cottages, tipis and benders (tents formed around a structure of branches).

A notable feature of the holistic communities in Scotland, which reflects the intense life experience in many communes, is their small size.

Findhorn, the largest, once possessed 300 fully incorporated members, but in 2001 the number is 100, with another 500 or so living in the neighbourhood as 'associates'. Most other Scottish communes number fewer than 30 members. Yet the modest numbers are compensated for by extensive communication networks. People from Monimail, for example, visit both Laurieston Hall and Talamh (an eco-anarchist community in Lanarkshire) to exchange ideas and experiences and have fun. Many groups publish pamphlets about themselves and about sustainable living. Findhorn is at the fore in this, with the Findhorn Press boasting some fifty book-length titles in print and an international network of distributors. Most important of all is the worldwide web which has numerous sites dealing with issues relating to sustainable living and helps holistic communes get and keep in touch with one another on a global basis. Findhorn has its own, impressive web pages.

FINDHORN

Findhorn, 30 miles (48 km) east of Inverness, occupies a former caravan site and the surrounding duneland, now reclaimed for agriculture.[4] Peter and Eileen Caddy, two of the three founders, came with their children to the site in 1962, and their caravan still stands within the commune's extensive complex of buildings and garden ground (Fig. 40.1). Near to the caravan is a large patch of ground on a shaded and woody bank, which remains

Figure 40.1 A treasured icon of the Findhorn commune is the caravan in which Peter and Eileen Caddy arrived with their family in 1962. The founders' first dwelling is parked in a tranquil location beside the flourishing organic vegetable garden. Source: D Riches.

Figure 40.2 'Today, the Community has diversified into more than 30 different businesses and initiatives, and provides a valuable insight into sustainable living. It is a place where everyday life is transformed by living, working and learning with spiritual values.' Source: extract from Findhorn Foundation brochure, 2001.

uncultivated. This area is given over to the devas, the plant spirits, who in the first days of the commune conveyed an understanding of the mystical nature of agricultural processes to the Caddy's friend and fellow founder, Dorothy MacLean. The recognition of these spiritual beings, along with Findhorn's location at the intersection of lines of special energy within the earth (ley lines) and its being a 'Light Centre' in direct connection with the cosmos, is viewed by the residents as the propitious combination responsible for the rapid emergence of the commune. Findhorn not only flourished but found fame quickly too, in part the product of extensive international publicity about the almost supernaturally sized vegetables grown on the site.

These days Findhorn has become an educational commune, whose members and associates – originating from all parts of the world, but especially England, other northern European countries and North America – provide, supervise and facilitate a vast range of courses, programmes and conferences, ranging from pottery workshops through to week-long deliberations on topics such as 'Sex and Spirituality'. Those who avail themselves of such educational provision again come from far and wide. At Findhorn in 2001 the gardens continue to exist, but the commune's emphasis is more on bearing witness to harmonious and spiritually attuned modes of living which can serve as a template for similar experiments elsewhere. As one resident put it, 'These days we grow people.' (Fig. 40.2).

The present-day Findhorn site, which the public is encouraged to visit, is criss-crossed by metalled and non-metalled paths, illuminated at night by subdued lighting and divided into distinct domains including communal and craft buildings and residential areas. There is motorised access to most parts of the site but car-parking facilities are located near the entrance, as are a large number of caravans, available both to people attending courses and workshops and for rent by the general public. Close by there is a 'village green' with play facilities and a shop which comprises a grocery (with a large range of organic produce), an extensive bookshop (concentrating mainly on 'alternative' themes) and a purveyor of remedies based on herbal and other 'traditional' knowledge. Across from the shop, which also sells bread made by the commune's bakery, a former toilet block is now used by a company manufacturing flower essences (Fig. 40.3). Receptacles for recycling glass, paper, aluminium and other refuse are also positioned in this zone.

Universal Hall, surrounded by trees, is the largest building on the site. It comprises a visitor centre, a café, a drama studio and a pentagonal-shaped auditorium decorated with huge murals. The auditorium can seat around 200 people and is available for use by both Findhorners and outsiders for theatre productions and musical events. The outer walls of the building have the sun, the moon and rivers sculpted in the stone work, while the entrance doors are embellished with stained glass and overlook a flat area decorated with mosaics. Another communal building, closed to the casual visitor, is the Community Centre. This is constructed in part from recycled telegraph poles and includes a large well-appointed communal refectory with solid wooden tables. The excellent food is appreciated by those who choose to eat there, but people are not prevented from eating in their houses if they wish (Fig. 40.4).

Figure 40.3 The former toilet block now houses one of the thriving 'alternative' businesses which have been set up within the Findhorn commune's precincts. Source: D Riches.

Figure 40.4 One of the few communal areas not accessible to the general public, the Community Centre is the focal point for casual relaxation within the Findhorn precinct. Source: D Riches.

Walking or cycling along the paths in Findhorn one encounters clusters of residential buildings of all imaginable architectural styles whose construction reflects the community's present emphasis on ecologically friendly living. Caravans are gradually being phased out and more prominent now are buildings with turf-covered roofs bearing solar panels and walls whose timber, the commune insists, comes from renewable forests (Fig. 40.5). The commune covers the cost of these developments principally by private donations, including some from members, and by its many and varied educational and experience programmes which attract large numbers of 'paying guests' from the outside, but this has not prevented the accumulation of quite large amounts of debt. The Field of Dreams, a number of large, custom-built family houses where both members of the commune and sympathisers live, is the latest area to be developed. It includes the first house to be built from straw bales, following a prototype which now serves as a large tool shed. Looming over the Field of Dreams is the wind generator, which supplies a good proportion of the commune's electrical power.

Another cluster of buildings in Findhorn houses facilities for youth, including a large wooden house which is the main home of up to ten young people of student age. There is also the turf-roofed Youth House, a prominent recreational building beside the village green whose structure incorporates recycled wood pulp; like many ecologically experimental buildings in Findhorn, this has breathable walls, fabricated so that internal water vapour can find egress. The most unusual residential area in the commune is a group of five family houses of differing sizes, each one created from wooden whisky vats obtained cheaply from nearby distilleries changing over to stainless-steel vessels (see Fig. 40.6).

Figure 40.5 Recently constructed accommodation for guests attending courses and conferences at the Findhorn commune is ecologically friendly in design, including turf-roofed houses for super-efficient insulation. Source: D Riches.

Figure 40.6 One from the cluster of whisky-barrel houses which are a distinctive feature of the Findhorn architectural landscape. When nearby distilleries changed to stainless steel vats the wooden structures could be bought cheaply and were converted into family accommodation. Source: D Riches.

Figure 40.7 One of the several 'sanctuary' buildings within the Findhorn commune, which is sited in a peaceful garden. Inside, seating is arranged in a circle, available for anyone to use in meditation. Source: D Riches.

Other intriguing buildings serve as 'sanctuaries' both for formal and informal meditation and for quiet reflection, and secluded outdoor spaces are likewise dedicated to peace, harmony and contemplation (see Fig. 40.7). Eileen Caddy, now in her eighties and still a resident, occupying an 'ecological house' built for her by her son, attends many of the commune's meditation events. Because the physical location of the Findhorn commune is significant, both historically and spiritually, there is no question of moving the community to another site. Findhorners wanting total seclusion, however, do have some options. There is an outpost of the commune on the Isle of Erraid, off the west coast of Mull, where around a dozen people live almost totally self-sufficient lives, supplementing their livelihood by candle making. Alternatively, the Trees for Life charity, which leases premises within the Findhorn commune, welcomes volunteers in its project to replant the Caledonian forest in parts of the Highlands.

MONIMAIL

Monimail, in north-east Fife, is smaller in numbers than Findhorn and also more rudimentary in its infrastructure and facilities. In the 15 or so years since its foundation the commune has seen a steady accretion and improvement of buildings, although many of these originally fulfilled other functions, such as the three big huts which were formerly part of a school. The commune also incorporates a tower house, part of the Bishop's Palace, which has been renovated and which now includes an exhibition room. There is more accommodation in the Walter Segal building, a kit house with energy-efficient design. Some members make their homes in caravans or garden sheds on the site. There are further large huts which provide communal space and accommodation for guests. Members generally take meals together.

Monimail describes itself as an Earth-based commune where individuals can connect to one another and to the land, which offers a healing place for people (including people with disabilities) to live. As with Findhorn, there are outsiders associated with the commune, particularly teachers and craftspeople who provide courses for the public on site. The dozen or so members, working in orchards and gardens, are conscious of the beauty of the place, with its walled garden and a driveway flanked by laurel bushes. In comparison to Findhorn, facilities are simple, but there are ample compensations, such as sitting around a log fire in the orchard at the end of the day, perhaps drinking wine and playing music.

Findhorn and Monimail, alike in some ways, unlike in others, share with many other 'intentional communities'[5] problems of social organisation. These make daily life in communes on the one hand exhilarating but on the other suffused with tension and potential conflict. Many people at Findhorn have been commune members for decades, but it is significant that most others come and leave within a fairly short period of time. Taking into account all members, five years is the average length of stay at Findhorn, and around four years in Monimail. The contradictions of social existence in holistic communes would appear to revolve around three problem areas. Firstly the commune represents a way of life alternative to that of mainstream society, yet commune members continually have to interact with the mainstream in order to keep their communities in existence. Secondly, communes attempt to put an alternative way of life into practice, yet nearly all adult members will have been socialised in the mainstream world and most will have taken up the commune life recently. Finally, the holistic commune usually celebrates the social ethics of egalitarianism and individual autonomy, yet some people can take these values too far and it is difficult to secure their co-operation in daily tasks and in consensual decision making.

The difficulties in having to interact with the mainstream world stem from the fact that holistic communities, for all their ambition, are never totally self-sufficient. In many communes members depend on social security payments, or take part-time work in the mainstream world, for example as teachers, craftspeople, tree surgeons, carpenters or assistants in 'alternative' retail shops. Some individuals are unhappy about being dependent on the mainstream in this way. As a strategy, a number of communes have commoditised their way of life. Findhorn, once based entirely on horticulture, now declares that its educational purpose is its *raison d'être*. To sample the 'Findhorn experience', by taking courses lasting for either a week or for three months, outsiders pay large sums of money. They are often not accommodated on the main site but instead at Cluny Hill College, a large former hotel that the commune owns in the nearby town of Forres. However, this smacks of materialism, so over the years participants in some courses have been permitted to pay only what they can afford – to the commune's financial detriment. In addition, running the courses, as well as keeping all the other aspects of commune life in order, is hard work, even though the visitors contribute vital labour, as gardeners or kitchen workers. One former commune member from Monimail likened the work regime there to the Protestant work ethic.

The additional difficulty for holistic communes is that without rigid social barriers prohibiting social intercourse with the outside, commune ideals can become contaminated by the mainstream. Many communes are quite selective concerning whom they admit to full membership, yet people periodically move back and forth between the commune and the mainstream world, as their social, economic, family and spiritual priorities change and mature. A paradoxical situation is that newcomers arrive full of idealism, and are disappointed that long-standing members have resorted, at least in part, to mainstream procedure. Thus one newcomer to a commune was dismayed that instead of total harmony he encountered political factionalism, and another felt that, in place of equality and consensus, hierarchical principles were employed to effect decisions. The fact of the matter, too, is that the life of self-sufficiency, as another commune member admitted, can be, 'lonely, tough and hard going'. This is the case, even at Findhorn. There members enjoy free food, accommodation and access to many facilities, but one person (interviewed in 2001) described the meagre £50 per week 'pocket money' he received from the commune as amounting, on his part, to a 'sacrifice' in relation to a chosen way of life.

In the holistic commune, egalitarianism and consensual decision making can become starkly opposed to another cherished social value, that of autonomy (of being permitted to do 'one's own thing'). Holistic communes generally provide a supportive framework within which people can explore their own individuality, and this type of community is often joined by those who seek to escape what they see as the rigid routines of mainstream life. The problem for many communes is that they have had to develop their own forms of routine (kitchen rotas, gardening details, etc) to deal with practical matters, and yet must also find a way to accommodate someone who, for example, wants 'time off' to meditate. Likewise consensual decision making is the ideal, yet how can decisions be made when every voice must be respected, and voices point in different directions? At Findhorn, one member (giving thanks for the wise heads who inspired the correct course of action) recalled the old days when it took two years to decide on the type of a new kitchen carpet.

Placing emphasis on the 'spiritual' is the chief means by which communes deal with these contradictions. The spiritual connotes another, far more valuable domain, which transcends the problems of ordinary existence and makes them bearable. Many holistic communes, accordingly, are known not just for their 'deep ecology', but also for their New Age religious commitment. This especially speaks of the divinity within each person and the oneness of this divinity with an embracing and purposeful cosmos. New Age spirituality is generally positive and optimistic. New Agers see themselves at the forefront in relation to an impending 'new age' – the Age of Aquarius – when people will make contact with the 'divinity within'. They draw on old wisdoms as inspiration for this quest. Peace, harmony and love mark the Age of Aquarius, and North American Indian and European pagan traditions, as well as Eastern philosophies, are deemed to bear witness to, and provide the practical guidance for the renunciation of hate and fear. A celebration of feminine essence, connoting nurturance and non-aggression, which is present in all people, is a major strand in New Age thinking. Findhorn is commonly described as a New Age commune, but spiritual ideas seem much less important in other communes in Scotland.

Another way in which holistic communes smooth over the problems of daily life is by emphasising the social value of tolerance. Egalitarian ethics shuns the idea of authority, which means that it is difficult to control the miscreant or the unorthodox, so let unorthodoxy be celebrated! Accordingly, in New Age communes each person is permitted to develop his or her own personal or spiritual path. In the less spiritual communes tolerance is valued as an end in itself. For example, the self-help manual for communal living, *Creating Harmony: Conflict Resolution in Community*, has a chapter focusing precisely on 'tolerance and solidarity'.[6] The volume as a whole stands as an honest recognition of the difficulties inherent in living communally, with some sections devoted to techniques of conflict resolution and others to examples of communities of various types throughout the world where harmonious co-existence is a reality. The metaphor which appears throughout the book is that of 'healing', with reference to both the community and the individual.

The respect which holistic communards accord to children perhaps encapsulates some of these values, especially as commune children are generally treated as belonging to the commune as a whole as well as to their particular parents. Young Findhorn children are lucky in that the school they attend alongside local 'straight' children is a Rudolph Steiner School which propounds social and spiritual values rather close to New Age type principles. Commune children elsewhere may be less fortunate, and their lifestyle and sometimes 'hippy' appearance can lead to mainstream children treating them unsympathetically. Spontaneity and emotionality, often especially marked in children, are important values in holistic communes, which generally see education as being the process of a child's self-development. For teenagers in particular, the tolerance and the lack of orthodoxy is something which makes commune life more attractive than the mainstream world, where, as they perceive it, more rigid routines apply. The general absence of hierarchy between adults and non-adults is also appreciated.[7] Most young people will eventually leave the commune and make their way in the mainstream world, for which they will be well enough prepared as, in Scotland, few commune children will not have experienced mainstream schooling at secondary level. But very many will spread a positive message about a way of life which proves that, in relation to prevailing mainstream routines, there are viable alternatives.

NOTES

1. Kanter, 1972, 2.
2. The name of the Greek earth goddess was invoked by James Lovelock to describe his theory of the Earth and its surrounding atmosphere as a self-regulating system. Lovelock did not in any way posit Gaia as a spiritual notion – this is something that the New Age beliefs associated with many holistic communes have added.
3. Diggers and Dreamers 2000–1, 2000.
4. See Riddell, 1990. Jeremy Tucker, the author's PhD student, carried out anthropological research at Findhorn in the mid 1990s. Jeremy died before completing his thesis, and this chapter is dedicated to his memory.
5. 'Intentional community' is the sociological term referring to communities which exist not through natural historical evolution, but as an act of will.
6. Jackson, 1999.
7. Diggers and Dreamers 2000–1, 2000, 30–1.

BIBLIOGRAPHY

Devall, B, Sessons, R. *Deep Ecology: Living as if Nature Mattered*, Utah, 1985.
Diggers and Dreamers 2000–1. *The Guide to New Communal Living*, London, 2000.
Jackson, H, ed. *Creating Harmony: Conflict Resolution in Community*, Holte (Denmark), 1999.
Kanter, R M. *Commitment and Community: Communes and Utopias in Sociological Perspective*, Cambridge (Mass), 1972.
Lovelock, J. *Gaia: A New Look at Life on Earth*, Oxford, 1979.
Pepper, D. *Communes and Green Vision: Counterculture, Lifestyle and the New Age*, London, 1991.
Riches, D, Prince, R. *The New Age in Glastonbury: The Construction of Religious Movements*, Oxford, 2000.
Riddell, C. *The Findhorn Community: Creating a Human Identity for the Twenty-First Century*, Forres, 1990.
Rigby, A. *Communes in Britain*, London, 1974.
Sutcliffe, S. A colony of seekers: Findhorn in the 1990s, *Journal of Contemporary Religion*, 15/2 (2000), 215–31.
Walker, A, ed and comp. *The Kingdom Within: A Guide to the Spiritual Work of the Findhorn Community*, Findhorn, 1994.

41 Religious Communities

MARK DILWORTH

Although monasticism probably originated in the East, it became established as a Christian way of life after St Anthony (251–356), in Upper Egypt in the late third century, began to live as a solitary, devoting his time to prayer and manual work. His influence grew and many young men imitated him. It was a natural development for groups of these hermits to begin meeting at regular intervals for the Eucharist and other purposes, and they were then organised by St Pachomius (c286–346) into eremitical communities.

Upper Egypt was an ideal region for a simple monastic life, for its climate allowed the monks to live all year round in primitive shelters and on a very sparse diet. Soon monasticism spread to other lands but met with difficulties in a less favourable ambience. In southern Italy it was a luxuriant but disordered growth, with some elements less than edifying. Accordingly St Benedict (c480–547) wrote a rule for the small monasteries he had founded.

Benedict placed the major emphasis on community life, stressing adherence to common practices and obedience to the superior, the abbot. Monastic life of its nature entailed poverty and chastity and in organised communities obedience was necessary. The vows taken by Benedict's monks were Stability (lifelong commitment to a particular community), *Conversatio Morum* (untranslatable but meaning the monastic way of life) and Obedience. Poverty and chastity were taken for granted. Perhaps most important of all, he counselled moderation in every area. So ordered and moderate was his rule that, over the centuries, it supplanted all others.[1]

Efforts were made, with varying success, to remedy the lack of co-ordination and form groups of monasteries with the same observance and ethos. There were also attempts at reform when laxity or abuses crept in. One reform, that at Cluny, achieved enormous success as countless houses adopted its observance, while other reforms, notably the Cistercian, formed new and fervent groupings. By about 1100, not only was St Benedict's rule almost universally observed, but the new centralised groupings were clearly in the ascendancy. The canons regular, priests living a monastic life and following the rule of St Augustine, had adopted Benedictine customs. In 1215 the 4th Lateran Council decreed that in each region all uncentralised monasteries were to hold general chapters together. Even the Carthusians followed the basics of Benedict's rule, although they differed in being uncompromisingly austere and living as hermits in community.[2]

Monasticism, reaching these islands in the fourth century, shared many characteristics of the Egyptian model, in particular austerity according to individual choice and including voluntary exile, *peregrinatio pro Christo*. The most important foundation in Scotland was that of St Columba on Iona.[3] These Scotic, Gaelic-speaking monks evangelised various parts of Europe and founded their 'Celtic' type of monasteries. Gradually, however, they adopted aspects of Benedict's rule and from about 1100 formed a centralised Benedictine congregation in south Germany known as the Schottenklöster.[4]

The new monastic groups (for example, Tironensians, Cistercians, Augustinians, Premonstratensians) at this time were at the peak of their fervour and vitality. Invited into Scotland by rulers and magnates, they made dozens of foundations and supplanted the less organised Celtic monasteries (although certain Celtic communities, such as that at Inchaffray, Perthshire, survived but adopted Augustinian ways). Later, in the 1420s, Carthusians also arrived. Until the Reformation put an end to monastic life, there were 40 monastic communities in Scotland.[5]

Scottish monasticism, however, did not disappear completely at the Reformation. The Irish Schottenklöster in Germany, taken over by Scots monks in 1515, survived until 1862, to be succeeded by Fort Augustus Abbey in Inverness-shire.[6] In 1946-8 two more monasteries were founded: Nunraw (Cistercian), 1946, near Haddington, East Lothian; and Pluscarden (Benedictine), 1948, near Elgin, Moray.[7]

Monastic life, however, is only one part of the spectrum of religious life. In the early Middle Ages, monasticism was, for all practical purposes, the only form, but in modern times – worldwide as well as in Scotland – it is a small fraction of the whole. The friars (chiefly Franciscans, Dominicans and Carmelites) arriving in Scotland in the thirteenth century were a different kind of religious, with an ethos which encouraged mobility.[8] Then, from the sixteenth century on, congregations devoted to active apostolates (education, care of the sick, foreign missions and so on) have proliferated.

This development is especially exemplified by religious women. The relatively few and small communities of nuns in the Middle Ages were strictly enclosed but, beginning with the French Daughters of Charity in the seventeenth century, women religiouses working outside their convents have greatly outnumbered men.[9] When religious life returned to Scotland with the arrival of the Ursuline sisters on the outskirts of Edinburgh in 1834, it was very different from the pre-Reformation scene, for it was predominantly female and almost entirely non-monastic.[10]

The number of Roman Catholic male religiouses in Scotland reached a peak in the 1960s, that of women religiouses perhaps later, and has since been declining.[11] A parallel development in the opposite direction has also been taking place, with diverse groups adopting a communal lifestyle. These vary from being tightly knit to merely having certain weekly exercises in common. The types include those with members united by their Christian faith (and not only Roman Catholics) or following the practices of an Eastern religion or devoted to aspects of the amalgam summed up as New Age.[12]

Notable examples are the staff and helpers at Iona Abbey, L'Arche communities at Inverness and Leith, Edinburgh, where mentally handicapped people and carers share an identical common life, and the Findhorn community, Moray.

The dwellings can vary from groups of caravans to clusters of small buildings to a single large building (see Figs 40.1, 40.2, 40.3, 40.4, 40.5, 40.6). The dwellings of non-monastic religious orders from the thirteenth century to the present day have likewise been extremely diverse, although a place for praying in common and one for members to eat together can be presumed. Even for the autonomous monastic communities it is difficult to describe the variations although all had certain features in common and there was an established norm.

MONASTERIES

In Europe the early monasteries probably followed the Pachomian pattern of an enclosure, within which there were various separate units, but the Benedictine emphasis on community brought about a more compact layout.[13] This was basically a hollow square with buildings on each side (Fig. 41.1). Christian churches traditionally had the sanctuary and high altar at the east end; one corner of the square was in the angle formed by the nave and transept of the church. In northern Europe the north side of the square

Figure 41.1 St Andrews priory church and precincts, Fife, as they might have been about 1550, drawn by Alan Sorrel in 1965. Source: HMSO booklet, *St Andrews Cathedral*, 1986, 25.

usually lay alongside the nave and got the most of the sun, but the slope or drainage could result in the church being built on the south side.

Along each side inside the square ran a covered walkway, the outer side of which was mostly open or at least pierced by windows. The enclosed open quadrangle together with the walkway, termed the cloister, was the most striking feature of monastic buildings. In fact it became almost synonymous with the monastery itself and it was not only where much activity took place but also the passageway from one building to another. As a modern architectural historian wrote: 'The more one studies the monastic plan, the more does one marvel at its sheer efficiency,'[14] which explains its adoption everywhere. The same pattern was followed by nunneries, as can be seen in the Augustinian nunnery and nearby abbey on Iona, and even by some friaries, such as Jedburgh (Franciscan), Roxburghshire and Linlithgow (Carmelite), West Lothian.

There were of course variations in each monastery but none departed far from the basic pattern. In the church, beginning at the east, were the high altar, the choir where the monastic offices took place and, separated by a screen, the nave for the laity. In St Benedict's rule, the choir office, the communal prayer of the monastic community, was given priority and shaped the monks' lives. It consisted mostly of the Psalms, together with readings from Scripture and the Fathers, and was divided into eight single offices. First came Matins, the night office, followed at intervals by Lauds and the short office of Prime. The three 'little hours' (Terce, Sext and None) were recited during the day. In the evening came Vespers and, at the end of the day, the short office of Compline.[15]

In the north cloister the monks read or wrote; from the fourteenth century the windows were glazed and from the fifteenth there were perhaps carrels provided, and not far away was the library. In the east wing, nearest the church, was the sacristy, where gear for the services was kept. Also near the church was the chapter house, to which the monks processed after Prime and where talks were given and meetings held. Above these was the dormitory, with the night stair at the north end descending towards the church – this was used for the night office – and, at the south end, the day stair leading to the cloister.

Near the foot of the day stair were the novices' quarters and the calefactory (warming room), with the rere-dorter (latrine block) running off to east or south. On the south side were the refectory and kitchen. The drainage constituted one of the most remarkable features of medieval monasteries, for running water was diverted to flow underneath kitchen and latrine, as well as the infirmary further east, and carry away waste. The drainage channels were well constructed, sometimes with dressed stone, and often compel the modern visitor's admiration. Water for cooking and washing was piped separately.[16]

The west wing was given over to more mundane things, such as storage and offices, with the abbot's quarters, guest rooms and parlours where monks and visitors could converse. In Cistercian houses, however,

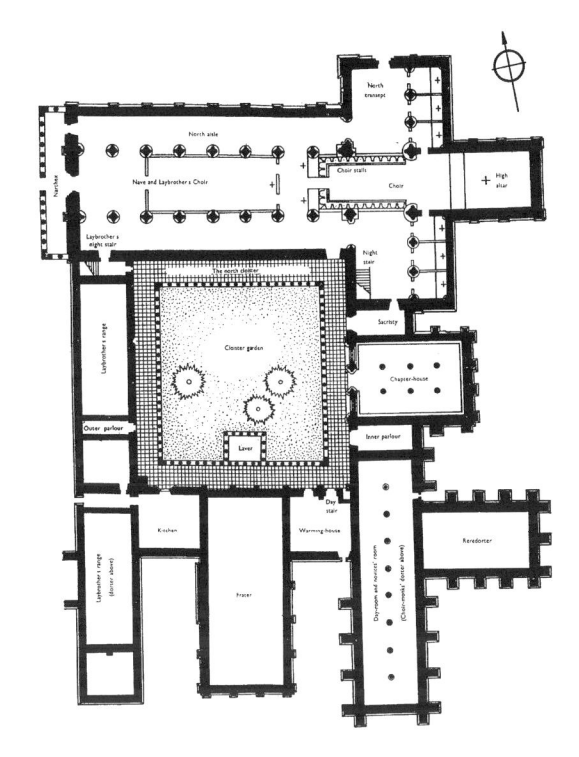

Figure 41.2 A typical monastic plan, based on Dundrennan (Cistercian), Kirkcudbrightshire. Source: Cruden, 1960, plan 1.

this wing provided quarters for lay brothers, who had their own refectory and dormitory with day and night stair. Outside the west wing was the gatehouse and between this and the monastic buildings was a space occupied by, for example, workshops. It was the place where the monastery and the outside world met. Surrounding the whole complex was the perimeter wall, often very impressive. Some gatehouses were fortified and sometimes the church tower as well (see Fig. 41.2).

One can see how the monks spent their day in a settled routine without having to go far for any of the essentials of living. In winter the calefactory was surely more of a necessity than a luxury and was perhaps the common room and where any writing was done. In the later period, when monks had their private 'yairds' (vegetable gardens), these were probably to the east or south of the main buildings.[17] And no doubt the amount of time spent on the west side would vary with each monk and his duties and inclinations.

The cloister lay at the heart of Carthusian monasteries too, but in a very different way. Along its external walls was a series of doors, each leading to a small two storeyed house. The ground floor comprised a workshop and woodstore and there was a covered walkway and small garden; the upper floor was the monastic cell, where the individual monk prayed, studied and slept. His meals were provided through a hatch, except on Sundays

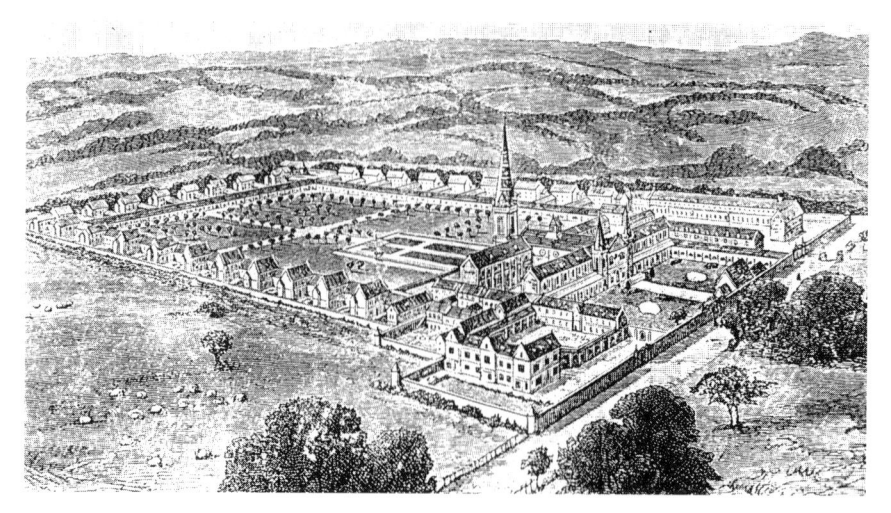

Figure 41.3 Bird's-eye view, St Hugh's Charterhouse, Parkminster, Sussex.
Source: M Dilworth.

and feast days, when all ate together. The night office, conventual Mass and Vespers were sung by all in the church, with other offices said privately. Once a week they took a long walk together and conversed. Carthusians were essentially hermits living together in community[18] (Fig. 41.3).

FORT AUGUSTUS ABBEY

Fort Augustus Abbey was originally a Hanoverian military fort, built by General Wade 150 years before its conversion to a monastery.[19] It was square shaped and symmetrical with a building along each side and a bastion jutting out at each corner (see Fig. 41.4). When work began to make it into a monastery in 1876, the south block and the south-east and south-west bastions formed no part of the plan; they had perhaps been partially destroyed in the siege by Jacobite forces in 1746. On the other three sides the existing low buildings were raised to a height of three or four storeys. A covered cloister was built round the perimeter of the barrack square enclosed by these buildings, creating a remarkable similarity to a medieval monastery, a similarity reinforced by there being entrances from the cloister to the buildings on all four sides (Fig. 41.5).

The abbey grounds were bounded by Loch Ness to the east, the River Tarff to the south and the Caledonian Canal to the north, so that access was from the west. The west wing was the hospice (guest house), the north wing was the abbey school and the east wing was the monastery proper, containing, amongst other things, library, refectory, monks' living quarters and common rooms. Opening off the south cloister were the sacristy and chapter house and beyond them was a space for a large church to be built. One fundamental change had taken place in monasteries since medieval

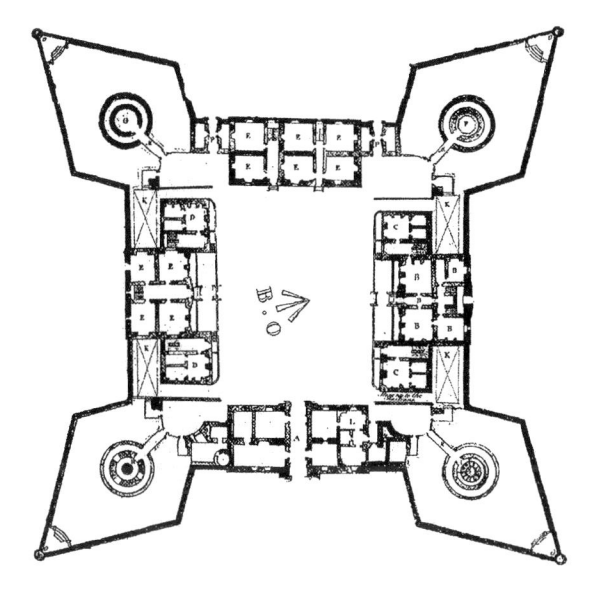

Figure 41.4 Outline plan of Fort Augustus, Inverness-shire. Source: Scottish Catholic Archives.

times, namely that each monk had his own cell (in modern parlance, a bedsit). The cloisters as a consequence were no longer a place of communal activity (Fig. 41.6).

Figure 41.5 Fort Augustus Abbey, Inverness-shire, aerial view. Source: Reproduced by kind permission of Lovat Estates.

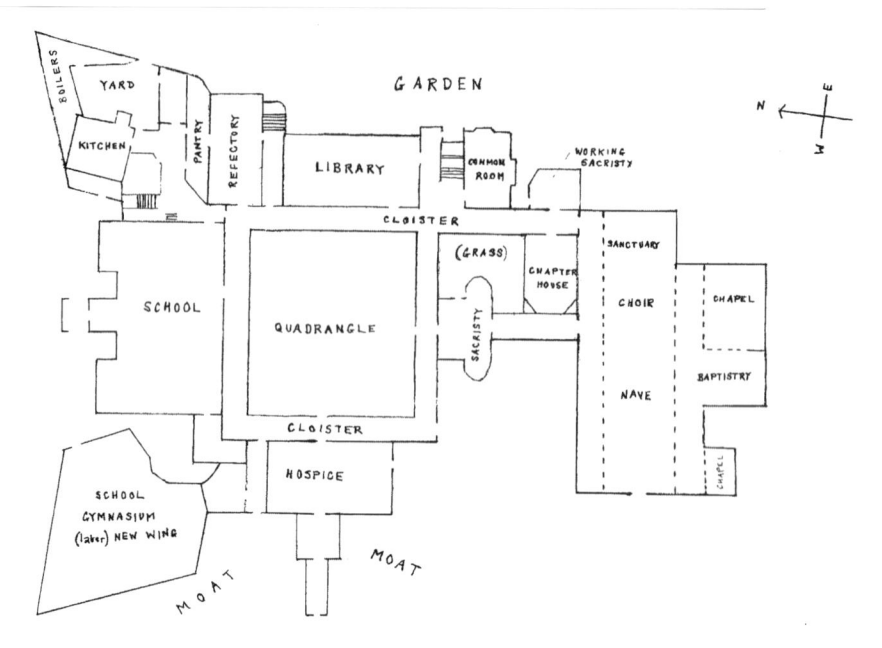

Figure 41.6 Fort Augustus Abbey, Inverness-shire, ground plan. Source: Made by Mark Dilworth from plan made for insurance purposes.

During the 120 years of the Fort Augustus Abbey's existence, the outside world naturally enough made its impact. The Victorian plumbing was still largely in use a century after its installation. A few bathrooms had been set up and wash-hand basins (cold water only) installed in most cells, and, when the whole system had to be replaced, hot water became available on tap. Fires in rooms gave way to central heating and in 1889 the abbey constructed its own system whereby a turbine engine, driven by water, supplied electricity to the neighbourhood until the coming of the hydro-electric grid.

Monastic life can hardly be compartmentalised. Life in the abbey was a structured whole, governed by the spiritual principles of St Benedict's rule, the modern and more detailed Constitutions of the Benedictine Congregation and the customs of an autonomous community. St Benedict's basic requirement was that the monk should 'truly seek God' and, having withdrawn from the usual world of family and workplace, he was to do so together with his fellow monks and in obedience to an abbot chosen by the community. The day was to be spent in prayer, reading and work, with priority given to the community's prayer in common, the choir office.

At Fort Augustus the day began with Matins and Lauds at 5 o'clock, followed by Mass, private prayer and Prime.[20] After an informal breakfast came the 'little hours' and conventual Mass, and the forenoon was then clear for work and activity. The midday meal was formal and in silence, the monks taking their turn to wait at table and read aloud. A short period of recreation in common followed and the day was again clear until Vespers at

6 o'clock. Supper, partly formal, was at 7 o'clock and the day ended with Compline shortly after 8 o'clock. It was an uncluttered horarium, into which each individual fitted his personal prayer and spiritual reading. Once or twice a week there was a formal meeting in the chapter house, usually after Vespers.

The abbey was a complex institution. The community looked after its own recruits, who had their living quarters somewhat apart but shared with their seniors in refectory and choir. Their formation lasted until they made their perpetual vows after four years and then completed their studies for the priesthood after another three. A community of monks in perpetual vows was a constitutional entity in canon law and formed a legislative body or chapter with canonical rights and duties. Monks had chapter rights both as a body and as individuals; they had to be consulted on important matters and they voted in secret ballot on accepting recruits, financial affairs and the like.

Almost every monk carried responsibilities: as monastic officials, such as prior (abbot's deputy) or novice master; in charge of particular areas, such as sacristy, library or garden; with responsibility in the school or teaching there or teaching the young monks. The horarium in the school was, however, basically very different from that of the monastery.[21] Although care was taken to avoid, if possible, clashes between monastic duties and the demands of teaching, supervision and pastoral care, an individual monk inevitably had at times to be absent from conventual duties. Officials such as the headmaster or the monastic bursar had to balance the demands of their work against their monastic commitments, as had been the case also in medieval monasteries.

There were also lay brothers, leading a life separate in many ways, with their own living quarters and a simple choir office in English but sharing formal meals with the choir monks. Many of the routine chores and much of the maintenance was undertaken by them and they were often skilled tradesmen (plumbers, electricians, tailors, for example) and in charge of areas such as the printing office, bakery, engine house or carpenter's shop. Choir monks and lay brothers were really two communities living under one roof, meeting together on occasion and enjoying cordial informal relations with each other.

Naturally the personal relations of monks with each other were important. Given the wide varieties of background, temperament and interests, some friction was inevitable. Two things, however, helped to deal with this. Firstly, care was taken during each recruit's formation to ensure that he was fitted for the community life. Normality, a sense of humour and the ability not to take himself too seriously were regarded as essential. Secondly St Benedict was fully aware of the friction caused by human beings living together, and in several places in his rule he stressed the need for mutual forbearance and submission and the need to be forgiving. If disagreement arose over work, the superior could deal with it, as of course he could in any more serious conflict.

The relationship between the individual and the community as an entity was also important. Anything the monk received, whether earned or as a gift, belonged not to him but to the community, and the community was obliged to provide each person with what he needed. The monk would make his needs known in such matters as clothes or footwear, although all wore the same habit, and there was a common store for basic items. An infirmarian was appointed to oversee health needs, and access to doctor or dentist offered no difficulty. The mutual commitment of individual and community was for life. Those who had grown elderly or infirm were looked after, unless they needed medical care not possible in the monastery, and took part in community exercises insofar as they were able.

It was obviously sensible for a community to keep its numbers well and happy and in return the individual would feel it correct to 'pull his weight' in the common enterprise. Indeed many were sent to universities or church colleges to gain qualifications. Although canon law required monasteries to have capital underpinning, this could hardly provide adequate financial support, especially in a period of inflation. It was also in accordance with monastic ethos that monks would live by their own labours. Each community therefore embarked on some enterprise which would provide financial support and also, if possible, benefit the wider community. In a sparsely populated district like Fort Augustus, the running of a boarding school brought pupils to the monastery and provided a pastoral service for the Scottish Catholic Church.

The ethos was very much that of a family, for each one had applied for entry, received a new 'religious' name and after due probation had been accepted into the community for life. The person's name day (feast day) was usually marked in some way. Jubilees (25th and 50th anniversaries) of taking vows or being ordained priest, as well as occasions such as an 80th birthday, were matters for celebration. The standard form would be to adjourn after the midday meal for, say, a glass of port in the common room, where the abbot would prepare a not-too-wordy toast and the jubilarian would reply. For notable occasions, such as a golden jubilee of priesthood, there would be a celebratory meal with monks and guests together.

Male guests were welcomed at the abbey. Usually they wanted to have a quiet time and attend choir office, but often they included relatives of monks, particularly the younger ones. Guests slept in the hospice and ate in the monastic refectory, sharing the same food as the monks with the addition of a cooked breakfast. The food was adequate, for the monks were expected to do a good day's work. Breakfast was simple, with porridge, very welcome in winter; the main meal at midday was the standard three-course meal of the period and was the same as that provided for the school; supper was a simple dish together with bread and jam. Tea and coffee were provided at breakfast, a cup of tea in late afternoon, tea and cocoa at supper. The feasts and fasts of the Church were of course observed, so that on major feast days the fare was more festive and water was replaced by wine or homemade beer.

It was prudent to avoid stress and overwork. The horarium included unscheduled time, with some afternoons free for walks and occasional longer periods allowing the individual to go further afield. Fort Augustus lies in spectacular highland scenery; most young monks walked the hills or went for cycle runs and many became proficient in managing a boat on the loch. Hobbies were not discouraged and some benefited the community: bookbinding, gardening, beekeeping or fishing for example. Often there was no dividing line between work and hobby in areas like playing the organ or serious reading. A more than adequate library and a range of periodicals was available to all. Nor were the arts banished. Reproductions of classical art, mostly religious, or of portraits, together with prints of various kinds, adorned communal rooms as well as the cloisters, staircases and corridors. As for music, some monks played the piano and occasionally other instruments and could use music rooms in the school, while records could be listened to in recreation rooms at suitable times. School holidays and the Christmas and Easter seasons brought greater flexibility and leave of absence was granted to allow contact to be maintained with one's natural family and prevent staleness.

There was, nevertheless, a large measure of withdrawal and solitude. Monastic 'enclosure', which excluded school and hospice but included a large garden, marked an area which the monk did not leave without a reason and was not usually entered by others. The privacy provided by individual cells had become a feature of monastic spirituality, giving rise to aphorisms like 'The monk keeps (French *garde*) his cell and the cell keeps the monk' and 'The good monk loves his cell'. Each monk looked after his own cell; the furniture was simple, solid and had survived decades of use, while anything which could be called ornamentation was fairly minimal. Apart from recreation, the general rule was to keep silence unless there was some reason for speaking, and the 'greater silence' between Compline and Morning Office was strictly enforced. Noise in the area of the cells was discouraged, even in daytime. Individual temperaments varied greatly but one essential element was willing acceptance of a large part of each day being spent alone and in silence.

An attempt should be made to assess the quality of life in the monastery. Although the horarium was so structured and so many areas were regulated by authority, permissions could always be sought (say, to speak to a friend arriving unexpectedly) and were usually granted. If one accepted the limitations of living in a highly regulated community of men only, the life was not burdensome and could be very fulfilling. The primary purpose of monastic life was to provide an ambience for the spiritual life to be taken seriously; it also gave freedom from the cares of family and work (although monastic officials often had to shoulder similar burdens). The danger to be guarded against was of lapsing into a passive acceptance of the limitations imposed by the life rather than keeping one's embracing of them active and purposeful.

The closure of the abbey was due to two general factors. The first was the decline in recruitment, a factor which had been operating in the Western world for some decades. In a period of greater recruitment, more houses would be founded such as the two monasteries established in Scotland in the 1940s, then, when recruitment diminished, not all could be sustained. The other factor was financial and practical. The abbey school, like so many other independent boarding schools, had been forced to close in 1993 by lack of pupils. Situated on a popular tourist route, the abbey had great potential as a heritage centre, with facilities for conferences and other events. This became in fact a successful enterprise, unfortunately hampered by a succession of poor tourist seasons. The decision to close was taken in November 1998.

NOTES

1. Ryan, 1931, 15–56; Knowles, 1969, 9–40.
2. Knowles, 1969, 44–97.
3. Ryan, 1931, 80–109; Knowles, 1969, 28–32.
4. Dilworth, 1974, 11–18; Dilworth, 1976.
5. Dilworth, 1995, 2–11; accounts of each house are in Cowan, Easson, 1976, 55–106, 143–56.
6. Dilworth, 1974, 18–19, 267–9.
7. Dilworth, 1978, 109.
8. Cowan, Easson, 1976, 107–42.
9. Addis, Arnold, 1960, 145–6.
10. *History of St Margaret's Convent, Edinburgh*, c1886; for Ursulines of Jesus see *Dictionary of Religious Orders, Congregations and Societies of Great Britain and Ireland*, c1953, 489–90.
11. Dilworth, 1978, 93, 100–1; Cameron, 1993, 706.
12. See Chapter 40 Communes.
13. Ryan, 1931, 285–94; for the enclosure (*vallum*) in Iona, see Ritchie, 1997, 36–9; medieval monastic buildings are described in Cruden, 1960, 41–8; Fawcett, 1994, 94–117; Knowles, 1969, 98–107.
14. Cruden, 1960, 47.
15. The choir office is regulated in detail in St Benedict's Rule, chs 8–18, with a summary in ch 16.
16. For the drain at Melrose, see Fawcett, 1994, 114; the remarkable drain at Paisley is described in Malden, 2000, 173–80, 212.
17. Dilworth, 1995, 29.
18. *The Carthusians*, 1924, 58–86.
19. The buildings are described in *Fort Augustus Abbey*.
20. The decades before the impact of the Second Vatican Council in the late 1960s are taken as the model.
21. School events are chronicled in Turnbull, 2000.

BIBLIOGRAPHY

Addis, W E, Arnold, T. *A Catholic Dictionary*, 17th edn, London, 1960.
Bulloch, J. *Adam of Dryburgh* (Church Historical Society), London, 1958 (a personal account of medieval monastic life in Scotland).
Cameron, N M de S, et al. *Dictionary of Scottish Church History and Theology*, Edinburgh, 1993.

Cowan, I B, Easson, D E. *Medieval Religious Houses: Scotland*, 2nd edn, London and New York, 1976.
Cruden, S. *Scottish Abbeys*, Edinburgh, 1960.
Dictionary of Religious Orders, Congregations and Societies of Great Britain and Ireland, Glasgow, ndp (c 1953).
Dilworth, M. *The Scots in Franconia*, Edinburgh and London, 1974.
Dilworth, M. The Schottenklöster. In Cowan, Easson, 1976, 240–4.
Dilworth, M. Religious Orders in Scotland, 1878–1978, *Innes Review*, 29 (1978), 92–109.
Dilworth, M. *Scottish Monasteries in the Late Middle Ages*, Edinburgh, 1995.
Fawcett, R. *Scottish Abbeys and Priories*, London, 1994.
Fort Augustus Abbey: Past and Present, Fort Augustus, various edns.
History of St Margaret's Convent, Edinburgh, Edinburgh, c 1886.
Hunter Blair, D O, ed and trans. *The Rule of St Benedict*, 5th edn, Fort Augustus, 1948.
Knowles, D. *Christian Monasticism*, London, 1969.
Malden, J. Archeological overview: The discovery of the drain. In Malden, J, ed, *The Monastery and Abbey of Paisley*, Glasgow, 2000.
Ritchie, A. *Iona*, London, 1997.
Ryan, J. *Irish Monasticism: Origins and Early Development*, Dublin and Cork, 1931.
The Carthusians: Origins, Spirit, Family Life, Partridge Green, Sussex, 1924.
Turnbull, M T R B. *Abbey Boys* (corbie.com), Perth, 2000.

42 Fishing and Homes

REGINALD BYRON

INTRODUCTION: THE ORGANISATION OF THE DOMESTIC ECONOMY

An often-heard remark in Scottish fishing villages is that fishermen marry other fishermen's daughters. Behind this light-hearted aphorism is a serious ethnological point: fishermen's wives have an uncommon way of life. Those women who have themselves grown up in a fisherman's home are already accustomed to what this unusual domestic work involves and, perhaps, are more willing to accept it. The absence of men during the daylight hours, or for weeks at a time during certain seasons, means that virtually all aspects of household organisation are the responsibility of their wives who, quite often, have no choice but to carry out these activities in collaboration with their female relatives or the women of other families connected through their husbands' joint involvement in a boat's crew.

A collaborative relationship between a man and a woman, and further sets of collaborative relationships between men, and between women, are the most convenient and frequent solutions to the resourcing problems that are peculiar to this way of life. In order that the men can pursue the fishery at all, they must team up with other men. In order that they can spend enough time at sea to make a reasonable living, a number of supporting roles must be played by those who stay ashore. A fisherman who does not have a wife, unmarried sister or mother who can act as his shore-side partner and is willing and able to do the female work that must be done to complement his own labour is at a grave disadvantage, and he is unlikely to make an adequate living from the fishery. Yet his wife, sister or mother cannot cope on her own; she will need the help of other women.

In some types of fishery, the daily or seasonal preparation and maintenance of fishing gear in the home is, or was, an essential part of the fishing effort. Domestic arrangements varied by the type of fishery pursued, whether it was predominantly pelagic (for example, herring, which feed in the upper layers of the sea) or demersal (fish such as cod or haddock, which feed on the sea floor), or a seasonal mixture of the two. The work of the domestic unit also varied by the range of other, supplementary income or subsistence-producing activities available to the household, including migrant labour, crofting and farming, knitting and craft work, fish processing, and marketing and catering for summer visitors. Different combinations of these variables were to be found in fishing communities round the coasts of Scotland in the nineteenth and twentieth centuries, and all of them had their own implications for the organisation of domestic life.

Deep-sea fishing is a form of hunting rather than of husbandry. Since fish swim about freely and are subject to variation in population levels, sometimes quite extreme variation, there is, typically, a good deal of uncertainty and unpredictability about where they are to be found, when, and in what amounts. These matters are largely beyond the control of any given fisherman. In contrast, a shepherd can count his flock and estimate the number of lambs that will be born next season with a measure of confidence, and a farmer knows how much land he has and what kind of harvest his inputs are likely to produce. Deep-sea fishing is much less predictable. People who have, over the generations, succeeded in making their living in this way typically organise their affairs along certain lines. Some of the organisational mechanisms that we find in Scottish deep-sea fishing societies have widespread ethnographic distribution and seem to have occurred to people as workable solutions to shared problems in widely scattered places round the North Atlantic region and beyond.

Household form and domestic organisation are a demonstration of these organisational principles. Any household, whether it is a fishing household or not, must meet a range of economic demands occurring over its lifecycle. As a young married couple set up a home, they enjoy a period of relative prosperity until children arrive and begin to make heavy demands upon the couple's earnings. After the children begin to earn and eventually move out, the parental couple may again enjoy years of relative prosperity until they retire and lose their earning capacity, when they return to a condition of dependency upon others. In fishing households, this standard lifecycle trajectory is complicated by the instability in the economic resource upon which they depend. The supply of fish and the level of fish prices rise and fall, not always in inverse ratios, from month to month, season to season, and year to year, in irregular and unpredictable rhythms, in response to vagaries in the weather, the complex array of ecological variables governing fish populations and the intensity of human predation. There is no telling when feast or famine may occur, how serious these variations might be, or whether they will catch a family at the 'wrong' point in its developmental cycle.

How fluctuations in the supply of fish affect individual households depends upon the household's particular circumstances in terms of its consumption requirements and its available labour supply. A household which happens to have – by virtue of its stage in the lifecycle – the labour power (several young adult children) to take advantage of a boom in the availability of fish is well placed to 'strike it rich' and provide the resources for each of these young adults to marry and set up their own households. Conversely, such a household may have young adults coming of working and marriageable age at a moment when the fishery goes into a deep and extended depression. In this scenario the children will have very reduced prospects of a decent local livelihood and of accumulating the resources to get married.

Success in bringing up children with good prospects, and still having any of them nearby in one's old age is, perhaps, a greater test of organisational ingenuity and good household management in a deep-sea fishing household than it is in a farming or urban household, where either labour and capital provide a more predictable return, or other means of livelihood are more readily available locally. Characteristically, in deep-sea fishing societies, elementary families – husband, wife and children – are drawn into flexible networks of alliance with others, jointly pooling their labour and capital within and across the generations in organisational forms which in some ways resemble extended households. The term 'collegial corporation' has been used to describe these domestic units.[1] They are composed of elementary families united in production and consumption, even though they do not all, at all times, share the same hearth or sleep under the same roof. Where the fishermen own their boats (as in Shetland and many other parts of Scotland), they usually do so by pooling their capital and drawing from it a common livelihood, as well as frequently providing a livelihood for their sons or sons-in-law. Their wives, daughters, daughters-in-law and other womenfolk team up for mutual support while the men are at sea. The important point to note is that the elementary or conjugal family is rarely a complete, independent, self-contained economic unit in deep-sea fishing societies. Making ends meet demands that each such unit is intimately interlinked with others through relationships of collaboration.

Analytically, these relationships of collaboration are of three distinct kinds:

Moral economies of resource allocation.
Where it is feasible (as in some types of inshore fishing), the allocation of fishing space or territories may be regularly redistributed and/or divided up on the basis of need. Who gets what will be largely a function of household lifecycles over time. At any given moment, some households will need more access to resources, some will need less, and the way these resources are allocated is, at least in part, a reflection of these needs. In riverine, lake, intertidal and shallow-water sea fisheries, re-allocating territory as the occasion demands is quite straightforward. In deep-sea fishing it is generally not possible to divide up fishing space. Even so, moral considerations may govern access to, and the distribution of, valued goods such as jobs for crewmen on boats or shore-side fish processing, or other livelihood-generating opportunities.

Moral economies of capital risk spreading.
Here, households pool some of their capital assets (money and/or equipment), and so spread some of the risk and even out the temporal disparities associated with the constituent households' lifecycles. The fundamental mechanism is some variation on a 'share' system, which is especially

common in Scotland and the Nordic countries. Boats may be capitalised in this way, as well as shore-side businesses. People within the group each contribute some capital to the start-up and operating costs of a joint venture and draw a cash/subsistence income from it in proportion (more or less) to their capital share. Each of these shareholders is a member of a different household, which is highly likely to be (and it is desirable that it is) at a different stage of its lifecycle from some of the others. Multigenerational shareholding enterprises have the most flexibility as internal adjustments and re-allocations are possible which can help those of their members whose households are at a strained point of development; those less in need can reallocate some of their share of income to those more in need. Should a depression occur, the risk can be spread flexibly to cushion the impact on the most vulnerable of the shareholding households. Single-generational share systems, where all the constituent households are at the same stage of the lifecycle, have much less flexibility and are much more risky. Members' consumption demands can all peak at the same time and drive the group to the limit of risk taking and beyond it, sometimes with tragic consequences, as when a boat is lost at sea.

Moral economies of collaboration.
Here, labour is pooled or shared between households. Labour-intensive operations may strain or exceed the capacities of individual households at certain points in their lifecycles. These collaborative operations may be of almost any kind. Inter-household collaboration may run along the lines of kinship, of neighbouring, or of involvement in the same shareholding enterprise (for example, the mothers, sisters, wives and daughters of men who own a fishing boat in common may collaborate in a range of non-fishing, shore-side domestic operations), or may involve various combinations of these organisational principles. Help contributed to one kind of collaborative task may be reciprocated in a different way, and these inter-household relationships may include a flow of material goods (favours and obligations 'in kind') as well as of labour.

The collegial corporation involves all three of these kinds of moral economy. As a form of local-level social organisation which is frequently found in Scotland (as elsewhere in the North Atlantic), it is related to the development of markets, trade and technology which brings about the intensification of production and gradually transforms deep-sea fishing from a subsistence activity into an industrial process. These corporate groupings of kinsmen and neighbours are a social-organisational response to a particular set of economic conditions associated with commodity production based upon an especially fluid resource where there are no, or few, limitations upon access. It is not necessary to suppose that these extended groupings have any great historical depth, and in most places they probably did not. Alan Macfarlane, Martine Segalen and A F Robertson, anthropologists of the European household, have shown that the independent elementary family was, in the past, as frequently found as it is today, but that certain situations stimulate the development of extended domestic organisation. The conditions imposed upon fishers' livelihoods by industrial-scale fishing in various places around the North Sea and the North Atlantic at various times during the eighteenth, nineteenth and twentieth centuries appear to have been responsible for the creation of just such specialised social groupings.[2]

THE EAST NEUK

The East Neuk provides a good illustration of the ways in which fishing families have organised their affairs, and the east coast is a locality where the topic of domesticity is well documented in the literature for those who wish to take their reading further.[3] The Fife fishing ports of the East Neuk (Crail, Cellardyke, Anstruther, Pittenweem and St Monans) are fairly typical of the communities which developed round the coasts of Scotland with the rise of the Great Herring Fishery during the nineteenth century, a phenomenon which brought about a dramatic expansion in the coastal population (Fig. 42.1). The development of this fishery was closely tied to Victorian industrial capitalism and, as in many other places, owed relatively little to the existence of earlier or other modes of fishing livelihood which might, or might not, have existed in the locality. In Pittenweem, the industrial nature of the herring fishery can be seen in the rows of small terraced cottages in which the fishers lived, houses which were little different from those which might have been built in a coal-mining village at around the same time. By contrast, in Shetland and Orkney, fishing with lines for cod during the winter season for subsistence purposes and to some extent for market had been a part of the annual round of activities for many households for centuries before the Great Herring Fishery arose.[4] These households normally had access to crofts and combined fishing with subsistence agriculture and animal husbandry, while those fishers who lived in communities which became established or greatly expanded during the Victorian era – such as Pittenweem – generally did not possess any land-based resources other than their cottages.

In the early part of the nineteenth century, herring fishing was emerging as a major activity along the Fife coast. The herring fishery had two seasons: the summer months, when the men fished in the Firth of Forth; and the early autumn, when they moved down the east coast of Britain to East Anglia, landing their catches at the ports of Lowestoft, Suffolk and Great Yarmouth, Norfolk.[5] The summer fishing season in the Forth, which was called the 'Lammas Drave', peaked in the 1860s and went into a gradual decline thereafter, but the autumn herring fishery continued until after World War II, enjoying its most successful period between the Wars. In addition to the herring fishery, some of the East Neuk men took part in the winter season of cod fishing, which took them northwards to the Faroe Islands on trips lasting up to three weeks; others fished closer to home for white fish.

To give some indication of the part played by the East Neuk fishers in the Scottish fishing industry as a whole during the late nineteenth century, the official figures for the Anstruther district in 1883 show that the combined

Figure 42.1 Low tide at the harbour. The villages of Anstruther and Cellardyke are continuous. The church spire to the left is in Austruther; the fishers' cottages on the right are in Cellardyke, 1180s. Source: Courtesy of the Scottish Fisheries Museum.

value of boats, nets and lines was one eighth of the total valuation for Scotland. In the same year, the following numbers of people were employed in fishing in the Anstruther district: men and boys as fishermen, 3,491; fish curers, 46; coopers, 76; and a catchall category of 'others connected', 2,362. It seems likely that most of the last were women and girls, whose categorisation in this way by no means did justice to their contribution to the economy of the East Neuk and to the welfare of their households. The women certainly were not idle while their menfolk were at sea. In addition to running the home and looking after the children, the fishermen's wives, mothers, sisters and daughters would have been involved in a number of unpaid activities connected with fishing and might also have had a waged job or other cash-producing work for at least part of the year.

The fishing people of the East Neuk were cottars, who lived in small houses along the shore in a cluster of isolated villages surrounded by farmland, yet they had no agricultural holdings to supplement their subsistence or to fall back upon in times of scarcity in the fishery. They relied exclusively upon cash earnings which were, for the most part, seasonal and unstable. Although earnings could be reasonably good at peak times during the year, there were other periods when little or no cash was brought into the household. The standard of living for the majority of East Neuk fishing families was poorer than that of the farming people in the immediate vicinity and the miners' and shipwrights' families in the more industrialised parts of Fife nearer the Firth of Forth.

Until the 1920s, some East Neuk women worked as gutters and packers for the herring boats landing their catches at Pittenweem or Anstruther during the Lammas Drave in the Firth of Forth, or what remained of it. Those women who stayed at home during the summer could also bring in some cash by renting out their houses to visitors from the cities. During the two-week Glasgow holiday, or 'Glasgow Fair', up to 2,000 industrial workers and their families arrived annually in the East Neuk in the years before World War II. The fishermen would be away and their wives and children lived in the attics, renting out the rest of the house and perhaps providing meals for the guests. At other times of the year, the women had a more direct involvement in fishing. When the men came home after the herring season, one of the tasks done mainly by women was net mending (Fig. 42.2). Each fisherman was responsible for providing a certain number of nets for the boat (as his contribution to the boat's 'fleet' of nets), and after each season's use the nets would need to be dried, mended and retarred (Fig. 42.3). A fisherman who had no wife, sister, mother or daughter to mend his nets would have to pay a net mender out of his share of the catch. At the turn of the century, net menders were paid 1s per day for twelve hours' work, plus meals, if they worked in the fisherman's home.

Figure 42.2 Net mending, Ferryden, Angus. One of the regular, seasonal tasks of a fisherman's wife until the 1950s. Note the children playing a game under the watch of the net mender. Source: Courtesy of the Scottish Fisheries Museum.

Figure 42.3 Two women (at the left) mend, while the men tie the individual herring nets to the common 'buss rope' to form a 'fleet' ready to be set at sea. St Monans, Fife. Source: SLAW 585227.

In the winter, the East Neuk women made a substantial contribution to line fishing. The lines for haddock fishing had to be baited daily, early in the morning, so the bait would still be fresh when the lines were set at sea. Each of up to 1,800 hooks had to be baited individually by hand with cut-up pieces of horse mussel or lugworms, then the hooks arranged and the lines coiled in baskets to prevent snags and tangles when the men set them. This messy, time-consuming and painstaking task was a part of the lot of most fishermen's wives and daughters during the haddock season, the length of the Forth coast, until the adoption of the seine net in the late 1930s and early 1940s did away with the hooks and lines and baiting:

> I hated the small lines for this meant so much more work for the adults shelling the mussels, baiting the lines ... I have seen both of my parents fall down with exhaustion at the end of a day, after my father had come in from the sea.[6]

Unmarried women, or those without small children to look after, were employed in small numbers at Robert Watson's oilskin factory in Cellardyke, founded in 1859. Fifteen of the 26 girls who left the Cellardyke school in 1920 went to work at Watson's. Other women, including widows with young or elderly dependants, made money from taking in dressmaking, tailoring and mending, washing and pressing, serving tea or meals, or by converting the front rooms of their terraced cottages into shops, selling groceries, clothing and other items.[7] Knitting was another occupation for women and also girls. Most was done to supply menfolk with warm clothing to wear at sea, but knitted goods could also bring in cash. The East Neuk fishermen wore many layers of knitted garments, including long drawers, sea-boot stockings, socks, jerseys ('ganseys') and mitts. The amount of knitting needed to equip a man for fishing was prodigious; he would need six or seven of each item for the long trips away from home, and his kit of seagoing clothes would have to be renewed every year. One advantage of knitting over other, fishing-related domestic tasks was that it was a portable activity, allowing women to gather with friends and socialise while they worked. Girls were able to knit by the age of six or seven, having learnt by copying their mothers, and would soon be knitting socks and mitts for their fathers and brothers, imitating their mothers in the way they sat with their friends, chatting while they knitted.

The most significant way in which East Neuk women were involved in the cash economy of fishing before World War II was as 'herring lassies'. Every year, during the season, from sometime around the middle of the nineteenth century until the late 1930s (and again for a few years on a reduced scale in the late 1940s and early 1950s), there was a complex pattern of migratory labour to and from the herring ports all round the coasts of Britain, involving young Scottish women, the wives, sisters and daughters of the fishermen of numerous communities, notably the East Neuk villages. There are still several women in Pittenweem, now in late old age, who, in their time, were herring lassies, travelling the length of Britain from Suffolk to Shetland, and remember this way of life.

To an unmarried woman, especially, the relatively high wages which were possible in the migratory work must have been attractive. While she would be expected to contribute some of her wages to her parents' household and to cover her own keep, she could set aside part of her wages for her 'drawer' (trousseau) or, not uncommonly, the deposit on the purchase of a house. Her fiancé might have been putting aside his savings to purchase a share in a fishing boat, to give himself and his family a greater security of income. Such jointly-capitalised marriages, and households, may have been relatively common; if so, this would account for the relatively late age of marriage before World War II, at about 26 or 27 for women and 30 for men. For a woman, then, it took ten or 12 years of working and saving in order to be in a position to marry and set up an independent household with sufficient resources to cover the costs of having at least the first child; for a man this period of accumulation was 12 to 15 years. Whether the wife needed to, or could, go back to work as a herring lassie after the arrival of children depended upon a greater number of variables, and how the household subsequently developed: how well 'set up' her husband was in the summer and winter fisheries; whether or not he had a share in a family boat with his father and brothers or brothers-in-law, and the luck of his crews from season to season and from year to year; whether or not the wife had a mother, sisters or sisters-in-law nearby upon whom she could call for help; whether the household contained any other dependants, such as an aged or infirm parent; whether there were children, how many, how they were spaced; and what aspirations the couple had for themselves and their children.

FISHING HOUSEHOLDS TODAY

Jane Nadel-Klein, writing of Ferryden, near Montrose, Angus, observes that so sharply defined and gender-specific were men's and women's roles, and so cramped their living conditions, that:

> It seems unlikely that men, women, and their families could have spent much time together. People in Ferryden remember that children would often be turned out of their beds to make room for their fathers and older brothers upon their return late at night from the fishing. Their houses were called 'but-and-ben'; one room front, and one back, and in these, fisherwomen raised families of ten or 12.[8]

There have been dramatic changes in the contribution of women to fishing, and in the living conditions of fisher families, over the last 50 or 60 years. The herring industry went into a decline after World War II. By the late 1950s, drift netting was rapidly on the way out and would be virtually extinct in the British Isles within a decade. Nowadays, it is nearly 40 years since an East Neuk fisherman's wife mended a herring net and 50 to 60 years since any local woman followed the fleets, living the life of a herring lassie. Only one woman was encountered during the author's research in the locality who had been still baiting lines into the 1950s; for the rest, baiting lines is a childhood memory of women in their seventies or eighties (see Fig. 9.3). It has been decades since an East Neuk fisherman went to sea with a complete outfit of woollens knitted by his womenfolk; his kit bag now contains only factory-made work clothes bought from the fisherman's co-operative store and charged against his income tax. And, of course, as throughout Britain, families are much smaller nowadays.

Although women no longer go out to work in the ways that they used to, and they no longer prepare fishing gear and make their menfolk's seagoing clothing in their homes, some elements of a specialised life remain. It is still the case that the men are away from home through the week and sometimes longer periods of several weeks at a time. Women frequently say that they are both mother and father to their children, meaning that they cannot call on their husbands to discipline the children. As one woman said, 'When the children are naughty, it is the mother who has to deal with it. When the father comes home at the weekend, he gets all the attention.'

The relationship between mother and daughter (and sometimes between mother-in-law and daughter-in-law, where a women has married into the community from another place) remains of considerable importance. Deirdre Chalmers, in her field research,[9] interviewed one in-marrying fisherman's wife who initially thought it strange that the locally born wives seemed to prefer the company of their mothers to that of their friends and age mates; soon, however, this woman found herself in the same situation when she began to have children of her own. The mother's mother in a fishing community fulfils a number of roles. Since it is likely that she does not have a paid job and lives close by, she is free to visit her daughter every day, help with the housework and shopping, and look after the children either in her daughter's house or her own. The presence of a mother is a way of sharing the workload of running a house and caring for a family in the absence of the husband. Chalmers noted, in 1988, that:

> One young Pittenweem woman whom I visited on a number of occasions, and who had several children including a baby, told me that if I were to call round and she was not there, then I should go to her mother's house in the same street. In fact, on almost every occasion that I visited, her mother was in the house, helping her daughter with housework. Both women are fishermen's wives, and this gives them a level of shared experience that clearly affects their relationship. The younger woman not only knew the lifestyle into which she was marrying, but she could turn to her mother for advice, in the knowledge that she would appreciate the situation. When asked about any problems or difficulties about this way of life, she replied: 'I am used to it as my dad was the same.' As a young mother with children, she was pleased to have her mother close at hand to help her run her home. On the topic of housework, she joked that her mother has had a great deal of influence on her, because 'she is a very tidy person and I'm not.' As she gets older, she may find that the relationship alters, and that she will be visiting her mother as the children get older and she is not tied to the home. Another Pittenweem woman said that as a young woman she would spend several hours every day with her mother, but that now she just phones her a couple of times a week, and may on occasion do her shopping for her. Now that her children are teenagers, she does not need her mother to babysit, or to keep her company in her home. The influence a fisherman's wife may have on her daughter can be substantial, as one skipper's wife, now in her sixties, said, 'You never forget the help and advice you get from your mother. I find myself doing the same things she did.'[10]

The last woman quoted above was herself now a grandmother. She looked after her grandchildren a great deal, and had a very close relationship with her daughters, whom she saw every day, 'My grandchildren practically live with me.' The tendency in this community is for the older generation's home to be the focal point of family life. Chalmers notes that the mother may visit her young married daughters in their homes during the week, especially if they have young children, but it is likely that on formal occasions the visiting pattern will be reversed; the grandparents' home will be the venue for Sunday lunch, for example. Interestingly, however, Sunday lunches tend to bring together a woman's sons – who are ashore for the weekend – and their wives and children, and not her daughters and daughters' children, who will be going to the house of their mothers-in-law with their husbands and children. Of course, this pattern depends upon the mothers of both husbands and wives living locally. In many Scottish fishing villages, endogamy was the norm. Jane Nadel-Klein, writing of Ferryden, says of this:

> [A] reason [for] the strong preference for village endogamy ... emerged from the nature of male and female work groups. Both male and female work groups were organised around core groups of male and female kin. Boat crews were organised agnatically [through the male line of descent]. Men said they felt more comfortable working with their own kin. Women felt the same way, however, and women's groups often consisted of mothers, sisters, and daughters, and the evidence suggests that women were extremely reluctant to break these ties. 'When the men were away, as they often were, for eight to ten weeks at a time then the women and children had each other for company and protection.'[11] Parish birth records from Ferryden reveal a strong pattern of women having given birth at least to their first and second children in their mothers' houses. Some women even returned from North America to do so.[12]

Where it is the case that either wife or husband has married into the community from outside, the nature of their incorporation into local social networks is rather different. It is more often the case that wives have married in than husbands, because the men have tended to be tied to the place through their boat and their relationships with other men – often their male kinsmen – with whom they co-own the boat or have long-standing obligations of mutuality. In-marrying, wives often expressed, and still express, regret that they do not have their mother or sisters nearby, upon whom they can call for help. Clearly, they regard having to rely on their mothers-in-law or sisters-in-law as a more difficult situation to be in.

For the oldest generation of women, who are the great-grandmothers of the youngest generation, to be a fisherman's wife was to be a herring lassie, a baiter of lines, a mender of nets, a knitter of seagoing woollens and to have grown to adulthood before, or shortly after, World War II, at a time when money was scarce and it was a struggle to set up an independent household. It is the experience of this generation of women which is portrayed in the Fisheries Museum in Anstruther, although perhaps a somewhat romanticised version of it, for it is difficult to appreciate from the tidy displays the conditions which drove young women to become migratory labourers, and to suffer the discomfort and stigma which went with it. These women taught their daughters that there was nothing romantic about the past, and in the changing world of the post-War period, this legacy of hardship, together with the prosperity of the times, encouraged and provided the means for their daughters to distance themselves from the past. No longer forced through circumstance to work as herring lassies, bait lines, or take any active role in the local fishing industry, few of them actually chose to do so. They preferred, rather, to stay at home minding the children and to 'make a home', luxuries their mothers never had the opportunity to enjoy. During this period, this younger generation of women began to move out of the small, terraced cottages down along the shore to newer and more spacious houses at the back of the villages. To these women, the daughters of the herring lassies, a life of quiet domesticity on the middle-class pattern was a break with the past.

The daughters of these women, the granddaughters of the herring lassies, have carried on much of this self-image and way of life. When discussing their roles, they most often describe themselves as being both mother and father to their children, and say that their place is to make a home for their husbands where they can relax in comfort on their rest days between fishing trips. It is rare for a fisherman's wife to have a paid job outside the home. This preference for domesticity is out of keeping with current cultural ideologies in Britain which emphasise the economic empowerment of women and the sharing of domestic responsibilities between husbands and wives. Indeed, this is one of the main ideas which creates and sustains a social separation between the remaining fishermen's wives and the other women in the East Neuk villages. Fishermen's wives are called 'diehards' because, to other women, they appear irrationally attached to an outmoded wifely role; the stay-at-home, the complacent domestic, still attached to her mother and content to be dependent upon her husband and unwilling to join in with other village women in 'community' activities. Ironically, of course, seen from the perspective of a fisherman's wife, her contemporary role is the very opposite. Being in sole charge of domestic decisions while their husbands are away, such women see themselves as having more independence and autonomy than other women, not less. Moreover, they see themselves as truly liberated, freed from the conditions their grandmothers and great-grandmothers were forced to endure.

CONCLUSION

In the East Neuk, as elsewhere in the North Atlantic region, the organisation of the domestic sphere, so vital to the survival and success of the fishing household and the fishing community as a whole, was almost exclusively the domain of women. In recent decades, women's lives in coastal communities appear, in general, to have changed much more than men's, and women themselves can be seen to have taken the initiative in many changes in the domestic sphere. The nature and direction of these changes might be understood as attempts to resolve the tension between the ideal of the independent, elementary, conjugal family, and the practical arrangements, made necessary by prevailing economic conditions, which compromised their idealised perceptions of themselves as the exclusive providers of domesticity to their husbands and children.

In fishing societies, the central persons within the fundamental units of social organisation – households – were women. Not only did women play a pre-eminent part in the socialisation of the next generation, they took a direct economic role in production and distribution which was complementary to that of their menfolk. Because they were so often absent, the men left the affairs of the household to their women, but the women had to rely on other women to perform a range of routine domestic tasks. In the East Neuk, this was extended further since the women were often away themselves, at the herring gutting for example, putting extra responsibilities upon those of their female relatives who stayed behind, including, sometimes, even the most fundamental daughterly, wifely and motherly duties concerned with the care of their children and husbands, younger siblings and aged parents.

Because of unpredictable variation in the supply of fish, the stresses of production and reproduction in deep-sea fishing households made it extremely difficult for an elementary family to cope on its own. There were regular occasions when the demand for labour exceeded the supply, and consumption exceeded production, and there were critical moments when both occurred simultaneously. Making ends meet required that women allied themselves with others to enlarge the pool of labour available to them; to work out ways to call in extra labour when it was in short supply and lend it out when it was in surplus, especially when there was a need to sell their own labour for cash. As long as the amount of labour which was required to bring in a sufficient household income or to perform routine domestic tasks exceeded the elementary family's capacity to deliver it unaided, women could not, in practice, achieve their ideal role as the exclusive providers of domesticity to their husbands and children. In peasant farming societies this sort of situation is merely a temporary phase in the developmental cycle of the domestic group. In deep-sea fishing societies, however, it is more often a permanent structural condition that is not ameliorated, all other things being equal, just by the passage of time. The household does not 'grow up' from a childlike condition of dependency to mature self-sufficient independence, but must continue to rely upon the women of other households to perform its complete repertoire of domestic functions.

NOTES

1. Byron, 1986.
2. Byron, 1994.
3. Byron, Chalmers, 1993; Chalmers, 1988; Nadel-Klein, 1988; Nadel-Klein, 2000; Nadel-Klein, 2003.
4. See, for example, Byron, 1986; Fenton, 1978; Goodlad, 1971; Nicholson, 1978.
5. See also Chapter 33 'Time Out' Work.
6. Sir David Fraser (Christian Watt) (1983). Quoted in Nadel-Klein, 1988, 195.
7. See also Chapter 19 Working from Home.
8. Nadel-Klein, 1988, 195.
9. Chalmers, 1988, 155ff
10. Chalmers, 1988, 155–7, with minor corrections.
11. Quotation attributed to an unpublished ms by Margaret Buchan in Nadel-Klein, 1988, 201.
12. Nadel-Klein, 1988, 201.

BIBLIOGRAPHY

Byron, R. *Sea Change: A Shetland Society, 1970–9*, St John's, 1986.
Byron, R. The maritime household in northern Europe, *Comparative Studies in Society and History*, 36/2 (1994), 271–92.
Byron, R, Chalmers, D. The fisherwomen of Fife: History, identity, and social change, *Ethnologia Europaea*, 23/1 (1993), 97–110.
Chalmers, D. 'Fishermen's Wives: The social roles of women in a Scottish community', PhD thesis, The Queen's University of Belfast, 1988.
Fenton, A. *The Northern Isles: Orkney and Shetland*, Edinburgh, 1978.
Goodlad, A. *Shetland Fishing Saga*, Lerwick, 1971.
Nadel-Klein, J. A fisher laddie needs a fisher lassie: Endogamy and work in a Scottish fishing village. In Nadel-Klein, J, Davis, D L, eds, *To Work and to Weep: Women in Fishing Economies*, St John's, 1988, 190–210.
Nadel-Klein, J. Granny baited the lines: Perpetual crisis and the changing role of women in Scottish fishing communities, *Women's Studies International Forum*, 23/3 (2000), 363–72.
Nadel-Klein, J. *Fishing for Heritage: Modernity and Loss along the Scottish Coast*, Oxford and New York, 2003.
Nicholson, J. *Traditional Life in Shetland*, London, 1978.

43 Living Alone

SINÉAD POWER

INTRODUCTION

Living alone is a state experienced by individuals at different stages in their lives, for a variety of reasons, and in a range of ways, and can be both positively and negatively viewed. The category of those living alone may encompass, for example, a young adult having left his or her parental home to undertake work or study, or a recently bereaved older person living alone for the first time. Not only can the experience of living alone vary across the lifecourse, but also in other ways, such as according to gender, ethnicity and socio-economic circumstances. Historically, cultural stereotypes have influenced attitudes and perceptions about those who live alone. Again these may be positive or negative, but certainly seem to be gendered. There is, for instance, the archetypal 'spinster' who has cared for but lost her aged parents, or the 'bachelor' avoiding marriage in favour of a carefree life.[1]

This chapter explores the increasing trend towards one-person households in Scotland and identifies some of the key characteristics of those who live within this household structure. Then, drawing upon a contemporary qualitative study which investigates the housing pathways of a group of older people in Edinburgh, some of the issues and experiences associated with living alone in later life are discussed. A sense of lifecourse passage emerges as children leave home and partners pass away. A particular focus is on the transition to living alone for those who have previously lived with another person or other people and on how this impacts upon the nature of their relationship with, and their experience of, their 'home'.

WHO LIVES ALONE?

Households have been variously described as the basic units of society, primary consumption units, and the units through which most people organise their lives.[2] Over the past 40 years a new household form has come to prominence: that of the non-elderly, single-person household. The growing importance of one-person households reflects wider changes in society such as a rising divorce rate and the increased purchasing power of younger generations.[3]

From 1971 to 2001 the proportion of households in Scotland consisting of one person increased from 18 per cent to 31 per cent. Most of this change

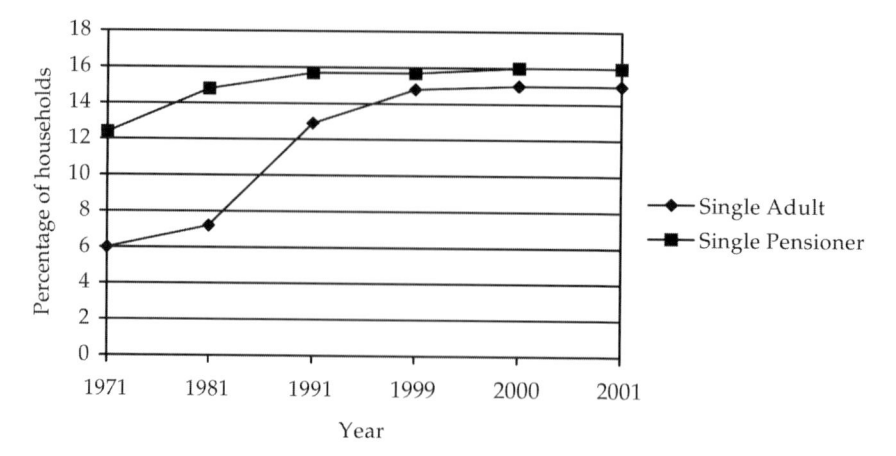

Figure 43.1 Increase in single-adult and single-pensioner households, Scotland, 1971–2001. Source: Scottish Household Survey, 1999; Scottish Household Survey, *Bulletin*, 8 (July 2002).

was accounted for by a rise in single-adult households[4] (6 per cent to 15 per cent) as opposed to single-pensioner households[5] (12 per cent to 16 per cent), as illustrated in Fig. 43.1, although the latter group continues to account for just over half of all single-person households.

The trend towards increasing numbers of one-person households, moreover, is expected to continue. Overall there were an estimated 732,790 one-person households in 2000. Household projections for Scotland suggest that this figure is set to rise to 969,380 by the year 2014, and could account for 39 per cent of all households. Of all household types, single adult male households are projected to increase by the largest percentage, although single adult female households are thought to continue to comprise the larger share of one-adult households throughout the period.[7]

Living alone in Scotland also has a distinctive distribution pattern. For single adults the experience is very much a 'big-cities' phenomenon with Edinburgh and Glasgow having a much greater proportion of their households comprising single adults (aged 16 to 59) than other areas of the country.[8]

Although the greatest increases in numbers of one-person households are accounted for by those under the age of 59, the likelihood of living alone does increase with age, with a greater proportion of older people living alone. In fact 55 per cent of those aged 75 and over live alone, compared with 14 per cent of those aged 25 to 34. Older women are more likely to live alone than older men; 76 per cent of single-pensioner households consist of a female compared with 24 per cent being formed of a male. The distribution of one-person households, particularly in later life, is clearly a very gendered one.[10] (Fig 43.2)

In this context, the following section explores the contemporary experience of living alone in later life in Edinburgh. A particular focus is upon

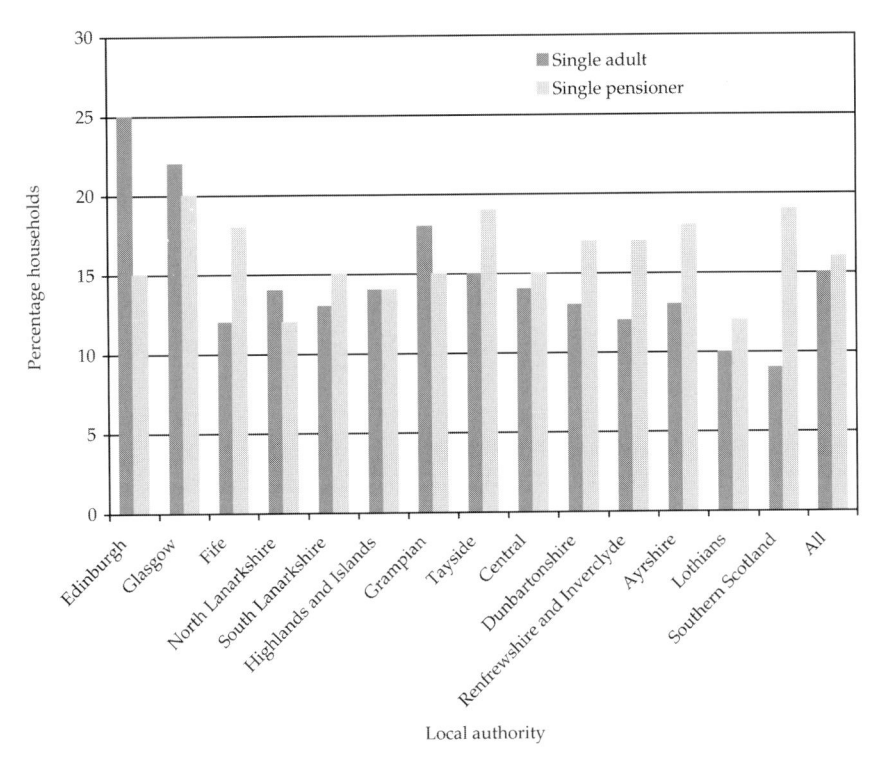

Figure 43.2 Single-adult and single-pensioner household types by local authority grouping. Source: Dudleston, et al, 2001.

how the nature of people's relationship with their home changes as they make the transition to living alone having been recently bereaved. The 'dwelling', and the meanings which are attached to it, emerge as interwoven with the relationships which are formed within it. A changing household structure, and in particular the change to living alone, plays an important part in shaping the experience of the home environment.

LIVING ALONE IN LATER LIFE IN EDINBURGH

Behind the above statistics lies a range of experiences of being, and reasons for living, in a single-person household. Those who live alone often choose to do so for positive lifestyle reasons as they seek independence and freedom.[11] In a survey of 538 older people living alone in four areas of England – Harrow (London), Northampton, Oldham (Lancashire) and South Norfolk – Tunstall found that the 'overwhelming majority' of those interviewed preferred to live alone. Tunstall, however, did acknowledge that this was a very specific question about a complex reality and that participants may have been putting a brave face on a situation over which they had little control.[12]

This section is based on doctoral research conducted by the author in

Scotland's capital city, on the housing pathways and experiences of a group of people over the age of 65. As part of this study, 34 in-depth qualitative interviews were carried out. Participants were asked to describe their housing options, decisions and pathways since they were aged 50 with a view to exploring how their housing pathways were affected by the wider macro-environmental context as well as events within their own lives.

Nearly half of those interviewed had gone from a situation of living with another person or other people at the age of 50 to living alone over the age of 65. Descriptions of experiences of what for many was their first experience of living alone varied. For some remaining in their current home was what was important; for others a new start and a new home was desired.

RESPONSES TO BEREAVEMENT AND THE EXPERIENCE OF LIVING ALONE: TO MOVE OR TO STAY PUT?

For those interviewed as part of this study, the transition to, and experience of living alone is imbued with an emotional significance beyond the numerics of changing household structure. The concept of home is overlaid with the domestic idyll, with family relationships being central to ideas about home life.[13] It has been argued that home and family are virtually interchangeable terms,[14] and that a home is only a home as long as the family is in it.[15] 'Home,' Jones argues, 'is thus not just about housing, but also about a household, and preferably a family household.'[16]

A number of the participants in this study had to negotiate a new relationship with their home centred on it no longer being a 'family' home. For younger households the transition to a single-person household usually involves moving from the family home to a single-person home. This is not generally the case for older people, for whom the same dwelling usually goes from being a family home to a single-person home (Fig. 43.3).

For a number of those interviewed, memories which associated their home with the recent sadness of bereavement and the loss of their loved one meant that for them moving to a new home was the preferred option. Helen talks about the impact that a declining household composition had on her feelings about her home:

> I felt that we had gone from four, and then to three and then to two and then to one. Em, I didn't particularly like living alone. He died in the house so I thought 'no, no I can't', so that is when I moved. So that kind of kept me going for a while, all this, and then I moved. I have never regretted moving. I wouldn't have been happy in the house.[17]

Connie explains why she decided to move home following the death of her husband:

> I just wanted away. I go back sometimes to see the neighbours but it

Figure 43.3 Mrs Nettie (Annette) Kellock in her Rodney Street home, Edinburgh, 1973. Source: SLA 128A(1) xvi/16/9.

takes an effort ... But I find that I see my husband going up that path for the last time and I don't like it.[18]

Helen, Connie and indeed all the participants in the study who moved following bereavement were homeowners. Being an owner seems not only to offer the economic possibility of making such a decision but also choice over the timing of the decision. For both Connie and Helen this decision was made quickly:

> The day after he was cremated. He was cremated on the Friday and I looked at a paper on the Saturday and I saw these flats advertised. And on the Saturday afternoon, and by the Wednesday I had put the deposit down.[19]

> Aha, I moved in the April. We went out for a walk and these houses were being built. And we never walked that way and we thought we will have a wee look at them. And I said, 'right, I am going to move here, I am going to put my name down for one being built'. And it was like that. A lot of people had said 'you are doing it too quick'. But na, I wasn't, I could not have lived in the house by myself.[20]

For Irene, it was a longer process of adjustment and negotiation, in terms of *when* she moved.

> The girls were always on to me to move you see. And I said no when I am ready to move. I was always aware of [husband's name] presence in the house ... I felt a presence for four years and then I said now I am ready to move.[21]

LIVING ALONE • 781

and *where* she moved to:

> So I looked at some houses in [existing area]. And [son-in-law] said no. He said if you are moving you are better to move up this way to the girls [her daughters].[22]

Those who rent their homes from the local authority do not seem to have the opportunity to adjust to their new situation in the same way or as easily. Following the death of her husband, Jean spent some time living with her daughter before moving back to her own home:

> SP: Can I ask you after your husband passed away was it hard for you living in the house?

> Well, I stayed with my daughter for about a month. But as each day was going on I was getting fidgety. So I says to her one day, I said, '[daughter's name] I should go back to my own house'. She said, 'do you feel like you are wantin' to go back'. I said, 'oh yes,' I said, 'I am not goin' to stay here. I am fine, you have done well and I have got myself a wee bit settled'. She had cleared everything away, you know.

> SP: That belonged to your husband?

> Cleared all his suits away and everything like that before I went there. So when I went in it was just myself and that was it. I would say it took me nearly a year to get myself mobile. But then the first year I was a fit woman then. It was no problem, ken. But eh otherwise I had to get back. I had to get back … [23]

Jean had previously spent five years asking the council for a move back to her native area, so it is unlikely that moving again after the death of her husband would have seemed a feasible option to her.

Not all of those interviewed who had experienced bereavement and subsequently lived alone wanted to move. As a homeowner Margaret had the choice to move to a new home and make a new start, and indeed was advised to do so by some friends. Her home, however, was still a place that held her, a place represented by something more than just the actual dwelling, and encompassed the wider environment:

> No, no. And I remember when my sister died one friend wrote and said, 'you must move, there will be too many sad associations'. But I never wanted to. Because I am outdoors and it is near the beautiful [name of local park] Park. And that was why my parents went there from Leith in the first place in 1917, because they liked the out of doors and [name of local park] Park. A great place. So you could say that the environment was pulling me to stay near. This beautiful environment which meant a great deal to our family.[24]

Her transition to living alone was aided by a strong social support network:

> SP: And after she passed away how did you feel living in the house

after she was gone? Was that something that was very difficult for you? I know that is quite a hard question.

No I tell you it wasn't difficult living because the church was a great source of help of, of what is the word, support. And my sister had been a great member. So that was, but I couldn't bear to have her photograph in my bedroom until four years had passed.[25]

MATERIAL DIMENSIONS OF HOME

In addition to the emotional relationships discussed above, the experience of living alone also interacted with the participants' relationship with the material dimensions of their home. For Mary, a council tenant, the transition to living alone was a reason to move to a smaller house. She states that:

No, I planned to stay in that house. And of course when the children got older and I was only there myself, I said well, if the town wants a bigger house and if they ask me to move into a wee one I will do so. But of course the town didn't. And I was in the big house on my own. And it was when I fractured my hip that I got the smaller house. Although I had a bedroom downstairs, the toilet was upstairs so they gave me the wee cottage that I am in now.[26]

Initially this move to a smaller dwelling 'was a load off [Mary's] mind' but subsequent experiences have unfortunately proved less positive as the step access into her home inhibits her mobility.

For Alice, the owner of a high-value property, the situation seems more complex. The size of her home constantly reminds her that she is now alone and reinforces feelings of isolation:

Now that my husband is gone and all my friends kind of thing I feel that I very seldom use the lounge, very seldom. And then I don't have very many people coming to stay.[27]

Alice talks about the possibility of selling and moving somewhere smaller. Her doctor, however, advised her to cordon off parts of her home and just to use what she needed. Her son would prefer her to go and live nearer him and his wife. For Alice, the experience of living alone, in her own home, although lonely, is preferable to losing her sense of independence:

And my son what he would like, he would rather if I didn't look for anywhere here but that I moved nearer him. He has got plenty of room, he could take me there, but I wouldn't want that. He could even, in the ground, he has got so many acres of grounds that he could even build a little granny house, you know. But I would be there by myself all of the time because they are out at work. Stuck in the garden as it were, not seeing anybody, can't get on a bus because there are no buses near, quite a way to walk, and the bus is once an hour. So it is no use going to my son's. He would prefer. When I say I am selling this time, I get them all worried. He says where are you going Mam? I say I don't know where I am going but I am selling. He is frightened that I will make a mistake by selling the house that I shouldn't have sold, you know. So he is trying to always keep in touch with me and see that I don't sell the house. So I don't know where I'll land.[28]

'COPING' WITH THE EXPERIENCE OF LIVING ALONE

Whilst most of those interviewed for this research were contented with their housing circumstances, the experience of living alone was not always a happy one. This applies to both owner occupiers and to local authority tenants. Irene and Sheila allude to the loneliness and fear of living alone:

> SP: You are happy?
>
> Well I've got accustomed to speaking to myself![29]
>
> SP: And how did you feel about living in the house in Craigour Road after your parents passed away?
>
> Well it was something that I had to get used to. I had some friends of my mother, elderly people, you know, and I would go round to them. I was like another daughter to her, you know. I kept her young and she enjoyed my company. And I enjoyed hers. But if I got frightened I could cut down the back ground, through to them. Oh it is not nice.
>
> SP: Did you find it difficult being alone?
>
> Oh aye. People knocking at the door at 12 o'clock at night … I am putting on a strong voice … I had a little bit of fear when you are living alone and it got worse. But it was semi-detached. I wasn't on my own as such.[30]

Living alone, however, does not necessarily mean being alone or indeed coping alone. Support in managing around one's home can be received from both informal and formal networks. Jean talks about the support that she receives from both her neighbours and her family:

> Oh aye. I have got good neighbours. I mean if they don't see you for a few days they come up to the door to see if you are all right. And when I was ill they were very good. Oh aye. They were up all the time. Asking if I wanted anything and that. But my daughter she is good. Oh aye. She comes up Wednesday and Friday. I write out the messages and she gets all the messages and she won't let me hoover. I can hoover but she'll not let me. She says, 'you leave that alone. I'll do the hoovering. You do your dusting and your washing'. I can do all of that. I do my own cooking. I've got a micro of course. It is easy![31]

For others it is the more formal support of a home help which facilitates their experience of living alone:

Oh, I didn't get out very much. I had to have a home help as well, because I couldn't deal with the things.[32]

SP: And you have a home help as well?

She comes two days a week, Tuesday and Thursday. She is very good, aye. She has been doing her job for nearly twenty years. I have had her for quite a while now. She does the housework. Although it is not dirty like. On Tuesday and then on Thursday she goes up to the store and gets my messages for the week like. She makes my dinner for one day and I get her to get readymade meals in the store. And I say, 'well get me three like.' And I get butcher meat from across the road. I have got to watch when I am cooking because I can hardly stand like.

SP: So how do you manage?

Well, when I say cooking, I mean boil an egg or something like that. My main meal I use ready meals and I can put that in the oven.[33]

SP: Do you have a home help?

Yes. And I have got someone to do the messages. The home help used to do the messages but I have changed it. It used to be £2.50 and then it went up to £3 and I pay for the home help because I get attendance allowance.[34]

This type of help can be combined with family support:

SP: And how do you manage about the house?

I have a help who comes on a Wednesday at four o'clock for two hours.

SP: And do you cook your own meals?

I don't cook them all. I get two meals here. And I cook the weekend's meals. And the girls bring me Marks and Spencers. Collectively you know. They fill up the freezer.[35]

Ultimately, however, it is a process of individual adaptation to a new living circumstance:

SP: Okay and how do you manage about the house? Are there any tasks that you find difficult?

[Shakes head]. I open the plugs up and have a look. It is surprising what you learn to do when you are on your own. Or what you try to do when you are on your own.

SP: You don't have the other option.

That is right, you try everything, and then you think well no I cannae manage it, you know. But I can manage myself. My daughter will not let me put up curtains now, she will not let me go up ladders.[36]

CONCLUSION

Over the past 40 years living alone in Scotland has become a less marginal household structure and as the projections above show it is likely to become even less so. Those who live alone are a varied group who occupy such a household structure for a variety of reasons. This chapter has explored how a group of older people in Edinburgh have negotiated the experience of living alone as a result of being bereaved. Household type, and in this instance the transition to living alone, emerges as an important factor in how one's 'home' is experienced.

NOTES

1. Jamieson, et al, 2001.
2. Hall, Ogden, Hill, 1999.
3. Marsh, Arber, 1992; for an overview of changes in households, especially family households, see Chapter 28 Hierarchy and Authority.
4. A single-adult household consists of an adult of non-pensionable age and no children, Scottish Household Survey. *Bulletin*, no 7 (2002).
5. A single-pensioner household type consists of one adult of pensionable age and no children. Pensionable age refers to men aged 65 and over and to women aged 60 and over, Scottish Household Survey, *Bulletin*, no 7 (2002).
6. Scottish Household Survey, 1999; Scottish Household Survey, *Bulletin*, 8 (July 2002).
7. Scottish Executive, 2000.
8. Dudleston, et al, 2001.
9. Dudleston, et al, 2001.
10. Hope, et al, 2000.
11. Jones, 1987; Hall, Ogden, Hill, 1999.
12. Tunstall, 1966, 55.
13. Allan, Crow, 1989.
14. Oakley, 1976.
15. Gilman, 1980.
16. Jones, 1995, 2.
17. Helen, interview 25.
18. Connie, interview 9.
19. Connie, interview 9.
20. Helen, interview 25.
21. Irene, interview 12.
22. Irene, interview 12.
23. Jean, interview 24.
24. Margaret, interview 19.
25. Margaret, interview 19.
26. Mary, interview 8.
27. Alice, interview 4.
28. Alice, interview 4.
29. Irene, interview 12.
30. Sheila, interview 16.
31. Jean, interview 24.
32. Mary, interview 8; for further information on the home care available to those who are infirm see Chapter 26 Infirmity.
33. William, interview 26.

34. Elizabeth, interview 33.
35. Irene, interview 12.
36. Helen, interview 25.

BIBLIOGRAPHY

Allan, G, Crow, G, eds. *Home and Family: Creating the Domestic Sphere*, Basingstoke, 1989.

Dudleston, A, Hope, S, Littlewood, A, Martin, C. *Scotland's People: Results from the 2001 Scottish Household Survey, vol 5: Annual Report*, Edinburgh, 2001.

Forrest, R, Murie, A. Housing market, labour markets and housing histories. In Allan, J, Hamnett, C, eds, *Housing and Labour Markets: Building the Connections*, London, 1991, 63–91.

Gilman, C P. The home: Its work and influence. In Malos, E, ed, *The Politics of Housework*, London, 1980, 72–83.

Hall, R, Ogden, P, Hill, C. Living alone: Evidence from England and Wales and France for the last two decades. In McRae, S, ed, *Changing Britain: Family and Households in the 1990s*, Oxford, 1999, 265–97.

Hope, S, Braunholtz, S, Playfair, A, Dudleston, A, Ingram, D, Martin, C, Sawyer, B. *Scotland's People: Results from the 1999 Scottish Household Survey, vol 1: Annual Report*, Edinburgh, 2000.

Jamieson, L, Wasoff, F, Cunningham-Burley, S, Backett-Milburn, K, Kemmer, D. *Solo Living in Scotland: Trends and Issues, June, 2001*, Edinburgh, 2001.

Jones, G. Leaving the parental home: An analysis of early housing careers, *Journal of Social Policy*, 16/1 (1987), 49–74.

Jones, G. *Household Formation amongst Young Adults in Scotland*, Edinburgh, 1990.

Jones, G. *Leaving Home*, Buckingham, 1995.

Marsh, C, Arber, S. *Families and Households: Divisions and Changes*, London, 1992.

Oakley, A. *Housewife*, Harmondsworth, 1976.

Pickles, A, Davies, R. The empirical analysis of housing careers: A review and a general statistical modelling framework, *Environment and Planning*, A/23 (1991), 465–84.

Scottish Executive. *Household Projections for Scotland*, Edinburgh, 2000 based, <www.scotland.gov.uk/stats/bulletin>

Tunstall, J. *Old and Alone: A Sociological Study of Old People*, London, 1966.

44 Communities on the Move

HUGH GENTLEMAN

INTRODUCTION

Most of this volume deals with conventional dwellings since that is where the majority of the population lives, whether in flats or houses in cities and towns, in smaller settlements or isolated farms and cottages. Even the more communal forms of living, covered in other chapters, are still centred on permanent buildings, firmly rooted to the ground – if their occupiers have to move for any reason, it is they who move, while the dwelling remains to be occupied by newcomers. Although this may be seen as the dwelling norm, it should not be forgotten that small groups within the Scottish population live in less conventional structures which do not possess foundations and in many instances go with their occupiers if they move from place to place, in the past even like the snail with their homes literally carried on their backs. This chapter, and the one which follows, look at some of these groups and the factors governing lifestyle in their dwellings.[1]

The groups include two which, while distinct, share certain similarities; Scotland's Travelling People, and Show or Fairgound People.[2] Brief mention is also made of a third group, the 'New Age' travellers. Although these three together amount to only a small part of the Scottish population, each illustrates an ongoing facet of Scottish life in which patterns and lifestyles have been adapted to the particular constraints of the structures in which people dwell.

TRAVELLING PEOPLE

The background
Travelling People have been a feature of Scottish life for at least 800 years, whether known by the old names of 'Tinkers', 'Egyptians' or 'Gypsies' or, more recently, as 'Travellers',[3] initially adopted as an inclusive term avoiding the more pejorative earlier names, although later, in turn, acquiring negative connotations. Their roots seem to lie in both an indigenous ancient Celtic tinsmith caste and Romany Gypsies, the groups coming in contact probably around the sixteenth century as the latter spread from continental Europe through England and northwards into Scotland (see Fig. 14.3). Whether there occurred an actual fusion or just a paralleling of two groups following similar ways, lifestyle and work is unclear. While other people may have

been absorbed later, the original two strands remained identifiable into the early twentieth century.[4] Borders Travellers, with their long history of movement between England and Scotland, tended to be known as Gypsies, particularly around the end of the nineteenth century, with Kirk Yetholm the home of their 'King'. While they adopted the term 'Traveller' for expediency in the 1960s, some members of this group have now reverted to their old title for similar reasons. Conversely, many Scottish Travellers, particularly in the north, who also accepted the Traveller title, feel quite distinct and reject the term Gypsy if it is applied to them, 'Dinna cry me ain o thae foreigners. Am no wan o yer ethnics. We've aye been here', being the strongest assertion from one Traveller.[5]

Travellers have always been seen as an out-group, never accepted, occasionally tolerated but often punished, sometimes severely, by the law and generally marginalised. At one time in Scotland to be, 'repute and haldin to be an egyptiani' was punishable by death.[6] Nineteenth- and early twentieth-century government committees began to look at the Travelling People more objectively, although they were sometimes lumped with insalubrious-seeming bedfellows.[7] Most reports concluded that it was necessary to turn Travellers away from their existing lifestyle through settlement into houses and education of children into settled ways.[8] Only about 30 years ago did government begin to accept the Travellers' right to continue their traditional semi-nomadic lifestyle and occupations with places where they could legally stop and park or erect their dwellings, whether for short or long periods.

Scottish Travellers are part of a wider Traveller community stretching not just through England and Wales but also Ireland, much of Europe and effectively worldwide. A common misconception is of a homogeneous community. In practice it is made up of many groups all sharing some common features of lifestyle, kinship and descendency-based membership and economic role. All share marginalisation and often discrimination or persecution by settled society, and lack equality of rights. Even among any one group within the Traveller community, social structure and wealth cover a spectrum from very poor to very rich and from those living in a small caravan by the roadside to those who reside in a large house and travel only occasionally. Similarly it covers those 'who live in quiet retreats, rubbing along with their neighbours and avoiding confrontation and also includes those who gain from the exertion of power and extortion within the Traveller group'.[9] Self-ascription as 'Travellers' is the essential link. Their adopted title has generated the fallacy that they are, or preferably should be, always travelling, whereas their pattern has always involved summer mobility and a more sedentary winter. Consequently any generalisation is a minefield and this chapter can hope to give only a flavour of some past and present conditions and some reasons for them.

Numbers
Estimates of numbers of Travellers in Scotland have varied over the past century or so, using differing criteria, and have not always been objective.

Better figures began to emerge from the 1960s. Two detailed censuses in 1969[10] estimated that there were around 450 households (c2000 persons) using caravans, tents or some form of hut shelter throughout the year on some type of camping place, usually illegal. A further national count in 1992 revealed more, but smaller, households (c750 with c3000 persons in total),[11] probably reflecting the growing availability of legal camping sites whose individual household pitches are not necessarily compatible with the earlier extended family groupings found on illegal encampments. Although each count from the 1890s until 1992 was essentially a time snapshot, those in 1969 and 1992 shared a common methodology, content and coverage. Counts by every Scottish council each January and July since 1998, covering official council sites, privately owned sites and unauthorised encampments, have begun to provide better time-series information; by the end of 2002, totals across the three kinds of stopping place had settled down to around 600 households (c2200 persons).[12]

Some early estimates included Travellers who had settled more or less permanently in houses but who still regarded themselves as part of the Traveller community and culture. The number of these people ranged up to 10,000–15,000, or even higher[13] but, while often quoted as fact in subsequent texts, the figures have never been successfully validated. No count since 1969 has attempted to assess this probably unmeasurable group.

The Travellers' role
Travellers have traditionally adopted a mobile lifestyle for most of the year, partly to follow chosen ways of making a living but also because of being unwelcome at one place for any length of time. Their accommodation generally moved with them or, in time of need (seasonal work or family reasons such as age, illness or children's education), they made temporary use of fixed accommodation (disused farm cottages or bothies or empty urban tenement flats), all more available than now, when accommodation may have to be found from day to day or may even involve the use of self-built shacks. This move between 'traditional' and more conventional dwellings was probably more common in Scotland than in the south. The increasing difficulty of finding camping places, pressure from the settled community and sometimes encouragement from those concerned with their welfare, have pushed many individuals into long-term or even permanent house dwelling, although without reducing their sense of being part of the Traveller community, and many people who have lived in houses for years still take to the road at some point in the summer.

The interplay of three main factors has always governed Traveller domestic life: work activity (involving most members of a family and also determining wealth or poverty); the nature of their dwelling itself (which depends in part on relative wealth); and where the dwelling happens to be.

Work, activity and lifestyle
The Traveller lifestyle has always been based on self-sufficiency and self-enterprise. As buyers and sellers of items at the doorstep, particularly in rural areas, a channel for news, dealers in horses and collectors of unwanted material and recyclers, Travellers identified a need and provided a service.[14] From the 1950s these more traditional activities began to be superseded by scrap-metal collecting, sorting and selling. As demand for and the suitability of this declined, new roles were identified and adopted such as tree cutting, fence and gate painting and small-scale tarmac work. Scotland has always offered varied opportunities, from the central belt with its metal-based industries and the agricultural areas with demand for short-term labour at fruit or potato harvests, to the coast with opportunities for shellfish gathering. The main key to survival over the centuries has been practical adaptability. However, occupations have always varied in profitability between or within years and their success or otherwise means substantial swings in families' relative wealth or poverty, in turn affecting standards of domestic life and prompting moves upwards or downwards within or between given types of dwelling to a greater extent than most non-Traveller households.

Dwellings and lifestyle
A transition can be identified through three main forms of traditional accommodation: tents, horse-drawn caravans and trailer caravans. Official reports over the past hundred years or so have noted other unconventional dwellings used from time to time including caves, huts or sheds, disused sail lofts, empty bothies, or even shelters made of branches.

The twentieth century as a whole and, even more, its second half, saw significant changes in the conditions for Traveller domestic life, both in terms of the dwellings themselves and where these could be located. Most marked was the shift from the tent as the principal mode, to the horse-drawn caravan from the late nineteenth century and later, mostly after World War II, to the trailer caravan, towed by a van or lorry. This shift can be seen in Fig. 44.1, summarising data from Scottish reports over the past 80 years with comparative figures from elsewhere in Britain (note that numbers may total more than 100 per cent owing to households having more than one form of accommodation)[15] (Fig. 44.1).

THE TENT
The most basic form of accommodation, probably for the greatest part of the Travellers' existence in Scotland, and throughout Europe, was the tent. By the end of the twentieth century use of tents in Scotland had almost entirely died out, although probably lasting longer than south of the border, reflecting a response to terrain and poorer households. In 1917 about a third of families in Scotland lived in tents, mostly as their only dwelling and, even 50 years later, the figure was not very different. Ireland in 1961 was similar, whereas in England and Wales only 4 per cent of Travellers were tent-dwelling in

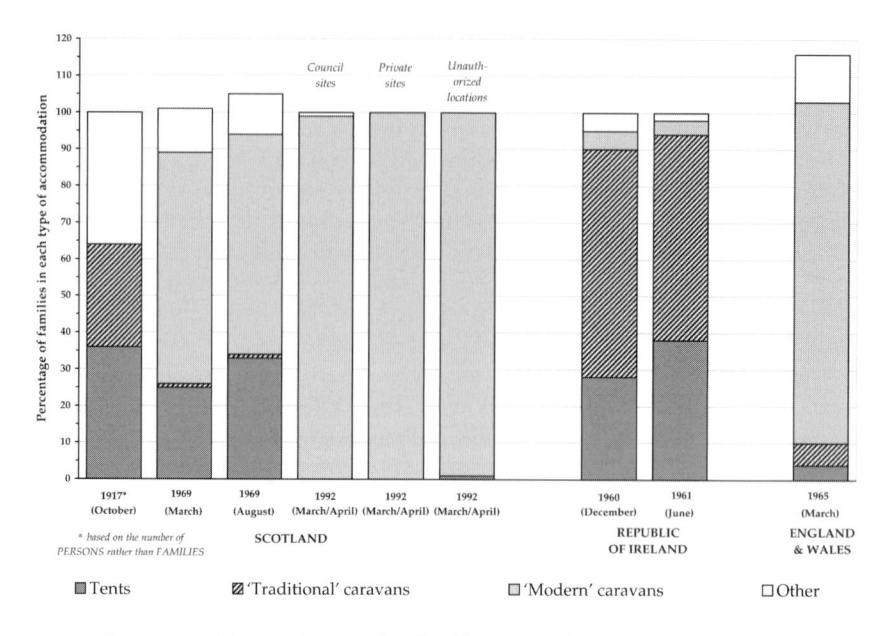

Figure 44.1 Changes in Traveller dwellings over the years. Source: H Gentleman.

1965. The marked decline to only a tiny proportion of Travellers in Scotland living in tents by 1992 probably reflects the fact that tents were not allowed on the new official sites being developed by local councils and probably not on privately-owned sites either. Only on unauthorised encampments did they appear.

Traveller tents did not conform to the conventional image of a tent, but rather were formed of a simple round or oval frame of flexible boughs cut from hedges or woodland (hence the common terms of 'bough', 'bow' or 'bender' tents and 'humpies') covered with some form of tarpaulin or canvas, or later sometimes with plastic sheeting, the edges being tied or pegged down or weighted with stones. Tents might vary in size and style from a small single or twin tent, perhaps only 5 or 6 feet (1.5–1.8 m) across and 4 or 5 feet (1.2–1.5 m) high, to much larger structures of 12 to 15 feet (3.7–4.6 m) by 8 to 10 feet (2.4–3 m) and up to 6 or 7 (1.8–2.1 m) high in the centre. Often only the cover travelled with the family, a frame being found at each stopping place, although this too might be carried in areas with little suitable woodland. Depending on its size and the family's means of transport, tents could have more permanent iron hoops to support the covering.

The 1971 study of Travellers in Scotland[16] gives details of different kinds of tents, including the occasional manufactured gateleg tent, found at least up to the 1970s (Fig. 44.2). It also emphasises that domestic life in a tent was very basic. A well-pitched and carefully built tent could form an extremely secure and weatherproof habitation and the heat level inside when the stove was lit was a cause for comment by both Travellers

Figure 44.2 Travellers camped illegally on a former bing in Bellshill, Lanarkshire in 1969. The camp includes trailer caravans, a large bow tent and a gateleg tent. Source: © H Gentleman.

Figure 44.3 Interior of a Traveller family's large bow tent in Kintyre, Argyll in 1970. Note the roof structure, the homemade oil tin stove and the hooked rod for the kettle. Source: © A Stewart.

Figure 44.4 Continuity of traditional construction styles; paired bender tents. The upper example is from Perthshire, 1893, the lower from Kintyre, Argyll, 1969. Sources: upper; Report of the Departmental Committee on Habitual Offenders, Vagrants, Inebriates and Juvenile Delinquents, Edinburgh, 1895), lower: © H Gentleman.

themselves and by outsiders on the rare occasions when they were invited in.[17] Such standards were, however, unusual and only possible when a household was in one place for a period, with time and effort spent on the structure. Most tents were much less secure, offering limited space, and were damp and dark and not proof against sustained bad weather. Some larger tents may have had wooden floors but others often had little or no flooring, save perhaps pieces of linoleum or old carpet (see Fig. 44.3). Similarly, while some larger long-term tents might have had a few pieces of furniture such as an old iron bedstead or a couple of battered armchairs, others would have none and their occupants little in the way of other belongings. Tents were recorded in 1969 in which 'the only belongings for an elderly couple were a few scraps of straw, a few piled up blankets and a cardboard box with a little food and a few cooking and eating utensils'.[18]

In good weather any cooking would be done on an open fire outside, particularly with smaller tents. Small, paired tents would generally have the fire between them[19] (Fig. 44.4). The fire provided warmth and a means of cooking and was a focus for socialising among large or multiple families. Larger tents might use an open fire for much of the time but have some form of stove inside for both heating and cooking. Some stoves were small cast-iron models but often they were homemade from old paint or oil drums. Fuelled by wood or coal, these stoves could be hazardous, leading to tents burning down, occasionally with fatalities, as happened in Argyll in 1969 and near Airdrie, Lanarkshire, in 1970. In 1969, rather more than half the families living solely in tents used solid-fuel stoves of some kind for all their cooking and heating, others using both stoves and open fires. Paraffin stoves were occasionally used, but only rarely was bottled gas to be found among tent dwellers in the 1960s.

Despite these problems, many Travellers have retained a fondness for tent dwelling and some tents may still be used in more remote areas during the summer months. As late as the 1970s, Travellers talked to the author about the pleasures of the tent, and some of the literature about and by Travellers paints a glowing picture of this way of life.[20] In practice, most reflects the summer story and probably is a nostalgic referral to times – perhaps pre-War – when camping places were easier to find, acknowledging that much wintering was in slightly more solid, if not necessarily conventional, dwellings.

THE HORSE-DRAWN CARAVAN
The horse-drawn caravan may be a development of the two-wheeled cart. While some Travellers transported tent and belongings on their backs, others might have a hand or horse-drawn cart, the latter more necessary for those with a large tent and a cast-iron stove. Carts probably then acquired a cover, allowing some members of the family to sleep there, with others using the tent. There followed a transition to more identifiable caravan structures, mainly from the late nineteenth century onwards, but often used in conjunction with tents. The Scottish pattern, involving a lighter structure than the English one, which was less suited to the terrain, was generally a flat wooden base on smallish wheels with a top of wooden hoops covered with canvas,[21] almost akin to a tent on wheels, a format also used in the south and common in Ireland at the same period (Fig. 44.5). Horse-drawn caravans have largely disappeared as regular dwellings in Scotland, although they occasionally appear at seasonal Traveller fairs. The advantages of a horse-drawn caravan included not having to dismantle it each time the family moved and it being usable in transit. It also was a more weatherproof type

Figure 44.5 A horse-drawn Traveller caravan near Wormit, Fife in 1953. Source: © D C Thomson & Co., Ltd.

of dwelling, being raised off the ground. Personal belongings could be stored more easily within it, in boxes or drawers under beds or in higher-level cupboards. It might also contain a small solid-fuel stove but consequently, as in tents, fire could be a hazard. Space, although more generous than in a tent, would still be very limited, so most daily life continued to take place outside.

THE TRAILER CARAVAN

Gradually, but particularly after World War II, trailer caravans (or trailers) superseded horse-drawn vans. Although providing more space, trailers need a motor vehicle to tow them and adequate financial resources for both. Initially the homes were mostly conventional smaller holiday or touring vans, bought new, or more often second-hand, and sometimes adapted to suit families' own particular needs. The variety of style and quality used over the past 50 or more years is illustrated in many of the official reports.[22] Compared with horse-drawn vans, but also depending on the trailer's age and the family's wealth or poverty, they could be reasonably spacious and comfortable or extremely cramped and dilapidated. While in general they provided better protection from the weather than earlier types of moveable dwelling, much depended on the age of the trailer and its maintenance, and in practice, many identified in the 1969 counts were very derelict.

At the wealthy end of the Traveller scale, specialised, custom-built trailers began to appear from the 1960s onwards. At first these were built to order by specialist firms in England, but imports from Germany subsequently became common. Extending to the maximum size towable on the road, they can be spacious and luxurious, with fitted kitchens, electric lighting and even central heating from a solid-fuel or bottled-gas stove. (see Fig. 44.6). Internal space has become more versatile, allowing for either specific bedroom areas or at least effective partitioning to allow some degree of personal privacy, a concept completely lacking in most of the earlier forms of Traveller dwelling.

Despite these improvements over time, all forms of moveable Traveller dwelling share a number of inevitable restrictions in the way in which the family can live within them. While newer trailers may be better constructed, few are designed for long-term accommodation and leaks and condensation can still be a problem, particularly if the van is not well maintained. One of the greatest constraints is cramped living space, with relatively large families occupying very small dwellings. Traveller families live at a higher density per unit of accommodation than the settled population, although, as will be seen later, household size has come down. The 1971 study[23] noted that:

> ... the average size of Traveller family or household is four to five persons (the 'mean', based on the total numbers of people recorded

Figure 44.6 Interior of a comfortable modern Traveller trailer in Lanarkshire in 1969. Source: © H Gentleman.

Figure 44.7 Children form an important part of Traveller life, as do horses. Camping on an old farm in Lanarkshire in 1970. Source: © H Gentleman.

divided by the number of households) and if the caravan, tent or hut is treated as the unit of accommodation, the great majority of Traveller families live at this kind of density.

Lack of space within the dwelling itself has major implications for privacy between family members in terms of generation and gender segregation, issues which are being increasingly recognised. Modern trailers may have internal partitions to divide up space but older and smaller trailer caravans, horse-drawn caravans and, most of all, tents, meant a single communal space having to suffice for all. Recently, some families have acquired a second trailer in order that age groups or genders can be segregated.[24]

A feature common to all trailers, and of course to earlier forms of home, has been the complete absence of toilet facilities within the dwelling. This is not only a function of lack of space, but also reflects inherited cleanliness customs and mores. Although some families now have portable camping toilets, for those who have to stop on unauthorised locations, use of 'all of Scotland' is more common, a frequent cause of complaint from any nearby house dwellers. Domestic life tends to be governed by other cleanliness customs relating to the separation of washing bowls for food, clothes or personal use, practical responses to the constraints of cramped and nomadic existence.

Although better trailers and improved facilities, both within dwellings and on recognised sites, mean more activities can take place within the dwelling, much of domestic life continues to be lived out of doors, whether within an individual family or between a group of families travelling together. Close integration of domestic life and work generally leads to the dwelling being surrounded by the paraphernalia of the household's economic activity, whether work vehicles or materials. Children play outside and from an early age are likely to get involved in some way in parental work activities – often acquiring a knowledge and practical skills which many settled children may not have at an equivalent age – even though their conventional education remains a significant problem (Fig. 44.7).

A Traveller's caravan is an almost instantly tradeable commodity, whether in time of need or by choice, as an upwards or downwards deal with another Traveller. If families' resources are adequate, their domestic expectations will have changed just as those of the settled community's have. Early on, many Travellers in their trailers acquired battery-powered televisions and a number of the other trappings of modernity, and these have tended to break down some of their isolation from the wider world. The author recalls the seemingly strange juxtaposition of values and technologies to be observed while sitting with Travellers in their modest but comfortable trailer on an unauthorised, muddy patch of derelict land with no facilities in North Lanarkshire, watching on their battery television the 'splash down' from the first moon-landing flight in 1969.

Places to stay and lifestyle
While domestic lifestyle depends in part on the nature of the dwelling itself, it is equally a function of its location. Throughout their history Travellers have found places to stay as and where they could while moving seasonally, generally in search of work, but often following fairly regular movement patterns. A farmer may have found them a corner of land when using their seasonal labour at fruit or potato harvest time. Another seasonal activity for some, shellfish gathering on beaches, generally led to finding a place on or near the foreshore. Traditional activities such as hawking meant stops on common or waste land on the fringes of small towns and villages, or roadside stops in more isolated rural areas.

Up to, and perhaps just after, World War II, the presence of Travellers was often tolerated as their stay was fairly brief. Tents or horse-drawn carts and caravans were not too visually intrusive, their activities might be seen as useful to local communities and a place used on a regular basis might be left reasonably clean and tidy on departure so as to permit return visits. Nevertheless, Travellers could be convenient scapegoats for accusations of theft or damage. They could also feel the consequences of appearing in villainous roles in folklore, particularly in relation to child stealing, mothers in the settled community calming their young children with, 'Hush, hush, dinna fret ye, the black tinkler willna get ye', or its southern counterpart, 'My mother said, I never should, play with the Gypsies in the wood ... '

At that time too, interfamily differences among groups sharing a stopping place could be resolved without too much conflict by one or other family moving away, and alternative stopping places could be found fairly easily. When it was necessary to find more solid winter living quarters, empty bothies or farm cottages could sometimes be rented or 'bought' for a few months. Nevertheless, significant practical difficulties accompanied this kind of lifestyle. The pleasures of moving from place to place were countered by suddenly being told to move on, or it might be difficult to find anywhere to stop, irrespective of weather or the seriousness of family circumstances. Stopping places might be unsuitable, on dangerous or derelict land, or by the side of a busy road. Whatever the stopping place, the greatest problem has always been that of access to water, the most basic commodity of all. In rural settings it might come from a spring or burn but this was not always available and so water had to be carried or obtained from a farmer, a garage or a householder, perhaps only grudgingly or with payment. With modern trailers and lorries, transport of water may be easier than in the past, but is a far cry from the constant and convenient water supply which the settled population takes for granted.

By the 1950s and 1960s the situation was beginning to change. Many conventional dwellings formerly available to Travellers at certain seasons were being demolished, being gentrified as second homes, or farmers and other owners were less willing for Travellers to occupy them. Continuing adaptation to new ways of making a living and new forms of accommodation and transport in turn led to difficulties with the settled community as some Travellers moved into the scrap-metal business, often concentrating in and around urban areas and tending to leave debris and damage behind them as they moved or were moved. Increasingly, Travellers were being moved on, sections of old road by-passed by improvements, and patches of empty or derelict land were being closed off and obscure Acts were being used to prevent any kind of unauthorised encampment. At the same time public antagonism towards the Traveller community was rising.

These difficulties were not limited to Scotland. Post-War planning controls everywhere, notably the Caravan Sites and Control of Development Act, 1960, plus increasing demand on land, began to limit the range of places where it was possible to camp. Old Highways Acts were increasingly being used to prevent Traveller encampments. Research into the nature and scale of the presence of Travellers in England and Wales,[25] together with pressure from a variety of bodies, led to the introduction of legislation in England and Wales in 1968, placing an obligation on local authorities to make appropriate provision of places where Travellers could stay legally.[26]

Neither the research nor the legislation extended to Scotland. Coupled with growing evidence of problems facing both Travellers and local authorities in Scotland and the increasing awareness by Scottish Travellers and those working on their behalf of the policy differences, The Scottish Office realised that it had a problem on its hands but little knowledge of its nature. In the light of this, in 1969, it commissioned its own research, initially a national count, which developed into a comprehensive study of the nature of Travellers' way of life and difficulties, producing recommendations aimed at helping them.[27] From this arose policies[28] based on creating a network of official serviced Traveller sites throughout Scotland, matched to local need but allowing for seasonal movement as needed or desired. Sites would be funded by the Scottish Office (initially at 75 per cent of capital cost but subsequently increased to 100 per cent) but developed by local authorities. This was coupled with appointment of a committee to advise the Secretary of State on implementation of site policy and on wider issues affecting Traveller welfare.[29] In the event this committee lasted for 28 years and while it had no executive powers it was an important catalyst for change. In its ninth report, in addition to summarising its final term of office, it also assessed much of the background, the extent to which the planned network of official sites was achieved between 1971 and 1999, and the fact that it clearly was not going to meet the needs or preferences of the whole Traveller community.[30]

The new policies brought the greatest potential change to the Travellers' domestic lifestyle since their arrival in Scotland. For the first time they could find places where they had a right to stay and call their own which also provided the basic necessities of life, primarily direct access to water but also to electricity. The official sites, to be provided by each local authority in which need could be identified, were based on a concept of individual fenced pitches for each household, providing a hard-standing for their trailer and vehicle(s) with, perhaps most importantly, their own chalet or 'amenity unit', usually in semi-detached pairs, with toilet and shower or bath facilities, a sink with hot and cold running water and space for their domestic appliances such as a washing machine. They also provided a storage area for household belongings. While the chalets were not intended as living space they took pressure off the limited space in a trailer (Fig. 44.8). Standards of construction, based on Advisory Committee guidelines,[31] were basic but were seen as appropriate to a stay which might vary in duration but was not intended to be long-term. One or two early sites had centralised blocks of toilet and washing facilities, but these were rapidly seen as unsuitable and replaced by individual facilities on each pitch. The Advisory Committee also produced guidelines for site management.[32] Development of sites was slow and a number of councils continued to be reluctant to provide them and raised many difficulties and reasons for not doing so, even where need clearly could be identified. Despite policies and grants being introduced in 1971, it was 1978 before the first official site was opened. The grant scheme was discontinued at the end of 1998, although applications in the pipeline would still be funded. By the end of 2002 a total of 37 sites had been developed in 26 of the 32 Scottish councils, theoretically providing almost 600 pitches although, for a variety of reasons, not all were always available.[33] Most council sites are open all year but two or three are for summer use only. Some early sites were later upgraded to provide better standards of build or facilities and one or two were expanded in size to meet additional local need.

Provision of sites has led to changes in lifestyle for many families,

Figure 44.8 A new council site for Travelling People in Scotland in the late 1990s. Source: © E Jordan.

some economic, some personal. Since the sites came into existence a number of ways of making a living have changed. Some can now be carried on from a settled base with working members of the family travelling out for shorter periods but, at the same time, site rules do not permit work being carried on within the site itself – in contrast to former life on unauthorised locations where work and domestic life were coterminous. While improvements in the quality and reliability of vehicles together with improved road networks should enable coverage of a greater potential work area from a base site, it has been suggested that in reality there has been a reduction in both the overall amount and distance of travel in recent years.[34] Personal factors include a better quality of life for some older members of families and also for children. Occasionally long-stay families have acquired larger trailers or a second small van which can be used for part of the family when temporarily away on work or provide the necessary extra accommodation to segregate children. Most of all, the possession of 'an address' which sites provide, helps with access to many essentials of modern life, such as medical or education services or access to state benefits. As council tenants paying rent on the sites, many families with limited financial resources can qualify for housing benefit to meet their rent. Together, these changes have affected the nature of seasonal movement. While it was expected that most families would continue to move around Scotland on a seasonal basis, for the first time the sites gave households security of tenure in return for rent. Those who wished to stay on a longer-term basis could leave for short periods, usually up to a certain number of weeks, without sacrificing their tenancy, although having to pay a retaining fee. However, the extent to which some families have chosen to base themselves on the same site for increasing lengths of time is noteworthy; by July 2002 almost half of those on official sites had been there for more than two years and a few families had been tenants for ten or more years.

Life on a council site is a balance between pros and cons and studies have looked at this from the perspective of both the site providers and the site users.[35] Benefits are seen as: fairly secure tenancy if wanted – now leading to increasingly long stays on site; clearly defined space, some of which households have enhanced with flower tubs or even small gardens; each family having its own 'amenity space' (that is a bath or shower, toilet, sink and utility space), storage space, hot and cold running water and electricity (with the possibilities of connecting both water and power to the trailer); better opportunities of access to social services such as schools and health services; and possibly, over time, improved relations with the surrounding community. Against these have to be set perceived disadvantages: rules and regulations and management styles; costs of rent and electricity; constraints on working on or from the site; and the loss of perceived 'freedom', however phantasmal this might be. Concentrating Travellers into larger groups on a small number of places enforces close proximity to neighbours, without the former opportunities to move away to avoid confrontations. Unfortunately the latter still occur and have become more serious and in the most extreme cases they have led to intimidation and violence and damage, to families themselves, to personal property, and ultimately, in some recent incidents, to the complete trashing of council sites.

Of families recorded in the twice-yearly counts since 1998, around half have been using council sites. But despite the site network, not all Travellers wish to use them, some because there are no vacancies, or no site where they want to be. Others may have been on a site but have either left or been evicted for non-compliance with site rules and are no longer allowed access, or they may feel that the cons listed above have outweighed the pros. Others again may have no wish to stay on a site under any circumstances, seeing them as 'reservations' and a barrier to freedom.

On average, about a quarter of the total number of Travellers has found space on the few privately owned sites which take them. These sites may have been developed by Travellers for themselves and their extended family groups but may also take on a few other people. The layout and facilities on these tend to be much less formal than on council sites, but while there may be fewer rules and regulations, occupants may be much more dependent on the whim of the owner as to who gets access and when they have to go. Private sites of this kind are taken into account in assessing the overall level of provision for Travellers within a given area, but in practice they are an uncertain resource. Other privately owned resources are occasional holiday or touring sites, some of which may take a few Travellers at particular times of the year, although often only specific groups and with restrictions on vehicles and activities.

Finally, for those who cannot, or choose not to use council or privately owned sites, about a quarter of the total, and those who have not moved into conventional housing as a short-term or permanent measure, the only alternative remains to stop on unauthorised locations, little different from those used for decades and still subject to the same difficulties and uncertainties. Some who choose to travel and stop illegally in this way are house-dwelling Travellers who come out onto the road for part of the summer months, others are the more mobile groups such as English and Irish Travellers who may travel the length and breadth of Britain at different times of the year.

The future
Travellers in Scotland have survived the vicissitudes of their chosen way of life for centuries. The period of time during which there has been recognition of their right to retain the essentials of their lifestyle and culture while also facilitating improvements in their conditions is very short within their history. Nevertheless, within that brief period there have been significant improvements in their position. Substantial physical provision has been made across much of Scotland, although it has not, and possibly could not have met the perhaps contradictory expectations either of those creating it or those using it. Public awareness of Traveller needs and culture has improved, but many attitudes, on both sides, have failed to change. When coming to the end of its final term in 1999, the Advisory Committee identified changes in working and domestic lifestyle which had taken place over 30 years and issues which would still have to be addressed if progress was to be made.[36] However, underlying these was the recognition that it would have to be a two-way process: acceptance by the settled community of Travellers' lifestyle, needs and rights would have to be matched by Travellers' recognition of their own responsibilities towards their settled neighbours.

The old way of life, romantic in idea but hedged round with practical difficulties and constraints for Scotland's Travelling People, can never return as it was. No doubt, as over past centuries, their ability to adapt to changing circumstances, constraints and lifestyles will continue, although the context, for them and for the settled community alike, will undoubtedly change.

FAIR OR SHOWGROUND PEOPLE

As noted earlier, there are both differences and similarities between traditional Travellers and the largely separate and distinct group of Fair and Showground families (Show People) who provide supporting entertainment, for example, for annual agricultural shows and charter fairs and who are part of the Occupational Travellers group as identified by the European Commission. There is little source material and no analytical research into their domestic life appears to have been undertaken on this group, certainly in Scotland. Published material on fairs tends to concentrate either on their historical aspects or on the development of the material culture (the 'rides' themselves, the vehicles hauling them or styles of caravan used as living accommodation) and most relates to England with little specifically on Scotland, although in practice there is probably little difference between the areas in these aspects. Some more recent information has emerged in the context of children's educational needs and the difficulties of a mobile existence in terms of access to services and of acceptance by the settled community. From these different kinds of source it is possible to glean something of the nature of domestic life in the travelling period of the Show People's year and, where practicable, their voices have been used as illustrations.[37]

The Travelling People and Show People form two separate groups and although intermarriage between the communities occurs, it is not common. Yet in addition to self-ascription, various similarities exist. These lie mainly in having been a feature of Scottish and wider British life for a long time, the dependence on kinship and inheritance, traditional custom and more within community and lifestyle, their movement around the country in pursuit of a living and, while travelling, their use of caravans as dwellings. Scotland has its own very long-established Show families and although the dominance of heredity in establishing membership and right to work within the group might point to a claim for ethnic status they have rejected such a claim, describing themselves as a 'business community'. Nevertheless comments by members of the community illustrate the importance of kinship and sense of 'belonging':[38]

> My mother was a show person, her grandparents, all her people, it goes way back 100 and odd years; My family are Show People generations back. My grandmother had these old hobby horses. I've got a photograph of them over 100 years old; I was born into the business, my mother and father were born into it. It's a way of life for us. You're born into the business and you just carry on; Both my mother and father were Show People and from an early age we were taught that we had to learn the business too.

The Show community differs from the Travelling People in being much more organised, not just in their seasonal travelling but with a necessary common voice to speak on their behalf which grew in response to increasing pressure on the lifestyle of those Travellers earning their living in shows and fairgrounds. In 1889, the Show People organised themselves into an association in order to protect their rights to mobility, an association which developed into The Showmen's Guild of Great Britain. While a Scottish section of the Guild gives Scottish Show People their own representation, its functions are limited to the establishment and maintenance of work permits for its members, negotiation with councils and public bodies which have health and safety responsibilities and mediation in disputes between its members over pitches and infringement of hereditary rights. On the other hand it has no locus in accommodation and health issues within the community itself.

The pattern of domestic life is largely pre-determined, both from year to year and also within the year, by the timing, often centuries old, of traditional fairs held in particular towns at a given season.[39] Typically Show People set out on the road in about March, moving from fair to fair for the next eight months or so before returning to a winter base:

> The first fairs are in March, the second week in March; There's Dumfries, that's a street fair that is quite busy. Then in the second week in March nearly all the shows go to Perth. Sometimes they go to Dunfermline.

While the travelling season used to end in October, it is now extending, partly a function of new demand for events in the autumn and even in winter, 'There is a big fair at the Scottish Exhibition Conference Centre in Glasgow that goes on for a month over Christmas and New Year. A lot of Show People are there.' Partly also economic pressures of investment in increasingly sophisticated fairground rides means that they have to be utilised for as long a time as possible:

> There are shows at bonfires [end October/early November] now and we never used to get that. A lot of people are travelling all the year now because the expenses are so much that you can't afford to sit idle and have your equipment closed.

Over-wintering facilities are a necessity for Show People but in recent years these have been associated with increasing difficulties. A winter base is needed both to store and carry out maintenance on equipment, whether this is a large and sophisticated ride, a smaller children's roundabout or just various kinds of stall. It is also necessary to store and maintain the caravans which will be home throughout the travelling months. In the past there were well-known grounds or 'yards' of this kind, particularly in the Bridgeton area, at Whiteinch and Partick (all in Glasgow), although some families base themselves in Edinburgh, in towns in the central belt or in Aberdeen. As well as the caravans they use for travel through the season, many people own their own houses while some rent houses privately, and others may live through the winter in their caravans on a piece of land which they own, or rent space from another Show Person, allowing them to access electricity and water supplies. The winter season is also the only opportunity to recharge the human batteries which have been almost continuously at work through the travelling season, erecting, running and dismantling rides or stalls and moving on to the next fair:

> Weddings, most of our functions, are on in the wintertime. Some people if they are having a wedding will maybe have it in November when they come in. January, February, that is when you will have a holiday if you can afford to go, that is if you have had a good season. As we work all summer we can't go for a holiday then. In January people will be painting up their equipment, getting ready for moving away for Easter, for 'starting out' as we call it. Or they might be

working at a job over the winter. At that time of the year everybody is getting ready for the move.

It is the Show People themselves who are responsible for making their own arrangements for winter quarters. Over the years the number of wintering grounds has dwindled and space has become increasingly difficult to find, only those able to pay for the full year being guaranteed a return to the same place the next winter, a system which discriminates against poorer members of the community. Members of the community tend to concentrate in particular towns. Of all European cities, Glasgow has the highest number of Occupational Travellers over-wintering in this way [40] but even though they pay local rates and taxes, the city allocates no wintering grounds. Many of the remaining yards are now of very poor quality with minimal or non-existent facilities.

Traditionally, once a family moved onto the road for the season they moved from fair to fair and were able to park their living van beside or near to their ride or their stall, enabling them to keep an eye on it at all times, although increasingly tight safety regulations may now prohibit such proximity. Many families would follow the same circuit through the year, functioning almost as a small village of their own with mutual support systems, 'We all look after one another. Never mind where you are, if there's trouble we're all there.' At the same time there are the inevitable frictions of a very close community incorporating different status levels:

> This place here, it's like another wee village inside a village and everybody knows everybody, and we all talk to one another. There are family feuds that happen from time to time, but in general you don't allow it to go on because you've got to keep living together.

As with traditional Travellers, the living accommodation itself covers a broad spectrum of size and quality. Just as Show families have taken vehicles originally built for other purposes and have adapted them for heavy haulage or for generators, some have converted buses into residential accommodation.[41] Most accommodation has been purpose built, however. Many caravans in the late nineteenth century would be horse drawn, but were soon superseded by steam traction engines and later by heavy lorries. At one end of the spectrum may come quite small touring caravans which can be towed by a van. At the other, show caravans can be very large, spacious and highly sophisticated, much more so than even the better modern Traveller trailers. Some of the bigger examples may have originated as large conventional caravans but subsequently have been modified in sub-structure and wheels to make them more stable and to match them up to the kind of heavy haulage vehicle which has to be used for fairground equipment. Others are modern custom-built caravans, some even articulated with the towing vehicle. Particularly in the case of earlier caravans dating from the first half of the twentieth century, there were close similarities in appearance to early railway coaches, sometimes as large and luxurious and with specialised coach building and painting, now the subject of much specialised study[42] (Figs 44.9, 44.10). A particular feature, which is not to be found in Traveller trailers, is the sophisticated 'pull-outs' allowing the sides or ends of the caravan to be extended to give additional living space and provide an entrance porch when on a fairground or a winter site, '... fetch the caravan in, level it off, push out the pull outs ... If the ground is soft it may take three quarters of an hour just to level the caravan.' (Fig. 44.11).

Although in the larger Show vans domestic life can be not just comfortable but even luxurious among the well-to-do, with most of the amenities of modern life, there are still constraints which are likely to impinge on everyone. The most basic of all requirements of life is access to water. For Show People this is perhaps easier to access than for Travellers, partly through their regular patterns of movement and pre-determined stopping places and partly because in most cases their caravans are larger with tailored construction. Water supplies vary depending on individual fair sites. Some have standpipe access while there may be possibilities of directly connecting a water supply to the caravans themselves:

> We know which grounds have electricity and which have water. We don't waste water if there is no tap. It's a way of life we are used to ... ; A box to supply electricity is supplied on the side of the site and a standpipe. We can connect to the local water supply. But that is only on certain grounds.

Where this is not possible, large caravans may incorporate stainless-steel storage tanks, 'The water tanks are inside the underboxes. They look just like wee square doors in the bottom of the wagon.' These have to be filled up whenever there is an opportunity:

Figure 44.9 An older-style Show family's trailer in Scotland. Source: © E Jordan.

Figure 44.10 All modern comforts. A modern trailer at a Scottish fair in the late 1990s. Note the two large pull-outs giving additional interior space. Source: © E Jordan.

> Most caravans have now got stainless steel water tanks underneath so that if we knew there wasn't any water supply at the next fair then we'd fill the tanks up at this fair. Then when we got to the next fair we'd just have to try to conserve the water.

Smaller caravans do not have these and water has to be collected in cans. But the larger specialist caravans might have showers and bathrooms. The nature of all the caravans is such that the best possible use has to be made of the available space, necessitating multipurpose rooms and carefully designed storage space:

> My mum's bed is in the sitting room and it folds into the wall. The tele [sic] and the settee are in there. In the middle is the kitchen. The cooker is calor gas. We have a bathroom and a shower. The shower is in my bedroom, so when anyone is getting a shower I have to move out. My bed folds into the wall too because if it stayed out I'd have no room to play with my toys.

Lighting in the early caravans was generally by paraffin lamps and there are references to the better off having more than one lamp and 'the really posh ones having a light outside', but lighting later progressed to bottled gas. In the early days cooking would have been on a solid-fuel stove:

> My mother had a caravan, an old fashioned square one and she had a Hostess stove alight in it when we were going along the road on a Sunday. My mum and me were in the caravan while it was moving

COMMUNITIES ON THE MOVE • 809

Figure 44.11 The interior of a Show family's luxurious trailer at a Scottish fair in the present day. Source: © E Jordan.

and my mum cooked a Sunday dinner on the road. We would pull into a lay-by and get our dinner which mum had cooked. The Hostess stove was a coal stove like a little range ...

Cooking in its turn also moved on in the later twentieth century to bottled gas and later, depending on the caravan, lighting improved:

Years ago the caravans were not as big as they are now. They didn't have pull out sides. We weren't always tapped on to the water supply. We used to have stainless-steel cans for the water and calor-gas light inside. Then we got flush toilets. Now we have washing machines in the caravans.

With the increasing range of domestic equipment a source of electricity becomes a necessity. Show People have to be prepared for a range of eventualities while at fairs if they are to enjoy the facilities which the average house dweller takes for granted and run their working lives satisfactorily. While on many fair sites there is now the opportunity to 'hook up' to an electrical supply, this is not always the case and so most families have a generator, either for their caravans alone or also to power fairground equipment, 'Every showman has their own generator. To hoover the floor or do the washing we might have to put the generator on. It's an expensive way of heating water.' However, the benefits of this can be offset by the noise caused, 'If there is no electricity we would have to put our own generators on. Then we would have to put them off early because of the noise.'

Another feature of Show People's domestic life which differs from that of the settled community is the need to pack carefully personal possessions almost on a weekly basis if they are not to be damaged on the move:

> Well everybody has their own way of packing their ornaments for travelling. All my wee bits and bobs I wrap them in towels and put them in a washing basket and just lay it in the bath so that I can pull them back out easily. The coffee table goes on the bed.

As with Travellers, Show People's domestic life and work life are intermingled, not just because the dwelling place and the workplace are generally adjacent during the working season, but also because the whole family is involved in the work. Older members are likely to travel with the rest of the family and undertake some activity appropriate to their years. While the menfolk of varying ages are involved in setting up, running and dismantling rides and stalls, women assist in the day-to-day running of these, in tasks such as selling tickets or handling much of the increasing amounts of paperwork. Older children may expect, or be expected, to follow in their parents' footsteps and take an interest in and help with the family's work, learning practical skills as they go, 'Well I do help my dad. I wouldn't say I'm always helping, but of course when I'm asked to.'

> I'm self taught. Then I went to night school and learned how to weld and just went from there. I used to play with Lego a lot when I was young! My youngest boy is really keen to help with that kind of work. If he can get a hammer and a piece of steel he's interested in it. If there's a nut to tighten up he'll do it. He just learns by my side.

At the same time, children need to be protected from the danger inherent in the fairground and also need to have as far as possible a normal opportunity to grow up and be educated. The difficulty of maintaining continuity of education is admitted by some:

> I've probably been to 32 schools in my life by going to different places up and down; I took a record book to each school so they would know where I was. There were two or three schools where I'd already done the work they were doing so they tried to get me to do the same work all over again; As a child I travelled all over the north of Scotland and dodged school as much as possible. I regret it now, but I used to be a great favourite for the pictures in the afternoon and tell the old man I'd just come from school.

As with the traditional Travellers these problems among Show People have been both a cause of concern and also of research and initiatives over recent years:

> When I go out in summer I take work from school, but I go to loads of different schools and they give me their own work to do. Going into different schools is OK because you get to have more friends than staying in one place. You get used to it anyway.

According to the Scottish Traveller Education Programme:

> The Glasgow winter-base schools welcome them and support their learning within the limits of their flexibility of their staffing and resources. Show pupils are now assured of remaining on the roll while travelling and in secondary schools subject choices are made before moving out in the early spring so that places are reserved for their return in October/November. Schools provide packs of materials to accompany the children, to be completed and returned for marking. But this is only a stop-gap measure; children need teaching and such packs are not true 'open or distance' learning.[43]

Despite these measures it seems that while most Show children can achieve results comparable with those of their peers, at least up to early stages of secondary education, some difficulties remain, with lower levels of achievement in the reading and writing elements of language. On the other hand, Show children may demonstrate high levels of achievement in more practically oriented subjects, such as business studies, maths and art, reported as 'favourite' and 'best', which perhaps both reflects their background and points to a future role in continuing the Show way of life.

NEW AGE TRAVELLERS

Brief mention should be made of a group with a minor and short-lived presence in Scotland. Known as 'New Age' travellers or sometimes as 'hippies' or 'alternative lifestylers', small numbers of these people were recorded in Scotland at the end of the 1980s and the early 1990s. In contrast there existed much larger groups with this lifestyle in certain parts in England, where they were a feature of many seasonal festivals. New Age travellers are regarded as forming the least cohesive of the nomadic groups with little or no community history or recognisable familial structures, although there appear to be a very small number of second- or third-generation members.

This group was neither regarded by traditional Travellers as part of their community nor recognised by central and local government as a distinct community with needs of its own. Over time, some members began to develop a voice, but many individuals regarded any attempts to do so as being antipathetic to their chosen way of life.[44] The 1992 count of the Traveller community[45] made some attempt to assess numbers and the characteristics of the group, although it was recognised that this was difficult to do given many members' reported lack of co-operation with, and sometimes overt antagonism to any form of authority. The most that could be learned at that time was that their presence was limited to a few areas – mainly Midlothian, Lochaber and Skye and Lochalsh, with a scatter in seven other districts – and that their lifestyle was different from that of the traditional Traveller community. In the main, New Age travellers were to be found in rural and often hidden locations. The 1992 count was able to estimate only around 60 households in Scotland which might be classified under this heading. Much of the evidence for their presence was anecdotal, for example in northern parts of Stirlingshire, although in one or two areas a closer watch was kept by the local council, such as that of Argyll and Bute where:

> The survey was cut short by the unprovoked attack of their dogs ... 'there seemed to be about 30 to 40 people and at least 12 dogs resident in the lay-by' ... at the time of the subsequent count ... 'following the incident on the last visit to this location, the survey was carried out from a safe distance'.[46]

More so than in the case of mainstream Travellers, information about New Age travellers' household characteristics and lifestyle was difficult to obtain, and again much had to be based on anecdotal information. Household size appeared smaller than in the main Traveller community and closer to that of the settled community, but people did not necessarily follow conventional family groupings. Household structure was generally younger than for traditional Travellers, with no old people and a substantial proportion of very young children.

Unlike other travellers, these people appeared to have no culture of self-sufficiency and self-employment and their accommodation tended to be a hotchpotch of run-down caravans, vans, trucks and conversions of, or simply use of other vehicles such as buses or ambulances. Most of those recorded in 1992 had been in the area for up to six months, perhaps moving from place to place within it, with a few having stayed in one spot for longer periods. Even though, by that date, official council sites were already available in some areas, New Age travellers never sought accommodation on them, nor would they have been eligible to stay there. While they were regarded as a periodic thorn in the flesh of the few local councils where they took temporary root, they were mostly moved on by the police on a fairly regular basis.

Despite occasional anecdotal references to their presence in the last few years in certain parts of Scotland, most recently from council Traveller-site managers, and a number of sources in the south-west, there remains little or no detailed information about these travellers. The author has, however, recently been told of one or two small groups which may have New Age origins and who have adopted some aspects of the traditional Traveller lifestyle of 60 or 70 years ago, using bow-topped, horse-drawn caravans to travel in the summer months and sell items of their own making, mainly paintings and craft goods. In the winter they appear to live in either their caravans or, as they were referred to, 'wigwams', creating a retrospective link with past styles of traditional Traveller life.

The New Age group have chosen a separate lifestyle which rejects much of conventional society but does not fit with either of the culture-based Travelling groups discussed here. They appear from the conventional population to live an unconventional lifestyle for a while and may disappear whence they came.

NOTES

1. With the exception of the section on Fair or Show People, this chapter draws on the author's own research within this field undertaken for and published by The Scottish Office (now The Scottish Executive) over the past 35 years. The section on Show People is drawn from secondary sources and particular thanks for assistance are due to the Scottish Traveller Education Programme in Moray House College of Education, University of Edinburgh.
2. While there are many practical differences between these two groups and the way in which they relate with the settled community, each is recognised by the European Parliament as a sub-set of a generic group of Travellers, the first sub-set covering Gypsies/Travellers and the second Occupational Travellers. The latter includes not just the Show/Fairground group but also Circus and Bargee groups (a feature of the large inland waterways of Europe).
3. Gentleman, Swift, 1971, 9; Throughout this chapter the terms 'Travellers' or 'Travelling People' are used as shorthand. Definitions have been changing in their acceptance and perceived correctness over the past 30 to 40 years. In the late 1960s, the above terms began to be adopted fairly widely throughout much of the UK as inclusive and non-pejorative and applicable to all those people otherwise known as Tinkers, Gypsies, Romanies, plus various other names. Problems of definition in Scotland were addressed in Gentleman, Swift, 1971. In 2001 the Scottish Executive said that in future it would be adopting the term 'Gypsies/Travellers' rather than simply 'Travellers'. However, in practice it is not universally accepted and many people within the community still prefer to call themselves Travellers, Scots Travellers or members of the Travelling People, while others may specifically refer to themselves as Gypsies. The term excludes the group who became known as 'New Age' travellers. In 2001 the Scottish Executive formally re-emphasised the need for the already commonly accepted initial capitalisation. This usage was further confirmed in a report on Gypsy and Traveller education, Learning and Teaching Scotland. *Inclusive Educational Approaches for Gypsies and Travellers*, Edinburgh, 2003.
4. Gentleman, Swift, 1971, 9; Report of the Departmental Committee on Tinkers in Scotland, 1918, 6; Gentleman, Swift, 1971, 10, 19; Report of the Departmental Committee on Habitual Offenders, Vagrants, Inebriates and Juvenile Delinquents, 1895, Cmnd 7753; Report of the Departmental Committee on Vagrancy, Edinburgh, 1906.
5. Quoted in evidence submitted by the Scottish Traveller Education Project to the Scottish Parliament Equal Opportunities Committee, 2000.
6. Report of the Departmental Committee on the Tinkers in Scotland, 1918.
7. Report of the Departmental Committee on Vagrancy in Scotland, 1936, Cmnd 5194.
8. Children's Act, 1908.
9. Evidence submitted by the Scottish Traveller Education Project to the Scottish Parliament Equal Opportunities Committee, 2000.
10. Gentleman, Swift, 1971.
11. Gentleman, 1993, 66.
12. Gentleman, 2002.
13. Rehfisch, 1958.
14. Another important aspect of Scots Travellers' contribution to wider society is their role in maintaining a large corpus of Scottish oral tradition.
15. In Fig. 44.1, categories have been simplified: 'Tents' covers tents used on their own; 'Traditional caravans' includes horse-drawn caravans, either alone or in combination with tents; 'Modern caravans' includes trailer caravans on their own, in

conjunction with tents or with miscellaneous 'other' kinds of dwelling; 'Other' includes huts and sheds, shelters of branches, caves, known squatters in more conventional dwellings and any other form of shelter. (Note that some proportions add up to more than 100 per cent since in some cases accommodation type was not known, while families also may have been using more than one type at the time of the count.)

16. Gentleman, Swift, 1971, 36–8, Plates I-V.
17. Gentleman, Swift, 1971, 37.
18. Gentleman, Swift, 1971, 37.
19. Gentleman, Swift, 1971, Plate II.
20. For example, Whyte, 1979.
21. Gentleman, Swift, 1971, Plate I.
22. Gentleman, Swift, 1971; Ministry of Housing and Local Government, Welsh Office, 1967.
23. Gentleman, Swift, 1971, 39.
24. Traveller age structure differs from that of the settled population with a greater preponderance of young people (44 per cent aged under 16 in 1992 compared to 20 per cent in the 1991 Scottish population census) and far fewer elderly people (4 per cent aged 60–5 and over as opposed to 18 per cent).
25. Ministry of Housing and Local Government, Welsh Office, 1967.
26. Caravan Sites Act, London, 1968.
27. Gentleman, Swift, 1971.
28. Scottish Development Department, 1971.
29. Although issues affecting Travelling People were broadly similar both north and south of the border, responses differed between the two areas. Scotland offered the carrot of funding plus the benefits of an ongoing Advisory Committee but without the stick of legislation. England and Wales adopted a legislative approach, placing a duty on local authorities to provide sites but initially without central-government funding (although this was later introduced) and without any form of Advisory Committee.
30. Advisory Committee on Scotland's Travelling People, 1998–9, 4–8.
31. Advisory Committee on Scotland's Travelling People, 1990a.
32. Advisory Committee on Scotland's Travelling People, 1998.
33. Gentleman, 2001; Gentleman, 2002; Gentleman 2003a; Gentleman, 2003b.
34. Lomax, Lancaster, Gray, 2000.
35. Douglas, 1997; Lomax, Lancaster, Gray, 2000.
36. Advisory Committee on Scotland's Travelling People, 1998–9, 18–19.
37. All the personal quotations used in this section have been drawn from three booklets produced co-operatively by the Scottish section of the Showmen's Guild of Great Britain and The Scottish Traveller Education Programme under the general editorship of Cathy Kiddle; Kiddle, et al, c2002a; c2002b; c2002c.
38. Kiddle, et al, c2002b.
39. Through the year, the pattern of a typical run of fairs might be:
    Dumfries Street Fair (March)
    Carlisle East Fair (April)
    Kirkcaldy Links Market Fair, Fife (April)
    Perth Spring Fair (April)
    Rosyth Fair, Fife (April/May)
    Edinburgh: Meadows Spring Fair (April)
    Edinburgh: Leith (April/May)
    Burntisland Summer Fair, Fife (May/August)
    Edinburgh: Meadows Festival (June)
    Hawick Fair, Roxburghshire (June)

Edinburgh: Leith Links Festival (June)
Appleby, Westmoreland (June/July)
Aikey Fair, Aberdeenshire (July)
St Boswells, Scottish Borders (July)
Musselburgh, Eastlothian (August)
Aberdeenshire Highland Games, Aboyne (August)
Perth Agricultural Fair (August)
Perth Fair (October)
Dumfries Road Fair (September)
Burntisland Bonfair, Fife (November),
Kiddle, et al, c2002b.
40. Estimates have been given of some 500 or so families based in the Glasgow area.
41. Slater, 1988.
42. Various books have been published on the styles and technical aspects of both caravans and haulage vehicles. One with good illustrations of styles and variety is Braithwaite, 1999. While most of the major builders of Showmen's vans were south of the border, a number of Scottish makers are identified within Braithwaite's main directory.
43. Jordan, 2000.
44. Evidence submitted by the Scottish Traveller Education Programme to the Scottish Parliament (Equal Opportunities Committee), 2000.
45. Gentleman, 1993a.
46. Gentleman, 1993a, Appendix I, 8–9.

BIBLIOGRAPHY

Acton, T. *Gypsy Politics and Social Change*, London, 1974.
Adams, B, Oakley, J, Morgan, D, Smith, D. *Gypsies and Government Policy in England*, London, 1975.
Advisory Committee on Scotland's Travelling People. *First Term Report, 1971–4; Second Term Report, 1975–8; Third Term Report, 1979–82; Fifth Report Report, 1986–8; Sixth Term Report, 1989–91; Seventh Term Report, 1992–4; Eighth Term Report, 1995–7, Ninth Report, 1998–9; Final Report, 1998–9*, Edinburgh.
Advisory Committee on Scotland's Travelling People. *Guidance Notes on Site Provision for Travelling People* (The Scottish Office), Edinburgh, 1990a.
Advisory Committee on Scotland's Travelling People. *Traveller Site: Guidance Notes on Upgrading* (The Scottish Office), Edinburgh, 1990b.
Advisory Committee on Scotland's Travelling People. *Guidance Notes on Site Management (Revised)* (The Scottish Office) (1991), Edinburgh, 1998.
Baird, W. *Memoir of the Revd John Baird, Minister of Yetholm: With an Account of his Labours in Reforming the Gypsy Population of that Parish*, London, 1862.
Bancroft, A, Lloyd, M, Morran, R. *The Right to Roam: Travellers in Scotland, 1995/96*, Edinburgh, 1996.
Braithwaite, P. *A Palace on Wheels: The History of the Showman's Living Wagon, with an A–Z of Manufacturers, 1860–1960* (Fairground Heritage Series, no 4), Maidenhead, 1999.
Brockie, W. *The Gypsies of Yetholm*, Kelso, 1884.
Department of Health for Scotland. *Report of the Committee on Vagrancy in Scotland* (Her Majesty's Stationery Office), Edinburgh, 1936.
Dorner, J. *Markets and Fairs* (1973), Hope, 1977.
Douglas, A (The Douglas Consultancy). *Local-Authority Sites for Travellers* (The Scottish Office Central Research Unit), Edinburgh, 1997.
Gentleman, H. *Counting Travellers in Scotland, the 1992 Picture: Estimates of the Number,*

Distribution and Characteristics of Travelling People in Scotland Based on a Count Undertaken for the Scottish Office (The Scottish Office Central Research Unit), Edinburgh, 1993.

Gentleman, H (Research Consultancy Services). *The Twice-Yearly Count of Travellers in Scotland: The First Three Years* (The Scottish Executive Central Research Unit), Edinburgh, 2001.

Gentleman, H (Research Consultancy Services). *Gypsies/Travellers in Scotland: The Twice-Yearly Count, January 2002* (The Scottish Executive Central Research Unit), Edinburgh, 2002.

Gentleman, H (Research Consultancy Services). *Gypsies/Travellers in Scotland: The Twice-Yearly Count, no 2, July 2002* (Scottish Executive Social Research), Edinburgh, 2003a.

Gentleman, H (Research Consultancy Services). *Gypsies/Travellers in Scotland: The Twice-Yearly Count, no 3, January, 2003* (Scottish Executive Social Research), Edinburgh, 2003b.

Gentleman, H, Swift, S. *Scotland's Travelling People* (Scottish Development Department), Edinburgh, 1971.

Government of the Republic of Ireland. *Report of the Commission on Itinerancy in Ireland*, Dublin, 1963.

Hethrington, K. *New Age Travellers: Vanloads of Uproarious Humanity*, London, 2000.

Jordan, E. Traveller Families from Home–School Partnerships: A challenge for school boards, Conference Paper – Section 4.1, 2000.

Kiddle, C. *Traveller Children: A Voice for Themselves*, London, 1999.

Kiddle, C, ed. *The Fairground: A Family Business* (Scottish Traveller Education Project, Moray House College of Education, University of Edinburgh), Edinburgh, c2002a.

Kiddle, C, ed. *The Fair: Past, Present and Future* (Scottish Traveller Education Project, Moray House College of Education, University of Edinburgh), Edinburgh, c2002b.

Kiddle, C, ed. *Our Families: Children from the Fairground* (Scottish Traveller Education Project, Moray House College of Education, University of Edinburgh), Edinburgh, c2002c.

Lomax, D, Lancaster, S, Gray, P. *Moving On: A Survey of Travellers' Views* (The Scottish Executive Central Research Unit), Edinburgh, 2000.

McCormick, A. *The Tinkler Gypsies* (1907), Wakefield, 1973.

McCraw, I. *The Fairs of Dundee*, Dundee, 1994.

Ministry of Housing and Local Government, Welsh Office. *Gypsies and Other Travellers*, London, 1967.

Okely, J. *The Traveller-Gypsies*, Cambridge, 1983.

Rehfisch, F. The Tinkers of Perthshire and Aberdeenshire, unpublished ms, SSS, University of Edinburgh, 1958.

Rehfisch, F, ed. *Gypsies, Tinkers and Other Travellers*, London, 1975.

Report of the Departmental Committee on Habitual Offenders, Vagrants, Inebriates and Juvenile Delinquents, Edinburgh, 1895.

Report of the Departmental Committee on Tinkers in Scotland, Edinburgh, 1918.

Report of the Departmental Committee on Vagrancy, London, 1906.

Report of the Departmental Committee on Vagrancy in Scotland, 1936

Scottish Development Department. Circular, 9 (1971).

Scottish Parliament, Equal Opportunities Committee. *First Report: Inquiry into Gypsy Travellers and Public-Sector Policies, vol 1; Report, vol 2 – Evidence*, Edinburgh, 2000, 2001.

Showmen's Guild. *All the Fun of the Fair* (Showmen's Guild of Great Britain), npp, 1987.

Slater, M. *Fairground Transport, vol 1*, York, 1988.

Werninck, J H A. *Woonwagenbewoners* [*The Dutch Caravan Dwellers: A Sociological Study of a Marginal Group, with English Summary*], The Netherlands, 1959.

Whyte, B. *The Yellow on the Broom*, Edinburgh, 1979.

45 Domestic Life in Temporary and Self-Built Dwellings

HUGH GENTLEMAN

The homes of ordinary Scots in the past were often characterised by their temporary nature, in some elements of their construction at least. They were also frequently self-built using local resources, by the household but usually with community help, and employing traditional knowledge and few or no specialist skills. The gradual adoption of more permanent materials, many brought in from beyond the local area, the imposition of controls on building and the development of specialist building trades changed this. In terms of non-finite, year-round dwellings, self-building and the use of ephemeral materials lasted longest (until the early twentieth century), in, for example, the blackhouses of the Western Isles.[1]

Today most people expect their homes to be constructed by qualified professionals and trades people and for the dwellings' major components to be permanent, or at least to outlast their own lives. Even so, elements of past patterns continue. People with appropriate skills can still build their own dwellings, albeit within the confines of planning policy and building control. Others use specialists for certain aspects of the construction of their homes but undertake the simpler tasks themselves, while some may not be directly involved in constructing a 'self-built' home but take an active role in its design.

As well as the initial construction, maintenance and often modification is vital in the continuity of a dwelling. All houses require repair or replacement of some components from time to time, whether outside or in. While people are no longer required periodically to renew walls of turf and are unlikely to have to rethatch roofs, installing new windows and doors, for example, is relatively common. Elements of older ways may also be seen in homes, where, whether by choice or through economic necessity, householders have to be 'handy' in order to maintain the building or personalise it to reflect preferences and fashions. In recent years this has been made very much easier by the growth of a huge do-it-yourself industry.

As well as 'permanent' types of dwelling, there are those whose role is not expected to outlive the inhabitants. They may be a feature of particular communities, as discussed in the previous chapter. Others may be used by the general population, usually in times of need, as, for example, the 'prefab' bungalows of the mid twentieth century, static caravans and mobile homes. In addition, accommodation for workers can be of a temporary nature,[2] for example 'portakabins' for construction or forestry workers, as can some holiday and second homes.

This chapter focuses on the strands of self-building and temporality in Scottish homes, and the effect of these on the domestic life of the households within them, by examining in detail two types of dwelling; mobile homes and self-built leisure huts.

MOBILE-HOME DWELLERS

Large holiday caravan sites, often visually intrusive on a section of coastline, are a common feature of parts of the Scottish landscape. Most of the serried ranks of dwellings (Figs 45.1, 45.2, 45.3, 45.4) on these are what are generally known as 'mobile homes' but are destined for seasonal holiday use rather than for permanent residence. At the same time there are still many sites, generally much smaller, which are licensed for year-round residential use and function differently from holiday sites.

Unlike the Travelling People discussed in the previous chapter, for whom a temporary structure as principal form of dwelling is a function both of the underlying need for mobility and of cultural heritage, most mobile-home dwellers are no different in background or lifestyle from the majority of the population who live in houses. The difference is that, for a variety of reasons, this is the kind of dwelling which they need to use at a particular time in their lives. It may be that they are working in an area only for a period of months or years and consequently do not wish to acquire a conventional house; or else it may be a case of a young couple needing somewhere to live on their own rather than with family while they save for a

Figure 45.1 A large mobile home site on the edge of a Scottish town in 1976. Source: © H Gentleman.

Figure 45.2 Modern twin-unit mobile homes on a Scottish site in 1976. Source: © H Gentleman.

Figure 45.3 Part of a small Scottish mobile-home site in 1976. Source: © H Gentleman.

Figure 45.4 A large mobile-home site developed for an incoming workforce in northern Scotland in 1976. Source: © H Gentleman.

conventional dwelling; or perhaps they are people with little likelihood of being able to buy a house or of gaining access to public-sector housing.

The only substantive research into mobile homes and their residents in Scotland now dates back over 25 years and was undertaken for the Scottish Office, paralleling similar work in England and Wales.[3] The context for these studies was, in part, the potential of this kind of accommodation to meet housing needs, but it was also a response to growing disquiet about matters on some privately owned sites, particularly security of tenure.

From a Scottish perspective it was felt that conditions in mobile homes might be different from those in England and Wales and also that this kind of accommodation might, in the short term at least, cope with demand from the growing influx of workers in areas affected by the developing North Sea oilfields and their related land-based industries. The study looked at aspects of mobile homes themselves: size, style and structural problems, perceived benefits and drawbacks and costs. Sites on which year-round mobile homes were located were assessed in terms of spread across Scotland, size, period at which they had been developed, the facilities available and levels of resident satisfaction. The study also assessed the size and structure of the mobile-home population, length of time living in this kind of home, the reasons for choosing it and the residents' experience of other housing, together with attitudes towards mobile-home living. Other aspects investigated included site owners, management, site development in general and the extent to which these homes might meet housing needs.

Mobile homes are generally much larger than conventional trailer caravans. Although the term 'mobile home' has been widely accepted for many years, in practice it is correct only in the sense that, if necessary, the structure can be moved. Unlike Traveller or Show People's caravans, mobile homes are not intended to be moved from place to place throughout their life and are usually located on special licensed mobile-home sites. Indeed, once located on a site, they are likely to remain on the same spot until their owners wish or need to replace them. Although the homes have wheels, these are only to allow them to be manoeuvred into place on a site and if they have to be moved from place to place they must be transported on a low-loader vehicle rather than being towed like a conventional caravan.

In the mid 1970s, 166 sites for mobile homes (licensed sites with more than three permanent residential units) were identified in Scotland, all but eight privately owned, the others owned by local councils. Sites were mainly concentrated in the central belt, the north east and around the Moray Firth, the latter two areas being major centres of North Sea oil development at this period. Throughout Scotland sites were generally small, most having fewer than 15 residential units and only three with 100 or more, while a number combined both residential and holiday use. Site development began mainly in the 1960s with only about one in ten having developed in the five years before the study.

As defined for the study, mobile homes include all dwellings which can be moved by road in no more than two sections (ie as defined in the 1960 and 1968 Acts),[4] though it was recognised that the term was often restricted to twin units only, the term 'caravan' referring to anything smaller. In practice, size and style varied greatly from small touring-style caravans of as little as 100 square feet (9.3 m^2) up to twin units with up to about 1000 square feet (93 m^2) of floor space, with appendages such as verandas and sun lounges. Nowadays the twin unit, moved as two sections and bolted together on site, has probably become the norm and can often be the size of a small bungalow. But at the time of the study more than four out of five homes were single-units, basically one long box, up to ten-foot (3 m) wide and varying in length from about 20 to 40 feet (6–12 m). Despite retaining their wheels for the few occasions on which they need to be moved, once on site mobile homes are supported either by feet or by more substantial supports, and the study revealed about half resting on a specially constructed tarmac or concrete base. Although probably now more common, the practice of 'bricking in' – bridging the gap between the ground and trailer with wooden panels or sometimes a more substantial semi-permanent structure – was then fairly rare on the Scottish sites.

Most of the mobile homes included a separate kitchen or a kitchen/diner and in addition had toilet and bathroom facilities, either separate or combined. Nearly all were attached to a mains cold water supply and consequently most also had running hot water, both connected to a drain or soak-away. Smaller conventional caravans, sometimes found on these sites, generally had fewer facilities and despite being used as year-round homes by their residents, clearly were regarded as unsuitable for permanent residence, despite the same dwelling type often being viewed as suitable for groups such as Travellers. Most of these caravans were connected to mains electricity, allowing conventional lighting and heating, although solid-fuel stoves were also used for heating.

At the time of the study, these sites provided a home for some 3800 households, a total of about 10,000 people. Allowing for the then rarity of the twin unit, space standards were therefore fairly minimal. Even though households were generally smaller than the Scottish average, with two thirds of the homes having floor areas of between 200 and 400 square feet (18.5 m^2–37 m^2) over a fifth of all residents were living at a density of more than one person per 100 square feet (9.3 m^2). More than half of the residents comprised younger married couples, sometimes with children, with a head of household aged 40 or less, while the remainder were generally older married couples on their own or older single adults. Four out of five households had an employed head, often a manual worker with a reasonably high income. Half had at some stage been owner occupiers of conventional housing, with most of the others having rented in the public sector, and almost half were on a waiting list for a council house.

People move into mobile homes from both necessity and choice and may move out of the sector and into a conventional home or trade up or down within it. In 1975 most often it was job change, followed by financial reasons and the need to avoid sharing accommodation with other households, which prompted the move into a mobile home. There might also be a wish to downsize one's home after retirement, for example. For two out of five households the mobile home was their first dwelling together and many were on a waiting list for a council house. The great majority of residents owned their own homes, although they had to pay rent for their pitch to the site owner, often together with other charges such as rates, car parking and, where appropriate, fees for certain communal facilities. Only about one in six of these dwellings had been purchased new but, while in theory a new mobile home came furnished, additional setting-up costs could still be very great. If to these costs was added the often substantial depreciation of the home itself, real living costs could be high. Few residents in fact saw this as a particularly cheap approach to housing. Despite the predominance of owner occupation, on a proportion of sites, all very small, the site owner also owned all the individual mobile homes and rented these out, but on the whole levels of satisfaction with such sites were lower than where owner occupied.

Turnover on sites was generally high. Four out of five residents had moved in within the previous nine months and only about one in ten had been there for more than five years. Nevertheless, residents were more or less evenly divided into those disliking the lifestyle and actively intending to move, those who were ambivalent in their attitudes and housing aspirations and those who were happy and contented to stay in this kind of dwelling. For each resident satisfaction with the nature of the accommodation depended on factors such as the type of household (expanding, static or contracting), space, facilities and even the type of site. In contrast to attitudes towards the lifestyle in general, most residents appeared to like their own home with only a few to any extent dissatisfied. Perceived advantages often related to size and layout in terms of convenience, compactness and ease of care, using phrases such as 'homely' or 'cosy' and 'absence of stairs'. Others found effective or easy heating an advantage. At the same time, what is favourable to one household can be regarded as a drawback by others. 'Cosy and compact' can translate into 'cramped and lacking in privacy', while 'airiness' can also be seen as draughty. Important but less tangible features were the independence and freedom offered by having a home of one's own, together with pleasant surroundings and an open-air life.

Satisfaction with the sites themselves was generally high, particularly on private and on smaller sites. Perceived site advantages included peace and quiet, a view, a rural atmosphere, recreational opportunities and the friendliness and helpfulness of neighbours.

Certain problems were inherent in the design and structure of mobile homes at that time, although improvements were being introduced into some newer and larger dwellings. Difficulties included condensation, leaks and rising damp. Poor insulation seemed worse in terms of under-heating in winter rather than overheating in summer. Particularly among households in conventional caravans, as opposed to mobile homes, the limited space and absence of adequate room division resulted in a lack of privacy, and even in smaller mobile homes this could be a problem for expanding families. At least a quarter of the households had no bath in their home and nearly one in five had no toilet, either in the home itself or on their pitch, having to use communal facilities somewhere on the site, a sizeable minority of residents complaining about having to share in this way.

Generally, satisfaction levels, particularly in respect of the freedom and independence of the lifestyle and the compactness, convenience and effective heating of the dwellings themselves, were greatest among 'static' or 'contracting' households. Least satisfied were the 'expanding' households, perhaps in rented accommodation on council-owned sites, who suffered most from lack of space and facilities in older homes and who saw the only advantage of a mobile home as being a relatively cheap way of having a home of their own.

Although over the past 30 years the average structural standard of most mobile homes has almost certainly improved and the dwellings are now much larger and better equipped, problems of security of tenure and conditions imposed by site owners seem to continue. Nevertheless this method of dwelling remains popular and meets a variety of needs and preferences.

HUTS AND HUTTERS

This group of temporary structures, which has been around for some 80 years, although until very recently there has been little awareness of its existence and even less known about it, illustrates a particular interplay between dwelling and occupant. Its future is uncertain and many examples may wither on the vine in the coming years, so they should not be lost without some record of their existence (Figs 45.5, 45.6, 45.7, 45.8).

Figure 45.5 A medium-sized hut site in the Scottish Borders in 1999. Source: © H Gentleman.

Figure 45.6 Assorted huts on a large, long-established site in Angus in 1999. Source: © H Gentleman.

Figure 45.7 Huts on a small site in Stirlingshire in 1999. Source: © H Gentleman.

Figure 45.8 Huts on a small site near Loch Lomond in 1999. Source: © H Gentleman.

As with other contexts, detailed knowledge of these structures and their occupiers has come to light only as a result of localised difficulties and concerns over security of tenure, leading to research being commissioned to assess the scale and nature of the problem. The results of that study[5] gave, for the first time, a picture of the presence of hut sites in Scotland, how they operate, who uses them and the nature of their current and possible future role.[6] Aspects examined in detail included: the location of sites; their origin, growth, character and changing ownership; the characteristics of the huts themselves and of their occupiers; site management and costs; patterns of hut use and the pros and cons of having a hut; finally, the report looked at what might happen in the event of changing legislative structures.

In 1999 Scotland still appeared to have some 650 huts fitting the study's criteria, described variously as huts, living huts or chalets (but excluding certain specialised categories such as fishing or climbing huts and bothies). Groups of these structures were spread across some 37 locations in Scotland, mostly on land under single ownership although there were some on which the occupiers had obtained ownership of their individual piece of land. The 'sites', if they could be so described, covered a belt broadly extending from Angus to Ayrshire, together with outposts in the Lothians and Borders and to the Solway coast. Coastal proximity was important for Angus and for Clyde-estuary sites, while many of the others were in fairly close proximity to major urban areas such as Glasgow, Edinburgh and Dundee. It seemed likely that there had been other such sites in the pre- and immediate post-War period which had faded away, perhaps through lack of interest, a change of land ownership or pressure from planning authorities. Numbers of huts at any one location varied widely. About half the sites had around two to nine huts and another third had c 10–20. Above this size, groups of c 20–30 were found on five sites, c 50 on two sites and finally the largest, and the first in origin, comprised over 150 huts, albeit in scattered groupings within a tract of estate land to the north of Glasgow. This first development began shortly after World War I, with a second in the mid 1920s, but most sites developed in the 1930s, and a few shortly after World War II. The advent of tighter planning controls on land use and the availability of alternative holiday locations probably account for no more starting up after this time. Certainly all seemed to have developed prior to there being any controls on this kind of land use.

Contrary to what initially had been assumed, hut sites proved not to have been 'set up' in any organised way but rather to have grown organically in response to requests to a farmer or estate owner from an individual or a group of people for a place to erect some kind of shelter for weekend recreational use. They seem to have coincided with a desire by city dwellers to enjoy fresh air and the countryside, coupled with 1930s ideas of healthy living. Certainly a few of the sites on the fringes of Glasgow were started off by cycling clubs in the 1930s, their members seeking an overnight base for weekend trips. Some grew up, particularly in Ayrshire and on the Angus coast, in places where people from the cities had been coming for their summer trades holidays, initially staying in tents but then later preferring rather more solid accommodation for this purpose and also for weekend breaks, the latter possibly made easier by the growth of rural bus services. Only two sites, one in East Lothian (now owned by the local council), and the other on the shores of Loch Lomond, appear to have been consciously 'developed' from the outset. In most cases, the only contribution on the landowners' part was to allow people to use a patch of his or her land, often of little use for anything else.

Perhaps the most intriguing aspect of the sites is the dwellings themselves and the ingenuity of their owners. As the study report indicated, 'Lifestyle in a hut seems to lie at some midpoint in a triangle between camping, a caravan and a house,'[7] but it depends largely on how often a hut is used, on the kind of use made of it and the nature of the individual 'hutters'. With a few exceptions, huts are fairly small but varied in structure and style. A small number may date back to the original structures of the 1930s while others may be completely new. Most are built of some form of timber but with additions of a variety of other materials such as tarred felt and corrugated metal, all of these often second hand or recycled from other purposes. In the early days a handful of structures appear to have incorporated parts of old bus bodies, but while these have now gone, one dwelling based on an old railway coach could still be seen on a site in the Borders in 1999. A few may have a brick, stone or concrete fireplace and a chimney. Since each hut has been self-built, its size and its external and internal appearance depend very much on what the original builders and successive owners could spend, their practical skills and their preferences in terms of how the hut was to be used and how it should look. In practice most huts have more than one room and some have been made into versatile miniature houses, fully carpeted and curtained, with stylish, if tiny, bedrooms and minute kitchens. At the other end of the spectrum some are no more than fairly bare shacks and may now be very run down. One essential feature of all huts is that they do not have conventional foundations but rather they sit on the ground and in theory can be removed completely, although in practice few could be without being destroyed.

Services such as mains electricity and running water are rare except on one or two sites and a few sites specifically prohibit any form of mains connection. However, the ingenuity of occupiers comes into play here. Bottled gas can be used to provide heating and lighting or for cooking, while fluorescent lights can be powered by batteries or a small generator. Traditional oil lamps or candles continue to be used, either for simplicity or perhaps to help create the 'feeling' of a hut. In a few isolated cases occupiers have even installed a small wind-powered generator for their lighting, although it is unlikely that this can produce sufficient output for cooking.

Water supply is often very primitive and although some sites may have a standpipe, only a few, mostly with owned or formally leased plots, have mains water plumbed in. Otherwise drinking water is generally brought in from home, obtained from a nearby hotel or cafe or bought as commercially bottled water. A few sites are located close to a spring or burn but these rarely provide suitable drinking water and are mostly used only for washing purposes, although in one instance a burn has been observed in use as a form of refrigerator, with bottles in a milk crate suspended in the water. Again there is scope for enterprise on the part of occupiers and a number of examples can be seen where quite complicated systems have been created to collect rainwater from roofs into storage vessels, either inside or outside the hut. A few occupiers have even gone so far as to pipe this source into the hut to provide water for a sink tap or even to supply a simple form of shower. Despite the limited scope for facilities most occupiers have provided their huts with some form of toilet, usually a chemical or earth closet. Once more it is only on the few sites in which individual plots are owned or on long leases that huts have any flushing toilet connection to a main sewer or to a centralised septic tank.

By the very nature of their structure, huts need a high level of maintenance if they are not to deteriorate. At the same time they have always been essentially organic structures which can be enlarged, reduced, changed inside, reclad outside or reroofed, not just of necessity but according to the tastes, interests and practical skills of the occupier. Certainly maintenance as both a necessity and a choice featured in occupiers' perceptions of the drawbacks and benefits of owning a hut. It is not just within the hut itself that a form of 'nesting' instinct can be seen since occupiers may create outdoor sitting areas, flower beds, displays of hanging flower baskets and occasionally vegetable patches. In contrast, some huts may be little more than a tumbledown shack in a bare field. One of the most important features of hut owning has been a greater freedom for this kind of individuality and self-expression than in forms of permanent housing or even other types of holiday home. While in most cases occupiers theoretically have to inform the landowner before making changes, in practice this seems not to be bothered about much by either side. Similarly, as happens in a number of cases, if an occupier loses interest in the hut or it becomes too much to maintain it can just be left to decay as it has little capital value and the responsibility can be shrugged off. None of the huts was envisaged as having real longevity, being without conventional foundations, although a small number have in fact survived for decades, certainly longer than many caravans and mobile homes.

By the time the phenomenon of huts was explored the hutters were already a predominantly ageing group, whose children might have enjoyed the dwellings when everyone was younger but now had other interests and opportunities, while they themselves found it less easy to maintain a hut or make sufficient use of it. Nevertheless, the children's interest is sometimes rekindled when they in turn have young families, and a hut may get a new lease of life. Landowners also tend to be older people, whose parents, perhaps, allowed the site to start up, and they may have little interest in it and be prepared to let matters drift on, provided the site creates no difficulties or requires no effort from them. Sites bring little in the way of rental returns but, on the other hand, require little or no input. Only in a few instances has there been any serious form of management or significant level of rent charged. While occupiers own their huts, they rarely have any formal right over the ground on which the huts rest and in theory are required to remove the structures at the landowner's request.

A variety of reasons for liking or disliking huts was put forward by their occupiers. Especially important among the perceived advantages were 'escape' and 'tranquillity', in effect representing the need to get away from everyday life. Feelings of health and relaxation also featured as benefits, sometimes because the locality offered the opportunity for activities such as walking or using the sea and boats. In the study, older people identified the benefits for their children of being able to use a hut, either because grandchildren were able to enjoy it or as nostalgia for their own times there as young people. Younger parents regarded the huts as a way of giving children access to a better and safer environment than the inner city and its routines and as a means of encouraging family links. Huts could be seen not only as a release from conventional daily lifestyles, with an absence of television and phones, but also as a return to a 'basic' existence with oil lamps and limited cooking and washing facilities. This lifestyle was also seen as encouraging shared activities within the family, whether outdoors or in, extending to a sense of camaraderie between families. Finally, the element of self-sufficiency and self-help in terms of maintaining or modifying a hut gave some hutters both the need and the opportunity to learn new skills, since huts are not the type of structure for which one calls in tradesmen when something needs to be done. Perhaps for the same reason, isolation was sometimes seen as a problem, in relation to actual or potential vandalism when the occupiers were away from a hut for a long period of time. Uncertainties about security of tenure and relationships with landowners could also cause worry, although this tended to be confined to only a few sites.

Huts developed to meet specific needs among a particular segment of the Scottish population. It is possible that similar little settlements existed, or continue, south of the border, although little seems to be known about them. Their future is uncertain. Some sites are declining, others are static. While most of the landowners are elderly, a few younger people have taken over and are content to let the huts stay but in return expect both a greater degree of return and a greater degree of control. From the other perspective, new occupiers may be prepared to pay rather more realistic rentals, but expect improved services in return. If left to their own devices, this type of dwelling as it has existed up to now will most likely continue for a while but then gradually disappear, witnessing the end of a peculiar form of Scottish dwelling. Whatever, if anything, replaces them will certainly be different and, while probably a lot tidier and more acceptable to landowners and to planning authorities, will fulfil different kinds of need and very likely be less picturesque.

AN OVERVIEW

These two chapters have dealt with unconventional dwellings. At the outset it was noted that the common factor linking the forms of home discussed is that, unlike conventional dwellings, they are not permanently rooted to the ground. However, the way in which these, and other types of home have been used depends on the interplay between the structure and the needs, or sometimes the preferences of its occupiers. Of the five groups examined, each has used its dwellings in its own particular way. None of the groups is exclusive to Scotland, although all have at times displayed some Scottish features. For two of the groups, the Travelling People and the Show People, the basis for their choice of dwelling and lifestyle is essentially cultural. Two others, mobile-home dwellers and the hutters, are essentially conventional members of the settled community whose choice of dwelling has been for reasons often pertaining to particular stages in their lifecycle. New Age travellers form a more anomalous group, lying somewhere between the other two pairs, with no historical or economic, and no specific cultural rationale for their chosen lifestyle. At times the boundaries between the five groups have not been hard and fast, and the dwellings used by one group have occasionally been used by the others.

There have been many other instances of Scotland's residents creating and living in unusual types of dwelling. Some are permanent year-round homes.[8] Others have been created for specific work or leisure purposes, such as the farm bothies for unmarried male farm workers or shooting or fishing shelters for seasonal use.[9] A further group, including the huts and perhaps even the humble allotment shed, provide a more general recreational experience, simply offering escape from the day-to-day round of conventional life. All illustrate different facets of the less conventional aspects of Scottish lifestyles.

NOTES

1. See Fenton, 1989; Fenton, Walker, 1981.
2. See Chapter 33 'Time Out' Work.
3. Gentleman, Wilkinson, 1977.
4. Caravan Sites and Control of Development Act, 1960; Caravan Sites Act, 1968.
5. Gentleman, 2000.
6. The terms 'huts' and 'hutters' were used collectively to cover a variety of local descriptions within the study's initial assumptions as set out below:
    The origin of huts is very uncertain. It appears that in some cases, for example during and after the two World Wars, some Scottish landowners made land available on lease for which ex-servicemen and other town and city dwellers were allowed to erect dwellings at their own cost, primarily to enjoy the benefits of the countryside and fresh air for holidays and at weekends. Such dwellings were generally of modest timber construction and, over the years, generically the name 'huts' has been applied to them.
    Generally sites comprise a number of dwellings within a specific area of land although in some cases they may be found as small clusters, or two or three

dwellings over a more widely dispersed area (NB this definition excludes beach huts and chalets used for holiday letting, huts on allotments, caravans and mobile homes).

'Hutters' occupy their plots as 'tenants' or 'licensees' generally paying an annual rental for their plot, although they may own the actual dwellings on the land. The nature of tenancy or licence arrangements is often uncertain.

Both on account of the nature of the structures and the extent, or lack, of services available to them, these dwellings were not intended to be used as permanent residences, although in practice some have come to be used for protracted periods.

7. Gentleman, 2000, 12.
8. See Chapter 46 Idiosyncratic Homes.
9. See Chapter 34 'Time Out' Leisure.

BIBLIOGRAPHY

Bird, B, O'Dell, A. *Mobile Homes in England and Wales: Report of Surveys* (Department of the Environment, Building Research Establishment, Urban Planning Division), London, 1977.

Fenton, A. *The Island Blackhouse*, Edinburgh, 1989.

Fenton, A, Walker, B. *The Rural Architecture of Scotland*, Edinburgh, 1981.

Gentleman, H (Research Consultancy Services). *'Huts' and 'Hutters' in Scotland* (The Scottish Office Executive Research Unit), Edinburgh, 2000.

Gentleman, H, Wilkinson, D. *Residential Mobile Homes in Scotland: A Study of Mobile Homes, their Sites and their Residents in 1975* (The Scottish Office Central Research Unit), Edinburgh, 1977.

46 Idiosyncratic Homes

DEBORAH MAYS

'There was an old woman who lived in a shoe ... ', so the well-known nursery rhyme begins, serving to open a child's imagination to a bizarre but comforting image. Idiosyncratic homes have always been the product of resourceful minds, occasionally of practicality or necessity and, where conversions, invariably result in a sustainable solution which perpetuates local character.[1] Many are part of an age-old tradition and answer modern demands of diversification and re-use. The range of subjects eligible to be defined as eccentric is wide, and the topic will be addressed in stages, through history, and through the two principal categories of 'purpose-built' (see Figs 45.6, 45.7, 45.8) and 'adapted', with some consideration of their presence in modern culture.

PURPOSE-BUILT HOMES

Historically, practicality has always been a prime mover in the creation of distinctive but functional homes; fertile imaginations have found solutions. The defensive brochs of early history with their honeycomb walls and enclosed courtyards, for example, give early witness. It is no surprise that their successors, the thatched longhouses or byre dwellings, which survived as late as the 1800s, should have evolved through practical economy of heat and material. While their ubiquitous form arguably disqualifies them from definition as idiosyncratic, to contemporary eyes they were certainly provident and prudent, with the residential area sharing mutual walls (and in the earliest examples, sharing doors) with the animal byres, all under one roof.[2]

The 'Colonies'
Similarly founded on an ergonomic principle were the compact and interlocked 'colonies' of the nineteenth century, the creations of an economic imperative by workmen's co-operatives (see Fig. 3.2). Economy of land and material dictated the ingenious form of, for example, the Glenogle Colonies in Edinburgh, built between 1861 and 1865 by the Edinburgh Co operative Building Association under the management of James Colville. These consist of ranked terraces where each 'flatted cottage' is part of a two-storeyed block, one dwelling sitting on another and boasting a garden and its own front door. Stone steps at right angles to the terrace access those at first floor level. James Begg, working with Colville, advocated the enterprising venture

which made the terraces possible in his seminal text on socially conscious architecture, *Happy Homes for Working Men*.[3]

Lighthouses
Coastline lighthouses required tailored and storm-proof habitations for the men who typically 'lived below the shop' in distinctive flat-roofed accommodation at the structures' feet. The Bell Rock or Inchcape Lighthouse off the Angus coast, completed in 1811 by Robert Stevenson and John Rennie, was an exception owing to its particular location. It was built in granite and sandstone on reefs in a triumph of masterful engineering. Lacking any surrounding terrain on which to site accommodation, it housed within it the keepers who maintained the light, while additional accommodation was built for their families in the nearby coastal town of Arbroath.[4] The bunk room and the kitchen-living room occupy the top tiers of the six-level accommodation, each chamber with a flat-vaulted ceiling. The Commissioners of the Northern Lighthouse Board explained that while homes provided for keepers were often remote and primarily functional in design, they were free, furnished and 'of a description greatly superior to that occupied by the same class in ordinary life'.[5] Such rock lighthouses were nonetheless uncomfortable for the keepers, allowing only a minimum of exercise, although Muckle Flugga, Shetland (1854), was highly rated for its relative spaciousness and outstanding views. The accommodation in lighthouses after the 1820s was always for three attendants at any time and this served to minimise the loneliness endured in the more remote locations. Ardnamurchan Lighthouse, Argyll (1848), designed by Alan Stevenson, boasts the only Egyptian-style houses for keepers, and breaks from the tradition of unadorned austerity.

Toll houses
The openly stated function of purpose-built toll houses, usually single-storeyed with canted or bow-ended elevations, was born of commercial and civic need; that is the collection of funds for the maintenance of roads and bridges created to improve transport and communication, usually as a result of local legislation and predominantly in the decades around 1800.[6] This occurred, for example, when a toll house accompanied a new suspension bridge at Boat of Brig, Moray, in 1830. Turnpike Trusts, authorised by parliamentary legislation, commissioned the accommodation for the toll keepers. One such trust arose from 'An Act for Repairing the Roads in the County of Forfarshire' in 1789, and led to a rash of toll houses in Angus. The tailored form of many toll houses, allied to their strategic positioning, provided a clear view of approaching travellers, with ticket windows fitted on occasion, and toll gates or barriers discouraging fee dodgers. The circular toll house at 1 Barrhead Road, Glasgow (c1800), and the octagonal example at Maryburgh Old Cononbridge, in Easter Ross, by Joseph W Mitchell (1828), illustrate the full application of conspicuous design to optimise on the view. Toll keepers on occasion seized the opportunity to sell victuals and ales.

Exceptionally, at Gretna, Dumfriesshire, and at Lamberton, Berwickshire, adjacent to the border, toll houses served as the venue for quick weddings.[7]

Hermitages
The impractical, small cottages adorning isolated locations in the Picturesque, Classical landscapes of wealthy estates in the eighteenth and nineteenth centuries were often referred to as hermitages, and examples of these can be found at Dalkeith Park, Midlothian, Acharn on the Taymouth Estate and Friars Carse, both Perthshire, Craigieburn (by Moffat), and Dunscore, both Dumfriesshire.[8] They do not boast the additional practical functions of lighthouses and toll houses, and it is uncertain, indeed, if any were inhabited long-term, serving as little more than punctuations in the landscape or tourist attractions. The Hermitage Folly, Dunkeld, Perthshire, is a bow-ended gazebo with octagonal entrance chamber, named Ossian's Hall, built for the second Duke of Atholl in 1758 and remodelled by George Steuart in 1783. Sited at the head of a dramatic waterfall on the River Braan, and known to have been occupied by a 'hermit' by day (if not by night) through the tourist season, it remains a popular tourist attraction (see Fig. 22.2).

The Gilmerton Cave
Particularly colourful are those idiosyncratic homes which gained their uniqueness purely from artistic or social curiosity. Only these aspirations can explain the underground cave in Drum Street, Gilmerton, Edinburgh.[9] It is a residential complex which lies 10 feet (3 m) below the ground and the rock ceilings of the rooms average about 6 feet (1.8 m) in height. Legend has it that the cave was carved out of the hill by George Paterson, a smith, in 1719–24.[10] The cave is approached by a flight of steps and contains several rooms leading off a central passage, including a forge and washhouse, apparently once lit by skylights. It is furnished with benches, bed recesses, a stone table and 'punch bowl', and also possesses chimneypieces and a well, all carved out of the rock. Paterson is said to have lived here for a long time with his family.

Idiosyncratic town and country houses
Occasionally individual patrons and their architects have wrought outstanding masterpieces of idiosyncracy for purely aesthetic or social pleasure. Most can be attributed to the wealthy patrons of the nineteenth century, although Robert and James Adams' 'castles' of the eighteenth century might qualify. These Adam dwellings display proto-Romantic taste, classically symmetrical in plan but castellated Gothic in elevation. They include such dramatic compositions as Culzean, Ayrshire, Seton, East Lothian, and Wedderburn, Berwickshire.[11] Internally, the decoration was of the highest calibre, but rather more mainstream than eccentric. Equally romantic in design were the 'priories' designed in the early nineteenth century; these are (or were) Gothick mansions, following the example of Horace Walpole's seminal, mid eighteenth-century Strawberry Hill in London, and typically

assumed the name 'priory' simply to conjure up the idea of medieval retreat and lineage. Examples include Crawford Priory (1810) and Inchrye Priory (1827), both Fife, and Cambusnethan Priory, Lanarkshire (1819).

Two further richly artistic houses can be considered in greater detail, representing extremes of taste, scale and patronage. Hospitalfield, Arbroath, Angus, is a unique property, commissioned by Patrick Allan Fraser between 1842 and 1870, enclosing an earlier fabric.[12] It is a fantastical red-sandstone Baronial villa, incorporating medieval fragments, with tailored interiors, created using the skills of local craftsmen. Fraser was a collector of art and antiques, and a patron of the arts. Appropriately, the house was bequeathed by the Frasers in 1890 to 'the promotion of Education in the Arts' and a trust now operates from the house offering international study retreats for students (Fig. 46.1).

Mount Stuart, Isle of Bute, was created by Sir Robert Rowand Anderson and a leading team of craft designers for the third Marquess of Bute, from 1878 to about 1905. The marquess was a leading patron of the arts in the two short decades of his adult life, and a key benefactor of the Roman Catholic Church, to which faith he was an ardent convert. This intensity of feeling imbues the architectural expression of his home. The house is a romantic Gothic casket on a palatial scale, rich in high art and, for its day, in technological innovation. It was inspired by medieval architecture on the continent of Europe and is imbued with mystical symbolism.[13]

Figure 46.1 Hospitalfield, Arbroath, Angus, a Baronial confection commissioned by the art collector, Patrick Allen Fraser, using the skills of local craftsmen. It is an extreme example of a Victorian genre of artistic houses. Source: D Mays.

The interior decoration was provided by a large team, notably Farmer & Brindley, William Frame, Charles Campbell and H W Lonsdale, and includes a marble hall and marble chapel, swimming pool, Turkish bath and horoscope room.

Such wholesale adherence to romantic revival styles and craftsmanship by architect and patron may also be seen in smaller projects. Frederick Thomas Pilkington, one of a number of 'romantic Goths' working in Scotland in the later nineteenth century, applied the style to comparatively humble terraced housing in Broomhill Road, Burntisland, Fife (1858) and in Park End, Bridge Street, Penicuik, Midlothian (c1862). He worked with greater funds to a richer degree, but with the same trademarks, in the Edinburgh suburb of the Grange in a double villa at 48–50 Dick Place, and his own house, Egremont, 38 Dick Place in 1864. The latter displays Pilkington's play with masonry in imaginative, perverse composition and his love of carved Ruskinian foliage, yet by contrast, the interior boasts a fine Louis XVI decorative scheme. Thornton & Shiells, James Gowans (Edinburgh) and Pirie & Clyne (Aberdeen) were other notable architects in the same camp, who similarly added colour to the Scottish townscape.

A comparatively recent idiosyncratic country house is that at Callernish, North Uist, designed by Martyn Beckett for Lord Granville in 1962. It comprises a circular courtyard plan inspired both by the ancient ring of standing stones at neighbouring Callanish, Lewis (playfully) and by historic precedents such as the Round Square at Gordonstoun, Moray (where Beckett had worked) or the round stables at Prestonfield, Edinburgh, and Errol Park, Perthshire.

Estate dwellings
Wealthy patrons over recent centuries have on occasion decorated their wider estates with imaginative and colourful properties for their employees, more functional than hermitages and follies although no less idiosyncratic. Lodges, in particular, lent themselves to this expression, declaring the patron's fortunes and architectural connections; such are the South (Rockville) and East Lodges to the Balgone estate, East Lothian (1859), by J Anderson Hamilton, for the Suttie family, at once Baronial *tours de force* and Picturesque landmarks. Another notable example of adornment to policies is the former Factor's House at Paxton, Berwickshire (1871), by William J Gray of Coldingham, latterly serving as a dower house. This is a Swiss Cottage in form, a cocktail of pink masonry, rusticated and vermiculated at the quoins, with decorative half-timbering, stencil work and filigree. The Retreat, Abbey St Bathans, Berwickshire, is a further example of 'estate' enrichment with a difference. It was designed originally for the Earl of Wemyss (Francis Charteris) in the later eighteenth century and is generally thought to have been his country seat and hunting lodge with separate flanking wings providing stabling, kennels and staff quarters.[14] Circular in form with a conical roof and Gothick windows, the planning of the interior is ingenious with a bow-ended drawing room and undulating ceilings (Fig. 46.2).

Figure 46.2 The Retreat, Berwickshire, a round house imbued with a sense of fun, is an example of estate adornment in the late eighteenth century. Source: D Mays.

ADAPTED BUILDINGS

Conversions of unusual buildings to residential use can constitute imaginative achievements and provide rich interest. They can also make continued use of otherwise redundant property and therein support the current appreciation of sustainable solutions. Public support for this is evidenced in such publications as Historic Scotland's *Passed to the Future* (2002), on the sustainable management of the historic environment. Conversions of listed buildings to residential use are given the added incentive of relief from Value Added Tax. As more than 50 per cent of the building trade in the late twentieth century was employed on old buildings rather than new, for which its members were predominantly trained, the redevelopment of old fabric can be seen to contribute significantly to the economy.[15] The demand (especially in statutorily listed property) for traditional materials and craftsmanship in the work provides further employment and preserves traditional skills. Government backing for re-use in Britain is escalating as Cadw's recent booklet on the conservation and conversion of chapels in Wales further demonstrates.[16]

The challenges facing architects and owners seeking to convert are numerous and varied. Compatibility with domestic conversion is a preferred starting point in the selection of a property, otherwise the task may be too problematic. Where residential use has not been the predominant element in the area the infrastructure and amenities may not be suited for domestic purposes. Listed Building status, which is probable for the more unusual properties, will bring some advantage (as above) but will require Listed Building Consent from the Scottish Executive prior to any alteration. Planning Permission will almost always be required for the change of use. Building regulations for residential purposes are necessarily strict and the need to satisfy these and domestic requirements without losing the character of a building can be demanding. Once completed, the level of public curiosity that the property will naturally attract, and consequent probability of uninvited visitors and lack of privacy, must be borne in mind.

While imaginative re-use of redundant agricultural buildings, discussed below, has long been commonplace, redeployment of other properties for residential use is a more recent phenomenon. This has been made possible by the diversity in building types arising in particular since the eighteenth century and their subsequent redundancy in the face of social and technological change. An increasing desire to utilise well-designed structures, notably for reasons of sustainability, historical continuity, preservation of artistic form and economy, has fuelled the practice of conversion. Of course a thirst for strikingly different and characterful homes also plays its role.[17] The Landmark and Vivat Trusts work to locate, restore and convert idiosyncratic gems across Great Britain, partially funding future projects by letting those completed. Their catalogues demonstrate the range and wealth of examples, such as Landmark's The Pigsty, Robin Hood's Bay, Yorkshire. Building Preservation Trusts across Scotland can fulfil a similar role. The Buildings at Risk Register operated by the Scottish Civic Trust, provides a source for available properties of character, potential imaginative homes of the future.[18]

Towers

Towers of all sorts have lent themselves most naturally to the search for eccentricity. It is no coincidence that the protagonist of Iain Banks' novel, *Espedair Street* (1987), should have been housed in a church tower. Three very different but wholly characterful 'vertical' home conversions can be found in the Forth Valley. The first of these, Bridgeness Tower, West Lothian, was converted in 1988 by W A Cadell Architects, and in 1989 won a Saltire Award for its architectural achievement. A converted seventeenth-century windmill, it contains five storeys and a roof-top patio with panoramic views of the Firth.[19] The Greigs, who occupied the tower from 1994 to 2002, reported on the unusual historical features which graced their home, such as the spiral stair joining each level, with ships' masts forming the newel, but pointed out that its verticality and quirky room plans would not suit everyone.

A late nineteenth-century water tower constructed by James Leslie for the Dalkeith Town Council, Midlothian (1879), is another remarkable domestic conversion. The octagonal form and polychrome brickwork of the five-stage tower was imaginatively recreated by Gerry Goldwyre in 1987, and remains a much-loved local landmark.[20] He surrounded the uppermost level with a jettied timber balcony and created a weather-boarded lantern eyrie at the site of the former water tank. Goldwyre described the logistical difficulties for those undertaking such an imaginative task. The most awkward element to incorporate was the stair, which had to comply with strict fire regulations but now adds to the mystery of the property in its enclosed, circuitous form. The local council was supportive of the conversion and worked with Goldwyre to overcome potentially inhibiting Building Regulations. While the tower provides a family home of considerable character, tailored to the residents' needs, with views of Penicuik, Arthur's Seat and, on a good day, North Berwick Law, Goldwyre points out that its beacon-like qualities and eccentric form serve as a magnet for inquisitive tourists (see Fig. 46.3).

The third distinctive tower conversion to be considered is very different again. A polygonal, two-stage signal tower with rooftop terrace, in North Queensferry, Fife, it lies in the shadow of the Forth Bridge and was built in 1917 as a Royal Naval property to guide warships along the Forth.[21] It ceased operation in 1995 and its original purpose is still warm in local

Figure 46.3 Dalkeith's Water Tower, Midlothian, is a prominent local landmark and an imaginative conversion, illustrating how even the most constrained of spaces can be utilised to effect. Source: D Mays.

memory. It is undergoing alterations by Euan Millar for conversion to his own residence. The former engine room at the top of the tower is now a bedroom, while a living room, kitchen and bathroom have been formed downstairs. Owing to its exposed position, triple glazing has been installed to counter the icy winds of winter.

Canal boats
Canal boats, that is barges, lighters and narrowboats, have become increasingly fashionable for conversion since World War II and the ensuing shortage of homes.[22] The deckhouses of the first narrowboats were very small, essentially single multipurpose spaces, with low ceilings, fitted cupboards, a stove and bed recess, serving as the home of the skipper and his family and were by modern standards often unacceptably compact.[23] Nowadays the transport function and cargo space of the boats has all but disappeared in the vessels' conversion to floating homes. In Scotland, the Union Canal at Ratho, West Lothian, is a notable centre for such residences, close to Edinburgh and such attractions as the Millennium Wheel, Falkirk, Stirlingshire.[24]

Some houseboats retain the livery of the commercial company under which they formerly operated. Most have steel hulls and require thorough insulation. Heating the small spaces is relatively inexpensive. Heaters are often part of a diesel-fired hot-air blown central heating system, while bottled gas is another common source of power. Electricity may be supplied by mooring operators, and batteries and mains generators serve otherwise. Water is usually supplied in Inland Waterways' sites by taps at the moorings, the vessels equipped with holding tanks and shower pumps. Sewage is contained in separate holding tanks and pumped periodically into the sewage system. Since December 1963 Planning Permission has been required to moor a houseboat as it forms a material change in the use of the land. Floating homes do not escape taxes; some are subject to Council Tax, while others may pay a tax on the mooring fee. Neil Robinson, a former houseboat resident in Ratho, reports from experience that the boats are vulnerable to theft and that there is occasional hostility from passers-by over what is seen as a privileged lifestyle. If a chandler's shop is not available, securing essential amenities can be awkward and time consuming.

Agricultural buildings
Redundant agricultural properties, as indicated above, present a wealth of rural buildings suitable for residential conversion, and those in commuting proximity to urban centres have invariably been seized by keen-eyed developers. Mills in particular, frequently with a wheel house and located by a river or extant mill race, offer a picturesque ensemble, and examples can be found all over Scotland. The converted, earlier nineteenth-century Mill of Ireland, Stenness, Orkney, for example, includes the former milling equipment and offers an open airiness within, while retaining a sense of its wholesome original purpose. By contrast in scale and location, the substantial mill group at Stanley Mills, Perthshire (1785–1850), has recently been

Figure 46.4 Crauchie Steading, East Lothian, now residential, retains the presence of an agricultural property. Source: D Mays.

converted to housing, while the early nineteenth-century West Mill in Dean Village, Edinburgh, has been inhabited since its conversion in 1972 by Philip Cocker & Partners.[25] While very popular, the character of steading conversions may be destroyed by the financial pressure to incorporate too many units. One of the most successful examples is that at Crauchie, East Lothian, where the character of the cartshed and granary has been respected, with recessed glazing to the cart arches and retention of the forestair. The old feeding openings of the former cattle courts have been unaltered and read as such, while the former horsemill has been carefully glazed to demonstrate its original function (see Fig. 46.4).

Stables and coach houses
Stables and coach houses similarly offer an interesting architectural form for conversion, and the mid eighteenth century coach house at Alderston, Haddington, East Lothian, has been saved from ruin by Bob Heath in a conversion for his own residence (see Fig. 46.5). On occasion, the owners of great houses have opted to downsize to smaller premises, converting the stables to use as their own domicile: for example at Saltoun, East Lothian (where the Great Hall has been converted into flats); and at Penicuik, Midlothian, and Dunglass, East Lothian (where the main houses were damaged by fire).

Dovecots
Dovecots may seem a bizarre choice for residential conversion as their dimensions are small, but on occasion they have proved an attractive subject. The lectern dovecot at East Morningside House, Edinburgh, for example, offers a cosy cabin-like interior, while Morris & Steedman's conversion of the larger eighteenth-century Humbie Dovecot, East Lothian, in 1971 has a more modern decorative scheme. Saved from decline, both continue to contribute recognisable profiles to their setting and a rich sense of local history.

Churches
Declining congregations across many of the main Christian denominations in Scotland have led to an ever-growing estate of empty churches, and a great variety of possible types of conversion. Residential use is not, invariably, the most compatible choice, but with care and imagination, interesting and adaptable homes can often be created. The former Episcopal Christchurch, 118 Trinity Road, Edinburgh, is one such. Built by John Henderson in 1854 for Walter Mitchell Goalen, it was originally a proprietary chapel. In 1980, it was converted by Gordon & Dey, and continues to proclaim its unusual history, nestling in the villa suburb of Trinity as a single dwelling.

Figure 46.5 Alderston Coach House, Haddington, East Lothian, prior to conversion to a family home and architectural practice by Bob Heath. Its inviting form captured the architect's imagination and it now stands as a skilful conversion. Source: D Mays.

Other buildings
Village schools are often under threat from declining pupil numbers and downturns in local economies, so that redundant examples reach the market from time to time. The Old School in Dean Village, Edinburgh, built by Robert Wilson in 1874, for instance, was flatted in 1985. This change of use preserved its central architectural place within the village community, recalling a former era still very much in living memory (see Fig. 46.6). Mews properties, which originally served the terraces of later Victorian Scotland, have in conversion provided many temporary and permanent homes. Thirlestane Lane Mews, Edinburgh, for example, was originally a terrace of two-storeyed design, with stalls, coach houses and harness room at ground level and hayloft and domestic quarters above but, with the ground-floor function now redundant, many are now fully domestic.

One of the most colourful and unusual conversions in Scotland, which would merit a place in an update of Lucinda Lambton's *Curious Houses*, is that of the Pineapple, Dunmore, Stirlingshire. Converted in the late twentieth century, it can now be enjoyed as a residence by all, available year round to rent from Landmark. Built for the Earl of Dunmore about 1771 at the centre of a range of bothies, it is said to mark his return from Virginia where he was Governor, echoing the tradition there of placing a pineapple on a

Figure 46.6 The Old School, Dean Village, Edinburgh, illustrates how careful conversion can retain a property's character, allowing a major player in local history to continue to declare its contribution. Source: D Mays.

Figure 46.7 The Pineapple, Dunmore, Stirlingshire, is an unusual example of international inspiration which is now, together with the flanking bothies, used for residential purposes. The fruit would have been unknown to the majority of the population at this date. Source: D Mays.

gatepier when in residence (see Fig. 46.7). The proposed conversion of the concrete World War II Control Tower at Fearn Airfield, Ross-shire, to a home and photographic studio demonstrates the level of enthusiasm, perseverance and breadth of imagination among those embarking on the creation of an idiosyncratic home through conversion. Neither water nor electricity is supplied to this remote site, but inside remains the layout, mural signage and atmosphere of war rooms bustling with activity.

When undertaking a conversion consideration must be given to a supporting infrastructure (drainage, water supply, the provision of gas and electricity, street lighting and road maintenance, in particular), Planning Permission, accordance with building regulations and, in the best, the survival of character. Residents in such properties must expect unorthodox internal planning and quirky spaces, they may need to be agile and to expect inquisitive visitors (see Fig. 40.6). While the conversion of redundant jute mills in Dundee or steadings in the commuter belt may be lucrative, conversion for single occupancy may not bring as much financial gain, but certainly the sustainable result is commendable in an age of accelerated change.

The pattern of eccentric design and adaptation of redundant subjects has been as widespread in other parts of Britain and abroad as it has been in Scotland. The conversion of the former dock buildings of Liverpool and London for residential purposes, for example, is well known. Sir Edwin Lutyens' sense of fun was applied in his designs, his 'vivreations', and his imaginative playfulness is evident in the windswept Lindisfarne Castle,

Holy Island, Northumberland (1902), and the romantic Castle Drogo, Drewsteignton, Devon (1910). Einstein's Tower, Potsdam, Germany (1917–24), a residential laboratory by Eric Mendelsohn for the great eponymous scientist, is one of the best known and most remarkable creations in the world.

NOTES

1. For the purpose of this chapter, a 'home' is defined as a year-round, permanent dwelling rather than a seasonal or holiday home (for these see Chapter 33 'Time Out' Work; Chapter 34 'Time Out' Leisure; Chapter 45 Domestic Life in Temporary and Self-Built Dwellings), with only the occasional illustrative exception. The Scottish Executive's lists of buildings of special architectural or historic interest provide invaluable information on many of the subjects detailed in this article.
2. For further information on longhouses see descriptions in Marshall, 1794; the various volumes of the *OSA*; the overview by Fenton, Walker, 1981. Early examples were built in drystone, had earthen floors and were heated by fires whose smoke escaped through holes in the open cruck-framed roof. Contemporary descriptions indicate that earth and turf were also used for the walls of longhouses. As vernacular housing, the type evolved considerably and demonstrated distinct regional variation. Generically, the longhouse included a byre and a kitchen-living area, with further divisions arising over time, plastered timber screens providing the slender divide. Later examples included enclosed box beds and separate family rooms. The use of glass or chimneys was rare and confined to late examples such as those in Angus before 1800.
3. Later ventures by the Edinburgh Co-operative Building Company (as the Association became) can be found off Dalry Road (Dalry Colonies), London Road (Abbeyhill Colonies) and Slateford Road ('Flower' Colonies). The Pilrig Colonies of 1849, by Patrick Wilson for the Pilrig Model Dwellings Company, pre-dated the Glenogle Colonies as flatted cottages, but sprang from a more philanthropic root.
4. See Bathurst, 1999, for an account of the feats of engineering by the Stevensons. Scotland's Lighthouse Museum at Kinnaird Head, Fraserburgh, Aberdeenshire, and the Stromness Museum, Orkney, provide a wealth of information on lighthouses and their residential components.
5. In 1857. See Bathurst, 1999, 237.
6. The fee for crossing a bridge was sometimes referred to as the 'pontage'. The board which showed the fees for crossing the toll gate at Wanlockhead is now housed at Leadhills Mining Museum, Lanarkshire.
7. The toll house built at Gretna in 1818 to accompany the new bridge over the River Sark is referred to as Scotland's 'first house'. The house at Lamberton provided the equivalent service for the east-coast wedding trade.
8. The term 'hermitage' was first applied to isolated settlements of communal monasticism where the monks dwelt in separate cells; and became synonymous with retreats. Idealised in the Picturesque Movement of the eighteenth century, it evolved again and by the early nineteenth century was the name for certain *cottage ornée* houses, indicating the romantic isolation of the dwellings. It became popular again later in the century for Arts and Crafts houses similarly designed and situated.
9. Gillon, 1990, 50–2, gives an account of this troglodyte's home; also Cant, 1999; Gillon draws on a 1792 account by the Rev Thomas Whyte; see also Chapter 1 Shelter from the Elements.
10. The eighteenth-century attribution has been challenged by later historians, such as

F R Coles, Assistant Keeper of the National Museum of Antiquities in Edinburgh, who in 1897 considered the Paterson legend to provide an inadequate explanation and suggested that a more sinister origin (linked to the Free Masons or smuggling) might in fact be the case.
11. Rowan, 1985, discusses the Adam designs in general and some in detail with illustrations.
12. Hospitalfield is now available as a retreat for artists.
13. The earlier house at Mount Stuart was destroyed by fire in 1877 and Rowand Anderson was commissioned to work with the marquess to create the magnificent home of his dreams, embodying his philosophy and giving a canvas to craftsmen.
14. For further information on The Retreat, see the Rev John Sked in the *OSA*, Berwickshire, 1791–9, 6; Lindsay, 1931.
15. This statistic on the building trade was supplied by Ingval Maxwell, Director of Technical Conservation, Research and Education at Historic Scotland.
16. Cadw (Welsh Historic Monuments), 1999.
17. The production of literature supporting re-use and conversion has accelerated in the decades after 1945 in the United Kingdom, and publications such as those by Strong, Binney, Harris, 1974; Scottish Civic Trust, Scottish Development Department, 1981; Department of the Environment, Scottish Development Department, Welsh Office, 1972–7 (including vol 3, *New Life for Old Churches*), have gone some way to encourage redevelopment.
18. For further information on the British Archaeological Reports, contact the Scottish Civic Trust, 42 Miller Street, Glasgow G1 1DT.
19. The windmill has a history of conversion, not always to residential use. It ceased operation as a mill in 1749 when bought by a shipmaster as an observatory. The Cadell family acquired it subsequently, letting it to a businessman in the 1800s as a factory for the manufacture of iodine and sulphuric acid made from local seaweed. By the late nineteenth century it had reverted to the Cadells and with the architect Hippolyte J Blanc, they added the top storey and patio with castellated effect in 1895. See *Press and Journal*, 13 June 2002; Jacques, 2001, 146, for further information.
20. The following information was gained courtesy of Gerry Goldwyre, in conversation. He now runs a successful restaurant from the tower.
21. See the article on its conversion in *Dunfermline Press*, 29 August 2002, for an illustration.
22. More than 15,000 people in Britain live permanently in floating homes, although converted canal boats are only a percentage. See www.liveaboardcruising.co.uk for further information on this domestic option.
23. There is a houseboat museum in Amsterdam, on the Prinsengracht, which provides useful archival material on such waterborne existence.
24. See the website for The Bridge Inn, Ratho.
25. The three former mills at Stanley were saved from demolition by Historic Scotland and the Phoenix Trust.

BIBLIOGRAPHY

Bathurst, B. *The Lighthouse Stevensons*, London, 1999.
Begg, J. *Happy Homes for Working Men and How to Get Them*, London, 1866.
Cadw (Welsh Historic Monuments). *Chapels in Wales: Conservation and Conversion*, Cardiff, 1999.
Cant, M. *Villages of Edinburgh: An Illustrated Guide*, Edinburgh, 1999.
Department of the Environment, Scottish Development Department, Welsh Office. *The*

Aspects of Conservation (Her Majesty's Stationery Office), London, 1972–7, including vol 3, *New Life for Old Churches.*
Fenton, A, Walker, B. *The Rural Architecture of Scotland,* Edinburgh, 1981.
Gillon, J. *Eccentric Edinburgh,* Edinburgh, 1990.
Jacques, R. *Falkirk and District,* Edinburgh, 2001.
Lambton, L. *An Album of Curious Houses,* London, 1988.
Landmark Trust. *The Landmark Handbook,* Maidenhead, annually.
Lindsay, I G. Selected precedents: Circular plans from Scotland, *Architects' Journal,* 29 (July 1931), 131–2.
Marshall, W. *General View of the Agriculture of the Central Highlands of Scotland,* Edinburgh, 1794.
Rowan, A. *Designs for Castles and Country Villas by Robert and James Adam,* Oxford, 1985.
Scottish Civic Trust, Scottish Development Department. *New Uses for Older Buildings in Scotland* (Her Majesty's Stationery Office), Edinburgh, 1981.
Scottish Executive, Historic Scotland. *Passed to the Future* (Operational Policy Paper), Edinburgh, 2002.
Scottish Minister's lists of buildings of special architectural or historic interest.
Strong, R, Binney, M, Harris, J. *The Destruction of the Country House, 1875–1975,* London, 1974.
Vivat Trust. *Catalogue,* London, annually.

PART THREE

•

Homelessness

47 Homelessness

PETER A KEMP

Dwellings provide much more than just shelter against the elements. They also constitute the locus of the 'home', a term which is imbued with many meanings and steeped in symbolism. Our domestic lives are played out within our dwellings and we invest much of our financial and emotional capital in them, perhaps even more when they are owned than when they are rented. For most people, the home is a sanctuary or haven from the outside world.[1] The phrase 'home is where the heart is' may be hackneyed but it still has a ring of truth. It follows that to be without a dwelling of one's own – to be *home*less – means to be without much more than just shelter from the storm. It also involves the loss of the locale within which we spend most of our time, keep most of our material possessions, and entertain family and friends. Furthermore, the very experience of being homeless can cause people to suffer from other adverse circumstances – such as poor physical and mental health, and unemployment – and make it difficult for them to obtain or exercise the rights which other citizens take for granted, such as voting in elections or claiming social security benefits.

And yet what we mean by homelessness is not straightforward or universally agreed. In fact, the situations which may be regarded as constituting homelessness are a matter of argument and political debate.[2] Definitions of homelessness can vary from being roofless and without any shelter at all, to being badly housed.[3] Where exactly one draws the line between these two positions depends partly upon one's values and political sympathies. And it also depends upon the period in which one lives, for perceptions of homelessness and what, if anything, should be done about it are not static, but have changed over time. But speaking generally, most people in Scotland today would probably regard those who are literally roofless or sleeping rough as homeless. Many would probably also agree that people using night shelters and hostels, or staying in bed-and-breakfast accommodation and not having a permanent address of their own, are in a real sense homeless too.

HOMELESSNESS AND THE POOR LAW

Although the term 'homelessness' is relatively recent, there has been concern about the homeless poor for centuries in Scotland, as elsewhere in

Great Britain. Worry about begging and vagrancy was a recurrent feature of public debates from the Middle Ages onwards and a frequent subject of legislation. For the most part, official responses to the homeless poor were punitive, the aim being to deter or punish rather than to comfort or accommodate. Laws were passed to prohibit both begging and vagrancy, but these appear to have had little success in preventing either phenomenon.[4]

A continuing feature of debates and practice under the Poor Laws was whether, in providing relief to vagrants or the 'casual poor', one was simply encouraging people to beg or making it possible for them to pursue their unsettled way of life. The homeless poor were not to be provided with alms lest it encourage others to become feckless and rootless. Moreover, the direct link under the Poor Law between providing help for the vagrant and paying for it via local taxes (the rates), added further to the desire to act punitively and with deterrence rather than hospitality in mind. Similar concerns were aired about helping the destitute more generally, but the anxiety was that much greater when it came to vagrants and beggars. Opposition to the levying of a Poor Rate was a powerful factor underlying the largely negative wider response to vagrants and the destitute. The 1844 Royal Commission on the Poor Laws in Scotland, for example, received evidence of lairds in the north east who had evicted tenants and sub-tenants from their land in order to avoid a heavy assessment under the threatened new Poor Law.

Nevertheless, public attitudes to vagrants were not straightforward. Generally speaking, two different stereotypes were reflected in official responses to them. On the one hand, there was the so-called 'sturdy beggar', who was considered to have opted for a life on the road, preferring idleness and begging to regular employment. On the other, there were those whose homelessness and destitution were due, not so much to wilfulness, but to infirmity or old age and who were incapable of doing paid work. While the former were to be discouraged and even punished, the latter were to be pitied and perhaps to be provided with alms, although preferably by charity rather than by the local ratepayers under the Poor Law. Although the Poor Laws were reformed in 1845, there remained no statutory provision for the able-bodied poor in Scotland until 1921, which meant that the sturdy beggar had still to rely largely on charity, casual work and perhaps theft in order to survive.

The recurrent public concern about vagrancy suggests that the scale of the problem must have been not inconsiderable. However, it is unclear exactly how many vagrants there were in Scotland and the numbers probably fluctuated over time. In 1879, when requesting public support to 'check the fearful evil of vagrancy', the Glasgow Charity Organisation Society estimated that there were 50,000 to 60,000 vagrants in Scotland. While this may be an overestimate (and it is difficult to verify the figure one way or the other), it does suggest that there were many thousands of homeless poor in Victorian Scotland.

Vagrants and others without a dwelling of their own often stayed in common lodging houses, which sprang up in Scottish cities and elsewhere during the nineteenth century. The Public Health (Scotland) Act 1897 defined common lodging houses as:

> ... a house, or part thereof, where lodgers are housed at an amount not exceeding 4d a night, or other such sums as shall be fixed under the provision of this Act, for each person, and shall include any place where emigrants are lodging and all boarding houses for seamen irrespective of the rate charged for lodging or boarding.

Sleeping accommodation within common lodging houses was usually shared with the other occupants and typically possessed bunk beds, cubicles and beds in dormitories (Figs 47.1, 47.2). Thus, what people were paying for was in effect a bed space rather than a dwelling. The residents of common lodging houses paid by the night and sometimes by the week. They had no security of tenure, but were not necessarily vagrants or regarded by either themselves or their contemporaries as homeless.[5]

Most common lodging houses were commercial premises seeking to make a profit by providing temporary accommodation, not just to homeless

Figure 47.1 Woman in bed in a dormitory, common lodging house, Glasgow.
Source: Reproduced by kind permission of Mitchell Library, Glasgow.

people but also to travelling workers and people otherwise in transit. However, the Corporation of Glasgow began to provide municipal common lodging houses as early as 1871. And during the late nineteenth and early twentieth centuries, organisations such as the Salvation Army and Church Army also began to provide common lodging houses, with missionary rather than commercial objectives in mind. The number of bed spaces in common lodging houses in Glasgow reached a peak of 13,844 in 1913, at which time there were 60 establishments (Fig. 47.3).

By the inter-War years, the homeless and rootless were given temporary relief under the Poor Law in so-called 'casual wards'. These shelters were sometimes kept separate from the poorhouse or located in a cottage or other building which was an annexe to it. Conditions within these casual wards were often more basic and even less attractive than in the poorhouse.[6] In some places the local policy under the Poor Law was to refer vagrants to common lodging houses rather than to provide assistance in casual wards. Nonetheless, the provision of common lodging houses began to decline quite sharply from the early 1930s.

Figure 47.2 Upper and lower bunk, common lodging house, Glasgow. Source: Reproduced by kind permission of Mitchell Library, Glasgow.

Figure 47.3 Men in a common lodging house dining room, Glasgow. Source: Reproduced by kind permission of Mitchell Library, Glasgow.

HOMELESSNESS AND THE WELFARE STATE

The 1948 National Assistance Act marked an important watershed in the development of legislation towards homeless people. The Act abolished the Poor Law, closed most of the casual wards, and transferred the relief of poverty to central government. The newly established National Assistance Board took over the few remaining casual wards and maintained them as 'reception centres' to meet its new obligation under the 1948 Act to provide shelter for those 'without a settled way of living'. A survey carried out in January 1950 found there were 88 men and 13 women and children in reception centres in receipt of a bed for the night in Scotland.[7] Thus, the number of homeless poor in the new, post-War welfare state was only a fraction of the supposed number of vagrants in Scotland a century earlier. Indeed, the Medical Officer of Health for the City of Glasgow noted in 1955 that, 'In Scotland today the genuine tramp is becoming a rarity.'[8]

The 1948 Act also required the new local authority welfare departments to provide permanent residential care for people 'who by reason of age, infirmity or any other circumstances are in need of care and attention which is not otherwise available to them'. In addition, local authorities were required, for the first time, to provide temporary accommodation for people 'who are in urgent need thereof, being need arising in circumstances which could not reasonably have been foreseen or in other such circumstances as the authority may in any case determine'.

The 1948 legislation was an important step towards a more humane approach to the way in which society tackled homelessness. Nevertheless, it was still a heavily circumscribed approach. The fact that the duty to provide accommodation was given to welfare rather than to housing departments reflected a prevalent view that homelessness was not so much a housing problem (a shortage of dwellings) as a social work one. Moreover, the need to be rehoused had to be 'unforeseen', an important qualification which was subsequently embodied in the later notion of 'intentionality' under the 1977 Homeless Persons Act (see below). The implication was that those who could have averted their loss of accommodation were not deserving of assistance. Furthermore, even where the applicant's homelessness could not reasonably have been foreseen, the assistance was to be temporary rather than the provision of permanent accommodation. This all amounted to a view that 'legitimate' homelessness was the result of an emergency (such as fire or flood) or a personal crisis which could be resolved by counselling or other social-work skills. In either case, the appropriate response was seen to be the provision of temporary accommodation to enable the family to get back on its feet.

The way in which local authorities interpreted the legislation varied, however, not least because the 1948 Act gave them considerable scope to exercise their discretion. Many local authorities took a punitive approach to those who sought their assistance and they seemed unable to break free from the harsh Poor Law tradition. Indeed, instead of using ordinary housing, many councils used former workhouses to provide temporary accommodation to homeless families, while others resorted to bed-and-breakfast establishments. It was also common for local authorities to restrict assistance to families and exclude single homeless people from the definition of persons in 'urgent need' of help under the Act. In fact, many councils also excluded fathers and provided temporary accommodation only to the mother and children (a practice many had operated when they were responsible for the Poor Law prior to 1948).

Immediately after World War II the extreme shortage of accommodation meant that homelessness hit the headlines for a brief period and led to a temporary growth in squatting in Scotland and elsewhere in Great Britain. In August 1946, for example, an estimated 1,500 families in Glasgow were squatting in empty properties including churches, hotels and mansion houses, which in turn led to legal action to evict them. The Labour Government agreed to allow homeless families to occupy former Army camps and by October 1946 about 7,000 people were doing so in camps across Scotland. In addition, the government requisitioned empty dwellings, converted properties, repaired badly damaged houses, and built temporary homes ('prefabs') in order to tackle the homelessness crisis.

During the 1950s, the number of households admitted to local authority temporary accommodation fell as the post-War house-building drive gathered momentum. The establishment of the NHS and the creation of a comprehensive social security system by the 1945 Labour Government, more or less full employment, and rising real incomes also seem to have reduced the number of applicants for temporary accommodation under the National Assistance Act.

In the 1960s, however, the number of families provided with temporary accommodation by local authorities increased. It was at this time, as part of a wider 'rediscovery' of poverty by academics and social commentators, that homelessness emerged as a new social problem and one that seemed to affect 'ordinary' families and not just down-and-out single people. The broadcast of the TV drama 'Cathy Come Home' helped to make homelessness a media issue, and subsequently the pressure group Shelter was established to campaign for the homeless and badly housed. This helped to bring homelessness to the attention of the public and politicians.

Concern about homelessness increased still further in the 1970s as a growing number of reports focused on the subject. Official recognition of the expanding problem of homelessness was heightened by the publication of the Morris Committee Report (1975) on the relationship between housing and social work services in Scotland. The report, along with others such as the Finer Committee Report (1974) on lone-parent families, saw homelessness as more of a housing than a social work problem and, accordingly, one that should be the responsibility of housing rather than social work departments. The problem was less that homeless people were inadequate or the unlucky victims of an emergency, but more that they simply were unable to find accommodation because there was a shortage of affordable, rented dwellings.

Meanwhile, common lodging houses – or hostels in more contemporary jargon – were a declining feature of Scottish urban life. In 1962, there were 2,087 bed spaces in common lodging houses in Glasgow and a further 788 in Edinburgh. By 1972 the figures were 812 and 656, a fall of 61 per cent and 17 per cent respectively. This suggests that, despite the growing public concern about family homelessness, the numbers of single homeless people were continuing to fall, especially in Glasgow.

HOUSING THE HOMELESS

The Housing (Homeless Persons) Act 1977, which came into force in Scotland on 1 April 1978, has proved to be the single most important legislative response to homelessness to date. Under this legislation, certain groups of homeless people were given a statutory right to be rehoused by their local authority. This was an important new social right, even if it did not apply to all who were homeless. In amended form, the right still applies in Scotland today. This new measure reflected the growing belief that the problem of homelessness was due to a lack of accommodation rather than to the mendacity or inadequacy of the homeless people themselves. However, notions of blame were not entirely absent from the legislation and, in addition, assistance was to be provided to the deserving rather than the undeserving homeless.

Under the 1977 Act, and successor legislation, the duty of local authorities to re-house homeless people applies only to those who are in 'priority need'. The latter is defined as being households with dependent children or in which a woman is pregnant, people who are deemed to be vulnerable (for example, as a consequence of their age, or physical or mental disability), and people who are made homeless by an emergency (such as fire or flood). However, even among priority need groups, the statutory duty to provide permanent housing is confined to those who are deemed to be 'unintentionally' homeless; people who are intentionally homeless have a right only to temporary accommodation. Finally, the right to be rehoused extends only to those (priority need, unintentionally homeless) people who can prove they have a 'local connection'. If the homeless applicant fails that test, they can be referred to a local authority with which they do have such a connection.

Thus, as well as drawing on the age-old Poor Law distinction between the deserving and the undeserving, as now embodied in the concepts of priority need and intentionality, the local connection rule under the 1977 Act also draws on the old Poor Law concept of the parish test. In effect, too, the law drew a clear distinction between two types of homeless people. On the one hand, there were 'priority need' or 'statutory homeless' people, who are mainly households with children and to a lesser extent vulnerable people who are pregnant, old or sick. Provided they were not intentionally homeless, these people were provided with accommodation and had a right to be re-housed. On the other hand, there were those people who did not qualify for rehousing, either because they were intentionally homeless or because, whatever the cause of their homelessness, they were not deemed to be in priority need. The latter mainly comprised single homeless people who were sleeping rough, staying in hostels or bed and breakfasts, or were among the 'hidden homeless' staying temporarily with friends and relatives. Policy and practice towards homeless people in Scotland has bifurcated along these lines since 1977.

RISING HOMELESSNESS

In the 1980s and 1990s the number of 'statutory' homeless people increased dramatically and the number of people staying in hostels or sleeping rough in Scotland also rose. Between 1987/8 and 1999/2000 alone, the number of applications to Scottish local authorities for rehousing on the grounds of homelessness increased from 24,668 to 46,023, a rise of 87 per cent in just 12 years (Fig. 47.4). In 1999/2000, 74 per cent of applicants were assessed as being homeless or potentially homeless, while 44 per cent were deemed to be both homeless and in priority need.

Although the reasons for the growth in homelessness in Scotland since the late 1970s are a matter of dispute, it is clear that a rise in unemployment over the same period has been a major factor. Other reasons include the reduction in the supply and affordability of rented homes, the growth in family breakdown, and the policy of releasing patients from long-stay psychiatric institutions into the community. These causes of the growth of homelessness are sometimes referred to as 'structural' factors because they are beyond the control of any one individual, rather than being the result of

April to March	Number of applications
1987/8	24,668
1988/9	24,741
1989/90	29,068
1990/1	35,061
1991/2	40,623
1992/3	42,822
1993/4	43,038
1994/5	41,495
1995/6	40,936
1996/73	40,989
1997/8	43,135
1998/9	45,723
1999/2000	46,023

Figure 47.4 Number of households applying to local authorities under the homelessness legislation. Sources: Pawson, 2000, Table 2.1; Homelessness Task Force, 2002a, Annex D.

personal inadequacy or mendacity. Ability to cope with adverse circumstances may sometimes explain why one person as opposed to another is homeless, but not the aggregate *level* of homelessness. The 'presenting' causes may be relationship breakdown, mortgage arrears or eviction by the landlord, but the underlying causes explaining why so many people are homeless are more structural factors such as the relationship between housing demand and supply, the level of unemployment and poverty.

By the end of the twentieth century, homelessness had become a visible and much-remarked feature of Scottish life. Rough sleepers and beggars had once again become a common sight in city centres. One study identified 126 hostels across Scotland with a total of 3,700 bed spaces, the majority of them located in Glasgow. According to a survey of Scottish households conducted in 1995, one in 20 respondents or their partners had personal experience of being homeless in the previous ten years. Although only a very small number had ever slept rough, a quarter had stayed at least one night in a hostel or bed and breakfast; and the majority had stayed temporarily 'care of' friends or relatives. Despite the welfare state, therefore, homelessness in Scotland has not withered away, and has in fact become more pervasive since the passage of the 1977 Homeless Persons Act.

WORKING THE ACT

In order to help local authorities interpret their duties under homelessness legislation, the Scottish Executive and its predecessors have published a Code of Guidance. This Code does not have the force of law but is meant to inform the way in which local authorities implement their duties under the Act. However, in practice there are significant differences in the way councils have implemented legislation. Decisions have to be made about whether a particular situation or set of events should be regarded as 'unintentional homelessness', or whether a particular applicant should be deemed to be 'vulnerable' in some way, and thus has a right to be rehoused. In order to decide on such matters, council staff interview applicants and may undertake investigations to check the veracity of the applicant's claims. Judgements have to be made and, perhaps inevitably, these are influenced by stereotypes and prejudice.

Although the homelessness legislation is meant to be gender neutral, in practice staff implicitly, and sometimes explicitly, draw on gendered interpretations of applicants' situations and of the validity of their stories about how their homelessness occurred. Research in one Scottish city found that officers would sometimes provide overnight bed-and-breakfast accommodation to non-priority need homeless women, but invariably expected men in identical circumstances to sleep rough. Similarly, non-priority need women might sometimes be given the tenancy of a flat while men are offered hostel accommodation, on the grounds that women are better able to cope on their own with a tenancy and less happy to stay in a hostel.[9] The point is not that these perceptions are necessarily erroneous, but rather that, as these examples show, they can influence who gets what and why.

For their part, homeless applicants seek to present their case in ways that they hope will persuade the council's staff and be sufficient for them to be seen as deserving of being rehoused. In effect, they have to 'construct' the story of their homelessness in ways that seem convincing to the person sitting at the other side of the desk in the homeless person's unit (HPU). The way in which homeless people present themselves – how they behave and the picture they paint of their circumstances – can have a critical bearing on the outcome of their application or the way in which they are dealt with by the council staff.

These negotiated encounters in the HPU are usually very stressful both for applicants and staff. The applicants may have lost their home, frequently under distressing emotional or financial circumstances. They may, for example, have been left by their partner, be fleeing domestic violence, or have struggled with the financial consequences of unemployment or of ill-health. Moreover, their children may also feel humiliated or ashamed, lose contact with their friends, suffer taunts from other children, or have their education disrupted. Meanwhile, the HPU staff may be struggling to get through a heavy and relentless caseload; they may have to face rudeness, threats or even physical violence from some desperate, emotionally disturbed or mentally ill applicants. It is not surprising that officers develop routines and stereotypes as ways of coping with the stress and the heavy workload that seem to be endemic in many local authority HPUs.

There is significant variation between local authorities in the number of priority-need or statutory-homeless households, even when differences in population size are taken into account. In 1997/8, for example, the number of 'priority need' homeless households per 1,000 resident households varied from 3.0 in East Renfrewshire to 13.0 in Glasgow. The rate of priority need homelessness is higher in urban local authorities than in rural ones; 8.0 compared with 6.5 per 1000 resident households in 1997/8. Although homelessness is lower in rural than urban areas, it is also less visible and more dispersed, making it more difficult to provide services for people who lose their accommodation.

SINGLE AND HOMELESS

For priority need applicants who are rehoused by their local authority, homelessness is invariably a very difficult experience. But for those who are not rehoused or who have not applied in the first place, it can be still more stressful. While some 'non-priority need' homeless people may be fortunate enough to be accommodated, at least for a while, by friends or relatives, the options for others may be to stay in a hostel or sleep rough; neither of which is attractive for most of those who endure them. Research has consistently found that, while a very small minority of people are homeless from choice, most would rather live in their own home (including a rented room) rather than a hostel or a bed and breakfast.[10]

Life in many hostels is by no means a hospitable existence, although conditions are slowly improving. Many hostels are large, unappealing buildings, sometimes poorly kept and often dirty and with only limited facilities. This is less true of the more modern hostels, which are often relatively small and targeted at particular categories of homeless people, such as women or adults under 25 years of age. In the older and larger hostels illicit drug use appears to be widespread. Violence and theft are not uncommon and security for personal possessions is quite limited. The conditions are so poor in the older and larger hostels that some homeless people opt for sleeping rough instead.

The overwhelming majority of people sleeping rough are young or middle-aged white men. The streets are dangerous enough places to live for them, but even more so for women and for people from a minority ethnic group. The lack of physical shelter which dwellings provide means that homeless people, especially those sleeping rough, suffer from poor health. In addition, they find it difficult to gain access to healthcare. Most are unemployed even before they become homeless, but the lack of a dwelling to call home makes it difficult for them to obtain, or to retain paid employment. Stuck in the 'no-home, no-job cycle', many have to rely on social security benefits in order to get by, although claiming benefits without an address and national insurance number (necessary to verify identity and prove that the individual is not trying to defraud the social security system) is difficult. Significant numbers of rough sleepers resort to begging. Some sell *The Big Issue*, a magazine which aims to provide homeless people with an alternative source of income to begging, although in fact it is common to combine both, especially among those with a drug dependency to fund.

In response to the growth of street living and the problems it brings, in the 1990s the government introduced a Rough Sleepers' Initiative, which has helped to reduce the scale of the problem but by no means eliminated it. Then in 1999 the new Scottish Executive set up a Homelessness Task Force in order to investigate rough sleeping and homelessness more generally and to make recommendations as to how it could best be tackled. The report of the Task Force set out a range of recommendations for better homelessness policy and practice. Some of these recommendations were incorporated into new legislation, while others are being taken forward by local authorities and other agencies working in the homelessness field. The Homelessness (Scotland) Act 2003 is particularly important in that it expands the categories of people who are defined as being in priority need (hence who have a statutory right to be rehoused). By 2012 the concept of priority need will be abolished altogether, so that everyone who is unintentionally homeless will be entitled to be rehoused. In addition, the Act removes the need for people to demonstrate that they have a local connection.

While this Act and the other initiatives arising from the report of the Task Force will undoubtedly help to minimise the extent of the problem and alleviate some of its worst effects, homelessness is likely to remain a significant feature of Scottish life for the foreseeable future.

CONCLUSIONS

Homeless people have been a recurrent feature of Scottish life for many centuries. The nature and scale of the problem of homelessness have altered over time, both in response to changing social and economic conditions, and also as perceptions of the causes of homelessness and what (if anything) should be done about it, and by whom, have developed over time. The largely punitive and deterrent approach characteristic of the Poor Laws was gradually replaced by more humanitarian and understanding responses. Although more sympathetic than in the past, the Scottish public remains ambivalent, especially about begging and rough sleeping, both of which have re-emerged in recent years when many had thought that they had been banished from Scottish society. Moreover, the distinction between the deserving and the undeserving homeless remains in public debate and in legislation, even if the language in which it is couched and the attitude of officialdom towards it have changed.

In response to the growth in homelessness, a range of measures has been introduced to tackle the problem. Yet more initiatives seem likely to emerge in the public policy context created by Scottish devolution. These measures extend well beyond the provision of dwellings for the homeless and reflect a changed perception of the nature and causes of the problem. There is a new consensus that, while the provision of better hostels and more permanent housing have helped, many of the most disadvantaged homeless people need more than just shelter. Some also need social work help, counselling, drug services, employability programmes and a range of other forms of social support. In other words, it is now believed that homelessness in Scotland is not just a housing problem, but reflects multiple disadvantages which may not be solved simply by providing accommodation. Homes are more than just dwellings, but many homeless people need more than a home in order to solve their problems.

NOTES

1. For a minority of people, the home may be a site of conflict, fear or even violence. Indeed, women's refuges have been set up in several Scottish cities to provide a safe haven for women and their children fleeing domestic violence.
2. Perceptions of homelessness vary not just within Scotland but also between different countries.
3. Some feminist authors have stretched the term to include women living in unsatisfactory relationships with their male partner, although most people would probably not consider that to constitute homelessness as such.
4. Legislation was introduced in Scotland as early as 1424 to control vagrants and beggars. Other enactments included a statute of 1449 which authorised the imprisonment of vagabonds, while laws passed in 1455 and 1477 ordered that beggars be put to death. Legislation in 1535 made parishes responsible only for the poor who were born within the parish and made begging within the parish lawful only for natives. The pre-1845 Poor Law system in Scotland rested largely on a statute of 1579, which was headed 'For the punishment of the strong and idle beggars, and relief of the poore [sic] and impotent'.
5. Many thousands more people avoided homelessness by living in highly overcrowded dwellings, sharing their living accommodation and sometimes even their beds with other families in conditions which would not be tolerated or allowed today. Laidlaw, 1956; see also Chapter 32 Paying for Dwellings.
6. See Chapter 37 Children's Homes for further information on the Poor Law.
7. Meanwhile, the number of common lodging houses continued to fall. By 1953, for example, there were only 19 left in Glasgow, occupied by around 3600 residents per night. At this time there were six common lodging houses in Edinburgh, and two each in Ayr, Dundee and Paisley, according to Laidlaw, 1956.
8. Laidlaw, 1956.
9. See Cramer, 2002b.
10. See Fitzpatrick, Kemp, Klinker, 2000.

BIBLIOGRAPHY

Anderson, I, Tulloch, D. *Pathways through Homelessness*, Edinburgh, 2000.
Ann Rosengard Associates, Scottish Health Feedback. *The Future of Hostels for Homeless People*, Edinburgh, 2001.
Brandon, D. *The Decline and Fall of the Common Lodging House*, mimeo, npp, 1972.
Carter, I. *Farm Life in North-East Scotland, 1840–1914*, Edinburgh, 1979.
Curruthers, A. *The Scottish Home*, Edinburgh, 1996.
Clapham, D, Kemp, P A, Smith, S J. *Housing and Social Policy*, Basingstoke, 1990.
Cloke, P, Milbourne, P, Widdowfield, R. *Rural Homelessness*, Bristol, 2002.
Cramer, H. *Homelessness. What's Gender Got to Do with It?*, London, 2002a.
Cramer, H. 'Engendering Homelessness: An ethnographic study of homeless practices in a post-industrial city', unpublished PhD thesis, University of Glasgow, Glasgow, 2002b.
Crowther, M A. *The Workhouse System, 1834–1929*, Athens (Georgia, USA), 1982.

Dean, H. *Begging Questions: Street-Level Economic Activity and Social-Policy Failure*, Bristol, 1999.
Fitzpatrick, S. *Young Homeless People*, Basingstoke, 2000.
Fitzpatrick, S. *Research for the Homelessness Task Force: A Summary*, Edinburgh, 2001
Fitzpatrick, S, Kemp, P A, Klinker, S. *Single Homelessness: An Overview of Research in Britain*, York, 2000.
Fitzpatrick, S, Kennedy, C. The links between begging and rough sleeping: A question of legitimacy?, *Housing Studies*, 16 (2001), 549–68.
Greve, J. *Homelessness in Britain*, York, 1991.
Homelessness Task Force. *Helping Homeless People: An Action Plan for Prevention and Effective Response*, Edinburgh, 2002a.
Homelessness Task Force. *Research for the Homelessness Task Force: A Summary*, Edinburgh, 2002b.
Humphreys, R. *No Fixed Abode: A History of Responses to the Roofless and the Rootless in Britain*, Basingstoke, 1999.
Johnstone, C. Housing and class struggles in post-War Glasgow. In Lavalette, M, Mooney, G, eds, *Class Struggle and Social Welfare*, London, 2000, 139–54.
Kemp, P A, Lynch, E, Mackay, D. *Structural Trends and Homelessness: A Quantitative Analysis*, Edinburgh, 2001.
Kennedy, C, Fitzpatrick, S. Begging, rough sleeping and social exclusion: implications for social policy, *Urban Studies*, 38 (2001), 1–16.
Kershaw, A, Singleton, N, Meltzer, H. *Survey of the Health and Wellbeing of Homeless People in Glasgow*, London, 2000.
Laidlaw, S. *Glasgow Common Lodging Houses and the People Living in Them*, Glasgow, 1956.
Lindsay, J. *The Scottish Poor Law: Its Operation in the North East from 1745 to 1845*, Ilfracombe, 1975.
Nichols, G. *A History of the Scotch Poor Law*, London, 1856.
Pawson, H. *A Profile of Homeless People in Scotland*, Edinburgh, 2000.
Pawson, H, Third, H, Tate, J. *Repeat Homelessness in Scotland*, Edinburgh, 2001.
Saunders, P, Williams, P. The constitution of the home, *Housing Studies*, 3 (1988), 81–93.
Somerville, P. Homelessness and the meaning of the home: Rooflessness or rootlessness?, *International Journal of Urban and Regional Research*, 16 (1992), 529–39.
Watchman, P Q, Robson, P. *Homelessness and the Law in Britain*, 2nd edn, Glasgow, 1989.
Watson, S, Austerberry, H. *Housing and Homelessness: A Feminist Perspective*, London, 1986.
Yanetta, A, Third, H, Anderson, I. *National Evaluation of the Rough Sleepers Initiative in Scotland*, Edinburgh, 1999.

Index

A

Abbey St Bathans, Berwickshire, 837
abbeys see Dryburgh Abbey; Fort Augustus Abbey; Iona Abbey; religious communities
Abbotsford, Roxburghshire, 84, 180, 183, 186, 384, 573
Aberdeen
    allotments, 413
    backlands, 408
    beds, 102–3, 111
    Bethany House, 693
    brewers, 344
    candle making, 76
    children's home, 693
    coal transport to, 85
    fish trade, 429–30
    hospice, 672
    married couples' freedom, 528
    maternity hospitals, 440
    meat trade, 429–30
    pottery, 65, 556
    prison, 656, 658
    Provost Skene's House, 202–3
    radio missions, 215
    rosaries, 203
    sheltered housing, 18
    Show People, 806
    Steiner school, 708
    street lighting, 70
    tower blocks, 10–11, 17–18, 35
    typhoid fever, 486
    weaving, 348
    women servants, 361–2
    see also Watson, Sheriff of Aberdeen
Aberdeen University student accommodation study, 717, 718, 728, 731–2
Aberdeenshire
    ale selling, 350
    Auld Yule customs, 168
    Balmoral Castle library, 384
    baskets, 333
    box beds, 100
    Catholic aids to worship, 208
    ceiling decoration, 180
    Corriemulzie Cottage, 187
    Craigievar overmantel, 179
    croft houses, 69, 71–2, 83, 100
    domestic medicine, 503
    dressers, 61
    fires, 83
    fishwives, 352, 621
    girnals, 58
    'hingin-lum' fireplace, 71–2
    Kildrummy Castle, 26
    kists, 55–6
    lighting, 69, 71
    meat offerings in tombs, 583
    midwifery, 443
    Peterhead prison, 656
    plaid production, 341
    spinning, 341, 345
    spirituality, 201
    women's employment, 346
    youth hostel, 646
    see also Balmoral Castle
Aberlour House, Banffshire, 182
Aberlour Orphanage, 685, 689, 692
Abernyte, Perthshire, fuel, 88
Abertay University, 171
Acharn, Perthshire, 835
acupuncture see complementary medicine
Adam, James, 94, 182, 835
Adam, John, 94
Adam, Robert, 94, 182, 835
Adam, William, 84, 179, 385
addresses, status, 552–3
adolescents see teenagers
advertisements, 377
after-dinner activities, 426, 432, 433–4, 435
after-school clubs, 475
afternoon teas, 430–1, 531
aftershave, 267
Aga cookers, 325, 428
age differences
    head coverings, 229, 230
    hierarchy and authority, 515
    personal appearance, 248–50, 252
    personal hygiene, 278
    prisoner classification, 658
    status and, 557–8
    technology use, 159
    see also children; old people
aggregation see incorporation
agricultural buildings
    conversion, 839, 841–2
    see also barns; bothies; farm steadings; mills; outbuildings; stables
agricultural tenancies, 30, 364, 403, 406, 543
agricultural work see farm work

agricultural workers *see* farm servants
AIDS, 487, 673, 674, 680
Airdrie, Lanarkshire, 347, 795
airing, laundry, 319
Airlie, earl of, 646
'AIROH' bungalows, 17
Aith, Shetland, 461–2
Aithbhreac Inghean Corcadail, 203
Aithsting, Shetland, 86
Albert, Prince, 433, 638
alcoholic drinks, 424, 425, 427, 631, 664
   see also ale; beer; brandy; whisky; wine
alcove beds, 100, 112
   see also bed recesses; box beds; corner beds; wall beds
alder trees, 404
Alderston, Haddington, East Lothian, 842
Aldous, Fred, 193
ale, 344, 350, 425
   see also beer; brewers and brewing
Alexander Morton & Co, 184
All Scotland Crusade, 216
Allan, David, 92
Alloa, Clackmannanshire, 295
Allotment Movement, 413
allotments, 400–1, 413–16, 831
Alston, Robin, 381, 382, 394
alternative medicine *see* complementary medicine
Alva, Clackmannanshire, 460
Alvie, Inverness-shire, 227
amber beads, 504
Amish commune, 737
Anaglypta, 184
Anderson, Jane, 367
Anderson, Sir Robert Rowand, 182, 836
Anderson, William, 208
Anderson's Royal Polytechnic, Glasgow, 184

anglers
   accommodation, 640–1
   see also fishermen
angling holidays, 640–1
Angus
   blue caum, 315
   bothies, 631
   box beds, 100
   Brown and White Caterthuns, 31
   cradle decoration, 56
   cycling, 648
   dung as fuel, 88
   Edzell Castle gardens, 402
   farm servants, 266, 424, 631
   fir candles, 72
   fishing communities, 771, 773
   flax mills, 349
   folk museum, 61
   footwear, 231
   Fowlis Easter church, 177
   fruit picking, 620
   head coverings, 230
   Hospitalfield, 836
   House of Dun, 84, 179
   huts, 827–8
   inn beds, 102
   Invermark Lodge, 640
   kist decoration, 56
   lairds' houses, 12
   Links Foundry, 287
   open prison, 657
   peat storage, 89
   plasterwork, 179
   seaside resorts, 648
   shelving, 61
   spinning, 346
   status, 542
   temporary lighting, 71
   toll houses, 834
   weaving, 77, 348
   windows, 79
   youth hostel, 645–6, 648
   see also Forfarshire
Angus Folk Museum, 60
animal fat, 76, 290
   see also lard; tallow
animal hair
   tethers made from, 336
   see also horsehair products

animal skins, 581, 584
animal stomachs, boiling meat in, 421
animals, 36–7, 581–96
   see also cattle; dogs; donkeys; horses; meat; pigs; sheep; vermin
Anna, duchess of Bedford, 430
Anstruther, Fife, 766–7, 768, 773
antenatal care, 443–4, 451
apothecaries, 255, 262
appearance *see* personal appearance
Appiehouse, Stenness, Orkney, 582
apple trees, 400, 401, 407
apprentices, 357, 528
   accommodation, 342, 343, 519
   garden lads, 409
   girls, 343
   holidays at fairs, 423
   nailers, 347
Approved Schools, 689
aprons, 235, 249, 364, 564, 623
Aquinas, St Thomas, 584
Arbroath, Angus, 834, 836
archery, 169
Ardclach, Nairn, 229, 231
Ardersier, Inverness-shire, 229
Ardnamurchan Lighthouse, Argyll, 834
Ardoyne, Insch, Aberdeenshire, 71–2
Ardverikie, Inverness-shire, 640
Ardvorlich, Dunbartonshire, 504
Argand, Aimé, 318
Argyll
   Ardnamurchan Lighthouse, 834
   climbing, 642
   Crarae Glen Garden, 410
   dressers, 59
   Dunadd fort, 25–6
   fuel, 75, 87, 88

866 • SCOTTISH LIFE AND SOCIETY

hospice, 672
hotels, 642
    Inveraray Debtors'
        Prison, 650
    Oban High School, 709,
        710
    Travelling People's
        tents burned, 795
Argyll and Bute, New Age
    travellers, 813
Ariés, P, 528–9
arks see girnals
Arkwright, Richard, 349
armed forces, 17, 141, 223,
    381–2, 632–3, 708
Armitage Shanks, 297
    see also Shanks
        (sanitary ware
        manufacturer)
Army camps, homeless
    people, 856
Arniston, Midlothian,
    283
Arniston House,
    Midlothian, 84, 385
Arnol, Lewis, 100
Arnot, William, 75–6
Arran, 410
art, 174–96, 583, 759
    see also decoration;
        drawing; murals;
        ornaments;
        paintings
artisans see craftsmen
Arts and Crafts Movement,
    175, 179
asbestosis, 486
ash, 83, 284, 286, 300, 314
ash closets, 284
ash collection and disposal
    ash pits, 83–4, 92, 300,
        715
    baskets, 333
    grates, 94
    middens, 83, 284, 286,
        333
    pails and tubs, 311
ash trees, 402, 403, 404, 406,
    504
asylum cookers, 32
    see also immigrants
asylums, 486
Atholl, 2nd duke of, 405,
    835

Atholl, 4th duke of, 405, 406
Atholl, Perthshire, 72, 340
    see also Blair Atholl;
        Forest of Atholl
Atkinson, Eleanor, 585
attics, 64, 80
    see also lofts
attractiveness, 245–6, 247–8,
    249
au pairs, 475
Auchinleck, Lord, 357–8
Auchinleck House,
    Ayrshire, 358
Augustinian order, 749, 750,
    752
Auld Cassidy, 633
Auld Yule, 168
autistic people, 496
auxiliary uses, baths, 292,
    469
auxiliary work, 330–7, 346
Ayr, 70, 85, 361, 366, 368
Ayr Bank, 366
Ayrshire
    Auchinleck House,
        358
    Bargeny, 369
    Catholic worship, 208
    caves, 15
    clothing, 226, 227
    coal transport from, 85
    Dumfries House lady's
        closet, 63
    farm house breakfast,
        366
    girnals, 58
    handball, 168
    huts, 827–8
    lavatory cleaning, 469
    Maybole Ragged
        School, 688, 692
    Ochiltree house fire, 426
    Penkill Castle murals,
        182
    plaids, 228
    poor relief, 368
    prison, 656
    Protestant family
        worship, 209
    reading by firelight, 72
    salt industry, 340
    sanitary ware
        production, 295, 296
    seaside resorts, 648

    servants, 226, 358, 364,
        365–9
    temporary lighting, 71
    textile furnishings,
        184
    water supply
        engineering
        products, 287
    weaving, 348
    see also Culzean Castle
Ayrshire needlework, 137

B
babies
    bathing, 279, 468–9
    beds, 466
    clothing, 139, 140, 445,
        450, 461–5
    death customs, 567–8
    health care, 461, 467
    layettes, 445
    pushcarts, 465
    pushchairs, 466
    saining, 564
    shawls, 445, 465, 466
    swaddling, 461
    umbilical cord
        dressings, 448–9
    see also birth customs;
        breastfeeding;
        childbirth; children;
        cradles; infant
        mortality
back gardens, 408, 413
backgammon, 164, 165
backlands, 342, 408
Badbea, Caithness, 14
Baden-Powell, Robert and
    Agnes, 473
bagnios, 280–1, 282, 837
bagpipe music, 121, 129,
    201, 363, 619
Baillie, Miss, 280
Baillie, Lady Grizell, 280,
    308, 311, 315, 320,
    321, 364
Baillie, Joanna, 122
Baillie, Robert, 166–7
bait and baiting, line
    fishing, 350, 352,
    621, 769, 771, 773
Baker, G P, 184
bakers, 339, 342, 344, 361
bakestones, 420

baking, 200, 344, 420–1,
    427–8, 431, 619
    see also bannocks;
        bread; confectioners
        and confectionery
Bàl Stocainn, 619
Baldev Singh, 575
Balgone, East Lothian, 837
ball games, 168, 169–70, 470,
    631
ballan, the, 502–3
Balloch, 673
ballrooms, 121
Balmacaan House,
    Inverness-shire, 640
Balmoral Castle,
    Aberdeenshire, 180,
    187, 384
Banff, 167
Banffshire
    Aberlour House, 182
    Aberlour Orphanage,
        685, 689, 692
    ash pits, 84
    baskets, 333
    bog fir digging, 72
    dressers, 61
    kists, 56
    lighting, 73, 84
    umbilical cord
        dressings, 448–9
    windows, 79
    women's head
        coverings, 229
Banks, Ella, 447
Banks, Iain, 839
bannocks, 200, 425, 449, 619
banquets, 426, 449
baptisms
    at home, 216
    children's clothing and,
        461
    christening pieces, 450
    Gude and Godlie
        Ballates, 205
    midwives, 450
    ministers' duties, 217
    personal appearance
        and, 247
    protection before, 450,
        564–5
barbers, 253, 254, 266, 268,
    281
bards, 120, 121, 127, 363

Bargeny, Ayrshire, 369
barley, 312, 366
    see also bere; grain
barley bannocks, 425
barley meal, 619
Barlinnie prison, Glasgow,
    656
barmkins, 36
Barnard, Lady Anne, 122
Barnardo's homes, schools,
    708
barns
    auxiliary work, 330, 331
    doors, 13
    farm servants' wives'
        work, 525
    rope-making, 335
    sleeping accommoda-
        tion, 628, 633, 643
Barra, 476, 709
barracks, 17
Barrhead, Renfrewshire, 295
Barrhead Road, Glasgow, 834
Barrhead Sanitary Ware, 297
Barrogill Castle, Caithness,
    382
Barrowfield pottery,
    Glasgow, 65
Barvas, Lewis, 65, 224, 231
basins, 257, 267, 279, 469
    see also wash-hand
        basins
baskets and basket-making,
    116, 333–4
    see also willows
Bass, Nick, 495
bastles, 36, 37
bath fats, 280
bath houses, 280, 281
bath stoves, 280
bath tubs, 61, 278, 279, 280,
    468
bathing, 259, 262, 277, 278,
    278–83, 291, 297,
    468–9
    frequency, 278, 279–80,
        291, 468, 547
    skin care, 263
    see also washing
bathing facilities
    portable, 283–4
    Travelling People's
        camping sites, 801,
        803

bathrooms, 258, 279, 280,
    288–98
    caravans, 809
    Fort Augustus Abbey,
        756
    infirm people, 493
    miners' houses, 283
    mobile homes, 822, 824
    North Queensferry
        signal tower
        conversion, 841
    prisons, 716
    privacy, 289, 290, 716
    shared flats and houses,
        716, 717, 721–2,
        728
    Show People's
        caravans, 809
    storage furniture, 67,
        291
    see also en-suite
        bathrooms; sanitary
        ware
baths, 279–80, 291, 292,
    468–9
    manufacture, 287
    mobile homes, 824
    reading in, 383
    storage in, 61, 469
    see also bath tubs;
        bathrooms; pit-head
        baths; public baths;
        Roman baths;
        Russian baths;
        Turkish baths
bathstools, 280
baxters *see* bakers
Bayne, Anne, 460–1, 463,
    495
BBC, 128–9
BBC Radio Scotland, 643
beadwork, 190
Bealtainn, 569, 570–1
    see also May Day
bean bags, 116
beans, 366, 403, 415
bear baiting, 585
beards, 265–6, 267
    see also shaving
Beaton family, 502
beauty, 244, 245, 254, 260–1
    see also cosmetics;
        personal
        appearance

868 • SCOTTISH LIFE AND SOCIETY

Beavers (scouting movement), 473
Beckett, Martyn, 837
bed and breakfast accommodation, 649–50, 856, 859
  see also boarding houses; lodgings
bed bugs, 102, 115, 317–18, 586–7, 588
  see also vermin
bed hangings, 178, 361, 586
bed recesses, 101, 112, 649, 835, 841
  see also alcove beds; wall beds
bed warming, 83, 493
bedding, 114–15
  blackhouses, 363
  cotton, 114, 232
  infirm people, 493
  knitted bedspreads, 135
  laundry, 236, 301, 318, 320
  linen, 102, 103, 114, 232, 361
  plaids as, 114, 227, 363
  prisons, 664
  quilts, 141, 191
  servants, 365
  Shetland, 86
  Travelling People, 445, 795
  vermin, 586
  youth hostels, 646
  see also blankets; mattresses; sheets
Bedford, duchess of, 430
bedroom furniture, 62–3, 111, 112, 189, 239, 256–9
  see also beds; dressing tables; wardrobes
bedrooms
  aristocratic households, 111
  bagnios, 281
  Balmacaan House, 640
  Bell Rock lighthouse, 834
  best rooms, 556
  catechism in, 206–7
  children, 383, 553, 714, 716
  cleaning, 317
  computers in, 154, 383, 472
  croft houses, 189
  decoration, 182, 185
  dressing rooms and, 258
  farm workers' houses, 80
  heating, 83, 189
  huts, 828
  infirm people, 494
  living and working spaces, 97
  location, 44
  personal hygiene activities, 256
  private spaces, 47–8
  reading and study in, 383, 384
  seating, 111, 112
  shared flats and houses, 714, 716, 720–2, 726, 727, 728
  status, 553, 556
  student residences, 715–6
  teenagers, 185, 383
  television sets in, 146
  Travelling People's caravans, 797
  wall hangings, 178
  window seats, 111
  see also boudoirs; garderobes; sleeping accommodation
Bedrule, Roxburghshire, 226
beds, 97–102, 111–15
  babies, 466
  blackhouses, 100, 363
  bothies, 110, 630, 631, 715
  caravans, 796, 809
  cats and, 588
  childbirth, 445
  children's, 111, 114, 383, 466, 716
  curtains, 103, 111
  farm houses, 86, 99, 103
  independent schools, 704
  infirm people, 493, 714
  inns, 102, 103, 115, 637, 638, 715
  lodging houses, 853, 854
  making, 317
  marriage customs, 567
  Orkney, 99, 100, 101, 564, 582
  prisons, 664
  reading in, 382, 383, 384–5, 388, 390
  salmon netters' lodges, 625
  sharing, 714, 715, 716
  Shetland, 86, 466
  status, 111, 556
  storage spaces, 61–2, 67, 103, 111
  teenagers, 383, 384–5, 388
  Travelling People, 795, 796
  vermin, 102, 113, 115, 317–18, 363, 637
  see also alcove beds; bed recesses; box beds; bunk beds; chaff beds; corner beds; feather beds; folding beds; hurly beds; iron bedsteads; orthopaedic beds; roofed beds; wall beds; waterbeds; wooden beds
bedsits, 717
  see also cells; sheltered housing
beech trees, 404
beef broth, 366
beef tea, 504
beer, 568, 758
  see also ale; brewers and brewing
beeswax, 317
Beeton, Mrs, 310, 369, 377, 385
Beeton, Samuel, 140
beetroot, 415
beets, 400
Begg, James, 833–4
Begg, Rev James, 210
beggars, 633, 852, 859, 861, 862
Behn, Aphra, 363
Beldam, Valentine, 360
Belhaven, Lord, 403
Bell, John Zephaniah, 182
Bell Rock lighthouse, 834
bellows, 86, 311

INDEX • 869

Beltane *see* Bealtainn
Ben Hough, Tiree, 13
Benbecula, 711
benches, 56, 99, 386, 835
   see also settles (seating)
benders (tents), 17, 110, 738, 792
Benedictine order, 749, 750, 751, 752, 756–7, 759
benefits *see* social security system
bere, 37, 340
   see also barley
Berlin woolwork, 140, 190
berry picking *see* fruit picking
Berwickshire
   box beds, 100
   clothing, 227, 232
   Dryburgh Abbey, 177
   houses, 79, 100, 837
   libraries, 382, 384
   toll house, 835
   see also Baillie, Lady Grizell
besoms, 311, 334
   see also brooms
best rooms
   bedrooms, 556
   book storage, 379
   childbirth, 567
   croft houses, 191
   death customs, 496, 567
   fires and fireplaces, 84, 96
   New Year celebrations, 567
   ornaments and furnishings, 78, 80, 191
   tenements, 546
   see also parlours
Bethany House, Aberdeen, 693
betrothals, 565–6
Beveridge, Rev John, 393
Beveridge Report, 604
Bewan, North Ronaldsay, Orkney, 100
Bhagwan, 737
Bhopal, Professor Raj, 485
Bible
   disguise for other reading, 383

House of Binns' copies, 382
   in birth customs, 564
   ownership of, 202, 204–5, 206, 208, 213
   sale by colporteurs, 627
   toothache cure, 504–5
   see also family Bibles
Bible reading, 204–5, 211, 213, 388, 391
   Iona Community, 214
   prevention of, 202, 204–5
   sermons and, 206
   Tell Scotland movement, 215
   temporary lighting for, 71
Bible stories, illustration, 140
bicarbonate of soda, 506
bicycles *see* cycling
bidets, 291, 294–5, 297–8
billiards rooms, 384
Billings, R W, 180
birch trees, 402, 404
birds, as pets, 588–9, 591
birth *see* childbirth
birth customs, 200, 449, 563–5, 575
   see also baptisms
birth rate, 458
birthmark removal, 263–4
Bishop's Palace, Fife, 744–5
Bissell, Mr , 313
black bun, 421, 434
Black Death, 586
Black, William S, 182, 184
blackberry juice, 321
blackhouses, 13, 15–16, 97, 100, 363, 818
Blackie, Walter, 182
blacklead, 314, 316, 317, 588
blacksmiths, 75, 340, 342, 351, 490, 502, 509
Blackwater River, 641
Blair, Anna, 248, 462, 463–4, 468
Blair Atholl, Perthshire, 404
Blair Drummond, Perthshire, 404
Blairgowrie, Perthshire, 620, 621
blankets, 114, 115, 365
   babies' wraps, 445, 466

Highland houses, 102, 103, 363
   laundry, 301, 318
   Shetland houses, 86, 99, 466
   Travelling People's tents, 445, 795
Blay, Ida, 647
bleaching, 232, 237, 268, 300
   see also whitening
bleeding, domestic medicine, 505
blemish removal, 264
blind people, 494, 708–9
blistering, 505, 508
blue, linen whitening, 320, 321
blue bonnets, 230, 231
blue caum, 315
blue kelt, 231
Bluecoat Boys, 703–4
Blunden, Edmund Charles, 704
board games, 76–7, 164–6, 170, 434, 471, 535, 631
Board of Trustees for Fisheries and Manufactures, 344–5
boarding houses, 648–50
   see also lodging houses; lodgings
boarding out
   children, 475–6, 683, 686–8, 709–11
   mentally infirm people, 486
   see also fostering
boarding schools *see* residential schools
Boat of Brig, Moray, 834
bobbin winding, 350
Bod of Gremista Museum, Shetland, 61
bodach *see* peermen
body clothes *see* clothing
body hair, 268
bog fir *see* fir candles
bogies, 470–1
boilers, 95, 237, 279, 291, 321, 427, 434
boiling, 421, 427
boils, 504, 505

bondagers, 264, 335
bonesetters, 490, 502
Bo'ness, West Lothian, 85
bonfires, 570–1
bonnets, 229, 230, 231, 239, 264, 321, 366
Book of Kells, 583
books, 377, 378–9
　blind people, 494
　bothymen's reading, 631
　children and, 473
　commune publications, 739
　Dollar Academy, 707
　furnishings, 378, 379, 380
　gardening, 409–10, 412, 415
　medical recipes, 503
　prisoners' reading, 664
　radio readings, 391
　storage, 379–81, 385
　see also Bible; Book of Kells; cookery books; Diggers and Dreamers Handbook; fiction; psalm books; reading
bookselling, 343, 627, 628, 741
bools, 470
booths, 37, 343
boots, 231, 311, 462, 463, 464, 557
borax, 314
Borders
　common ridings, 33
　huts, 827, 828
　music, 125
　plaids, 228
　reading, 378, 389, 391
　Travelling People, 789
　weaving, 348
　youth hostels, 644
　see also Berwickshire; Roxburghshire; Selkirkshire
Boston, Thomas, 202
Boswell, James, 102–3, 301, 308, 358, 522
botanical gardens, 400, 410

bothain àiridh see huts, shielings
bothies, 110, 630–1, 715, 831
　clothing storage, 238, 631
　deer watchers, 640
　kists, 55, 56, 110, 631
　migrant workers, 518
　music, 125, 631
　salmon netsmen, 624–5
　sleeping accommodation, 110, 630, 631, 643, 648, 715
　Travelling People, 791, 800
　see also bunkhouses
Bothwell Castle, Lanarkshire, 26
bothymen, 266, 518, 630–1
bottled food, 430
boudoirs, 379, 431
bouets see lanterns
Bourtreehill Coal Co, 296, 297
bowatts see lanterns
bowling, 169
box beds, 61–2, 100–1, 102–3, 111–12, 461, 556
　see also alcove beds; corner beds
box irons, 323
box making, 352
box mangles, 323
box rooms, 64, 721, 728
Boy Scouts, 473
Boyd, Alice and Spencer, 182
Boyd Orr, John, 1st Baron, 467
boys
　bathing, 468
　clothing, 230, 461, 462, 463–4
　employment, 308, 341, 346, 347, 361, 533, 534
　games, 152, 168, 169, 171
　girls subservient to, 534, 537
　hair, 463
　handcrafts, 170

　housework, 308, 469
　knitting stockings, 341
　personal appearance, 248
　reading, 388, 392
　servants, 308, 361
　toys and play, 170, 469–71, 472
　weaving, 346, 347
　see also Bluecoat Boys; children; education; Life Boys; teenagers; young people
Boys' Brigade, 473
Bracadale, Skye, 238
braces
　canvas work, 140
　fireplaces, 91, 94
brae o' the broth, 504
Braemar, Aberdeenshire, 646
braid bonnets, 230
brambles, 321, 468, 510
bran, 263, 320, 427
Branch Davidians, 737
Brand, Colin, 645
branders, 12, 72
brandy, 267
brass candlesticks, 75
brass fenders, 189
brass fireside equipment, 86
brass headboards and footboards, 113
brass kettles, 211
brass mantelpiece rods, 85
brass plant pots, 190
brass polishing, 315, 317
brats (aprons), 235
braziers, 91–3
bread
　afternoon teas, 430
　birth customs, 564–5
　bothymen, 631
　farm house meals, 366
　Findhorn community, 741
　jeely pieces, 467–8
　lykewakes, 568
　plate cleaning, 427
　shieling workers' food, 619
　spirituality and, 200
　see also bakers; baking; bannocks

INDEX • 871

bread ovens, 420
bread poultices, 504, 505
Breadalbane, Lord, 12, 121
breakfasts, 211, 366, 421, 424, 758
breastfeeding, 118, 466–7, 564
    see also wet nurses
breeches, 366
    see also trews; trousers
Brereton, Sir William, 308, 361
brewers and brewing, 342, 344, 361, 639
    see also ale; beer
brickdust, 314
bridal dresses, 559
Bridge of Weir, Renfrewshire, 689, 708
Bridge Street, Penicuik, Midlothian, 837
Bridgeness Tower, West Lothian, 839
brimstone, 508
Bristol board, 140–1
British Broadcasting Corporation see BBC
British Linen Company, 345
British Telecom, 494
broadcasting, 120–1, 128–9, 214–15, 216
    see also radio; television
Broadmeadows, Selkirkshire, 644
brochs, 9, 31, 833
Brodick Castle, Arran, 410
bronchitis, 510
broom, as fuel, 85, 87, 88
Broomhill Road, Burntisland, Fife, 837
brooms, 307, 311
    see also besoms
brose, 424, 619, 631
broth, 366, 425, 504
brothels, 343
Broughty Ferry, 351
Broun, Issobell, 359–60
Brown Caterthun, Angus, 31
Brownies, 473
Bruce, Andrew, 631

Bruford, Alan, 633
brushes, 263, 265, 311, 312
    see also brooms; hair brushes
Brussels sprouts, 415
Buccleuch, duke of, 543
Buccleuch Estate, Dalkeith, Midlothian, 406, 543
Buchan, Anna, 186
Buchan, David, 503, 505, 507–8
Buchan, Dora, 587, 588, 589
Buchan, John, 641
buckets see pails
bucking, 320, 321
Buckney, Andrew, 393
Buckquoy, Orkney, 583
Buddhism, 218
budgerigars, 588, 591
Buick's of Alloa, 295
building conversions, 838–46
building materials, 9–10, 17
    see also slate; stone; thatched roofs; tiles; wood
Building Preservation Trusts, 839
Buildings at Risk Register, 839
built-in furniture, 64, 67, 112, 238, 239, 546
    see also fitted kitchens; wall beds
bun breists, 100
bungalows, 550–1, 556–7
bunk beds, 114, 383, 466, 625, 853
bunkers, coal, 64
bunkhouses, 650
    see also bothies
bureaux see desks
burgages, 408
burials, 25, 38, 99, 575, 583
    see also death customs; funerals
Burke, Edmund, 180
Burn, William, 180
Burnet, Gilbert, bishop of Salisbury, 206
burns, 510
Burns, Rev A R G, 641
Burns, Robert, 167, 202, 208, 390, 544, 586

Burntisland, Fife, 837
Burrafirth, Aithsting, Shetland, 86
Burrell, Sir William, 179
Burt, Capt Edward, 90, 102, 235, 308, 311–12, 319–20
Burton, Mary Hill, 182
Burton of Trent, Lord, 639
Bute, 3rd Marquess of, 182, 836
Bute, 4th Marquess of, 84
Bute, 182, 225, 229, 836–7
    see also Argyll and Bute
butlers, 310, 361, 367, 427
    see also stewards
butler's pantries, 316
butter
    afternoon teas, 430
    breakfasts, 424
    domestic medicine, 506
    farm house meals, 366, 425
    housekeepers' duties, 364
    making, 341, 364, 618–19
    oil lamps, 74
    selling, 331, 341
    transporting, 333
buttermilk, 263, 425
butterwives, 344
button boots, 464
buying
    books, 378
    see also bookselling; fairs; hawkers and peddlers; markets; shops
byre dwellings, 36, 97, 110, 330, 363, 582, 833
byre lanterns, 70
byres
    as toilets, 299
    entry through, 86
    mucking out, 332
    protective charms, 504
    waste disposal, 582
    work of farm servants' wives, 525
    see also middens; springhouses

C

cabbages, 403, 409, 415
  see also kail
caber tossing, 631
cache (ball game), 167, 168
Caddy, Eileen, 739, 740, 744
Caddy, Peter, 739, 740
Cadell, W A, Architects, 839
Cadogan, William, 458
Cadw, 838
caich, 167, 168
'Càin Domnaig', 199–200
Cairngorm Club, 641
Cairngorms, 642, 648
cairties, 471
caisies, 333, 335
Caithness
  Barrogill Castle library, 382
  cruisie lamps, 73
  heather ropes, 350
  kitchen ranges, 95
  plantiecrues, 403
  seating arrangements, 117
  straw work, 333
  tethering of children, 14
Calbost, Lewis, 89
calchens, 72
Caledonian Horticulture Society, 409
calendar customs, 185, 569–74
callendering, 318
Callernish, North Uist, 837
Calton Hill, Edinburgh, 322, 424
Calton Jail, Edinburgh, 656
Calvin, John, 523–4
Calvinism, 199, 246
  see also Church of Scotland
Cambuslang, Lanarkshire, 207
Cambusnethan Priory, Lanarkshire, 836
Cameron, Catherine, 207–8
Cameron, Hugh, 180–1
camomile, 510
Campbell, Åke, 618
Campbell, Archibald, 383
Campbell, Charles, 837
Campbell, Sir Duncan, 63
Campbell, Sir George, 410

Campbell, J & W, 184
Campbell, John, 2nd marquis of Breadalbane, 12, 121
Campbeltown, Argyll, 88
camphor, 314
camping, 650
  see also caravans; tents; Travelling People
camping sites
  New Age travellers, 813
  Travelling People, 621, 789, 790, 792, 795, 799–804
  see also mobile homes
campus accommodation see student accommodation
camstone, 314, 315
canal boat conversions, 841
canaries, 588, 591
cancer, 487, 491, 672, 674
Candlemas, 569
candles, 75–7, 138, 344, 448–9, 744, 828
  see also fir candles
candlesticks, 72–3, 75, 76
Canna, 618
canned food see tinned food
Canongate, Edinburgh, 14, 408, 628
canvas products, 140–1, 181–2, 192
caps, 229–30, 366
caravans
  communes, 738, 739, 741, 742, 745
  holiday caravans, 819
  Monimail commune, 745
  New Age travellers, 813
  religious communities, 751
  Show People, 805, 806, 807–10
  Travelling People, 621, 791, 795–9, 805, 807, 808, 822
  see also camping sites; mobile homes
card games, 166–7, 361, 388, 426, 434, 471, 631

carding, 77, 84, 191, 335, 341, 346
care in the community see home care
carles, 76
Carluke dressers, 61
Carlyle, Alexander, 167
Carmelite order, 750, 752
Carmichael, Alexander, 199–201, 206
Carmylie, Angus, 230
Carnoustie, 351
Carpenter, Mary, 688
carpets, 178–9, 184, 185, 189, 313, 316, 317
Carron Iron Company, 94, 287, 347, 349
carrots, 415
carrycots, 466
Carse of Gowrie, 12, 424, 587, 627
Carswell, Catherine, 174, 379
Carthusian order, 749, 750, 753–4
carving, 425
cash see money
casks, 64, 65
cast iron products, 184, 287
Castle Drogo, Devon, 846
Castle Huntly, Perthshire, 657
castles, 26, 37, 79, 93, 96
  see also names of individual castles
casual wards, 854, 855
catechisms, 206–7, 209, 391
Catholic children's homes, 689
Catholic pupils, boarding, 709
Catholic teacher training college, Glasgow, 711
Catholicism, 38, 198–205, 206
  post-Reformation, 208–9, 213, 217, 529, 569, 711
cats
  depictions of, 583
  games with, 585
  gardens and, 591
  meat and skins, 581, 584

INDEX • 873

Orkney, 582
    pets, 588, 591–2, 593
    vermin control, 586, 587, 591
    witches and, 584–5
cattle, 581, 582
    Beltane custom, 570
    housekeepers' duties, 364
    mucking out, 332
    Orkney houses, 582
    prayer and, 200
    security of, 36–7
    servants' tasks, 363, 364
    status and, 36, 542–3
    see also byre dwellings; byres; dairy products; droving; shielings
cauliflowers, 415
caum, 314, 315
caustic soda, 290
caves, 15, 405, 628, 643, 648, 650, 791
    see also Fingal's Cave; Gilmerton Cave
CDs (compact disks), 67, 129, 144, 149, 383
ceilidh houses, 125, 127
ceilings
    cottage kitchens, 64
    decoration, 176, 177–8, 179, 180, 181, 182, 202
    houseboats, 841
    Ossian's Hall, 405
    The Retreat, Abbey St Bathans, 837
    status, 554
celebrations see festivals
Cellardyke, Fife, 766, 769
cellaret cupboards, 62
cellars, 37, 64, 318, 343
cells
    prisons, 654, 660, 661, 664
    religious communities, 753, 755, 756, 759
Celtic spirituality, 199–202
Celtic year, 569
central heating, 16, 83, 96, 104, 185
    children's clothing and, 462–3

cleaning and, 325
Fort Augustus Abbey, 756
furniture arrangement, 188
houseboats, 841
infirm people's homes, 493
open plan arrangements, 97
shared flats and houses, 728
Travelling People's caravans, 797
villa outbuildings, 553
youth hostels, 646
Central Midwives Board for Scotland, 442, 443, 448
ceramic bricks, 291, 296
ceramic door furniture, 184
ceramic ewers and basins, 279
ceramic ornaments, 187, 189–90
ceramic sanitary ware, 295, 297
ceramic storage utensils, 65–6
ceramic tiles, 95, 184, 291, 551
cereals see grain
certification, deaths, 484
ceruse, 264, 265
cesspools, 294
chaff beds, 101, 114, 365, 625
chairs, 115–16
    bedrooms, 111, 112
    bothies, 110
    canvas work, 140
    Corriemulzie Cottage, 187, 188
    croft house parlours, 189
    fireside, 86
    gender differences in use of, 117
    kitchens, 78
    reading in, 382, 383, 385
    spinning in, 117
    status and, 115, 714
    student accommodation, 720, 728

study in, 383
Travelling People's tents, 795
see also pushchairs; rocking chairs; wheelchairs
chalets see huts
Chalmers, A K, 442
chamber pots, 62, 102, 299, 300, 317, 318
chambermaids see housemaids
chaplains, 217–18
    aristocratic households, 202
    hospitals and hospices, 217, 675, 680
    independent schools, 707
    prisons, 217–18, 666, 668, 688
    see also ministers
Chapman, Anne, 445
chapmen see hawkers and peddlers
charcoal irons, 323
charity, 572
    see also beggars; poor relief
Charles Star Cycling Club, 647
Charleston, Dundee, 589
Charlotte Square, Edinburgh, 84, 182
charms see protective charms and customs
Charteris, Francis, earl of Wemyss, 837
chaulmers, 125, 630
    see also bothies
cheese
    birth customs, 449, 564–5
    farm workers' food, 366, 425
    housekeepers' duties, 364
    lykewakes, 568
    selling, 331, 341
    shieling work, 619
chemicals, skin care, 263
chemists' shops, 255, 262
cherries, 406

cherry trees, 402, 407
chess, 164, 166, 170
chest conditions, 486, 503–4
chestings, 495, 568
chestnut trees, 402
chests *see* kists; tea chests
chests of drawers, 63–4, 239
chickenpox, 508
chicky-mellie, 168
chilblains, 506
child migrants, 692
childbirth, 200, 440–57, 460–1, 563–5, 567, 575
    see also babies; baptisms; howdies; midwives
childcare, 458–81
    aristocratic households, 361
    childminders, 339, 475
    fathers, 477–8, 528, 537
    neighbours, 325, 490
    nurses, 361, 432, 530
    rescue movement, 689, 692–3
    servants, 373, 432, 475, 531
    Shetland, 490
    social workers, 697
    widows and widowers, 686
    women's work, 324–5, 528, 537
    see also babies; boarding out; children's homes; fostering; nannies; wet nurses
childhood, 458, 528–9, 688
childminders, 339, 475
children
    as prisoners, 688–9
    attractiveness, 245
    bathing, 279, 280, 283–4, 468–9
    bathrooms, 289–90
    bedrooms, 383, 553, 714, 716
    beds, 111, 114, 383, 466, 716
    boarding out, 475–6, 683, 686–8, 709–11
        see also fostering
    books and, 473

church attendance, 463, 693
communes, 747–8
computers and video games, 152–3, 472
cots, 114, 461, 465–6
crafts and hobbies, 170, 171
dental health, 468
discipline, 528–33, 703–4
    see also parent-child relationships
emigration, 692
family households, 518–19
fishing communities, 763–4, 770–1, 772
food, 432–3, 443, 466–8, 475, 476, 533, 693
fostering, 69, 475–6, 518, 532, 686, 697, 698
    see also boarding out
gardens, 413
homelessness, 857, 858, 860
hospice visiting, 678
households with sick or infirm people, 489, 492
huts, 830
latch-key children, 475
meals, 432, 467–8, 475, 531, 716
migration, 692
ministers and, 530
mobile phones, 472–3
music, 121, 471–2
New Age travellers, 813
personal appearance, 248, 261
poorhouses, 684, 697
privacy, 280, 694
reading, 388, 389–90, 392, 473
rhymes and songs, 120, 471
seating, 116, 117
security, 14, 24, 27
Show People, 811
sleeping accommodation, 111, 114, 383, 461, 465–6, 716

special needs, 491, 492
status, 553, 557
television, 149–50, 152, 472, 716
tethering, 14
Travelling People, 445, 789, 790, 799, 802
vermin, 468, 686, 687–8
videos, 150
washing, 278–9
workhouses, 686, 687–8, 689
    see also babies; boys; education; girls; illegitimate children; orphans; parent-child relationships; play; teenagers; young people
children's clothing, 461–4
    boys, 230, 461, 462, 463–4
    children's homes, 687, 693, 694
    footwear, 462, 463, 464, 557, 686
    Highland dress, 228
    knitted garments, 139, 462, 464
    laundry, 321
    school uniforms, 248, 703
    Sunday clothing, 227
children's comics, 392
Children's Fair, Grangemouth, 573–4
children's homes, 683–700
children's hospices, 673, 674, 680
children's work, 459–60, 528, 533–4, 684
    as carers, 489
    boys, 308, 341, 346, 347, 361, 533, 534
    gender differences, 533, 537
    girls, 343
    home industries, 339
    housework, 308, 459, 469, 532, 533–4, 537
    wages, 476, 526, 533, 534

INDEX • 875

water carrying, 318
weaving, 347
chimneys, 83, 84, 96, 104
   cleaning, 311, 317, 633
   coal fires, 420
   farm workers' houses, 80
   Gilmerton Cave, 835
   huts, 828
   see also hingin lum fireplaces; roof openings; round-about firesides
china, 59, 61, 64, 189, 427
   painting, 191
   sanitary ware, 297
   see also crockery; dressers; sideboards
Chincough Well, Glasserton, Dumfriesshire, 507
Chinese communities, 511
chiropody, 488–9
chiropracty see complementary medicine
cholera, 286, 485, 486, 587
Christchurch, Trinity Road, Edinburgh, 843
christening see baptisms
christening pieces, 450
Christianity, 197–222, 584, 692
   see also Catholicism; Protestantism; Sunday
Christie, Jimmy, 647
Christmas, 569
   courtiers' clothing, 223
   food, 419, 434
   Fort Augustus Abbey, 759
   hospices, 677
   meals, 434
   prisons, 664
   shared flats and houses, 730
   toilette for, 261
   see also Yule
Christopher, F J, 193
Christ's Hospital, 703–4
chromolithographs, 190
Church Army, 854
church attendance, 109
   children, 463, 693

clothing, 226–7, 229, 231, 238, 463
   personal appearance, 247
   shieling workers, 619
   see also Sunday observance
church buildings
   conversion to houses, 843
   religious communities, 751–2
   squatting in, 856
   street lighting, 70
   windows, 79
   see also Fowlis Easter church
Church of Jesus Christ of the Latter Day Saints, 218
Church of Scotland
   care homes, 681
   General Assembly, 211–12, 213
   views on personal appearance, 247
   see also Calvinism; kirk sessions; ministers; Sunday observance
Churchill, Winston S, 703
churching, 564, 575
cinnabar, 265
circulatory problems, 510
Cistercian order, 749, 750, 752–3
Clach Dearg, Ardvorlich, Dunbartonshire, 504
Clackmannanshire, 79, 226, 295, 460, 656, 704–8
clan chiefs, 121, 228, 357, 363, 521
clans, 357, 521–2, 531
claret, 425
Clarke, Edward, 97, 98
cleaning, 310–18
   bothies, 631
   central heating and, 325
   chimneys, 311, 317, 633
   cooking utensils, 315, 325, 421, 427
   floors, 307, 311–12, 314, 316–17
   grates, 311, 316, 317, 588
   hearths, 85, 311, 316

kitchen ranges, 317, 325
outside lavatory, 469
ovens, 420
prisons, 666
silverware, 310, 427
stairs, 317, 727
windows, 307, 315, 317, 373
   see also dry cleaning; housework; laundry; polishing; washing; washing up
cleaning equipment and materials, 291, 310–17, 325, 427
cleanliness
   box beds, 101
   crowded spaces, 461
   hospitals, 482
   inns, 102, 308, 637
   kitchen sinks, 362
   kitchens, 325
   shared flats and houses, 720, 727
   silverware, 308
   stairs, 102, 286, 288
   Travelling People, 798
   see also housework; laundry; personal hygiene; skin care; washing
clearances, 14, 638
Cleish Castle, Kinross, 180
clergymen, 351, 381
   see also chaplains; ministers
Clerk, Sir John, of Penicuik, 12, 122, 181
climbing see hill walking and mountaineering
Clinical Resource and Audit Group, 452
clivvies see peermen
cloakrooms, 288, 290, 297
cloaks, 229
clocks, 189, 568
cloisters, 752, 753, 754, 755
close stools, 284
cloth, 231–2
   see also linen; weavers and weaving; woollen cloth

clothing, 223–42, 243
　babies, 139, 140, 445, 450, 461–5
　brushing, 311
　canvas work, 140
　church attendance, 226–7, 229, 231, 238, 366, 463
　colours, 228–9, 231, 232, 559, 664
　death customs, 568
　fashion, 223, 224, 559
　fishermen, 770, 771
　games, 139
　heating and, 83, 462–3
　herring gutters, 623
　hospice patients, 676
　hospice staff, 675–6
　knitted, 135, 139, 331, 349, 462, 464
　leisure activities, 139
　linen, 228, 229, 231–2, 361
　lodgers, 343
　mending, 235, 769
　needlework, 134–5, 137, 139, 140, 142
　plaids, 225, 227–9, 366
　prisoners, 664
　religious communities, 758
　remnants as dusters, 311
　servants, 223, 225, 226, 251, 366
　smoky smell, 227, 235, 420
　status, 223, 224, 542, 559
　storage, 55, 62, 111, 116, 238–9, 631
　Sunday clothing, 227, 463, 468
　tacksmen, 363
　weddings, 559
　women servants, 226, 366
　see also children's clothing; footwear; graveclothes making; head coverings; laundry; layettes; plaids; shirts; stockings; toilette; trousers; underwear; uniforms; woollen clothing
cloths
　dressers and sideboards, 556
　medical uses, 508
　see also dusters; floorcloths; neck cloths; tablecloths
Cluanie Inn, Kintail, Ross and Cromarty, 642
club chairs, 116
Cluny, 749
Cluny Hill College, 746
Clyde Report, 695
Clyne, Sutherland, 229
Co-operative Movement, 429
Co-operative Wholesale Society, 184, 290, 291
coach houses, 553, 640, 842
coal, 64, 72, 85, 86, 88, 292, 341
　see also Bourtreehill Coal Co
coal bunkers, 64
coal fires, 84–5, 94, 96
　chimneys, 420
　cooking and baking, 420, 421
　housework and, 311
　lighting, 72
　salmon netters' lodges, 625
　smooring, 89–90
　see also grates
coal industry, 282–3, 339, 684
　see also miners
coat hangers, 238
coats, 231, 463
Coats and Clark, 192
cobles, 624, 625
Cochrane, Sir John, 426
cock fighting, 585
Cockburn, Lord Henry, 656
Cocker, Philip, & Partners, 842
cockles, 418
cod fishing, 766
coffee, 424, 426, 758
coffee houses, 425
coffinings see kistings
Coirechatachan, Skye, 103
Colclough, Stephen, 394
cold cream, 263, 264
cold cures, 503
Cole, Robert, 360
Coleridge, Samuel Taylor, 637, 704
Colin Du, 363
Coll, 522
Collace, Perthshire, 255
collars, 137, 140, 235, 239, 320
　see also neck cloths
College Street, Glasgow, 182
colleges, gardens, 400
collegial corporations, 764, 765–6
collieries see coal industry; miners
collies see cruisie lamps
'colonies', 833–4
colouring, hearths, 628
colours
　bonnets, 230
　clothing, 228–9, 231, 232, 559, 664
　decoration, 182–3, 193
　hair, 244, 502
　laundry, 320–1
　plaids, 228
　skin, 264–6
　see also dyes and dyeing
colporteurs, 627
Coltness, Lanarkshire, 402
Colville, James, 833
combinations, 462
combs, 230, 255, 259
comedy, 558–9
comics, children's reading, 392
commodes, 257, 284, 298, 493
common brick, 314
common lodging houses see lodging houses
common ridings, 33
communal dwellings see bothies; huts; shielings; military accommodation; shared flats and houses; student accommodation

INDEX • 877

communal reading, 391
    see also Bible reading
communes, 737–48
Communion, 202, 205–6, 207–8
Community Centre, Findhorn, 741
community security, 30–5
compact disks, 67, 129, 144, 149, 383
complaints procedures, prisons, 670
complementary medicine, 401, 489, 500, 501, 505, 510, 741
    see also domestic medicine
computer games, 145, 152–3, 167, 170–1, 387, 472
    see also video games
computer tables, 383
computer workstations, 383
computers, 153–7
    children, 152–3, 472
    compact disks and, 144, 149
    education, 152, 153, 159
    gender differences in use of, 146, 151, 152, 153, 157, 158
    location in home, 154, 383, 472
    old people's use of, 154, 159
    ownership, 145–6, 153
    play and, 469
    research on use of, 144
    television sets as, 146
    work at home, 353
    see also home offices; Internet access; technology
Comrie, Perthshire, 507
confectioners and confectionery, 424, 426
confinements see childbirth
Congrès Internationale des Gouttes de Lait, 458
conifers, 406, 410
    see also fir trees; larch trees; Scots pine trees

conservatories, 16, 190, 406, 413, 554
consumption see tuberculosis
contraception, 460
Control Tower, Fearn Airfield, Ross-shire, 845
conversions, 838–46
cookers, 325, 427–8, 434, 494, 727, 795, 809–10
    see also kitchen ranges; microwave cookers; stoves
cookery books, 307, 325, 392, 421–2, 424, 433, 503
cooking, 420–1, 427–8, 434–5
    bothies, 630, 715
    cyclists, 648
    fireplace accessories, 86–7
    hearths, 86, 420
    huts, 828
    lorry drivers, 629
    prisons, 664, 666
    salmon netsmen, 625
    shared flats and houses, 727–8, 730
    Show People's caravans, 809–10
    Travelling People, 795
    see also Edinburgh School of Cooking; food preparation
cooking utensils, 420, 421, 428, 434
    candle making in, 76
    cleaning, 315, 325, 421, 427
    shielings, 618
    youth hostel, 645–6
cooks
    children's meals, 432
    Duke of Hamilton's household, 361
    men, 308, 427
    recruitment, 369
    training, 370, 371, 372
    women, 308, 309–10, 427
coppers see boilers
cord ties, umbilical cords, 449

còrdadh, 565–6
corn mills, 339
corner beds, 99, 100, 564, 582
    see also alcove beds
corner cupboards, 61
Cornton Vale Prison, Stirlingshire, 657, 658
coronary artery disease, 487
Corriemulzie Cottage, Deeside, 187, 188
cosmetic surgery, 254, 264
cosmetics, 254, 251, 261, 262, 264–5, 557
    see also personal appearance
cots, children's, 114, 461, 465–6
cottage gardens, 400, 403, 406–7
cottages, 64, 546, 551, 588, 639, 738
    see also Corriemulzie Cottage; croft houses; fishermen's cottages; hermitages
cotton bedding, 114, 232
cotton clothing, 224, 232
cotton mills, 349
    see also spinning mills
cough cures, 503
council houses, 17–18, 597, 602–3, 604, 605–7, 610
    Edinburgh, 97, 603
    Glasgow, 18, 547
    heating and double glazing, 97
    location, 35
    old people, single-person households, 781, 783, 784
    right to buy, 608
Council of Illiberis, 199
council tax, 722, 841
country houses, 835–7
    bathing arrangements, 280
    conservatories, 16
    decoration, 180, 181–2
    deer stalking, 638, 639–40
    gardens, 402–3, 404–6, 410

icehouses, 424
kitchens, 96
larders, 424, 430
libraries, 381, 385
location and setting, 11–12
oratories, 202
safe rooms, 38
servants, 55, 369
Shetland, 86
storage furniture, 62–4
storerooms, 37
youth hostels, 644
Coupar Angus, 349
Couper, James, 181
courses
dinners, 425, 431
see also desserts
courtiers, clothing, 223
courtyards, tower houses, 36
covenants, 208
Cowal, Argyll, 672
Cowan, Dr Nairn, 488
cowls (head coverings), 230
cradle songs, 123
see also lullabies
cradles, 56, 114, 466
crafts and hobbies, 76–7, 170, 171, 175, 190–4, 759, 762
see also Arts and Crafts Movement; needlework
craftsmen, 342, 351–2, 373
building trades, 369
gardens, 407–8
itinerant, 633
religious communities, 757
workplaces, 340–1, 342–3, 351, 357
see also apprentices; gardeners
Craig, J & M, 295–6
Craig, Ross and Cromarty, 645
Craigentinny-Telferton, Portobello, 415
Craigieburn, Dumfriesshire, 835
Craigievar, Aberdeenshire, 179
Craiglockhart, Edinburgh, 711

Craigmillar Health Project, Edinburgh, 492
Crail, Fife, 766
Cramond, 418
cramp, 508
crannogs, 37
crans (fireplace equipment), 420
Crarae Glen Garden, Argyll, 410
Crauchie, East Lothian, 842
Crawford, John, 378
Crawford, John C, 394
Crawford Priory, Fife, 836
cream, butter making, 618–19
creams, skin care, 262, 263, 264
creels, 333, 335, 466
Crichton-Stuart, John Patrick, 3rd Marquess of Bute, 182, 836
Crieff, Perthshire, 60
crime, 631, 637, 861
see also security
crochet, 134, 140, 556
crockery, 633, 646
see also china; washing up
croft houses, 16, 30
ash pits, 84
bed and breakfast accommodation, 650
bedrooms, 189
box beds, 100
carpets, 189
chaff beds, 114
children's clothing, 464
fireplaces, 71–2, 189
fires, 83, 96, 189
furniture, 189
lighting, 69, 70–1
milk, 543–5
ornaments and furnishings, 191
paraffin-fired ovens, 96
parlours or best rooms, 189, 191
provision of, 598
sanitation, 84, 300
tables, 189
see also blackhouses

crofting, fishing communities, 762, 766
crogain, 65, 74
Cromarty
women's work, 108
see also Ross and Cromarty
Cromarty Firth, coal transport to, 85
Crompton, Samuel, 349
Cromwell, Oliver, 290
crooks and links (fire equipment), 72, 86–7, 92
cruisie lamps, 73–5, 138, 336
cryin bannocks, 449
cubbies see kubbies
cucumbers, 406
Culross, Fife, 404
Culzean Castle, Ayrshire, 182, 835
caves, 15
clothing, 226
servants, 226, 364, 365, 367, 369
Cumbernauld, 18
Cunningham, William, 165
Cupar, Fife, 58
cupboards, 61–2
bathrooms, 291
beds stored in, 112, 114
built-in, 64, 67, 238
cleaning equipment and materials, 315
cleaning of, 317
houseboats, 841
kitchens, 427
laundries, 322
Travelling People's caravans, 796
see also display cabinets; dressers; larders
curds, 425
Curran, James, 169
Currie, Janet, 199
Currie, John, 559
curtains, 77–8, 184
beds, 103, 111
best rooms, 78, 80
cleaning, 318
Corriemulzie Cottage, 187

INDEX • 879

cradles, 114
death customs, 568
dyeing, 318
farm workers' houses, 80
Curtis Report, 695
cushions, 115, 116, 178, 361
customs, 75, 168, 562–80
see also birth customs; death customs; protective charms and customs
cutlery, 191, 312–13, 431, 646
cycling and cyclists, 644, 645, 647–8, 759, 827
Cyclists' Touring Club (CTC), 647

D
D R Simpson & Sons, 95
Daiches, David, 124, 384, 392, 575
Daiches, Lionel, 124
Dailly, Ayrshire, 364
dairy maids, 308, 361, 364
see also milkmaids
dairy products, 37, 331, 341, 364
see also butter; cheese; curds; milk; milk processing; whey
Dale, David, 349
Dalgairns, Mrs, 421
Dalgliesh and Forrest, 62
Dalkeith, Midlothian, 84, 169, 300, 406, 423, 543, 839–40
Dalkeith Park, Midlothian, 835
Dalnacardoch, Perthshire, 637
Dalry, Ayrshire, 227
Dalry Meadows, Edinburgh, 78
Dalrymple, Sir John, 405
Dalyell, General Thomas, 382
dam brods, 165
damp proofing, cradles, 114
dampness, 8, 18, 547, 797, 824
see also drainage; waterproofing

dams, the see draughts (board game)
dancing, 201, 260, 619, 631
dancing classes, 472
Darlo, Mary, 361
darning, 235
Darnton, Robert, 390
Darvel, Ayrshire, 184
Daughters of Charity, 750
Dava Moor, Moray, 230
David, Elizabeth, 434
day centres, 492
day hospices, 677
days of rest see holidays; Sunday observance
De La Rue, Thomas, & Co, 193
De Voe, Msr, 121
dead people
   laying out, 201, 260, 490, 495, 567–8
   use for birthmark removal, 263–4
deaf people, 494, 709
Dean Village, Edinburgh, 842, 844
death customs, 201, 260, 495–6, 567–9, 575, 679
see also burials; funerals
deaths, 200, 217, 484, 494–6, 574
see also hospices; infant mortality
Debtors' Prison, Inveraray, Argyll, 650
decency, personal appearance, 246–8, 265
deck-access schemes see tower blocks
decoration, 174–85
   bathrooms, 290
   books as, 378, 379
   calendar customs, 185, 573–4
   ceilings, 176, 177–8, 179, 180, 181, 182, 202
   colours, 182–3, 193
   cradles, 56
   doorsteps, 315
   drawing rooms, 182, 187, 190

dressing rooms, 258
embroidery, 178–9, 189, 190, 192
fireplaces, 84, 315
floors, 315–16
girnals, 57–8
hearths, 85, 315
herring gutters huts, 623
kists, 55–6, 57
kitchen ranges, Glasgow, 95
libraries, 182, 385
paint, 176, 177–8, 183, 318
shared flats and houses, 720, 728
shelving, 61
status, 174, 176, 177, 178, 180–4, 553
studies, 385
tea chests, 62
textiles, 176, 178–9, 184, 361
windows, 181, 193
windowsills, 315
wood carving, 176, 191
wood panelling, 176, 178, 193
see also furnishings; ornaments
deep freezes, 435
deep-sea fishermen, 624
see also fishing communities
deer stalking, 638–40
Delacour, William, 180
delivery packs, childbirth, 445–6
Dempster, John Austin, 283
Denmark, geese in kitchens, 331
Denny, Stirlingshire, 672
dental care
   children, 468
   domestic medicine, 504–5, 510
   married women, 249
   personal appearance, 249, 254, 256
   personal hygiene, 279
   prisoners, 667
   religious communities, 758
toothpastes, 291

Denton, Howard, 575
Department of Agriculture for Scotland, 415
Departmental Committee of Inquiry into Allotments, 414–15
desk lamps, 387, 728
desks, 383, 721, 728
dessert knives, 312
desserts, 426, 427, 431, 432
detergents, 291, 300, 436
Deutsche Gesellschaft für Volkskunde, 330–1
Dewar, Sir John, 484
diabetes, 487
Dick Place, Edinburgh, 837
Dickens, Charles, 704
diet *see* food
dieting groups, 254
Digby, Sir Kenelme, 421
Diggers and Dreamers Handbook, 738
dining halls, 715, 716
dining rooms
    cleaning, 310, 316
    closets for prayers, 209
    Corriemulzie Cottage, 187
    decoration, 180, 181, 182
    Findhorn, 741
    furniture, 62
    libraries, 382
    living and working spaces, 97
    parlours and, 186, 426
    religious communities, 751, 753, 754
    status, 556, 557
    see also food; meals
dinners, 424, 425–7, 430, 431–2
    after-dinner activities, 426, 432, 433–4
    farm houses, 366
    schools, 467, 468
    servants' duties, 425, 426–7, 431, 434
    Western Isles, 418
disabled people, 26–7, 373, 701, 708, 751
    see also infirm people
discipline
    children, 519, 528–33, 703–4

    see also parent-child relationships
    prisons, 666, 670
diseases, 486–7, 507
    epidemics, 485, 486
    food and, 419
    maternity hospitals, 440, 441
    sanitation and, 286, 287
    vermin and, 586
    workhouses, 687–8
    see also Black Death; bronchitis; cancer; chest conditions; chickenpox; cholera; eye diseases; health; lung diseases; measles; rose (erysipelas); smallpox; tuberculosis; typhoid fever; typhus; whooping cough; workhouse disease
dish racks, 191
    see also dressers
dishes *see* crockery; washing up
dishwashers, 436
display
    books, 379
    china, 59, 61, 64, 189
    glassware, 380
    ornaments, 188, 189, 380, 556
display cabinets, 189
disposable nappies, 464–5
distilling, 341
    see also whisky
District Nursing Associations, 490–1
District Schools, 688
divination, 574
divorce, 477, 520, 523–4, 608
DMA Design, 171
doctors, 500, 503
    bonesetters and, 490
    care of infirm people, 488, 490, 494
    childbirth, 440, 441, 442–3, 444, 450, 451, 461

    education, 440, 442
    Highlands and Islands Medical Scheme, 484
    home visits, 483, 489, 511
    hospices and nursing homes, 675, 676, 680
    infirm people, 488, 490, 494
    libraries, 381
    postnatal care, 448
    prisons, 666
    relationship with patients, 511
    religious communities, 758
    status, 559
    vocational reading, 390
    work at home, 351
    World War I, 485
document storage, 38
Dods, Meg, 425
dog fighting, 585
dogs, 581, 584, 587–8, 591–2, 593
    deer herd control, 640
    Orkney houses, 582
    prehistoric evidence, 583
    shieling work, 61
    see also Greyfriars Bobby; gun dogs
Dollar Academy, Clackmannanshire, 704–8
dollies (laundry equipment), 321
domestic medicine, 264, 300, 500–14
    see also complementary medicine; protective charms and customs
domestic servants *see* servants
domestic violence, 524
Dominican order, 750
Donald Brothers, 192
Donaldson's School for the Deaf, 709
donkeys, 507
door furniture, 184, 316, 547

doors
　barns, 13
　blackhouses, 15
　decoration, 181
　farm houses, 78–9
　farm workers' houses, 79
　lighting and, 80
　space division, 100
　status, 547, 551, 552
　symbolic importance, 562, 565, 568–9
doorsteps, decoration, 315
dormitories
　children's homes, 693
　hostels, 644, 646, 715, 716
　lodging houses, 853
　religious communities, 752, 753
　student residences, 716
Dornoch, Sutherland, 226
double glazing, 16, 83, 96, 97, 104, 552, 728
double villas, 549
'doubling up', 600, 715
Douglas, David, 410
Douglas fir seeds, 410
dovecots, 843
Dowanhill, Glasgow, 211
drainage, 8, 288, 307, 318, 752, 822, 845
　see also dampness; sewerage; waterproofing
draught protection, 16, 111, 114, 333
draughts (board game), 164, 165
drawers
　babies' beds, 466
　see also chests of drawers
drawing, 118
drawing rooms
　after-dinner activities, 426, 433
　afternoon teas, 431
　book storage and display, 379
　cleaning, 310, 316
　Corriemulzie Cottage, Deeside, 187, 188
　decoration, 182, 187, 190

　fireplaces, 84
　furniture, 62, 187, 188
　interior design, 174, 181
　lighting, 385
　music, 122
　parlours and, 186
　Retreat, The, Abbey St Bathans, 837
　status, 553
dress see clothing
dressers, 59–61, 93, 191, 546, 554–6
dresses, 231
　brides, 559
　children's, 461, 462, 463–4
　crochet trim, 140
　embroidery, 137
　knitting, 139
　storage, 239
dressing rooms, 63–4, 259, 640
dressing tables, 112, 239, 257, 258
dressmaking, 118, 136–7, 142, 226, 233, 372, 769
dried fish, 619, 621
drinks see alcoholic drinks; milk
Dripping Well, Tillichuil, Perthshire, 507
droving, 228, 633, 637
drugget, 366
druggists, 255
　see also chemists' shops
drugs, 450, 861
Drumchapel, Glasgow, 547
Drummond Castle, Perthshire, 410
Drummond, James, 383
dry cleaning, 237–8
dry larders see pantries
dry laundries, 319
Dryad (company), 191, 193
Dryburgh Abbey, Berwickshire, 177
drying
　firewood, 421, 588
　food preservation, 424
　laundry, 85, 86, 319, 322, 413
　rags, 421
drying cupboards, 322

dullers, 342
dulse, 506–7
Dumbarton, 64
Dumfries
　brothel, 343
　fair, 806
　hospice wards, 672
　linen smoothing, 322–3
　prison, 656, 658
　sewerage, 284–5
　water supply, 284
　young offender institution, 657
Dumfries House, Ayrshire, 63
Dumfriesshire, 12, 228, 430, 507, 644, 647, 835
Dunadd, Argyll, 25–6
Dunbar, East Lothian, 362, 423
Dunbartonshire, 64, 182, 504
Dunblane, Perthshire, 708
Duncan, John, 182
Duncan, Joseph F, 300
Dundee
　allotments, 413
　bonnet-making, 230
　coal bunkers, 64
　cottages, 588
　cycling, 647
　flax mills, 349
　hospice, 672
　houses, 351, 589
　huts, 827
　jam factories, 620
　kitchen ranges, 588
　marmalade making, 344, 351
　milk depots, 467
　pets, 587–91
　public baths, 287
　restaurants, nursing mothers, 467
　sack sewing, 352, 353
　salmon distribution, 626
　selling from homes, 343
　servants' trade union, 373
　tenements, 587–8, 589
　tower blocks, 10–11
　water supply, 287
Dundee Bicycle Club, 648
Dundee Western Cycling Club, 647

Dunfermline, Fife, 348, 806
dung, 88–9, 91, 263, 320, 332, 502
  see also manure; middens
Dunglass, East Lothian, 842
Dunkeld, Perthshire, 405, 835
Dunmore, earl of, 844
Dunmore, Stirlingshire, 406, 844–5
Dunscore, Dumfriesshire, 835
Durham, Dr James, 484
Durness, Sutherland, 229–30, 231
dusters, 307, 311
DVDs and DVD players/recorders, 151
dwellings
  design and construction, 15–18
  location and setting, 11–15, 35, 552
  see also caravans; caves; houses; huts; military accommodation; shelter; tenements; tents; tower blocks; tower houses
dyers, 342
dyes and dyeing
  cosmetic uses, 265, 266
  curtains, 318
  laundry, 320
  textiles, 232, 238
  urine as mordant, 301
  see also colours
dykes see garden walls
Dysart, Fife, 347

E
Earlshall, Fife, 410
Earn (river), salmon netsmen, 624–5
earnings see money; wages
earth
  from graves, 503
  see also fullers' earth
earthenware, 64, 65–6, 74, 83, 421, 427

East Lothian
  backlands, 408
  bathing house, 280
  coach house conversion, 842
  fairs, 423
  farm house windows, 79
  farm workers' shoes, 231
  fortifications, 31
  housekeepers' duties, 364
  Humbie Dovecot, 843
  huts, 828
  lady's closet, 63
  laundry, 362
  lodges, 837
  married women, 249
  meat offerings in tombs, 583
  monastery, 750
  painted panels, 180
  seaside resorts, 648
  Seton, 835
  stables, 842
  Traprain Law fortifications, 31
  woollen manufactory, 339
East Morningside House, Edinburgh, 843
East Neuk see Fife
East Renfrewshire, homelessness, 861
Easter, 223, 677, 759
Easter Ross see Ross-shire
Eastfield Farm, Lanarkshire, 91–2
eating out, 425, 435
ecclesiastical buildings
  decoration, 177
  see also church buildings; religious communities
Edinburgh
  allotments, 413, 415
  backlands, 408
  bagnios, 281
  bakers, 339
  building conversions, 843, 844
  canvas decorative panels, 182

ceiling decoration, Holyrood, 180
cleanliness, 308, 362
clothes drying, 322
coal bunkers, 64
council houses, 97, 603
Craiglockhart, 711
Craigmillar Health Project, 492
dovecot, 843
drainage, 318
drawing room furniture, 62
fairs, 423
fireplace, Charlotte Square, 84
gardens, 322, 408–9, 410
Gilmerton Cave, 15, 835
graveclothes making, 344
Greyfriars Bobby, 585
hawkers and peddlers, 429, 628
home ownership, 42–54, 604, 720
homeless person, 681
hospices, 672, 673, 674–5, 681
houses, 62, 548, 603, 833–4, 842, 843, 844
huts, 827
icehouses, 424
Lauriston Home, 440
lawyers, 342, 343, 357–8
lodging houses, 857
maternity hospitals, 440, 450
mews conversion, 844
midwifery, 444
milk supply, 430
mill conversion, 842
Old Tolbooth, 656
Orphan Hospital, 58
People's Story Oral History Archive, 102
postnatal care, 440
prayers, 201–2
praying apartments in tenements, 209
prisons, 656, 660–1
sanitary ware production, 295
sanitation, 285, 286, 318

schools, 686, 701, 708, 709, 844
selling from homes, 343
servant recruitment, 366, 367
sewerage, 318
shared dwellings, 715
Show People, 806
single-person households, 777, 778–86
stables, 837
status, 110–11, 542
street lighting, 70
Sunday clothing, 227
tea tables, 62
tenements, 14, 42–54, 78, 110–11, 209, 210
tower blocks, 10–11
town houses, 62, 63, 110–11, 178, 837
Tynecastle Company, 180
Tynecastle Tapestry, 184
water supply, 285, 287
women property owners, 42–54, 344
women's privacy, 42–54
see also Leith; Ramsay, Edward Bannerman, dean of Edinburgh; Scott, Walter (Edinburgh goldsmith); Steel's of Edinburgh
Edinburgh City Museums and Art Galleries, 102
Edinburgh Co-operative Building Association, 833
Edinburgh Florist Society, 409
Edinburgh New Town, 15, 62, 322, 548
Edinburgh Ragged School, 686
Edinburgh Royal Infirmary, 281
Edinburgh School of Cooking, 371
Edinburgh University student accommodation study, 717

Edington, Thomas, 287
education, 377, 473–4
at home, 121, 361, 373, 384
boarded-out children, 476
buildings, 384
children's homes, 684, 688, 689, 692
communes, 747
computers, 152, 153, 159
cooks, 370, 371, 372
doctors, 440, 442
Findhorn community, 740, 742, 745–6
hospice and nursing home staff, 673
housekeepers, 372
housework, 370–2
knitting, 134, 136, 372
laundry work, 237, 370, 371
midwives, 440, 442
millinery, 372
Monimail commune, 745
music, 121, 361, 471–2
needlework, 134, 135–6, 343, 372
prisoners, 666–7
Reformation and, 458
religious communities, 758
servants, 309, 370–2
Show People's children, 805, 811–12
status and, 542, 544–5
teacher training, 711
television, 148, 149
Travelling People's children, 789, 790, 799, 802, 803
women and girls, 134, 135–6, 309, 370–2
see also books; homework; schools; study; training
Edzell Castle gardens, Angus, 402
eelie dollies *see* cruisie lamps
eggs, 264, 333, 419, 468, 631
Egremont, Dick Place, Edinburgh, 837

Eilean Mór, Islay, 582
Einstein's Tower, Potsdam, Germany, 846
electric appliances, 268, 323, 325, 427–8, 430, 434, 435
electric boilers, 321
electric heating, 96, 291
electric lighting, 70–1, 76, 138, 385, 386–7
huts, 828
mobile homes, 823
Travelling People's caravans, 797
electricity supplies
building conversions, 845
effect on housework, 307, 325, 434
Findhorn community, 742
Fort Augustus Abbey, 756
houseboats, 841
huts, 828
mobile homes, 822–3
prison cells, 654, 660, 661, 664
rural areas, 299
Show People's caravans, 806, 808, 810
Travelling People's camping sites, 801, 803
Elgin, Moray, 750
Elie, Fife, 648
Eliot, Simon, 394
elm trees, 403, 404, 409
email, 153–4, 155–6, 353
embarrassment, status and, 558
embroidery, 134, 140–1
aristocratic households, 361
dresser and sideboard cloths, 556
Fairistytch, 193
footstools, 86
forbidden on Sabbath, 199
interior decoration, 178–9, 189, 190, 192
lighting for, 138

884 • SCOTTISH LIFE AND SOCIETY

professional, 136, 137, 349
raffia, 191
training, 136
emigrant children, 692
employment
  baking, 344
  bobbin winding, 350
  brewing, 344
  children's homes, 692
  commune members, 745
  factories, 460, 684
  fishing communities, 350, 352, 767, 769–70
  flax spinning, 345–6
  gender differences, 343–4, 353, 525–6, 534
  homeless people, 851, 858, 859, 861
  knitting, 137, 191, 339, 340, 341, 348, 349
  muslin flowering, 349
  nailers, 347
  oil-rig workers, 629
  old people, 341, 350
  prisoners, 666
  Show People, 807
  single women, 525, 769
  spinning, 349
  teenagers, 460
  weaving, 346, 347
  young people, 534
  see also children's work; farm servants; holidays; miners; servants; wages; women's work; work at home
en-suite bathrooms, 289, 468, 716
Enchantress stoves, 96
English Landscape style, gardens, 404–5, 410
engravings, 189, 190
entertainment *see* leisure activities
Enzer, Joseph, 179
Enzie, Banffshire, 73
Episcopal Christchurch, Edinburgh, 843

Episcopalianism *see* Scottish Episcopal Church
Erraid, Isle of, 744
Errol Park, Perthshire, 837
Errol, Perthshire, 77
erysipelas, 300
escritoires, 383
estate workers' accommodation, 640, 837
ethnic minorities
  in hospices, 679
  see also Chinese communities; immigrants
European Court of Human Rights, 670
evangelical Christianity, 207, 214–16, 692
evictions, 599, 601, 603
ewers and basins, 257, 279
exercise
  prisoners, 666, 667
  see also leisure activities
extended families, 477, 519–20
exteriors, status, 545–53
eye diseases, 504, 688
  see also Glasgow Eye Hospital
eye infections, 506, 510
eyebright, 510

F
face packs, 263
face powder, 265
facial hair, 266–8
  see also barbers; shaving
factors *see* house factors
Factor's House, Paxton, Berwickshire, 837
factors' lines, 599
Faculty of Physicians and Surgeons, 442
fairies, 450, 504, 564
Fairistytch art embroidery, 193
fairs, 122–3, 806
  food supply, 340, 422, 429
  personal appearance and, 247

  servants' engagement at, 309, 364
  travelling traders and, 627
  see also Children's Fair; markets; Show People
Fairygreen, Collace, Perthshire, 255
Falkirk *see* Carron Iron Company
Falkirk Iron Company, 184
Falkland Palace, 167–8, 558
family Bibles, 211, 213
family households, 518–19, 520
family sizes, 444, 460–1, 477, 534, 536, 537, 797–8
  see also household sizes
family worship, 71, 209, 211
  see also Bible reading
famine, 418
farm houses
  ash pits, 84, 92, 300
  bed and breakfast accommodation, 650
  beds, 86, 99, 103
  doors, 78–9
  food, 366, 425
  gardens, 403, 407
  hawkers and peddlers, 628
  hingin lum fireplaces, 91
  kists, 56
  kitchens, 91–3, 94, 99
  lighting for reading, 73
  location and construction, 16, 598
  meals, 366, 425
  outbuildings and, 16
  reading, 73, 378, 389
  round-about firesides, 91
  sanitation, 84, 103, 298, 301
  seating, 91
  water barrels, 64
  windows, 78–9
  youth hostels, 644
  see also byre dwellings; croft houses

INDEX • 885

farm servants, 357
　bedrooms, 80
　box beds, 100, 111
　children as, 459, 684
　engagement, 364, 630
　food, 366, 424, 425, 631
　footwear, 231, 334-5
　gardens, 403, 406-7
　houses, 79, 80, 100, 299-300, 597, 598
　kists, 55, 56, 110, 631
　leisure activities, 631, 648
　meals, 630
　reading, 378
　Scottish Farm-Servants' Union, 299-300
　seasonal, 617-18
　sleeping accommodation, 80, 100, 110, 111
　wages, 364, 406, 631
　washing and shaving, 266
　wives' work, 137, 349, 525
　women, 341, 364
　work hours, 387
　see also bothies; bothymen
farm steadings, 16, 330, 598, 842
　see also barns; bothies; mills; outbuildings; stables
farm work
　days of rest, 108, 199
　gender differences, 332, 364
　women, 137, 332, 341, 349, 364, 525
　see also harvest work; shielings
farm workers see farm servants
farmed-out accommodation, 599
Farmer & Brindley, 837
farmers
　vocational reading, 390
　see also house farmers
farming, fishing communities, 762
Farrier, Reginald, 410
fashion, 174, 223, 224, 252, 265, 559

fast food, 425, 435
fasts, 206
fathers
　childbirth, 447-8, 449, 563-4
　childcare, 477-8, 528, 537
　homelessness, 856
　marriage customs, 565, 566
　see also parent-child relationships
Fearn Airfield, Ross-shire, 845
feather beds, 101
feather dusters, 311
Federation of Edinburgh and District Allotments and Garden Associations, 413
feet, 278, 506
　see also footwear
females see girls; women
fenders, 85-6, 189
fennel, 510
Fenton, Professor Alexander, 634
Ferguson, Donald, 587
Ferguson, Edith, 587
Ferrier, Susan, 186
Ferry Inn, Inver, Perthshire, 637-8
Ferryden, Angus, 771, 773
Festival of Britain, 193
festivals, 109, 677, 679, 697, 730, 758
　see also calendar customs; Christmas; Easter
festive decoration, 185
　see also decoration, calendar customs
Fetlar, Shetland, 99
feu duties, 37
feu superiors, 543
fevers, 489, 506-7, 586
　see also puerperal fever; typhoid fever
fiction, 392-3
fiddle music, 120, 121, 124, 125, 201, 472, 567
Field of Dreams, Findhorn community, 742

Fiennes, Celia, 308, 420
Fife
　birthmark removal, 264
　bonesetters, 502
　coal transport from, 85
　cradle decoration, 56
　domestic medicine, 300, 503, 504
　dung as fuel, 88
　fair, 806
　fishing communities, 766-74
　fruit picking, 620
　gardens, 404, 410
　hospice wards, 672
　houses, 836, 837
　kists, 56
　linoleum production, 291
　mangling, 350
　Monimail commune, 738, 739, 744-5, 746
　nailers, 347
　pit-head baths, 283
　salmon distribution, 626
　salmon netters' lodges, 624, 625
　seaside resorts, 648
　signal tower conversion, 840-1
　status, 542
　weaving, 348
　William Rodger & Son, 58
figs, 505
Findhorn commune, 738, 739-44, 745-6, 747, 751
Findhorn Press, 739
Finer Report, 857
Fingal's Cave, Staffa, 637
Finlaggan, Islay, 37
Finlay's (Glasgow ironmonger), 386
Finnish baths, 280
Fintry, Dundee, 589
fir candles, 71, 72-3, 75, 84, 138, 564
fir trees, 403, 410
fire
　divination by, 574
　house fire at Ochiltree, 426
　protective customs, 570, 571

fire coal, 72
fire hazards, 71–2, 86, 795, 796
fire screens, 86
fireguards, 494
firelight, 71, 72, 385, 386
fireplaces, 83–4
    best rooms, 86
    braces, 91, 94
    castles, 93
    cleaning, 316
    cooking and baking, 86–7, 420
    crans, 420
    croft houses, 71–2, 189
    decoration, 84, 315
    farm workers' houses, 80
    hingin lum fireplaces, 71–2, 91–3, 420
    huts, 828
    mirrors, 84, 86, 189
    ornaments and accessories, 85–7, 189, 316
    salmon netters' lodges, 625
    status, 193
    see also braziers; grates; hearths; round-about firesides
fires
    best rooms, 84, 96
    calendar customs, 570–1
    cooking and baking, 420
    croft houses, 83, 96, 189
    crooks and links, 72, 86–7, 92
    footmen's duties, 316
    Fort Augustus Abbey, 756
    kitchens, 78, 83, 89, 91–2, 96
    prayer and, 200
    smooring, 89–90, 200, 570
    Travelling People, 795
    waste disposal in, 83
    water heating, 291
    see also ash collection and disposal; braziers; chimneys;
    coal fires; grates; hearths; ovens; peat fires; smoke; stoves; wood fires
fireside accessories, 85–7
fireside chairs, 86
firewood, 64, 85, 87, 88, 421, 588
Firrhill School, 709
First National Conference on Infant Mortality, 458
Firth, John, 332
Firth of Forth, herring fishing, 766, 768
fish
    dried, 619, 621
    farm house food, 366
    preparation, 422
    selling, 341, 352, 429–30, 621
    shieling workers' food, 619
    smoked, 88, 619
    storage, 37
    tavern, alehouse and coffee house food, 425
    see also shellfish
fish larders, 424, 430
fish-liver oil, 74, 75
fish processing, 762
fish storage, 37
fisher girls see herring gutters
Fisheries Museum, Anstruther, 773
fishermen, 56, 469, 622, 624, 640–1, 770, 771
    see also deep-sea fishermen; salmon netsmen; seamen
fishermen's cottages, 12, 766
fishing, 199, 641, 759
    see also Board of Trustees for Fisheries and Manufactures; cod fishing; herring fishing; line fishing; salmon fishing
fishing booths, Shetland, 37, 99
fishing communities, 213, 350, 352, 621–3, 762–76
    see also deep-sea fishermen; fishwives; herring gutters
fishing lodges, 624, 625, 640, 831
fishing nets, 350, 768, 771, 773
fishing stations, salmon netting, 624
fishwives, 352, 429, 621
fit wazzies, 332
fitness, personal appearance, 254
fitted kitchens, 185, 554–5
flackies, 333
flannel dusters, 311
flat irons, 226, 323
flats, 546, 547, 552, 553–4, 718
    see also shared flats and houses; sheltered housing; tenements; tower blocks
flax mills, 349
flax spinning, 341, 345–6, 349
    see also harn; linen
Flaxman, John, 180, 181
fleas, 586, 588
    see also vermin
fleshers, 342
Fletcher, Andrew, of Saltoun, 364
flets, 333
floor coverings, 86, 316–17, 427
    see also carpets; linoleum
floorcloths, 291, 317
floors
    bathrooms, 291
    cleaning, 307, 311–12, 314, 316–17
    decoration, 315–16
    farm workers' houses, 80
    kitchens, 317, 427
    sand, 314–15
    Travelling People's tents, 794

INDEX • 887

florists' societies, 409
flour, 57, 263
    see also meal
flour bags, 466, 623
flowers
    allotments, 401, 415
    gardens, 400, 401, 402, 403, 407, 410
    see also honeysuckle; sulphur flowers; violets
flues see chimneys; hot walls
flush toilets see WCs
folding beds, 112
folding tables, 116
folk healers, 502
folk medicine see domestic medicine
food, 225, 418–39
    baptism celebrations, 450
    birth celebrations, 449
    calendar customs, 571–2
    children, 432–3, 443, 466–8, 475, 476, 533, 693
    children's homes, 693
    communes, 738
    farm houses, 366, 425
    farm servants, 366, 424, 425, 631
    Findhorn community, 741
    free for mothers and children, 443
    garlic, 512
    health and, 419, 512
    Highlands, 363
    holiday homes, 649
    hospices and nursing homes, 676, 679
    infirm people, 433
    lorry drivers, 629
    meals on wheels, 491
    oil-rig workers, 629
    personal appearance and, 249
    personal hygiene, 278
    prisons, 661, 664
    religious communities, 753–4, 758
    salmon netsmen, 625
    servants, 418

shared flats and houses, 722, 727
shieling workers, 619
storage, 37, 58, 64, 66–7, 86, 116
tinned, 330, 419
see also bread; breastfeeding; brose; broth; dairy products; eggs; fish; fruit; game; grain; jam; meals; meat; obesity; offal; porridge; potatoes; poultry; tinned food; vegetables
food preparation, 362, 421–2, 428
    crowded spaces, 461
    herring gutters, 623
    holiday homes, 649
    infirm people, 493
    prisons, 664
    shared flats and houses, 727–8
    see also baking; cooking
food preservation, 407, 424, 430, 435, 468
    see also refrigerators
food storage, 37, 58, 64, 86, 116, 631
    see also presses (cupboards)
food supply, 331, 340, 419, 422–4, 425, 429–30, 435
    see also meals on wheels
football, 169–70, 470, 631
footmen, 309, 310, 316, 361, 365, 427
footstools, 86, 140
footwear
    Angus gravestone, 231
    children, 462, 463, 464, 557, 686
    farm servants, 231, 334–5
    religious communities, 758
    straw, 334–5
    see also boots; shoes; slippers

Forest of Atholl, Perthshire, 88
Forfar, Angus, 631
Forfarshire, 56, 345
    see also Angus
Forgandenny, Perthshire, 88
formal gardens, 404, 406, 410
Formica, 193
forms of address, in hospices, 679
Forres, Cluny Hill College, 746
Forret, Thomas, 205
Fort Augustus Abbey, Inverness-shire, 750, 754–60
Fort William, Inverness-shire, 17, 709
forts, 25–6, 31
    see also crannogs
fostering, 69, 475–6, 518, 532, 686, 697, 698
    see also boarding out
Foulis, John, of Ravelston, 121
foundations, 12, 828, 829
four-in-a-block villas, 547, 549
Fourth Lateran Council, 749
Fowlis Easter church, Angus, 177
Fox and Geese, 166
Frame, William, 837
Framework for Action Working Group on Maternity Services in Scotland, 452
Franciscan order, 750, 752
Fraser, Eugenie, 248
Fraser, Patrick Allan, 836
Fraserburgh, Aberdeenshire, 621
freckles, 263
Fred Aldous Co, 193
free personal care, 491, 493, 496, 681
freezers, 435
French irons, 323
French style dinners, 425
Friars Carse, Perthshire, 835
Froggatt, Katherine, 675
frogs, 505
front gardens, 413, 547

888 • SCOTTISH LIFE AND SOCIETY

fruit, 400, 401, 402, 407, 410, 415, 468
  see also brambles; cherries; grapes; jams; jellies; melons; peaches; pineapples; strawberries
fruit picking, 618, 620–1
fruit trees, 400, 401, 402, 407, 408
  see also orchards
fruitwives, 344
Fuaran a' Mhinisteir, Petty, Inverness, 507
fuels, 84–5, 87–9
  children foraging for, 533
  cooking and baking, 420
  effect on housework, 307, 311, 325, 434
  irons, 323
  lamps, 74, 75
  storage, 64, 86, 89, 292
  Travelling People's stoves, 795
  see also coal; electricity; firewood; gas; oils; peat
fullers' earth, 262, 314
fulling mills, 339
Fulton, Rikki, 527
fumigation, 318
functional music, 122–5
funerals, 201, 202, 217, 226, 261, 496
  see also burials; death customs
furnaces, hot walls, 406
furnishings, 175
  best rooms, 78, 80, 191
  books as, 378, 379, 380
  canvas, 140, 192
  crafts and hobbies, 175, 190–4
  croft houses, 191
  huts, 828
  libraries and studies, 385
  linen, 178, 184, 190, 193, 361, 365
  metalwork, 184, 190, 191

shared flats and houses, 720, 728
  textile, 184, 192–3, 361
  see also curtains; cushions; decoration; linen; ornaments
furniture, 111, 193–4
  bothies, 110
  central heating and, 188
  croft houses, 189
  dining rooms, 62
  drawing rooms, 62, 187, 188
  dressing rooms, 63–4
  Fort Augustus Abbey, 759
  polishing, 307, 310, 317
  positioning, 186, 188
  reading and study, 382–3
  shared flats and houses, 721, 728
  Travelling People's tents, 794–5
  washing, 317
  see also bedroom furniture; beds; built-in furniture; desks; door furniture; kitchen furniture; seating; storage furniture
furze, 85, 87, 88
futons, 114

G
Gaelic ideals, personal appearance, 244
Gall, George, 332
Gall, Mary, 332
Galloway, 76, 167, 266, 333
'Galoshins, The', 572–3
Galston, Ayrshire, 168
gambling, 166, 167
game, 422
  see also grouse
game larders, 424, 430
game of the goose, 166
gamekeepers, 640, 646
games, 163–73, 471
  after dinner activities, 426, 432, 433–4
  bothymen, 631

boys, 152, 168, 169, 171
cat-in-soot game, 585
clothing for, 139
prisons, 667
Sunday observance, 108, 199
  see also ball games; billiards rooms; board games; card games; computer games; play; quoits; role-playing games; video games
Garden, Dr George, 362
Garden History Society, 411
garden lads, 409
garden walls, 193, 400, 402, 403, 405, 406
gardeners, 401, 407, 409, 410, 502
gardens and gardening, 400–17
  cats and, 591
  clothes drying in, 322
  Edinburgh, 322, 408–9, 410
  Findhorn community, 740
  flats, 546
  front gardens, 413, 547
  huts, 829
  Monimail commune, 745
  play in, 471
  prisons, 666
  privacy, 193, 249
  radio and television programmes, 412
  religious communities, 400, 401, 409, 753, 757, 759
  rented housing, 546, 547
  sports, 169
  tenements, 547
  Travelling People's camping sites, 803
  villas, 169
  see also allotments; orchards; vegetables
garderobes, 284
garlic, 508, 510, 512
Garnett, T, 637
Garve Hotel, Ross-shire, 641
Garvock, Kincardineshire, 227

gas-fired appliances, 325, 828
   boilers, 321
   cookers, 325, 427, 795, 809, 810
   irons, 323
   stoves, 797
gas heating, 96, 291, 841
gas lighting, 71, 76, 138, 385, 640, 809
gas supply, building conversions, 845
gas works, 503
Gattonside, Melrose, Roxburghshire, 236
Gaugain, Jane, 138
geese, 331, 582
   see also gioco dell' oca (game of the goose)
Geikie, Walter, 56, 61, 64, 253, 391
gellies (tents), 110
gender differences
   afternoon teas, 431
   auxiliary work, 331
   bathing, 468
   children's chores, 533, 537
   children's clothing, 462
   computer ownership and use, 146, 151, 152, 153, 157, 158
   cooks, 308, 427
   decoration, 182
   employment, 343–4, 353, 525–6, 534
   experience of home, 24, 43, 338–9
   farm work, 332, 364–5
   hierarchy and authority, 515, 517, 522–8
   homeless persons' treatment, 860
   horticultural society membership, 409
   housework, 307, 526–7, 528
   Internet access, 146, 153
   leisure activities, 534
   personal appearance, 244, 248, 249, 251
   personal hygiene, 278
   prisoner classification, 658

radio and television audiences, 129
reading, 388–9, 392
seating arrangements, 117, 118
servants' work, 308, 309–10, 359
single-parent households, 477
single-person households, 778
social freedom, 534
toys and play, 469–71
wages, 308, 427, 526–7, 528
see also men; women
gender equality, access to divorce, 523–4
gender role reversals, 570
General Assembly of the Church of Scotland, 211–12, 213
general practitioners see doctors
Gennep, Arnold van, 562, 563, 576
George III, king of Great Britain and Ireland, 365
geriatric medicine, 488–9
   see also old people
German Society for Folk-Life Study, 330–1
Gerrard, Betty, 588, 589
Giles, James, 187
Gillette, King Camp, 268
Gillham, M, 144, 153, 154, 157
Gillies, Mrs, 519–20
Gilmerton Cave, Edinburgh, 15, 835
Gilpin, William, 637
gioco dell' oca (game of the goose), 166
girdles (baking equipment), 420
Girl Guides, 473
girls
   bathing, 468
   chaperoning of, 534, 535
   clothing, 462
   cosmetics, 557
   education, 134, 135–6, 309, 370–2

employment, 343
fishing communities, 770
hair, 248, 249, 250, 559
housework, 469, 533–4
independent schools, 703, 704
laundry work, 469, 533, 534
personal appearance, 248, 260, 557
reading, 388, 392
Scouts membership, 473
servants, 308, 309
shieling work, 618–20
subservient to boys, 534, 537
toys and play, 169, 469, 470, 471, 472
see also children; herring gutters; teenagers; women; young people
Girls' Guildry, 473
girnals, 57–8
Glamis, Angus, 179
Glasgow
   allotments, 413, 415
   Anderson's Royal Polytechnic, 184
   bishop's palace gardens and orchards, 409
   boarding out of children, 476
   Boys' Brigade, 473
   cholera, 485
   churches, squatting in, 856
   coal bunkers, 64
   council houses, 18, 547
   death customs, 567–8
   decoration, 95, 178, 180–1, 183–4
   drawers as babies' beds, 466
   evictions, 599
   family worship, 211
   farmed-out accommodation, 599
Finlay's (Glasgow ironmonger), 386
hill walking and mountaineering, 643

Holmwood House, decoration, 180–1
homelessness, 854, 855, 856, 857, 859, 861
hospice, 672
hostels, 859
hotels, squatting in, 856
huts, 827
International Exhibition (1901), 183
iron industry, 287
kitchen ranges, 95
lodgers, 600
lodging houses, 854, 857
maternity hospitals, 440, 451
Melfort House, 553–4
midwifery, 442, 444–5
milk depots, 467
North Woodside Mission, 473
overcrowding, 600
population and housing, 285
portable bathing facilities, 283–4
pottery, 65
prayers, 201–2
prisons, 656–7
privately rented housing, 597, 599, 601
public baths, 281–2
religious pictures, 202
rent strikes, 601
restaurants, nursing mothers, 467
sanitary ware production, 295, 297
sanitation, 285, 286, 287, 297, 298
Scottish Exhibition Conference Centre, 806
servants' training, 370, 371
shared ownership of homes, 720
shaving routine, 266–8
Show People, 806, 807, 812
single-person households, 778
soap making, 290
squatters, 856
status, 542
street lighting, 70
Sunday clothing, 227
swimming baths, 282
teacher training colleges, 711
Templeton of Glasgow, 184
tenements, 182, 210–11, 237, 286, 297
toll house, 834
tower blocks, 10–11, 17, 547
town houses, decoration, 178
Training School for Servant Girls, 370–1
United Secession Church, 211
vagrants, 855
washhouses, 237, 281, 282, 325
water supply, 281, 285, 287, 485
weaving, 347
women's employment, 352, 353
Glasgow Cathedral, 177
Glasgow Charity Organisation Society, 852
Glasgow Diocese, 442
Glasgow Eye Hospital, 483
Glasgow Garden Suburb Tenants, 546
Glasgow Green, 281, 325
Glasgow Incorporation, 409
Glasgow School of Art, 191
Glasgow School of Cookery, 371, 372
Glasgow University, 489, 711, 715
glass porches, 16
Glasse, Hannah, 315, 320
Glasser, Ralph, 267–8
Glasserton, Dumfriesshire, 507
glassware, 315, 380, 427, 431
glazed doors, 80
glazing, 79, 406, 841
  see also double glazing; windows
glazing irons, 323
Glen Affric, Inverness-shire, 645
Glen Clova, Angus, 645–6, 648
Glen Doll, Angus, 646
Glen Dorchaidh, South Uist, 618
Glencoe, Argyll, 642
Glenelg, Inverness-shire, 103, 628
Glenesk, Angus, 61, 89, 100, 102
Glenesk Hall, Saughton Prison, Edinburgh, 660–1
Glenfield Kennedy, 287
glengarries, 230
Glenochil, Clackmannan-shire, 656
Glenogle Colonies, Edinburgh, 833–4
Glenorchy and Innishail, Argyll, 87
Glensheil, Ross-shire, 231
Glenshirra Lodge, Inverness-shire, 640
Glentress, Peeblesshire, 650
gloves, 239, 464
glue, laundry work, 321
glycerine soap, 264
Goalen, Walter Mitchell, 843
goffering irons, 226, 323
goffering tongs, 323
gold wedding rings, 506
Goldwyre, Gerry, 839–40
goose, game of the, 166
gooseberries, 407
Gorbals, Glasgow, 267–8, 547
Gordon & Dey, 843
Gordon, duke of, 314
Gordon, James, 408
Gordon, Seton, 129, 618
Gordonstoun, Moray, 837
Gorgie, Edinburgh, 444
Gow, Ian, 385
Gowans, James, 837
Gowthens Muir, Perthshire, 621
graces (prayers), 201–2, 205
Graham, Billy, 215–16
grain, 37, 86, 333, 118, 586
  see also barley; bere; bran; harvest work; meal

INDEX • 891

gramophones, 120–1, 129, 433, 535
Grange, the, Edinburgh, 837
Grangemouth, 291
Grangemouth Children's Fair, 573–4
Grant, Mr (Baron of the Exchequer Court), 409
Grant, Elizabeth, 244, 278, 280, 423, 432
Grant family, Monymusk, Aberdeenshire, 345
Grant, I F, 58, 522
Grant, James, 409
Granville, Lord, 837
grapes, 406
grass
    gardens, 400, 402
    see also lawns
grass chairs, 116
grass ropes, 335
grates, 84, 94, 95–6, 103, 420
    cleaning, 311, 316, 317, 588
    cooking accessories, 86
    croft houses, 83, 189
    see also braziers
graveclothes making, 344
graves, 503, 507, 583
gravestone, footwear depicted on, 231
Gray, John, 585
Gray, John Miller, 385
Gray, William J, 837
grease paint, 264
great halls, music, 121
Great Yarmouth, 623
Greenock, Renfrewshire, 285, 287, 656, 672
Gretna, Dumfriesshire, 835
Greyfriars Bobby, 585
grooming see toilette
group reading, 391
grouse, 638
guarduvines, 62
Gude and Godlie Ballates, 205
guesthouses, 649, 650, 716
guests see visitors
Guides, The, 473
guising, 571–3
    see also role-playing games

gun dogs, 589, 591
    see also hunting dogs
Gurness, Orkney, 31
guse nebs see cruisie lamps
Guthrie, Thomas, 686
gutters see herring gutters
gyms, 254
gypsies, Travelling People and, 788–9

H
Haddington, East Lothian, 408, 423, 750, 842
haggis, 421
hair, 244, 259, 262, 268, 279
    boys, 463
    children in institutions, 686, 687
    colours, 244, 502
    girls, 248, 249, 250, 559
    married women, 249–50
    see also animal hair; facial hair
hair brushes, 256
hair powder, 309
hairdressers, 252–4, 260, 373
hairdressing equipment, 256
    see also combs
half lades, 333
Hall of Ossian, Penicuik House, Midlothian, 181
Halliday, Wull, 633
Halloween, 569, 571, 574
halls, 546, 721
    see also great halls
Hamilton, duchess of, 360–1
Hamilton, duke of, 360–1, 426
Hamilton, Gavin, 167
Hamilton, J Anderson, 837
Hamilton, John, 2nd Baron Belhaven, 403
hammermen, 342
hamsters, 591
handball, 168
handcrafts see crafts and hobbies
hanging lums see hingin lum fireplaces
Hanley, Bob, 633
Harley, William, 281

harn, 366
Harper, Mrs, 645
Harris, 544–5, 586
harvest work, 332, 525, 618, 627, 791, 799
    see also fruit picking
hats, 239, 321, 463, 464
    see also millinery
hawkers and peddlers, 422, 425, 429, 627–8, 633, 791, 799
hay
    laundry work, 321
    see also harvest work
Hay, David Ramsay, 183
Hay, George, bishop and Vicar Apostolic, 208
haylofts, 553, 844
Hayward & Son, 184
head coverings, 229–30, 239, 259, 264, 366, 373, 559
    see also bonnets; caps; glengarries; mutches
healing wells, 504, 507, 509–10
health, 419, 485–6, 512, 851, 861
    see also dental care; diseases; infirm people; mental health
health benefits, pets, 593
health care, 290, 461, 467, 667, 861
    see also medical services; National Health Service
health cures, bagnios, 281
health visitors, 443, 488, 491, 494
heart conditions, 487, 510, 672
hearth bisoms, 311
hearth outshots, 93–4
hearths, 103, 225
    calendar customs, 570
    cleaning, 85, 311, 316
    colouring, 628
    cooking and baking, 86, 420
    decoration, 85, 315
    hospitality and, 554

peat fires, 89–91
symbolic importance, 562, 566
see also fireplaces; fires; mantelpieces
Heath, Bob, 842
Heath Robinson, W, 556
'Heather Jocks', 627–8
heather ranges (brushes), 311
heather ropes, 335, 350
Heatherbell (Ladies') Cycling Club, 647, 648
heating, 83–107
  bathrooms and, 259–60, 291
  bedrooms, 83, 189
  bread ovens, 420
  clothing and, 83, 462–3
  council houses, 97
  houseboats, 841
  huts, 828
  infirm people's homes, 493
  mobile homes, 823, 824
  outbuildings, 553
  toilette and, 268
  Travelling People's accommodation, 792, 794, 795, 797
  youth hostels, 646
  see also central heating; fires; stoves; water heating
Hebrides see Western Isles
heckling, 342
hedges, 402
Helensburgh, Dunbartonshire, 182
hemp fibres, 71
Henderson, Hamish, 621
Henderson, John, 843
henna, 265–6
Henryson, Robert, 584, 585, 586
hens, 582, 631
henwives, 361
herb poultices, 505
herbal preparations
  cosmetic uses, 260–1, 262, 264, 265
  Findhorn community, 741

medical uses, 401, 501, 505
skin cleansing, 263
see also complementary medicine
herbalists, 255
herbs, 66, 317, 400, 401, 402, 407
Hermitage Folly, Dunkeld, Perthshire, 405, 835
hermitages, 835
herring, 366, 619
herring fishing, 226, 622, 766, 768, 771
herring gutters, 55, 622–4, 768, 770, 771, 773
Hick, Alice, 563–4
hide and seek, 471
hides see animal skins
hierarchy and authority, 515–41
  Bealtainn fire, 571
  bothy seating, 631
  clans, 357, 521
  communes, 738, 746, 747
  Highland betrothal procedure, 566
  parent-child relationships, 526, 528–37, 557
  salmon netsmen, 625
  shared flats and houses, 724–5
  women, 357, 515, 521, 522–8
  see also status
High Street, Perth, 584, 585
high teas, 431
Highland and Agricultural Society of Scotland, 406
Highland dancing classes, 472
Highland dress, 228
  see also kilts; plaids
Highlands
  bedding, 102, 103, 363
  beds, 102, 103, 111
  betrothals, 565–6
  blue bonnets, 230
  boarding out of children, 476
  clearances, 14, 638

clothing, 223, 226, 228, 363, 463
dairy products, 341
death customs, 201
distilling, 341
dressers, 60–1
dyeing of cloth, 232
food, 363
fuel, 87–8
gardens, 404
guising, 571
health, 485–6
hospitality and status, 554
houses, 110, 598
  see also blackhouses; byre dwellings
infirm people, 483–4
inn beds, 102
medical services, 484
medicine, 502, 503
ministers' libraries, 382
music, 121, 123, 125
New Year customs, 571
religion and play, 164
sanitation, 102
seating, 115–16
servants, 309, 363
spirituality, 201
stalking and shooting holidays, 638
status, 542–3, 554
tourism, 637
tuberculosis, 485–6
women's employment, 341
see also Caithness; clans; Ross-shire; shielings; Sutherland; Western Isles
Highlands and Islands Development Board, 194
Highlands and Islands Medical Scheme, 484
Highlands and Islands Medical Service, 113
Hill Farm, Banffshire, 73
Hill House, Helensburgh, Dunbartonshire, 182

INDEX • 893

Hill of Aithsness, Vementry, Shetland, 86
hill walking and mountaineering, 641–4, 646, 647
hillforts, 25–6, 31
Hilton Fireclay Works, 295
hinds *see* farm servants
Hindu communities, 679
Hinduism, 218
hingin lum fireplaces, 71–2, 91–3, 420
hippens *see* nappies
Historic Scotland, 838
HIV/AIDS, 487, 673, 674, 680
hobbies *see* crafts and hobbies
hobs, 94, 95, 420, 427
Hoddom Castle, Dumfriesshire, 644
hoeing, 332
Hogg, James, 102, 543, 637
Hogmanay, 419, 421, 434, 569, 570, 573
see also New Year
holiday accommodation, 819, 827–8
see also bed and breakfast accommodation; boarding houses; camping sites; hotels; youth hostels
holiday homes, 649, 650, 819
holidays, 108–9
angling, 641
apprentices, 423
fishing communities, 768
Fort Augustus Abbey, 759
servants, 373, 423
Show People, 806
stalking and shooting, 638
see also Christmas; fairs; New Year; Saturday working; Sunday; tourism
Holmwood House, Glasgow, 180–1
Holy Island, 846

holy wells *see* healing wells
Holyrood, Edinburgh, 180
home
computer location in, 154, 383, 472
education at, 121, 361, 373, 384
gender differences in experience of, 24, 43, 338–9
selling goods from, 343
weddings and baptisms at, 216
see also work at home
home births, 440, 441–8, 449, 450–1, 452
home care, infirm people, 482, 483–4, 488, 489, 490–4, 496, 519
home healers, 501–3
see also domestic medicine
home helps, 491, 493, 785
Home, Henry, Lord Kames, 405
home offices, 97, 145, 154, 353, 384
see also computers
home ownership, 597, 603–4, 605, 608–11, 719
decoration and, 176
Edinburgh, 42–54, 604, 720
mobile homes, 823
old people in single-person households, 781, 783, 784
security and, 24–5, 30, 38
shared flats and houses, 718, 720, 725, 726
Show People, 806
status, 551, 554
see also mortgages; property ownership
Home, Patrick, Baron Polwarth, 489
home visits, doctors, 483, 489, 511
homeless people, 440, 592, 681, 716, 851–64
homeopathy *see* complementary medicine

Homes for the Dying, 672
homework, 84, 388, 475
honey, 424, 510
honeysuckle, 504
Hoover, W H, 314
hopscotch, 169
Hornby, Frank, 170
horse dealers, 791
horse races, 164
horsehair products, 114, 336
horses, 582, 584
see also stables
horticultural societies, 409
horticulture *see* gardens
hosiery *see* knitting
hospices, 217, 672–82, 754, 758, 759
Hospitalfield, Arbroath, Angus, 836
hospitality, 546, 554–6, 568, 571
hospitals, 217, 482, 483, 486, 495
see also doctors; maternity hospitals; nurses; Orphan Hospital
hostels
dormitories, 644, 646, 715, 716
homeless people, 716, 851, 858, 859, 860, 861, 862
secondary school pupils, 710
see also lodging houses; student accommodation; youth hostels
hot air baths, 280
hot tar, medical uses, 503
hot walls, 406
hot water bottles, 83, 493
hot water supply *see* boilers; water heating
hotels, 641, 642, 647, 649, 716, 856
see also inns
Hough, Tiree, 13
Hounam, Roxburghshire, 225
house factors, 598, 599, 600, 601

house farmers, 599
House of Binns, West Lothian, 382
House of Dun, Angus, 84, 179
house plants, 190
houseboats, 841
household sizes, 797–8, 813
    see also family sizes
households, 518–21
housekeepers, 309
    Duchess of Hamilton's household, 361
    duties, 310, 311, 361, 367, 427
    training, 372
housekeepers' store rooms, 315
housekeeping costs, 527, 722
housemaids, 308, 315–16, 361
houses, 9, 109–11
    Berwickshire, 79, 100, 837
    communes, 738
    construction, 17–18
    conversions to, 838–46
    Dundee, 351, 589
    Edinburgh, 62, 548, 603, 833–4, 842, 843, 844
    environmental adaptation, 7–9
    estate workers, 837
    farm servants, 79, 80, 100, 299–300, 597, 598
    Fife, 836, 837
    Findhorn community, 742, 744
    Highlands, 110, 598
    lighthouse keepers, 834
    manufacturers, 351
    Midlothian, 837, 839–40, 842
    miners, 283
    Monimail commune, 745
    Orkney, 27, 582, 841
    poultry and animals in, 331, 582
    shops in, 343, 769
    Show People, 806
    Travelling People, 790, 799

weavers, 77
see also blackhouses; byre dwellings; ceilidh houses; cottages; council houses; country houses; croft houses; farm houses; flats; lodges; prefabricated houses; rented houses; roundhouses; shared flats and houses; summerhouses; toll houses; tower houses; town houses; wheelhouses
housework, 307–29, 362
    children, 308, 459, 469, 532, 533–4, 537
    fishing communities, 772
    fuels and, 307, 311, 325, 434
    gender differences, 307, 526–7, 528
    herring gutters' huts, 623
    hospices and nursing homes, 676
    howdies, 491
    reading and study and, 387
    salmon netsmen, 625
    servants, 307, 308
    shared flats and houses, 720, 727–8
    Show People's caravans, 810
    training, 370–2
    water supplies, 307, 311, 318, 319, 325, 427
    see also cleaning; laundry
housing, environment and, 7–9, 12, 13, 14–16, 18
housing associations, 605, 606, 607, 608, 610

housing benefit, 606–8, 609
Howard, Henry, 421
howdies
    childbirth help, 442, 443, 444, 447, 448–9, 490, 491
    housework, 491
    laying out of dead people, 490, 495, 568
    medical advice, 502
    see also midwives
Howie (sanitary ware manufacturer), 296
Humanist Society, 217
Humbie Dovecot, East Lothian, 843
Hume, David, 209
Hume, Lady Grisell, 121
hunting see stalking and shooting
hunting dogs, 592
    see also gun dogs
hurlies, 470–1
hurly beds, 62, 103, 111, 546
huts, 824–30, 831
    climbing clubs, 642
    fruit pickers, 620–1
    herring gutters, 623
    Monimail commune, 744, 745
    shielings, 17, 99, 109–10, 618
    Travelling People's accommodation, 791
    youth hostels, 644, 647
    see also bothies; Nissen huts
Hutterite commune, 737
hygiene see cleaning; cleanliness; personal hygiene; vermin
hymn singing, 123–4
hypogea, 15
hypothec, law of, 599, 601

I
ice chests, 430
icehouses, 424, 430, 627
    see also refrigerators
Icolmkill see Iona
Ikea, 380, 381

INDEX • 895

illegitimate children, 361–2, 368, 477, 520, 685–6
illnesses *see* diseases; health; infirm people
Imbolc, 569
immigrant workers, 353, 621, 684
immigrants
    allotments, 415
    care of infirm people, 485
    children's clothing, 464
    customs, 574–6
    group reading, 391
    security, 24, 29, 32
    see also Chinese communities; Hindu communities; Muslim communities
Inchaffray, Perthshire, 750
Inchcape lighthouse, 834
Inchrye Priory, Fife, 836
incorporation, rites of, 563, 565, 576
Incorporation of Surgeons and Barbers, 281
independent schools, 701–8, 754, 757, 758, 759, 760
indigestion, 506, 510
indoor head coverings, 230
indoor planting, 190
Industrial Schools, 688
infant mortality, 458, 467, 476, 477, 487, 567
Infant Welfare movement, 458
infants *see* babies
infirm people, 26–7, 324, 368, 482–99, 855
    beds, 493, 714
    care at home, 482, 483–4, 488, 489, 490–4, 496, 519
    food, 433, 493
    poorhouses, 468, 486, 488
    religious communities, 758
    see also disabled people; old people

information and communication technologies *see* technology
Ingoldsby, Colonel, 360
inheritance tax, 610
Inland Waterways, 841
inns, 637–8, 641, 642, 672
    beds, 102, 103, 115, 637, 638, 715
    cleanliness, 102, 308, 637
    see also hotels; rooms, letting
insanity *see* mental health
Insch, Aberdeenshire, 71–2
insulation, 16, 104
interior decoration *see* decoration
International Congress on Milk Depots, 458
International Exhibition, 1901, 183
Internet, 412, 435, 739
Internet access, 145–6, 157
    Dollar Academy, 707
    gender differences, 146, 153
    Girl Guides, 473
    reading and study and, 387–8
    shared flats and houses, 722
    work at home, 353
    youth hostels, 646
invalids *see* infirm people
Inver, Perthshire, 637–8
Inveraray, Argyll, 650
Inverewe, Ross and Cromarty, 410
Invermark Lodge, Angus, 640
Inverness, 249, 287, 507, 656, 658, 672, 751
Inverness-shire
    Balmacaan House, 640
    bathing arrangements, 279–80
    Bealtainn fire, 571
    clothing, 227
    deer forest, 640
    Fort Augustus Abbey, 750, 754–60
    Fort William barracks, 17

gamekeepers and hill walkers, 646
hawkers and peddlers, 628
inn beds, 103
local government reorganisation, 710
Lochaber High School, 709
lodges, 640
tartans, 229
WCs, 279
youth hostel, 645
Invershiel, Kintail, Ross-shire, 103, 637
Iona, 345, 750
Iona Abbey, 401, 752
Iona Community, 213–14, 751
iron bedsteads, 113, 115
iron industry, 287
    see also Carron Iron Company; cast iron products; Falkirk Iron Company
ironing, 236, 319, 321, 323, 353, 432, 769
irons, 226, 323
ironwork, 179, 184
Irvine, Ayrshire, 85, 295
Irvine, John, 567
Irvine valley, Ayrshire, 348
Islam, 218, 575
    see also Muslim communities
Islay, 37, 504, 582
Isle of Erraid commune, 744
Italian irons, 323
itinerant craftsmen, 633
    see also Travelling People
itinerant musicians, 633
itinerant traders *see* hawkers and peddlers
ivy leaves, laundry work, 321

J
J & M Craig, 295–6
J & W Campbell, 184
jackets, 231, 462
Jacob, Ida, 647
Jaeger, Dr , 139

jam factories, 620
James I, king of Scots, 164, 167, 182
James IV, king of Scots, 120, 339
James V, king of Scots, 120, 167, 339, 402
James VI, king of Scots, 166, 285, 360, 558, 584
jams, 407, 426, 631
    see also jeely pieces
jars, food storage, 64
Jarvis, Tom, 626
Jedburgh, Roxburghshire, 752
jeely pieces, 467–8
Jeffrey & Co, 184
Jehovah's Witnesses, 218
Jekyll, Gertrude, 410
jellies, 426, 468
    see also jams; jeely pieces
jerkins, 366
jewellery, 38, 248, 504
    see also wedding rings
Jewish community, 124–5, 575, 680
    see also Judaism
jigsaws, 433–4
John Smith & Co (Wools), 192
Johnson, Dr Samuel, 102–3, 308, 418, 420, 424, 637
joiners, 369, 373
Judaism, 218
    see also Jewish community
jugging, 421
juices, 262, 263, 321
Junior Mountaineering Club of Scotland, 642
Jupp, Rev, 689
jupps, 366
juveniles see children

K
kail, 76, 100, 403, 409
    see also cabbages
kallyards, 298, 400, 402, 406
    see also gardens
Kames, Lord, 405
Kant, Immanuel, 194
Kaufman, Paul, 394

Kay, Billy, 643
kazies, 333
kegs, food storage, 64
Keillor family, 344, 351
Keir, Thomas, 343
Keith, Banffshire, 72, 84
Kellie Castle, Fife, 404, 410
Kells, Kirkcudbrightshire, 87
Kelso, Roxburghshire, 102, 226, 585
kelt, 231
Kelvinhaugh, Glasgow, 283–4
Kelvinside, Glasgow, 415
Kennedy, Lewis, 410
Kenneth I, king of Scots, 531
kerls, 76
Kesson, Jessie, 245, 249
Kilbarchan, Renfrewshire, 348
Kilbride, Bute, 225
Kildrummy Castle, Aberdeenshire, 26
Kilkenneth, Tiree, 13–14
Kilmacolm, Renfrewshire, 549
Kilmarnock, Ayrshire, 72, 287, 295, 656
Kilmarnock bonnets, 230
Kilmartin, Skye, 509
Kilmory, Bute, 229
Kilravock, laird of, 382
Kilsyth, Stirlingshire, 347
kilts, 228, 463
Kincardineshire, 58, 88, 179, 227
kindling see firewood
King, Mary Truby, 458, 464–5, 466
Kingshouse Hotel, Glencoe, Argyll, 642
Kinkhost Well, Comrie, Perthshire, 507
Kinross, 180, 673
Kinross House, 84
Kinross-shire, 84, 180, 583, 673
Kintail, Ross-shire, 103, 637, 642
Kirbister, Orkney, 99, 101
Kirk Close, Perth, 284
kirk sessions, 206, 362, 368

Kirk Yetholm, 789
Kirkcaldy, Fife, 291, 542, 672
Kirkcudbright, 71
Kirkcudbrightshire, 87, 738, 739
Kirkinner, Wigtownshire, 87
Kirriemuir, Angus, 77
kishies, 333, 334
kistings, 495, 568
kists, 55–8, 116
    clothes storage, 238
    dressing rooms, 258
    farm servants, 55, 56, 110, 631
    seating, 56, 110, 111, 116, 631
    servants, 55, 238
    work surfaces, 257
    see also Scotch chests
kitchen boys, 361
kitchen fires, 78, 83, 89, 91–3, 96
kitchen floors, 317, 427
kitchen furniture, 57–61, 78, 427
    see also kitchen tables
kitchen gardens, 400, 402, 406, 408
    see also herbs; vegetables
kitchen maids, 308, 427, 432, 531
kitchen ranges, 94–6, 103, 169, 325, 427, 460
    boilers, 95, 291, 427
    ceramic tiles, 95, 184
    cleaning, 317, 325
    Dundee, 588
    tenements, 546
    water heating, 95, 291
    youth hostel, 645
    see also stoves
kitchen sinks, 78, 279, 286, 362, 460, 469, 546
kitchen tables, 61, 78, 315, 382
kitchens, 127, 434, 567
    auxiliary work in, 330, 331, 335–6
    Bell Rock lighthouse, 834
    bungalows, 556–7
    caravans, 797, 809

INDEX • 897

castles, 96
cleanliness, 325
cottage ceilings, 64
country houses, 96
croft houses, lighting, 69, 70
Dollar Academy, 707
farm houses, 91–3, 94, 99
fitted, 185, 554–5
geese in, 331
herbal preparations, 262, 263
holiday homes, 649
huts, 828
infirm people, 493–4
location, 44
mobile homes, 822
North Queensferry signal tower conversion, 841
pantries and, 424
Perthshire, 77–8
religious communities, 96
round-about firesides, 91
shared flats and houses, 721, 727–8
sleeping accommodation, 99, 101, 112, 556, 649
status, 554–5, 556–7
television sets in, 146
tenements, 78, 286, 546
villas, 554, 556, 557
windows, 77–8
youth hostels, 646
see also living rooms
Knap of Howar, Orkney, 55
Knapdale, Argyll, 75
knife boxes, 312
Knight, Mrs, 588
knitting, 134, 138–40
bedspreads, 135
clothing, 135, 139, 331, 349, 462, 464
education, 134, 136, 372
employment, 137, 191, 339, 340, 341, 348, 349
fishing communities, 762, 769–70, 773
herring gutters, 623
lighting for, 84

men and boys, 341
old people, 341
Shetland, 191, 335, 340, 341, 349
shieling workers, 619
stockings, 341, 346, 619, 770
women's employment, 339, 340, 341, 349
women's work, 339, 340, 341, 349
knitting sheaths, 55
knives
cleaning, 312–13
see also cutlery
knot gardens, 402, 408
Knox, John, 199, 205–6
Knoydart, Inverness-shire, 646
Koresh, David, 737
kringle stuils, 332
kubbies, 333, 334, 335

L
lace, laundry of, 321
lace making, 134, 137, 138, 140, 331, 344, 556
Ladies' Scottish Climbing Club, 641
lady's closets, 62–3, 256, 257
lady's maids, 308, 361
lairds, 121–2, 559
see also landlords
lairds' houses *see* country houses
Lamb, Ann, 444, 448
Lamb, Charles, 704
Lamb, Patrick, 421
Lamberton, Berwickshire, 835
Lambton, Lucinda, 844
Lammas, 569
Lammas Drave, 766, 768
Lammermuirs, the, 9
lamps, 74, 75, 386
see also cruisie lamps; desk lamps; lanterns; paraffin lamps; Tilley lamps
Lanark, 689
Lanarkshire
Bothwell Castle, 26
Cambusnethan Priory, 836

candle making, 76
Carluke dressers, 61
clothing, 226
cotton mills, 349
evangelical Christianity, 207
farm house kitchen fires, 91–2
fruit picking, 620
garden, 402
orphanage, 689
prison, 656
Talamh community, 739
Travelling People's tents burned, 795
weaving, 347
women's employment, 345
Landess, Robert, 505
landlords
farm steadings, 598
huts, 827, 828, 829–30
mobile home sites, 823, 824
power of, 543, 601
shared flats and houses, 717–8, 721, 723–4, 726, 729, 731
see also lairds; rented houses; tenancies
Landmark Trust, 839
landscape gardens, 404–6
Landseer, Sir Edwin, 585, 639
Langalour, Orkney, 99
lanterns, 70
larch trees, 406
L'Arche communities, 751
lard, 74, 262, 263
larders, 424, 427, 430
latch-key children, 475
Lateran Council, 749
Lath a' Mhàil, 108
latrines
religious communities, 277
see also sanitation
launderettes, 237, 282, 325, 547
laundry, 235–8, 312, 319–24, 362, 547
bedding, 236, 301, 318, 320
blankets, 301, 318

898 • SCOTTISH LIFE AND SOCIETY

camping sites, 801
cotton, 232
detergents, 291
drying, 85, 86, 319, 322, 413
education, 237, 370, 371
fishing communities, 769
girls' work, 469, 533, 534
lodgers, 343
nappies, 465
place mats, 432
prisons, 664
Show People's caravans, 810
soap, 320, 325
sports clothing, 169
storage of equipment and materials, 316
student residences, 716
Travelling People, 798, 801
tubs, 236, 279, 292, 311, 319, 320, 321
work at home, 352
see also washhouses
laundry maids, 308, 321
see also washerwomen
Lauriston Hall commune, 738, 739
Lauriston Home, Edinburgh, 440
lavatories, 84, 284, 299, 300, 469, 546, 553, 704
outbuildings, 284, 294, 479
see also latrines; privies; sanitation; WCs
lavatory basins see wash-hand basins
Law of The Lord's Day, 199–200
lawns, 405
see also grass, gardens
Lawson, John, 366
Lawson's Intelligence Office, 366
lawyers
Duke of Hamilton's household, 360
libraries, 381, 393
location in Edinburgh, 342

mortgage arrangement, 598
servants, 357–8
vocational reading, 390
work at home, 343
lay baptisms, 450
lay brothers, 753, 757
layettes, 445
laying out, dead people, 201, 260, 490, 495, 567–8
lead, 263, 264, 265
see also blacklead
Leadhills, Lanarkshire, 345
Leblanc process, 290
Lee Penny, 504
leeches, 505, 510
leeks, 400, 415
Leigh Hunt, James Henry, 704
leisure activities
allotments, 415
armed forces, 632–3
clothing for, 139
Dollar Academy, 707
farm servants, 631, 648
Findhorn community, 741
gender differences, 534
herring gutters, 623
home-based, 535
homeless person, 681
hospices and nursing homes, 677
lace making, 140
lorry drivers, 629
needlework, 134, 137–42
oil-rig workers, 629
prisoners, 661, 664, 666, 667
religious communities, 754, 756, 759
salmon netsmen, 625
seasonal workers, 618
shared flats and houses, 729
Sunday observance, 109, 199
temporary accommodation, 637–53
textile workers, 648
women, 134, 137–41, 648

see also computers; crafts and hobbies; cycling; games; gardens and gardening; hill walking; holidays; mountaineering; music; play; sport; television; tourism; video
Leith, 85, 281, 467, 469, 751
see also Water of Leith
lemon juice, 262, 263
lemon water, 264
lemonade, 504
Lennoxlove, East Lothian, 63
Leslie, James, 839
Lesmahagow, Lanarkshire, 226
Letterewe, Wester Ross, 646
letters, 118, 155–6, 669
lettuce, 400
Lever Brothers, 283
Lever, William Hesketh (Viscount Leverhulme), 290
Lewis
box beds, 100
clothing, 224, 231
croft house provision, 598
earthenware pots, 65
extended families, 519
hospice, 672
hymn singing, 123–4
peat-burning hearths, 89
plantiecrues, 403
shielings, 618
women's dress, 224
Lewis, C S, 703
liberty bodices, 462
libraries, 381–2, 384, 385, 391, 393, 394
decoration, 182, 385
Fort Augustus Abbey, 754, 757, 759
prisons, 664, 667
public, 378, 391
religious communities, 752
talking books, 494
websites, 388

lice, 586, 686
Life Boys, 473
life expectancy, 487
light slots, 79
lighter conversions, 841
lighthouses, 834
lighting, 69–82, 138
   Balmacaan House, Inverness-shire, 640
   caravans, 797, 809, 810
   ceramic bricks and, 296
   drawing rooms, 385
   housework, 311
   huts, 828
   lace making, 138
   living rooms, 96–7
   mobile homes, 823
   needlework, 76, 84, 138
   outbuildings, 553
   privacy, 256
   reading, 71, 72–3, 76, 80, 382, 385–7, 664
   rope-making, 335
   spinning, 76, 77, 84
   work at home, 77, 138
   see also candles; electric lighting; firelight; gas lighting; lamps; street lighting
Lillie, Helen, 384, 388
limpets, 418
Lincrusta Walton, 184
Lindisfarne Castle, Northumberland, 846
Lindsay, Sir David, 204
line fishing, 350, 352, 621, 766, 769, 771, 773
linen
   aristocratic households, 361
   bedding, 102, 103, 114, 232, 361
   bleaching, 232, 237
   clothing, 228, 229, 231–2, 361
   furnishings, 178, 184, 190, 193, 361, 365
   laundry, 236, 320, 321, 362
   making, 226, 350
   mending, 137
   samplers, 190
   smoothing, 322–3
   storage, 239
   umbilical cord dressings, 448–9
   whitening, 320, 321
   see also flax spinning; harn
linen dusters, 311
Liniclete, Benbecula, 711
links (fire equipment), 72, 86–7, 92
Links Foundry, Angus, 287
Linlithgow, West Lothian, 423, 752
linoleum, 291, 317
linseed oil, 317
Linton, Roxburghshire, 171
liquor cases, 62
listed buildings, 838, 839
Lister, Joseph, 1st Baron, 483
literacy, 204, 377, 378, 382
   see also reading; writing
livestock see animals; poultry
Living Memory Project, 144
living rooms, 96–7, 110, 193
   auxiliary work, 330, 331
   Bell Rock lighthouse, 834
   computers, 154, 383
   libraries, 382, 384
   North Queensferry signal tower conversion, 841
   painting of Perthshire, 77–8
   shared flats and houses, 721, 727, 728
   sleeping accommodation, 96, 109, 110
   status, 556–7
   study areas, 383
   villas, 557
   see also kitchens
Livingston Development Corporation, 411–12
Lloyd George, David, 601
lobsters, 469
local authorities
   homeless people, 855–8, 859–61
   see also council houses; council tax
Local Nursing Associations, 490–1
local prisons, 656, 658
Loch Awe Hotel, Argyll, 642
Loch Lomond, 673, 828
Loch Maree, Ross and Cromarty, 509
Loch Tay, Perthshire, 641
Lochaber, 812
Lochaber High School, Fort William, Inverness-shire, 709
Lochhead, Marion, 378–9
Lockerbie Aircraft Bomb Trial, 657
lodgepole pine seeds, 410
lodgers, 343, 519, 599–600, 623, 716
lodges, 187, 188, 837
   see also bothies; fishing lodges; shooting lodges
lodging houses, 628, 715, 853–4, 857
   see also hostels
lodgings, 623, 715–16, 717, 718, 719
   see also bed and breakfast accommodation; boarding houses; rooms, letting
lofts, 343, 384, 791
   see also attics; haylofts
Logie Wester, Ross-shire, 226
London Board Schools, 297
long lets, privately rented houses, 600–1
long-term prisons, 656, 658
longhouses see byre dwellings
Longmate, Norman, 703
looking glasses see mirrors
Lop Ness, Orkney, 583
Lorimer, Sir Robert, 179, 297, 410
lorry drivers, 629
Lothian Incorporation, 409
Lothians, 56, 64, 91, 525, 827
   see also East Lothian; Midlothian; West Lothian
Loudon, Jane, 410

900 • SCOTTISH LIFE AND SOCIETY

Loudon, John Claudius, 401, 409, 410
Louis XVI, king of France, 299
lounges *see* living rooms
Low Moss prison, Glasgow, 656–7
Lower Taes, Fife, 625
Lowestoft, 623
Lowlands, 37, 60–1, 522
luckenbooths, 343
lullabies, 120, 123
lum chests *see* Scotch chests
lumbago, 508
lunacy *see* mental health
Lunasdal, 569
lunch clubs, 492
lunches, 425, 430, 467–8, 475, 758
lung diseases, 487, 672
Lutyens, Sir Edwin, 846
lye, 300, 320
lying-in *see* maternity hospitals; postnatal care
lykewakes, 201, 568
Lyons, Martyn, 387, 392

M
Macbeth, Ann, 192
McCaskill, Mary, 446
MacColl, Ewan, 129
MacDiarmaid, J M, 640
MacDonald, Annie, 618, 619, 622
MacDonald, Colin, 108, 114
MacDonald, Finlay J, 297, 544, 557
MacDonald, John, 367
MacDonald, Patrick, 123
McDowall, William, 284
Macewen, Sir William, 483
Macfarlane, Alan, 765
Macfarlane, Walter, & Co, 184, 287
MacFisheries Ltd, 626
McGettrick, Professor Bart, 712
McGibbon, William, 122
MacGregor, Alasdair Alpin, 531
Macgregor, Forbes, 585
MacGregor, Margaret, 201
MacGregor, Rob Roy, 531

McIntosh, Dr Farquhar, 712
Mackay, Jessie, 725
Mackay, John, 382
MacKenzie, John Alexander, 279
MacKenzie, Rev Neil, 99
McKenzie, Osgood, 410
Mackenzie, W L, 441, 476
Mackie, Charles, 182
Mackintosh, Charles Rennie, 182–3
Macklin, Mr, 280
Maclagan, R C, 163
MacLean, Calum, 618
MacLean, Rev Dr Donald, 199
MacLean, Dorothy, 740
Maclellan, Angus, 489
MacLeod, George, 213, 214, 215
McLintock, Mrs, 421
Macmillan nurses, 672–3
McMurtrie, Elizabeth, 368
McNabb, Capt John, 705
MacNeacail, Iain, 127
MacNeil of MacNeil, 203–4
MacRae, Peigi, 301
McVeigh, Patrick, 249, 460
magazines, 380, 387, 391–2, 409, 412
Magdalene Institutes, 689
maggots, 510
magico-religious medicine *see* protective charms and customs
Maiden, The, Sutherland, 644
maids *see* dairy maids; housemaids; kitchen maids; lady's maids; laundry maids; milkmaids; nursery maids; parlour maids; scullery maids
maid's rooms, 546, 554
maises, 333
Mallaig, Inverness-shire, 279–80
malting, 342
Manderston, Berwickshire, 384

mangle bats, 322–3
mangling, 236, 282, 323, 324, 350
Manguel, Alberto, 390
manicures, 254, 262
mantelpieces, 85, 187, 188, 189, 193, 596
see also overmantels
mantuas, 233
manure, 83, 284, 286, 300, 525
see also dung; middens
Mar, earl of, 26
marble floors, 317
Marie Antoinette, queen of France, 299
Marie Curie centres, 672, 675
marigolds, 403
markets, 331, 419, 422, 429
see also fairs
Markham, Gervase, 421
marmalade, 344, 351, 421, 424, 426
marriage, 520, 521, 524, 574, 770, 773
see also betrothals; divorce; weddings
marriage customs, 124, 565–7
married couples, hierarchy and authority, 522–8
married women
day of rest, 108
employment, 324–5, 339, 343–4, 352–3, 526, 527, 537
needlework, 137, 352
head coverings, 229, 230
landed families, 357
legal rights, 522–3, 528
personal appearance, 249–50
reading, 389
seating, 117
technology, 158
see also housework; mothers; widows
Marshall, David, 212, 495
Martin, John, 100
Martin, Martin, 508, 509, 586
Mary of Guise, 402

Mary, queen (consort of George V), 141
Mary, queen of Scots, 584
Maryburgh Old Cononbridge, Easter Ross, 834
masonry see stone buildings
masons, 369
masseurs, 281
matchmakers, 566
maternity hospitals, 440–1, 450–2
Maternity Services Schemes in Scotland, 442
mats see place mats; rugs; straw mats; table mats
mattresses, 114
   salmon netters' lodges, 625
   sprung, 113, 114
   straw, 101, 114, 365, 466, 645
   vermin, 101, 586
   see also chaff beds
Maxwell, Gavin, 628
May Day, 573
   see also Bealtainn
May, Robert, 421
Maybole, Ayrshire, 368
Maybole Ragged School, Ayrshire, 688, 692
Meadows, Edinburgh, 322
meal, 37, 116, 263, 333, 631
   see also barley meal; flour; girnals; oatmeal
meals, 424–6, 427, 430–3, 435
   birth celebrations, 449
   children, 432, 467–8, 475, 531, 716
   Christmas, 434
   farm houses, 366, 425
   farm servants, 630
   Findhorn commune, 741
   hospices and nursing homes, 676
   in tombs, 583
   infirm people, 493
   Monimail commune, 745
   religious communities, 756, 757, 758
   salmon netsmen, 625
   shared flats and houses, 727–8
   toilette before, 261
   see also banquets; breakfasts; dinners; food; lunches; suppers
meals on wheels, 491, 493
means testing, 604, 606
measles, 508
meat
   boiling in animal stomachs, 421
   carving, 425
   cats, 581, 584
   farm house dinners, 366
   offerings in tombs, 583
   preparation, 422
   roasting, 420
   storage, 37
   tavern, alehouse, and coffee house food, 425
   transporting, 429–30
   see also animals; beef broth; beef tea
meat larders, 424, 430
medical clubs, 483
medical practitioners
   Chinese communities, 511
   see also doctors; nurses
medical recipes, books, 503
medical services, 483–5, 500, 511
   deaths and, 495
   ministers, 490, 502
   old people, 487, 488
   prisons, 661, 666, 667
   religious communities, 758
   travelling craftsmen, 633
   Travelling People, 802, 803
   see also bonesetters; doctors; geriatric medicine; health care; hospitals; midwives; National Health Service; nurses
Medici, Francesco de, 166
medicine, 300, 500–14
   see also complementary medicine; geriatric medicine
Mein, James, the Elder, 63
Melfort House, Glasgow, 553–4
Mellerstain, Berwickshire see Baillie, Lady Grizell
melodeons, 120, 125–6
melons, 406
Melrose, Roxburghshire, 227, 236, 322–3
Melville, Andrew, 558
men
   after-dinner activities, 426
   allotments, 415
   auxiliary work, 331
   care of infirm people, 489
   childbirth, 447–8, 449, 563–4
   cooks, 308, 427
   facial hair, 265–8
   farm servants, 364
   guising, 571
   hierarchy and authority, 515, 521, 522–8, 533
   Hogmanay fire smooring, 570
   homelessness, 856, 860
   knitting, 341
   needlework, 141
   personal appearance, 244, 248, 249, 250, 251, 252, 263, 265
   reading, 388, 392
   seasonal work, 618
   Shetland marriage customs, 567
   Show People's tasks, 811
   single-parent households, 477
   single-person households, 778
   suicides, 487
   weaving, 346, 347
   see also fathers; gender differences
Mendelsohn, Eric, 846

mending
　　clothing, 235, 769
　　crockery, 633
　　fishing nets, 768, 771, 773
　　household linen, 137
menservants, 308, 309–10, 360–1
　　clan chiefs', 363
　　clothing, 366
　　cooks, 427
　　interviews, 367
　　mangling work, 323
　　recruitment, 367
　　shoes, 366
　　taxes on, 309, 365
　　wages, 308, 361, 364, 367, 427
　　World War I, 359
　　see also butlers; farm servants; footmen
mental health, 449, 486, 487, 509–10, 558–9, 666, 667
mentally infirm people, boarding out, 486
Merchant Company of Hospitals or Schools, 686
Mercier, Louis-Sébastien, 298
mercury, 263, 264, 265
meridians (refreshment breaks), 425
merry meht, 449
metals, 315, 316, 325, 564
　　see also brass; iron; steel
metalwork, 184, 190, 191, 421, 788
　　see also iron industry; ironwork
metalworkers, 342
mews
　　conversions, 844
　　see also stables
mice, in art, 583
Michael, Fife, 283
microwave cookers, 434, 435, 646
middens, 83, 284, 285, 286, 299, 333, 587
　　see also manure
Middleton, Cecil H, 412
Midhowe, Orkney, 31

Midlothian
　　Arniston House, 84, 385
　　conservatories, 406
　　conversions, 839–40, 842
　　fairs, 423
　　farm house sanitation, 84, 300
　　hermitage, 835
　　houses, 837
　　New Age travellers, 812
　　Newhailes House, 62, 369
　　paraffin production, 96
　　Penicuik House, 181
　　pickies (hopscotch), 169
　　pit-head baths, 283
　　servants, 369
　　water tower conversion, 839–40
midwives, 442–7, 451
　　baptisms, 450
　　care of infirm people, 490
　　district nurses, 490
　　drug administration, 450
　　laying out of dead people, 495
　　postnatal care, 448–9
　　prayer and, 200
　　training, 440, 442
　　see also childbirth; howdies
migrant workers, 17, 518, 617–36, 762, 854
　　see also herring gutters; seasonal workers; Travelling People
migrants, children, 692
Milestone Hospice, 673
military accommodation, 17
　　see also armed forces
milk
　　children's food, 466–7
　　croft houses, 543–5
　　cryin bannocks, 449
　　Edinburgh supply, 430
　　farm house food, 366, 425
　　farm servants' food, 366, 424, 425
　　floorcloth cleaning, 317
　　housekeepers' duties, 364

　　selling, 331
　　shieling workers' food, 619
　　sieves, 618
　　skin care, 263, 264
　　Western Isles, 418, 420
　　see also breastfeeding; buttermilk
milk depots, 467
　　see also International Congress on Milk Depots
milk processing, 332, 618
　　see also dairy products
milking, 332, 364, 618, 619
milking cogs, 618
milking songs, 582–3
milkmaids, 333
　　see also dairy maids
Mill of Ireland, Stenness, Orkney, 841
Millais, Sir John Everett, 704
Millar, Euan, 841
Miller, Christian, 382–3, 384, 388, 389, 390, 530
Miller, Hugh, 380, 386, 387
millinery, 372
　　see also hats
millipedes, 508
mills, 75, 841–2, 845
　　see also corn mills; cotton mills; flax mills; fulling mills; spinning mills
miners, 282–3, 486
　　see also coal industry
Miners' Welfare Fund, 283, 296
mines
　　children's employment, 684
　　domestic medicine, 504, 507
　　Parys Copper Mine, 421
ministers
　　children and, 530
　　days of rest, 108
　　duties, 217
　　hospices, 680
　　libraries, 381, 382
　　medical services, 490, 502
　　servants, 357
　　status, 559

INDEX • 903

visits to parishioners, 206, 210–11, 216–18
see also chaplains; clergymen; Spring of the Minister
Ministry of Defence, 708
Ministry of Labour, 372
minorities *see* ethnic minorities
Minto, John, 394
mirrors
   bathrooms, 291
   cleaning, 310
   death customs, 568
   dressing rooms, 257–8
   fireplaces, 84, 86, 189
   in painting of Perthshire kitchen/living room, 78
   lady's closets, 62
   Ossian's Hall, 405
   personal grooming, 255, 257, 258, 259
   polishing, 317
   shaving, 267
missions, 214–16, 473, 623
Mitchell, John, 209
Mitchell, Joseph W, 834
Mitchison, Naomi, 460
mixed planting, gardens, 410
Moaney, Pat, 633
mobile homes, 818, 819–24, 831
   see also caravans
mobile phones, 472–3
model ships, 191
Molloy, Jake, 634
monasteries *see* religious communities
money, 340, 342, 345, 346, 527
   see also wages
Monimail commune, 738, 739, 744–5, 746
monkey-puzzle trees, 410
Monopoly, 166, 434, 471
Monteviot, Roxburghshire, 382
Montrose, Angus, 287, 348, 349, 542
Monymusk, Aberdeenshire, 345

Monzie House, Crieff, Perthshire, 60
Moonies, 737
mops, 311–12
Moray, 16, 230, 750, 834, 837
   see also Findhorn commune
Moray, earl of, 408
Moray Place, Edinburgh, 63
Morgan, Johnny, 628
Mormons, 218
morning rooms, 379, 382
Morris & Co, 184
Morris & Steedman, 843
Morris Report, 857
Morris, William, 181
Morrison, Angus, 625
mortgages, 597–8, 604, 606, 608–9, 610, 718, 720
Mortlake workshops, 178
Morton, Alexander, & Co, 184
Morton, William Scott, 180
Moryson, Fynes, 230
Mossetter, Orkney, 99
mothers, 466–7
   average age, 477, 519
   folk healers, 502
   free food, 443
   knitted garments for children, 139
   mortality rate, 458
   needlework teaching, 134
   personal appearance, 250, 251
   relationships with children, 530, 531–2, 537, 771–3
   rocking chairs, 116
   saining, 564
   see also childbirth; childcare; parent-child relationships; pregnancy; unmarried mothers
mother's helps, 475
Mount Stuart, Bute, 182, 836–7
mountain ash trees, 504
mountaineering *see* hill walking and mountaineering
Mountaineering Council of Scotland, 646

moustache cups, 266
moving house, 599, 600
Mowat, Ian, 393
Muchalls, Kincardineshire, 179
Muckle Flugga lighthouse, 834
Mugdrum Island, Newburgh, Fife, 624
mugs, 65
Muir, Molly, 444, 447, 466
Muirhouse, Edinburgh, 182
Mull, 97–8, 744
Mulrine, Stephen, 168
multi-storey buildings *see* tenements; tower blocks
multiple occupation *see* shared flats and houses
mummers' plays, 572–3
Mundell, Martha, 343
municipal baths *see* public baths
Munro, Mrs, 556
Munro, Rev George, 123
Munro Kerr, J, 441
Munro, Neil, 189
Munro, William, 58
Munroists, 641–2
murals, 182, 202
Mure, Elizabeth, 530
museums, 60, 61, 102, 152, 773
   see also National Museums of Scotland
music, 120–33
   after-dinner activities, 426, 433
   aristocratic households, 121, 361
   bothies, 125, 631
   children, 121, 471–2
   education, 121, 361, 471–2
   family activities, 535
   Fort Augustus Abbey, 759
   herring gutters, 623
   itinerant musicians, 633
   James I, king of Scotland, 164
   lyke wakes, 201

play and, 164
salmon netsmen, 625
Shetland, 124, 125, 567, 623
Travelling People, 127, 621
washboards, 321
see also bagpipe music; fiddle music; recorded music; rough music; songs
Muslim communities, 575, 664, 679
see also Islam
muslin clothing, 224
muslin flowering, 349
Musselburgh, Midlothian, 62, 369, 423
mussels, 418
mustard, 263
mutches, 229, 230
Muthesius, Hermann, 183
mutton broth, 366
mutual improvement societies, 377
Mylne's Court, Edinburgh, 715

N
NAAFIs, 632
nail care, 254, 256, 262
nailers, 347, 349
Nairn, 229, 231
Nairne, Lady Carolina, 122
Nairnshire, 16, 55, 229, 231
Naismith, Robert, 360
names
  dwellings, 552
  in hospices, 679
  on storage utensils, 65–6
  streets, 552–3
nannies, 308, 474–5
Napier Commission, 639
nappies, 464–5
narrowboat conversions, 841
National Assistance Board, 855
National Childbirth Trust, 451
National Cyclists' Union (NCU), 647
National Domestic Servants' Union, 373

National Health Service
  domestic medicine and, 503, 511
  home care, 493
  homelessness and, 856
  hospital service, 483
  nursing associations replaced by, 491
  old people's care, 488–9
  paying for, 484
National Induction Centre, 656
National Museums of Scotland, 57, 61, 77–8, 142, 223
National Trust for Scotland, 411
National Union of Allotment Holders, 413
National Year of Reading, 388
natural lighting, 77–81
  see also windows
natural remedies *see* complementary medicine
nausea, 504
navvies, 383, 391
neck cloths, 231, 236
needlework, 134–43, 192, 233
  after dinner activity, 426
  aristocratic households, 361
  education, 134, 135–6, 343, 372
  forbidden on Sabbath, 199
  lighting for, 76, 84
  seating for, 118
  women's employment, 136–7, 138, 339, 343, 352, 353
  work at home, 137, 138, 142, 352
  see also dressmaking; embroidery; knitting; mending; sack sewing
neighbourhood-watch schemes, 33
neighbours
  bereavement help, 260, 447

care of infirm people, 26, 324, 483, 490
childbirth help, 446, 447, 448
childcare help, 325, 490
nervous complaints *see* mental health
NetGuides, 473
Nether Benzieclett, Orkney, 99
nets, 350, 768, 771, 773
netting stations, 624
nettles, 508
neuk beds *see* corner beds
New Age spiritualities, 218, 747, 750
New Age travellers, 812–13, 831
New Lanark, 349
New Pitsligo, Aberdeenshire, 350
New Testament *see* Bible
new towns, 18, 35, 547
New Year, 261, 434, 567, 569, 571, 664
  see also Hogmanay
Newbigging, Glen Clova, Angus, 645–6, 648
Newburgh, Fife, 624, 625, 626
Newhailes House, Musselburgh, Midlothian, 62, 369
Newmills, East Lothian, 339
newspapers
  as fuel, 88
  childbirth uses, 445, 446–7
  Dollar Academy, 707
  reading, 378, 386, 388, 391, 392, 631, 664
  sports reporting, 169
  Stamp Duty, 377
NHS *see* National Health Service
NHS Direct, 511
Niddrie Fireclay Works, 295
night caps, 230
night soil, 286, 287–8, 587
Niles, John, 127
Nimlin, Jock, 643
Nissen huts, 17
Nissen, Lt Col Peter, 17

INDEX • 905

NMS (National Museums of
    Scotland), 57, 61,
    77–8, 142, 223
Noise Abatement Society, 126
noise nuisance, 126
Noranside, Angus, 657
Norie, James, 180
Norse houses *see* Viking
    houses
North Berwick, East
    Lothian, 648
North Knapdale, Argyll, 75
North Queensferry, Fife,
    840–1
North Ronaldsay, Orkney,
    100
North Uist, 25, 230, 231, 837
North Woodside Mission,
    Glasgow, 473
Northern Isles, 56, 73,
    115–16, 334, 403,
    502–3
    see also Orkney;
    Shetland
Notre Dame College,
    Glasgow, 711
novels, 392–3
nuclear families, 477
nuncheon, 425
    see also lunches
nunneries, 752
Nunraw monastery,
    Haddington, East
    Lothian, 750
nuns, 750
nursery gardens, 401, 402
nursery maids, 308, 474
nursery meals, 432
nurserymen, 409
nurses
    care of old people, 488,
        490, 494
    childcare, 361, 432, 530
    hospices, 672–3, 675–6,
        680
    laying out of dead
        people, 495
    see also home care;
        midwives; nannies;
        wet nurses
nursing associations, 490–1
nursing homes, 496, 673–4,
    675, 676, 677, 679,
    680–2

nursing mothers, 466–7
    see also wet nurses
nut gardens, 402
nutrition *see* food
nuts, 574

O

oat straw, 466
oatcakes, 344, 366, 425
oatmeal, 263, 403, 418, 424,
    449, 619, 631
    see also haggis;
        porridge
oats, 333
Oban High School, Argyll,
    709, 710
obesity, 487
Ochiltree, Ayrshire, 226, 426
offal, 422
offenders *see* prisoners
offices (religious services),
    752, 754, 756–7, 759
offices (workplaces) *see*
    home offices;
    outbuildings;
    servant registry
    offices
oil-fired heating, 841
oil lamps, 73–5, 138, 318,
    828
    see also paraffin lamps
oil of vitriol, 263
oil-rig workers, 628–9, 821
oilcloths *see* floorcloths
oils
    beauty preparations,
        262, 265
    cleaning substances,
        314, 317
    fuels, 74, 75, 325
    soap making, 290
    see also paraffin
Old Cumnock, Ayrshire, 228
old people
    care of, 373, 487–9, 490,
        494, 519, 681, 855
    church representatives'
        visits, 217
    computer use, 154, 159
    diseases, 487
    employment, 341, 350
    extended families,
        519–20
    free care, 491, 681

health visitors, 488, 491,
    494
hierarchy and authority,
    515, 519
medical services, 487,
    488
personal appearance,
    245, 250–1
pets, 593
poorhouses, 468, 488,
    684
religious communities,
    758
security, 24, 26–7
sexual attractiveness,
    245
single-person
    households, 777,
    778–86
television watching,
    148–9, 159
    see also home care;
        infirm people;
        nursing homes;
        sheltered housing
Old School, Dean Village,
    Edinburgh, 844
Old Tolbooth, Edinburgh,
    656
Oldwood Croft,
    Aberdeenshire, 100
one-parent families *see*
    single-parent
    families
one-person households *see*
    single-person
    households
onions, 415, 506, 510
online shopping, 435
open fires *see* fires
open prisons, 657, 658
open razors, 266, 267–8
oratories, 202
orchards, 401, 402, 405, 409,
    627, 745
    see also fruit trees
Ordlay Croft,
    Aberdeenshire, 100
Orkney
    animals, 582, 583, 586
    auxiliary work, 332,
        333, 336, 346
    beds, 99, 100, 101, 564,
        582

906 • SCOTTISH LIFE AND SOCIETY

brochs, 31
childbirth, 564
converted mill, 841
fishing communities, 766
flax spinning, 345, 349
houses, 27, 582, 841
kists, 55
linen making, 350
mattresses, 101
peat-burning hearths, 89
poultry, 331, 582
prehistoric burial sites, 38, 583
prehistoric woollen hood, 223
St Magnus Cathedral, 177
shoemaking, 231
Skara Brae, 38, 55, 100, 555
souterrains, 37
stove, 96
straw work, 349–50
see also Northern Isles
ornaments, 174, 175, 185–90
  bathrooms, 290
  croft houses, 191
  fireplaces, 85–7
  storage and display, 188, 189, 380, 556
  see also ceramic ornaments; decoration
Orphan Homes of Scotland, 689
Orphan Hospital, Edinburgh, 58
orphanages, 685, 689, 692
orphans, 475–6, 518, 701
  see also children's homes
Orr, Willie, 638
Orthodox Christian communities, 218
orthopaedic beds, 113
Orwell, George, 703
Ossian's Hall, 105, 835
Oswald, James, 122
outbuildings
  cruisie lamps, 75
  farm houses and, 16
  sleeping accommodation, 628
  status, 549

toilets, 284, 294, 479
villas, 553
see also barns; bothies; byres; farm steadings; mills; stables; storerooms
outshots, 100
oven-to-table ware, 428, 434–5
ovens, 83, 420
  firewood and cats in, 588
  hot walls, 406
  kitchen ranges, 94, 95, 427
  paraffin-fired, 96
  see also microwave cookers
overcrowding, 115, 285, 419, 485, 600, 728, 797–8
  see also privacy; single ends
overmantels, 179, 187
  see also mantelpieces
ox gall, 314

P
packmen see hawkers and peddlers
pails, 299, 307, 311, 315–16, 317, 318, 333
painted panels, East Lothian, 180
painters, 369, 373
painting, 176, 177–8, 183, 318
  china, 191
  hanging shelves, 61
  kists, 55–6, 57
paintings, 77–8, 189, 202, 203, 585
Paisley, Renfrewshire, 64, 285, 295, 348
  see also Tannahill, Robert
palm trees, 410
pans see cooking utensils
pantries, 110, 316, 424, 430
pantry boys, 361
Paolozzi, Eduardo, 180
paper bag making, 352
Paper Duty, 377
paraffin-fired appliances, 96, 323, 325, 795

paraffin lamps, 69, 70, 385, 461
  candles and, 76
  cruisie lamps and, 75
  Show People's caravans, 809
  youth hostel, 645
  see also oil lamps
paraffin production, 96
Paré, Ambroise, 510
parent-child relationships, 526, 528–37, 557, 771–3
  see also discipline
Park End, Bridge Street, Penicuik, Midlothian, 837
parlour maids, 432
parlours, 110
  after dinners, 426
  book storage, 379
  croft houses, 189, 191
  dining rooms and, 186, 426
  drawing rooms and, 186
  house plants, 190
  music, 122
  status, 556
  tenements, 186
  see also best rooms
parrots, 588–9
parsley, 400
parsnips, 400, 403, 415
parterres, 400, 402, 403, 408, 410
Partick, Glasgow, United Secession Church, 211
Parys Copper Mine, Anglesey, 421
pasta storage, 66–7
patches, smallpox scar concealment, 264
patchwork, 141
Paterson, George, 835
Patey, Dr Tom, 644
Pathhead, Fife, 347
Paton, Walter Hugh, 186
Patons and Baldwins, 192
Paxton, Berwickshire, 837
Paxton, Margaret, 236
peaches, 406
pear trees, 400, 401, 407, 408
peas, 403, 415

INDEX • 907

pease, 366
peat, skin care, 262
peat fires, 89–91, 311, 420
peat hooses, 99
peats, 85, 87, 88, 89–91
    seasonal workers, 617
    storage, 64, 89
    temporary lighting, 71
    transporting, 333
    work of farm servants' wives, 525
Peeblesshire, 91, 225, 232, 650
Peel Report, 451
peels (baking equipment), 420
peenies, 235
peermen, 72–3, 75
peevers, 169
peggy sticks, 321
pen wipes, 140
Penicuik House, Midlothian, 181
Penicuik, Midlothian, 837, 842
Penkill Castle, Ayrshire, 182
Pennant, Thomas, 201, 504, 582
penny wabble, 350
People's Story Oral History Archive, 102
People's Temple commune, 737
Pepys, Samuel, 266
perfumers, 254
periodicals *see* magazines
permaculture, 738
Perrot, Michelle, 174–5
personal appearance, 243–52
    children, 248, 261
    decency, 246–8, 265
    dental care, 249, 254, 256
    dieting groups, 254
    fashion, 252, 265
    fitness, 254
    girls, 248, 260, 557
    men, 244, 248, 249, 250, 251, 252, 263, 265
    nail care, 254, 256, 262
    sports clubs, 254
    status and, 246, 247–8, 251–2, 256–7, 265
    tanning centres, 254
    weddings, 247, 261, 278

    women, 243–5, 246, 249–50, 251–2, 259, 263, 264, 265
    see also beauty; clothing; hair; jewellery; skin care; toilette
personal care, old and infirm people, 491, 493, 496, 681
personal grooming *see* toilette
personal hygiene, 243, 277–306
    basins, 257, 259, 279, 469
    see also wash-hand basins
    bathrooms, 258, 260, 277, 280, 282, 288–98
    bedrooms, 257
    childcare, 468–9
    infirm people, 491, 493
    prayer and, 200
    prisons, 654
    washing, 257, 266, 277, 278–9
    see also bathing; skin care
personal libraries *see* libraries
personal security, 23–8
personalised storage utensils, 66
Perth
    bagnio, 281
    candle works, 75–6
    dog skeleton, 584
    fair, 806
    flooding hazard, 12
    games, 165, 167, 585
    High Street, 584, 585
    prison, 656
    salmon distribution, 626, 627
    status, 542
    toilet in outbuildings, 284
Perthshire
    Bealtainn fire, 571
    camping site, 621
    craftsmen, 351–2
    dressers, 60, 61
    farm servants' food, 424

fir candles, 72
flax spinning, 341
fruit picking, 620, 621
fuel, 88
gardens, 404, 410
healing waters, 507, 509–10
Hermitage Folly, 405, 835
hermitages, 835
inns, 637–8
kitchen presses, 58–9
kitchen/living room, 77–8
lairds' houses, 12
meat in tombs, 583
monastery, 750
Monzie House, 60
music, 121
open prison, 657
Queen Victoria School, 708
reforestation, 404
rural crafts, 351–2
salmon fishing, 624, 641
shoemaking, 340
souterrains, 37
stables, 837
Stanley Mills, 841–2
symbol stones, 255
Taymouth Castle, 12
weaving, 77, 351
windows, 77–8, 79
Peter, Alexander, 58
Peterhead prison, 656
pets, 581–2, 583, 584, 585, 587–93, 679
    see also cats; dogs
Petty, Inverness, 507
pewter ware, 308, 315, 362, 427
pharmacies, 255, 262
Philip II, king of Spain, 166
Philip Cocker & Partners, 842
Phillips, Alistair, 212–13
Phoenix Foundry, 287
physical size, status and, 557–8
physicians *see* doctors
piano stools, 116
pickies, 169
Pictish stones, 255, 583
pictures, 189
    see also art; paintings

pieces *see* christening pieces; jeely pieces
pies, 425
pigs (animals), 581, 582, 583
pigs (earthenware), 65, 83
Pigsty, The, Robin Hood's Bay, Yorkshire, 839
pilers, 470–1
Pilkington, Frederick Thomas, 837
pinafores, 235, 462
pine needles, 262
Pineapple, the, Dunmore, Stirlingshire, 406, 844–5
pineapples, 406
pipe music, 121, 129, 201, 363, 619
pipeclay *see* camstone
piped water supplies, 279, 284, 285, 287
  housework, 307, 318, 325, 427
  huts, 828, 829
  mobile homes, 822
  religious communities, 752, 756
  shared, 715
  Travelling People's camping sites, 801, 803
Piper Alpha disaster, 634
Pirie & Clyne, 837
pirnies, 230
pit-head baths, 283, 296
Pitcur, Perthshire, 37
Pittenweem, Fife, 766, 768, 770, 772
place mats, 432
plaid production, Aberdeenshire, 341
plaids, 114, 225, 227–9, 363, 366, 559
  see also Highland dress
plain sewing, 134, 135, 136, 137, 141
plane trees, 402
plantiecrues, 403
plants *see* complementary medicine; gardens and gardening; herbal preparations; house plants
plaster, dung as, 91

plasterers, 369
plasters, 505, 508
plasterwork, 179, 311
plates *see* crockery; washing up
play, 163–73, 199, 249, 469–73, 475, 535
  see also games; leisure activities; sport; toys (playthings)
play areas, 707, 741
Playfair, W H, 180
playing cards *see* card games
plum trees, 401
plumbing, 291, 296
  see also drainage; piped water supplies; sanitation
Pluscarden monastery, Elgin, Moray, 750
Plymouth Brethren, 218
pneumoconiosis, 486
poaching, 543, 641
poets *see* bards
polished stones, 320, 322
polishing, 310, 313, 314, 315, 316, 317, 325
polishing irons, 323
Polkemmet, West Lothian, 283
Poll-ewe, Ross-shire, 633
Polmont, Stirlingshire, 657
Polwarth, Lord and Lady, 489
ponds, 403
poor relief, 362, 368, 476, 488, 684–5, 852
  see also boarding out; charity; children's homes; homeless people; poorhouses; social security system; workhouses
poorhouses, 468, 486, 488, 684, 697, 854
  see also workhouses
Pope, Alexander, 404
porches, 16, 551, 552, 554, 808
porridge, 366, 421, 424, 468, 619, 631
portable bathing facilities, 283–4

Portobello, 65, 415
Portpatrick, Wigtownshire, 168
Portree High School, Skye, 709
possers, 321
postnatal care, 440, 448–9
potato picking, 618, 620
potato water, 321
potatoes
  allotments, 415
  domestic medicine, 506, 508
  farm house meals, 366
  food, 418, 419
  gardens, 403, 407
  shieling workers' food, 619
  storage, 58
pots
  food storage, 64
  see also cooking utensils; crogain
pottage, 366
potteries, 65, 342, 556
poufs (seating), 116
poultices, 504, 505
poultry, 331, 422, 425, 582
  see also cock fighting; geese; hens
powder *see* face powder; hair powder; soap powders
powder closets, 257
prams, 465–6
prayers
  Catholic worship and rituals, 200, 201–2, 205, 208–9, 213, 216
  Iona Community, 214
  Protestant worship and rituals, 206, 209, 211–13, 217
  Scottish Episcopal Church, 211
  see also graces (prayers); rosaries
prefabricated houses, 17, 818, 856
pregnancy, 361–2, 368, 460, 477, 857, 858
  see also antenatal care; childbirth; postnatal care

INDEX • 909

Premonstratensians, 750
preservation *see*
    conversions; food
    preservation
presses (cupboards), 58–61, 100, 238
pressing *see* ironing; smoothing
Preston Gaol, Lancashire, 688
Prestonfield, Edinburgh, 837
Prestonpans, East Lothian, 423
printer tables, 383
'priories', 835–6
prison cells, 654, 660, 661, 664
prison chaplains, 217–18, 666, 668, 688
Prison Visiting Committee, 670
prisoners
    children, 688–9
    classification, 657–8, 660
    clothing, 664
    dental care, 667
    education, 666–7
    employment, 666
    exercise, 666, 667
    food, Glenesk Hall, Saughton Prison, 661
    hair, 268
    leisure activities, 661, 664, 666, 667
    letters, 669
    mental health care, 666, 667
    reading, books, 664
    responsible prisoner, 659
    wages, 664, 666
    see also remand prisoners
prisons, 296, 650, 654–71, 688, 716
    see also tolbooths
privacy, 42–54
    allotments, 415
    bathing, 280, 468
    bathrooms, 289, 290, 716
    children, 280, 694
    children's homes, 694

crowded spaces, 461, 567
curtained beds, 111
employers of servants, 256–7, 309
gardens and gardening, 193, 249
hospices, 676, 680, 681
independent schools, 704
lighting and, 256
mobile homes, 824
oil-rig workers, 629
personal grooming, 253, 256–7
public baths, 468
reading, 382–3, 384–5, 393
rural toilet arrangements, 299
shared flats and houses, 714, 715, 716, 717, 721, 726–7, 730
single ends, 567
smoke and, 256
tenements, 42–54, 567
Travelling People's caravans, 797, 798
women, 42–54, 185
    see also overcrowding
private mortgages, 598
privately rented houses *see* rented houses
privies, 284, 286, 287, 294, 298
    see also lavatories; WCs
Proctor & Gamble, 291
property ownership, 42–54, 344, 542
    see also home ownership
protective charms and customs, 29, 38, 450, 501, 504–5, 564–5, 570–2
    see also birth customs; death customs
protective clothing, 235
Protestantism, 202, 205–8, 209, 529
Provost Skene's House, Aberdeen, 202–3
psalm books, 206
psalm singing, 208

psychiatric disorders *see* mental health
public baths, 280–4, 287, 468
public libraries, 378, 391
public schools, 702
public washhouses *see* washhouses
puerperal fever, 440, 441
pullin the swingletree, 631
pullishees, 322
puppies *see* dogs
purgatives, 505
purses, 140
Purves, Andrew, 378, 389, 391, 392
pushcarts, 465
pushchairs, 466
pyes, 471

Q
Quanterness, Orkney, 583
Quarrier, William, 689
Quarrier's Homes, 689, 708
quarterers, 633
quarters, Celtic year, 569
Queen Victoria School, Dunblane, Perthshire, 708
Queen Victoria's Jubilee Institute, 490
Queen's Park, Edinburgh, 322
querns, 200
quilts, 141, 191
quoits, 631

R
race games, 166, 170
    see also horse races
Rachel House, Kinross, 673
radio, 146, 433, 471
    aids for deaf people, 494
    gardening programmes, 412
    gender differences in audiences, 129
    music, 120–1
    prisons, 654, 661, 664
    reading and, 387, 388, 391
    religious programmes, 214–15, 216
    see also BBC Radio Scotland

910 • SCOTTISH LIFE AND SOCIETY

Radway, Janice, 392
raffia, 191
Rainbows, 473
rainwater, 263
rainwater cisterns, 284
Rajneeshees, 737
rambling, 647
Ramsay, Miss, 280
Ramsay, Edward Bannerman, dean of Edinburgh, 211
Ramsay Garden, Edinburgh, 182
Ramsden, Sir John W, 640
Rangers (scouting movement), 473
ranges *see* kitchen ranges
Rannoch, 570–1
rantle trees, 86
rape, 524
raspberry picking, 620
Ratho, West Lothian, 841
rats, 76, 586, 587, 591, 624
Raven, James, 382, 394
Ray, John, 362
razors, 256, 266, 267–8
reading, 71, 377–99
    bothymen, 631
    children, 388, 389–90, 392, 473
    farm houses, 73, 378, 389
    Fort Augustus Abbey, 759
    lighting for, 71, 72–3, 76, 80, 382, 385–7, 664
    newspapers, 378, 386, 388, 391, 392, 631, 664
    prisoners, 664
    private time and, 49
    religious communities, 752
    seating for, 118, 382
    time for, 349
    see also Bible reading; books; literacy; study
Reading Experience Database, 394
reading groups, 391
readymade clothing, 233–4
real tennis, 167–8
reception centres, homeless people, 855

recipes
    beauty preparations, 260–1, 262
    medicines, 502, 503
    see also cookery books
recorded music, 121, 126, 128–30, 380, 472
    see also gramophones
recreation *see* leisure activities
recruitment, servants, 309, 364, 366–7, 369–70
red deer, 638
red grouse, 638
Red House, Bexleyheath, Kent, 181
Red Stone, Ardvorlich, Dunbartonshire, 504
redcurrants, 407
refectories *see* dining rooms
reforestation, 406, 744
Reformation, the, 205, 419, 458, 523, 529, 750
reformatory schools, 688–9
refrigerators, 427, 430, 434, 435
    see also icehouses
Reid, John, 404, 410, 421
rèiteach beag and rèiteach mór, 565–6
religions
    dress codes, 224
    hierarchy and authority and, 529
    personal appearance and, 246–7
    personal security and, 25
    see also Christianity; Hinduism; Islam; Judaism; offices (religious services); Sikhism; spirituality
religious broadcasting, 214–15, 216
religious communities, 749–61
    hairdressing, 259, 268
    hospices as, 674–5, 679, 680
    kitchens, 96
    personal hygiene, 277, 278
    sanitation, 277, 752, 756

sports, 167
water supplies, 277, 752
religious festivals, 109
    see also Christmas; Easter
religious music, 123–4
religious observance, 164, 463, 664, 668, 693, 707, 708
    see also Bible reading; church attendance; family worship; Sabbath, Jewish; Sunday observance
religious pictures, 202, 203, 213
religious revivals
    fishing communities, 213
    see also evangelical Christianity
religious texts
    reading, 391–2
    see also Bible
remand prisoners, 656, 657, 658, 659, 660, 669
remedies *see* domestic medicine
Renfrewshire
    coal bunkers, 64
    four-in-a-block villas, 549
    homelessness, 861
    hospice, 672
    prison, 656
    Quarrier's Homes, 689, 708
    sanitary ware production, 295
    temporary lighting, 71
    water supply, 285, 287
    weaving, 348
Rennie, John, 834
rent day, 108
rented houses, 597, 598–608, 610–11, 719, 720
    cottages, 546
    fishing communities, 768
    gardens and gardening, 546, 547
    maintenance obligations, 317
    migrant workers, 617

INDEX • 911

mobile homes, 823
Show People, 806
student accommodation, 718
see also council houses; farm servants, houses; shared flats and houses; tenancies; tenements
rents
houses, 602–3, 604, 605–7, 609, 610
Travelling People's camping sites, 802, 803
reputation, status and, 544
rescue movement, childcare, 689, 692–3
residential homes, 217, 496
see also children's homes; nursing homes
residential schools, 532, 701–13, 754, 757, 758, 759, 760
respite care, 492
rest, 108–19
see also sleeping accommodation
Retreat, The, Abbey St Bathans, Berwickshire, 837
Retreat, The, Glen Esk, Angus, 61
revivals *see* religious revivals
rheumatic complaints, 508
rhododendrons, 410
rhubarb, 415
rhymes and songs, children's, 120, 471
Rice, Marjorie Spring, 441
rice water, 320
Richards, Frank, 702
ridicule, status and, 558
Riga bowls, 191
right to buy, council houses, 608
rings, eye infection treatment, 506

rites of passage, 250, 557, 562–9, 576, 714
see also birth customs; death customs; marriage customs
rivlins, 230
roasting, 420, 427, 585
Robbins Report, 711
Robert I, king of Scots, 299
Robertson, Rev A E, 641
Robertson, Helen, 349
Robertson, Jeannie, 127
Robertson, John, 707, 712
Robertson, Stanley, 127
Robin's House, Balloch, 673
Robinson, Neil, 841
Robinson, Thomas, 94
rocking chairs, 116, 189
Rockville Lodge, Balgone, East Lothian, 837
Rodel, Harris, 586
rodents *see* mice; rats
Rodger, William, & Son, 58
role-playing games, 171, 470
see also guising
Rolland, Margaret, 362
Roman baths, 277, 280
Roman Catholicism *see* Catholicism
Roman gods, 198
romance fiction, 392–3
roof openings, 79
see also chimneys; smoke extraction
roof spaces, storage in, 64
roofed beds, 111, 112
roofs, 9–10, 15, 625
see also thatched roofs
room-and-kitchen accommodation, 96, 101, 112, 122, 587
rooms, letting, 343
see also inns; lodgings; shared flats and houses
Rooney, Charles (Chic), 587, 588, 589, 591
rope stools, 116
ropes and rope-making, 332, 334–5, 336, 350
Rorie, David, 263, 443, 503
rosaries, 202–3, 213
rose (erysipelas), 300
Rose, Jonathan, 391, 394

Rosebery, Countess of, 77
Ross and Cromarty, 410, 509, 642, 645, 646, 710
see also Cromarty; Ross-shire
Ross, bishop of, 166–7
Ross, Willie, 129
Ross-shire
clothing, 226, 231
curtains, 78
dresser and sideboard, 556
Fearn Airfield Control Tower, 845
gas lighting, 76
hill walking and mountaineering, 642, 646
inn beds, 103, 637
library, 393
salmon fishing, 641
shoemaking, 231
story telling, 633
toll house, 834
windows and ventilation, 79–80
women's head coverings, 559
see also Ross and Cromarty
Rosskeen, Ross-shire, 559
rotstone, 314, 315, 317
rouge, 264
rough music, 524
Rough Sleepers' Initiative, 862
round-about firesides, 91
Round Square, Gordonstoun, Moray, 837
roundhouses, 31
see also brochs; wheelhouses
rowans, 504
Rowling, J K, 473
Rows, The, Great Yarmouth, 623
Roxburgh, Lord, 362
Roxburghe House, 672
Roxburghshire
Abbotsford, 84, 180, 183, 384, 573
cat-in-soot game, 585

clothing, 225, 226, 227
fireplace, 84
friary, 752
inn beds, 102
libraries, 382, 384
linen smoothing, 322–3
role-playing games, 171
round-about firesides, 91
Sunday clothing, 227
washdays, 236
Royal Blind School, 708–9
Royal College of Physicians, 281
Royal Commission on Long-Term Care of the Elderly, 491
Royal Commission on the Housing of the Industrial Population of Scotland, 598, 602–3
Royal Commission on the Poor Laws in Scotland, 852
Royal Highland Show, 194
Royal National Institute for the Deaf, 494
Royal Society, 382
rubbers (brushes), 311, 312
Rudolph Steiner schools, 708, 747
ruffies, 71
rugs, 184, 191
rullions, 230
Runciman, Alexander, 181
rushes, 71, 74, 75, 114, 138, 336
Russian baths, 280, 282
Russian style dinners, 430
Rutherglen, public baths, 281
Rutter, J O N, 385

S
Sabbath
    Christian see Sunday
    Jewish, 575
sack sewing, 352, 353
sad irons, 323
safe rooms, 38
safety razors, 268
sail lofts, 791
saining, 564

St Andrews, Fife, 58, 648
St Andrews College of Education, Glasgow, 711
St Anthony, 749
St Augustine, 749
    see also Augustinian order
St Benedict, 749, 750, 757
    see also Benedictine order
St Brendan, 200
St Bride, 200
St Bride's Day, 569
St Christopher's Hospice, London, 672, 674
St Columba, 750
St Columba's Hospice, Edinburgh, 672, 674–5
St Dominic, 202
St Fillan's pool and chapel, Strathfillan, Perthshire, 509–10
St Kilda, 15, 98–9, 112, 448, 529
St Madoes, Carse of Gowrie, 587
St Maelrubha, 509
St Magnus Cathedral, Orkney, 177
St Monans, Fife, 766
St Ninians, Stirlingshire, 347, 349
St Pachomius, 749
St Rollox Works, Glasgow, 290
St Thomas Aquinas, 584
saithe, 619
salmon distribution, 626–7
salmon fishing, 624, 641
salmon netsmen, 624–7
salt, 424, 574, 623
salt industry, 339–40
Saltcoats, Ayrshire, 340
salted herring, 619
Saltoun, East Lothian, 842
Saltoun Hall, East Lothian, 364
Salvation Army, 854
Samhainn, 569
samplers, 76, 134, 190, 235
sand, 314–15, 317
Sanderson, Arthur, 180

Sandison, Chrissie, 447, 461–2, 490
sanitary ware, 291–8
    see also baths; bidets; showers; wash-hand basins; WCs
sanitation, 277, 284
    ash pits, 84, 300
    byres, 299
    croft houses, 84, 300
    diseases and, 286, 287
    Edinburgh, 285, 286, 318
    farm houses, 84, 103, 298–301
    farm servants' houses, 299–300
    Glasgow, 285, 286, 287, 297, 298
    Highlands, 102
    houseboats, 841
    huts, 829
    infirm people, 493
    overcrowding, 285
    prisons, 654, 660, 661, 664
    religious communities, 277, 752, 756
    rural areas, 298–301
    shared, 715
    Show People's caravans, 810
    tenements, 284, 286, 288, 297, 546, 587
    Travelling People's camping sites, 801, 803
    Travelling People's caravans, 798
    see also chamber pots; commodes; latrines; lavatories; privies; sewerage; WCs
saponification, 290
sardines (game), 471
Saturday working, 109, 387
saucepans see cooking utensils
sauch yards, 400
Saughton Prison, Edinburgh, 656, 660–1
saunas, 280
Saunders, Dame Cicely, 672, 674

INDEX • 913

sausages, 631
Savoch, Aberdeenshire, 168
Scalpay, salmon netsmen, 626
scalps, shaving, 268
scar concealment, 264
Schama, Simon, 703
school building conversions, 844
schoolrooms, 384
schools, 473–4, 475
   art, 191
   Barnardo's homes, 708
   Benbecula, 711
   blind people, 708–9
   boarded-out children, 476
   cookery, 371, 372
   deaf people, 709
   dinners, 467, 468
   discipline, 519, 533
   Edinburgh, 686, 701, 708, 709, 844
   Edinburgh Ragged School, 686
   Fort Augustus Abbey, 754, 757, 758, 759, 760
   London Board Schools, 297
   Maybole Ragged School, 688, 692
   music, 471–2
   portable bathing facilities, 283–4
   Quarrier's Homes, 708
   Queen Victoria School, 708
   servants' training, 370–1
   status and, 545
   workhouse children, 688
   see also Approved Schools; homework; independent schools; Merchant Company of Hospitals or Schools; public schools; reformatory schools; residential schools; secondary schools; Steiner schools; teachers

Schottenklöster, 750
sciatica, 508
Scotch chests, 57, 63–4, 239
   see also kists
Scotland, Mr, 626
Scotland's Garden scheme, 411
Scots pine trees, 404
Scott, Lady Alicia, 122
Scott, Walter (Edinburgh goldsmith), 166
Scott, Sir Walter, 183, 230, 393, 637, 639
   see also Abbotsford
Scott, William Bell, 182
Scott, Willie, 125
Scottish Allotment and Garden Society, 413–14
Scottish Arts Council, 378, 387
Scottish Board of Health, 485
Scottish Civic Trust, 839
Scottish Co-operative Wholesale Society, 184, 290, 291
Scottish Council of Independent Schools, 701
Scottish Development Agency, 194
Scottish Episcopal Church, 211, 529, 569, 843
Scottish Executive
   broadband, 387–8
   homelessness, 859, 862
   listed building conversions, 839
   maternity services, 452
   old and infirm people, 491, 493
   social inclusion of pupils, 709
Scottish Exhibition Conference Centre, Glasgow, 806
Scottish Farm-Servants' Union, 299–300
Scottish Health Management Efficiency Group, 452
Scottish Home and Health Department, 451

Scottish Home Industries Association, 77
Scottish Household Survey, 518
Scottish Ideal Homes Exhibition, 194
Scottish Institute for Residential Childcare, 695
Scottish Mountaineering Club, 641, 642
Scottish Office, 598
Scottish Prison Service, 656, 657, 658–9, 668, 670, 671
Scottish Society for Autism, 496
Scottish Traveller Education Programme, 812
Scottish Youth Hostels Association, 644, 646, 647
scouring see cleaning
scouting movement, 473
scrofula, 502
Scroggie, Jack, 645
Scroggie, Sydney, 643, 645–6, 648
sculleries, 179, 299, 553, 556
scullery maids, 308, 427
scutching, 342
Seafield, Lord, 639
seamen, 56, 618, 853
   see also fishermen
seamstresses, 233, 344
   see also dressmaking; needlework
seaside resorts, 648–50
seasonal workers, 518, 617–27, 715
   see also herring gutters; migrant workers; shielings
seasons, quarter days, 569
seating, 115–18
   bedrooms, 111, 112
   beds as, 112, 115
   bothies, 631
   canvas work, 140
   farm houses, 91
   hospitality and, 554
   kists as, 56, 110, 111, 116, 631
   reading, 118, 382

round-about firesides, 91
settles, 91, 99, 116
status and, 86, 115, 116–18, 566, 631, 714
straw, 332–3, 334
workers, 109
youth hostels, 644, 645
see also benches; chairs; sofas; stools; window seats
Seaton, Margaret, 523
Seaton Pottery, Aberdeen, 65, 556
seawater, 281
seaweed, 262, 506–7
seaweed mattresses, 101
secondary schools, 709–11
secret compartments and rooms, 38
security, 23–41
  children, 14, 24, 27
  hospices, 676
  hostels for homeless, 861
  infirm people, 494
  residential schools, 707, 709
  shared flats and houses, 725–6, 730
  see also crime; social security system
security categories, prisoners, 658
seeing coal, 72
Segal building, Monimail commune, 745
Segalen, Martine, 765
self-built dwellings, 818–32
  see also blackhouses
Selkirkshire, 91, 644
selling
  ale, 344, 350
  books, 343, 627, 628, 741
  coal, 341
  dairy products, 331, 341
  fish, 341, 352, 429–30, 621
  from homes, 343
  poultry, 331
  urine, 288
  see also fairs; hawkers and peddlers; markets; shops

separation, rites of, 563–4, 565, 576
sequestration, 599
sermons, 206
servant registry offices, 367, 370, 373
servants, 308–10, 357–76, 427, 474, 536
  bathroom use, 289
  care of infirm people, 482, 489
  chaperoning of girls, 535
  childcare, 373, 432, 475, 531
  children as, 460, 528
  clothing, 223, 225, 226, 251, 366
  country houses, 55, 369
  Culzean Castle, 226, 364, 365, 367, 369
  dinner duties, 425, 426–7, 431, 434
  employers' privacy, 256–7, 309
  employers' toilettes, 256–7, 259–60
  family households, 519
  food, 418
  holidays, 373, 423
  kists, 55, 238
  personal appearance, 251
  seating, 118
  security, 26
  shaving by, 267
  sleeping accommodation, 365–6, 556, 715, 716
  taxes on, 309, 365
  toilets, 553
  wages, 308, 361, 364, 365, 367, 373, 427
  water carrying, 278
  Western Isles, 363
  see also farm servants; menservants; women servants
servants' quarters
  beds, 111
  clothes storage, 238
  flats, 546
  kists, 55

status, 546, 548, 551
villas, 554, 556
see also bothies
Seton, East Lothian, 835
settles (seating), 91, 99, 116
Seventh Day Adventists, 218
sewerage, 277, 284–5, 287–8, 294, 298
  Edinburgh, 318
  houseboats, 841
  tenements, 286, 288
sewing *see* needlework
sexual attractiveness, 245–6, 247, 248, 249
sexual intercourse, 567
Shakers, 737
shallots, 415
Shanks (sanitary ware manufacturer), 295–6, 297
shared flats and houses, 714–36
  see also rooms, letting
shared sleeping accommodation, 714, 715, 716
  see also dormitories
sharpening, knives, 313
shaving, 256, 262, 266–8, 557
  see also barbers
Shaw, Margaret Fay, 301
shawls, 191, 226, 229, 348, 445, 465, 466
sheds, 791, 831
sheep, 581, 582, 583, 638
  see also clearances; mutton broth
sheets (bedding), 86, 114, 115
  Ferry Inn, Inver, 638
  Highlands, 103
  making, 134, 361
  peat hoose beds, 99
  prisons, 664
  servants' bedding, 365
  torn for candle wicks, 75
  Travelling People, 115
shellfish, 418, 425, 469, 791, 799
shelter, 7–22
  see also dwellings
Shelter (organisation), 857

sheltered housing, 18, 27, 217, 496, 717
shelving, 61–2, 67, 189, 315, 317, 380–1
see also storage, books
Shetland
  auxiliary work, 332, 333, 335–6, 346
  beds and bedding, 86, 99, 466
  betrothals, 567
  Bod of Gremista Museum, 61
  byres, entry through, 86
  candles, 76
  carding, 346
  ceramic storage utensils, 65
  childbirth, 443, 444, 447
  childcare, 490
  children's clothing, 461–2
  country houses, 86
  croft houses, 191
  cruisie lamps, 75
  cupboards, 61
  farm houses, 86
  fiddles and fiddlers, 124, 125, 567
  fishing booths, 37, 99
  fishing communities, 764, 766
  flax spinning, 345
  furnishings, 191
  grain storage, 86
  herring gutters' huts, 623
  hingin lum fireplaces, 91
  infirm people, care of, 490
  kists, 56
  knitting, 191, 335, 340, 341, 349
  linen making, 350
  marriage customs, 124, 567
  medical services, 484
  midwives, 443, 444
  Muckle Flugga lighthouse, 834
  music, 124, 125, 567, 623
  ornaments, 191
  peat hooses, 99

peat-burning hearths, 89
playing cards, 167
rugs, 191
seating arrangements, 86
shelving, 61
storerooms, 37
straw work, 349–50
windows, 79
see also Northern Isles
shielings, 17, 61, 99, 109–10, 617, 618–20
shifts (clothing), 231
shinty, 164
shirts, 231, 235, 236, 239, 366
shoemakers, 340, 342, 633
shoemaking, 231, 340
shoes, 230–1, 239, 364, 366, 464, 557
shooting holidays, 638
shooting lodges, 640, 646, 831
shop workers, 343, 344
shopping, shared flats and houses, 727
shops, 343, 429, 435, 741, 769
see also chemists' shops; fairs; markets; online shopping
short-term prisons, 656–7, 658
shortbread, 421, 434
shot putting, 631
shottles, 55
Shotts, Lanarkshire, 656
Show People, 788, 804–12, 831
shower rooms, 288
showers, 291, 292
  canal boats, 841
  huts, 829
  infirm people, 493
  miners, 283
  oil-rig workers, 629
  Show People's caravans, 809
  Travelling People's camping sites, 801, 803
  youth hostels, 646

Showmen's Guild of Great Britain, 805
shutters, windows, 193
Sibbald, Sir R, 418
sideboards, 62, 179, 188, 554–6
sideburns, 266
sieves, 336, 618
signal tower conversion, 840–1
Sikh birth customs, 575
Sikhism, 218
silk, 232, 238, 321
silk dusters, 311
silver candlesticks, 75
silver knives, 312
silverware, 111, 308, 310, 315, 427
simmons, 335
Simpson, D R, & Sons, 95
Simpson, Sir James Y, 450
Sinclair family, Mey, Caithness, 382
Sinclair, Henry, bishop of Ross, 393
Singer Manufacturing Company, 184
Singh, Baldev, 575
singing *see* songs and singing
single ends, 30, 285
  childbirth in, 460–1
  grates and kitchen ranges, 96
  lodgers, 600
  privacy, 567
  Scotch chests, 64
  sleeping accommodation, 102, 461
single mothers *see* unmarried mothers
single-parent families, 477, 518, 608, 857
  see also unmarried mothers
single people
  homelessness, 856, 857, 858, 861–2
  shared flats and houses, 717, 719, 720
  see also single women
single-person households, 24, 777–87

916 • SCOTTISH LIFE AND SOCIETY

single women, 482, 522, 525, 528, 769
sinks, 78, 715, 728, 801, 803, 829
   see also kitchen sinks
Sitka spruce seeds, 410
sitting rooms, 116, 187, 188
   see also best rooms; drawing rooms; parlours
size
   laundry work, 321
   see also family sizes; household sizes; physical size
Skara Brae, Orkney, 38, 55, 100, 555
skeelies, 502
skimmed milk, 425
skin care, 262–6, 268
   see also cosmetics
Skinner, James Scott, 472
Skinner Report, 698
skinners, 342
skins, 581, 584
skipping, 469, 470
skirts, 235, 238
Skye
   bed curtains, 103
   brochs, 31
   climbing, 642
   clothing, 238
   domestic medicine, 506–7, 509
   farm house sanitation, 103, 301
   Portree High School, 709
Skye and Lochalsh, New Age travellers, 812
skylights, 80
Slains, Aberdeenshire, 352
slate floors, 317
slate roofs, 10
slaves, 260
sledging, 471
sleeping accommodation, 97 107, 111–15, 714
   apprentices, 343
   barns, 628, 633, 643
   bothies, 110, 630, 631, 643, 648, 715
   caves, 628, 643, 648, 650
   children, 111, 114, 383, 461, 465–6, 716

cyclists, 648
dressing rooms, 258
farm servants, 80, 100, 110, 111
fishing communities, 771
hawkers and peddlers, 628
herring gutters, 623
hill walkers and mountaineers, 643
holiday homes, 649
homeless people, 716
hostels, 644, 646, 715, 716
independent schools, 704
kitchens, 99, 101, 112, 556, 649
lodging houses, 853
lodgings, guest houses and hotels, 716
lorry drivers, 629
outbuildings, 628
religious communities, 752, 753
salmon netsmen, 625
servants, 365–6, 556, 715, 716
shared, 714, 715, 716
single ends, 102, 461
studies, 384
teenagers, 185, 383, 384–5, 388, 556
travellers, 628, 633
Travelling People, 795, 797
visitors, 384
youth hostels, 628, 644, 646
   see also bedrooms; beds; cradles; dormitories; prams
slickstones, 322
Sligachan Inn, Skye, 642
slippers, 140
Sloboda, John, 130
slop pails, 299, 317
slugs, 508
smallpox, 264, 119, 486
Smith & Wellstood, 96, 291
Smith, George, 347
Smith, John, & Co (Wools), 192–3

Smith, William, 473
smiths *see* blacksmiths
smoke, 90, 227, 235, 256, 420
smoke extraction, 83, 91, 363, 420
   see also chimneys; roof openings
smoked fish, 88, 619
smoking
   food preservation, 424
   in hospices, 676
   see also tobacco
smoking rooms, 433
Smoogrow, Orkney, 96
smooring, fires, 89–90, 200, 570
smoothing, laundry, 322–3
Smyllum Orphanage, Lanark, 689
snoods, 229, 259
Snow, John, 286
snuff, 568
soap, 262–3, 278, 290–1
   household cleaning, 317
   laundry work, 320, 325
   shaving, 267
   skin smoothing, 264
   taxes on, 290, 320
   urine as, 300–1
   washing up, 436
soap powders, 325
social security system, 604–5, 609
   commune members, 745
   homeless people, 851, 856, 861
   housing benefit, 606–8, 609
   salmon netsmen, 627
   shared flats and houses, 719, 723, 731
   Travelling People, 621, 802
Society for the Relief of the Destitute Sick, 367
Society of Antiquaries, 382
sofas, 189
solicitors *see* lawyers
soliman, 264
Sollas, North Uist, 25
Somerville, Alexander, 389–90
Somerville, Dr William Francis, 473

INDEX • 917

song poets *see* bards
songs and singing, 121, 471
   after dinners, 426
   children's, 120, 471
   fruit pickers, 621
   Gude and Godlie Ballates, 205
   herring gutters, 623
   play and, 163, 164, 249
   shielings, 619
   soul-peace, 200
   Travelling People, 621
   women, 122, 123, 249, 582–3
   see also hymn singing; lullabies; music; psalm singing; work songs
songwriters, 122
sookans, 334
soot, 255, 585
   see also sutiemen
sore feet, 506
sore throats, 506
soul-peace, 200
soup *see* broth
sour dook *see* buttermilk
souterrains, 37
South Queensferry, West Lothian, 85, 423
South Uist, 61, 489, 566, 618–20, 709
Spangler, J Murray, 313–14
special needs, children, 491, 492
Spence, Sir Basil, 547
Spence, Isabella, 308
Spey (river), 624, 641
spices, 66
spinners, 342
spinning, 232, 331, 335, 339, 341, 342, 352
   Aberdeenshire, 341, 345
   Angus, 346
   chairs for, 117
   clothing, 224, 227
   employment, 349
   forbidden on Sabbath, 199
   heating and, 91
   laundry maids, 321
   lighting for, 76, 77, 84
   Shetland, 191

spinning mills
   Forfarshire, 345
   see also cotton mills; flax mills
spinning songs, 346
spinning wheels, 76, 78, 92, 226, 227, 232, 345
spirewort, 508
spirituality, 197–222
   communes, 738, 740, 744, 746–7
   hospices, 679–80
   New Age communities, 218, 747, 750
   see also religions
Spock, Dr, 458
spoons, 191
   see also cutlery
sport, 163–73
   see also games
sports areas, Dollar Academy, 707
sports clubs, 254
Spottiswoode House, Berwickshire, 382
spring cleaning, 318
Spring of the Minister, Petty, Inverness, 507
springhouses, 36
sprouts, 415
sprung mattresses, 113, 114
squatting, 721, 856
stable lanterns, 70
stables, 332, 553, 626, 640, 837, 842, 844
   see also horses
Staffa, 637
stained glass, 182, 184, 546, 741
Stair, earl of, 166
stairs
   Bridgeness Tower, West Lothian, 839
   cleaning, 317, 727
   cleanliness, 102, 286, 288
   inns, 102
   monasteries, 752, 753
   tenements, 284, 317, 546
   toilets on, 284, 546
   town houses, 317
   water tower conversion, 840
stalking and shooting, 638–40

Stamp Duty, 377, 610
stamp holders, 140–1
Stamp, Linda, 466
Standing Maternity and Midwifery Advisory Committee, 451
Stanley, Perthshire, 624
Stanley Mills, Perthshire, 841–2
starching, 320, 321, 325, 432
status, 542–61
   bathrooms, 288–9
   Bealtainn fire, 571
   beds, 111, 556
   book-buying, 378, 379
   cattle ownership, 36, 542–3
   clothing, 223, 224, 542, 559
   decoration, 174, 176, 177, 178, 180–4, 553
   Edinburgh, 110–11, 542
   fireplaces, 193
   Highland dress, 228
   libraries, 381
   personal appearance, 246, 247–8, 251–2, 256–7, 265
   personal hygiene, 278, 279, 280
   personal security, 25–6, 27
   seating, 86, 115, 116–18, 566, 631, 714
   servants, 310, 357
   shared flats and houses, 724
   skin colour, 264
   stone buildings, 16
   tenements, 110–11, 546
   see also hierarchy and authority; home ownership
steam baths, 280
steamies *see* washhouses
Steel's of Edinburgh, 295
Steiner schools, 708, 747
Stenness, Orkney, 582, 841
Sterne, Jonathan, 128
Steuart, George, 835
Steven, Alasdair, 628
Stevenson, Alan, 834

Stevenson, Robert, 834
stewards, 310, 361
    see also butlers
Stewart, D Y, 287
Stewart, James, earl of
        Moray, 408
Stewart, Martha, 634
Stewart, Sandy, 621
Stewart, Thomas, 402
Still, Ann, 96
still rooms, 258, 262
Stirling, 70, 285, 402, 692
Stirlingshire
    gardens, 406
    hospice, 672
    nailers, 347, 349
    New Age travellers, 813
    Pineapple, the, 406,
        844–5
    prison, 657, 658
    soap manufacture, 291
    young-offender
        institution, 657
    see also Carron Iron
        Company
Stocking Ball, 619
stockings
    children's, 462, 464
    darning, 235
    knitting, 341, 346, 619,
        770
    menservants, 366
    Ross-shire, 231
stomach complaints, 510
    see also indigestion
stomachs, boiling meat in
        animals', 421
stone, polishing, 315
stone buildings, 9, 16, 625
stone floors, 317
stone stairs, 317
stone storage utensils, 65
Stonehaven,
        Kincardineshire, 88
stones
    protective charms, 504
    see also bakestones;
        polished stones;
        slickstones; symbol
        stones
stools, 111, 115, 116, 117,
        257, 332, 334
    see also close stools;
        footstools

storage, 55–68
    books, 379–81, 385
    butter and cheeses, 619
    candles, 75
    china, 59, 61, 64, 189
    cleaning equipment and
        materials, 315–16
    clothing, 55, 62, 111,
        116, 238–9, 631
    cobles, 625
    compact disks, 67, 383
    documents, 38
    food, 37, 58, 64, 66–7,
        86, 116
    fuels, 64, 86, 89, 292
    glassware, 380
    grain, 86
    jewellery, 38
    laundry equipment and
        materials, 316
    linen, 239
    lobsters, 469
    magazines, 380
    meal, 37, 116, 631
    ornaments, 188, 189,
        380, 556
    recorded music, 380
    religious communities,
        752
    silverware, 111
    tea, 38, 62, 85
    toilette equipment, 255,
        256
    tools, 116
    urine, 232, 300, 311
    videos and video
        players/recorders,
        380
storage furniture, 62–4, 67,
        291
    see also cupboards;
        dressers; girnals;
        kists; presses
        (cupboards); Scotch
        chests; shelving;
        sideboards
storage spaces
    baths, 61, 469
    beds, 61–2, 67, 103, 111
    best rooms, 379
    bothies, 238, 631
    camping sites, 801, 803
    caravans, 796, 809
    castle fireplaces, 93

    cellars, 343
    drawing rooms, 379
    lofts, 343
    parlours, 379
    roof spaces, 64
    servants' quarters, 238
    shieling huts, 619
    Travelling People's
        accommodation,
        796, 801, 803
    see also storage
        furniture;
        storerooms
storage utensils, 64, 65–6
store tea chests, 62
storeage spaces, see also
        barns
storerooms, 37
Stornoway, Lewis, 672
Story, Mrs, 370
story telling, 127, 633
stoups (water pails), 318
stoves, 83, 94, 96
    caravans, 796, 797,
        809–10
    gas-fired, 797
    houseboats, 841
    mobile homes, 823
    paraffin-fired, 96, 795
    tents, 792, 794, 795
    villa outbuildings, 553
    water heating, 280
    see also bath stoves;
        kitchen ranges
Strachan, Fr Robert Francis,
        208
Strange, Lady, 529
Stranraer, Wigtownshire, 85
Strathcarron Hospice, Denny,
        Stirlingshire, 672
Strathfillan, Perthshire,
        509–10
straw, plate cleaning, 427
straw beds, 466
straw cradles, 114
straw footwear, 334–5
straw mats, 333
straw mattresses, 101, 114,
        366, 466, 645
straw ropes and rope-
        making, 332, 334–5
straw seating, 332–3, 334
straw work, 332–3, 346,
        349–50

strawberries, 402, 425, 620
Strawberry Hill, London, 835
streakin boards, 495
street layouts, 14–15
street lighting, 70
street names, 552–3
stucco, 179
student accommodation
  furniture, 721, 728
  halls of residence, 711, 716, 718
  home ownership, 718
  lodgings, 715–16, 717, 718
  rented houses, 718
  shared flats and houses, 717–18, 719, 720, 721, 722, 723, 726, 728, 731
  studies of, 717, 718, 728, 731–2
  see also independent schools
students
  fruit picking, 621
  salmon netsmen, 627
studies (rooms), 384, 385
study, 377–99, 728–9, 753
  see also reading; writing
Stulaval, South Uist, 618
styes, 506
sugar, 407, 424, 430, 449, 468, 506, 510
sugar poultices, 504
suicides, 487, 569
sulphur, 264
sulphur flowers, 508
summerhouses, 409
sun tanning and protection, 254, 264
Sunday observance, 108, 109, 199
  herring gutters, 623
  salmon netsmen, 626
  toilette, 199, 261, 278
  water caddies, 318
  see also church attendance
Sundays
  clothing for, 227, 463, 468
  duration of, 199

reading on, 388, 391
religious instruction, 209
superstitions see customs
suppers, 366, 424, 425, 431, 758
surgeons see doctors
Sutherland, 226, 229–30, 231, 637, 644
Sutherland, Mima, 444
Sutherland Report, 491
sutiemen, 317
Suzy Lamplugh Trust, 709
swaddling, 461
sweating, 506
sweating balnes, 280
swedes, 409, 415
Sweeney, Dr Dorothea, 712
sweet oil, 314, 317
sweetmeats see confectioners and confectionery
sweir tree, 631
sweys/swees, 86–7, 91–2, 420
swimming baths, 282
swingletree pulling, 631
sycamore trees, 404
symbol stones, 255
Synod of Whitby, 199
syphonic WCs, 293

T
table mats, 140
  see also place mats
tablecloths, 426, 431–2, 638
tables (board game) see backgammon
tables (furniture)
  carpets, 178–9
  Corriemulzie Cottage, 187
  croft houses, 189
  folding, 116
  Gilmerton Cave, 835
  kists as, 56
  reading at, 382
  salmon netters' lodges, 624
  setting, 310
  symbolic importance, 566

washing, 317
see also computer tables; dressing tables; kitchen tables; tea tables; toilet tables; writing tables
taigheirm, 585
tailors, 134, 141, 233, 340, 342, 364, 633
Tain, Ross-shire, 76, 78, 79–80, 393
Talamh community, 739
talc, 265
talking books, 494
tallow, 71, 74, 290
tallow candles, 75, 76, 138, 344
tally irons, 323
Tannahill, Robert, 525
tanners, 342
tanning see sun tanning and protection
Tansley, Kevin, 712
tape lace, 140
tapestries, 178–9, 182
tar, 503
tartans, 226, 229, 231
tattooing, 264
taverns, 425
taxes
  hair powder, 309
  home ownership, 609–10
  houseboats, 841
  listed building conversions, 838
  reading material, 377
  servants, 309, 365
  soap, 290, 320
  windows, 79, 552
  see also council tax; Stamp Duty
Tay (river), 12, 624–5, 626, 641
Tay Salmon Fisheries Company, 626
Tayfield, St Madoes, Carse of Gowrie, 587
Taylor, John, 102, 230, 363
Taymouth Castle, Perthshire, 12
Taymouth Perthshire, 121
Tayvallich, Argyll, 75

tea
	after dinners, 426
	breakfasts, 424
	domestic medicine, 505
	Fort Augustus Abbey, 758
	lykewakes, 568
	storage, 38, 62, 85
	see also afternoon teas; beef tea; high teas
tea chests, 62
tea leaves, 316, 321
tea tables, 62
teachers, 390, 559, 703–4, 707, 708, 709, 711
	see also education; schools
technology, 120–1, 126, 128–9, 144–62, 170–1, 353
	see also computers; Internet; radio; telephones; television; videos and video players/recorders
teenagers
	children's homes, 696
	communes, 747
	employment, 460
	personal appearance, 248–9
	reading, 384–5, 388, 392
	sleeping accommodation, 185, 383, 384–5, 388, 556
	technology, 159
	see also young people
teeth see dental care
tein-eigin (Bealtainn fire), 570–1
telephones, 153–4, 155, 159, 494, 661, 669, 707
	see also mobile phones
television, 144, 145, 146–50, 158
	after-dinner watching, 435
	aids for deaf people, 494
	children and, 149–50, 152, 472, 716

craft and home improvement programmes, 193, 194
Dollar Academy, 707
gardening programmes, 412
gender differences in audiences, 129
homelessness awareness, 857
in hospices, 676
music, 120–1
old people and, 148–9, 159
play and, 171, 469
private time and, 49
reading and, 387
religious activities and, 216
shared flats and houses, 722
status and, 559
television lounges, 646
television sets, 146, 188, 654, 660–1, 667, 799, 809
Tell Scotland movement, 215–16
Templeton of Glasgow, 184
temporary accommodation, 617–53, 818–32, 855–6
	see also bothies; holiday accommodation; huts; lodging houses; lodgings; youth hostels
temporary lighting, 71–2, 74, 75, 138
tenancies
	agricultural, 30, 364, 403, 406, 543
	allotments, 413
	peat fuel and, 89
	shared flats and houses, 724, 726
	status, 542, 543
	Travelling People's camping sites, 802–3
	see also rent day; rented houses; rents
tenements, 10, 17
	back greens, 413

best rooms, 546
care of infirm people, 490
ceramic tiles, 184
childbirth in, 445
chimney sweeping, 317
clothes drying, 322, 413
Dundee, 587–8, 589
Edinburgh, 14, 42–54, 78, 110–11, 209, 210
gardens, 547
Glasgow, 182, 210–11, 237, 286, 297
kitchens, 78, 286, 546
leisure and, 167
music, 122, 126
parlours, 186
play, 169
prayer spaces, 209
privacy, 42–54, 567
sanitation, 284, 286, 288, 297, 546, 587
Scotch chests, 64
security, 34–5
social problems, 28
stairs, 284, 317, 546
status, 110–11, 546
vermin, 586–7
water supply, 284–5, 286
women's privacy, 42–54
	see also flats; rented houses; room-and-kitchen accommodation; single ends; washhouses
Tennants' St Rollox Works, 290
tennis, 167–8
Tennyson, Alfred, 1st Baron, 180
tents, 284, 445, 621, 791–2, 794–5
	see also benders (tents); camping; camping sites; tepees
tepees, 738
terraces, 402, 403, 410
tethers, 14, 336
TextDirect, 494
textile artists, 142
textile furnishings, 184, 192–3, 361

INDEX • 921

textile industries, 342,
    345–9, 352
  clothing, 224
  home furnishings,
    192–3
  lighting and, 77
  soap, 290
  urine, 232, 288, 300
  water supplies, 342
  waulking songs, 123
  see also needlework;
    spinning; weavers
    and weaving
textile workers, leisure
    activities, 648
textiles
  dyes and dyeing, 232,
    238
  interior decoration, 176,
    178–9, 184, 361
  personal hygiene, 278,
    290
  see also embroidery;
    linen; tapestries
thatched roofs, 9–10, 335
theft see crime
therapets, 679
thigging, 572
Thirlestane Lane Mews,
    Edinburgh, 844
Thomas De La Rue & Co,
    193
Thompson, Alistair R, 394
Thomson, Alexander
    'Greek', 180–1, 182
Thomson, Christian, 362
Thomson, D C, 392
Thornton & Shiells, 837
Thorpe, Professor Harry,
    414
Thorpe Report, 414–15
thresholds, 565
  see also doors
Throughcare Centre,
    Saughton Prison,
    667, 668
Thrower, Percy, 412
thrush, 300
tic and tac, 471
tied accommodation, 598,
    640
tig, 471
tiles, 95, 184, 291, 551
Tilley lamps, 70–1

Tillichuil, Perthshire, 507
Tillicoultry, Clackmannan-
    shire, 226
timber floaters, 624
time off see holidays
'time out', Jewish Sabbath,
    575
'time out' leisure, 637–53
'time out' work, 617–36
Tingwall, Shetland, 484
tinkers see Travelling
    People
tinned food, 330, 419
tipis, 738
Tiree, 13–14, 15–16
Tironensians, 750
Titchmarsh, Alan, 412
tobacco, 568
  see also smoking
Tobar nan Dileag, Tillichuil,
    Perthshire, 507
Tod and Lambs, 166
toilet tables, 257
toilets see lavatories;
    privies; sanitation;
    WCs
toilette, 199, 243–76, 278,
    288
tolbooths, 366, 367, 656
toll houses, 834–5
tombs see graves
tonics, 263
tool storage, 116
toothache, 504–5, 510
  see also dental care
toothpastes, 291
topiary, 410
torches, 71
Torridon, Ross and
    Cromarty, 646
Torthorwald Castle,
    Dumfriesshire, 12
Tour of Europe, 166
tourism, 618, 637, 638, 641,
    760, 762, 768
  see also bed and
    breakfast
    accommodation;
    boarding houses
towel rails, 279, 291
tower blocks, 10–11, 17–18,
    28, 34–5, 547
tower conversions, 839–41,
    845

tower houses, 9, 10, 36, 179,
    744–5
town gardens, 408–9,
    411–12, 413
town houses, 62–4, 110–11,
    178, 180–2, 317,
    835–7
townit, 335–6
toys (head coverings), 229
toys (playthings), 170,
    469–72, 557
  see also play
trade unions, 299–300, 373
tradesmen see craftsmen
Trail, Betty, 365
trailers see caravans
Traill family, 585
Traill, Nellie, 236–7
trainers, 557
training see education
Training School for Servant
    Girls and
    Temporary Home
    and Free Registry
    Office, 370–1
transhumance see shielings
Traprain Law, East Lothian,
    31
Traquair, Peeblesshire, 232
Traquair, Phoebe, 182
travellers, 381, 383, 628, 633
  see also cycling and
    cyclists; hawkers
    and peddlers; hill
    walking and
    mountaineering;
    hotels; inns;
    itinerant craftsmen;
    itinerant musicians;
    New Age travellers;
    tourism; visitors;
    youth hostels
Travelling People, 788–804,
    831
  bedding, 445, 795
  camping sites, 621, 789,
    790, 792, 795,
    799–804
  caravans, 621, 791,
    795–9, 805, 807, 808,
    822
  childbirth, 445
  children, 445, 789, 790,
    799, 802

music, 127, 621
security, 32
Show People and, 805
social security system, 621, 802
stories, 127
tents, 445, 621, 791–2, 794–5
work, 621, 791, 799, 800, 802, 803
travelling traders *see* hawkers and peddlers
Treaty of Union, 409
trees, 400, 401, 402–3, 404
see also ash trees; elm trees; fir trees; fruit trees; larch trees; monkey-puzzle trees; orchards; palm trees; reforestation
Trees for Life charity, 744
trews, 228
see also breeches; trousers
Trinity Road, Edinburgh, 843
trivets, 420
Trivial Pursuit, 166, 471
Trossachs, the, 637
Trotter, William, 62, 63
see also Young and Trotter
trough closets, 286
trousers, 231, 239, 557
see also breeches; trews
truckers, 629
tuberculosis, 485–6, 489, 502
tubs
bathing, 61, 278, 279, 280, 468
candle making, 76
fish oil processing, 74
house cleaning, 311
laundry, 236, 279, 292, 311, 319, 320, 321
urine collection, 232, 300, 311
see also baths
tunics, 228
turf, 85, 88
turf stools, 116

Turkish baths, 280–1, 282, 837
Turnbull, James, 13
turnips, 403, 409
Turnpike Trusts, 834
Tweed (river), 641
Tweedsmuir, Peeblesshire, 225
Tynecastle Company, Edinburgh, 180
Tynecastle Tapestry, Edinburgh, 184
typhoid fever, 286, 486
typhus, 419, 586, 587, 687

U

Uist *see* North Uist; South Uist
umbilical cords, 448–9
umbrellas, 230, 352
underground dwellings, 15
see also caves
undertakers, 496
underwear, 231–2, 233, 235, 236, 238, 321, 462
Unification Church, 737
uniforms
armed forces, 141, 223
hospice staff, 675–6
patchwork from, 141
schoolchildren, 248, 703
servants, 223
workhouse children, 687
Unilever, 291
Union Canal, Ratho, West Lothian, 841
United Secession Church, Glasgow, 211
Universal Hall, Findhorn community, 741
universities, 400, 708, 715–16
see also Aberdeen University; Abertay University; Edinburgh University; Glasgow University; student accommodation
unmarried mothers, 361–2, 368, 440, 477, 520, 685–6
see also single-parent families

Unst, Shetland, 443, 444
untried prisoners *see* remand prisoners
upholstery, 115
uplighters, 386–7
urban layouts, 14–15
urinary problems, 510
urine, 232, 262, 288, 300–1, 311, 320, 506
Urquhart, Hector, 633
Urquhart, Ross-shire, 226
Ursuline sisters, 750
utility rooms, 288
UTOPIA project, 144–5, 159

V

vacuum cleaners, 313–14
vagrants, 852–3, 854, 855
see also beggars
valets, 310, 361
vegetables, 400
farm servants' gardens, 403, 407
Findhorn community, 740
food, 468, 512
hut gardens, 829
mixed planting, 403, 410, 415
monastery gardens, 753
Stirling royal gardens, 402
town gardens, 409, 413
washing, 312
see also allotments; beans; cucumbers; kail; leeks; lettuce; parsnips; peas; potatoes; swedes; turnips
veils, 321
Vementry, Shetland, 86
ventilation, 16, 79–80, 291
see also smoke extraction
vermilion, 265
vermin, 586–7
beds, 102, 113, 115, 317–18, 363, 637
cats and, 586, 587, 591
children and, 468, 686, 687–8
clothes storage, 238

food storage in roof
    spaces, 64
girnals, 58
head-shaving, 268
mattresses, 101, 586
salmon netters' lodges,
    624
see also bed bugs; rats
Versailles, 299
Victoress stoves, 96
Victoria, Queen, 121, 187,
    410, 467, 639, 640
video games, 145, 152–3,
    158, 707
see also computer
    games
videos and video players/
    recorders, 145,
    150–1, 380, 469, 707
Viking houses, 99
village gardens, 407–8
villas, 338, 369, 837
    gardens, 169
    status, 546, 548–9, 552,
        553–7
vinegar, 264, 320, 505
violets, 507
visitors and visiting
    calendar customs, 570,
        571
    communes, 746
    death customs, 496, 568
    hospices, 676, 677,
        678–9
    ministers, 206, 210–11,
        216–18
    prisons, 666, 669–70
    religious communities,
        752, 758
    shared flats and houses,
        716, 730
    shielings, 619
    studies as sleeping
        accommodation,
        384
    see also hawkers and
        peddlers; health
        visitors; home
        visits; tourism
vitrines, 189
vitriol, oil of, 263
Vivat Trust, 839
vocational reading, 390
Voe, Msr De, 121

voles, 586
voluntary workers, 676, 677,
    678, 679, 692
see also Women's Royal
    Voluntary Service
vulnerable people, 24, 26–7,
    857–8, 860
see also children;
    disabled people; old
    people

W

W A Cadell Architects, 839
Waco commune, 737
Wade, General George, 754
wages, 339, 340
    children, 476, 526, 533,
        534
    farm servants, 364, 406,
        631
    gender differences, 308,
        427, 526–7, 528
    herring gutters, 770
    housing costs and, 597,
        598, 599
    in kind, 231, 340, 364
    net menders, 768
    prisoners, 664, 666
    servants, 308, 361, 364,
        365, 367, 373, 427
    shoes, 231, 364
    tailors, 364
    weavers, 348
    women, 346, 350, 352,
        353, 770
    women servants, 308,
        361, 364, 367
waistcoats, 231
Walker, Catherine, 634
walking aids, 494
wall beds, 98–9, 101, 112,
    445
see also alcove beds;
    bed recesses
wall hangings, 178
wall paintings, 182, 202
Wallace, John, 559
Wallace, Neville Barnes, 704
Wallis, John, 166
wallpapers, 176, 180, 184,
    187, 586, 638
walls
    bathrooms, 291
    blackhouses, 15

decoration, 176, 177,
    178, 179
farm servants' houses,
    80
Findhorn community
    buildings, 741, 742
Ossian's Hall, 405
St Kilda houses, 99
see also garden walls;
    hot walls; murals;
    wood panelling
wally closes, 184, 546
wally dugs, 189–90, 191
walnut trees, 402
Walpole, Horace, 835
Walter Macfarlane & Co,
    184, 287
Walter Segal building,
    Monimail
    commune, 745
Wanlockhead,
    Dumfriesshire, 647
wardrobes, 112, 238, 239,
    259
warm figs, 505
Warriston, Archibald, 207
warts, 264, 506
wash balls, 262
wash-down WCs, 294
wash-hand basins, 258, 287,
    291, 293, 297, 298,
    756
wash-out WCs, 293
washboards, 236, 321
washerwomen, 308, 321,
    323, 361
see also laundry maids
washhouses, 237, 281–2,
    320, 321, 325, 547,
    715
    Gilmerton Cave, 835
    villas, 553
washing
    floors, 311–12, 314, 317
    furniture, 317
    personal hygiene, 256,
        266, 277, 278–9, 798
    vegetables, 312
    windows, 307, 317
see also bathing;
    cleaning;
    cleanliness; laundry;
    skin care
washing boynes, 64

washing machines, 235, 237, 282, 321, 801, 810
washing up, 427, 434, 436, 618, 664, 798
washstands, 112, 256, 257, 279
wassocks, 333
waste disposal, 284–8
    byres, 582
    disposable nappies, 465
    in household fires, 83
    religious communities, 752
    shared flats and houses, 727
    tenements, 286, 587
    Travelling People, 791
    see also ash collection and disposal; sanitation; sewerage
water
    divination by, 574
    domestic medicine, 506
    protective customs, 570
    skin care, 263
    see also healing wells; lemon water; potato water; rainwater; rice water; seawater
water barrels, 64, 284, 285
water caddies, 285, 318
water carrying, 236, 278, 279, 284–5, 286, 318, 319
water closets see WCs
water heating, 95, 267, 279–80, 291, 321, 427, 428
Water of Leith, 318
water pails, 318
water supplies, 277, 278, 279, 284–7, 298
    bagnios, 281
    building conversions, 845
    farm servants' houses, 299–300
    Gilmerton Cave, 835
    Glasgow, 281, 285, 287, 485
    houseboats, 841
    housework, 307, 311, 318, 319, 325, 427
    huts, 828–9
    mobile homes, 822
    potteries, 342
    religious communities, 277, 752
    rural areas, 299
    shared, 715, 716
    shielings, 618
    Show People's caravans, 806, 808–9, 810
    textile industries, 342
    Travelling People's camping sites, 800, 801
    see also piped water supplies
water supply engineering products, 287
water tower conversion, 839–40
waterbeds, 113
waterproof flooring, 291
waterproofing
    houses, 8, 363
    see also damp proofing; drainage
Watson, Sheriff of Aberdeen, 688
Watson, baillie of Banff, 167
Watson, Robert, 769
Watt, Christian, 621
waulking songs, 123, 164
wazzies, 332
WCs, 277, 288, 291, 293–4, 296, 297, 298
    farm servants' houses, 300
    huts, 829
    Inverness-shire, 279
    manufacture, 287, 297
    mobile homes, 822, 824
    tenements, 286
    Versailles, 299
weavers and weaving, 77, 331, 342, 346–9, 364
    artist, 551
    clothing, 224, 227, 231, 232
    industrialisation, 224, 339, 346–7, 348–9, 352
    lighting for, 77, 84
    Perthshire, 77, 351
    plaids, 227–8
weavers' caps, 230
work at home, 339, 340, 341, 342, 347–8, 349, 351, 352
    see also plaids; tartans
weaving, see also tapestries
weaving galleries, 77
web sites, 388
    see also Internet access
Webb, Philip, 181
Wedderburn, Berwickshire, 835
wedding rings, 506
weddings
    at home, 216
    clothing, 559
    graces, 202
    ministers' duties, 217
    music, 124
    personal appearance and, 247, 261, 278
    Show People, 806
    toll houses, 835
    see also betrothals
Weir, Molly, 237, 248, 249
Weir, Tom, 643–4, 646, 647–8
Welsh Historic Monuments, 838
Wemyss, earl of, 837
Wemyss-ware, 190
West Lothian
    Bridgeness Tower, 839
    coal transport from, 85
    fairs, 423
    House of Binns library, 382
    houseboats, 841
    paraffin production, 96
    pit-head baths, 283
    religious community, 752
West Mill, Dean Village, Edinburgh, 842
Wester Hailes, Edinburgh, 97
Wester Ross see Ross-shire
Western Isles
    boarding out of children, 476, 709–11
    care of infirm people, 483–4
    cats, 585
    crogain, 65, 74
    cycling, 648

dairy products, 341, 418, 420
dinners, 418
distilling, 341
domestic medicine, 502–3, 509
dressers, 60
dung as fuel, 88
herring gutters, 622
medical services, 484
music, 123, 125
religion and play, 164
salmon netsmen, 626
seating, 115–16
secondary school pupils, 709–11
servants, 363
shielings, 99
sleeping accommodation, 98, 99, 112
souterrains, 37
spirituality, 199–200
taigheirm, 585
underground dwellings, 15
waulking songs, 123
women's dresses, 224, 231
women's employment, 341
woollen cloth making, 228–9
youth hostels, 648
see also Benbecula; blackhouses; byre dwellings; Coll; Harris; Iona; Islay; Lewis; Mull; North Uist; Skye; South Uist; Staffa; Tiree
Westruther, Berwickshire, 232
wet larders, 424
wet laundries, 319
wet nurses, 344, 361, 362, 467
see also breastfeeding
Whatman, Susannah, 236
wheelchairs, 494
wheeled goffers, 323
wheelhouses, 25
whey, 425, 507
whins, 85
Whinwell Children's Home, Stirling, 692

whisky, 341, 434
White Caterthun, Angus, 31
white lead, 263, 264
White Stripe, Enzie, Banffshire, 73
whitecurrants, 407
whitening
  cleaning substance, 314, 315, 317
  linen, 320, 321
  skin, 264
  see also bleaching
Whithorn, Wigtownshire, 79, 198
whiting (cleaning substance), 314, 315, 317
whooping cough, 504, 507–8
Wick, Caithness, 350
wicker chairs, 116
wicks
  candles, 75, 76
  cruisie lamps, 75, 336
widows, legal rights, 522, 525
widows and widowers, childcare, 686
wigs, 257, 265, 266, 361
Wigtownshire, 79, 85, 87, 168, 198
Wilkie, David, 253, 391
William III, king of Great Britain and Ireland, 79
William Rodger & Son, 58
Williams, Lawrence, 377
willows, 400
  see also baskets and basket-making
Wilson, John, 367
Wilson, Robert, 844
windmill conversion, 839
window boxes, 190
window hangings see curtains
window seats, 111
window shutters, 193
window tax, 79, 552
windows, 78–80
  bathrooms, 291
  cleaning, 307, 315, 317, 373
  decoration, 181, 193
  flats, 546

kitchens, 77–8
  religious communities, 752
  status, 547, 551–2, 554
  symbolic importance, 562, 568
  taxes, 79, 552
  villas, 552
  weavers' houses, 77
  see also curtains; glass; glazing; natural lighting; skylights
windowsills, decoration, 315
wine, 262, 263, 426, 431, 502, 508, 758
  see also claret
Winksetter, Orkney, 99
Winning, Cardinal Thomas, 216
winshoos, 336
winter, quarter days, 569
winter dykes, 86
wiskers, 55
witches, 504, 584–5
wives see married women
woad, 264
Wolf Cubs, 473
women
  after-dinner activities, 426
  afternoon teas, 431
  allotments, 415
  bathing, 468
  clothing, 224, 226, 228, 231, 232, 233, 234, 235
    see also dresses; dressmaking; skirts
  days of rest, 108
  dental care, 249
  domestic violence, 524
  education, 134, 135–6, 309, 370–2
  evangelical Christianity, 207
  facial hair, 268
  fishing communities, 350, 429, 621–3, 762, 764, 765, 767, 768–74
  hair, 249–50
  head coverings, 229–30, 264, 366, 559

hierarchy and authority, 357, 515, 521, 522–8
homelessness, 440, 856, 857, 858, 860, 861
horticultural society membership, 409
landed families, 357
legal rights, 522–3, 528
leisure activities, 134, 137–41, 648
long lets of rented houses, 600
merchants' and manufacturers' wives, 358
personal appearance, 243–5, 246, 249–50, 251–2, 259, 263, 264, 265
privacy, 42–54, 185
property owners, 42–54, 344
radio and television audiences, 129
reading, 388–9, 392–3
religious communities, 750
rent arrears management, 599
rent strikes, 601
seating, 116, 117
security, 24
single-parent households, 477
single-person households, 778
songs and singing, 122, 123, 249, 582–3
straw footwear, 334–5
student residences, 716
suicides, 487
teacher training college, 711
technology and, 146, 158
see also gender differences; girls; married women; mothers; single women; widows; witches
women servants, 308–10, 343, 357, 358, 359–60, 364–5, 373

Aberdeen, 361–2
clothing, 226, 366
Duke and Duchess of Hamilton's household, 361
farm servants, 341, 364
head coverings, 366
male followers, 370
numbers, 372–3
pregnancy, 361–2, 368
recruitment, 364, 367
single women forbidden other work, 525
training, 370–2
wages, 308, 361, 364, 367
World War I, 359
see also au pairs; cooks; dairy maids; housekeepers; housemaids; kitchen maids; lady's maids; laundry maids; milkmaids; mother's helps; nannies; nursery maids; parlour maids; scullery maids; washerwomen
women's prisons, 657, 658
Women's Royal Voluntary Service, 669
women's work, 324–6, 339, 341–2, 343–4, 345–6, 349–50, 352–3, 525–8, 534, 537
agricultural work, 137, 332, 341, 349, 364, 525
see also bondagers
auxiliary work, 331, 335
care of infirm people, 482, 489
childcare, 324–5, 528, 537
children's homes, 692
cooks, 308, 309–10, 427
days of rest, 108
employment opportunities, 310, 338

factory work hours, 460
farm servants, 341, 364
farm servants' wives, 137, 349, 525
fire smooring, 570
fishing communities, 350, 352, 767–70
folk healers, 502
laying out of dead people, 495
needlework, 134–41, 142, 339, 343, 352, 353
oil-rigs, 629
seasonal work, 618
shielings, 618–20
Show People, 811
single women, 525, 769
wages, 346, 350, 352, 353, 770
weaving, 347
see also childcare; housework; howdies; knitting; laundry; midwives; nurses; spinning; washerwomen; women servants; work at home
wood *see* firewood; timber floaters
wood, *see also* firewood
wood ash, 314
wood carving, 176, 191
wood fires, 311, 420
wood graining, 183
wood lice, 508
wood panelling, 176, 178, 193
wooden beds, 111, 112, 318
wooden buildings, 9
wooden cradles, 114
wooden dressers, 60, 61, 555
wooden kists, 55
wooden porches, 16
wooden seating, 115–16
wool, 581, 584
see also Berlin woolwork
wool industry, 339
see also knitting; spinning; weavers and weaving

INDEX • 927

woollen cloth manufacture, 228–9, 232, 300, 339
woollen clothing, 224, 227, 231–2, 238, 321
   see also knitting; plaids; stockings
Wordsworth, Dorothy, 12, 308, 311, 322–3, 325, 637
work at home, 331–2, 338–56
   canvas work pattern-making, 140
   craftsmen, 340–1, 343, 351, 357
   home offices, 97, 145, 154, 384
   lighting, 77, 138
   linen making, 226
   needlework, 137, 138, 142, 352
   see also women's work
work hours
   factories, 460
   farm servants, 387
   herring gutters, 622
   reading and study and, 387
   salmon netsmen, 625
   shieling work, 619
work-related diseases, 486
work songs, 122–3, 164, 346, 582–3
workers' housing, 598
   see also estate workers accommodation; farm servants, houses; shielings
workhouse disease, 687
workhouses, 368, 684, 686, 687–8, 689, 856
   see also poorhouses
workplaces
   craftsmen, 340–1, 342–3, 351, 357
   other people's homes, 357–76
   seating in, 109
   see also work at home; workshops
workshops, 338, 342, 349
   homes as, 331, 332
   luckenbooths as, 343
   Marquess of Bute's, 182
   Mortlake, 178
   prisons, 666
   religious communities, 753
   weaving, 346–7
workstations, 383
World Wide Web see Internet access
worms, 505
worship see religions; spirituality
wounds, 510
Wright, Sgt Major Jimmy, 633, 634
wrights, 495
writing, 118, 155–6, 377, 378, 382, 752, 753
   see also songwriters; study
writing tables, 383
wrought iron see ironwork
Wylie & Lochhead, 183

Y
yards see courtyards; gardens; kailyards; sauch yards
Yarmouth, 623
yellow ochre, 265
Yester, East Lothian, 180
Young and Trotter, 62
young offenders' institutions, 657, 658, 688–9
young people
   communes, 742, 747–8
   employment, 534
   Findhorn community, 742
   fishing communities, 763–4
   hierarchy and authority, 515
   housing benefit, 608
   lodgings, 719
   shared flats and houses, 717–20
   see also children; students; teenagers
youth hostels, 628, 644–6, 647, 648, 650, 716
Youth House, Findhorn community, 742
Yule, 168, 569

Z
Zeist, Netherlands, 657
Zoltie, Hyman, 575